CONCRETE TECHNOLOGY AND PRACTICE

Fourth edition

CONCRETE TECHNOLOGY AND PRACTICE

Fourth edition

W. H. TAYLOR
MCE, CEng, FICE, FIE Aust.
Consultative Executive and Technical Service Engineer
Formerly of CSIRO Division of Building Research

McGRAW-HILL BOOK COMPANY Sydney
New York St Louis San Francisco Auckland Bogotá Düsseldorf
Johannesburg London Madrid Mexico Montreal New Delhi
Panama Paris Sao Paulo Singapore Tokyo Toronto

123456/MP/210987

A93343

National Library of Australia
Cataloguing-in-Publication data:
Taylor, Walter Harold.
    Concrete technology and practice.
    Index.
    Includes bibliographical references.
    ISBN 0 07 093343 x.
    1. Concrete. 2. Concrete construction—
    Handbooks, manuals, etc. I. Title.
620.136

Printed in Australia by Macarthur Press, Parramatta

Contents

vi

Foreword

The writing of the first edition of *Concrete Technology and Practice,* published in 1965, was a prodigious undertaking. To update it in three further editions was equally demanding, particularly the production of this metricated version of a monumental work which won the first Building Science Forum of Australia book award shortly after being published.

These publications have been of outstanding value internationally to designers, constructors, manufacturers, researchers, educators and students. They have become the prescribed text used by creative faculties in tertiary and post-tertiary education, a field in which the author has made a significant contribution to community service in a part-time capacity.

Mr Taylor is well qualified to have produced this oft-quoted "Koran of concrete". Graduating in Civil Engineering, as a Simon Fraser Scholar at the University of Melbourne, he obtained First Class Honours and won the Argus Scholarship at the Final Honours Examination. After extensive civil engineering experience in Government, semi-Government and Municipal services, as well as in private industry, he joined the staff of the Division of Building Research of the Commonwealth Scientific and Industrial Research Organisation in 1945: specialising in research and development of the composition, properties and uses of both normal and special concretes.

In the twenty-five years he spent with the Division, he made valuable contributions to the technology of concrete and extremely strong calcium silicate products; and his advice was widely sought by professional engineers, architects, industrialists, builders and lawyers on the more difficult features that face formulators and users of concrete from time to time. On retiring from the CSIRO in 1970, the author joined a company of consulting engineers as a technical service engineer, a role with plenty of scope for the progressive use of his unique knowledge, talents and initiative in getting things done.

The foregoing contributions to creative achievement have led to the author being signally honoured with personal profile entries in such international biographies as *The Dictionary of International Biography, The Writers Directory, The International Authors and Writers Who's Who, Who's Who in Community Service, Men of Achievement, Who's Who in the World, The Blue Book (Leaders of the English-Speaking World)* and *The International Register of Profiles.* A bronze award for distinguished achievement in creative service was bestowed on the author in 1976 by the International Biographical Centre, Cambridge, England.

The reputation established by the earlier editions will surely guarantee a large demand for this first metricated text, the contents of which have been rewritten to meticulously meet the many advancing requirements of professional, educational and practising personnel.

The scope of the text forms an invaluable source of information for persons engaged in advancing their knowledge and applying their special skills in creative concrete services. Its straightforward, well-integrated, complete coverage and format facilitate the practical conversion of essential basic concepts and specific data on concrete technology into highly rewarding and satisfying results.

Sir John Holland
Chairman, John Holland Construction Group, Australia.
Ian Langlands
Deputy Chancellor, Monash University, Australia.
Former Chief, Division of Building Research, CSIRO.
John Connell
Principal, John Connell Group of Consulting Engineers, Australia.

Preface

This fully revised, updated and extended metric version of the text, first published in 1965, is designed to fulfil a continuing demand for advanced knowledge of concrete technology and practice. The technical format is orientated to give concise, comprehensive information and data in SI units to professional consultants, industrialists, practitioners and research educationists.

To this end, the factual resources of the text are systematically arranged to provide a direct means of ready reference, review, retrieval and research. These objectives are realised through the following media: an introductory subject classification in the Contents supported by a complete Index; a schematic diagram on requirements and procedures (Appendix III); eight composite Parts subdivided into chapters and sections; a cohesive linkage of topics and Standards with expeditious cross-referencing; a select Glossary of Terms (Appendix VI); and a convenient presentation of Bibliographic Abbreviations (Appendix VII) and Conversion Factors (Appendix VIII), which serve as further facilities in equipping the text with well-documented information that conveys prerequisite meaning for practical professional usage.

The basic principles, component materials and operational procedures employed in every phase of concrete work are expounded in considerable detail, extending from the research laboratory and design office to the production industry or constructional site concerned. Standard Specifications are tabulated for English-speaking countries, and large bibliographies are sectionally tied to the text with the aid of small keynumbers. A well-integrated gamut of references, including their own literature surveys, extends not only the scope of the volume but its effective use in different countries.

Cordial appreciation is recorded here of the valued services rendered by book reviewers and referees. Also, to the donors of photographs and the following industrial donors of annual concrete honour awards, which have greatly helped to advance the knowledge and expertise of numerous persons to national advantage:

ARC Industries Ltd,
Humes Ltd,
John Holland Construction Group,
Monier Brick and Concrete Industries (Monier) Ltd,
Rocla Concrete Pipes Ltd,
The Readymix Group.

These awards have been conferred on candidates who have earned the right to them, by emerging progressively on merit from the 5000 registrants of my qualifying training courses in Concrete Technology and Practice, which have been held since 1945. This programme has been carried out mainly through the Departments of Civil Engineering, Architecture, Technisearch and External Studies of the Royal Melbourne Institute of Technology.

Technical change creates an increasing demand for greater knowledge, and the purpose of recasting this work of world repute is to help satisfy that demand. In addition, it is to bring established principles into proper perspective, factually promote forward and flexible constructive thinking, and provide guided inspiration for those concerned with the materialisation of proposed concrete projects into gratifying structural realities.

This publication, which has some forty per cent greater capacity than its precedents, has arisen primarily as the result of an earlier motivation to make a worthwhile contribution of diligent development in community service, over and above that normally earned through vocational enterprise and endeavour. Over thirty thousand hours of voluntary additional service have thus been privately contributed for useful public purposes.

W. H. Taylor,
May 1977.

Acknowledgements

The author acknowledges with thanks the valued support given to the publication of this book by the following persons:

K. J. CAVANAGH, Director, Cement and Concrete Association of Australia.

R. W. SEDDON, Secretary, Australian Portland Cement Ltd.

A. W. LAWSON, Metric Engineer, Australian Department of Housing and Construction.

E. J. TAYLOR, the author's wife.

Knowledge is power (Socrates)

Build on Knowledge (*The Age,* Melbourne)

Progress through Knowledge (American Concrete Institute)

Out of much study comes much thought,
Out of inspired effort comes growth,
Good thought and good growth go together (Maxim)

The end is to build well (D. Hardy, Britain)

LIST OF PLATES

Note: The concrete illustrations given in the text typify building procedures of recent origin or concerning the preceding few decades.

PART 1

TRADITIONAL MATERIALS AND PRACTICES

CHAPTER 1

HISTORICAL

The cement used by the Egyptians was calcined gypsum, and both the Greeks and Romans used a cement of calcined limestone. Roman concrete was made of broken brick embedded in a pozzolanic lime mortar. The cementitious matrix of the latter consisted of lime putty mixed with brick dust or volcanic ash. Hardening was produced by a prolonged chemical reaction between these components in the presence of moisture.

Dead weight was reduced by casting hollow pots into the concrete, as in the tomb of the Empress Helena near Rome, and metal bars were sometimes incorporated, as was done in the roof of the Baths of Caracalla in Rome. Marble mosaics were embedded in the surface for decorative effect.

With the decline of the Roman Empire, concrete fell into disuse. The first step towards its reintroduction was in about 1790, when an Englishman, J. Smeaton, found that when lime containing a certain amount of clay was burnt, it would set under water. This cement resembled that which had been made by the Romans. Further investigations by J. Parker in the same decade led to the commercial production of natural hydraulic cement, which was widely used early in the nineteenth century, both in England and later in France. The first (unreinforced) concrete bridge was built in 1816 at Souillac in France.

An important advance toward the manufacture of a dependable hydraulic cement was made by Joseph Aspdin, an English mason, in 1824. His product was called portland cement because it resembled a building stone that was quarried on the Isle of Portland off the coast of Dorset. Until the end of the nineteenth century, large quantities of this cement were exported to many parts of the world. The first factories for portland cement outside the British Isles were opened in France in 1840, Germany in 1855, and the United States in 1871. In 1882 William Lewis established a plant at Brighton, South Australia, and produced a cement similar to imported portland cement.

Reinforced concrete was evolved and initially developed by such Englishmen as W. B. Wilkinson of Newcastle-on-Tyne in 1854, and the following Frenchmen: J. L. Lambot in 1855, F. Coignet in 1861, and Joseph Monier in 1867. By the turn of the century, forty-three patents had been taken out on reinforced concrete, and most of its basic properties and principles of design had been determined.

By 1911, probably the biggest structure of its kind in the world was the Melbourne Public Library with a dome 35 m in diameter. The value of mechanical vibration was discovered in 1917 by E. Freyssinet, who used it in the construction of an airship hangar at Orly in France. In 1918, Duff Abrams[1] established the relationship between the compressive strength of concrete and the ratio of mixing water to cement. Confirmatory records that have since been determined are indicative of the fifty-year properties of concrete.[2]

The growing acceptance of reinforced concrete as a material of construction is reflected in the world's demand for portland cement, which reached about 250 megatonnes per annum near the middle of this century.* With increasing demand, and continuing improvements in technology, the annual output should be more than doubled by the year 2000. Advanced creative expertise will then make it possible to expedite the much-needed construction of reinforced concrete thoroughfares, floating airports, and systems-built structures which would be 100 or more storeys tall or have over 500 m suspended span.[4]

---

*Transistorised computers and television cameras are now installed in certain plants for the automatic control of cement manufacture.[3]

CHAPTER 2

GENERAL REQUIREMENTS

New developments in the use of concrete are constantly taking place, and its wide application in the present will be exceeded by new and greater applications in the future. A fundamental requirement for making concrete structures and products is good concrete, which is a carefully controlled mixture of cement, water, and fine and coarse aggregate, combining admixtures as needed to obtain the optimum in quality and economy for any use.

Good concrete, whether plain, reinforced or prestressed, should be strong enough to carry superimposed loads during its anticipated life. Other essentials (as circumstances may require) are impermeability, durability, and a minimum amount of shrinkage, cracking, cavitation, efflorescence, spalling and surface wear. The following factors contribute to these ends.

A knowledge of the properties and fundamental characteristics of materials and the principles of design.

Reliable estimates of site conditions and costs.

Component materials of satisfactory quality.

The minimum or optimum quantity of mixing water.

A careful measurement or weigh-batching of cement, water and aggregate.

Proper transport, placement and compaction of the concrete.

Early and thorough curing.

Competent direction and supervision.

The tremendous investments in concrete work more than justify the modest cost of the last item for effective and economical results. Although good concrete costs little more than poor concrete, its performance is vastly superior. A project specification for concrete properties should make reference to certain selected prerequisites, such as the following.

5

Grade designation for characteristic strength (see Table 3.25 and Appendix VIII, 5).

Maximum size, properties, single-size, and gap or continuous grading limitations of coarse, fine and combined aggregates.

Slump value for mechanical or manual handling, compaction and finishing.

Cement classification, certification and content for special performance, durability and impermeability criteria.

Pretested pozzolans or chemical admixtures.

Plant or project control testing requirements.

Design density and type of lightweight concrete.

Efficient (both early and effective) curing facilities.

Acceptable crack widths, shrinkage and permeability or absorption coefficients, specific creep, elastic modulus, strengths at various ages, variation in quality, and serviceability criteria of hardened concrete.

CHAPTER 3

MAIN COMPONENTS

3.1 CEMENT

3.1.1 GENERAL INFORMATION

In the manufacture of portland cement, limestone and clay or shale are intimately ground and mixed (wet or dry process) and subsequently clinkered at about 1450°C in either a rotary or a shaft kiln.[5] The final grinding is done with a 4–6 per cent addition of gypsum, which retards the hydration of the aluminate component of the cement and renders it fit for use; raw cement otherwise would have a tendency to flash set on the addition of water (Plate 2).[6, 7]

Portland cement is composed of four main oxides (in addition to others) within the following limits: lime (CaO), 60–66 per cent; silica (SiO_2), 19–25 per cent; alumina (Al_2O_3), 3–8 per cent; and iron oxide (Fe_2O_3), 1–5 per cent. Magnesium oxide (MgO) is limited to 4 per cent. Portland cement is varied in type by changing the relative proportions of its four predominant chemical compounds and by the degree of fineness of the clinker grinding. The finer the grinding, the more rapid is the rate of hydration. Oil must be excluded at all stages.

Loss of weight on ignition is an indication of prehydration and carbonation, and is generally a measure of the freshness of cement. The permissible maximum loss of weight is 3 per cent in temperate climates and 4 per cent in tropical climates. Sodium and potassium oxides, in amounts greater than 0·5–0·6 per cent, cause destructive expansive reaction between the cement and certain aggregates. Cements of high alkali content are associated with increased crazing, checking, cracking and efflorescence of concrete. Free alkali content is associated with the composition of raw materials and the burning process. [8, 9, 97, 103]

The theoretical percentage compound composition of portland cement can be calculated from its percentage content of raw materials by means of the following formulae.

Tricalcium silicate ($3CaO \cdot SiO_2$) $= 4 \cdot 07 \times CaO - 7 \cdot 60 \times SiO_2 - 6 \cdot 72 \times Al_2O_3 - 1 \cdot 43 \times Fe_2O_3 - 2 \cdot 85 \times SO_3$

Dicalcium silicate ($2CaO \cdot SiO_2$) $= 2 \cdot 87 \times SiO_2 - 0 \cdot 754 \times 3CaO \cdot SiO_2$

Tricalcium aluminate ($3CaO \cdot Al_2O_3$) $= 2 \cdot 65 \times Al_2O_3 - 1 \cdot 69 \times Fe_2O_3$

Tetracalcium aluminoferrite ($4CaO \cdot Al_2O_3 \cdot Fe_2O_3$) $= 3 \cdot 04 \times Fe_2O_3$

Small amounts of gypsum, magnesium oxide, alkali, free lime and silica in the form of glass may be present in portland cement. The relative proportions of the main compounds and the fineness to which the clinker is ground give rise to several types of this cement. A small variation in the composition or proportion of its raw materials leads to a large variation in compound composition. [10, 11, 12, 80]

In ordinary portland cement, the tricalcium silicate content is high (about 55 per cent), but it is kept below 40 per cent in low-heat portland cement. The dicalcium silicate content is about 17 per cent in ordinary portland cement and 40 per cent (minimum limit, 35 per cent) in low-heat portland cement (see Table 3.2).

Most of the strength-developing characteristics of portland cement are controlled by these compounds. The tricalcium silicate reacts very rapidly with water and is mainly responsible for high early strength; the dicalcium silicate reacts very slowly with water and provides most of the ultimate strength. Their combined calculated content, which may differ from the actual, usually averages about 72 per cent.

The tricalcium aluminate content is 10 or 11 per cent in ordinary portland cement, 5 or 6 per cent in low-heat portland cement, and is limited to 5 per cent in sulphate-resisting cement. It promotes high early strength, but is the most vulnerable constituent of portland cement toward possible chemical attack. The tetracalcium aluminoferrite content is 9 or 10 per cent in ordinary portland cement and 12 or 13 per cent in low-heat portland cement. It serves as a useful flux during manufacture and hydrates rapidly with a relatively small development of strength. The compounds tricalcium aluminate and tricalcium silicate develop the greatest heat, then follows tetracalcium aluminoferrite, with dicalcium silicate developing the least heat of all.

Low-heat portland cement, therefore, is not only low in heat generation but it also develops strength much more slowly than does ordinary portland cement. During the setting and hardening of portland cement, silicates with a high calcium content are transformed into silicates with a reduced calcium content. Calcium hydroxide is liberated as a by-product, and the amounts present in set cement are approximately 15–16 per cent in high-early-strength portland cement, 14 per cent in ordinary portland cement, and 12 per cent in low-heat portland cement.

The amount of calcium hydroxide is determined more readily petrographically (by counting the crystals in prepared sections) than chemically. The foregoing amounts of calcium hydroxide, which were determined in this way, may be increased by up to 5 per cent so as to allow for the presence of minute crystals that would not be discerned in the count. Electron micrographs, tracer techniques, spectrophotometry (AS 1378, BS 3875, ASTM E275), atomic absorption spectroscopy, X-ray diffraction and ethylenediamine-tetraacetate (EDTA) methods [13] are being increasingly used for the analysis of minerals in set cement. The setting and hardening of portland cement is dealt with further in Chapter 31.[131]

Although a great deal of change in cement composition has been caused by manufacturers being requested to meet demands for strength, almost to the exclusion of all other desirable characteristics, cements can be produced or are obtainable with a wide range of properties for particular uses. Their properties at the time of use should comply with relevant standard specifications, which also describe cement-testing procedures. Those given in Table 3.1 form part of a family of standards that are listed in Appendix I, including AS K54, BS 1014 and ASTM Part 28 on pigments.[121]

TABLE 3.1   STANDARDS ON CEMENT

Type of Cement	Standard Specification		
	Australian	British	ASTM
Blended	1317	—	C595
High-alumina	—	915	—
Masonry	1316	5224	C91
Natural	—	—	C10
Portland			
Air-entraining	—	—	C175; C226
Blastfurnace	1317	146 Pt 2; 4246	C595
High-early-strength	1315; MP27	12	C150
Low-heat	1315; MP27	1370; 4246	C150
Ordinary	1315; MP27	12; 4550	C150; C114
Pozzolan	1317; A181	—	C595
Sulphate-resisting	1315; MP27	4027	C150
Slag	—	—	C595
Supersulphate	—	4248	—

Relevant standards of the American Society for Testing and Materials include: autoclave expansion, C151; calcium sulphate content, C265; chemical analysis, C114; compressive strength, C109, C349; false set, C359, C451; fineness determination, C115, C184, C204, C430; flexural strength, C348; heat of hydration, C186; normal consistency, C187; optimum SO_3, C563; processing additions in manufacture, C465; sampling, C183; setting time, C191, C266; specific gravity, C188; sulphate resistance, C452; tensile strength, C190; and volume change, C490, C596 (Tables 16.1 and 17.1).

General data are summarised in Tables 3.2–3.6.

TABLE 3.2   POTENTIAL COMPOSITION (PER CENT) OF
                        PORTLAND CEMENTS

Compound	Type of Portland Cement			
	High-early-strength	Low-heat	Ordinary	Sulphate-resisting
Tricalcium silicate (C_3S)	61	36	55	45
Dicalcium silicate (C_2S)	11	40	17	35
Tricalcium aluminate (C_3A)	11	5	11	3
Tetracalcium aluminoferrite (C_4AF)	9	13	9	11
Miscellaneous constituents	8	6	8	6

TABLE 3.3   FINENESS INDEX

Portland Cement	Specific Surface Area (m²/kg)*
Blastfurnace	225–360
High-early-strength	400–460
Low-heat	280–360
Ordinary	300–400
Sulphate-resisting	280–360

* Air-permeability method (Britain) which gives results 1·6–1·8 times those obtained with the Wagner turbidimeter (United States).

TABLE 3.4   TOTAL HEAT OF HYDRATION OF CEMENTS

Cement	Heat of Hydration (kJ/kg)					
	1 day	2 days	3 days	7 days	28 days	90 days
High-alumina	395	—	440	440	—	—
Portland						
Blastfurnace	105	185	255	300	350	370
High-early-strength	280	—	370	405	440	465
Low-heat	—	—	185	230	280	315
Ordinary	185	230	280	325	370	395
Supersulphate	—	—	130	175	200	—

TABLE 3.5   CHARACTERISTICS OF CEMENTS

Cement	Chemical Resist-ance	Cracking Resist-ance	Drying Shrink-age	Rate of Evolution of Heat of Hydration	Rate of Strength Develop-ment
High-alumina	Very High	Low	Medium	Very High	Very High
Portland					
Blastfurnace	Medium	Medium	Medium	Medium	Medium
High-early-strength	Low	Low	Medium to High	High	High
Low-heat	Medium	High	Medium	Low	Low
Ordinary	Low	Medium	Medium	Medium to High	Medium
Pozzolan (e.g. Fly-ash)	Medium to High	High	Medium to High	Low to Medium	Low
Sulphate-resisting	High to Very High	Medium	Medium	Low to Medium	Low to Medium
Supersulphate	High	Medium	Medium	Very Low	Medium

TABLE 3.6   **SETTING TIME AND STRENGTH**

Type of Cement	Standard	Setting Time (h)		Minimum Compressive Strength, Cubes (MPa)*			
		Initial not less than	Final not more than	1 day	3 days	7 days	28 days
High-alumina	BS 915	2	8	41	48	—	—
Masonry	AS 1316	1·5	24	—	—	3·4	6·2
	BS 5224	0·75	10	—	—	4	6
	ASTM C91	2†	24†	—	—	3·4	6·2
Portland							
Air-entraining	ASTM C175	1†	10†	—	6·2	10·5	19·5‡
Blastfurnace	AS 1317	1	12	—	19	31	45
	BS 146	0·75	10	—	11	21	34
Blastfurnace (Type IS)	ASTM C595	0·75	7	—	8·3	14·5	24‡
High-early-strength	AS 1315	1	12	—	28	38	52
	BS 12	0·75	10	—	21	28	—
High-early-strength (Type III)	ASTM C150	1†	10†	11·5	21	—	—‡
Low-heat	AS 1315	1	12	—	17	24	31
	BS 1370	1	10	—	17·6	14	28
Low-heat (Type IV)	ASTM C150	1†	10†	—	—	5·5	14‡
Ordinary	AS 1315	1	12	—	19	31	45
	BS 12	0·75	10	—	15	23	—
Ordinary (Type I)	ASTM C150	1†	10†	—	8·3	14·5	24‡
Pozzolan	AS 1317	1	12	—	19	31	45
Type IP	ASTM C595	0·75	7	—	8·3	14·5	24‡
Type P	C595	0·75	7	—	—	10·5	21‡
Sulphate-resisting	AS 1315	1	12	—	19	28	38
	BS 4027	0·75	10	—	15	23	—
Moderate, Type II	ASTM C150	1†	10†	—	6·9	12·5	24‡
High, Type V	ASTM C150	1†	10†	—	—	10·5	21‡
Slag (Type S)	ASTM C595	0·75	7	—	—	4·1	10·5‡
Supersulphate	BS 4248	0·75	10	—	14	23	34

* 1 part cement : 3 parts sand, by weight.

† Gillmore needle used instead of Vicat needle.

‡ 1 part cement : 2·75 parts sand, by weight.

Note: Concrete and mortar strengths may not compare for a given cement.

While a high early strength is frequently sought in practice, it is worth noting that portland cement with a high rate of hardening has a strong tendency to effloresce and to crack under restraint at an early age.[5] This characteristic is brought about by high contents of tricalcium silicate (C_3S) and tricalcium aluminate (C_3A) and a high degree of fineness of the cement grinding. A relatively slow-hardening cement, through plastic deformation at early ages, will minimise the effect as shrinkage increases (see Section 31.2.1). With the fast-hardening cement, early surface cracking is induced by a high rate of heat of hydration, by differential thermal volume and moisture changes, and by an early reduction of workability and ambient humidity (see Sections 3.1.2(e)(iii) and 15.2.2(b)).

The temperature rise in mass concrete (assuming heat is not lost) can be calculated from the specific heat of the concrete and its cement content per unit weight, once the heat of hydration of the cement is known (Table 3.4). For this purpose, the specific heat of concrete may be taken as 1·05 kJ/kg·°C for 1:6 or richer mixes and 0·96 for leaner mixes, such as 1:9 by weight. Typical calculated values of adiabatic temperature rise (when no heat is lost) for 1:9 portland cement concrete are illustrated by the curves in Figure 3.1. In practice, the contours and dates of maximum temperature rise and gradient depend not only on the proportion and heat of hydration of the cement, but also on losses of heat at increasing

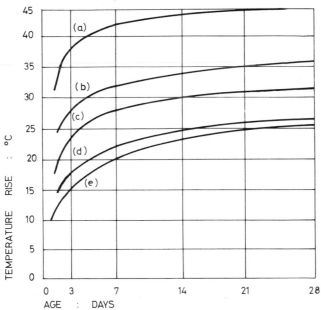

Fig. 3.1 Temperature rise. Mix, portland cement concrete 1:9 by weight. Conditions: adiabatic. (a) High-early-strength cement. (b) Ordinary cement. (c) Blastfurnace cement. (d) Sulphate-resisting cement. (e) Low-heat cement.

ages. The latter are related to the rate of placing the concrete, its bulk and thermal conductivity, the formwork used and external conditions (see Sections 24.4, 30.1, 32.1.2, 32.2.2 and References 11–15, 19, 61, 113 of Part 6).[15, 80, 129, 144]

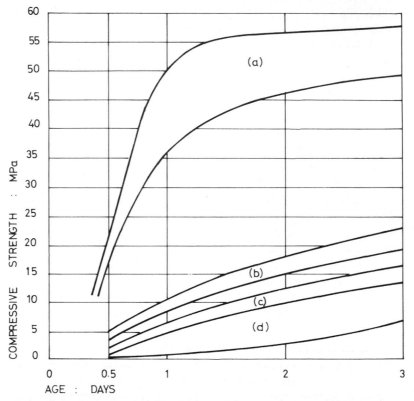

Fig. 3.2 Compressive strength range at early ages. Mix, 1:2:4 concrete, water/cement ratio 0·6 by weight. Temperature: 18°C. Specimens: 150 mm diameter × 300 mm cylinders. (a) High-alumina cement. (b) High-early-strength portland cement with calcium chloride. (c) High-early-strength portland cement. (d) Ordinary portland cement.

3.1.2 INDIVIDUAL AND BLENDED CEMENTS

Individual and blended cements are alphabetically summarised below, the information given being supplementary to that in Sections 3.1.1, 17.1 and 29.1. The properties of these cements are likely to be affected by handling and storage facilities, the type and condition of containers, and the age factor at the time of use (Sections 3.1.3(a) and (b)). It is essential that these considerations be taken into account with imported cements (see Reference 60 of Part 2).[89]

The generic term *hydraulic cement* is defined in the Glossary, Appendix VI.

(a) Expansive and stressing cements.
Expansive cement is a mixture of portland cement, an expanding agent of calcium sulphoaluminate cement (calcium oxide, aluminate and sulphate) or ettringite, and a stabilising element in the form of slag cement. Calcium sulphate is the principal reagent causing expansion, and the stabiliser has the effect of absorbing an excess of it after the desired expansion has taken place.[17-20]

The more finely ground the expanding agent, the sooner the expansion starts. Its amount and duration are controlled by adjusting the proportions of the three components and the period of effective hydration of the cement. For water-cured, neat-cement specimens, the expansion is normally between 0·2 and 2·0 per cent at ages of 1–4 weeks. With cement of suitable composition, it is possible to make a mix apparently nonshrink.[92-94]

A neat mix of 1 part portland cement to 4 parts supersulphate cement has an expansion of 0·15 per cent when cured in water for 7 days. Expansion is increased by an admixture of an optimum amount of high-alumina cement and it is decreased by introducing aggregate into the mix. Expansion by the hydration of magnesia takes place when portland cement is mixed with 5–7 per cent of calcined dolomite.

The early compressive strength of expansive cement is less than that of ordinary portland cement, but at 28 days the former may be as strong as, or stronger than the latter. Expansion is greatest when the water/cement ratio is low and curing is done under water. In thick sections, blind holes are temporarily formed so as to promote hydration of the interior of the mass. Expansion ceases between 1 and 2 days after the removal of curing water. Warm moist curing increases early strength but reduces the amount of expansion.

Expansive cement has a high degree of impermeability and affords a good measure of protection to reinforcement. It is very sensitive to humidity and sulphate attack, but its freeze–thaw durability is improved with air-entrainment. Highly expansive cement, which contains an excess of sulphur trioxide, has reduced strength and low resistance to seawater or sulphate solutions. Some of the expansive effect is lost by subsequent drying shrinkage and creep under load. The moisture movement of dried-out expansive-cement mixes is somewhat greater than that of ordinary cement mixes.

With restrained expansion, expansive cement is suitable for underpinning and structural repair work. It is also used for joint-filling purposes, connecting precast units and making impervious plaster. Only partial prestressing is possible with expansive cement.

Stressing cement consists approximately of 75 per cent portland cement, 15 per cent high-alumina cement, 10 per cent gypsum, and a setting retarder.[21] Whereas with expansive cement an increase in volume takes place mostly during the hardening process, with stressing cement this occurs with hydrothermal treatment of the hardened product. The linear expansion is 3–4 per cent, and the concrete attains high strength and a high degree of impermeability at pressures of up to 5 MPa. It can be used for making thin shell-roof units and high-pressure pipes of prestressed concrete, the prestress being achieved by the expanding action of the concrete. The expansion of portland cement mortar by means of admixtures is described in Chapter 25.[104]

(b) High-alumina cement.

High-alumina cement is a pulverised fused mixture of limestone and bauxite. It is essentially anhydrous monocalcium aluminate which, on setting, becomes dicalcium aluminate with a release of aluminium oxide. Concrete that is made with high-alumina cement and suitable aggregate develops great strength very quickly. Much heat is generated during setting and hardening, the total heat of hydration at 1 day being as much as that of ordinary portland cement at 90 days (see Table 3.4).

Pores can thus be formed in concrete at an early age by the rapid evaporation of water. Even at very low temperatures, the concrete gains strength rapidly with a high rate of heat evolution, provided that there is adequate bulk and protection against frost action for the first 4–6 hours. Where finishing operations are likely to be delayed (e.g. in industrial concrete floors) the concrete should be placed soon after it is mixed (see Section 20.1.5(g)).

In mortar and concrete, a compressive strength of 40 MPa or more can be attained in a day or two (see Table 3.6 and Fig. 3.2). The shear strength is about one-quarter of the compressive strength, and Young's Modulus of Elasticity (at stresses of 3·4–20 MPa) varies from 35×10^3 MPa at 1 day to $36 \cdot 5 \times 10^3$ MPa at 28 days. Increased water/cement, aggregate/cement and fine-aggregate/coarse-aggregate ratios are permissible, but the aggregate should be free from stonedust. Although the cement is stored in airtight drums, fresh cement will give better results than cement which has become stale with age.

High-alumina cement concrete, if either mixed or subsequently kept at temperatures above 29°C, may lose considerable strength and durability through undesirable changes in crystal structure. The loss arises from the conversion of hydrated monocalcium or dicalcium aluminates into the tricalcium compound with a release of hydrated alumina. This effect is minimised by using ice-water for mixing, keeping the temperature of the constituents at batching below 24°C, densely precasting low-slump mixes, and placing either when shade temperatures are not above 29°C or at night in some climates.[69, 142, 143]

The heat of hydration must be dissipated by limiting the depth of concrete below coolant surfaces to not more than 200, 300 or 450 mm for mixes containing 390, 330 or 270 kg of cement/m^3 of concrete, respectively. Curing is necessary as soon as the concrete is hard enough to withstand defacement, starting lightly not more than about 6 hours after casting and increasing to continuous and copious spraying or flooding with flowing cold water for 24 hours or more. Temporary cooling ducts may be required to prevent overheating of the interior of the mass. Deferred or ineffective curing is detrimental; it is essential to strip side forms and apply cold water as early as possible (see Section 9.2). The concrete can then be allowed to dry and be put into service. High-alumina cement concrete piles should be driven within a week, preferably a few days after being cast.[22]

Note: Recommendations on the use of concrete made with high-alumina cement were removed from BS CP110 and CP114 to CP116 in 1974.

Mixtures of high-alumina and portland cements have quick-setting and reduced-strength characteristics.[23] With 15–60 per cent of portland cement in the mixture, a flash or false set occurs. Such mixtures have minor uses, such as plugging leaks in honeycombed concrete and quick-setting repair work. Normally, as these cements are very sensitive to each other, special care must be taken to use clean tools and plant and to avoid any trace of contamination.

Admixtures of calcium chloride, lime, or integral waterproofing agent should not be used with high-alumina cement, nor should aggregate which liberates lime or alkalies (such as blastfurnace slag) be used. Air-entraining, cement-dispersing agents, wetting agents and colour-lightening titanium dioxide (5 per cent by weight) may be used. Under ordinary temperature conditions, high-alumina cement mixes can be bonded to hardened portland cement concrete that is free of calcium chloride and at least 7 days old.

"La Arania" Overpass, Caracas,
Venezuela

Cement Manufacture
PLATE 2

Kiln and heat exchanger
(dry process)

Clinker-storage rill towers
with slotted sides
and floor-gate discharge
to the grinding mill

Kiln control console
with mimic indicator diagram

Grinding mill

Bag-filling bay

The cement is highly resistant to sulphate attack and to some dilute acids in nontropical climates, but not to attack by either caustic alkalies or sulphuric acid which forms in main sewers. As a warm, moist environment causes progressive deterioration in high-alumina cement concrete, it must not be used for structural or pavement purposes under warm, moist conditions. Nevertheless, a mixture of high-alumina cement and crushed firebrick aggregate withstands temperatures of up to 1300°C. It is therefore widely used in refractory concrete for furnace and flue linings, where heat and corrosion resistance are more important than strength (see Section 29.1).

Sulphide occurs as a minor constituent of high-alumina cement.[12] If this cement is used in prestressed concrete and significant pores develop therein, carbon dioxide from the atmosphere is able to penetrate and release small quantities of hydrogen sulphide. In the presence of moisture near tendons, this initiates embrittlement of the steel (see Reference 133 of Part 4). Aggregates containing feldspar and mica have an accelerating effect on the rate of conversion of the cement. Structural mix design, placement and usage of the concrete require special professional knowledge, experience, judgement and technique (see Reference 24 of Part 2).

If high-alumina cement concrete is used in building construction, it can be tested on-site with an instrument devised by John Laing (Britain). The procedure employed consists of irradiating the concrete with energy from a plutonium-238 isotope source and measuring the emitted radiation with a portable X-ray fluorescence isotope analyser. The high iron content of the cement is a fundamental factor in the metering operation.*

(c) Masonry cement.

Masonry cement is a selected fine blending of such materials as portland cement clinker, natural or slag cement, pozzolan, lime, marl, gypsum and air-entraining agent. In manufacture, the cement clinker is finely ground and the softer or porous components are coarsely ground before they are blended equally together. The specific surface area of this cement is about twice that of ordinary portland cement. It produces a highly workable or "fatty" mortar with a strength that is suitable for masonry work and brickwork. Such mortars, containing washed concrete sand, are suitable also for external rendering and internal plastering, and their capillary

---

* High-alumina cement is available under such trade names as "Ciment Fondu", "Lightning", "Lumnite" and "Rolandshutte".

moisture characteristics facilitate integral curing (see Section 16.2).[24, 25]

Masonry–cement sand mixes vary from 1:3 to 1:5 by volume according to the conditions of loading and exposure in service. The average compressive strength of the 1:3 mix is from 7–10 MPa at 28 days and shrinkage corresponds approximately to that of a relatively strong cement–lime mortar (see Sections 27.2.2(b), (k) and 33.4.3). If a setting accelerator is used in cold weather, the quantity required is about one-half of that which would be used with ordinary portland cement (see Sections 10.1 and 16.2). White or tinted masonry cement mortar has aesthetic uses in building construction (see Sections 27.2.2, 33.4.3 and 36.4). Physical requirements and tests (as specified in AS 1316) relate to soundness, setting time, water retentivity, air content and compressive strength.

(d) Natural cement.
This cement is made by finely pulverising either calcined, argilla-ceous limestone or natural cement rock of suitable composition. The temperature of calcination is somewhat lower than that used for the production of portland cement clinker, being no higher than that necessary to drive off carbon dioxide. A small amount of gypsum is incorporated in the end-product, possibly with an air-entraining agent and up to 2 per cent of calcium chloride by weight. A saving in fuel is effected when the rock is impregnated with crude oil. Material of this kind is used in Sweden for the manufacture of autoclaved, foamed concrete (ASTM C10).

(e) Portland cement.
(i) Air-entraining portland cement is made by adding 0·01–0·03 per cent by weight of a permissible air-entraining agent to the clinker at the time of grinding with gypsum. With this cement, the content of air that is entrained in concrete is less readily controlled under varying conditions than when an air-entraining agent is added to concrete at the mixer. The mixing time must be kept constant, and the mix adjusted for the requisite slump and cement factor (see Chapter 16).

(ii) Blastfurnace portland cement is an intimate mixture of 1 part portland cement and between 0·5 and 2 parts of granulated basic blastfurnace slag. The slag, which has hydraulic properties, is produced by quickly quenching slag with water while in the molten state. Although the slag is harder to grind than cement clinker,

each must be be brought by milling to the normal fineness of cement. A small amount of gypsum is added to regulate the setting, which is most active in fresh products (see Section 3.1.2(f)).

A suitable cement of this kind, containing not more than about 50 per cent of slag, can be used in much the same way as ordinary portland cement. The former develops strength somewhat more slowly than the latter at early ages, but their ultimate strengths are similar. An increased rate of strength development and heat evolution is obtained with increased fineness of grinding. The curing time should be extended, particularly in cold weather (see Sections 15.2.3(c), 32.1 and 32.2).

Portland blastfurnace cement has been widely used for marine construction. Its sulphate resistance is between that of ordinary portland and sulphate-resisting portland cements, while its resistance to heat, aggressive ground water, dilute acid and damp acidic gas is higher than that of ordinary cement. Owing to its low heat of hydration, this cement is suitable for use in large masses of concrete. Its high permissible content of sulphuric anhydride (BS 146) militates against its use in prestressed concrete. Blast-furnace cement reduces possible efflorescence (see Section 36.6) and expansive alkali–aggregate reaction (see Section 15.2.3 (a)).[115, 140]

(iii) High-early-strength portland cement has more tricalcium silicate and less dicalcium silicate than ordinary portland cement and it is ground more finely (see Tables 3.2 and 3.3). This cement reaches a normal 28-day strength in approximately half the time required by ordinary portland cement, the initial setting and mix characteristics being the same. Its medium to high drying shrinkage and tendency to cause cracking are associated with the water requirement of a high specific surface area, a rapid heat evolution and a low plastic deformation under restraint at an early age.[6] The cement is suitable for speedy construction work and projects where early strength is required. Early and adequate moist curing are essential for the first 3–5 days, a longer period being beneficial.

Extra-high-early-strength portland cement is made by grinding high-early-strength portland cement with 2 per cent by weight of calcium chloride. Concrete that is made with this cement should normally be placed, compacted, and finished within 15–20 minutes of mixing. The admixture increases early strengths by 50–100 per cent during the first week in cold weather, but has no effect upon the ultimate strength of the concrete. This cement should be stored in a dry place and used within 1 month.

An increase in the rate of hardening and strength development of concrete is obtained by increasing the fineness of grinding of portland cement. If its specific surface area is doubled, the 3-day strength of concrete is doubled and the 28-day strength is increased by 50 per cent. As superfine cement is liable to air set and its properties will not last for more than seven days, ordinary cement must be stored in a silo or bin and ground in a vibratory mill at the site before being fed by a screw conveyor to the mixer. These cements have increased shrinkage and they have low autogenous healing properties when superfine grades are used in concrete (see Section 31.3.1).[21, 113] Calcium chloride, if used, must be considered in relation to information in Sections 10.1, 15.2.2(a), 22.3.1, 34.1.2 and 34.2.2.

(iv) Hydrophobic portland cement is made in the same way as ordinary portland cement, except that during final grinding of the clinker a small amount of film-forming, water-repellent material is sprayed into the mill. Grinding agents consist of 0·15 per cent of polyethylsiloxane or 0·35–1·0 per cent of pentachlorophenol, chlorinated cresylic acid, oleic or stearic acid. The additive forms a protective coating around each particle of cement, which inhibits hydration until the coating is broken by mechanical abrasion in the presence of water during mixing.

The cement does not deteriorate when stored under humid conditions. Batches of hydrophobic and ordinary portland cements have been subjected to air-setting conditions for periods up to 28 days. The treated material developed its full compressive strength when used in a 1:3 mortar mix, while the strength of similar mixes which were made with untreated cement fell to approximately one-fifth of the normal value.

The grinding agents facilitate grinding and greatly increase the workability and cohesiveness of concrete mixes. A reduced water/cement ratio can therefore be used, with its attendant advantages, provided the concrete is fully compacted and cured. The cement gives an excellent surface finish and a slightly reduced early strength. Setting and hardening characteristics are the same as for ordinary portland cement. Hydrophobic cement can be used to advantage in soil stabilisation and pumped concrete.

(v) Low-heat portland cement has less tricalcium silicate and aluminate and more dicalcium silicate than ordinary portland

cement, so as both to reduce and delay the generation of heat of hydration. It hardens rather slowly and the hardening of the concrete made with this cement is particularly delayed by cold weather. At ordinary temperatures the concrete has approximately two-thirds the strength of a similar mix with ordinary portland cement at 7 days, three-quarters at 28 days and an equal strength at 3 months. Low-heat portland cement is a little less vulnerable to chemical attack than ordinary and high-early-strength portland cements.

· Low-heat portland cement is used in construction that requires large masses of concrete, where ordinary cement would develop much heat, expand the concrete and cause cracking. Low-heat portland cement creeps more under tension than does ordinary portland cement. This characteristic, coupled with an extended time for plastic yield of the concrete, minimises the effect of thermal movements and stresses that arise through restraint, such as in large footings and dams (see Section 30.1).[129]

For concreting in very cold weather, use may be made of a modified cement consisting of 60 per cent low-heat portland cement and 40 per cent ordinary portland cement. Large masses of concrete have a minimum of cracks, however, if they contain 100 per cent low-heat cement. Manufacturers do not hold stocks of low-heat cement, and special arrangements must be made for a supply when it is to be used.

(vi) Ordinary portland cement has been dealt with in some detail in Section 3.1.1. Variations in its composition may produce variations of up to ±20 per cent in the strength of concrete that is made with it, but uniform results are obtainable by drawing cement from one source of supply and, where practicable, from a reserved silo or bin (see Section 3.1.3(b)).

Bagged cement loses potential strength under ordinary conditions of covered storage. The rate of loss for the 28-day strength is 4 per cent for each month of storage after the first 3 months, the average relative humidity being 70 per cent. The loss after 2 years is about 60 per cent. Aged, unlumpy cement has less potential shrinkage than new cement. Loss of strength is minimised or prevented by using hydrophobic portland cement, bulk silos for storage, vapourproof containers, freshly made cement, and proper facilities for all periods of storage. The last requirement is dealt with in Section 3.1.3, which includes the storage of cement. Provided that the cement complies with its appropriate standard specification at the time of use, the 7-day compressive strength of ordinary portland cement concrete is

0·6–0·8 of its 28-day strength under temperate conditions. Early-age strengths are shown in Figures 3.2, 4.3 and 9.2.[10, 91, 108]

(vii) Pozzolan portland cement is a mixture of ordinary portland cement and a finely ground pozzolan, the proportion of the latter being usually from 10 to 30 per cent by weight. Pozzolans are discussed in Chapter 17. This cement has a slow rate of hardening and a heat of hydration which is similar to that of low-heat portland cement. The pozzolanic action involved requires a prolonged period of moist curing for the attainment of a normal ultimate strength. Economy is effected where a suitable natural or artificial pozzolan is readily available (see Sections 17.1 and 17.2).

Portland-pozzolan cement is used in large-mass concrete structures, such as dams, where low heat evolution is more important than a normal rate of strength development. Well-made concrete containing this cement has an increased measure of resistance to mildly severe conditions of chemical attack. Pozzolans with a high content of opal, such as diatomite and opaline chert or shale, effectively inhibit any alkali–aggregate reaction in concrete. A high-magnesia cement can be volume stabilised with a siliceous pozzolanic admixture, such as fly-ash.[10]

(viii) Sulphate-resisting portland cement has a low content of tricalcium aluminate (see Table 3.2), which is the compound in portland cement that is least able to resist chemical attack.

The most promising cement of this type, as recommended by the United States Bureau of Reclamation, has less than 50 per cent tricalcium silicate (C_3S) and less than 12 per cent tricalcium aluminate (C_3A) + tetracalcium aluminoferrite (C_4AF), in which less than 4 per cent is C_3A. It has, after curing, a low content of hydrated lime. A requirement of ASTM Standard C150 for Type V cement is that the content of C_4AF plus twice the amount of C_3A shall not exceed 20 per cent. The value of C_3A given for the recommended cement is based on chemical analysis rather than on X-ray diffraction, which gives a lower result. Portland cements approaching this composition, particularly with respect to C_3A, give excellent service in cement-rich, densely compacted, properly cured concrete.[26, 111, 147]

This cement is designed to resist attack by mineral sulphates in ground waters and subsoils, which contain sulphur trioxide in amounts up to 0·1 per cent and 0·5 per cent, respectively. The characteristics of the sulphate-resisting cement (as shown in Table 3.5) indicate its suitability for use in aggressive environments, such as heavily polluted or humid industrial atmospheres, seawater, factories and sewers. A very low content of free alkali in the cement minimises possible efflorescence and alkali–aggregate reaction. Where the composition and fineness of the cement are designed for a low cracking propensity, with a corresponding increase of dicalcium silicate (C_2S) to about 40 per cent, the strength of concrete should be considered at an age of 2 months (see Reference 145 of Part 4).

If cement containing 5 per cent of C_3A were mixed with ground gypsum (sulphur trioxide content brought to 7 per cent), then cast into bars (ASTM C452) and kept moist for 1 year, the resultant expansion would be only about one-quarter of that of ordinary portland cement (C_3A, 11 per cent). For moderate sulphate resistance (e.g. C_3A, 8 per cent) this value would be about one-third. By comparison, BS 4027 sets an upper limit of 3·5 per cent for C_3A in sulphate-resisting portland cement.

Owing to the low content of C_3A, an admixture of calcium chloride should not be used in concrete that is made with sulphate-resisting portland cement. This precaution is necessary in order to minimise the amount of uncombined chloride that could arise and lower the pH value and durability of the concrete and its inhibition of corrosion of steel reinforcement. Reference should be made to Chapters 15 (Section 15.2.3), 19 and 22 (Section 22.3) for factors affecting durability. See also Plate 14.

(ix) White or coloured portland cement. White portland cement is, in practice, ordinary portland cement with an iron oxide content below 1 per cent. It is used in the same way as ordinary portland cement in concrete projects. Calcination is done with oil fuel at 1700°C, the burning temperature being high because of the absence of iron and so of its fluxing effect during sintering.

Coloured portland cements are made by grinding white cement clinker with 5–8 per cent of chemically inert pigment (AS K54, BS 1014) and a dispersing agent. These cements have nonstaining and uniform tinctorial properties, and they are obtainable in dry-batched form with specially selected aggregate. They are used in cement paint, precast stone, terrazzoconcrete floors, internal and external renderings, and for setting and pointing work. To compensate for a slightly reduced cementing power, the content of coloured cement in mixes is increased by 10 per cent.[112]

(f) Slag cement.
Slag cement is a finely divided material with hydraulic properties.

It consists essentially of water-quenched, granulated blastfurnace slag and hydrated lime, the proportion of the former being at least 60 per cent by weight. The slag is a carefully selected, nonmetallic product, consisting of calcium silicates and aluminosilicates. It is developed simultaneously with iron in a blast furnace and subsequently quenched rapidly with water. Small amounts of gypsum and cement-dispersing agent may be added, but the content of calcium sulphate should be limited where there is a danger of corrosion of steel reinforcement.

This cement is used as a blend with portland cement in making concrete and with hydrated lime in making masonry mortar. The two components are blended in a mill or in a mixer on the site. For the latter purpose, the specific surface area (air-permeability method) should be at least 420 m²/kg (ASTM C595). Adequate mixing, moist curing and control testing are required for best results.[27]

(g) Supersulphate cement.

Supersulphate cement is a very finely divided mixture of granulated basic blastfurnace slag (80–87 per cent), calcium sulphate (10–15 per cent) and portland cement or cement clinker (up to 5 per cent). It normally complies with the requirements of standard specifications for ordinary portland cement in connection with setting time, soundness and compressive strength. Differences between it and ordinary portland cement include a specific surface area of 350–500 m²/kg, a low heat of hydration, immunity from false set, an increased proportion of mixing water taken up in hydration, and a bulk density that may be only 1200 kg/m³ as compared with 1500 kg/m³ for portland cement. The shrinkage and permeability of the two cements are about the same.

The cement is so named because the equivalent content of sulphuric anhydride exceeds that for portland blastfurnace cement.[28] As supersulphate cement is dependent upon the small amount of portland cement that it contains for early setting and hardening, dry transport and storage conditions are essential for early strength development of the freshly mixed concrete. There must be no additions to it of another cement, hydrated lime, setting accelerator or integral chemical waterproofer.

Concreting practice generally is the same as with portland cement: the water/cement ratio law, full compaction and careful curing govern strength. Supersulphate cement gives a more workable mix than portland cement for the same water/cement ratio. The workability and strength of the mix, however, are adversely affected by the use of too much sand. An increased cement content should be used and lean mixes or dirty aggregate avoided, as the cement is mainly used where chemical resistance is required.

Initially, the strength of supersulphate cement concrete is similar to that of portland blastfurnace cement concrete, but after 3 days it becomes progressively higher. Table 3.7 shows a typical range of compressive strengths of supersulphate cement concrete at normal ambient temperature, using different brands of cement. The mix is 1:1·5:3 of cement, sand and clean gravel, with a water/cement ratio of 0·5 by weight. The specimens are water-stored cylinders of 150 mm diameter and 300 mm height, or are 100 mm cubes. The tabulated cylinder strengths are converted from cube strengths derived at the Building Research Station, Britain.[70] The variation in early strength is related to the composition, quality of slag and fineness of grinding of the cement (BS 4248).

TABLE 3.7   STRENGTH DEVELOPMENT

Age (days)	Compressive Strength (MPa)	
	Cylinder	Cube
3	8·3–15	12–21
7	15–28	21–34
28	29–34	36–41
90	45–48	52–55

In cold weather the development of strength is slower than with portland cement but the ultimate strength is unaffected. Strength increases with curing temperatures up to 50°C, but above this temperature it decreases. The surfaces of freshly placed concrete should be covered with plastic sheet and kept moist for at least 3 days, otherwise atmospheric carbonation and lack of hydration cause a soft skin to form. Early and adequate moist curing (e.g. by wet burlap or underfelt (see Sections 9.1 and 9.2)) is essential to stop evaporation and to ensure effective hardening.

Supersulphate cement concrete can be bonded to mature portland cement concrete without causing interaction. During the early life of reinforced concrete that is made with this cement there is slight corrosion of mild-steel reinforcement. This effect is not progressive and it is of no significance if the concrete is well made and there is at least 40 mm cover to the steel. Reduced cover is permissible if the steel is primer-coated against rust or is embedded in very rich and densely compacted mixes. If this cement is used for prestressed concrete, steam curing should not be used. The low heat of hydration is advantageous in mass concrete construction where thermal stresses must be kept low.

Supersulphate cement concrete, when well made and impermeable, resists attack by the highest concentrations of sulphates that are normally found in ground waters and soils. In highly concentrated solutions, magnesium and ammonium sulphates cause attack more rapidly than other sulphates. The resistance to seawater is very high and this concrete stands up to attack by weak organic and inorganic acids with pH values above 3·5. This concrete is particularly suitable for sewers in contact with peaty acids or which carry certain trade effluents. Some measure of protection is provided also against the action of sulphur-oxidising bacteria.

This concrete is suitable for floors and tanks which must resist attack by dilute acids, vegetable oils, sugar solutions and phenols. For corrosion and abrasion resistance, a 1:1:2 mix of cement, sand and stone is used in the surface course of composite concrete floors. The wearing properties of supersulphate cement concrete floors in general are improved by trowelling-in neat cement before the concrete has set, a practice which is not recommended for portland cement work. A 1:2 or 1:2·5 cement and sand mortar mix (without lime) may be used for jointing chemically resistant bricks or floor tiles and as a surface rendering. The rendering may be applied by a cement gun and subsequently moist cured. Further information on special cements is contained in Sections 3.1.1, 15.2.3, 17.1 and 19.1.2. The advice of the makers should be obtained for special applications.

3.1.3 HANDLING, STORAGE AND BATCHING

(a) Handling.
Cement is generally transported, off-loaded, stored and withdrawn from storage before it can be used. Methods of handling are different for building contracts, civil engineering projects and the manufacture of concrete products. Road vehicles for transporting cement in bulk include tipping lorries, air-assisted discharge lorries, lorries fitted with spiral-type discharge conveyors and vehicles fitted with bottom-emptying hoppers or pressure containers.

Mechanical and pneumatic plant is designed for moving and elevating cement for gravity feed in a dustless way. Plant includes lift trucks and cranes; belt, spiral, chain-type and air-activated conveyors; enclosed bucket and chain-type elevators; and high-pressure or low-pressure pneumatic pumps. Various containers and handling systems enable bulk cement to be transferred from one form of transport to another.

Overhead silos, hoppers or bins are used for the storage or dispensing of each type of cementitious material. The withdrawal of cement is facilitated by the use of pneumatic, vibratory and gyratory activators. Air pads, for loosening cement which has settled tightly, are actuated by a large volume of dry, low-pressure air (20–35 kPa), and feeder devices to automatic weigh-batchers are provided with a precise cutoff. These items of equipment are illustrated in some detail in the trade literature and in References 29, 60 and 98 (see also Reference 134 of Part 3).

(b) Storage.

At all stages up to the time of use, cement must be kept dry so as to prevent or minimise deterioration from the effects of moisture, atmospheric humidity and carbonation (see Section 3.1.2(e)(vi)). Airtight drums and internally coated bulk silos (see Sections 20.2 and 21.2) are ideal storages. Cement in multiwall paper bags should be stored in a weatherproof building with close-fitting doors, which should be kept closed as much as possible. The floor should be placed well above the ground and incorporate a vapourproof membrane so that it is dry. If this is not possible, the bagged cement should be stacked on pallets or on a dry board platform, some 150 mm clear of the surface. The roof should be pitched at a slope of 1 in 3, so that condensation drips will not fall. In regions of high relative humidity, insulation and a storage temperature of 45°C are advisable.

Bags of cement should be stacked close together, so as to restrict the circulation of air around them. The stacks should be kept 150-300 mm clear of the walls (with access ways 900 mm wide) and be arranged so that stored cement can be removed in the order in which it is received: that is, first in to be first out. Stacks that are more than eight bags high should be placed in header and stretcher formation, so as to increase their stability when laid up to the maximum height (for economic handling) of 4·2 m. Tall stacks are stepped back for stability when cement is being withdrawn. The bottom bags of stacks eight or more bags high show "warehouse pack", but can be loosened easily by rolling on removal.

Cement which is 4 months old should be classified as "aged" and be retested before use. The capacity of bulk-storage containers should exceed the requirements of cement for 1 week and they should be cleaned out at least once each 4 months. On small projects where storage without a shed is required for a few days, the cement should be placed on a raised platform and covered over with tarpaulins, polyethylene film (0·2 mm thick) or waterproof building paper. The cover must be tucked in around the edges and weighted down overnight or when cement is not being taken from the stack. For full strength development, the cement should be used within 3 months (see Section 3.1.2(e)(vi)).

(c) Batching.

For manual batching, cement should be measured by the bag unless fractional bags are weighed. (A 40 kg bag is assumed to contain 26·5 litres of cement.) At central mixing plants, cement is measured by weight to an accuracy of 1 per cent. If cement is weighed on the scales used to weigh aggregate, it is weighed first and placed in a separate compartment. On very large projects, bulk cement is weighed for each batch with separate, automatic, weigh-batching equipment. Cement batchers are made of smooth metal with rounded corners and are equipped with a vibrator as an aid to their complete discharge before they can be reloaded. There should be means of access for testing, inspection and sampling.

For placing bulk cement into a batch hopper for a stationary mixer, a pipe of suitable size to hold the cement should extend from the batcher discharge to a level near the bottom of the hopper. After covering the end of this pipe with aggregate, the cement may be discharged. It then enters the mixer without loss of dust and is well distributed through the entering aggregate. In portable dry-batch compartments, a separate container is provided for the cement. Loss during loading is minimised by the use of a kinked, canvas, drop chute or a telescopic, flexible hose.[60]

3.1.4 ACCELERATED STRENGTH TEST

Ordinary portland cement must be stored initially until concrete specimens or mortar cubes are tested in compression at an age of 7 or 28 days. This waiting period for acceptance can be reduced to within 1·5 days by using accelerated strength tests (e.g. ASTM C684, and those described here).[30–34, 92] They are not a substitute for standard strength tests (see Section 14.4), but are useful for making early adjustments to the proportions of concrete mixes in quality control procedures. Typical testing systems are as follows:

Test 1. The concrete is placed in cube moulds (in the standard way) and cover plates are placed over them. The moulded specimens are placed in a thermostatically controlled curing tank, where they remain at $23 \pm 2°C$ for 24 hours. While still in the moulds, specimens are then plunged into boiling water and boiled for 3·5 hours. On removal from the boiling water, the specimens are cooled, stripped, weighed and tested in compression after a further time period of 1 hour. For ordinary portland cement, the accelerated-cured strength bears the same relationship to the 7-day strength as the 7-day strength does to the 28-day strength. The future strength can be estimated (as an average based on three specimens) from Figure 3.3 or from actual test results.

Test 2. The concrete is placed in cube moulds (in the standard way without overfilling) and enclosed, so that the base and cover plates are in full contact with the greased bearing faces of the moulds. Two accelerated-test procedures may be used. In a 7-hour test, the moulded specimens are placed in a cold, fanless oven with all vents closed. When the age of the concrete is 0·5 hour, the temperature of the oven is raised to $88 \pm 3°C$ in 1 hour and 94°C in less than 2 hours. It is controlled at this temperature until the total heating time is 6 hours. The specimens are then removed, stripped and tested after 0·5 hour. The average compressive strength of three

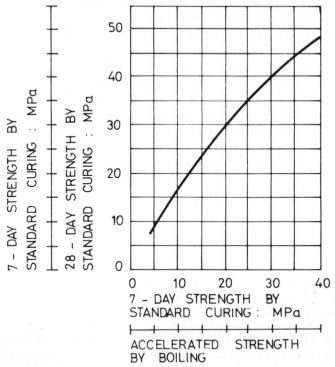

Fig. 3.3 Compressive strength relationship

specimens is used to predict the 7-day and 28-day strengths by means of Figure 3.4.

In a 23·5-hour test, the moulded specimens are stored in moist air at 23 ± 2°C for 19·5 hours after mixing. They are heated for the first 3·5 hours of the 6-hour regime described and tested after 0·5 hour. As with the first procedure, the 7-day and 28-day compressive strengths are predictable to within about ± 1·75 MPa. Where materials from constant sources are used on a particular site, the strength curves may be drawn to fit the local findings as testing proceeds; errors in prediction can thereby be approximately halved. In both procedures, the results are affected to a limited extent by the number and size of cubes cured in the oven.

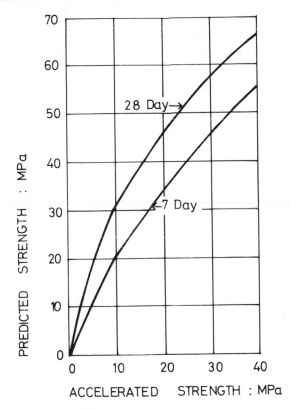

Fig. 3.4 Compressive strength relationship

Test 3. Concrete cylinders of 150 mm diameter and 300 mm height are made in the standard way (AS 1012 Part 8). After 0·5 hour, the specimens (in closed moulds) are placed in hot water, which is thermostatically controlled so that its temperature remains at a constant value of 74–77°C. At an age of 22 hours, the specimens are cooled to room temperature. They are stripped, weighed and tested in the standard manner at an age of 24 hours. The average accelerated strength of three specimens is equivalent to the 7-day strength of standard-cured specimens. A coefficient is therefore used to obtain the strength at 28 days. On 95 per cent of occasions, the deviation from the mean value of the ratio of the strength of quickly cured specimens to the strength at 28 days does not exceed 12 per cent for a given type of cement from one source (ASTM C684).

Test 4. Mortar cubes are made as specified for the testing of ordinary portland cement. During the first 24 hours after casting, the specimens are stored in air at a temperature of 23 ± 2°C and a relative humidity of 95 per cent. They are then put into an autoclave and covered with water. The bottom heating elements are switched on and the steam outlet is kept open for about 1 hour. The steam outlet is then closed and the side heating elements are switched on until, in about 1 hour, the gauge pressure is 1200 ± 50 kPa (192 ±2°C).

This condition is maintained for 3 hours by automatic controls, then the heat is switched off, the steam outlet is opened and the system is allowed to cool to 95°C in about 3 hours. The specimens are then removed from the autoclave, placed in hot water and then allowed to cool to room temperature for about 1 hour. They are then ready for crushing. The average compressive strength of three specimens obtained in this way is approximately the same as that attained by water-cured specimens of similar composition at an age of 28 days.

Test 5. For the 3-day RMIT† Technisearch test, demoulded, standard (age 24 hours) concrete cylinders are placed in strong plastic bags that are capable of retaining the moisture in a specimen while being hydrothermally cured for 48 hours in an oven at 74–77°C. From the accelerated-test strength (f_c MPa), the predicted 28-day compressive strength is estimated to be $1·23 f_c + 4·55$ MPa.

Test 6. For a 2-day autogenous curing test, sealed moulded cylinders are used in a calibrated insulated container. This test is specified with working details in Standard CSA A23.2.26, issued by the Canadian Standards Association. Its efficacy depends on the heat generated during setting and hardening of the cement. Predicted and actual 28-day test results are influenced by the proposed usage of materials, mixes, admixtures and curing regimes.

---

†Royal Melbourne Institute of Technology, Melbourne, Australia.

3.2 WATER

3.2.1 QUALITY

(a) Mixing water.
Water that is acceptable for drinking (except in respect of bacteriological requirements) is suitable for making concrete. Where a supply is drawn from natural sources, preliminary treatment may be required, such as settlement or filtration to remove suspended matter. The water should be free of materials that significantly affect the rate of hardening, or the strength and durability of concrete, or which promote efflorescence or the rusting of steel reinforcement.[36] Seawater (3·5 per cent of salts, mostly sodium chloride) is a promoter of the two latter defects, although qualitatively it does not reduce the compressive strength of the concrete. Experimental studies, including the use of an electrical-resistance corrosion-rate detection probe, have shown that higher salinities are likely to cause more adverse effects (see Sections 19.1.1(b) and 19.1.3).[81]

Mixing water should be free of even very small amounts of various sugars or sugar derivatives, and also of appreciable amounts of dissolved sodium or potassium salts where alkali–aggregate reaction is possible. Water with up to 1000 parts sulphur trioxide or 500 parts chlorine per million is usually acceptable for gauging purposes.[84]

Methods of examining water are specified in the standards of Appendix I. For water to be used in making concrete, BS 3148 requires that similar specimens be made in two groups, a sample of the water to be tested being used for one group and distilled water for the other. It is required that the initial setting times of the cement comply with the specified minimum and not differ in each instance by more than 30 minutes. Also, the average compressive strength of concrete test specimens at 28 days (or 24 hours, if high-alumina cement is used) shall not be less than 80 per cent of the average strength of test specimens of control concrete. The water is satisfactory if it meets these requirements and produces concrete with the desired 28-day strength. It is believed that the strength does not subsequently decrease by reason of the quality of the original mixing water.

Impure water gives approximately the same strength ratio with different proportions of mix. The strength of concrete that is mixed with impure water can be brought up to normal with additional cement. An amount of 1 per cent of additional cement is required to increase the strength of ordinary concrete approximately 1 per cent.

(b) Curing water.
Curing water should be free of materials that significantly affect the
hydration reactions of the cement, or promote possible alkali–
aggregate reaction, or produce unsightly stains or deposits on the
surface. Objectionable staining may be produced by organic
materials (e.g. tannic acid) or iron compounds (e.g. ferric chloride).
Staining tests can be carried out, on a comparative basis, by
evaporating a predetermined quantity of water from the surface of
white portland cement mortar or plaster-of-Paris specimens.[37]

3.2.2. QUANTITY OF MIXING WATER

One of the most common causes of poor-quality concrete is the use
of too much mixing water. Mixing water includes all free water in a
mix, but it excludes the moisture which is absorbed within the
surface of aggregate that is in a *saturated, surface-dry state* (see
Section 3.3.4(j)). In the manufacture of concrete, the least amount
of water should be used for purposes of manipulation and full
compaction. Excess mixing water causes segregation, settlement
and porosity, low strength and durability, poor wearing and bonding
qualities, laitance formation, increased shrinkage and cracking.

Portland cement, for adequate hydration, must combine with
about 20 per cent of its own weight of water and at least twice this
quantity of water is normally required in a mix to ensure workability. If a water/cement ratio of less than about 0·4 by weight is
used, the cement does not become fully hydrated. This characteristic
applies at a higher water/cement ratio with coarse particles of
cement.[24] It does not appear to affect the strength of the concrete,
however, as hydration of the cement particles begins at the outside
and progresses inwards. If hydration is incomplete, the cores of the
particles remain unaffected, while the cementing value of the outer
layers is fully developed. This state of affairs holds for water/cement
ratios as low as 0·2 by weight, or probably even lower.

As excess water evaporates or segregates from a mix, it forms
capillary ducts or small voids which reduce the quality of the cement
matrix in hardened concrete. The resulting voids/cement ratio is a
critical factor which governs the strength and durability of the
concrete. For purposes of convenience in practice, however, the
voids/cement ratio is replaced by the water/cement ratio, provided
that the concrete is otherwise fully compacted. From this concept

arises Duff A. Abrams' water/cement ratio law, which states that fundamentally *the strength of concrete is governed by the ratio of the volume (or weight) of water to the volume (or weight) of cement in a mix, provided that it is plastic and workable, fully compacted, and adequately cured* (Chapter 4). The water content in this expression refers to free water (i.e. the total content of water in the mix less that absorbed by the aggregate). Original and subsequent curves of the relationship, which incorporate the effect of successive changes in cement characteristics, are shown in Figure 3.5.[1, 2]

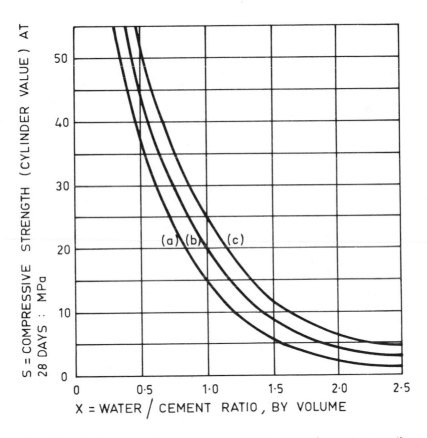

Fig. 3.5 Ordinary portland cement concrete. Water/cement strength relationship. (a) Primary curve: $S = \dfrac{100.}{7^x}$ (b) Secondary curve: $S = \dfrac{100.}{5^x}$ (c) Tertiary curve: $S = \dfrac{100.}{4^x}$

The curves in Figure 3.5 represent a family of very closely spaced curves for different portland cements, fineness indexes, water/

cement ratios, aggregates and degrees of curing. For brevity, each
is presented as a single, smooth curve for all mixes at an age of 28
days. Ordinary portland cement has changed since Duff Abrams'
law was formulated. The primary relationship, shown as (a) in
Figure 3.5, is

$$S = \frac{100}{7^x}$$

where S is the cylinder compressive strength (in MPa) of moist-
cured concrete at 28 days and x is the water/cement ratio by

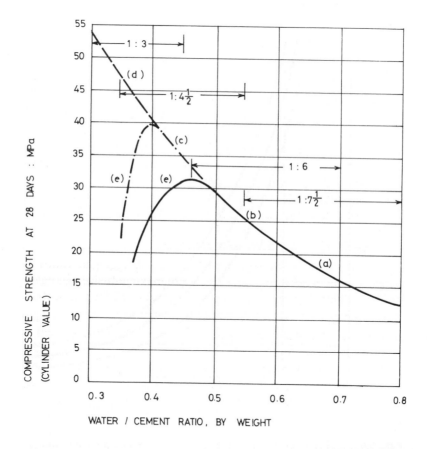

Fig. 3.6 Ordinary portland cement concrete. Proportions signify cement
and combined aggregate. (a) Sloppy mix. (b) Manual compaction.
(c) Vibratory compaction. (d) Vibropressure compaction. (e) Partial
compaction.
Note: High-early-strength, low-heat and ordinary tertiary-strength
relationships for different cement parameters are indicated in Section 3.1.2
(e) (iii) and (v), in Figures 3.5, 4.3, 4.4 and in References 5–8, 30–33 of Part 2.

apparent volume (on the basis that 40 kg of cement in a bagged state has a volume of 0·0265 m³ or 26·5 l).

The secondary and tertiary relationships, shown graphically as (b) and (c) in Figure 3.5, are indicative of successive values of

$$S = \frac{100}{5^x} \text{ and } S = \frac{100}{4^x}.$$

The potential limit is normally

$$S = \frac{100}{3 \cdot 5^x} \text{ for ordinary portland cement concrete.}$$

A conversion of the water/cement ratio by volume to a ratio by weight can be readily made on the basis that

1 m³ of cement weighs $\dfrac{40}{0 \cdot 0265}$ or 1500 kg

and, if $x = 1$ by volume, then

$$x = \frac{1 \text{ m}^3 \text{ water}}{1 \text{ m}^3 \text{ cement}} = \frac{1000}{1500} = \frac{2}{3} \text{ by weight.}$$

Therefore a volume to weight conversion of this ratio can be made by taking two-thirds of the former. A ratio by weight can be changed into a ratio by volume simply by increasing it by 50 per cent. As the water/cement ratio is reduced, the strength, durability and wear-resistance of fully compacted concrete are increased. With a low ratio, the actual and requisite workability are determined by the richness of the mix, the nature of the work and the means of compaction (Fig. 3.6). Compressive strengths of up to 380 MPa are possible with very high vibropressure.[100]

Typical ranges of water/cement and cement/aggregate ratios by weight are indicated for mixes with 19 mm maximum-size aggregate. The stiffer the mix, the greater is the energy required for compaction.

A small increase in the quantity of mixing water (e.g. of 1 per cent) reduces the strength of concrete by about the same amount as that which would occur if the quantity of cement were similarly reduced (i.e. by 1 per cent). With a small increase in water content, the consistence of the cement matrix and the workability of concrete are increased to a marked extent. Great care must be taken to see that the water/cement ratio is kept to a predetermined value within close limits. A small increase in the ratio can weaken concrete to a greater extent than that which would be caused by the most polluted mixing water that is ordinarily encountered.

A simple field test for estimating the water content of a fresh concrete mix is as follows. Fresh concrete (1–1·5 kg) is weighed in a shallow pan, mixed with 0·5 l of methylated spirits and ignited. The concrete is mixed with a further 0·5 l of spirits, ignited and reweighed. The water driven off is then determined. Desiccation by evaporation is dealt with further in Section 3.2.3(a).

3.2.3 WATER IN AGGREGATE

The moisture content of aggregate affects the amount of mixing water that is required for uniform concrete. The water in fine aggregate is an important factor in this respect and it also causes bulking of the aggregate. For purposes of adjustment in mix control, moisture determinations should be made frequently on representative samples of aggregate and related to particular batches of concrete. Changes in moisture content of fine aggregate are much greater than those in coarse aggregate, with the result that the former should be tested more frequently than the latter (BS 812, ASTM C70 and C566).

Methods of sampling, determination of specific gravity, and the preparation of *saturated, surface-dry aggregate* are given in Section 3.3.4(j), AS 1012 Part 2 and 1141. Specific gravity varies from 2·6 to 2·7 for sand and from 2·3 to 3·0 for stone (see Section 3.3.4(b)). With careful moisture-testing procedure, 98 per cent of the results may be expected to lie within ± 1·0 per cent.[38] The methods and appliances used for the determination of free moisture content, which can be measured in terms of the weight of either dry or moist aggregate, are described here.[101]

(a) Desiccation.
This method depends on evaporation or the use of calcium carbide. Desiccation by *evaporation* may be carried out by drying a 1000-g sample of aggregate to constant weight, either in a ventilated oven at 105 ± 5°C or over a stove, and the aggregate is stirred at intervals to ensure uniform drying. The difference between the original and final weights is a measure of the moisture content. A quick and simple field procedure is to mix a weighed sample of aggregate with a water-miscible, inflammable liquid in a shallow pan. About 0·15 l of methylated spirits, alcohol or acetone is required to 1 kg of sand. The liquid is ignited and the aggregate is stirred while it is being dried. The operation is repeated so as to ensure complete drying. The total percentage of moisture, which is determined on reweighing, is based usually on the weight of sample after drying (ASTM C566).

When using either of these methods, the water of absorption (which is lost during drying) should be deducted from the total moisture content in order to obtain the free moisture content. A correction value may be determined by bringing the dry aggregate to a saturated, surface-dry condition and reweighing, or the value may be taken arbitrarily as approximately 0·7 per cent for fine aggregate and 0·4 for coarse aggregate.

Desiccation by the interaction of moisture with *calcium carbide* is done in a sealed metal flask. The pressure produced by the liberation of acetylene gas is measured on a dial gauge, which is calibrated to show the percentage of moisture in a standard weight of fine aggregate. The test is simple and quick in operation, and it gives consistent results for uniform samples. A typical apparatus is the "Speedy" moisture tester (Plate 3).

(b) Displacement.
The determination of free moisture content by displacement is made with buoyancy moisture tester, measuring cylinder, pycnometer, steelyard moisture tester and siphon-can moisture tester. In this method, the volume of a given weight or amount of moist, dense aggregate is determined by measuring the quantity of water that it displaces when completely submerged. Similar tests are made with saturated, surface-dry aggregate of the same type, so that specific gravity, absolute volumes and moisture content can be calculated. Greater accuracy is obtained if, when inundating a sample of aggregate with water, the aggregate is gradually added to the water, rather than the reverse, as entrapped air is thereby more readily removed.

The method is practicable only because it has been found that the specific gravity of a particular type of aggregate, from a given source, does not vary appreciably. Three initial tests should be made on dry samples, which are taken from separate, similar consignments. Check tests should be made occasionally during the course of a project, so as to ensure accuracy of calibration and determination.

A *buoyancy moisture tester* is essentially a balance suitably assembled to enable a 1000-g sample of fine aggregate to be weighed in air and weighed again after being inundated and immersed in a large receptacle of water. The aggregate is held in a suspended, wide-mouthed container. The predetermined buoyancy loss in weight of the sample container is added to the result of the second weighing and the moisture content of the aggregate (by weight) is then read from a straight-line graph. For a range of specific gravities and moisture contents, this graph gives percentage value on a basis of dry or moist aggregate. Figure 3.7 shows a range of moisture contents on both bases of measurement for sand with a specific gravity of 2·65.

The graph has been prepared on the principle that
Weight in

$$\text{water of 1000 g dry sand} = 1000 - \frac{1000}{\text{Specific gravity of sand}}$$

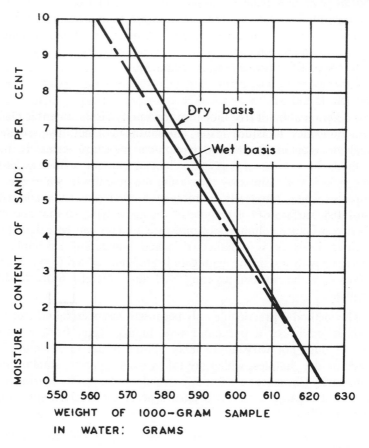

Fig. 3.7 Buoyancy analysis

For moist sand containing 10 per cent of moisture on a dry basis

$$\begin{aligned}\text{Weight in water of}\atop\text{1000 g moist sand} &= \left\{909{\cdot}1 - \frac{909{\cdot}1}{\text{Specific gravity of sand}}\right\} \\ &+ \left\{90{\cdot}9 - \frac{90{\cdot}9}{\text{Specific gravity of water } (= 1{\cdot}0)}\right\}\end{aligned}$$

A similar measurement on a moist basis

$$\text{Weight in water of} \atop \text{1000 g moist sand} = \left\{900 - \frac{900}{\text{Specific gravity of sand}}\right\}$$

$$+ \left\{100 - \frac{100}{\text{Specific gravity of water } (= 1{\cdot}0)}\right\}$$

For practical purposes of analysis, a variation of 0·02 in the specific gravity of the aggregate causes a variation in the weight of the immersed sample of approximately 3 g. Evaluations with a deviation within 0·5 per cent can be made at a mixing plant in about 3 minutes (BS 812).

A *measuring cylinder* is a convenient means of measuring moisture content. For determinations with fine aggregate, water is brought to the 300-ml mark of a 1-litre measuring cylinder, then the aggregate is gradually added and compacted until its surface is at the 600-ml mark. Compaction is effected by a shearing motion, which is brought about by rotating the cylinder backwards and forwards between the palms of the hands. The level of the water is then noted. The moisture content is represented by a difference in water level after tests have been made on dry and moist aggregates, respectively (Fig. 3.8).

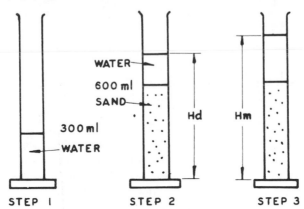

Fig. 3.8 Measuring cylinder

The percentage moisture content is calculated on a basis of dry weight from the expression

$$\frac{Hm - Hd}{(Hd - 300)G} \times 100$$

where Hm is the water level for moist aggregate, Hd is the water level for dry aggregate, and G is the specific gravity of the aggregate.

Values of this expression can be obtained alternatively from a nomogram.[41] The value of Hd remains steady with aggregate of constant grading and compaction. The measuring cylinder is of glass and a thick rubber ring placed around the top affords some measure of protection against breakage.

The moisture content of coarse aggregate may be measured in a metal tube, of 200 mm diameter and 750 mm height, with a graduated gauge glass attached to it (Fig. 3.9). The tube is filled with water to a standard level, the final adjustment being made by means of a stopcock. From readings with 20-kg samples of moist and saturated, surface-dry aggregate, the percentage moisture content is calculated or taken directly from a prepared table.[42]

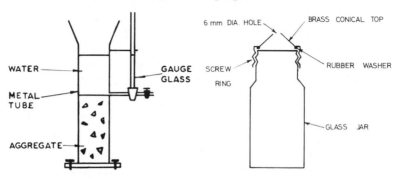

Fig. 3.9   Measuring cylinder   **Fig. 3.10**   Fruit-jar pycnometer

A *pycnometer* (for aggregate up to 10 mm gauge) consists of a 1-kg fruit-preserving jar fitted with a brass cone with a hole of 6 mm diameter at the apex (Fig. 3.10). It is used for finding the volume of a known weight of moist sand, by measuring the difference in the weight of the pycnometer when it is either filled with water or with water containing the sample of sand. In the latter instance, after expelling air from the pycnometer by rolling it, the water is again brought to the top of the cone. All weighings of the pycnometer, which should be kept dry on the outside, are made to an accuracy of 0·5 g.

Similar weighings to those mentioned are made with the sand in a saturated, surface-dry condition, in order to determine its specific gravity (see Section 3.3.4(j)). The free moisture content of 0·5–1 kg of the test sample is either read from a prepared chart[43] or calculated in the manner described in AS 1480 or Reference 134 of Part 3. In an alternative testing procedure, a pressure-type, air-entrainment meter (Fig. 3.11) may be used as a pycnometer when large samples of either fine or coarse aggregate are to be tested for moisture content.[44]

A *steelyard moisture tester* is a device for measuring the amount of moisture displaced from a can (containing a fixed amount of water and suspended at one end of a steelyard) by a constant weight of inundated sand. The higher the initial moisture content of the sand, the more water there is to be displaced for measurement.

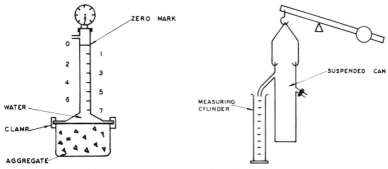

Fig. 3.11 Air-meter pycnometer   **Fig. 3.12** Steelyard moisture tester

The can has two spouts, one near the centre and fitted with a stopcock and the other near the top (Fig. 3.12). The can is filled with water up to the lower spout, which is then closed, and the moist sand under test is introduced until the steelyard is balanced by 2 kg of inundated sand. The surplus water which flows from the upper spout into a measuring cylinder represents the amount of free moisture in the test sample.

The zero position on the measuring cylinder is determined by putting an equal weight of saturated, surface-dry aggregate through the apparatus. Moisture content may be read directly off a transparent calibrated scale, and measurements can be made in 2 or 3 minutes with a deviation in value within 0·5 per cent.

A *siphon-can moisture tester*[46] is a can of double-conical shape. It is fitted with two siphon outlets, which are located at the bottom and top of the throat, respectively, and are fitted with stopcocks (Fig. 3.13). A wide inlet facilitates the entry of aggregate, which may be up to 40 mm nominal size. A wide base facilitates the removal of entrapped air, as the aggregate can be spread in a thin bed. The siphons enable water to be discharged quickly with a definite end point. Their inlets are centrally located for accuracy of performance, while, in plan, their outlets diverge at 30° for discharge into separate vessels.

Sample splitter

Quartering

"Speedy"
moisture tester

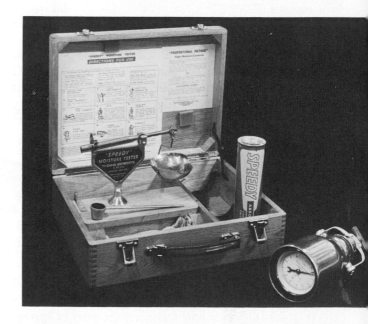

Sieve analysis of sand

Sieve analysis of coarse aggregate

Sand Processing
PLATE 4

(Plant operations include screening, washing, separation and fractional classification. Plant capacity: 40 to 80 m³/h)

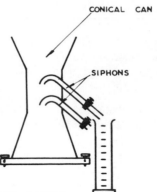

Fig. 3.13 Siphon-can moisture tester

In operation, the can is filled with water up to the lower outlet, which is then closed. A 2000-g sample of moist aggregate is then admitted and stirred. The excess water is discharged by the upper siphon into a measuring cylinder, from which the free moisture content of the aggregate is determined. The zero position on the measuring cylinder is determined similarly by using the same weight of saturated, surface-dry aggregate. Direct readings can be taken from a transparent scale (BS 812).

(c) Electrical resistance.
This method involves the use of a moisture meter, which measures the current that flows through moist, fine aggregate when a direct current is applied across two electrodes.[47, 48, 99, 119]

A moisture meter of this type shows instantaneous and progressive variations on a recorder chart. It is calibrated carefully on being installed in the batching plant, so as to include the effects of compaction and proximity of metal in the hopper. A wide range of moisture conditions must be set up artificially and sufficient samples are taken for the determination of average values throughout the range. Aggregate which differs in mineralogical composition or content of dissolved salt requires further calibration.

The electrodes should be placed in the stream of fine aggregate as it flows into the hopper. The abrasive action thus produced keeps the electrodes free from corrosion and dirt, which would produce a high contact resistance and low moisture reading. The electrodes should be placed sufficiently far from the sides and metal braces of the hopper to cause the current to flow through the aggregate between the electrodes, rather than through the metal of the bin.

The flow of current is a nonlinear function of the free moisture present, and this nonlinearity increases with increasing moisture content. Within a normal working range, a linear-scale chart may be set reasonably close to a calibration curve by means of the rheostat of the meter. Beyond 10 per cent free moisture, however, the variation becomes excessive for reliable reading. Reasonable accuracy is maintained by making a calibration check on a sample of aggregate at the beginning of each shift.

An electric moisture meter must be properly installed as part of a control panel and insulated from plant vibration. Ancillary requirements are the education of operating personnel in its use, periodic maintenance checks and servicing, easy recalibration by regular operating personnel, and a satisfactory repair or replacement plan. A radiation test is given in Section 14.9.1.

3.2.4 MEASUREMENT OF WATER AND LIQUID ADMIXTURES

Water should be measured either by volume or by weight to an accuracy of 1 per cent on well-controlled work. Measuring devices consist of either manually operated or automatic meters, weigh-batchers, and vertical cylindrical tanks with a centre-siphon discharge. The type of equipment used depends upon the type and size of mixing plant and the degree of control to be exercised; its accuracy of measurement should be verified periodically.

Truck mixers should have automatic water-measuring equipment which injects the water deeply into the drum. If wash water is to be used as a portion of the mixing water for succeeding batches, the wash water should be measured accurately in a separate tank provided for the purpose. An allowance for this water is then made when measuring the additional mixing water required.

Means should be provided for quickly measuring the amount of water in aggregate (particularly sand) and adjusting the quantities required so as to compensate for variations in moisture content. Where the moisture content is expressed as a decimal, the requisite batch weight of moist aggregate is determined from the following expressions.

Dry weight required $\times$ (1 + Moisture content)
where the moisture content is based on the weight of dry aggregate.

$$\frac{\text{Dry weight required}}{1 - \text{Moisture content}}$$

where the moisture content is based on the weight of moist aggregate.

Liquid admixtures are measured by volume or weight to an accuracy of within 5 per cent. When mechanical dispensing equipment is used, it must be maintained in a clean and accurate working condition. A daily verification of the batch quantity should be made by diverting a measured dosage into a small calibrated container. Uniformity in the measurement of total mixing water, with a given proportion and grading of materials, is indicated by the uniformity of consistence of the concrete (Plate 5).[48, 49]

The quantity of water to be added to a mix can be controlled instrumentally in certain types of mixing plant (e.g. those containing a pan mixer), where mixed material can be kept in intimate contact with an electrode in a waterproof housing.[79] The metallic part of the mixer into which the electrode is fixed is electrically earthed. The working principle is that the electrical resistance of a concrete mix decreases from infinity to a few ohms as its water content is increased from zero to that of a very wet mix. This variation is utilised instrumentally, so that when the resistance drops to a preset value, the current flowing to a solenoid-operated water valve is interrupted and the water is turned off.

The procedure is independent of initial water in the mix and can be supplemented with a valved by-pass of water for the intake of large quantities. Conductive admixtures to be used should be added last and each particular mix must be considered individually, as calibrated metering appliances are reliably automatic only when used within specific boundary conditions.

3.3 AGGREGATE FOR CONCRETE

3.3.1 GENERAL INFORMATION

Aggregate consists of uncrushed or crushed gravel, crushed stone or rock, sand, or artificially produced inorganic materials. Gravel may be excavated from pits by draglines, mechanical shovels or suction pumps, or it may be dredged from river beds. A screening plant separates gravel into various sizes and removes dirt by washing, and oversize particles are crushed and rescreened. Crushed stone in various gauges or sizes is produced by passing quarried material through primary and secondary crushers and screens. All gauges of coarse aggregate are retained on a 4.75 mm sieve, only fine aggregate passes it.

Aggregate for concrete should be hard, durable, and clean, with only limited amounts (if any) of flaky or elongated particles, adherent coatings, or harmful materials that may affect adversely the strength or durability of concrete or reinforcement. It may be classified according to its unit weight (as heavy, ordinary or light), or by particle shape and surface texture, or in mineralogical or petrographical terms. Ordinary aggregate, being the commonest, is considered first. The heavyweight and lightweight varieties are dealt with in Section 29.3 and Chapter 33, respectively.

Although systems of identification (see Section 3.3.4(k)) are a convenient means of classification, they cannot provide a reliable basis for the prediction of behaviour of an aggregate in concrete. Other factors must be considered, such as the physical and chemical properties of the cementitious matrix; the type of destructive agents that may come in contact with the concrete; and the effects of brittle fracture,[105] hydration, and possible physicochemical changes or weathering. Basic requirements on aggregates and their selection for use in concrete are contained in the standard specifications (see Table 3.8) and Reference 51.

TABLE 3.8   STANDARDS ON AGGREGATE

Title	AS*	BS*	ASTM
Abrasion resistance by Los Angeles machine	1141; 1465	—	C131; C535
Absorption	1012·2; 1141; 1465; 1467	812; 882	C127; C128
Aggregate abrasion value	—	812	—
Aggregate crushing and impact values	1141; 1465	812; 882	—
Aggregates for granolithic concrete floor finishes	—	1201	—
Aggregates for masonry mortar and grout	—	—	C144; C404
Alkali–aggregate reactivity mortar bar	1141; 1465	—	C227; C586
chemical method	1141; 1465	—	C289
Angularity number	1141; 1465	812	—
Blastfurnace slag aggregate	1466	877; 1047	—
Building sands from natural sources	—	1198-1200	—
Bulk density or unit weight	1141; 1467; 1480	812; 3681	C29
Clay and fine silt	1141; 1465; 1467	812	C142
Coarse and fine aggregates from natural sources	—	882	—
Concrete aggregates	1465	882; 1201	C33
Freezing and thawing tests	—	—	C290; C291
Friable particles	1141; 1465	812	C142
Lightweight concrete aggregates	1467	877; 1165; 3681; 3797	C35; C330; C331; C332
Light particles	1141; 1465	—	C123

Title	AS*	BS*	ASTM
Material finer than 0·075/0·002 mm in aggregate	1141; 1465; 1467	812	C117
Metallurgical furnace slag aggregate	1466	877; 1047	—
Moisture content of aggregate	1141; 1467; 1480	812	C70; C566
Mortar-making properties of fine aggregate	—	—	C87
Organic impurities other than sugar	1465; 1467	812; 882; 1198-1200	C40; C87
Particle shape	1141; 1465	812	—
Petrographic examination of aggregates	1141; 1465	812	C294; C295
Polished-stone value	—	812	—
Potential volume change of cement–aggregate combinations	1141; 1465	—	C342; C441
Sampling and testing of aggregates	1141; 1465; 1467	812	D75
Scratch hardness of coarse aggregate particles	—	—	C235
Sieve analysis of fine and coarse aggregates	1141; 1465	812; 1198-1200	C136; C330
Sieve analysis, density and voids of filler	1141; 1465	812	C238; D546
Soft particles in coarse aggregates	1141; 1465	—	C235
Soundness of aggregates by use of sodium sulphate or magnesium sulphate	1141; 1465	—	C88
Specific gravity and absorption of coarse aggregate	1141; 1465; 1467	812; 3681	C127
Specific gravity and absorption of fine aggregate	1141; 1465	812	C128
Sugar	1141; 1465	—	—
Surface moisture in fine aggregate	1141; 1465	812	C70
Ten-per-cent fines value	1141; 1465	812; 1201	—
Toughness of rock	—	812	—
Voids in aggregate for concrete	—	812	C30
Water absorption	1141; 1465; 1467	812	C127; C128; C566; D1864

* Australian (AS) and British Standards (BS) are designated in References 71-75 and Table 3.25; and Appendix I.

3.3.2 SHAPE AND TEXTURE

Aggregate may be classified on visual examination in terms of particle shape and surface texture (AS 1465 and BS 812), as is shown in Table 3.9.

TABLE 3.9   SHAPE AND TEXTURE OF AGGREGATE

Character- istic	Classifica- tion	Description	Examples
Particle shape	Angular	Well-defined edges at the intersection of rough faces, and three dimensions nearly equal.	Crushed rocks of all types; rubble; crushed slag.
	Elongated	Particles, usually angular, having a length/width ratio greater than 3.	—
	Flat or flaky	Particles having a width/ thickness ratio greater than 3.	—
	Flat and elongated	Particles having width/ thickness and length/ width ratios greater than 3.	Crushed or laminated rock.
	Irregular	Naturally irregular or partly shaped by attrition, having rounded edges and three dimensions nearly equal.	Pit gravel, land or dug flint.
	Misshapen	Either flat or elongated, or flat and elongated particles.	Crushed or laminated rock.
	Rounded	Fully water-worn or shaped by attrition, and three dimensions nearly equal.	River or seashore gravel; desert, seashore and wind-blown sand.
Surface texture	Crystalline	Easily visible crystalline constituents.	Granite, gabbro, gneiss.
	Glassy	Conchoidal fracture.	Black flint, vitreous slag.
	Granular	Fracture showing more or less uniform rounded grains.	Sandstone.
	Honey- combed and porous	Visible pores and cavities.	Pumice, scoria, clinker, foamed slag, expanded shale or clay, brick.
	Rough	Rough fracture of fine-grained or medium-grained rock with no easily visible crystalline constituents.	Basalt, felsite, porphyry, carboniferous limestone.

Character-istic	Classifica-tion	Description	Examples
	Smooth	Water-worn or smooth due to fracture of laminated or fine-grained rock.	Gravel, chert, slate, marble, some rhyolites.

3.3.3 MINERALOGY AND PETROGRAPHY

Rocks are naturally occurring crystalline, cemented or consolidated materials which form the immediate crust of the earth. They are classified according to origin in three major groups: igneous, sedimentary and metamorphic. They are subdivided into types according to mineralogical, petrological and physical characteristics. Most rock particles are composed of mineral grains of more than one type, although a rock (e.g. quartzite) may be composed of grains of only one mineral. The following minerals and rocks are dealt with in alphabetical order for identification purposes (see Section 3.3.4(k); AS 1141, 1465 and 1726; BS 812 and CP 2001; ASTM C294 and C295).[95]

(a) Minerals.
(i) Carbonate minerals. The most common carbonate mineral is calcite (calcium carbonate). The mineral dolomite consists of calcium carbonate and magnesium carbonate in equivalent chemical amounts, namely 54·3 and 45·7 per cent by weight, respectively. Both calcite and dolomite are relatively soft (the hardness of calcite is 3 and that of dolomite is 3·5–4 on the Mohs scale) and they are readily scratched by a knife blade. They have a rhombohedral cleavage and, on breaking, they form fragments with smooth, parallelogram-shaped sides. Calcite is soluble with effervescence in cold dilute hydrochloric acid; dolomite is soluble with effervescence only if the acid or the sample is heated, or if the sample is pulverised.

(ii) Clay minerals. The term "clay" refers to a rock or other natural material which contains particles of a specific size range and appreciable quantities of clay minerals (hydrosilicates of aluminium or magnesium, or both). Clay minerals are formed by the alteration of feldspars, other silicate minerals and volcanic glass. Most clayey materials are soft and porous, while some clay minerals of the montmorillonite and illite (hydromica) groups undergo large volume change with wetting and drying. Clay minerals are important constituents of shales: they occur in seams and pockets of limestone, and they are disseminated through sedimentary and weathered igneous rocks.

(iii) Feldspars. The minerals of the feldspar group are the most abundant rock-forming minerals. Since all feldspars have good cleavage in two directions, particles of feldspar usually show several smooth surfaces and frequently the smooth cleavage surfaces show fine parallel lines. All feldspars are softer than quartz and can be scratched by it. The various members of the group are differentiated by chemical composition and crystallographic properties. The potash feldspars (i.e. orthoclase, sanidine and microcline) are silicates of aluminium and potassium. The plagioclase feldspars include those that are silicates of aluminium and sodium, aluminium and calcium, or aluminium and both sodium and calcium.

The feldspar group is frequently referred to as the "soda-lime" group. It includes a continuous series (of varying chemical composition) from albite (the aluminium sodium feldspar) to anorthite (the aluminium calcium feldspar), with intermediate members of the series known as oligoclase, andesine, labradorite and bytownite. Feldspars containing potassium or sodium occur typically in granitic and rhyolitic rocks, whereas those containing calcium are found in rocks of lower silica content, such as diorite, gabbro, andesite and basalt.

(iv) Ferromagnesian minerals. The various types of igneous rock contain characteristic dark green to black minerals, which are generally silicates of iron or magnesium, or both, and include the minerals of the amphibole, pyroxene and biotite (black mica) groups. The commonest amphibole is hornblende and the commonest pyroxene is augite, and these three groups occur also in marble. Olivine (usually olive-green in colour) is a characteristic mineral of igneous rock of very low silica content.

(v) Iron oxides. The common iron oxide minerals may be grouped in three classes: black and magnetic (magnetite); red or reddish when powdered (haematite); and brown or yellowish (limonite). Magnetite is an important accessory mineral in many dark, igneous rocks. Limonite, as a term, is applied loosely to a variety of brown or yellowish minerals, some of which are hydrous; they include the iron minerals in many ferruginous sandstones, shales and clay ironstones. Ilmenite, which is iron titanate, is a commonly occurring black mineral having a specific gravity of 4·5–4·7.

(vi) Micaceous minerals. The micaceous minerals characteristically have a perfect cleavage, and particles of such minerals can be split into extremely thin flakes. The true micas are usually colourless or light green (muscovite), or dark green, or dark brown to black

(biotite), and they have elastic flakes. The green micaceous material which is found in schists represents minerals of the chlorite group. It can be distinguished from true micas because it has comparatively nonelastic flakes.

(vii) Silica minerals.

Chalcedony. Chalcedony has been considered both as a distinct mineral and as a variety of quartz, but is now believed to be composed of a submicroscopic mixture of fibrous quartz with a smaller but variable amount of opal. The properties of chalcedony are intermediate between those of opal and quartz, from which it can be distinguished only by laboratory tests. It frequently occurs as a constituent of the rock chert, and is reactive with the alkalies in portland cement.

Opal. Opal is a hydrous form of silica which occurs as an amorphous mineral, and therefore it is without characteristic external shape or internal crystalline arrangement. It has a variable water content, ranging from 2 to 10 per cent, the specific gravity and hardness are always less than those of quartz, and the colour is variable and the lustre resinous to glassy. Opal is found in sedimentary rocks and is the principal constituent of diatomite; it is found also as a secondary material, filling cavities and fissures in igneous rock. As a constituent of mineral aggregate it is of particular importance, because of its reactivity with the alkalies in portland cement.

Quartz. Quartz is a hard mineral composed wholly of silica (silicon dioxide). It scratches glass and is not scratched by a knife blade. When pure, quartz is colourless with a glassy (vitreous) lustre and a shell-like (conchoidal) fracture. It lacks a visible cleavage and, when present in massive rocks such as granite, it usually has no characteristic shape.

Tridymite and cristobalite. These minerals are crystalline forms of silica which are sometimes found in volcanic igneous rock. They are metastable at ordinary temperatures and pressures, so unless they occur in well-shaped crystals, they can only be distinguished from quartz by laboratory tests. These are rare minerals and they are included here only because of their reactivity with cement alkalies.

(viii) Sulphides. Many sulphide minerals are important ores of metals, but only pyrite and marcasite (both sulphides of iron) are frequently found in mineral aggregates. Pyrite is found in igneous,

sedimentary and metamorphic rocks. Marcasite is much less common and is found mainly in sedimentary rock. Pyrite is brass yellow in colour and has a metallic lustre; it is often found in cubic crystals. Marcasite is also metallic, but lighter in colour; it often oxidises with the liberation of sulphuric acid and the formation of iron oxides, hydroxides and (to a much smaller extent) sulphates. Pyrite does so less readily. Both minerals are known as "fool's gold" (see Section 19.1.1).

(ix) Zeolites. The zeolite minerals comprise a large group of soft, hydrous silicates, usually white or light-coloured, which are formed as a secondary filling in the cavities or fissures of rocks. Some zeolites, particularly laumontite, natrolite and heulandite, have produced deleterious effects in concrete, the latter two minerals being reactive with cement alkalies.

(b) Igneous rock—Natural and metamorphic types.
Natural igneous rocks are those that have been formed by cooling of a molten mass. They may be divided into two groups: *(i)* coarse-grained rocks and *(ii)* fine-grained rocks. The former are deep-seated and intrusive, and the latter are shallow-intrusive, extrusive or volcanic in origin. The coarse-grained rocks cooled slowly within the earth, while the fine-grained rocks were formed as rather quickly cooled lavas and they frequently contain natural glass. A composite variety of igneous rock, known as a porphyry, is characterised by the presence of large mineral grains in a fine-grained groundmass. This texture is the result of a sharp change in the rate of cooling or other physicochemical conditions during the solidification of the rock.

Within these two groups, rocks may be classified and named on the basis of their mineral content, which in turn depends very largely on their chemical composition. Rocks in the intrusive group generally have chemical equivalents in the extrusive group. Salamander is weathered basalt or decomposed igneous rock.

(i) Coarse-grained, intrusive igneous rocks (plutonic rocks).
Diorite. Diorite is a medium-grained to coarse-grained rock which is composed essentially of plagioclase feldspar and one or more ferromagnesian minerals, such as hornblende, biotite or pyroxene. The plagioclase is intermediate in composition, usually of the variety that is known as andesine. Diorite is darker in colour than granite or syenite and lighter than gabbro. If

quartz is present, the rock is called quartz-diorite, while an extremely fine grainsized diorite is classified as andesite.

Gabbro. Gabbro is a medium-grained to coarse-grained, dark-coloured rock, consisting essentially of ferromagnesian minerals and plagioclase feldspar. The ferromagnesian minerals may be pyroxenes, amphiboles, or both. The plagioclase is one of the calcium-rich varieties, such as labradorite. Ferromagnesian minerals are usually more abundant than feldspar.

Diabase and dolerite are similar in composition to gabbro and basalt but are intermediate in mode of origin. They usually occur in smaller intrusions than gabbro and have a medium-grained texture. The term "trap rock" is a collective term for dark-coloured, fine-grained to medium-grained igneous rocks, such as diabase, dolerite and basalt.

Granite. Granite is a medium-grained to coarse-grained, light-coloured rock, which is characterised by the presence of quartz and feldspar. The characteristic feldspars are orthoclase, microcline and albite; feldspar is usually more abundant than quartz. Dark-coloured mica (biotite) is common and light-coloured mica (muscovite) is frequently present. Other dark-coloured minerals, especially hornblende, may be present in amounts less than those of the light-coloured constituents.

Quartz-monzonite and granodiorite may be classed as rocks similar to granite, but containing more plagioclase feldspar (ASTM C615). Microgranite has an average grainsize less than 0·5 mm but greater than 0·05 mm. Rhyolite, which is a finer variety of granite, has an upper average grainsize limit of 0·05 mm, assessed on the greater dimension of grains in the groundmass.

Pegmatite. Extremely coarse-grained varieties of igneous rock are known as pegmatites; they are usually light-coloured and are generally equivalent to granite or syenite.

Pyroxenite and peridotite. Pyroxenites are composed almost entirely of pyroxene, while rocks composed almost entirely of olivine, or of both olivine and pyroxene, are known as peridotites. These rocks are relatively rare, but their metamorphic equivalent (serpentine) is more common.

Syenite. Syenite is a medium-grained to coarse-grained, light-coloured rock, which is composed essentially of feldspar, commonly of orthoclase type. Quartz is below 10 per cent or absent and dark ferromagnesian minerals (such as hornblende, biotite and pyroxene) may be present. Trachyte, which is a variety of syenite, has an extremely fine grainsize.

(ii) Fine-grained, extrusive igneous rocks (volcanic rocks).
The fine-grained equivalents of those coarse-grained igneous rocks described have similar chemical composition. The extrusive rocks are so fine-grained that the individual mineral grains are usually not visible to the naked eye; they may contain the same constituent minerals, or the rocks may be partially or wholly glassy.

Basalt. Basalt is the fine-grained, extrusive equivalent of gabbro. When basalt contains natural glass, the glass is generally lower in silica content than that of the lighter-coloured extrusive rocks. Hence it is less likely to be reactive with cement alkalies.

Felsite. Light-coloured, fine-grained igneous rocks are known collectively as felsite; the group includes rhyolite, dacite, fine-grained andesite and trachyte. These rocks are the equivalents of granite, quartz-diorite, diorite and syenite, respectively. They are usually light-coloured, but may be dark red or even black, but when they are dark they are more properly classed as "trap rocks" (see description of gabbro). When felsite contains natural glass, the glass frequently has such a high silica content that it is reactive with cement alkalies.

Obsidian, pumice and perlite. Igneous rocks composed wholly of glass have been named on the basis of their texture. A dense natural glass is called obsidian, while a glassy froth, filled with bubbles, is called pumice. Perlite is a siliceous or glassy lava with an onion-like structure and a pearly lustre, containing 2–5 per cent water. When heated quickly to the softening temperature, perlite puffs so as to become virtually an artificial pumice. These rocks may be reactive with the alkalies in portland cement.

(iii) Metamorphic igneous rocks.
Metamorphism is a process whereby igneous and sedimentary rocks are changed from their original state by heat, pressure or the introduction of extraneous chemical substances, or by a combination of these causes.

Gneiss. Gneiss is formed by the metamorphism of schists (see the description of argillaceous rocks) and granite. It is characterised by a layered structure, which arises from approximately parallel lenses or bands of platy minerals (e.g. mica and granular materials, or quartz and feldspar). Gneisses are generally coarser grained than schists and usually contain an abundance of feldspar. Intermediate varieties of rock, between gneiss and schist or granite, may be found in the same area.

Serpentine. Serpentine is a relatively soft metamorphic rock, which is light to dark green or almost black in colour. It is usually formed from silica-poor igneous rocks, such as pyroxenites and peridotites. Although some of the original pyroxene or olivine may be present, serpentine is composed largely of softer hydrous minerals; very soft, talc-like material is often a constituent of it.

(c) Sedimentary rock—Natural and metamorphic types.
Sedimentary rocks are stratified arenaceous or granular rocks that have been laid down for the most part under water, although occasionally they are formed by wind action. They may be composed of particles of pre-existing rocks (derived from them by mechanical agents), or they may be of chemical or organic origin.

(i) Argillaceous rocks. These rocks are largely composed of, or derived from, sedimentary silts and clays. When in soft and massive form they are known as claystones or siltstones, depending upon the particles of which they are composed. When they are hard and platy they are known as shales. When they are metamorphosed they become (through mineralogical change) slates, hornfels, phyllites and schists. These metamorphic rocks are usually characterised by a laminated structure and a tendency to break into thin pieces. Pyrophyllite, either in compact or foliated form, is hydrous aluminosilicate that arises largely from a hydrothermal alteration of feldspar. Calcination of the compact form gives rise to whiteware ceramics, shock-resistant refractories and white surfacing aggregate for bituminous pavements.

(ii) Carbonate rocks. Carbonate rocks are known as limestones, unless more than 50 per cent of the carbonate constituent consists of the mineral dolomite, in which case they are called dolomites. If 50–90 per cent of the carbonate content is of calcite, the rock may be called dolomitic or magnesian limestone; if 50–90 per cent is of dolomite, the rock may be called calcitic dolomite. Most carbonate rocks contain some noncarbonate impurities, such as silica minerals, clay, organic matter and hydrous calcium sulphate (gypsum). Carbonate rocks containing 10–50 per cent of sand are arenaceous or sandy limestones or dolomites; those containing 10–50 per cent of clay are argillaceous, clayey or shaly limestones or dolomites. Marl

is a clayey limestone which is fine grained and commonly soft, very soft carbonate rocks are known as chalk, while limestone that is recrystallised by metamorphism is known as marble (ASTM C503, C568 and C586).

(iii) Chert. Chert is a very fine-grained siliceous rock which is characterised by hardness, a conchoidal (shell-like) fracture in the dense varieties and a variety of colours. It scratches glass and is not scratched by a knife blade and the fracture is splintery in the porous varieties. The dense varieties are very tough and grey to black in appearance, but sometimes they are green, yellow, red, brown or blue in colour and have a waxy to greasy lustre. The grey to black varieties are known as flint and the coloured ones as jasper. The porous varieties have a chalky surface and are either white or stained yellowish, brownish or reddish in appearance. Chert is composed of silica in the form of chalcedony, cryptocrystalline quartz or opal, or a combination of them. Their identification requires a careful investigation of optical properties and absolute specific gravity. Chert occurs as nodules or bands in limestones and as particles in sand and gravel of suitable origin.

(iv) Conglomerates, sandstones and quartzites. These rocks consist of gravel, rock fragments and sand particles, cemented together in a matrix of finer material. If the rock fragments are predominantly rounded or gravel, the rock is a conglomerate. If the fragments are angular, it is a breccia. If the particles are of sand size, it is a sandstone or a quartzite. If when fractured the rock breaks around the sand grains, it is a sandstone. If the grains and the cement are largely quartz and the fracture passes through the grains, it is quartzite. Conglomerates and sandstones are sedimentary rocks, while quartzite may be either sedimentary or metamorphic sandstone. The cementing material of sandstone may be quartz, opal, calcite, dolomite, clay, iron oxides or other materials. If the nature of the cementing material is known, the designation of the rock may include a reference to it, such as "opal-bonded sandstone" and "ferruginous conglomerate" (ASTM C616).

Arkose is a coarse-grained sandstone, which contains conspicuous amounts of feldspar, and is derived from granite. Greywacke, which resembles shale or slate, contains dark grains of such rocks as chert, slate, phyllite and schist.

3.3.4 REQUIREMENTS AND TESTS

The requisite properties and tests for ordinary aggregate, which are specified in the standards listed in Table 3.8, are dealt with in detail here, priority being given to correct sampling. When using updated regional standards, variant details of procedure are taken into account in practice. For instance, the Standards Association and Quarry Industries of Australia are supplementing designated maximum sieve sizes with preferred nominal gauge sizes of quarry products (see Section 3.3.4(h), Tables 3.20 and 3.21).

Aggregates (coarse and fine) are defined in the Glossary (Appendix VI), separated on the 4·75-mm sieve, tested separately, broadly specified in Tables 3.11 and 3.12, and graphically mix-proportioned by the procedures illustrated in Section 4.6.

(a) Sampling.

Samples should show the true nature and condition of the materials which they represent. They should be drawn from points known to be representative of the probable variations in the material. A bulk sample is obtained by combining a sufficient number of increments that are drawn from different parts of the bulk, preferably during handling, by cross-sectioning the flow on conveyor belts or discharge from chutes at regular intervals.

The sampling of processed aggregates, supplied from established sources, is done on the basis that stock piles, built in sizes from 250 tonnes in six doubling-up stages to 16 000 tonnes, can be apportioned into at least one to seven equivolume sections according to these respective capacities. Five sample increments (Table 3.10) are taken from evenly distributed positions in each section, after removing outer segregated material to a depth of 0·3 m.

TABLE 3.10   INCREMENT AND BULK SAMPLES

Nominal gauge size (mm)	75	40	28	20	14	10	5
Sample increment (kg)	50	30	20	20	20	10	5
Bulk sample (kg)	125	75	50	50	50	25	10

Sampling from loaded conveyances is usually not truly representative of their contents. A small sample (e.g. 50 litres) is taken from each of five, evenly distributed positions within each rail truck or set of up to five road trucks. For fine aggregate, a sampling tube, 40 mm by 2 m with a tapered end, may be used at regular intervals in low stock piles or shallow bins.

Each set of five, successive sample increments is heaped on a clean, impervious floor, dampened if dry to minimise segregation, thoroughly mixed together and reduced by manual quartering or riffling with a sample splitter (AS 1141 and BS 1796, see also

Plate 3). Two bulk samples of relevant size (Table 3.10) are thus formed and seven sample increments are taken initially if a third, "referee" bulk sample is required. When quartering, the material is formed into a conical heap three times, flattened and quartered, and the process repeated with a pair of diagonally opposite quarters to produce the requisite size of sample.

Each bulk sample is packed in robust containers not exceeding 25 kg when filled. The sample is identified with the name of the supplier and source of supply, origin of sample and number of containers; the type, nominal gauge, location and amount of material sampled and represented; the date of sampling, name and status of the sampler, and the despatch address and nominated project. A duplicate label is placed on top of the material in each container; and material from all containers of each bulk sample, on arrival at the laboratory, is thoroughly mixed until quite uniform for storage and testing purposes.

(b) Bulk density.

The bulk density of aggregate, which is its average weight per unit volume (including voids) in kg/m^3 (see Glossary, Appendix VI), is affected by its moisture content and degree of compaction. For testing purposes, therefore, the aggregate is classified as oven-dry, saturated and surface-dry, or of certain moisture content (which has a significant effect on the bulking of sand, see Section 3.3.6). For conversion purposes in mix design, the bulk density should be determined on aggregate that has been sampled to represent that at the site. An average value should be based on at least three determinations.

Calibrated metal containers for testing aggregates (which are over 20 mm, 5–20 mm and under 5 mm nominal size) are of 25, 10 and 5 litre capacity. Their relevant minimum internal diameters are 350 mm, 250 mm and 200 mm, while their minimum internal depths are 250 mm, 220 mm and 170 mm, respectively (AS 1141 and 1480, BS 812 and 3681, ASTM C29).

For loose sampled aggregate, a container is filled with a scoop or shovel held approximately 50 mm above the top. The surface of the aggregate is levelled with a straight-edge, and the bulk density is determined from the net unit weight of the aggregate. For compacted sampled aggregate, the container is filled in three layers, each one being tamped with 25 strokes of a 15-mm diameter, round-ended rod that is 600 mm long. After striking off the surplus aggregate, the bulk density is determined as before. Its value for dense natural aggregate, dried to constant weight, should not be less than 1200 kg/m³. For comparative purposes, water has a bulk density of 1000 kg/m³ or 1 kg/litre and a specific gravity of 1·0.

$$\text{Bulk specific gravity of aggregate} = \frac{D}{1000}$$

$$\text{Percentage of voids in the aggregate} = \frac{(G \times 1000) - D}{G \times 1000} \times 100$$

where D = bulk density of oven-dry material and G = specific gravity (see Sections 3.2.3, 3.3.4(g)(iii) and (j)). Relevant information is included in Appendix VIII, Sections 4.3.2 and 23.1.5, in Table 3.8, BS 648 on weights of building materials and in Reference 1 of Part 3.

(c) Fineness modulus.

The grading of an aggregate can be expressed by a single figure known as the fineness modulus. This is the sum of the cumulative percentage of aggregate retained on each of a series of sieves (each sieve having a clear opening that is half that of the preceding one) and the total is divided by 100. The sieves used are 37·5 mm, 19·0 mm, 9·5 mm, 4·75 mm, 2·36 mm, 1·18 mm, 0·60 mm, 0·30 mm and 0·15 mm. Smaller sieves are not included, but coarser ones are used if necessary (Tables 3.21 and 27.4).

Practical limits are from 2 to 3·5 (±0·2 tolerance) for fine aggregate, 5·5 to 8 for coarse aggregate, and 4 to 7 for combined or all-in aggregate. The fineness modulus of any mixture of two or more aggregates is the weighted average of the fineness moduli of the separate materials.[102, 117]

(d) Grading.

Divided aggregate is preferable to all-in aggregate for consistent grading and the practical control of quality in concrete manufacture. The various sizes of particles of which an aggregate is composed should be uniformly distributed (Plate 3).

Tables 3.11 and 3.12 show permissible deviations in grading of aggregates supplied under any one contract. A composite sand grading should lie between the lower limit of the coarse and the upper limit of the fine sand classifications; but coarse aggregate gradings should remain within the applicable limits of nominal sizes. The mean of at least two samplings of a delivery is necessary before a rejection can be made.

TABLE 3.11   NATURAL AND CRUSHED FINE AGGREGATES

Sieve Size (mm)	Natural Fine: Percentage Passing			Maximum Deviation	
	Coarse	Medium	Fine	Daily	Total
4·75	90-100	90-100	95-100	± 5	± 5
2·36	60-95	75-100	95-100	± 5	± 5
1·18	30-70	55-100	90-100	± 10	± 10
0·60	15-35	35-80	80-100	± 10	± 15
0·30	5-20	10-40	15-50	± 5	± 15
0·15	0-10	0-10	0-15	± 3	± 5
0·075	0-5	0-5	0-5	± 2	± 5

Sieve Size (mm)	Crushed Fine: Percentage Passing	Maximum Deviation	
		Daily	Total
4·75	90-100	± 5	± 5
2·36	60-100	± 10	± 10
1·18	30-80	± 10	± 15
0·60	15-60	± 10	± 15
0·30	5-40	± 10	± 10
0·15	0·25	± 5	± 5
0·075*	0·10	± 3	± 3

* See footnote to Table 3.13.

Fine aggregate should not be used in reinforced concrete structures when it consists predominantly of one-size particles or conforms closely throughout with the finer limits of grading specified. In some regions, where only the latter type of very fine aggregate is available, mix redesign and testing must be carried out to achieve selected characteristics designated in Appendix III. Mix redesign means attention to such items as (a) reducing the proportion of fine aggregate as its grading becomes finer; (b) including coarser material (e.g. 3 mm crushed aggregate in an adjusted grading); and (c) controlling related increments in the water/cement ratio factor (with or without a water-reducing agent) for practical minimum workability with mechanical compaction (see Sections 3.3.5, 4.2.2, 4.3, 4.5, 4.6, 6.4, 20.1.2(b) and 23.1). Immediate membrane curing, subsequent moist curing and adequate jointing are essential precautions against plastic shrinkage cracking (see Sections 9.1–9.3, 20.1.4, 20.1.5 and 31.2.1–31.2.4).

TABLE 3.12

COARSE AGGREGATE

Sieve Size (mm)	Graded Aggregate				Single-size Aggregate							Total Deviation (± per cent)*	
	40	30	20	14	75	40	28	20	14	10	7	G†	S‡
75·0	100				100							—	—
53·0		100			40-70	100						—	10
37·5	90-100	90-100	100		0-30	90-100	100					10	10
26·5				100	0-5		90-100	100				10-15	10
19·0	30-70	25-60	90-100		0-5	0-20		90-100	100			10-15	10
13·2				90-100	0-5		0-20		90-100	100		10-15	5-10
9·5	10-35		25-55	40-85	0-5	0-5		0-20	0-45	85-100	100	10-15	5-10
6·7									0-10	0-20	85-100	5-15	5-10
4·75	0-5	0-10	0-10	0-10	0-5	0-5	0-5	0-5	0-5	0-5	0-40	5	5-10
2·36	0-2	0-5	0-5	0·5	0-2	0-2	0-5	0-5		0-5	0-2	—	10
0·075	0-2	0-2	0-2	0-2	0-2	0-2	0-2	0-2	0-2	0-2	0-2	—	—

Percentage Passing: Nominal Size (mm)

* Certain standards, which incorporate sieve analyses of aggregates, contain specific fractional limits of deviation within the ranges indicated (e.g. AS 1465, see Table 3.8).

† G = graded aggregate.

‡ S = single-size aggregate.

Various limits of grading are contained in AS 1465,[71] BS 882 and 1201,[74] BS 1198, 1199 and 1200,[75] and ASTM C33 (see Table 3.8). The method of test for sieve analysis is dealt with in Section 3.3.4(h). The results are plotted on charts, with sieve sizes evenly spaced along the base, so that different gradings can be easily compared.

(e) Harmful materials.
Aggregate is to be free of the following materials in amounts that may affect adversely the strength or durability of concrete or reinforcement.

Structurally weak substances (e.g. some shales, sandstones, limestones, mica particles)[33] and partly weathered rocks of volcanic origin (e.g. salamander). Friable particles must be limited also in architectural or fair-face concrete.
Clay, dust and silt coatings, which increase water requirement and impair bond between the cement paste and aggregate.[114]
Colloidal clay and fine crusher dust, when present as discrete particles in excessive quantity.[114]
Organic materials, superphosphate and traces of lime-soluble lead or zinc compounds that delay the setting of cement.[45] Other chemically active materials (e.g. some sulphides).[120] Iron compounds cause rust stains on concrete surfaces.
Water-soluble salts (e.g. chlorides, sulphates and some sulphides) which promote efflorescence, steel corrosion and surface staining. Calcium chloride may occur in flocculating agents used at sandwashing plants.

Certain alkali-reactive siliceous materials sometimes occur in aggregate. They consist of chalcedony, some cherts and flints, opal-containing shales and limestones, cryptocrystalline silica, and some glassy volcanic rocks (such as rhyolite, obsidian, andesite, dacite, phyllite, tridymite, and pitchstone). These expand in concrete because of their reaction with alkalies derived from the cement. Expansion caused by alkali–aggregate reaction may be delayed for a considerable time, but eventually becomes apparent as cracks, spalls or pop-outs (see Section 15.2.3).

The harmful effect of these substances in aggregate is related to the type of cement used and the conditions to which the concrete is exposed. Therefore, the failure of an aggregate to comply with all the requirements of Table 3.13 (AS 1465) may not give cause for its rejection, provided that special provisions based upon tests are observed regarding its use.

TABLE 3.13 PERCENTAGE LIMITS FOR HARMFUL MATERIALS

Material	Limit in Fine Aggregate	Limit in Coarse Aggregate	Test Method
Alkali-reactive materials*	Expansion not greater than 0·05 per cent in 6 months or 0·10 per cent in 12 months, when tested with cement containing more than 1·0 per cent alkali expressed as sodium oxide.		(i)
Clay, loam and silt	10		(ii)
Friable particles	1	0·25	(iii)
Light particles (density below 2 g/ml)††	1	1	(iv)
Material finer than 0·075-mm sieve**	Natural: 5 Crushed: 10	2	(v)
Organic impurities other than sugar†	Colour to be not darker than a standard colour in colorimetric test.		(vi)
Sugar‡	0	0	(vii)

* When the measured expansion ceases before 12 months have elapsed, the severity of the expansive reaction may be minimised by using a cement that contains less than 0·6 per cent total alkalies expressed as sodium oxide, or by the addition of a material (e.g. pozzolan) and air-entrainment that inhibits expansion.

** The increased limit for crushed (manufactured) fine aggregate may be permitted if its total percentage in the whole of the fine and coarse aggregate does not exceed 5 per cent. Material finer than 0·002 mm is limited to 0·8 per cent.

† A darker colour is permissible if attributable to small quantities of coal or charcoal, and the compressive strength of mortar at 7 and 28 days is reduced not more than 5 per cent, as compared with one made with a portion of the sample that has been washed in a 3 per cent solution of sodium hydroxide and rinsed in water.

†† Light particles (e.g. wood, coal and charcoal) tend to rise during vibration and produce poor-quality surface finishes. Coal residues cause disruption by swelling.

‡ Sugar, humic acid and urine retard the setting of cement.

(i) Alkali reactivity of cement–aggregate combinations.

Chemical test. An accelerated chemical test is described in AS 1141, 1465 and ASTM C289 and Reference 56. In this test, the ratio of silica release to reduction in alkalinity should be less than 1.

Mortar bar test. Under suitable conditions of temperature, moisture and concrete proportions, the minor alkalies of cement (sodium and potassium oxides) react with the amorphous silica which is present in certain aggregate to form reaction products (commonly called silica gel) which, through osmotic action, create hydrostatic pressures exceeding the tensile strength of the concrete. The method of test is specified in AS 1141, 1465 and ASTM C227, C342. Natural or crushed fine aggregate, unless specified otherwise, is graded as indicated in Table 3.14, each fraction being washed and dried after removing coarser particles.

TABLE 3.14   **MORTAR BAR TEST**

Designation	Sieve Sizes (mm)				
Passing-Retained	4·75-2·36	2·36-1·18	1·18-0·60	0·60-0·30	0·30-0·15
Percentage	10	25	25	25	15

Note: Crushed aggregate may give accentuated expansion owing to increased surface area. Therefore, coarse aggregate may not be classed as objectionably reactive with alkali, unless tests of concrete specimens confirm the findings of the tests of the mortar.

Steel specimen moulds, assembled either singly or in duplicate, are used for casting mortar bars 25 mm square by 285 mm long, which are provided with stainless-steel studs at their ends with a clear gauge distance between the studs of 250 mm. Ancillary equipment includes a mixer, paddle and bowl, medium-hard rubber tamper, steel trowel, length comparator and covered containers for storing the test specimens. The containers accommodate not more than thirty-six specimens each and are kept moisturetight. The mortar bars are supported vertically on their shoulders, which are kept about 25 mm above water in the containers, and the specimens are separated from each other by spacers. A separate container is provided for specimens that are made with a different brand or type of cement. (See ASTM C227 for illustrations of the equipment.)

The temperature of the moulding room and dry materials is maintained at 20–27°C, that of the mixing water, moist cabinet and measuring room being $23 \pm 2°C$. The relative humidities of the laboratory and moist cabinet should be not less than 50 per cent and 90 per cent, respectively.

Four test specimens, two from each of two batches, are made for each cement–aggregate combination. Moulds, but not stud inserts, are thinly covered with mineral oil. For watertightness, a mixture of 3 parts of paraffin to 5 parts of rosin by weight, heated to between 110°C and 120°C, is applied to the outside contact lines of the moulds.

Test mortar consists of 1 part of cement to 2·25 parts of graded aggregate by weight. The quantities of dry materials mixed at one time in a batch for making two specimens are 300 g of cement and 675 g of aggregate. The amount of mixing water should produce a flow of 105–120 (as determined in Section 9 of ASTM C109), except that the flow table (ASTM C230) is given ten 12 mm drops in 6 seconds instead of the usual twenty-five 12 mm drops in 15 seconds.

The mixing sequence consists of placing the mixing water in the mixer bowl, adding the cement, mixing at 140 ± 5 rev/min for 30 seconds and adding the fine aggregate slowly while mixing for 30 seconds. Mixing is continued at 270 ± 25 rev/min before allowing the mortar to stand for 1·5 minutes. During this interval, the mortar is quickly scraped down within the bowl and covered with a lid. The mixing is finished at 270 ± 25 rev/min for 1 minute. Mortar used in making a flow test is immediately returned to the mixing bowl and the entire batch is then mixed for a further period of 15 seconds.

In an allowable time of 2 minutes 15 seconds, the moulds are filled with freshly mixed mortar in two layers, each layer being compacted with the tamper until a homogeneous specimen is obtained. On trowelling the top surfaces smooth, the filled moulds are placed immediately in the moist cabinet for 24 ± 2 hours. Specimens are then demoulded, protected against loss of moisture and measured for length to the nearest 0·002 mm at 23 ± 2°C. (They should always be placed in the comparator with the same end uppermost.) They are then stood on-end, clear of water in a metal container maintained at 38 ± 2°C. At least 16 hours prior to measuring specimens, the container and its contents are transferred to the measuring room, where the comparator and reference bar are stored. The container is cleaned and the water changed each time that length measurements are made. Specimens are replaced in it in an inverted position as compared with the previous period of storage, and the container is returned to storage at 38 ± 2°C.

Readings are taken at 1, 2, 3, 6, 9 and 12 months, and at 6-month intervals thereafter. A determination is then made of the average percentage expansion of four specimens, of a given cement–aggregate combination, at a given period.[55] The effectiveness of mineral admixtures in preventing excessive expansion of concrete due to alkali–aggregate reaction may be determined where necessary by the test method described in ASTM C441 (see Table 17.1).

Rapid test. A rapid test is described in Section 14.7.1 and in References 12, 57 and 58.

(ii) Clay, loam and silt in fine aggregate. This field settling test, which is used for comparative guidance purposes, is not generally applicable to crushed stone sands. About 50 ml of a 1 per cent solution of common salt in water (2 teaspoonfuls to a litre of water) is placed in a 250 ml stoppered measuring cylinder. Aggregate is added gradually until the measured volume of the sand layer is 100 ml. Salt solution is added to bring the volume to 150 ml. The mixture is shaken vigorously until adherent clayey particles are dispersed and the cylinder, in an upright position, is gently tapped until the sand layer is at a uniform level. The cylinder is left to stand for 3 hours, when the height of clay and silt visible above the sand layer is expressed as a percentage by volume of the height of the sand below.

(iii) Friable particles. Dried test portions of fine and coarse aggregates, retained on a 0·60-mm sieve, are of 500 g and 5 kg minimum size, respectively. The sieving time is minimised in order to avoid breakage of the friable particles and an intermediate 4·75-mm sieve is used with coarse aggregate to prevent damage to the separating sieve. Material passing the 0·60-mm sieve at the initial stage is removed.

The selected test portion is then spread in a thin layer in a suitable dish or on a strong, impermeable sheet (e.g. polyethylene) and the friable particles are crushed by finger pressure. The test portion is again separated on the 0·60-mm sieve and, from test weighings, the percentage passing is calculated for the required characteristic under test.

(iv) Light particles (density below 2·0 g/ml). The method of test, which is specified in AS 1465, 1141, and ASTM C123, covers a procedure for determining the proportion of lightweight particles in fine and coarse aggregates by a sink–float means of separation in a heavy liquid. Test portions for the purpose (Table 3.15) are dried to constant weight at 105–110°C.

TABLE 3.15   **TEST PORTION**

Nominal size (mm)	7	10	14	20	28	40	75
Minimum portion (g)	200	500	1000	3000	4000	5000	10 000

The heavy liquid which has a density of 2·0 g/ml may consist of a mixture of liquids, such as acetylene tetrabromide, perchlorethylene, tribromomethane, monobromobenzene, carbon tetrachloride and benzene. Frequent checks are made to ensure that any evaporation of volatile components does not cause the density to increase to 2·01 g/ml or more. These chemicals are highly toxic and should be used only in a hood with precautionary care.

In connection with fine aggregate, the test portion is poured into at least three times its apparent volume of test liquid, the minimum volume of which is 300 ml. After freeing light particles by stirring the mixture and allowing it to become quiescent, the floating particles are removed by decanting the upper part through a 0·30-mm sieve. The decanted liquid is returned to the test vessel and the operation is repeated until light particles cease floating to the surface of the test liquid. The light particles are washed clean on the sieve with carbon tetrachloride, dried to constant weight at 105–110°C, cooled and weighed.

With coarse aggregate, a similar procedure is followed, except that coarse particles which rise to the surface are removed by tongs or by skimming with a piece of 0·03-mm sieve cloth. The percentage of light particles in the test portions is then calculated.

(v) Material finer than 0·075-mm sieve (decantation test). The method of test is specified in AS 1141, BS 812 and ASTM C117. A test portion is obtained from that despatched for testing by splitting or quartering (Table 3.16). After being dried and weighed, the test sample is placed in a pan, covered with water and vigorously agitated.

TABLE 3.16   **TEST PORTION**

Nominal size (mm)	Over 40	20-40	7-14	Under 7
Minimum portion (g)	5000	2500	1500	500

The wash water is poured immediately over a nest of two sieves, the upper being a 1·18-mm sieve and the lower a 0·075-mm sieve. Agitation and decantation are repeated until the wash water is clear. The washed aggregate, including that retained on the sieves, is dried and weighed. The difference in weight is expressed as a percentage of the original. (The result should agree with a residue obtained from the wash water by evaporation to dryness or filtration and drying.)

Material finer than 0·002 mm is determined from the wash-water residue derived after wet-sieving aggregate down to the 0·075-mm sieve. A hydrometric method of procedure is described in AS 1141 and A89.

(vi) Organic impurities or colorimetric test for sand. A test portion of approximately 250 g is obtained by splitting or quartering and a 3 per cent sodium hydroxide solution is prepared with 30 g of the alkali in 970 ml of water. A 50-ml portion of the solution is placed in a stoppered, 250-ml measuring cylinder or calibrated, 350-ml rectangular, clear-glass bottle. Fine aggregate is added to the 125-ml mark and the sodium hydroxide level is brought to the 200-ml mark using additional solution, while shaking the container to remove bubbles.

The mixture is vigorously shaken for at least 30 seconds to produce thorough wetting and dispersion, and is then allowed to stand for approximately 24 hours. Then 100 ml of the supernatant liquid is compared in colour with 100 ml of a reference colour solution, when each is contained in a similar rectangular, clear-glass bottle.

The standard reference solution is prepared by dissolving 2 g of tannic acid in 10 ml of ethanol (ethyl alcohol), diluting this liquid to 100 ml with distilled water and adding 2·5 ml of the resultant solution to 97·5 ml of the 3 per cent sodium hydroxide solution. The mixture is shaken vigorously, allowed to stand in subdued light for 1 hour and used within 2 hours of its preparation.

Colour comparisons may be made alternatively with a "Lovibond" comparator, or a triple combination of coloured acetate sheets‡ designated as No. 33 (deep amber), No. 50 (pale yellow) and No. 55 (chocolate tint). If the colour of the supernatant liquid is darker than that of the standard reference solution, the sand under consideration should be washed to reduce organic impurities to a harmless amount (see Plate 4). Otherwise, their effect on concrete making and aggregate properties should be determined. Information on sand-washing appliances is given in References 49 and 51–54; also in *Pit and Quarry,* 62, 5 (November 1969).

(vii) Sugar in aggregate. A sample weighing about 100 g is obtained from the sample despatched for testing by splitting or quartering. Fehling's solution is freshly made by mixing equal volumes of stock Solutions A and B prepared as follows.

Solution A. An amount of 17·3 g of powdered copper sulphate ($CuSO_4 \cdot 5H_2O$) is dissolved in water and made up to 250 ml.

‡Supplier: Strand Electric Aust. Pty Ltd, 19 Trent Street, Burwood, Victoria 3125.

Solution B. An amount of 86·5 g of crystalline sodium potassium tartrate (Rochelle salt) is dissolved in warm water, mixed with 30 g of sodium hydroxide and made up to 250 ml.

A 250-ml beaker containing the test sample, which is covered with water and 50 ml of 1 N hydrochloric acid, is placed in boiling water for 5 minutes. The suspension is filtered and 5 ml of the filtrate is neutralised with 1 N caustic soda (using litmus paper indicator) then 3 ml of the Fehling's solution are added. The mixture is placed in the boiling water bath for 5 minutes.

The solution remains blue if sugar is not present, but becomes decolourised, with the formation of a reddish-brown precipitate, if sugar is present. If the original filtrate is not acidic, a further 50 ml of 1 N hydrochloric acid are added and the procedure is continued as described. The test shows 1 part of sugar in 1000 parts of aggregate, but not 1 part in 10 000 parts of aggregate.

(f) Physical properties.
Coarse aggregate should comply with the following characteristics.

(i) Abrasion-resistance. A method of determining resistance to wear is specified in Standards AS 1141[71,72] and ASTM C131, C535. A Los Angeles abrasion machine (in the form of a cylinder of 710 mm diameter and 510 mm length internally) is provided with stub shafts, a removable, dust-tight cover and a 90 mm steel shelf. An

TABLE 3.17   TEST PORTION GRADING AND BALL CHARGE

Sieve Size (mm)		Weight and Grading of Test Portion (kg)									
Passing	Retained	A	B	C	D	E	F	G	H	J	K
75·0	63·0					2·5					
63·0	53·0					2·5					
53·0	37·5					5	5				
37·5	26·5	1·25					5	5			
26·5	19·0	1·25						5			
19·0	13·2	1·25	2·5						5		
13·2	9·5	1·25	2·5							5	
9·5	4·75										5
9·5	6·7			2·5							
6·7	4·75			2·5							
4·75	2·36				5						
Number of revolutions		500	500	500	500	1000	1000	1000	500	500	500
Number of balls		12	11	8	6	12	12	12	12	10	7
Weight of charge (g)		5000 ±25	4584 ±25	3330 ±20	2500 ±15	5000 ±25	5000 ±25	5000 ±25	5000 ±25	4165 ±25	2915 ±20

Note: A tolerance of ±2 per cent is permitted on all test portions.

abrasive charge consists of steel balls (with approximately 48 mm diameter) each weighing 417·5 ± 27·5 g. The size of charge and the number of revolutions of the machine, at 31·5 ± 1·5 rev/min depend upon the weight and grading of the test sample (see Table 3.17).

Unless otherwise specified, the proportion of misshapen particles in the fraction of coarse aggregate, retained on the 9·5-mm sieve, should not exceed 10 per cent when tested by a proportional calliper with a 3:1 ratio setting (see Section 3.3.4(g)(ii)). At the completion of the test, a preliminary separation of the sample is made on a sieve coarser than 1·7-mm gauge; the finer portion is then sieved on a 1·7-mm sieve.

The material coarser than a 1·7-mm sieve is washed, dried at 105–110°C to constant weight, cooled, resieved and weighed. The mean percentage of wear derived from two test results should not exceed the limits shown in Table 3.18.

Note: An abrasive-disc method of test is described in BS 812.[73]

TABLE 3.18   ABRASION AND CRUSHING VALUE PERCENTAGE LIMITS

Aggregate Type*	Los Angeles Value	Aggregate Crushing Value
Hornfels	25	20
Andesite; Basalt	30	25
Breccia; Dolerite	30	25
Microdiorite; Microgranite	30	25
Microsyenite; Rhyolite	30	25
Limestone; Quartzite	35	30
River Gravel; Slate	35	30
Granite; Quartz	45	40

* See Section 3.3.3(b) and (c).

(ii) Aggregate crushing value. Aggregate passing a 13·2-mm sieve and retained on a 9·5-mm sieve is compacted in a metal cylinder of 115 mm diameter and 180 mm height internally. Each of three layers is tamped twenty-five times with a round-ended steel rod of 15 mm diameter and 600 mm length. The net weight of the filled measure is then determined.

The test sample is dried at 105 ± 5°C for 4 hours, then cooled, weighed and similarly compacted in a cylinder (of 150 mm diameter and 135 mm height) to a depth of 100 mm over a base plate. Surfaces in contact with the aggregate are of machined, case-hardened steel (hardness No. 650 in BS 427). A load of

400 kN is applied through a plunger at the rate of 40 ± 4 kN per minute. On releasing the load, the fraction passing a 2·36-mm sieve is determined. The mean of two test results should not exceed the percentage limits shown in Table 3.18.

The ten-per-cent fines value (Glossary, Appendix VI) of the aggregate fraction passing the 13·2-mm sieve and retained on the 9·5-mm sieve should be not less than 50 kN and, for use in concrete for wearing surfaces, not less than 100 kN (AS 1141 and 1465, BS 812 and 1201).[71-74] Unsatisfactory particle shapes in the sample aggregate (see Section 3.3.4(g)) tend to increase the Los Angeles and aggregate crushing values and reduce the ten-per-cent fines value. In some areas, the acceptance of available aggregate may be based on satisfactory past experience or trial-mix results, these being related to the expected serviceability of the concrete.

(iii) Aggregate impact value. The aggregate, which is of 14–10 mm nominal size and prepared as in the previous test, is compacted in the cylindrical steel cup (100 mm diameter and 50 mm height) of an impact test machine (BS 812). The machine is placed on a firm, level and rigid base. A drop hammer (13·5–14 kg) is allowed to fall fifteen times on to the test sample at intervals of not less than 1 second from a height of 380 ± 6 mm. In duplicate tests, the fraction passing a 2·36–mm sieve is weighed to an accuracy of 0·1 g and the mean ratio of the weight of fines formed to the sample weight is determined. The aggregate impact value should not exceed 30 per cent for aggregate for use in concrete for wearing surfaces, and 45 per cent for aggregate for use in concrete for other purposes.

(g) Particle shape.
(i) Thickness and length gauges. A test sample of sufficient size is separated on the sieves designated here, in order to provide at least 200 pieces for each sieve-size fraction (which constitutes more than 15 per cent of the sample) and at least 100 pieces for each size fraction (which constitutes between 5 and 15 per cent of the sample). Size fractions below 5 per cent are not tested (BS 812). Sieved fractions of dry aggregate, in size (mm) sequence, are 63–53, 53–37·5, 37·5 and 31·5–26·5, 26·5–19·0, 19·0–13·2, 13·2–9·5 and 9·5–6·7.

Particle shapes and sizes are classified in Table 3.9. The openings of steel thickness and length-measuring gauges are 0·6 times and 1·8 times the respective mean sieve sizes calculated from the sizes designated here. When using updated regional standards, variant details of procedure are taken into account in practice (AS 1141 and 1465). For the first and fourth scheduled sizes, the thickness-gauge test only is carried out.

Having examined each size fraction separately by these gauges, the following percentage indexes are calculated and reported with the aggregate sieve analysis.

$$\text{Flakiness Index} = \frac{B}{A} \times 100$$

$$\text{Elongation Index} = \frac{C}{A} \times 100$$

where A = total weight of the sample gauged, B = total amount of material passing the thickness gauge, and C = total amount of material retained on the length gauge.

An assessment of particle shape of fine aggregate may be made by dividing it into fractions coarser and finer than a 0·30-mm sieve and separating the particles in each fraction into shape groups. Particles in the first group are separated and weighed, using a low-powered binocular microscope for small fractions. Those in the second group are counted under a microscope. Regular-rounded, regular-angular, flaky and elongated shapes may thus be identified.

Particle shape and size distribution influence the water content necessary to obtain a mix of suitable consistence, and thereby affect the compressive strength, drying shrinkage and durability of the resulting concrete. The sum of the flakiness and elongation indexes should not exceed 40 per cent.

(ii) Proportional calliper. A test sample is taken of sufficient size to yield not less than 100 particles in each size fraction, these comprising 10 per cent or more of the sample retained on a 9·5-mm sieve, that lies between two successive sieves of the size series 53, 37·5, 26·5, 19·0, 13·2 and 9·5 mm. After drying and quantitatively adjusting the sample, the particles of each relevant fraction are tested for flatness and elongation with a 3:1 ratio setting of the proportional calliper illustrated in AS 1141.

The weighted percentages of misshapen particles (Table 3.19) are calculated as the products of particle-type percentages and the proportion of material in each size fraction. Sieve sizes not tested are assumed to have the same percentages of misshapen particles as those of the next smaller or larger sizes present in 10 per cent or more of the sample. Table 3.19 shows a sample calculation for determining the weighted percentages of misshapen particles.

TABLE 3.19

PERCENTAGE OF MISSHAPEN PARTICLES

Sieve Size (mm)	Grading			Size Fraction Tested		Sorting Results			Weighted Percentages			
	Percentage Passing	Percentage Retained in Each Size Fraction		Weight D (g)	Number of Particles	Flat E (g)	Elongated F (g)	Flat and Elongated G (g)	Flat	Elongated	Flat and Elongated	Misshapen
		As Received B	Calculated on Retained 9·5 mm $C = \dfrac{B \times 100}{A}$						$H = \dfrac{E \times C}{D}$	$J = \dfrac{F \times C}{D}$	$K = \dfrac{G \times C}{D}$	$L = H + J + K$
53·0	10											
37·5	98	2*										
26·5	91	7*										
19·0	60	31	43·7	295·1	102	15·6	7·7	12·1	2·3	1·1	1·8	5·2
13·2	20	40	56·3	168·2	148	17·8	10·8	14·1	6·0	3·6	4·7	14·3
9·5	15	5*										
Totals		A = 71	A = 100									19

* These size fractions are less than 10 per cent of the sample as received and are not included in the total, A.

(iii) Angularity number. The voids remaining in aggregate, after compaction in a prescribed manner, are a measure of particle angularity. As rounded aggregate has about 33 per cent voids, the relative angularity of aggregate in general is the amount by which the percentage of voids exceeds 33 per cent. The angularity number ranges from 0 to about 12, but it does not apply to aggregate that breaks down during the test (AS 1141, BS 812).

A 10-kg test sample is obtained by screening aggregate with an appropriate pair of the following test sieves: 19·0 and 13·2, 13·2 and 9·5, 9·5 and 6·7, 6·7 and 4·75 mm. The predominant gauge-size sample is dried at 105–110°C and compacted specifically with a standard rod in a 5-litre calibrated metal container (see Section 3.3.4(b)).

Each of four layers is gently formed and evenly tamped with 125 strokes of the rod at the rate of about two blows per second. Without applying force, each blow is applied by a 50-mm free-fall of the rod. After striking off surplus aggregate, individual particles are added and lightly rolled into surface interstices without causing the rod to lift off the top of the container. Three determinations are made and if one of them differs more than 25 g from the mean, the procedure is repeated and the mean calculated on the six tests for insertion in the following formula.

$$\text{Angularity Number} = 67 - \left\{ \frac{100A}{W} \times \frac{1000}{D} \right\}$$

where A = mean weight (kg) of aggregate in the container, W = weight (kg) of water required to fill the container and D = bulk density (kg/m³), on an oven-dried basis, of the aggregate (see Section 3.3.4 (b)). The angularity number is reported to the nearest whole number.[85, 86]

(h) Sieve analysis.
Sieves and sieving media are specified in AS 1141 and 1152, BS 410, ASTM E11 and STP 447. Nominal gauge sizes of quarry products (see Section 3.3.4) are interrelated with designated sieve sizes in Tables 3.20 and 3.21.

Sample and sieve weights for lightweight aggregate are prescribed in AS 1467, BS 3681 and ASTM C330. For dense aggregate, a selected minimum test portion is obtained from the sample despatched for testing, using a sample splitter or by quartering (AS 1141, BS 1796), as follows.

TABLE 3.20   QUARRY PRODUCTS

Classification	Size (mm)										
Designated sieve size (mm)	75	63	53	37·5	26·5	19·0	13·2	9·5	6·7	4·75	3·35
Nominal gauge size (mm)	75	60	50	40	28	20	14	10	7	5	3
Superseded gauge size (in)	3	2½	2	1½	1	¾	½	⅜	¼	³⁄₁₆	⅛

TABLE 3.21   GENERAL SIEVE DATA

AS BS Sieves (mm)*	General Sieves ASTM	Dimension of Square Opening (mm)	Tyler Sieves	Dimension of Square Opening (mm)
75·0	3	76·2		
37·5	1½	38·1		
19·0	¾	19·05	¾	18·85
9·5	⅜	9·53	⅜	9·42
4·75	No. 4	4·76	No. 4	4·70
2·36	No. 8	2·41	No. 8	2·36
1·18	No. 16	1·20	No. 14	1·17
0·60	No. 30	0·599	No. 28	0·589
0·30	No. 50	0·295	No. 48	0·295
0·15*	No. 100	0·152	No. 100	0·147
0·075	No. 200	0·076	No. 200	0·074
0·053	No. 270	0·053	No. 270	0·053

* Smallest sieve opening for determinations of Fineness Modulus (see Section 3.3.4(c)).

Nominal gauge (mm)	75	40	28	20	14	10	7	3
Minimum weight (kg)	25	16	12	8	6	2	0.5	0.1

The test sample is dried to constant weight at a temperature not exceeding 110°C. When separating a sample containing a wide range of sizes into fine and coarse components, while still on the 4·75-mm sieve, the coarse components should be brushed to remove adhering fine material. Material finer than a 0·075-mm sieve is removed from the fine aggregate, in accordance with the instructions given for the test in Section 3.3.4(e)(v).

Coarse and fine aggregates are weighed and passed successively through appropriate sets of clean sieves, starting with the largest. Sieving is done mechanically or manually for a period of not less than 2 minutes, in such a way (with frequent jarring) that the material is kept moving· over the sieve surface in frequently changing directions. Sieve openings may be cleared by light brushing with a soft brush on the underside of sieves. On completion of sieving, the material retained on each sieve, together with any material cleaned from the mesh, is weighed.

The previously suggested sample sizes seldom lead to overloading of the sieves. However, if the weight of the material retained on any one sieve exceeds that specified in Table 3.22 for that sieve, the material on each overloaded sieve should be divided into suitable lots and sieved separately.

TABLE 3.22 MAXIMUM WEIGHT RETAINED ON EACH SIEVE

Sieve Size (mm)	450-mm Sieve (g)	300-mm Sieve (g)	200-mm Sieve (g)
63·0	5000	2200	1000
37·5	5000	2200	1000
26·5	4000	1800	800
19·0	3000	1300	600
13·2	2000	900	400
9·5		500	250
4·75		400	200
2·36			200
1·18			100
0·60			75
0·30			50
0·15			40
0·075			25

From the amounts retained on each sieve, expressed as a percentage of the weight of the total test sample, the cumulative percentage by weight passing each of the specified sieves is calculated and reported to the nearest whole number. Size fractions in fine aggregate are readily seen in a grading gauge, which is shown in Figure 3.14 for a 500-g sample.[59] Lines are drawn across the compartments to represent the maximum and minimum amounts of size fractions that may be retained on a series of sieves. After weighing, the size fractions of a sieved 500-g sample are poured into the appropriate compartments of the gauge. Their several heights enable a visible comparison to be made between an actual grading and its specified limits.

(i) Soundness.
The disintegrative resistance of an aggregate to sulphate solution is a means of estimating its soundness to weathering action, particularly where relevant service records are inadequate or unavailable (AS 1141 and 1465, ASTM C88 and Reference 76).

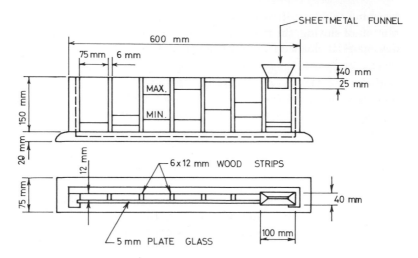

Fig. 3.14   Grading gauge

In a typical test, dry, washed samples of coarse and fine aggregates are separated into as many of the respective fractions of Table 3.23 as may be present. The only fractions then selected are those which constitute 5 per cent or more of the relevant sample, and the sample gradings are recalculated on this basis.

TABLE 3.23   **COARSE AND FINE AGGREGATES**

Fraction		Minimum	Sieve used to
Passing (mm)	**Retained (mm)**	**Test Portion (g)**	**determine Loss (mm)**
75·0	53·0	3000	
53·0	37·5	2000	26·5
37·5	26·5	1500	16·0
26·5	19·0	1000	13·2
19·0	13·2	750	8·0
13·2	9·5	500	6·7
9·5	4·75	300	3·35
4·75	2·36	100	1·70
2·36	1·18	100	0·85
1·18	0·60	100	0·425
0·60	0·30	100	0·212

A test solution of sodium sulphate is freshly prepared by dissolving 215 g of the anhydrous salt or 450 g of clear decahydrate crystals (sold as "Glauber's Salts") per litre of clear tap water. While being heated and cooled, the solution is stirred. It is then allowed to stand at 23 ± 2°C for at least 24 hours. This temperature is also used during the covered storage and testing operations. The density of the solution, as measured at 23°C by a 20/20 hydrometer, should be between 1·155 and 1·170 g/ml.

Each sample, in a perforated container or wire basket, is immersed in the sulphate solution for a period of 16–18 hours. The solution is at least five times the apparent volume of the test portion and covers it to a depth of at least 15 mm. Weighted wire grids are used to facilitate the coverage of lightweight aggregate.

On its removal from the solution, each sample is drained for 15 ± 5 minutes, dried at 105–110°C, cooled and again immersed in the prepared solution. The drying operation is continued until successive weighings (after additional drying for not less than 30 minutes), differ by no more than 1 per cent of the previous loss by drying.

After the completion of five cycles, the cooled sample is washed free from sulphate solution, as determined by a reaction of the wash water with barium chloride. Each fraction of the sample is dried, weighed and sieved as indicated in Table 3.23 for the appropriate size of particle.

The calculated weighted average loss, determined on the basis of the recalculated grading, should not exceed 12 per cent for coarse or combined aggregate or 16 per cent for fine aggregate, lower limits being specified for severer conditions of exposure. Owing to the empirical nature of the test, aggregate failing to meet its requirements may be accepted provided that concrete of comparable properties, made with similar aggregate from the same source, has given satisfactory service when exposed to weathering conditions similar to those expected.

(j) Specific gravity and water absorption.
Standard procedures of determination are given in AS 1141, BS 812 and 3681, ASTM C70, C127 and C128.

(i) Fine aggregate. A test sample, weighing approximately 1000 g, is obtained by splitting or quartering (see Glossary, Appendix VI). It is dried on a pan at 105 ± 5°C to constant weight, covered with water and allowed to stand for 24 hours. Entrapped air is removed by gentle agitation with a rod and the water is drained off. The sample is then dried to a surface-dry condition as described.

The sample is spread on a flat surface, exposed to a gently moving current of warm air and stirred frequently to secure uniform drying. When it approaches a free-flowing condition, it is placed loosely in a conical mould, which has a diameter of 38 mm at the top, 90 mm at the bottom and a height of 73 mm. The surface is tamped twenty-five times with a metal rod weighing 350 g and with a flat circular tamping face of 25 mm diameter.

If the cone of aggregate retains its shape when the mould is lifted vertically, free moisture is present, and drying with constant stirring is continued. Tests are made at frequent intervals, and when the cone first slumps upon removal of the mould, the aggregate is considered to have just reached a *surface-dry condition* while in a *saturated state*. An alternative procedure is to roll the fine aggregate in a dry glass jar, the particles being considered surface-dry when they do not cling to the glass surface, or to pass saturated air over wet aggregate.[101]

Approximately 500 g of the saturated and surface-dry material are weighed (B) and introduced immediately into a 500-ml flask, which is then filled almost to the 500-ml mark with water at $23 \pm 1°C$. The flask is rolled to eliminate air bubbles, placed in a bath maintained at $23 \pm 1°C$ for 1 hour and filled with water to the 500-ml mark. The flask is removed, dried externally and weighed (C). The aggregate is removed, dried to 110°C, cooled in a desiccator and weighed (A).

The flask is then filled almost to the 500-ml mark with water, placed in the constant temperature bath and, after 1 hour, is filled with water to the 500-ml mark. After removal and external drying, it is weighed (D).

Calculations

A = weight in g of oven-dry sample.
B = weight in g of saturated and surface-dry sample.
C = weight in g of flask plus water plus aggregate.
D = weight in g of flask plus water.

$$\text{Apparent specific gravity} = \frac{A}{A - (C - D)}$$

$$\text{Specific gravity on a dry basis} = \frac{A}{B - (C - D)}$$

$$\text{Specific gravity on a saturated and surface-dry basis} = \frac{B}{B - (C - D)}$$

$$\text{Water absorption} = \frac{B - A}{A} \times 100 \text{ per cent.}$$

(ii) Coarse aggregate. A test sample retained on a 4·75 mm sieve and weighing approximately 2 kg for sizes up to and including

19 mm, and 5 kg for larger sizes, is taken from the bulk sample. After being thoroughly washed, it is dried to constant weight at 105–110°C and immersed in water at approximately 23°C for 24 hours. On being removed from the water, the sample is rolled in a large absorbent cloth until all visible films of water are removed, although the surfaces of the particles still appear damp (B).

After weighing, the saturated surface-dry sample is placed immediately in a wire basket (4·75 mm mesh, 200 mm diameter and 400 mm height) and its weight (C) in water is determined.

Calculations
A = weight in g of oven-dry sample in air.
B = weight in g of saturated and surface-dry sample in air.
C = weight in g of saturated and surface-dry sample in water.

Apparent specific gravity $= \dfrac{A}{A - C}$

Specific gravity on a dry basis $= \dfrac{A}{B - C}$

Specific gravity on a saturated and surface-dry basis $= \dfrac{B}{B - C}$

Water absorption $= \dfrac{B - A}{A} \times 100$ per cent.

The specific gravity of dense aggregate should be not less than 2·3 (see Section 3.2.3). For site-placed and precast concretes, the water absorption of dense aggregates should not exceed 5 per cent and 2·5 per cent, respectively; otherwise, the aggregates must meet the sulphate soundness test for serviceability. Aggregates with more than 1 per cent water absorption within 30 minutes should be prewetted before being mixed into concrete. A water-absorption test for lightweight aggregate is included in AS 1467.

(k) Petrographic analysis and number.
Petrographic analysis enables coarse aggregate to be qualitatively appraised and numerically classified. The test apparatus includes the following items: hand lens of 10× magnification; Alnico magnet and 5·5–6·0 Mohs-scale pocket-knife; aggregate-breaking hammer and anvil; 5 per cent solution of technical-grade hydrochloric acid (20° Baumé); stereoscopic microscope with 6× to 25× final magnifications; 400-ml polyethylene squeeze-bottle with spout; and a 2000-g readable scale, accurate to the nearest g.

A 1000-g representative sample of oven-dry aggregate, passing 19·0 mm and retained on 9·5-mm sieves, is spread over a flat working surface. A quantitative examination is then made of clay-coated, cemented and encrusted particles; also clay, friable and misshapen particles (see Table 3.9; Sections 3.3.4(e)(ii)–(vi) and (g)). After separating out weak particles, the sample is washed to remove any clay or dust coatings and visually sorted into the rock types listed in Appendix V (see Section 3.3.3; CSA A23.2.30). This step in classification is aided with scratch, acid and microscopic tests (ASTM C294 and C295).

Particle identification is related to shape, surface, grain size, texture and colour, mineral composition, significant heterogeneities, chemical and physical properties, including weathering and fracture characteristics. From the gravimetric percentage of each rock type, the petrographic number (PN) is calculated as the sum of the products of each rock-type percentage, times the appropriate factor given in Appendix V. Excellent to very poor sample numbers range from 100 to 600, the requisite value for aggregate in concrete pavements and structures being limited normally to below 125. Upper limits of 135–150 apply to bituminous pavements and road mixes, with 200 for granular base courses.

A petrographic examination of metallurgical slag, for the determination of possible disintegration by the presence of dicalcium silicate, is described in AS 1141 and ASTM STP169A.

(l) Testing.
Practical procedures and policies in testing are indicated in Section 14.1.

3.3.5 HANDLING

(a) Fine aggregate.
Uniform concrete requires a consistent source, character, grading and moisture content of fine aggregate. The fineness modulus of a well-graded sand should not vary by more than $\pm$ 0·2 from that of a preliminary representative sample.

The fines, which are essential for good grading and workability, should be retained during washing operations and kept uniform in content. Sand, which must be well-graded to produce a plastic, workable mix, should have 15–20 per cent of material finer than a 0·30-mm sieve, of which not more than one-third should pass a 0·15-mm sieve. Lean mixes require larger amounts of these fine sizes than rich mixes. Coarse and fine sands are batched separately for consistently uniform results. For grading tests and specifications, see Sections 3.3.4(d), (h) and 20.1.2.

When dry sand is dropped from the end of an elevating conveyor, a chute or a chimney (with exit openings) should be installed so as to prevent a segregation of sizes by wind action. The segregation of aggregate moving down a sloping surface can be prevented by a retaining baffle. The bottom 0·5 m of wet sand in a stock pile is best allowed to serve as a drainage layer for the upper part. Stock pile drainage may take 2 days or more to reduce the moisture in sand to a uniform and stable amount (within 8 per cent). The moisture in sand, as batched, should be kept within a practical range of stability. Covered storage is helpful in reducing adjustments through variations in moisture content. Sand-washing plant is referred to in Section 3.3.4(e)(vi), iron elements (e.g. ilmenite) being separated magnetically or by settlement in sluiced pits.

(b) Coarse aggregate.

For uniform concrete, coarse aggregate is separated into several size fractions or gauges. The maximum size in each fraction is about twice the minimum size (see Table 3.21). Screens are stationary, sloping type (with slotted openings for damp aggregate) or horizontally operating, vibrating type. The latter type is more accurate than the former, owing to the tendency to overfeed screens. Unless coarse aggregate is placed directly from screening operations into the bins of batching plant, finish screening over the bins is recommended for cleanliness and size control. Undersized particles in any fraction should be limited in size to four-fifths of the designated minimum and in quantity to 3 per cent (below 26·5 mm).

Finish screening at a batching plant eliminates the main difficulties with coarse aggregate: namely, segregation and contamination in stock piles, careless loading to and from stock piles, and a variable overlap of fractional sizes. When washing is done during finish screening for regular batching, the free moisture content remains uniform. It may average about 1·3 per cent for 20 mm and 0·8 per cent for 40 mm nominal-size gauge aggregate. Some illustrations of finish screening over the bins of batching plants are contained in References 49 and 60.

Bulk storage should be on hard ground or a thin slab of weak concrete that is graded for drainage. A space or dwarf walls should be placed between different materials. When stock piles are required without finish screening, they should be built up in horizontal or gently sloping layers and not by end-dumping methods. Cranes should be operated so as to avoid swinging buckets of one aggregate over another. When stockpiling large-sized aggregate from an elevating conveyor, a rock ladder may be used to minimise breakage.[51, 60] The excessive fines which accumulate in large stock piles of coarse aggregate can be removed only by washing and finish screening at a well-equipped batching plant.

Storage or supply bins should permit all material to flow readily, directly, and evenly to the discharge opening. Each bin should have equal horizontal dimensions and a base which slopes (all round) towards a central outlet at an angle of 50° or more from the horizontal. Filling is done axially over the outlet and the aggregate, as batched, should be sampled and tested periodically. The purpose of careful handling and control, in proper sequence, is to reproduce satisfactorily a selected batch assembly for the continuous

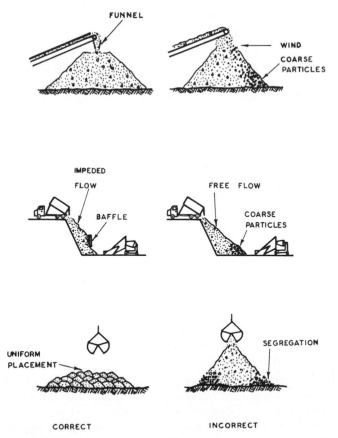

Fig. 3.15   Methods of handling aggregate

production of homegenous concrete. Bin design and rate of discharge are dealt with in Reference 61 and Appendix IV, 2, and a typical plant establishment is shown in Plate 5.

Ventilating silos, which may be constructed of precast prestressed staves, are a convenient means for bringing stored, wet aggregate to a drip-dry condition. The walls contain horizontal slots, in which the aggregate adopts a natural angle of repose without discharge, and the drying period required varies usually from 8 days for coarse aggregate to twelve or more days for fine aggregate. Some correct and incorrect ways of handling aggregate are illustrated in Figures 3.15 and 3.16.

DRAINAGE   SLAB   CONTAMINATION

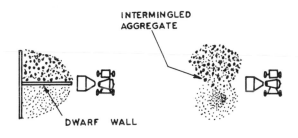

INTERMINGLED
AGGREGATE

DWARF   WALL

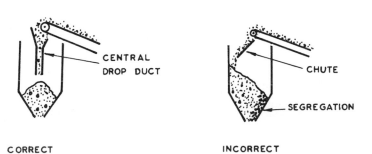

CENTRAL
DROP DUCT   CHUTE

SEGREGATION

CORRECT   INCORRECT

Fig. 3.16   Methods of handling aggregate

3.3.6 BATCHING

(a) Gravimetric.

Each type or size of aggregate should be stored and weighed separately, weighing being to a tolerance of 2 per cent individually or 1 per cent cumulatively on a saturated, surface-dry basis. Weigh-batching equipment is selected to suit the size of project, rate of production, and the requisite standard of batching performance, provision being made for the removal of samples for testing. Batching equipment should be insulated from plant vibration, such as by independent support, and charged from easily operated clamshell or undercut, radial-type bin gates. Power-operated gates should enable small increments to be added towards the end of a weighing operation. The equipment falls into three general categories.

(i) Manual batching. Wheelbarrows and platform scales are satisfactory and economical for projects that require up to 1000 m^3 of concrete. Thereafter, a manually operated weigh-batching installation for cumulative weighing is suitable for projects up to 5000 m^3 capacity. A stationary, sloping screen may be used to remove dirt and small undersized particles in coarse aggregate that is taken from stock piles.

(ii) Cumulative, semi-automatic batching. Equipment in this category automatically closes power-operated gates which control the supply of aggregate to a cumulative weigh-batcher. Manual operations include the setting of batch weights, release of the weighed batch to the mixer and push-button starting of each automatic weighing cycle. The cement is weighed separately and recorded on equipment which is clearly visible to the operator.

Installations of this kind are used on projects ranging in size from 5000–100 000 m^3 capacity, or in ready-mixed concrete plants which produce 30 000 m^3 or more per year. Horizontally operating, vibrating screens are used for finish screening. A typical plant with its control console is shown in Plate 5.

(iii) Individual, fully automatic batching. Equipment in this category has the following innovations.

A separate scale and batcher for the cement and each size of aggregate.

Push-button selection of the mix from a number of preset proportions.

Automatic batching that can be started automatically or manually, as desired, as soon as the individual batchers are discharged (but not before).

Simultaneous batching of all ingredients (which saves time).

Batcher discharge gates that close as soon as the batchers are empty (and not before).

Automatic recording of all aspects of the batching operation, including quantities and a continuous record of batch-to-batch consistence.[48, 49, 109]

The aggregate bins (preferably of cylindrical shape) are arranged concentrically around the cement bin. The individually weighed materials are discharged simultaneously with the cement, through a collecting cone, directly into the mixer.[60] Such installations are used for heavy constructional projects of more than 100 000 m^3 capacity, or in ready-mixed concrete plants of the same capacity per year. Finish screening is done on elevated, horizontally operating vibrating screens and closed-circuit television is installed for plant observation.

(b) Volumetric batching.
Volumetric batching on small-job sites is done with cuboid gauge boxes, which are screeded off level for uniformity in filling. Batches should be so arranged in size that whole bags of cement are employed. If fractional bags are used, an allowance must be made for the increase in volume of the cement, due to aeration by disturbance. The transfer of cement from bags into gauge boxes, in one or two operations, can lead to an inflation of its volume by some 10–20 per cent. Therefore unless the cement is weighed or collars are placed on the gauge boxes so as to allow for this effect, mixes may be undercemented to this extent.

Nominal proportions of mix (when specified) refer to dry materials. When gauging moist sand by volume, a correction is necessary for bulking due to moisture content, which causes sand (on being filled loosely into a container) to occupy a larger volume than it would if it were dry. The bulking of sand depends on its moisture content, grading and degree of compaction, and it varies from about 20–40 per cent with a moisture content of around 5–6 per cent. It is caused by the collective surface tension of water films on the particles and is eliminated when the sand is either completely dry or saturated. While the moisture content of wet sand may vary up to about 20 per cent, that of stone seldom exceeds 2 per cent by weight. Some typical bulking values are indicated in Figure 3.17, the volume of coarser aggregate staying fairly constant whether it be dry, moist or wet.

The gauge boxes for measuring sand must be increased in depth, proportionately to the bulking, if the nominal amount of fine aggregate is to be incorporated in the concrete. Otherwise, volumetrically gauged mixes are undersanded, difficult to work and below size for the requisite yield of concrete per bag of cement. Although volumetric measurement with poor control is susceptible to large variation, a predetermined correction of the right order facilitates the manufacture of concrete of uniform quality.

The bulking of a given sand may be determined from its equivalent loose dry volume (see Sections 3.2.3(a) and 3.3.4(j)), or from a prepared graphical relationship between bulking and moisture content. Alternatively, it may be determined approximately on the premise that the volume of inundated fine aggregate is the

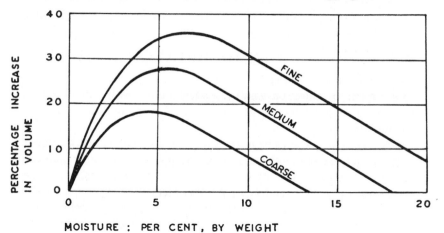

Fig. 3.17   Bulking of loose sand

same as if the aggregate were dry. It is possible, however, for the former to be smaller than the latter and give an apparent increase in bulking value.

An inundation test is made by placing damp sand loosely in a parallel-sided container until it is about two-thirds full (AS 1480). The surface is levelled without being pressed down and the average height *(H)* is measured by the insertion of a steel rule. The sand is emptied into another container and the original one is half-filled with water. The sample of sand is poured into the water in two halves, each being rodded or stirred vigorously and allowed to settle, so that its volume is reduced to a minimum. On measuring the new average height (h) of the levelled surface, the percentage of bulking is estimated from the expression

$$\frac{H - h}{h} \times 100.$$

Increased accuracy is obtained by batching by weight. No adjustment is required for bulking in this instance, but a correct allowance must be made for moisture in the combined aggregate (see Section 3.2.4). Batches should be protected from rain (see Section 9.3) and, where feasible, consideration should be given to the use of well-controlled, ready-mixed concrete or centrally batched, dry mixes. The size of the latter should suit the maximum capacity of the mixer at the site (see Section 7.1.2). Dry pigmented mortar and refractory concrete mixes are available in bagged form, ready for mixing with water. Lightweight concrete is dealt with in Section 33.2.2(d).[107]

3.4 STEEL AND FIBRE REINFORCEMENT

3.4.1 STEEL REINFORCEMENT

(a) Bar and fabric.

Quality and design requirements for steel reinforcement are dealt with in the standard specifications (see Tables 3.24 and 3.25) and in

TABLE 3.24   **QUALITY OF REINFORCEMENT**

| Type | Standard Specification | | |
	Australian	British	ASTM
Cold-worked, stress-relieved	1302	4461; 4486	A29; A311; A331; A434; A500
Deformed	1302	4449	A496; A615-7
Hard-drawn or high-tensile	1303; 1310 to 1313	2691; 3617; 4482; 4486	A82; A227; A416; A421; A431; A615-7
Hard-drawn fabric	1304	4483	A82; A185; A497
Lightweight (for brickwork)	Int. 325	CP 121, Pt 1	—
Plain	1302	4449	A184; A185; A615-7
Structural sections	1131; 1163; 1204; 1205; 1227; 1250; 1256; 1405; 1443; 1511	4; 449; 2994; 3500; 4360; 4848	A6; A36; A53; A242; A283; A306; A370; A500-1

Appendix IV and the Glossary (Appendix VI), while welding is referred to in Appendix I. Steel bars, wires and wire fabric, prepared and delivered from stockists, should be identifiable as to material history, type, tests and strength specifications.[132, 141]

Steel reinforcement, as suitably illustrated and grade designated on working drawings, may be bent cold or at a cherry-red heat not exceeding 850°C without subsequent quenching. Bends and hooks should comply with the requirements of relevant standards (e.g. AS 1480, BS 1478, 4466, CP110 Part 1, CP114 and CP116, or ACI Committee 318). By stacking reinforcement systematically in covered racks, it is protected, kept clean, straight and true to shape, also it is accessible for withdrawal in a desired order. Bundles of bent rods are labelled with diagram numbers for easy identification. At the time of use, reinforcement must be free from loose mill-scale, loose rust, mud, oil, grease, frost and other coatings which would reduce its bond strength. Slight rust is acceptable and steel analysis is cited in BS *Handbook 19*.

The reinforcement must be accurately located, securely fastened and be maintained in position with the aid of fastening hooks, ligatures, stirrups, supports or jigs during the placing of concrete. Fixing devices include soft-iron tying wire (diameter 1·60 and 1·25 mm), spring-wire clips, concrete or asbestos-cement cover blocks and ring spacers, metallic or plastic bar-chairs at 1000 mm maximum centres (both ways), and spot welding (if specified).

Cover is measured as the minimum distance between the outside of reinforcement (including stirrups, ties and helices) and the nearest permanent surface of the concrete. The latter excludes subsequent plaster, render or other finishing coat (if any). It varies with the type of member, requisite fire-resistance rating and condition of exposure such as are specified in AS 1480 and 1481, BS CP110 Part 1 and CP114–CP116, also by ACI Committees 318 and 324.[62, 63]

In ordinary reinforced concrete members, the cover should be not less than 40 mm in columns, 25 mm in beams and 20 mm in walls and slabs in sheltered locations. These values must be increased by 10–15 mm in locations with external exposure. They must be increased by 35–40 mm for main members or 45 mm for wall and floor slabs when the ground is used as formwork. A 65–100 mm cover of 30–40 MPa concrete, with protective treatments in highly aggressive environments, is required when reinforced concrete is exposed to corrosive liquids or seawater, the greatest hazard of the latter being in the intertidal zone (see Sections 19.1.2 and 19.1.3).

TABLE 3.25

DESIGN CRITERIA

Subject	Specification, Code or Committee*			
	AS	BS	ASTM	ACI
Aluminium structures, building components and alloys	1664; 1665; 1734; 1865; 1866; CA53; H46; H49-H52; H58; H61; H72	1161; 1470-7; 2032; 2033; 4300/15; 4872, Pt 2; CP118	B150; B169; B209-11; B221; B308; B483	—
Bending dimensions of steel reinforcement	1480, Suppl. 1 and MP28	1478; 4466; CP114	—	—
Composite steel-concrete construction	1480, Suppl. 1 and MP28	CP117, Pts 1 and 2	—	—
Loading code	1170, Pts 1 and 2	CP3(v), Pts 1 and 2	—	318; 318C; 437; 442‡
Metallic arc-welding in building construction	1554, Pts 1-3	499; 588; 1140; 1856	—	—
Oxyacetylene welding of mild steel	B29	693	—	—
Precast concrete drainage pipes	1342; 1597	556	C14; C76; C118	—
Prestressed concrete	1481	CP110; CP115; CP116; 5337	—	318; 344; 423‡
Schedule of unit weights of building materials	—	648; CP3(v), Pt 1	—	—
Serviceability and durability	1480, Suppl. 1 and MP28	CP3(ix),CP110,Pt1,App.A	—	—
Steel structures[116]	1163; 1250; 1538; MA1	153; 449; B157; CP113; CP2008	—	—
Structural recommendations for load-bearing walls†	1475; 1480; 1640	CP111; CP114	—	—
Structural use of reinforced concrete and precast concrete	1480, Suppl. 1 and MP28	CP110, Pts 1-3; CP114; CP115; CP116	—	317; 318; 318C; 324; 326; 333; 512‡
Tests on members, structures, panels and truss assemblies	1012.11; 1480	CP114; 1881	E72; E73; Pt 14	—

* See Abbreviations, Appendix VII.
† Standards on masonry construction are given in Table 27.2.
‡ See Appendix IV, 2 and Section 34.1.4.

Where precast concrete units have a cement content of at least 385 kg/m³ and have a water absorption not greater than 8 per cent for normal-weight concrete or 10 per cent for lightweight-aggregate concrete, the described covers to reinforcement may be reduced by 5 mm. The cover in wall panels, exposed to weather but not prolonged water inundation, may be minimised at 20 mm. The requisite cover not only excludes the thickness of applied heterogeneous finishes, but it must be increased to provide for surface loss by acid-etch cleansing. It should not be less than the diameter of reinforcement, pipe or conduit being protected (AS 1480, BS CP110 Part 1).[62, 63]

For concrete culvert, drainage, sewer and water pipes, the requisite cover is specified in relevant standards listed in Appendix I. In spun-concrete poles, the minimum watertight cover is 15 mm (BS 607 and 1308). Fire-resistant requirements are considered in Section 15.2.4 and the covers for lightweight and prestressed concretes are dealt with in Sections 33.3.2(e) and 34.1.3.

Permissible tolerances of cover are ± 5 mm for 25 mm or more, +5 mm to − 3 mm for less than 25 mm, and ± 2 mm for 15 mm or less. Reinforcing rods should be spaced or grouped so as to accommodate the coarse aggregate in a concrete mix (see Section 4.1) and internal vibrators if these are used. In structural concrete, the permissible tolerance in the finished size of members or thickness of slabs is + 5 mm to − 0 mm. In long precast concrete members, the permissible deviation in alignment is 5 mm in 6 m. In abraded finishes, cover must allow for the removal of matrix.

Reinforcement may be used as part of a lightning protective system, in accordance with BS CP326, and reinforcement connectors are available for making free-standing, full-strength splices to short starter bars at construction joints.[39] Lime wash applied to projecting reinforcement, or temporary plastic tubes placed around bars, prevents the spillage of rust-carrying water onto finished surfaces during construction in damp weather.

Selected references relating to steel, reinforcement and reinforced concrete generally include the following items.

Abrasive blast cleaning: AS 1627 Part 4; and BS 4232.
Bar-bending, dimensions and scheduling: BS 4466.
Bar-splicing systems: Reference 39 of Part 1.
Bond stress: Reference 137 of Part 1.
Brittle fracture: Reference 90 of Part 1; Reference 141 of Part 4; and Glossary, Appendix VI.
Cold working and strain hardening: Glossary, Appendix VI; and

ASTM A143.

Corrosion inhibition: Section 22.3; AS MA1.5; and BS CP2008.

Crack width and corrosion of reinforcement: Sections 15.2.2(a), (b), 22.3.2, 31.2 and 34.1.1.

Detailing of structures: Section 27.1.7; References 68, 88 and 130 of Part 1; and Park and Paulay, Appendix IV, 2(a)(i).

Estimating reinforcement quantities: References 67, 106 of Part 1.

Expanded metal reinforcement: Reference 148 of Part 1.

Galvanised reinforcement: Table 22.1.

High-strength reinforcing steel: Table 3.24; Sections 34.1.1 and 31.1.4(b); Glossary, Appendix VI; and References 16, 77 and 138 of Part 1.

Hydrogen embrittlement: Sections 3.1.2(b), 15.2.2(a) and 34.1.4(b); and Glossary, Appendix VI.

Placing, fixing and jigging: References 40 and 66 of Part 1; and Reference 58 of Part 5.

Prestressing tendons: Chapter 34.

Priming and protection: Section 22.3.2; Table 22.1; and Sections 33.2.3(d)(ii), 33.3.2(e), 33.3.3(f) and 33.4.4.

Specifications: Tables 3.24 and 3.25; Appendix I; BS 5328, CP110; References 64, 78, 82, 87, 96 and 110 of Part 1; and Reference 136 of Part 3.

Strain ageing and stress relieving: Glossary, Appendix VI.

Stress corrosion: Glossary, Appendix VI; Reference 19 of Part 4.

Testing: Chapter 14; Appendixes I and IV, 4(b).

Workmanship: Reference 118 of Part 1.

(b) Steel deck.

Cold-formed steel decks of galvanised, fluted, embossed, cellular or trapezially ribbed structural units (600 mm wide prior to joining) not only serve as permanent formwork but act as positive, soffit-bonded reinforcement in suspended, composite concrete slabs. For 2–3-hour fire-emergency conditions, when the steel sheathing loses structural value, supplementary steel fabric serves to resist bending moments induced in slabs of minimum thickness 120 mm. For design purposes, Supplement 1 to AS 1480 stipulates that the overstress potentiality of this reinforcement to meet unrelieved positive moment should be limited to 25 per cent.

Details of fire-rated structural systems and fire-tested prototypes are given in statutory and proprietary literature to meet the typical fire-resistance ratings of Table 34.1. Structural, modular and installation details are subject to approval by the appropriate local authorities concerned. They include flexural continuity in multiple spans, job scheduling, early-fixing, fusion welding and temporary propping prior to proper concreting. Also included are utility ducts, low sound transmission and vibration, built-in fire emergency provisions, and durability precautions against corrosion promoters and moist corrosive atmospheres (see Sections 15.2.2(a), 15.2.4 (Table 22.1) and 34.1.3(b)).[139]

3.4.2 FIBRE REINFORCEMENT

Fibre reinforcement, as typified in Table 3.26, is incorporated in cementitious composites primarily for crack arresting and ductility improvement purposes. Its effectiveness is dependent on such factors as fibre configuration, orientation, alignment and content, fibre aspect ratio (length/diameter), modular ratio (Elastic modulus, E of fibre/E of matrix), moist alkali attack at pH values of up to 13, matrix formulation and fabrication (see Section 7.1.2), fibre coating and admixtures (see Section 6.4 and Appendix I), vibratory and pressure or roller compaction (see Section 23.1), vacuum dewatering (see Section 23.4), injection or extrusion moulding, working environment, and appropriate means of moist or steam curing (see Sections 9.2 and 9.4). In this category, high-pressure steam curing is highly beneficial with asbestos-reinforced, silica-enriched, portland cement products (see Sections 9.4.6, 15.2.3(f) and 33.2.3(a)).[35, 50, 65, 83, 122-128, 133-136]

Depending on fibre orientation and the efficiency of fabrication of the composite, the volume of fibre that can be effectively accommodated to carry the precrack sustained load varies from 2–4 per cent (for random-solid arrays of steel fibre in concrete) to 5–10 per cent (for two-dimensional fibreglass-reinforced cement sheets). For ease of handling (see Section 7.1.2), the fibre aspect ratio of chopped steel fibres is typically 100–150, the higher limit or a deformed or specially coated profile being preferable for improved stress transfer and crack inhibition through bond. In premix projects, the fibres used are shorter than those employed in spray applications for optimum results.

Table 3.27 indicates that steel-fibre-reinforced concrete (proportioned with 2 per cent fibres by volume, 10 mm coarse aggregate and a 0·5 water/cement gravimetric ratio) has an impact strength and failure strain some three times those of the basic plain concrete. Other properties are improved also at more than

TABLE 3.26

FIBRE MATERIALS

Type	Diameter (mm)	Length (mm)	Specific Gravity	Tensile Strength (MPa)	Elastic Modulus, E (MPa)	Strength-loss Temperature (°C)
Steel	0·25-0·65	25-50	7·8	3150	210×10^3	300-750
AR Glass*	0·01-0·02 200/roving	25-50	2·7	2500	80×10^3	800
Asbestos						
Chrysotile "white"	20×10^{-4} (min.)	1-19 (max. 100)	2·6	3100	68×10^3	300-1450
Crocidolite "blue"	$0·1 \times 10^{-3}$	10-25 (max. 35)	3·4	3500	196×10^3	300-1100
Polypropylene	4×10^{-3} (min.)	25-50	0·9	400	8×10^3	160 mp
Portland cement	—	—	2·0	4	21×10^3	400

* "Cem-FIL" alkali-resistant, zirconia fibreglass.

proportionate cost and, while some rust spotting of exposed fibres can occur, its effect is limited to within 2 mm depth of the concrete (see Section 22.3 and Reference 180 of Part 5).

TABLE 3.27 CONCRETE PROPERTIES

| Type of Concrete | Strength | | | | | Elastic Modulus (MPa) | Failure Strain (per cent) |
	Tensile (MPa)	Com-pressive (MPa)	Flexural (MPa)	Shear (MPa)	Impact Ratio		
Steel-fibre	2·5	49	8	5	3	17×10^3	0·045
Plain	2·1	44	5	3	1	16×10^3	0·015

Built-up sheet and shell cementitious composites are expeditiously fabricated by spraying chopped, alkali-resistant fibreglass and cement slurry into a suction or shallow mould. A product of this type, made with 5 per cent of fibre by weight and with a bulk density of 2100 kg/m³, has a predictable 5-year-exposure strength in flexure of about 20 MPa.[136]

Alternatively, reinforcing tape of cement or latex-cement-impregnated fibre-fabric is drawn between suction rollers for integral use in panel production. Or, in conjunction with dewatering technique, the composite is wound onto a mandrel in connection with pole, pipe or conduit manufacture. Wet-usage durability of the product is improved by adding a pozzolan to the cement matrix (see Sections 17.1 and 17.2).

In fibrous cementitious composites, an inherent crack-arrest mechanism mitigates the severity of effect of possible impact, change-of-strain and crack propagation in building elements, thereby preserving the structural integrity of sections, particularly those containing concealed flaws. For fibreglass-reinforced cement paste, a fracture-toughness/crack-growth ratio, which expresses this factor, is four times that of cement mortar and over twice that of ordinary concrete. Polypropylene fibres, when incorporated in shell piles, are similarly beneficial in contributing to improved handling properties against damage by impact (see Reference 162 of Part 4).[133-136]

Appropriately designed and pretested fibrous cementitious composites provide suitable alternatives to timber members, scantlings or studs. They have comparable properties, in addition to improved fire resistance and rot resistance for building construction. Their wide range of potentially useful applications include lightweight veneered insulating panels and shell cladding, folded-plate and tubular sections, tied-back-coffer reinforced earth revetments, sewer and tunnel linings, industrial pallets, permanent forms, steel and reinforced concrete envelopes, and boat hulls built either alone or in conjunction with ferrocement (see Section 35.1). Mortarless concrete masonry walls can be strongly constructed in about one-third of the normal time with a fibreglass cement mix, which is sprayed or trowelled onto the surface of blocks that are laid without mortar.

The serviceability and ultimate strength of flexural members can be substantially improved by casting them within permanent forms of high-modulus fibrous cement, 6–9 mm thick. The main reinforcement may be placed directly in contact with the cementitious casing which contains, for example, 5 per cent of glass or 10 per cent of asbestos fibres, these being randomly distributed as two-dimensional reinforcement in the cement matrix. A strong bond preserves the integrity of this composite construction.

Fibre-reinforced concrete is applicable to prestressed concrete flanges, post-tensioned anchorage zones, tilt-up panels, blast and thermal-gradient-resistant fabrications, shock and vibration-loaded foundations, and pavement and bridge-deck overlays, where service durability and crack control are important prerequisites. A tough facing of the material, applied to hydraulic structures and marine breakwater units, is distinctly beneficial where surfaces may be subjected to waterborne cavitation, erosion, impact and wave damage.[145, 146]

Microcracks formed during the testing of fibre-reinforced cement specimens can be studied by a resin impregnation, vacuum treatment and etching technique.[149] Fibrous sprayed mortar and test guidelines are considered in Section 24.2.7, Chapter 14 and Appendix IV, 4(b).

REFERENCES*

1. ABRAMS, D. A. "Design of Concrete Mixtures". *Bulletin No. 1*, Structural Materials Research Laboratory, Lewis Institute (Chicago, December 1918), pp. 1-20.
2. WASHA, G. W. and WENDT, K. F. "Fifty Year Properties of Concrete". *ACI Journal*, 72, 1 (January 1975), pp. 20-28; 7 (July 1975), pp. 369-70.
3. "Computer Control at an Italian Cement Works". *Cement and Lime Manufacture*, 42, 2 (March 1969), pp. 19-26.
4. KESLER, C. E. *et al.* "Concrete–Year 2000". *ACI Journal*, 68, 8 (August 1971), pp. 581-89.
5. CZERNIN, W. *Cement Chemistry and Physics for Civil Engineers*, Crosby Lockwood, London, 1962.
6. FELDMAN, R. F. and RAMACHANDRAN, V. S. "Influence of $CaSO_4.2H_2O$ Upon the Hydration Character of $3CaO.Al_2O_3$". *Magazine of Concrete Research*, 18, 57 (December 1966), pp. 185-96.
7. HANSEN, W. C. "Quick and False Set in Portland Cements". *ASTM Materials Research and Standards*, 1, 10 (October 1961), pp. 791-98.
8. GOTTLIEB, S. "Vertical Kiln Operation in Australia". *Cement and Lime Manufacture*, 34, 6 (November 1961), pp. 83-90.
9. KUHL, H. *Zement Chemie*, I, II and III, Veb Verlag Technik, Berlin, 1965.
10. REHSI, S. S. and MAJUMDAR, A. J. "Use of Small Specimens for Measuring Autoclave Expansion of Cements". *Magazine of Concrete Research*, 19, 61 (December 1967), pp. 243-46; 20, 64 (September 1968), pp. 187-88.
11. BOGUE, R. H. *The Chemistry of Portland Cement*. Reinhold, New York, 1955.
12. LEA, F. M. *The Chemistry of Cement and Concrete*, Edward Arnold, London, 1970.
13. KANTRO, D. L., COPELAND, L. E. and ANDERSON, E. R. "An X-ray Diffraction Investigation of Hydrated Portland Cement Pastes". *ASTM Proceedings*, 60 (1960), pp. 1020-35.
14. AMERICAN PUBLIC WORKS ASSOCIATION. *Standard Specifications*. Building News Inc., Los Angeles, 1966.
15. KLEIN, A., PIRTZ, D. and ADAMS, R. F. "Thermal Properties of Mass Concrete During Adiabatic Curing". *ACI Publication SP-6, Paper 10*, 1963.
16. ACI COMMITTEE 439. "Uses and Limitations of High Strength Steel Reinforcement". *ACI Manual of Concrete Practice, Part 2*, 1976.
17. "Klein Symposium on Expansive Cement Concretes". *ACI Special Publication SP-38*, 1973; *ACI Journal*, 70, 8 (August 1973), pp. 590-95.
18. KLEIN, A. and TROXELL, G. E. "Studies of Calcium Sulphoaluminate Admixtures for Expansive Cements". *ASTM Proceedings*, 58 (1958), pp. 986-1008.
19. MIKHAILOV, V. V. "Stressing Cement: The Mechanism of Self-stressing Concrete Regulation". *Proceedings, Fourth International Symposium on the Chemistry of Cement*, 1960.
20. BERTERO, V. V. "Curing Effects on Expansion and Mechanical Behaviour of Expansive Cement Concrete". *ACI Journal*, 64, 2 (February 1967), pp. 84-96.
21. LEWICKI, E. "Building Research and Development Work at the Central Research Institute for Industrial Architecture, Moscow". *Library Translation No. 67*, C & CA, London, 1957.
22. ROBSON, T. D. *High-alumina Cements and Concrete*, Contractors Record, London, 1962.
23. ROBSON, T. D. "The Characteristics and Applications of Mixtures of Portland and High-alumina Cements". *Chemistry and Industry*, 1 (January, 1952), pp. 2-7.

---

*Titles corresponding to the abbreviations used herein are given in Appendix VII. References to certain technical documents, which are subject to periodical change, are intended to relate to their latest editions, which may be determined from technical institutions, publishers, retailers or libraries.

24. GOTTLIEB, S. "Production of Blended Cement". *Rock Products*, 53, 8 (August 1950), pp. 174-78.
25. WUERPEL, C. E. "Masonry Cement". *Proceedings, Third International Symposium on the Chemistry of Cement*, London, 1952. Illinois, 1976.
26. UNITED STATES BUREAU OF RECLAMATION. "Effect of Cement Type on Resistance of Concrete to Sulphate Attack". *Laboratory Report C-828*, Denver, Colorado, 1958.
27. KIEL, F. "Slag Cements". *Proceedings, Third International Symposium on the Chemistry of Cement*, London, 1952.
28. BAXTER, D. J. and BOARDMAN, J. M. "Manufacture, Properties and Use of Supersulphated Cement in Civil Engineering Works". *Civil Engineering and Public Works Review*, 57, 669 (April 1962), pp. 472-74; 670 (May 1962), pp. 627-29; 671 (June 1962), pp. 778-80; 672 (July 1962), p. 879; 673 (August 1962), p. 998; 674 (September 1962), p. 115.
29. ORCHARD, D. F. "The Handling of Cement". *ICE Proceedings*, 2, 5 (September 1953), pp. 616-42.
30. RYAN, W. G. "Accelerated Testing of Concrete—The State of the Art". *IE Aust. Journal*, 43, 9 (September 1971), pp. 6-10; 12 (December 1971), p. 9.
31. ERNTROY, H. C. "Variation of Works Tests Cubes". *Research Report No. 10*, C & CA, Britain, 1960.
32. KING, J. W. H. "Accelerated Testing of Concrete". *ICE Journal*, 40 (May 1968), pp. 125-29; 45 (March 1970), pp. 535-41.
33. LEECH, T. J. D. "The Snowy Mountains Scheme and the Application of Scientific Services". *IE Aust. Journal*, 30, 3 (March 1958), pp. 99-112.
34. MALHOTRA, V. M. *et al.* "Accelerated Method of Estimating 28-Day Strength of Concrete". *ACI Journal*, 66, 11 (November 1969), pp. 894-97; 67, 5 (May 1970), pp. 424-34; 68, 12 (December 1971), pp. 963-67.
35. PARRATT, N. J. *Fibre-reinforced Materials Technology*. Van Nostrand, London, 1972.
36. STEINOUR, H. H. "Concrete Mix Water—How Impure Can It Be?". *PCA R & DL Journal*, 2, 3 (September 1960), pp. 32-50.
37. UNITED STATES ARMY CORPS OF ENGINEERS. "Method of Test for the Staining Properties of Water". *Handbook for Concrete and Cement*, Vicksburg, Mississippi, 1949.
38. KIRKHAM, R. H. H. and HIGGINS, G. E. "Control Methods for Measuring the Moisture in Concrete Aggregates". *Civil Engineering and Public Works Review*, 50, 593 (November 1955), pp. 1221-24.
39. *Reinforcing Bar Splices*, Concrete Reinforcing Steel Institute, Chicago, Illinois 1976; *Concrete* (London), 7, 12 (December 1973), p. 55.
40. DISNEY, L. A. and REYNOLDS, C. E. *Reinforcement for Concrete*, C & CA, Wexham Springs, Slough, 1973.
41. BUTCHER, W. S. "Simple, Reliable Method for Determining Moisture in Sand". *Constructional Review*, 31, 10 (October 1958), pp. 28-29.
42. PROUDLEY, C. E. "Rapid Determinations of Free Moisture in Aggregates". *Concrete*, 56, 8 (August 1948), pp. 40-41.
43. RUFFLE, N. J. "Concrete Control for Small Works". *Structural Engineer*, 29, 8 (August 1951), pp. 211-20.
44. VAIL, P. G. "Use of Pressure Type Air Entrainment Indicator for Aggregate Moisture Tests". *ACI Journal*, 21, 3 (November 1949), pp. 221-24.
45. MIDGLEY, H. G. "Effect of Lead Compounds in Aggregate upon the Setting of Portland Cement". *Magazine of Concrete Research*, 22, 70 (March 1970), pp. 42-44.
46. McINTOSH, J. D. "The Siphon Can Test for Measuring Moisture Content of Aggregates". *Constructional Review*, 26, 1 (May 1953), pp. 20-26.
47. VAN ALSTINE, C. B. "Mixing Water Control by Use of a Moisture Meter". *ACI Journal*, 27, 3 (November 1955), pp. 341-47; 28, 6, 2 (December 1956), pp. 1209-13.
48. *Concrete Industries Year Book*. Pit and Quarry Publications, Chicago,

49. UNITED STATES BUREAU OF RECLAMATION. *Concrete Manual*, United States Government Printing Office, Washington, 1975.
50. MAJUMDAR, A. J. *et al.* "Glass Fibre Reinforced Cement". *Current Papers CP79/74, CP80/75, CP94/75*, Building Research Establishment, Garston, Watford, Britain.
51. ACI COMMITTEE 621. "Selection and Use of Aggregates for Concrete". *ACI Journal*, 58, 5 (November 1961), pp. 513-42; 59, 2 (February 1962), N.L. 28.
52. RUNDQUIST, W. A. "Washing Plants". *Construction Methods and Equipment*, 39, 5 (May 1957), pp. 237-64; 39, 6 (June 1957), pp. 178-210.
53. RUNDQUIST, W. A. "Sand Washing, Classifying Equipment. Which Type is Best for You?". *Rock Products*, 61, 11 (November 1958), pp. 82-86, 126, 131-32.
54. HILLESS, A. J. "The Production and Handling of Concrete Materials, Tinaroo Falls Dam". *IE Aust. Journal*, 31, 3 (March 1959), pp. 69-80.
55. ALDERMAN, A. R., GASKIN, A. J., JONES, R. H. and VIVIAN, H. E. "Studies in Cement–Aggregate Reaction". *CSIRO Bulletin*, 229, 1947.
56. MIELENZ, R. C., GREENE, K. T. and BENTON, E. J. "Chemical Test for Reactivity of Aggregates with Cement Alkalies; Chemical Processes in Cement–Aggregate Reaction". *ACI Journal*, 19, 3 (November 1947), pp. 193-221; 20, 4, 2 (December 1948), p. 224.
57. SCHOLER, C. H. and SMITH, G. M. "Rapid Accelerated Test for Cement–Aggregate Reaction". *ASTM Preprint*, 89, 1954.
58. WOOLF, D. O. and SMITH, T. R. "Rapid Test for Alkali–Aggregate Reaction". *Concrete*, 57, 2 (February 1949), pp. 18-22.
59. BOOTH, W. J. "Visual Determination of Size Fractions in Sand Sieve Analysis". *ACI Journal*, 32, 2 (August 1960), p. 226.
60. ACI COMMITTEE 304. "Measuring, Mixing, Transporting, and Placing Concrete". *ACI Journal*, 69, 7 (July 1972), pp. 374-414; 70, 1 (January 1973), pp. 55-60.
61. TANAKA, T. "What Do You Know About Bin Design?". *Rock Products*, 64, 2 (February 1961), pp. 115-20, 124-25.
62. ACI COMMITTEE 318. *Building Code Requirements for Reinforced Concrete.* See Appendix IV, 2(a)(i).
63. ACI COMMITTEE 324 and ACI–ASCE COMMITTEE 512. See Appendix IV, 2(a)(i).
64. *The Concrete Year Book.* Concrete Publications, London, 1976.
65. MONFORE, G. E. "Review of Fibre Reinforcement of Portland Cement Paste, Mortar and Concrete". *PCA R & DL Journal*, 10, 3 (September 1968), pp. 43-49.
66. *CRSI Handbook* and *CRSI Structural Lightweight Concrete Design Supplement.* Concrete Reinforcing Steel Institute, Chicago, Illinois, 1975.
67. BAKER, J. C. "Estimating Reinforcement Quantities in Concrete Building Construction". *Constructional Review*, 36, 1 (January 1963), pp. 24-29.
68. ACI COMMITTEE 315. *Manual of Standard Practice for Detailing Reinforced Concrete Structures*, American Concrete Institute, Detroit, Michigan, 1974.
69. NEVILLE, A. M. "A Study of Deterioration of Structural Concrete made with High-alumina Cement". *ICE Proceedings*, 25 (July 1963), pp. 287-324; 28 (May 1964), pp. 57-84; *High Alumina Cement Concrete*, Construction, Lancaster, 1975; *New Civil Engineer*, 135 (March 20 1975), pp. 12-13.
70. DEPARTMENT OF SCIENTIFIC AND INDUSTRIAL RESEARCH. "Supersulphate Cement". *BRS Digests*, 130 and 135, 1960.
71. STANDARDS ASSOCIATION OF AUSTRALIA. *Dense Natural Aggregates*, AS 1465; *Dense Slag Aggregates*, AS 1466; *Lightweight Aggregates*, AS 1467.
72. STANDARDS ASSOCIATION OF AUSTRALIA. *Sampling and Testing Aggregates*, AS 1141, 1465.

73. BRITISH STANDARDS INSTITUTION. *Sampling and Testing of Mineral Aggregates, Sands and Fillers*, BS 812.
74. BRITISH STANDARDS INSTITUTION. *Concrete Aggregates from Natural Sources*, BS 882, 1201.
75. BRITISH STANDARDS INSTITUTION. *Building Sands from Natural Sources*, BS 1198, 1199 and 1200.
76. WOOLF, D. O. "An Improved Sulfate Soundness Test for Aggregates". *ASTM Bulletin*, 213 (April 1956), pp. 77-84.
77. FREDMAN, R. "High Tensile Steel as Concrete Reinforcement". *Prestress*, 12 (June 1963), pp. 15-23.
78. ACI COMMITTEE 301. "Specifications for Structural Concrete for Buildings". *ACI Publication* SP-15 (73) and *Manual of Concrete Practice*, 2, 1976.
79. LEVITT, M. "An Appreciation of the Hydrobot". *Civil Engineering and Public Works Review*, 59, 690 (January 1964), pp. 59-63.
80. GUINIER, A. *et al. Chemistry of Cement*. Fifth International Symposium, Cement Association of Japan, Proceedings I–IV, Tokyo, 1969.
81. GRIFFIN, D. F. and HENRY, R. L. "Effect of Salt in Concrete on Compressive Strength, Water Vapor Transmission, and Corrosion of Reinforcing Steel". *ASTM Proceedings*, 36 (1963), pp. 1046-78.
82. CONCRETE INSTITUTE OF AUSTRALIA. *National Specification for Concrete*. Sydney, 1976.
83. SHAH, S. P. "New Reinforcing Materials in Concrete". *ACI Journal*, 71, 5 (May 1974), pp. 257-62; 11 (November 1974), pp. 582-84; 68, 2 (February 1971), pp. 126-35; 8 (August 1971), pp. 626-33.
84. DEPARTMENT OF SCIENTIFIC AND INDUSTRIAL RESEARCH. "Analysis of Water Used or Encountered in Construction". *BRS Digest*, 90, II, 1964.
85. LEES, G. "The Measurement of Particle Elongation and Flakiness: A Critical Discussion of British Standard and Other Methods". *Magazine of Concrete Research*, 16, 49 (December 1964), pp. 225-30; 17, 52 (September 1965), pp. 161-62.
86. MACKEY, R. D. "The Measurement of Particle Shape". *Civil Engineering and Public Works Review*, 60, 703 (February 1965), pp. 211-14.
87. WATSON, D. A. *Specification Writing for Architects and Engineers*, McGraw-Hill, London, 1964.
88. BRITISH STANDARDS INSTITUTION. *Bending Dimensions and Scheduling of Bars for the Reinforcement of Concrete*, BS 4466.
89. *World Cement Directory*. Cembureau, Malmo, Sweden, 1976.
90. DAVIES, S. R. "Failure of Reinforced Concrete due to Brittle Fracture of Reinforcement". *Civil Engineering and Public Works Review*, 60, 710 (September 1965), pp. 1291-95; 61, 715 (February 1966), p. 177.
91. DAVIS, D. E. "Strength of Concrete at Very Early Ages". *Civil Engineer in South Africa*, 7, 6 (June 1965), pp. 133-38.
92. LEA, F. M. *et al. Symposium on Concrete Quality*, Cement and Concrete Association, Britain, 1964.
93. LILJESTROM, W. P. "Shrinkage-compensating Cement Concrete". *Concrete Construction*, 21, 2 (February 1976), pp. 63-66.
94. BERTERO, V. and POLIVKA, M. "Effect of Degree of Restraint on Mechanical Behaviour of Expansive Concrete". *ASTM Proceedings*, 64 (1964), pp. 797-815.
95. DOLAR-MANTUANI, L. "Petrographic Examination of Natural Concrete Aggregates". *Highway Research Record*, 120 (1966), pp. 7-17.
96. BRITISH STANDARDS INSTITUTION. *Methods for Specifying Concrete*, BS 5328.
97. CARLSEN, H. "Behaviour of Alkalies during the Burning Process". *Rock Products*, 69, 5 (May 1966), pp. 87-8, 157.
98. LUNT, B. G. and MEULEN, G. J. R. VAN DER. "Bulk Storage of Cement for a Concrete Laboratory". *Magazine of Concrete Research*, 17, 51 (June 1965), pp. 103-5; 18, 54 (March 1966), p. 46.
99. LOVERN, J. D. "Important Variables Affecting Moisture Control". *Modern Concrete*, 30, 4 (August 1966), pp. 44-47.

100. LAWRENCE, C. D. "Properties of Cement Paste Compacted under High Pressure". C & CA Britain, *C & CA Research Report*, 19, 1969.
101. HUGHES, B. P. and BAHRAMIAN, B. "An Accurate Laboratory Test for Determining the Absorption of Aggregates". *Materials Research and Standards*, 7, 1 (January 1967), pp. 18-23.
102. POPOVICS, S. "Use of Fineness Modulus for the Grading Evaluation of Aggregates for Concrete". *Magazine of Concrete Research*, 18, 56 (September 1966), pp. 131-40.
103. BAILEY, R. B. and PAUL, J. F. "Rotary Cement Kiln Experimentation Utilizing a Hybrid Computer Simulation". *Minerals Processing*, 8, 3 (March 1967), pp. 16-21.
104. ACI COMMITTEE 223. "Expansive Cement Concretes—Present State of Knowledge". *ACI Journal*, 67, 8 (August 1970), pp. 583-610; 68, 4 (April 1971), pp. 293-96; *ACI Publication* SP-38, 1973.
105. ENDERSBEE, L. A. "Brittle Fracture in Concrete and Rocks". *IE Aust. CE Transactions*, CE9, 2 (October 1967), pp. 217-34.
106. BLACK, W. C. "Automatic Data Processing for Reinforcing Bar Estimates, Detailing, and Shop Order Entry". *ACI Publication*, SP-16 (1967), pp. 109-20.
107. TOBIN, R. E. "Lightweight Ready-Mix: A New Approach". *Concrete Products*, 70, 10 (October 1967), pp. 47-50, 71.
108. SWENSON, E. G. "Estimation of Strength Gain of Concrete". *Engineering Journal*, 50, 9 (September 1967), pp. 27-32.
109. PERRIN, L. "Automatic Control of Concrete Weigh Batching". *Contracting and Construction Engineer*, 22, 6 (February 1968), pp. 39-43.
110. DEPARTMENT OF THE ENVIRONMENT. *Specification for Road and Bridge Works*, Britain, 1976.
111. HANSEN, W. C. "Chemistry of Sulphate-resisting Portland Cements". *Performance of Concrete*, National Research Council, Canada. Toronto University Press, 1968.
112. FREEDMAN, S. "White Concrete". *Modern Concrete*, 32, 8 (December 1968), pp. 30-34; 9 (January 1969), pp. 30-35.
113. BENNETT, E. W. and COLLINGS, B. C. "High-early-strength Concrete by Means of Very Fine Portland Cement". *ICE Proceedings*, 43 (July 1969), pp. 443-52.
114. BUTH, E., IVEY, D. L. and HIRSCH, T. J. "Dirty Aggregate, What Difference Does It Make?" US Highway Research Board, *Highway Research Record*, 226, 1968.
115. RYAN, W. G. J. "Use of Fine Ground Granulated Blast Furnace Slag in Concrete". *IE Aust. CE Transactions*, CE11, 1 (April 1969), pp. 88-96.
116. AUSTRALIAN INSTITUTE OF STEEL CONSTRUCTION. *Safe Load Tables for Structural Steel*, AISC, Sydney, 1975.
117. LECOMPTE, P. "Fineness Modulus and Its Dispersion Index". *ACI Journal*, 66, 6 (June 1969), pp. 474-80; 12 (December 1969), pp. 1023-24.
118. SILBER, R. M. "Workmanship and the Draft Unified Code". *Concrete* (London), 3, 11 (November 1969), pp. 449-52.
119. ERNST, K. "Moisture Measurement for Concrete Production". *DOE BRS Library Communication*, 1506, 1969.
120. CEMENT AND CONCRETE ASSOCIATION. "Impurities in Aggregates for Concrete". *C & CA Advisory Note*, CZ 18, 1970.
121. RAEDE, D. "Color Ingredient in Building Products". *Building Materials and Equipment*, 13, 6 (April/May 1971), pp. 60-63, 76.
122. RILEM Symposium, London, 1975. "Fibre-reinforced Cement and Concrete". *Proceedings BRE*, Garston, Britain: Compendium by Prof. A. M. Neville, Construction, Lancaster, Britain, 1975.
123. SWAMY, R. N. "Technology of Steel Fibre Reinforced Concrete for Practical Applications". *ICE Proceedings*, 1, 56 (May 1974), pp. 143-59; (August 1974), pp. 235-56; 2, 57 (December 1974), pp. 701-707; 1, 58 (May 1975), pp. 229-48.
124. KENNY, R. A. "Patterns in Aerated Fibre Concretes". *International Construction*, 11, 5 (May 1972), pp. 81-84.

125. McCURRICH, L. H. and ADAMS, M. A. J. "Fibres in Cement and Concrete". *Concrete* (London), 7, 4 (April 1973), pp. 51-53.
126. SATHER, W. R. and WILSON, J. R. "Fibrous Concrete Topping of Bridge Deck". *Concrete Construction,* 18, 7 (July 1973), pp. 321-24.
127. EDGINGTON, J. *et al.* "Steel Fibre Reinforced Concrete". *BRE Current Paper,* 69/74, 1974.
128. ACI COMMITTEE 544. "Fibre Reinforced Concrete". *ACI Journal,* 70, 11 (November 1973), pp. 729-44; *ACI Publication,* SP-44, 1975.
129. HUGHES, B. P. "Temperature Rises in Low-Heat Cement Concrete". *Proceedings ASCE Structural Division,* 97, ST12 (December 1971), pp. 2807-23.
130. *Reinforced Concrete Detailing Manual.* Aust. CIA Sydney; and C & CA Britain, 1975.
131. ALEXANDER, K. M. "Relationship Between Strength and the Composition and Fineness of Cement". *Cement and Concrete Research,* 2, 6 (June 1972), pp. 663-80.
132. ELEY, W. G. "Reinforcing Steel—A Contractor's View". *Concrete* (London), 7, 6 (June 1973), pp. 22-24.
133. DEMPSTER, D. P. "Cement-coated Tape and Its Possibilities". *Concrete* (London), 7, 12 (December 1973), pp. 34-35.
134. BROWN, J. H. "Failure of Glass Fibre Reinforced Notched Beams in Flexure". *Magazine of Concrete Research,* 25, 82 (March 1973), pp. 31-38.
135. CORNELIUS, D. F. and RYDER, J. F. "New Fibrous Composites as Alternatives to Timber". *BRE Current Paper,* 66/74, 1974.
136. EDITORIAL. "Characteristics of Fibres and What They Do for Concrete". *Concrete Construction,* 19, 3 (March 1974), pp. 103-109; 20, 4 (April 1975), pp. 137-39; 7 (July 1975), pp. 281-83; 8 (August 1975), pp. 345-47; 10 (October 1975), pp. 443-45.
137. ACI COMMITTEE 408. "Bond Stress—The State of the Art". *ACI Journal,* 63, 11 (November 1966), pp. 1161-90; 64, 6, 2 (June 1967), pp. 1569-70.
138. ACI COMMITTEE 439. "Uses and Limitations of High Strength Steel Reinforcement". *ACI Manual of Concrete Practice,* 2, 1976.
139. RESEVSKY, C. G. *et al.* "Steel Floors". *AISC Steel Construction Journal,* 7, 3 (July 1973), pp. 1-16; *Concrete Construction,* 19, 9 (September 1974), pp. 445-47.
140. ATWELL, J. S. F. "Some Properties of Ground Granulated Slag and Cement". *ICE Proceedings,* 2, 57 (June 1974), pp. 233-50.
141. DISNEY, L. A. and REYNOLDS, C. E. *Reinforcement for Concrete.* C & CA, London, 1973.
142. SWAMY, R. N. and ANAND, K. L. "Behaviour of High Alumina Cement Concrete Under Sustained Loading". *ICE Proceedings,* 2, 57 (December 1974), pp. 651-71; 59 (June 1975), pp. 293-307; 59 (December 1975), pp. 853-65.
143. CUSENS, A. R. and JACKSON, N. "Properties of Converted HAC Beams". *Concrete* (London), 9, 5 (May 1975), pp. 30-32.
144. "Mass Concrete Pours". *ICE New Civil Engineer,* 146 (5 June 1975), pp. 22-23; 153 (24 July 1975), pp. 24-25.
145. HOFF, G. C. "Hydraulic Structures and Marine Environments". *RILEM Symposium on Fibre-Reinforced Cement and Concrete,* C & CA, Britain, 1975.
146. ZOLLO, R. F. "Wire Fibre Reinforced Concrete Overlays for Orthotropic Bridge Deck Type Loadings". *ACI Journal,* 72, 10 (October 1975), pp. 576-82.
147. MEHTA, P. K. "Evaluation of Sulphate-resisting Cements by a New Test Method". *ACI Journal,* 72, 10 (October 1975), pp. 573-75.
148. HANSON, J. M. *et al.* "Use of Expanded Metal as Reinforcement in Concrete Slabs". *ACI Journal,* 73, 2 (February 1976), pp. 97-103.
149. JARAS, A. C. "A Technique for Showing Microcracks in Fibre Reinforced Cement". *Cement and Concrete Research,* 6, 3 (May 1976), pp. 377-80.
150. ORCHARD, D. F. *Concrete Technology.* Volume 3. Applied Science, Barking, Essex, England.

PART 2

PROPORTIONING AND WORKABILITY

CHAPTER **4**

DESIGN OF CONCRETE MIXES

4.1 GENERAL INFORMATION

Many procedures have been advanced for the proportioning of concrete mixes, the trend being towards increased detail because of the complexity of the factors involved. Rene Feret[1] determined in 1892 that the compressive strength of concrete and mortar is proportional to the amount of cement in the binding paste, and that it increases as the volume of voids in the aggregate decreases. In 1907, Fuller and Thompson[2] introduced a method for grading aggregate and selecting proportions to produce concrete of maximum density. For a mixture of cement and aggregate, they represented ideal grading by a curve which approached an ellipse for the small-particle fraction and a tangent for the remainder (Fig. 4.1). Mixes thus produced tended to be harsh and to require vigorous compaction.[57]

Duff Abrams (see Chapter 3) demonstrated in 1918 the dependence of strength on the amount of water per unit volume of cement in the concrete. A relationship was derived between the water requirement of concrete and its proportions, including the grading of aggregate as expressed by its fineness modulus. Bolomey[3] proposed in 1926 a modified theoretical curve for guidance with continuous gradings. In the lower third, it contained a sufficient amount of graded small particles to ensure a plastic or workable mix that could readily be compacted manually. It is represented in Figure 4.1 by a parabola with the formula

$$P = A + (100 - A)\sqrt{\frac{d}{D}}$$

where P = percentage of *cement and aggregate* passing a given sieve of size d. D = maximum size of aggregate. A = a coefficient, varying from 12 to 14 for rounded to angular coarse-aggregate particles, respectively.

A modified version of Duff Abrams' water/cement ratio law (see Section 3.2.2) is taken as the basis of current procedures for the proportioning of concrete mixes. In this regard, when designing mixes for strength and workability, consideration must be given not only to the water/cement ratio, but to such factors as the aggregate/cement ratio and the grading, surface texture, stiffness, shape and largest size of the aggregate to be used. Additional precautions in mix design are necessary where the concrete must resist aggressive conditions during its service life (see Chapters 15 and 19).

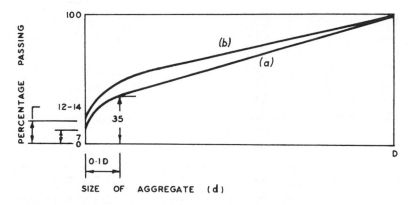

Fig. 4.1 Grading curves. (a) Fuller and Thompson. (b) Bolomey.

In the selection of aggregates for dense, strong concrete, the fine aggregate should be well-graded but without an excess of fines, while the coarse aggregate should contain moderately large particles, but should be deficient in medium-size particles (Plate 3). A high degree of subdivision of stone in an aggregate increases its surface area and, for a given degree of workability, increases the water requirement and shrinkage of the concrete (Fig. 30.1). To ensure effective placement and compaction, the largest particles should not exceed one-third of the narrowest dimension of a reinforced precast member or one-half when the concrete is compacted by high-frequency vibration or pressure, respectively (see Section 27.1); and should not exceed three-quarters of the cover over reinforcing rods or the clear distance between them (see Section 3.4 and ACI Committees 324 and 512 listed in Appendix IV, 2(a)(i)).

These maximum sizes should be adjusted where it is necessary for them to meet certain specified requirements of standard codes for concrete in buildings (e.g. AS 1480; BS CP110 Part 1 and CP114 to CP116). The maximum particle gauge relationship to cover may be waived in solid slabs. They may be made equal where steel reinforcement is widely spaced or high-quality concrete is compacted thoroughly (by mechanical means) around the reinforcement and into the corners of formwork or moulds. A maximum nominal size of 20 mm, however, is usually satisfactory for general purposes in building construction.

Typical water/cement ratio strength relationships (refer to Figs 4.3 and 4.4) are used for preliminary design, and adjustments are made subsequently for variations in strength that may arise from various causes, including variations in cements produced at different mills or at different times in the same mill (see Table 13.5). When making trial mixes for testing, the volume of the mix should be at least the cube of eight times the size of the largest particles. Test specimens should be made at least in triplicate, but statistical methods for sampling and design procedures should be applied in a large programme.

While strength development at ages up to 28 days may be regarded as a criterion of desirable quality, the effect of long-line restraint on early cracking should be taken into account when using a quick-hardening cement. In some instances it may be advantageous to use a slower-hardening cement, which, through early plastic deformation, has a reduced tendency to crack irrespective of the extent of the free-shrinkage characteristic (Sections 3.1 and 31.2).

Various units of measurement are used in different countries (including litres per bag or gallons per bag for water/cement ratio), with the result that international data may have to be adjusted before being used universally. In America, for instance, a US gallon of water (five-sixths of an imperial gallon) weighs 3·78 kg, a ton is equivalent to 0·907 tonne, and sieves are in ASTM designations (see Table 3.21). A bag of cement in Australia and Japan nominally weighs 40 kg, in America 42·6 kg and in Britain it is 50 kg (see Section 5.1 and Appendix VIII). The numbers of bags per tonne are 25, 23·5 and 20, respectively.

Compressive strengths in America and Australia, unless stated otherwise, are based on capped standard cylinders (150 mm diameter and 300 mm height), whereas in Britain they are based on cubes that are tested on accurately cast sides (refer to Fig. 4.15). Pressure, strength and stress conversion data are given in Appendix VIII, 3.

The various methods of mix design described later in some detail are an effective means of proportioning various types of concrete to obtain the requisite strength (Section 3.2.2), workability (Sections 6.1 and 6.2) and normal durability (Section 15.2). They enable available materials to be used advantageously and economically with good concreting procedures in practice.

4.2 FIELD METHODS OF DESIGNING PLASTIC CONCRETE MIXES

4.2.1 APPROXIMATE METHOD

The objective is, firstly, to proportion a mix so that fractional-size particles of aggregate fill the voids between those next larger in size and, secondly, to coat the particles with adequate cement paste of stiff to plastic consistence, so as to form a suitable cementing matrix. An approximate procedure is as follows.

1. Place damp, coarse aggregate in a receptacle.
2. Determine the quantity of water required to fill the voids and appear at the top surface of the aggregate. The volume of dry sand required to fill the voids in the coarse aggregate is equal to this volume of water.
3. Pour water onto this volume of sand (dry or adjusted for bulking) until it is fully inundated. The minimum volume of cement required is equal to the volume of water necessary to fill the voids in the sand.
4. Select a suitable water/cement ratio for the proposed class of concrete as indicated in Table 4.1 (i.e. 0·5–0·7 by weight or 0·75–1·05 by volume; see Sections 3.2.2 and 3.3.6(b)).
5. By means of a trial mix, determine the quantity of cement and water with a requisite ratio for quality or strength that gives to the mix the requisite workability and cohesive properties. The proportions of coarse aggregate and sand are adjusted when necessary in a subsequent mix to improve these characteristics.

Typical dry-batching plant and transit mixer
(Capacity: 50 to 100 m³/h)

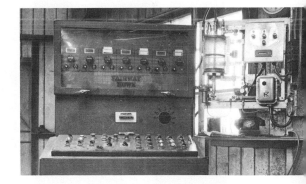

Control console
with an admixture dispenser

Concrete mixing plant at Lakes Dam,
Canberra
(Capacity: 40 m³/h)

1

2

3

4

5

Sequence of Operations

If stone aggregates are to be blended for the first step, select the heaviest combination (kg/m³) in a series of blendings, giving preference to combinations in which coarse particles predominate. The mortar matrix of concrete should be a little greater in volume than the voids in the coarse aggregate, which are about 45 per cent of the gross volume. Thus, for manual compaction, the volume of sand is about one-half that of stone in a mix.

4.2.2 PRACTICAL METHOD

In localities where ideally graded aggregates are not available locally, the Australian Government Department of Housing and Construction uses the following method[4] to obtain the best concrete mix from any given clean sand and stone or gravel aggregate.

1. Take samples of the materials to be used.

2. Examine the coarse aggregate and classify it as either "one size" or "graded". (For example, a 40 mm nominal-size crushed basalt with very few stones less than 19·0 mm would be classed as "one size", whereas a "graded" stone would have an appreciable quantity of all sizes.)

3. Decide on the class of concrete to be prepared and select a suitable water/cement ratio, as indicated in Table 4.1, which is based upon cylinder strengths.

TABLE 4.1   WATER/CEMENT RATIO AND COMPRESSIVE STRENGTH

Class	Water/Cement Ratio	Cylinder Compressive Strength (MPa)	
	by Weight	7 days	28 days
A	0·5	18·5	30·0
B	0·6	14·0	22·0
C	0·7	9·7	16·5

4. Dry about 5 kg of the sample of sand and determine its moisture content according to the methods described in Section 3.2.3.

5. Make a mortar using this sand in the way described here for Class A concrete.
    Weigh the dried amount of sand.
    Weigh out 1 kg of dry cement and 0·5 kg of water.
    Mix the cement and water to a paste.
    Gradually add and mix the sand into this paste until the

resulting mortar has stiffened sufficiently to satisfy the
conditions shown in Figure 4.2 and described here.

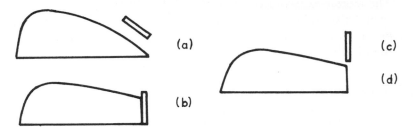

Fig. 4.2   Preparation of mortar. (a) Initial position of trowel blade. (b) Trowel
rotated into vertical position. (c) Trowel lifted away vertically. (d) Mortar
does not slump.

The mortar "sits up". This condition is reached when the
mortar is stiff enough to maintain a vertical face formed by
turning the trowel through 90°, pressing it horizontally
against the side of the mortar and lifting it away vertically.
The mortar is still fluid enough for cuts made in it to be
"self-healing" (i.e. the cuts tend to fill). They are made with a
vertical motion of the trowel at 20–25 mm intervals and
penetrate the mortar to a depth of 15–20 mm. The surface of
the mortar is not prepared in any way before this test is
applied.
When the mortar satisfies these conditions, the unused sand
is weighed. The weight of sand used is then obtained by
subtraction.

6. Determine quantities for a trial concrete mix in two stages.
    Stage 1. Table 4.2 shows batch weights of cement, water and
    dry sand for a 0·10 m³ mixer and the three classes of concrete
    referred to. Using the data in this table, weights of cement,
    water and sand are selected for a trial mix.
    An allowance must be made for the water in the sand. When
    the moisture content in about 5 kg of sand has been found, the
    amount of water in the weight of sand given in Table 4.2 can
    be calculated. For example, if the sand contained about 5
    per cent of water (on a dry basis of measurement) and 30 kg
    of dry sand were required in a batch, the total quantity of
    water that must be added with the sand would be about
    1·5 kg. Consequently, the batch weight of moist sand
    containing this amount of water would be 31·5 kg (see Section
    3.2.4).

TABLE 4.2   QUANTITIES FOR A TRIAL BATCH
                              (0·1-m³ mixer)

Class of Concrete	Sand in Mortar Mix (kg)	Sand for Trial Concrete Mix (kg)	
		With "One-Size" Coarse Aggregate	With "Graded" Coarse Aggregate
A			
Mortar Mix: 1 kg cement and	Less than 1·6	20·5	18·0
0·5 kg water	1·6 -1·75	22·0	19·5
Trial Concrete Mix: 13 kg	1·75-1·9	23·5	21·0
cement, 6·5 kg water and	1·9 -2·1	25·0	23·0
sand tabulated	More than 2·1	27·0	24·5
B			
Mortar Mix: 1 kg cement and	Less than 2·3	28·5	25·0
0·6 kg water	2·3-2·4	28·5	27·0
	2·4-2·6	30·0	28·5
Trial Concrete Mix: 13 kg	2·6-2·9	31·5	30·0
cement, 7·8 kg water and	2·9-3·5	33·5	31·5
sand tabulated	More than 3·5	31·5	30·0
C			
Mortar Mix: 1 kg cement and	Less than 2·6	33·5	30·0
0·7 kg water	2·6-2·8	35·0	31·5
	2·8-3·0	36·5	33·5
Trial Concrete Mix: 13 kg	3·0-3·2	38·0	35·0
cement, 9·1 kg water and	3·2-3·5	40·0	36·5
sand tabulated	3·5-4·1	41·5	38·2
	More than 4·1	40·0	36·5

The weight of mixing water to be added is obtained by subtracting the quantity of water in the sand from the total weight of water required. For example, if 1·5 kg of water were contained in the sand and a total of 8 kg were required in a batch, the quantity to add would be 6·5 kg.

Stage 2. Determine the quantity of coarse aggregate required as follows.

Weigh out 80 kg of stone. (If the sample is very wet, it should be dried.)

Next, "butter" the mixer by mixing a small amount of the selected mortar in it. Empty the mixer and prepare a trial concrete mix as follows.

Mix the cement, water and sand with about half the stone. Gradually add stone until the slump (see Section 6.2.1) is slightly greater than that required for the particular work. Add more stone, remix and check the slump, repeating this process until the required slump is obtained.

Weigh the remaining amount of stone and, by subtraction, determine the weight of course aggregate used.

7. If partial drying of the mix during preparation is suspected, repeat the procedure with the same weights of cement, water and sand, using initially the weight of coarse aggregate previously determined. Record the results finally obtained.

8. As water in a field mix is usually measured by volume, the weight of water determined should be converted into litres. (1 kg of water equals 1 litre.)

9. Adjust the batch quantities to suit the capacity of the mixer to be used, by multiplying them by one of the factors given in Table 4.3.

TABLE 4.3   **QUANTITY FACTOR**

Mixer Capacity in Bags of Cement	Factor by which Trial Batch Weights are Multiplied
0·5	1·5
1	3
2	6

4.3 ESTIMATION METHOD OF DESIGN OF CONCRETE MIXES

4.3.1 ORDINARY CONCRETE

Concrete mixes are usually designed by first selecting the water/cement ratio required to achieve a given compressive strength and durability, and then (using a type and grading of aggregate available) selecting an aggregate/cement ratio which will produce a mix sufficiently workable to be compacted by the means available. This is equivalent to preparing a cement paste with a water/cement ratio selected to ensure good-quality concrete, and then determining how much well-chosen aggregate (available locally) can be mixed into this paste, so as to produce concrete that can be fully compacted. The following data are required.

Anticipated service conditions and type of concrete structure, with an indication of sections and spacing of reinforcement.

Minimum compressive strength required at 28 days, and the durability requirement.
Suitable maximum size of aggregate and the characteristics of local supplies.
Particulars of equipment to be used for transporting, placing and compacting the concrete and the degree of job control anticipated.

British and American design procedures are given in References 5 to 8, 30-34, 43, 59 and 70, also in Reference 96 of Part 1. Certain basic principles and data contained in Reference 5(i), and incorporated in the design system given below, can now be applied through Reference 5(ii) to an extended range of acceptable aggregates with increased-strength cement. A typical design procedure (British), which is adaptable to the strength categories of curves (b) and (c) indicated in Figure 3.5 (3.2.2), is as follows:

1. Estimate a target average strength for the specified characteristic, nominal minimum characteristic, or grade-designated strength (Appendix VIII, 5) by dividing the latter by a suitable control factor (see Table 4.4 and Section 13.2). Normally, 95 per cent of test results are required above the minimum, but 99 per cent may be required for increased quality and target strength, or curtailed safety factor in design.[52, 71]

2. Find the requisite water/cement ratio to give this strength, using predetermined relationships or recent test data (see Figure 4.3 or 4.4).

3. Check the suitability of this ratio for known conditions of exposure (Table 4.5) and adopt the lower of the values required.

4. Estimate the minimum workability that is necessary for the concrete to be fully compacted, having regard to sections, reinforcement, and equipment available for handling and compaction (Table 4.6; and Sections 6.2.1 and 6.2.4).[38]

5. Choose an aggregate that is satisfactory with regard to maximum size, shape, texture and procurability. Select possible combinations of aggregate/cement ratio and overall grading of this aggregate to give the requisite workability with the nominal water/cement ratio. Tables 4.7 to 4.13 and Figures 4.5, 4.6 and 4.7 may be used for this purpose,[5-8] the respective ratios being by weight.[34, 66, 67]

From a choice of four gradings a suitable combination is chosen by striking a balance between two extremes or selecting either one of them. They are, respectively, as lean a mix as possible with a minimum permissible proportion of fine aggregate, and as cohesive a mix as possible with an upper limit in the proportion of fine aggregate. In vibrated and pumped concretes, Gradings No. 1 and No. 4 are respective choices, considerations being given generally to economy and minimal segregation (see Sections 4.3.4, 4.5, 23.1.5 and 24.1).

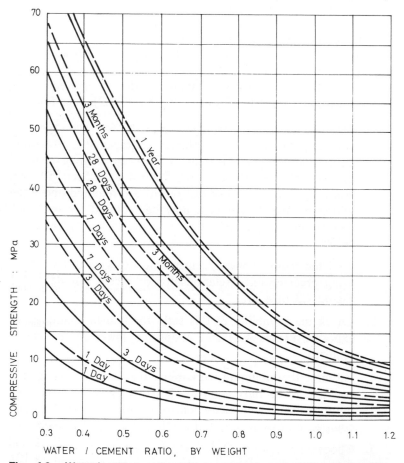

Fig. 4.3 Water/cement ratio and compressive strength relationship. ——, ordinary portland cement. – – –, high-early-strength portland cement. Specimens: 150 mm diameter × 300 mm cylinders (Australia and United States).

Note: See Figure 3.5 and footnote of Figure 3.6 for cement parameter influences on potential strength.

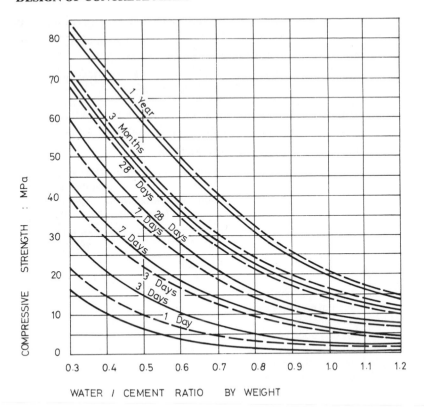

Fig. 4.4 Water/cement ratio and compressive strength relationship. ——, ordinary portland cement. – – –, high-early-strength portland cement. Specimens: 100 mm cubes (Britain).

Note: See Figure 3.5 and footnote of Figure 3.6 for cement parameter influences on potential strength.

6. Determine a fine-aggregate/coarse-aggregate ratio, using suitable local materials, so that their combined grading will approximate the selected overall grading (see Sections 3.3.4 (d) and 4.6 on proportioning). Adjust the mix to suit certain requirements of serviceability. These include such items as: special surface finish (see Sections 8.1, 27.2, 28.8–28.10 and 36.4), impermeability (see Sections 11.1, 11.2 and 14.10), durability (see Sections 12.1, 15.2.2, 15.2.3, 19.1 and 20.1.5), and acceptable volumetric changes (see Sections 30.1 and 31.1–31.4; Figs 3.1 and 33.7; and Reference 68 of Part 6).

7. Estimate the net quantities of water, cement and saturated

surface-dry aggregate that would be required in a trial mix, and make corrections for variations in the materials or the moisture content of aggregate (see Section 5.2).

8. Make trial laboratory mixes and effect adjustments, as necessary, to suit actual operating conditions.[30, 46, 50]

9. Conduct full-scale trial mixes under field conditions and establish a testing programme (see Sections 13.1 and 13.2).

TABLE 4.4   CONTROL FACTOR

Conditions of Production	Degree of Control	Coefficient of Variation (per cent)	Strengths Above Control Strength (per cent)		
			99	95	90
			Control Factor (per cent)		
Best-operated central plant	Excellent	6	86	90	93
	Range	9	79	85	89
Weigh-batching of all materials, determination of moisture content of aggregate, control of workability, very good supervision on site	Very good	12	72	80	85
Weigh-batching of all materials, control of workability, good supervision on site	Good	15	65	75	81
Volume-batching of aggregate, weigh-batching of cement, control of workability, good intermittent supervision	Fair	18	58	70	77
Volume-batching of all materials, little supervision on site	Poor	21	51	65	73

TABLE 4.5   MAXIMUM WATER/CEMENT RATIO BY WEIGHT
FOR REASONABLE DURABILITY

Condition of Exposure	Maximum Water/Cement Ratio	
	Plain Concrete	Reinforced Concrete
Internal, subject to heavy condensation	—	0·60
Alternate wetting and drying	0·60	0·60
Freezing and thawing*	0·55	0·50
Seawater or salt spray	0·50	0·45
Water-retaining structures	—	0·50

* Air-entrainment, recommended for this condition, is described in Sections 4.3.3, 15.2.3(c) and 16.1.

TABLE 4.6   WORKABILITY AND USAGE OF ORDINARY CONCRETE

Compactive Conditions and Usage	Degree of Workability	Slump (mm)	Compacting Factor Aggregate 10 mm	Compacting Factor Aggregate 20 mm
Intensive vibration of simple sections. Mechanical compaction of roads.	Very low	0-20	0·75	0·78
Vibration of mass concrete and simply reinforced sections. Manually operated vibration of slabs and roads.	Low	5-50	0·83	0·85
Vibration of heavily reinforced sections. Manual compaction of simply reinforced sections, slabs and roads.	Medium	20-100	0·90	0·92
Manual compaction of sections with congested reinforcement.	High	50-150	0·95	0·95

Note: The minimum slump of ready-mixed or transit-mixed concrete is 15-25 mm.

TABLE 4.7   10 mm ROUNDED AGGREGATE*/CEMENT RATIOS FOR FOUR DEGREES OF WORKABILITY WITH DIFFERENT WATER/CEMENT RATIOS AND GRADINGS

Water/ Cement Ratio by Weight	Very low Workability Grading number 1	2	3	4	Low Workability Grading number 1	2	3	4	Medium Workability Grading number 1	2	3	4	High Workability Grading number 1	2	3	4
0·35	4·0	3·6	3·0	2·3	3·4	2·9	2·4	1·9	—	—	—	—	—	—	—	—
0·40	5·6	5·0	4·2	3·2	4·5	3·9	3·3	2·6	3·9	3·5	3·0	2·4	3·5	3·2	2·8	2·3
0·45	7·2	6·4	5·3	4·1	5·5	4·9	4·1	3·2	4·7	4·3	3·7	3·0	4·2	3·9	3·4	2·9
0·50	—	7·8	6·4	4·9	6·5	5·8	4·9	3·8	5·4	5·0	4·3	3·5	4·8	4·5	4·0	3·4
0·55	—	—	7·5	5·7	7·4	6·7	5·7	4·4	6·1	5·7	4·9	4·0	5·3	5·1	4·5	3·9
0·60	—	—	—	6·5	—	7·5	6·4	5·0	6·7	6·3	5·5	4·5	5·8	5·6	5·0	4·3
0·65	—	—	—	7·2	—	—	7·1	5·6	7·3	6·9	6·1	5·0	X	6·1	5·5	4·7
0·70					—	—	7·7	6·2	7·9	7·5	6·7	5·5	X	6·6	6·0	5·1
0·75					—	—	—	6·7	—	—	7·2	5·9	X	7·1	6·5	5·5
0·80					—	—	—	7·2	—	—	7·7	6·3	X	7·6	6·9	5·9
0·85									—	—	—	6·8	X	—	7·3	6·3
0·90									—	—	—	7·2	X	—	7·7	6·7
0·95													X	—	—	7·0
1·00													X	—	—	7·3

* See Figure 4.5.

Note: X, mix would segregate. —, mix is outside the range tested. Indicated sizes of aggregate are their maximum in continuous gradings. The tabulated proportions are for aggregate with a specific gravity of 2·5-2·7. Values are based on total water added to air-dry aggregate.

TABLE 4.8   10 mm IRREGULAR AGGREGATE*/CEMENT RATIOS FOR FOUR DEGREES OF WORKABILITY WITH DIFFERENT WATER/CEMENT RATIOS AND GRADINGS

Water/Cement Ratio by Weight	Very low Workability				Low Workability				Medium Workability				High Workability			
	Grading number				Grading number				Grading number				Grading number			
	1	2	3	4	1	2	3	4	1	2	3	4	1	2	3	4
0·40	4·1	3·8	3·3	2·8	3·3	3·1	2·8	2·3	—	—	—	—	—	—	—	—
0·45	5·1	4·8	4·3	3·6	4·1	3·9	3·5	3·0	3·5	3·4	3·2	2·8	3·2	3·1	3·0	2·7
0·50	6·1	5·8	5·2	4·4	4·8	4·6	4·2	3·7	4·2	4·1	3·8	3·4	X	3·8	3·6	3·2
0·55	7·0	6·7	6·1	5·2	5·5	5·3	4·9	4·3	X	4·7	4·4	4·0	X	4·4	4·2	3·7
0·60	7·9	7·6	7·0	6·0	X	6·0	5·6	4·9	X	5·3	5·0	4·5	X	4·9	4·7	4·2
0·65	—	—	7·8	6·8	X	6·6	6·2	5·5	X	5·9	5·6	5·0	X	5·4	5·2	4·6
0·70					X	7·2	6·8	6·1	X	6·4	6·1	5·5	X	5·9	5·7	5·0
0·75					X	7·8	7·4	6·7	X	6·9	6·6	6·0	X	6·4	6·1	5·4
0·80					X	—	8·0	7·3	X	7·4	7·1	6·4	X	6·8	6·5	5·8
0·85									X	7·9	7·5	6·8	X	7·2	6·9	6·2
0·90									X	—	8·0	7·2	X	7·6	7·3	6·6
0·95													X	X	7·7	6·9
1·00													X	X	8·0	7·2

* See Figure 4.5.

Note: X, mix would segregate. —, mix is outside the range tested. Indicated sizes of aggregate are their maximum in continuous gradings. The tabulated proportions are for aggregate with a specific gravity of 2·5-2·7. Values are based on total water added to air-dry aggregate.

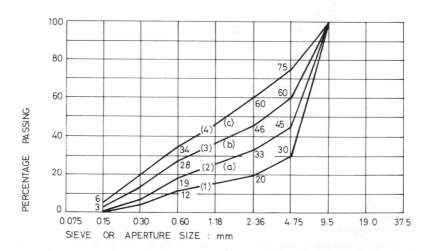

Fig. 4.5 Grading curves for 10 mm aggregate. (a) Zone A. (b) Zone B. (c) Zone C.

TABLE 4.9   10 mm ANGULAR AGGREGATE*/CEMENT RATIOS FOR FOUR DEGREES OF WORKABILITY WITH DIFFERENT WATER/CEMENT RATIOS AND GRADINGS

Water/ Cement Ratio by Weight	Very low Workability				Low Workability				Medium Workability				High Workability			
	Grading number				**Grading number**				**Grading number**				**Grading number**			
	1	2	3	4	1	2	3	4	1	2	3	4	1	2	3	4
0·40	3·7	3·3	2·8	2·0	—	—	—	—	—	—	—	—	—	—	—	—
0·45	4·5	4·1	3·5	2·6	3·8	3·6	3·0	2·2	3·3	3·1	2·7	2·1	—	—	—	—
0·50	5·2	4·9	4·2	3·2	4·4	4·2	3·6	2·7	3·8	3·7	3·2	2·6	X	3·2	2·9	2·4
0·55	5·9	5·6	4·9	3·8	4·9	4·8	4·2	3·2	X	4·2	3·7	3·0	X	3·7	3·4	2·8
0·60	6·6	6·3	5·5	4·3	X	5·3	4·7	3·7	X	4·7	4·2	3·4	X	4·2	3·8	3·2
0·65	7·3	7·0	6·1	4·8	X	5·8	5·2	4·2	X	5·1	4·6	3·8	X	4·6	4·2	3·6
0·70	7·9	7·6	6·7	5·3	X	6·3	5·7	4·6	X	5·6	5·1	4·2	X	5·0	4·6	4·0
0·75	—	—	7·3	5·8	X	6·8	6·2	5·0	X	6·0	5·5	4·6	X	5·4	5·0	4·4
0·80	—	—	7·8	6·3	X	7·2	6·6	5·5	X	6·4	5·9	5·0	X	5·8	5·4	4·7
0·85	—	—	—	6·8	X	7·6	7·1	6·0	X	6·7	6·3	5·4	X	6·1	5·8	5·1
0·90	—	—	—	7·3	X	—	7·5	6·4	X	7·1	6·7	5·8	X	6·4	6·1	5·4
0·95					X	—	7·9	6·8	X	7·5	7·1	6·1	X	6·7	6·4	5·7
1·00					X	—	—	7·2	X	7·8	7·5	6·5	X	7·0	6·7	6·1
1·05									X	—	7·8	6·9	X	7·3	7·0	6·4
1·10									X	—	—	7·2	X	7·6	7·3	6·7
1·15													X	X	7·6	7·0
1·20													X	X	7·9	7·3

* See Figure 4.5.

Note: X, mix would segregate. —, mix is outside the range tested. Indicated sizes of aggregate are their maximum in continuous gradings. The tabulated proportions are for aggregate with a specific gravity of 2·5–2·7. Values are based on total water added to air-dry aggregate.

TABLE 4.10   20 mm ROUNDED AGGREGATE*/CEMENT RATIOS FOR FOUR DEGREES OF WORKABILITY WITH DIFFERENT WATER/CEMENT RATIOS AND GRADINGS

Water/ Cement Ratio by Weight	Very low Workability				Low Workability				Medium Workability				High Workability			
	Grading number				**Grading number**				**Grading number**				**Grading number**			
	1	2	3	4	1	2	3	4	1	2	3	4	1	2	3	4
0·35	4·5	4·5	3·5	3·2	3·8	3·6	3·2	3·1	3·1	3·0	2·8	2·7	2·8	2·8	2·6	2·5
0·40	6·6	6·3	5·3	4·5	5·3	5·1	4·5	4·1	4·2	4·2	3·9	3·7	3·6	3·7	3·5	3·3
0·45	8·0	7·7	6·7	5·8	6·9	6·6	5·9	5·1	5·3	5·3	5·0	4·5	4·6	4·8	4·5	4·1
0·50	—	—	8·0	7·0	8·2	8·0	7·0	6·0	6·3	6·3	5·9	5·4	5·5	5·7	5·3	4·8
0·55	—	—	—	8·1	—	—	8·2	6·9	7·3	7·3	7·4	6·4	6·3	6·5	6·1	5·5
0·60	—	—	—	—	—	—	—	7·7	—	—	8·0	7·2	X	7·2	6·8	6·1
0·65					—	—	—	8·5	—	—	—	7·8	X	7·7	7·4	6·6
0·70					—	—	—	—	—	—	—	—	X	—	7·9	7·2
0·75													X	—	—	7·6
0·80													X	—	—	—
0·85													X	—	—	—
0·90																

* See Figure 4.6.

Note: X, mix would segregate. —, mix is outside the range tested. Indicated sizes of aggregate are their maximum in continuous gradings. The tabulated proportions are for aggregate with a specific gravity of 2·5–2·7. Values are based on total water added to air-dry aggregate.

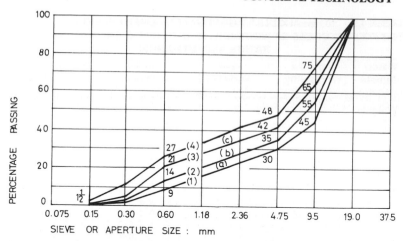

Fig. 4.6 Grading curves for 20 mm aggregate. (a) Zone A. (b) Zone B. (c) Zone C.

TABLE 4.11   20 mm IRREGULAR AGGREGATE*/CEMENT RATIOS FOR FOUR DEGREES OF WORKABILITY WITH DIFFERENT WATER/CEMENT RATIOS AND GRADINGS

Water/Cement Ratio by Weight	Very low Workability				Low Workability				Medium Workability				High Workability			
	Grading number				Grading number				Grading number				Grading number			
	1	2	3	4	1	2	3	4	1	2	3	4	1	2	3	4
0·35	3·7	3·7	3·5	3·0	3·0	3·0	3·0	2·7	2·6	2·6	2·7	2·4	2·4	2·5	2·5	2·2
0·40	4·8	4·7	4·7	4·0	3·9	3·9	3·8	3·5	3·3	3·4	3·5	3·2	3·1	3·2	3·2	2·9
0·45	6·0	5·8	5·7	5·0	4·8	4·8	4·6	4·3	4·0	4·1	4·2	3·9	X	3·9	3·9	3·5
0·50	7·2	6·8	6·5	5·9	5·5	5·5	5·4	5·0	4·6	4·8	4·8	4·5	X	4·4	4·4	4·1
0·55	8·3	7·8	7·3	6·7	6·2	6·2	6·0	5·7	X	5·4	5·4	5·1	X	4·8	4·9	4·7
0·60	9·4	8·6	8·0	7·4	6·8	6·9	6·7	6·2	X	6·0	6·0	5·6	X	X	5·4	5·2
0·65	—	—	—	8·0	7·4	7·5	7·3	6·8	X	X	6·4	6·1	X	X	5·8	5·6
0·70	—	—	—	—	8·0	8·0	7·7	7·4	X	X	6·8	6·6	X	X	6·2	6·1
0·75					—	—	—	7·9	X	X	7·2	7·0	X	X	6·6	6·5
0·80					—	—	—	—	X	X	7·5	7·4	X	X	X	7·0
0·85									X	X	7·8	7·8	X	X	X	7·4
0·90									X	X	X	8·1	X	X	X	7·7
0·95									X	X	X	—	X	X	X	8·0
1·00													X	X	X	X

* See Figure 4.6.

Note: X, mix would segregate. —, mix is outside the range tested. Indicated sizes of aggregate are their maximum in continuous gradings. The tabulated proportions are for aggregate with a specific gravity of 2·5-2·7. Values are based on total water added to air-dry aggregate.

TABLE 4.12 20 mm ANGULAR AGGREGATE*/CEMENT RATIOS FOR FOUR DEGREES OF WORKABILITY WITH DIFFERENT WATER/CEMENT RATIOS AND GRADINGS

Water/ Cement Ratio by Weight	Very low Workability				Low Workability				Medium Workability				High Workability			
	Grading number				Grading number				Grading number				Grading number			
	1	2	3	4	1	2	3	4	1	2	3	4	1	2	3	4
0·35	3·2	3·0	2·9	2·7	2·7	2·7	2·5	2·4	2·4	2·4	2·3	2·2	2·2	2·3	2·1	2·1
0·40	4·5	4·2	3·7	3·5	3·5	3·5	3·2	3·0	3·1	3·1	2·9	2·7	2·9	2·9	2·8	2·6
0·45	5·5	5·0	4·6	4·3	4·3	4·2	3·9	3·7	3·7	3·7	3·4	3·3	3·5	3·5	3·2	3·1
0·50	6·5	5·8	5·4	5·0	5·0	4·9	4·5	4·3	4·2	4·2	3·9	3·8	X	3·9	3·8	3·5
0·55	7·2	6·6	6·0	5·6	5·7	5·4	5·0	4·8	4·7	4·7	4·5	4·3	X	X	4·3	4·0
0·60	7·8	7·2	6·6	6·3	6·3	6·0	5·6	5·3	X	5·2	4·9	4·8	X	X	4·7	4·4
0·65	8·3	7·8	7·2	6·9	6·9	6·5	6·1	5·8	X	5·7	5·4	5·2	X	X	5·1	4·9
0·70	8·7	8·3	7·7	7·5	7·4	7·0	6·5	6·3	X	6·2	5·8	5·7	X	X	5·5	5·3
0·75	—	—	8·2	8·0	7·9	7·5	7·0	6·8	X	X	6·2	6·1	X	X	5·8	5·7
0·80	—	—	—	—	—	—	7·4	7·2	X	X	6·6	6·5	X	X	6·1	6·0
0·85					—	—	7·8	7·6	X	X	7·1	6·9	X	X	6·4	6·3
0·90					—	—	—	—	X	X	7·5	7·3	X	X	X	6·7
0·95									X	X	8·0	7·6	X	X	X	7·0
1·00									X	X	—	—	X	X	X	7·3

* See Figure 4.6.

Note: X, mix would segregate. —, mix is outside the range tested. Indicated sizes of aggregate are their maximum in continuous gradings. The tabulated proportions are for aggregate with a specific gravity of 2·5-2·7. Values are based on total water added to air-dry aggregate.

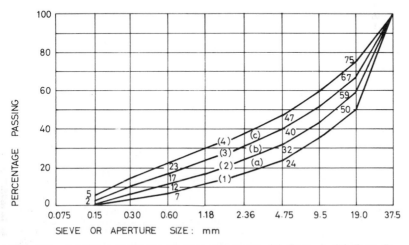

Fig. 4.7 Grading curves for 40 mm aggregate. (a) Zone A. (b) Zone B. (c) Zone C.

TABLE 4.13 40 mm IRREGULAR AGGREGATE*/CEMENT RATIOS FOR FOUR DEGREES OF WORKABILITY WITH DIFFERENT WATER/CEMENT RATIOS AND GRADINGS

Water/Cement Ratio by Weight	Very low Workability Grading number				Low Workability Grading number				Medium Workability Grading number				High Workability Grading number			
	1	2	3	4	1	2	3	4	1	2	3	4	1	2	3	4
0·35	4·0	3·9	3·5	3·2	3·4	3·3	3·2	2·9	2·9	2·8	2·6	2·5	2·7	2·5	2·3	2·3
0·40	5·3	5·3	4·7	4·3	4·5	4·5	4·2	3·8	3·8	3·8	3·7	3·4	3·5	3·5	3·3	3·1
0·45	6·5	6·5	5·9	5·3	5·6	5·6	5·3	4·8	4·6	4·7	4·6	4·3	4·1	4·4	4·3	4·0
0·50	7·7	7·7	7·1	6·3	6·7	6·6	6·3	5·7	5·4	5·7	5·5	5·1	4·8	5·2	5·1	4·8
0·55	—	—	8·1	7·3	7·6	7·6	7·2	6·6	6·2	6·5	6·3	5·8	X	5·9	6·0	5·5
0·60			—	—	—	—	—	7·4	7·0	7·3	7·1	6·6	X	X	6·7	6·2
0·65								8·1	7·8	8·1	7·8	7·2	X	X	7·3	6·9
0·70								—	—	—	—	7·9	X	X	—	7·4
0·75													X	X	—	8·0
0·80													X	X	—	—

* See Figure 4.7.

Note: X, mix would segregate. —, mix is outside the range tested. Indicated sizes of aggregate are their maximum in continuous gradings. The tabulated proportions are for aggregate with a specific gravity of 2·5-2·7. Values are based on total water added to air-dry aggregate.

It is relevant to note here that the grading curves of large aggregate, used in mass concrete,[15, 44] should contain more than sufficient small aggregate to fill the voids of the large aggregate and reduce the risk of segregation. Furthermore, for a given cement factor and degree of workability, there is an optimum maximum size of aggregate which, with least water requirement, produces the highest compressive strength. This maximum size decreases as the cement factor increases, and is usually in the region of 19·0–37·5 mm for medium to high-strength grades of ordinary concrete. Its selective use is influenced by the practical prerequisites of effective placement and reproduction of strength-test results.[16, 25, 29, 51]

An improved quality and strength, which results from removing excess water that collects under large particles of aggregate, may be obtained by effective reconsolidation (see Section 23.2). Further aspects of workability in the design of concrete mixes are dealt with in Sections 4.3.4 and 6.3.3, while factors affecting shrinkage and modifications of procedure in making concrete with lightweight aggregate are described in Sections 31.2.1 and 33.2.2(d).

The design of high-alumina cement concrete mixes is dealt with in Reference 24, and in Reference 134 of Part 3. If this type of concrete is used for long periods in hot moist conditions, it shows a marked reduction in strength and durability. The effect is greater for porous concrete than for dense concrete, and it must be taken into account in structural projects (see Section 3.1.2 (b)).

Mixes of portland cement concrete which are to be mixed and placed in a tropical climate should have a 15 per cent margin in target strength unless a cement is used with additional strength properties. The effect of high initial temperature on strength is illustrated in Sections 9.2 and 10.1.

The procedure described may be illustrated in the design of an ordinary mix for a concrete pavement which is to be constructed under fair supervision with 20–5 mm gravel and natural sand. The minimum (cylinder) compressive strength is to be 30 MPa at 28 days and power-operated equipment will be used for compaction.

With a control factor of 70 per cent (Table 4.4), the target average strength is

$$\frac{30}{0 \cdot 70} = 43 \text{ MPa}$$

at 28 days, for which a water/cement ratio of 0·4 by weight is required (Fig. 4.3). References to Tables 4.5 and 4.6 show that this value is satisfactory for ordinary durability and a very low workability is suitable for mechanically compacted roads. A group of four aggregate/cement ratios is listed in Table 4.11 for 20 mm irregular aggregate, a water/cement ratio of 0·4 and very low workability. Values of the four ratios correspond to the four gradings given in Figure 4.6, as follows.

4·8 : 1 for Grading No. 1.

4·7 : 1 for Grading No. 2.

4·7 : 1 for Grading No. 3.

4·0 : 1 for Grading No. 4.

It will be most economical if the grading of the aggregate can be made to approximate roughly to Grading No. 1. If, as determined in Section 4.6, 1 part of suitable sand should be combined with 3 parts of irregular gravel for this requirement, the mix proportions would be 1 part cement : 1·2 parts sand : 3·6 parts gravel : 0·4 part water/cement ratio by weight. The cement content, by calculation (see Section 5.2), would be 400 kg/m³ of concrete.

It is worth noting that differences in grading from Curve No. 1 to Curve No. 3 have little effect on the proportions of mix. The grading in this example may be adjusted, therefore, over a comparatively wide range without appreciably altering the aggregate/cement ratio. If very good control were exercised in all phases of the work, the control factor could be 80 per cent and the proportions would be 1 part cement : 6·0 parts combined aggregate : 0·45 part water/cement ratio by weight. The cement content would be 330 kg/m³, which means that improved control could effect an estimated saving of about 70 kg of cement per m³ of concrete. The economies to be effected by good control and supervision of concrete construction, as described in Appendix II, are thus clearly demonstrated.[34, 35]

Mix designs connected with flexural strength, fly-ash concrete, and surface area factor are given in References 36, in 42, 53 and 63. Mass concrete mix design is outlined in Section 30.1 and in the references given for this section.

4.3.2 NOMINAL BAG-UNIT MIX

The proportions of a nominal bag-unit mix can be estimated on the basis that the sum of the absolute volumes of the components is equal to the absolute volume of the concrete, the volume of any residual air in ordinary concrete being neglected. The absolute volume of a material is the actual total volume of solid matter in all of its particles. For purposes of calculation, the absolute volume is equal to the weight of the material divided by its specific gravity multiplied by the bulk density of water. (Specific gravity is dealt with in Sections 3.2.3, 3.3.4, 23.1.5 and the Glossary, Appendix VI.)

The procedure is best illustrated by an example, such as the determination of proportions by volume (dry) of a 9 bag/m³ mix (40 kg in Australia; see Section 4.1), with 21 litres of water per bag, and a fine-aggregate/coarse-aggregate ratio of 0·5. The specific gravities of the cement and aggregate may be taken as 3·15 and 2·65, respectively. The bulk density of water is 1000 kg/m³ and the bulk densities of loose dry aggregate may be regarded as 1410 kg/m³ for 20 mm stone screenings and 1520 kg/m³ for sand with a fineness modulus of 2·2 (see Section 3.3.4 (b), (c), and Appendix VIII).

Example. The absolute volume of cement paste per m³ of concrete

$$= 9 \left(\frac{40}{3 \cdot 15 \times 1000} \right) + \frac{21}{1000} = 0 \cdot 303 \text{ m}^3.$$

The absolute volume of combined aggregate per m³ of concrete
$$= 1 - 0 \cdot 303 = 0 \cdot 697 \text{ m}^3.$$

The absolute volume of 1 m³ of sand
$$= \frac{1520}{2 \cdot 65 \times 1000} = 0 \cdot 574 \text{ m}^3.$$

The absolute volume of 1 m³ of stone screenings

$$= \frac{1410}{2 \cdot 65 \times 1000} = 0 \cdot 532 \text{ m}^3.$$

The fine/combined aggregate ratio by absolute volume

$$= \frac{0 \cdot 574}{0 \cdot 574 + 0 \cdot 532 \times 2} = 0 \cdot 35.$$

In 0·697 m³ of combined aggregate, the absolute volume of sand = 0·697 × 0·35 = 0·244 m³ and the absolute volume of stone screenings = 0·697 − 0·244 = 0·453 m³.

The requisite volume of sand

$$= \frac{0 \cdot 244}{0 \cdot 574} = 0 \cdot 425 \text{ m}^3.$$

The requisite volume of stone screenings

$$= \frac{0 \cdot 453}{0 \cdot 532} = 0 \cdot 852 \text{ m}^3.$$

Assuming that a 40-kg bag of cement has a volume of 0·0265 m³,

the water/cement ratio by volume

$$= \frac{21}{1000 \times 0 \cdot 0265} = 0 \cdot 79.$$

The aggregate/cement ratio by volume

$$= \frac{0 \cdot 425}{9 \times 0 \cdot 0265} = 1 \cdot 8 \text{ for sand and } \frac{0 \cdot 852}{9 \times 0 \cdot 0265} \text{ or } 3 \cdot 6 \text{ for stone.}$$

Therefore, the nominal proportions (dry volume) are 1 part cement : 1· 8 parts sand : 3·6 parts stone screenings, and the water/cement ratio is 0·8 by volume. Mix design for bulk densities is set out in Reference 48.

4.3.3 AIR-ENTRAINED CONCRETE

Mix proportions for trial batches of air-entrained concrete are estimated by designing an ordinary concrete mix and modifying it to allow for the entrained air. As entrained air increases the workability of the mix and behaves apparently like particles of fine aggregate, the water/cement ratio and proportion of aggregate can be reduced for a given degree of workability and cement content per m³ of concrete.

Although a reduction in water/cement ratio increases the potential strength of the concrete, the entrained air may cause a net reduction in strength of about 10 per cent. Therefore the "average" target strength (corresponding to a minimum requirement) must be increased by this amount. If advantage were not taken, by reducing the water/cement ratio, of the increase in workability due to air-entrainment, the strength of the concrete would be reduced by about 5 per cent for each 1 per cent of entrained air.

A design procedure developed by Wright is as follows.[9]

1. The mix is designed as for ordinary concrete, but with an increase of 10 per cent in the estimated target strength.

2. The mix proportions thus obtained by weight are converted to proportions by absolute volume, by dividing them by the specific gravities of the individual constituents.

3. The volumes so obtained are expressed as percentages of the whole.

4. The percentage of entrained air is subtracted from the percentage of aggregate. If the former, which simulates fine aggregate, were all subtracted from the fine aggregate, a harsh mix would tend to form. In practice, therefore, 1 per cent is deducted from the coarse aggregate and the remainder from the fine aggregate.

5. The percentage of water is reduced by an amount which is taken from Table 4.14 and multiplied by the number of units of entrained air. (This table has been prepared for 19 mm maximum-size aggregate.) The amount thus removed is replaced by an equal volume of combined aggregate, which is split approximately in the same proportion as that of the fine to the coarse aggregate.

**TABLE 4.14   PERCENTAGE OF WATER TO BE REMOVED
FOR EACH 1 PER CENT OF ENTRAINED AIR**

Mix Proportions	Rounded Gravel Aggregate	Irregular Gravel Aggregate	Crushed Rock Aggregate
1 : 6	0·325	0·375	0·425
1 : 7·5	0·40	0·45	0·50
1 : 9	0·45	0·50	0·55

6. The resulting mix proportions (by absolute volume) are converted to proportions by weight, by multiplying each constituent by its specific gravity.

7. The proportions are converted to a nominal mix with a water/ cement ratio by dividing each respective weight by the weight of cement.

If a nominal, ordinary mix which is specified by volume is to be adjusted for air-entrainment, its proportions should be made gravimetric by multiplying them by the respective bulk densities (see the test described in Section 3.3.4(b)). When converting the proportions by weight to proportions by absolute volume, the respective specific gravities need be known only approximately, as the process is reversed subsequently and the changes involved are small. In the absence of actual values, the specific gravities of portland cement and aggregate may be taken as 3·15 and 2·6–2·7, respectively.

The procedure is best carried out in tabular form, as illustrated by the example in Table 4.15 for 17.5 MPa concrete. The mix resulting from Procedure 1 described is 1 : 2·5 : 5, with 20 mm nominal-size crushed-rock aggregate and a water/cement ratio of 0·65 by weight. The amount of entrained air to be incorporated is 5 per cent.

TABLE 4.15   MODIFICATION OF MIX FOR AIR-ENTRAINMENT

Procedure	Cement	Fine Aggre-gate	Coarse Aggre-gate	Water	Air	Total
1. Mix proportions by weight	1·00	2·50	5·00	0·65	—	—
2. Divide by specific gravities	3·15	2·65	2·65	1·00	—	—
Absolute volumes	0·32	0·94	1·88	0·65	—	3·79
3. Per cent absolute volumes	8·40	24·80	49·60	17·20	—	100·00
4. Add air and subtract it from aggregate	—	4·00	1·00	—	5·00	—
	8·40	20·80	48·60	17·20	5·00	100·00
5. Reduce water and increase aggregate (Table 4.14)	—	0·80	1·70	2·50	—	—
	8·40	21·60	50·30	14·70	—	100·00
6. Convert to weights	26·40	57·20	133·30	14·70	—	—
7. Divide by weight of cement	1·00	2·17	5·05	0·56	—	—

The reduction in strength due to air-entrainment can be estimated from an example.

Example. The estimated target strength (i.e. cylinder compressive strength) is 19 MPa at 28 days. The modified mix has a water/cement ratio of 0·56, which corresponds to a strength of 24·5 MPa for air-free concrete (Fig. 4.3). Since each 1 per cent of air reduces the strength by 5·5 per cent, the reduction in MPa in this instance would be

$$\frac{24 \cdot 5 \ \times 5 \cdot 5 \ \times \ 5}{100}$$

which equals 6·74, say, 6·8 MPa. The compressive strength, therefore, equals 24·5–6·8 or 17·7 MPa. This represents a reduction of approximately 1·3 MPa below the estimated target strength for ordinary concrete, which is due to air-entrainment.

The entrainment of air in concrete is dealt with further in Reference 69 and Chapter 16.[34]

4.3.4 HIGH-STRENGTH CONCRETE

The properties of rich mixes depend to a considerable extent upon the quality of constituent materials. The well-established relationship between compressive strength and water/cement ratio requires some qualification for design purposes, so as to allow for the type and maximum size of the coarse aggregate to be used, and the anticipated stiffness of mix that may necessitate vigorous means of compaction. High-strength rich mixes and vibratable lean mixes (see Section 23.1.5) are made with either coarse or gap gradings. Coarse gradings are identified by Grading No. 1 in Figures 4.5 and 4.6 and gap gradings are dealt with in Section 4.5.

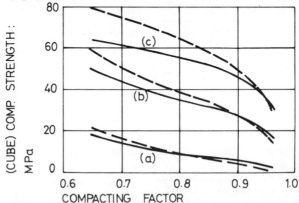

Fig. 4.8 Effect of type of coarse aggregate and workability on strength. ———, irregular gravel. – – –, crushed granite. (a) 1 day. (b) 7 days. (c) 91 days.

The following factors indicate that trial mixes should be made with the proposed materials and proportions in order to verify their suitability in practice. With similar rich mixes of stiff consistence, workability is greater when using crushed coarse aggregate (e.g. granite) than when using natural gravel. This property is reversed when the same coarse aggregate (with natural sand) is used with a water/cement ratio above 0·4 by weight.[10, 26, 32, 68]

With similar rich mixes, using either 10 mm or 20 mm nominal-size coarse aggregate, crushed granite gives a higher strength than irregular gravel in a mix of equal workability, the difference becoming progressively smaller as the workability increases (Fig. 4.8). The compressive strength of concrete, when made with any particular type of coarse aggregate, tends to reach a ceiling value, which is higher for crushed stone than for natural gravel. These results are related to an improved interlock and bond strength between the aggregate and matrix. (Workability and its measurement are dealt with in Chapter 6.)

The effect of the aggregate/cement ratio on the compressive strength of high-strength concrete is indicated in Figure 4.9. If the aggregate/cement ratio is reduced by 1·0 by weight, for purposes of workability, the water/cement ratio must be reduced by about 0·02 to maintain the strength. Figure 4.10 shows that the compressive strength of very stiff, rich mixes is virtually independent of the aggregate/cement ratio. For very high strength, therefore, it is preferable to use an intensive compactive effort with a very stiff mix, than to use a very rich mix in order to preserve workability for less vigorous means of compaction. Figures 4.9 and 4.10,[10, 35] which are purely illustrative, were derived from 7-day strengths using ordinary portland cement and irregular gravel.

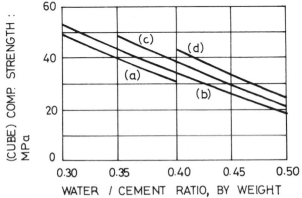

Fig. 4.9 Effect of aggregate/cement and water/cement ratios on strength. Aggregate/cement ratio by weight: (a) 2·5, (b) 3·5, (c) 4·5, and (d) 5·5.

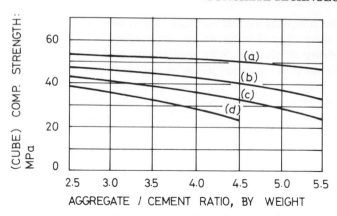

Fig. 4.10 Effect of workability and aggregate/cement ratio on strength. Compacting factor: (a) 0·68 (very stiff mix), (b) 0·78, (c) 0·85, and (d) 0·92 (plastic mix).

A design procedure developed by Erntroy and Shacklock is as follows.[10]

1. The average compressive strength at a particular age (e.g. 28 days) is estimated to meet a specified minimum value (see Section 4.3.1).

2. The type and maximum size of the aggregate and the requisite degree of workability are decided upon. The latter is selected from Table 4.16 (see Section 6.2 for workability tests), intermediate values being interpolated if required.

3. The "reference number" and related water/cement ratio are determined from Figure 4.11 or 4.12 and Figure 4.13 or 4.14. Although the former has been derived from test data for cube compressive strengths, it can be readily used in conjunction with Figure 4.15 for cylinder compressive strengths.

4. The corresponding aggregate/cement ratio is chosen from Table 4.17.

5. Fine and coarse aggregates are combined so that their grading approximates Curve 1 in Figures 4.5 and 4.6. Alternatively, a suitable gap grading may be used. Graphical means for blending aggregate are described in Section 4.6.

6. Trial mixes are made with the proposed materials and proportions, the moulded concrete being compacted by an electric vibrating-hammer and subjected to standard moist curing (AS 1012 Parts 2, 8 and 9; BS 1881; ASTM C31 and

C192). Adjustments are made to either or both factors until the desired test results are obtained. This operation is expedited by making additional test specimens that are 10 per cent richer and 10 per cent leaner than the designed mix.

TABLE 4.16   **WORKABILITY AND USAGE OF HIGH-STRENGTH CONCRETE**

Compactive Conditions and Usage	Degree of Workability	Slump (mm)	Compacting Factor	
			Aggregate 10 mm	Aggregate 20 mm
Vibropressure compaction of simple sections.	Extremely low	Nil	0·65	0·68
Intense vibration of simple sections. Mechanical compaction of roads.	Very low	0–30	0·75	0·78
Vibration of mass concrete and simply reinforced sections. Manually operated vibration of slabs and roads.	Low	5–50	0·83	0·85
Vibration of heavily reinforced sections. Manual compaction of simply reinforced sections, slabs and roads.	Medium	30–100	0·90	0·92

Note: The minimum slump of ready-mixed or transit-mixed concrete is 20 mm–30 mm.

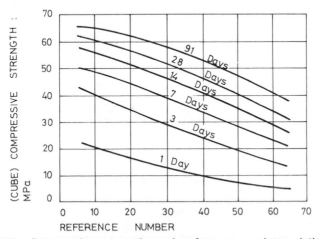

Fig. 4.11 Compressive strength and reference number relationship. Materials: ordinary portland cement with natural sand and *irregular gravel* as coarse aggregate. Specimens: 100-mm cubes.

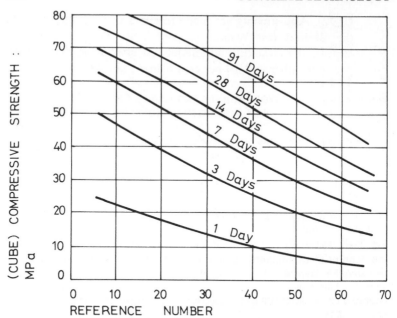

Fig. 4.12 Compressive strength and reference number relationship. Materials: ordinary portland cement with natural sand and *crushed granite* as coarse aggregate. Specimens: 100-mm cubes.

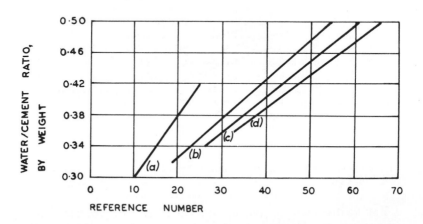

Fig. 4.13 Reference number and water/cement ratio relationship. Coarse aggregate: 10 mm nominal size. Degree of workability: (a) extremely low, (b) very low, (c) low, and (d) medium.

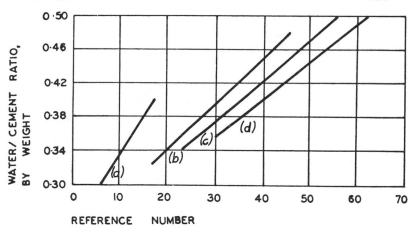

Fig. 4.14 Reference number and water/cement ratio relationship. Coarse aggregate: 20 mm nominal size. Degree of workability: (a) extremely low, (b) very low, (c) low, and (d) medium.

TABLE 4.17   AGGREGATE/CEMENT RATIO REQUIRED TO GIVE FOUR DEGREES OF WORKABILITY WITH DIFFERENT WATER/CEMENT RATIOS USING ORDINARY PORTLAND CEMENT

Water/ Cement Ratio by Weight	Aggregate/Cement Ratio by Weight															
	Workability*															
	EL	VL	L	M	EL	VL	L	M	EL	VL	L	M	EL	VL	L	M
	Irregular gravel† nominal size								Crushed granite† nominal size							
	10 mm				20 mm				10 mm				20 mm			
0·30	2·4	—	—	—	3·0	—	—	—	2·9	—	—	—	3·3	—	—	—
0·32	3·2	—	—	—	3·8	2·5	—	—	3·6	2·3	—	—	4·0	2·6	—	—
0·34	3·9	2·6	—	—	4·5	3·0	2·5	—	4·2	2·8	2·3	—	4·6	3·2	2·6	—
0·36	4·6	3·1	2·6	—	5·2	3·5	3·0	2·5	4·7	3·2	2·7	2·3	5·2	3·6	3·1	2·6
0·38	5·2	3·5	3·0	2·5	—	4·0	3·4	2·9	5·2	3·6	3·0	2·6	—	4·1	3·5	2·9
0·40	—	3·9	3·3	2·7	—	4·4	3·8	3·2	—	4·0	3·3	2·9	—	4·5	3·8	3·2
0·42	—	4·3	3·6	3·0	—	4·9	4·1	3·5	—	4·4	3·6	3·1	—	4·9	4·2	3·5
0·44	—	4·7	3·9	3·3	—	5·3	4·5	3·8	—	4·8	3·9	3·3	—	5·3	4·5	3·7
0·46	—	5·1	4·2	3·6	—	—	4·8	4·1	—	5·1	4·2	3·6	—	—	4·8	4·0
0·48	—	5·4	4·5	3·8	—	—	5·2	4·4	—	5·5	4·5	3·8	—	—	5·1	4·2
0·50	—	—	4·8	4·1	—	—	5·5	4·7	—	—	4·7	4·0	—	—	5·4	4·5

* EL = extremely low, VL = very low, L = low, M = medium. See Table 4.16.
† Natural sand used in combination with both types of coarse aggregate.

For *high-strength* in this section (as in British practice), *cubes* are more convenient to test than cylinders, as they can be tested on their sides without capping. Strength values can be transposed for different types of specimens by the use of *conversion factors* which are derived from actual test results or taken from a general relationship, such as that given in Section 4.4. Table 4.17 gives data for an overall aggregate grading with 30 per cent passing a 4·75-mm BS sieve. Minor adustments in the fine-aggregate/coarse-aggregate ratio may be made to suit various aggregates with different gradings and properties and to meet job requirements. The richness of mix may be adjusted to suit different cements and attain strengths up to 110 MPa.[58, 62, 72, 74]

A pan or paddle mixer with a positive stirring action is preferable, for high-strength mixes, to a drum mixer with a cascading action. Several suitable mixes with different degrees of workability may be designed with given materials. From their cement content, the cost of using a plastic, rich mix may be compared with the cost of compacting a stiff, leaner mix. A similar comparison may be made of mixes with a given degree of workability but with different types of coarse aggregate. A mix design should take into account, for effectiveness and economy, the possible use of revibration, high-speed slurry mixing and suitable admixtures.[64]

The following examples illustrate the procedure of design.

Example 1. Estimate the design of an ordinary, high-strength concrete mix with an average compressive strength of 35 MPa (100 mm cubes) at 7 days, using several degrees of workability and 10 mm crushed granite with natural sand.

The appropriate "reference number" (which is 43) and values of water/cement ratio are determined from Figures 4.12 and 4.13, respectively. The corresponding aggregate/cement ratios are obtained from Table 4.17 and the cement contents are calculated as in Section 5.2. The results are shown in Table 4.18 for purposes of comparison and selection.

TABLE 4.18   HIGH-STRENGTH MIX DESIGN

Degree of Workability	Water/Cement Ratio by Weight	Aggregate/Cement Ratio by Weight	Cement Content (kg/m³)
Very low	0·44	4·8	390
Low	0·42	3·6	480
Medium	0·40	2·9	550

Example 2. Estimate the design of an ordinary, high-strength concrete mix with an average compressive strength of 55 MPa (100 mm cubes) at 28 days, using very low workability and two types of coarse aggregate (20 mm nominal size) with natural sand. The results are shown in Table 4.19.

TABLE 4.19   HIGH-STRENGTH MIX DESIGN

Type of Coarse Aggregate	"Reference Number"	Water/Cement Ratio by Weight	Aggregate/Cement Ratio by Weight	Cement Content (kg/m³)
Irregular gravel	22	0·35	3·2	520
Crushed granite	35	0·42	4·9	390

A high-strength mix which is used for the manufacture of prestressed concrete (Chapter 34) should have one or other of the following minimum (*cube*) compressive strengths, these being converted to equivalent cylinder strengths where necessary by the use of Figure 4.15.

1. 27·5 MPa at 28 days where the concrete is to be used in certain hydraulic structures, such as circular tanks in which prestressing is desirable as a principle rather than as a direct means of reducing structural cross-section.

2. 30–50 MPa at 28 days for ordinary live-load requirements on structural sections, the first-mentioned limit pertaining to work with post-tensioned steel (see Section 34.1.10).

3. 55–70 MPa at 28 days for highly stressed applications. This compressive strength can be obtained with richer-than-usual mixes, carefully chosen aggregate, high-class workmanship, quality control and skilled supervision. Economy in mix design is possible, however, where post-tensioning is to be done 2–3 months after the concrete has been cast.

4. 30 MPa at 3–4 days, when post-tensioning is to be done at an early age or curing for several days is practicable, such as in the long-line, precasting, pretensioning process.

5. 27·5 MPa at 12–24 hours when the concrete is to be stressed at an early age, such as in the single-mould precasting process where moulds are used once per working shift (see Section 9.4.3). This value may be reduced to 20 MPa for partial prestressing (e.g. concrete piles).

6. 30–40 MPa at 28 days when coated, expanded-shale or expanded-clay aggregate concrete is used with a unit weight (dry) of 1750–1920 kg/m³. It is made with 9·5–12·5 40-kg bags of cement per m³ and a net water/cement ratio of

TABLE 4.20   SUGGESTED PROCEDURE FOR MAKING CONCRETE WITH SPECIFIED HIGH STRENGTHS AT VARIOUS AGES

Ordinary and High-Early-Strength Portland Cements

Nominal Minimum Cube Strength and Age* (MPa)	Degree of Control**	Approximate Average Cube Strength (MPa)	Method and Time of Curing†	Method of Compaction††	Approximate Mix Proportions by Weight‡	Water/Cement Ratio by Weight	Cement‡‡	Remarks
40 at 28 days	Good	60	Ordinary damp curing	1	1 : 3	0·40	OP	Steam curing is not generally applicable. The 7-day strength of dense-aggregate concrete is about two-thirds of the 28-day strength.
				2	1 : 4	0·40	OP	
				3	1 : 5	0·40	OP	
	Very good	50	Ordinary damp curing	1	1 : 5	0·45	OP	
				2	1 : 6	0·45	OP	
				3	1 : 7	0·45	OP	
60 at 28 days	Very good	75	Ordinary damp curing	1	1 : 2·2	0·30	OP	
				2	1 : 2·5	0·30	OP	
				3	1 : 3	0·30	OP	
30 at 3 days	Good	40	Ordinary damp curing	2	1 : 3	0·35	HES	Hand compaction is not usually practicable.
				3	1 : 3·5	0·35	HES	
			Steam curing for at least 24 hours	2	1 : 4	0·40	HES	
				3	1 : 5	0·40	HES	
	Very good	35	Ordinary damp curing	2	1 : 3·5	0·375	HES	
				3	1 : 4	0·375	HES	
			Steam curing for at least 24 hours	2	1 : 6	0·45	HES	
				3	1 : 7	0·45	HES	
25 at 24 hours	Very good	35	Ordinary damp curing	3	1 : 2·5	0·275	HES	Excellent control and intense vibration. Intense vibration and steam curing are usually essential
	Good	40	Steam curing for at least 12 hours	2	1 : 3	0·35	HES	
				3	1 : 3·5	0·35	HES	
	Very good	35	Steam curing for	2	1 : 3·5	0·375	HES	

					Approximate Mix Proportions by Weight‡	Water/Cement Ratio by Weight	Cement‡‡	Remarks
	Very good	35	at least 8 hours / Steam curing for at least 8 hours	3} 2} 3}	1:2·5 / 1:2·5 / 1:3	0·2·5 / 0·30 / 0·30	HES / HES / HES	
25 at 3 hours	Very good	35	Hot-water curing for at least 2 hours at 95°-100°C	3	1:2·6	0·30	HES	Excellent control and intense vibration are essential.

High-Alumina Cement

Nominal Minimum Cube Strength and Age* (MPa)	Degree of Control**	Approximate Average Cube Strength (MPa)	Method and Time of Curing†	Method of Compaction††	Approximate Mix Proportions by Weight‡	Water/Cement Ratio by Weight	Cement‡‡	Remarks
40 at 28 days	Good	60	Copious watering for first 24 hours	1} 2}	1:6 / 1:7	0·50 / 0·50		Mixes richer than 1 : 6 are not recommended except for very small members. Vibration is usually required.
60 at 28 days	Good	75	Copious watering for first 24 hours	1} 2}	1:5 / 1:6	0·45 / 0·45		
40 at 24 hours	Good	55	Copious watering for first 24 hours	1} 2}	1:5 / 1:6	0·45 / 0·45		
25 at 12 hours	Good	40	Copious watering for first 24 hours	1} 2}	1:5 / 1:6	0·45 / 0·45		

* The nominal minimum strength is taken as being approximately equal to the 1 per cent minimum for groups of cubes.

** "Good" means weigh-batching or very careful volume-batching with an allowance for the bulking of sand, determinations of moisture in the aggregate and continuous general supervision.
"Very good" means weigh-batching with frequent determinations of the moisture content of aggregate (obtained from specially chosen sources) and excellent general supervision.
Vibrated specimens are tested at the termination of steaming and cooling periods for purposes of control.

† Steam curing is carried out usually at 70-80°C.

†† Hand punning = 1. Ordinary vibration = 2. Intense vibration (sometimes with pressure) = 3.

‡ Mix proportions are given as the ratio of cement to total aggregate. The aggregate would be divided into various sizes to give a satisfactory overall grading.

‡‡ OP, ordinary portland. HES, high-early-strength portland.

0·35–0·4 by weight. The 1-hour absorption of dry aggregate necessitates an apparent increase in water content of 10–20 per cent. The slump would be zero.

7. 35–55 MPa at 1 day when lightweight calcium silicate hydrate or lime-silica concrete is used with a unit weight (dry) of 1120–1280 kg/m³. It is cured by autoclaving in high-pressure, saturated steam at 185°C for about 8 hours or more and tested in a dry state. The minimum compressive strength requirements 6 and 7 are dealt with in Sections 33.2.2 and 33.2.3.

Several means for obtaining high-early strength are summarised.

Rich mixes with a low water/cement ratio, vigorous compaction and thorough curing.

Warm materials, moulds, and mixing water, with or without a setting accelerator. If a setting accelerator is used, it should be compounded so as not to reduce the ultimate strength of the concrete or cause corrosion of steel wires in steam-cured, prestressed concrete.

High-early-strength or superfine portland cement, usually without a setting accelerator where it would cause corrosion (see Figs 3.2, 4.3 and 4.4).

Moist curing at an elevated temperature (above 23°C), adiabatic or mass curing, electric curing, steam curing and autoclaving with high-pressure, saturated steam. The latter is particularly suitable for the formation of a calcium silicate hydrate structure, in mixes containing lime and finely divided silica.

High-alumina cement under suitable hydrothermal conditions of use and with both immediate and special 24-hour curing.

The suggested procedures in Table 4.20 are for making concrete with specified high strengths at various ages.[11] They may serve as a useful guide in the determination of trial mixes and methods of making concrete with high-early and high-ultimate strengths. Only a full knowledge of the circumstances of a particular project will enable the relative merits of alternative mixes, methods of manufacture and standards of control to be evaluated satisfactorily.

4.4 COMPRESSIVE STRENGTH RESULTS
(Nonstandard Specimens)

If test specimens differ in size and shape from standard cylinders (with 150 mm diameter and 300 mm height), the compressive strength can be converted approximately to that of the standard cylinder from the data in Table 4.21, and Figures 4.15 and 4.16 for comparative purposes. Concrete acceptance data, however, must be based on standard specimens and tests (see Section 14.4).

TABLE 4.21   NONSTANDARD CYLINDERS TO STANDARD CYLINDERS

Diameter of Specimen (mm)	Multiplying Factor
75	0·94
100	0·96
125	0·98
150	1·00
175	1·02
200	1·04
250	1·07
300	1·10

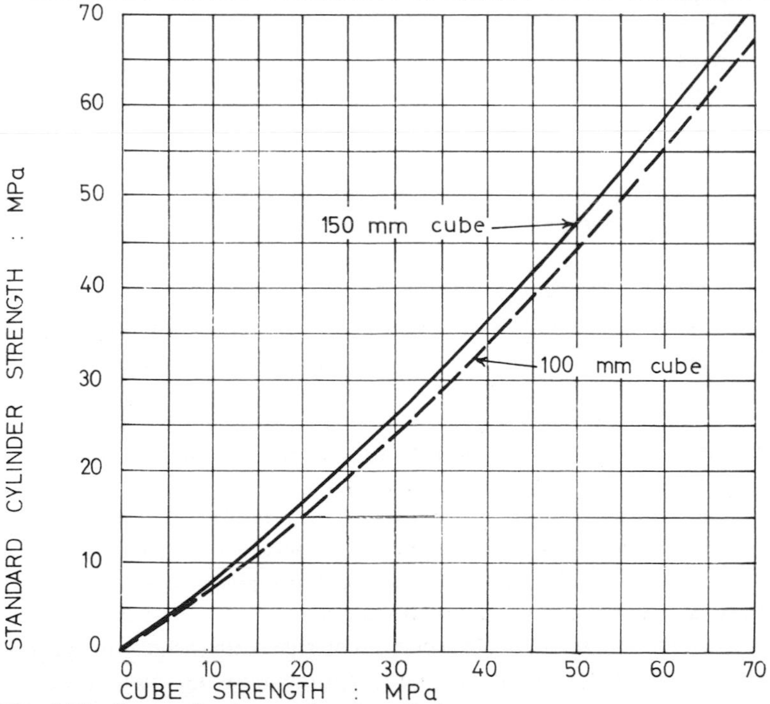

Fig. 4.15 Conversion from cubes to cylinders

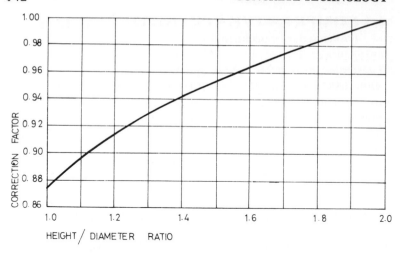

Fig. 4.16   Height/diameter ratio factors

4.5 GAP AND CONTINUOUS GRADINGS OF AGGREGATE

When designing a concrete mix with a high proportion of large-size particles, such as for vibratory compaction, the sizes from 4·75 mm to 9·5 mm may be omitted from coarse aggregate of 20 mm nominal size without causing segregation. With aggregate of 40 mm nominal size, the sizes from 9·5 mm to 19·0 mm may be omitted (or reduced in quantity) for an increased interlock of coarse particles in the mix.[12, 34, 45]

The greatest strength may be obtained with a gap grading but it is difficult to determine exactly the correct grading without preliminary tests. The advantage of using empirical grading formulae, such as those of Fuller, Bolomey and others, is that in every case a good concrete can be made, even if it is not the best, and it is better to have a good continuous grading than an indifferent gap grading (see Sections 3.3.4(d), 4.1 and 4.3).

The fineness-modulus method of grading, which is now superseded by other methods, is a link between the two ideas, and shows that an infinite number of solutions give a concrete having very nearly the greatest strength.

It is possible to depart to a certain degree from the grading judged to be the best without reducing appreciably the quality of the concrete. On the site it is necessary to consider not only the mechanical properties of the concrete, but freedom from segregation and workability in relation to the method of placing and consolidating the concrete, as well as the practicability of obtaining an aggregate having the grading desired.

Investigations by Feret (interpreted by Bolomey) have shown that the ideal grading is that in which the sizes of 38 per cent by weight of the particles are from 0 to 0·1D (D being the size of the largest particles) and 62 per cent are from 0·4D to D, the sizes from 0·1D to 0·4D being omitted.

The particles smaller than 0·1D must also be graded, so that 38 per cent of them (including the cement; i.e. 0·38 × 38 per cent = 14·5 per cent of the whole) are smaller than 0·01D and the remainder (38 − 14·5 = 23·5 per cent) are from 0·04D to 0·1D. These results are not conclusive, and gap gradings with ratios of fine to coarse particles other than 38 : 62 may be warranted under certain conditions.

It is apparent that an aggregate having a gap grading, if closely studied, will produce a concrete that is slightly stronger than one based on an aggregate with a continuous grading determined once for all cases. A good gap grading can be deduced from a continuous grading. It is sufficient to retain the same fineness modulus and to be assured that the requisite workability is obtained. This is so if the proportion of particles from 0·01D to 0·1D is not less than that in the continuously graded aggregate. Microconcrete in models is dealt with in Appendix IV, 2(b).

The choice between continuous and gap gradings depends on local conditions. If natural aggregate (e.g. gravel) is available, continuous grading is generally the more economical. If the fine and coarse aggregates are supplied separately, a gap grading may be more suitable for use, such as in low-slump, slip-formed and exposed-aggregate concretes.

For pump and manual placements, Bolomey's formula has an advantage in its simplicity and the fact that it gives a good concrete of the desired workability by adjustment of its empirical coefficient (see Section 4.1).

4.6 GRAPHICAL PROPORTIONING OF AGGREGATE

When two or more varieties of aggregate are available, they may be combined so as to approximate to some extent a specified grading. Of several methods which may be used for the graphical proportioning of aggregate, [5, 13, 14] two methods will be illustrated with examples. Reference data on nominal gauge sizes of quarry products and their interrelationship with designated sieve sizes, are given in Tables 3.20 and 3.21.

Method A. The following procedure shows how a fine and a moderately coarse aggregate (Table 4.22) may be combined, so that their proportions give a joint grading which approximates or approaches Grading No. 1 in Figure 4.6.

1. A square diagram is prepared with percentage scales along three sides (Fig. 4.17).

2. For fine aggregate, the sieve size and percentage passing are marked along the left-hand axis.

3. The right-hand axis represents these factors for the coarse aggregate.

4. Each point on the left-hand axis is joined by a straight line to the point with the same sieve size or number on the right-hand axis.

5. A vertical line is drawn through the point where the 5 mm size sloping line intersects a horizontal line which represents the percentage of requisite grading that passes a 4·75 mm sieve (30 per cent in this instance).

6. The vertical line thus drawn represents the combined-aggregate grading. The percentage of combined aggregate passing each sieve is the scale value of each point of intersection, which this line makes with the sloping, sieve-size lines. The percentage proportion of fine aggregate to total aggregate is read on the top scale, where it is intersected by the combined-aggregate line (25 per cent in this instance).

A comparison between the requisite and combined gradings in Table 4.22 shows a small discrepancy of up to 3 per cent. If the difference were large, however, it could be reduced by adjusting the proportions slightly. That is, by moving the combined-aggregate line a little to the left or right, or by using alternative or additional aggregate for the purpose.

TABLE 4.22   **GRADINGS OF AGGREGATE**

Sieve Size (mm)	Requisite Grading	Percentage passing		
		Fine	Coarse	Combined
19·0	100		100	100
9·5	45		31	48
4·75	30	100	7	30
2·36	23	92	0	23
1·18	16	76		19
0·60	9	48		12
0·30	2	20		5
0·15	0	3		1

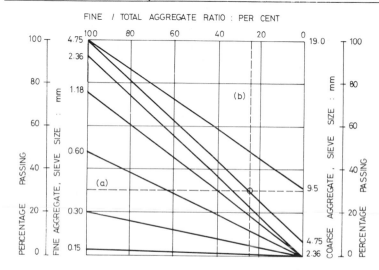

Fig. 4.17 Proportioning of aggregate. (a) 30 per cent. (b) Combined-aggregate line.

Where three varieties of aggregate are to be combined, two should be combined first and the resulting grading combined with the third. For instance, if two sizes of coarse aggregate (say, 40 mm nominal size and 20 mm to 5 mm nominal size) are to be combined with sand, the first two are combined initially, using the percentage passing a 19-mm sieve as a criterion of agreement or approximation. If two sands are to be combined, the criterion is the percentage passing a 0·60-mm sieve.

Method B. This procedure consists essentially of the following steps.

1. A graph of the required aggregate grading is prepared, using

the usual linear ordinates for percentage passing, but choosing the scale of sieve aperture size so that the grading plots as a straight line. This is readily done by drawing an inclined straight line and marking on it the sizes corresponding to the various percentages passing.

2. The gradings of the aggregates to be mixed are plotted on this scale. It is generally found that they are not straight lines.

3. With the aid of a transparent straight-edge, the straight lines which most nearly approximate to the grading curves of the single aggregates are drawn. This is done by selecting for each curve a straight line, in such a way that the areas enclosed between the straight line and the curve are a minimum and are balanced about the straight line.

4. The opposite ends of these straight lines are joined together. The proportions for mixing purposes are read off from the points where these joining-lines cross the straight line representing the required grading.

The actual procedure will be apparent from the following two examples.

Example 1. Table 4.23 shows certain limits of grading for a requisite mixture of aggregate. The average of these limits is the required grading.

1. The requisite grading is represented by the diagonal OO′ of a rectangle (Fig. 4.18). The vertical ordinates of the rectangle are graduated in per cent from 0 to 100 on a linear scale. The horizontal scale for sieve aperture size is graduated by drawing for each sieve a vertical line, which cuts the diagonal at a point where the ordinate equals the per cent passing that sieve: i.e. 100 per cent for 26·5 mm, 92 per cent for 19·0 mm, 82 per cent for 9·5 mm and so on.

2. The aggregates to be mixed are 26·5 mm coarse-graded stone, a normal concrete sand passing 4·75 mm and fine silica passing a 0·30-mm sieve. Their gradings are given in Table 4.23 and they are plotted on the same scale of sieve size as the required grading. They are represented in Figure 4.18 by the jointed lines CBA, CED and OF, respectively.

3. The nearest straight lines to these gradings are drawn with the aid of a transparent straight-edge, by the *minimum balanced areas* method described. They are the dotted lines HO′, CJ and OF.

4. The opposite ends of these lines are joined, giving the

dot-and-dash lines HJ and CF. The points where these lines cross the average required grading line are marked with circles, and the mixture which these points represent is shown on the right-hand side of the diagram.

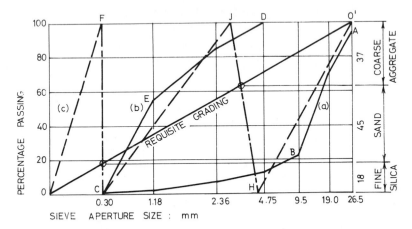

Fig. 4.18 Proportioning of aggregate. (a) Coarse aggregate. (b) Sand. (c) Fine silica.

The grading which results from mixing the aggregates in these proportions is given in the right-hand column of Table 4.23. Although it departs considerably from the average desired, it is within the specified limits. If other proportions are tried, it is found that they lie further from the average, and therefore this is the best mixture that can be obtained with the particular materials blended.

TABLE 4.23 BLENDING OF AGGREGATES—PERCENTAGES PASSING SIEVE SIZES

Sieve Size (mm)	Requisite Grading		Aggregates to be Combined			Mixture 37% (a) 45% (b) 18% (c)
	Limits	Average	(a) Coarse	(b) Sand	(c) Fine silica	
26·5	100	100	95	100	100	98
19·0	85-100	92	70	100	100	89
9·5	65-100	82	21	100	100	71
4·75	55-85	70	11	100	100	67
2·36	40-70	55	7	85	100	58
1·18	25-45	35	2	55	100	43
0·30	10-25	18	Nil	Nil	100	18

Example 2. Table 4.24 and Figure 4.19 show how a slightly different problem can be dealt with. It is required to find what mixture of nominally single-sized aggregates will give the required

grading of Table 4.24. No limits are set, because it is known that it is possible to get very close to the average. Inspection shows that 20 mm, 14 mm and 7 mm nominal-size aggregates will be required.

TABLE 4.24 BLENDING OF AGGREGATES—PERCENTAGES PASSING SIEVE SIZES

Sieve Size (mm)	Average Requisite Grading	Average Gradings of Aggregates to be Combined			Mixture 65% (a) 20% (b) 15% (c)
		(a) 20 mm	(b) 14 mm	(c) 7 mm	
26·5	100	100	100	100	100
19·0	95	92	100	100	95
13·2	45	20	92	100	46
6·7	15	1	3	92	15
3·35	5	1	1	15	3

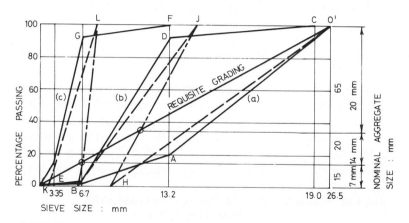

Fig. 4.19 Proportioning of aggregate. (a) 20 mm. (b) 14 mm. (c) 7 mm.

The average gradings of these aggregates are given in Table 4.24. In Figure 4.19, the diagonal OO′ is the required grading, from which the sieve-aperture size scale is determined as in Example 1. BAO′, EDC and OGF are the 20 mm, 14 mm and 7 mm gradings, respectively, drawn to the same scale. The dotted lines HO′, BJ and KL are the nearest straight lines approximating to these gradings. The dot-and-dash lines HJ and BL joining the ends of the dotted lines cross the required grading line at the points marked by circles. The relevant mixture is shown on the right-hand side of the diagram. Finally, the right-hand column of Table 4.24 shows how closely the grading resulting from this mixture approximates to that required.

CHAPTER 5

QUANTITIES OF MATERIALS

5.1 QUANTITIES AND USES FOR NOMINAL MIXES

Typical quantities per m³ and uses of concrete and mortar under average conditions are indicated in Tables 5.1 and 5.2. The amount of water includes that in the aggregate and, in each instance, the requisite volumes (m³) of sand and stone are obtained by multiplying the indicated number of 40-kg bags of cement by 0·0265 and the nominal proportions of the mix. The quantities, under average site conditions, are based on unit weights of 1410 kg/m³ for coarse aggregate and 1440 kg/m³ for slightly damp

TABLE 5.1   **QUANTITIES AND USES OF CONCRETE**

Mix Proportions Cement : Sand : Stone by Volume	Portland Cement (bags per m³)	Water (litres per bag)	Water/ Cement Ratio by Weight	28-Day Cylinder Compres- sive Strength (MPa)	Uses for Mixes
1 : 1 : 2	14	16	0·4	40	Heavy-duty floor surfacings. Thin sections. Prestressed concrete.
1 : 1·5 : 3	10·6	20	0·5	30	High-strength structural concrete.
1 : 2 : 3·5	9·1	22	0·55	—	Heavily trafficked roads.
1 : 2 : 4	8·4	23	0·57	25	Ordinary, reinforced concrete construction.
1 : 2·5 : 4	7·7	24	0·6	—	General structural purposes.
1 : 2·5 : 5	6·9	26	0·65	20	Footings and lean-mix roads.
1 : 3 : 6	5·8	30	0·75	15	Mass walls.

149

sand (see Section 3.3.6(b)). Prescriptive and grade-designated mixes for structural concrete are tabulated in BS CP110 Part 1 (1972, amended 1974) and Appendix VIII, 5.

TABLE 5.2 **QUANTITIES AND USES OF MORTAR**

Mix Proportions Cement : Sand by Volume	Portland Cement (bags/m³)	Water (litres/bag)	Water/ Cement Ratio by Weight	Uses for Mixes
1 : 1	27·2	15·4	0·38	High-strength grout mixture.
1 : 2	17·8	17·5	0·44	Watertight rendering (if carefully cured).
1 : 2·5	15·3	19·2	0·48	Early rendering for paths. Strong mortar.* Grout for prepacked macadam.
1 : 3	13·4	20·9	0·52	Wall plaster. Rough casting and stucco.†
1 : 5	8·8	26·4	0·66	Building bricks (see Section 27.2.2(b)).

* Wear resistance is 0·4 times that of heavy-duty granolithic concrete (see Sections 12.1 and 20.1.5).

† Cracking is minimised in ordinary rendering and jointing by substituting hydrated lime for one-half to two-thirds of ordinary portland cement in this mix. Alternatively, masonry cement may be used as recommended by its manufacturer.

5.2 CALCULATION OF QUANTITIES

From nominal mix proportions, the quantities of materials per m³ of concrete can be calculated on the basis of absolute volume. The procedure may be illustrated with a mix of 1 part cement : 1·75 parts sand : 3·5 parts stone screenings (dry volume) with a water/cement ratio of 0·55 by weight; and with the data given in Section 4.3.2, for loose dry aggregate. Assuming that a 40-kg bag of cement has a volume of 0·0265 m³ (in Australia; see Section 4.1 and Appendix VIII), the absolute volume of each component in m³ per bag of cement is as follows.

Cement: $\dfrac{40}{3150}$ = 0·0127 m³.

Sand: $\dfrac{0\cdot0265 \times 1520 \times 1\cdot75}{2650}$ = 0·0266 m³.

Stone: $\dfrac{0\cdot0265 \times 1410 \times 3\cdot5}{2650}$ = 0·0494 m³.

Water: $\dfrac{40 \times 0\cdot55}{1000}$ = 0·0220 m³.

Total = 0·1107 m³.

The number of bags of cement per m³ $\quad = \dfrac{1}{0 \cdot 1107} = 9 \cdot 0.$

Dry sand required $= 9.0 \times 0.0265 \times 1.75 = 0 \cdot 42$ m³

Dry stone required $= 9 \cdot 0 \times 0 \cdot 0265 \times 3 \cdot 5 = 0 \cdot 83$ m³.

Water required $= 9 \cdot 0 \times 0 \cdot 0220 \qquad\qquad = 0 \cdot 198$ m³ $=$
198 litres
or 22 litres/bag.

Should the sand have a bulking factor of 25 per cent, the requisite volume of damp sand would be $0.42 \times 1 \cdot 25 = 0 \cdot 53$ m³. If the free moisture contents of sand and stone screenings were 5 per cent and 1 per cent by weight, respectively, the amount of mixing water would be reduced accordingly.

The water in the sand would be $0 \cdot 42 \times 1520 \times 0 \cdot 05 = 31 \cdot 9$ kg. The water in the stone would be $0 \cdot 83 \times 1410 \times 0 \cdot 01 = 11 \cdot 7$ kg, which would be equivalent to 43·6 litres. The net quantity of mixing water required, therefore, would be $198 - 43 \cdot 6$ or 154·4 litres, i.e. 17·2 litres/bag.

Where quantities are required quickly for tendering purposes, they can be directly estimated from several design charts that are illustrated in Reference 47 for different concrete mixes.

5.3 GRAVIMETRIC CALCULATION OF ENTRAINED AIR IN CONCRETE

A formula for determining the air content of concrete (see Section 16.3) can be derived in the following manner.[28]

$$P_1 = \frac{W_0 - W_1}{W_0} \times 100 \text{ per cent}$$

where P_1 = design percentage of air, W_0 = theoretical bulk density of concrete on an air-free basis, and W_1 = bulk density of concrete containing the design percentage of air.

$$P_2 = \frac{W_0 - W_2}{W_0} \times 100 \text{ per cent}$$

where P_2 = actual percentage of air, W_0 = theoretical bulk density of concrete on an air-free basis, and W_2 = bulk density of concrete by test.

Solving for W_0:

$$W_0 = \frac{100W_1}{100 - P_1} = \frac{100W_2}{100 - P_2}.$$

Hence $\dfrac{W_1}{100 - P_1} = \dfrac{W_2}{100 - P_2}$ and $P_2 = 100 - \dfrac{W_2}{W_1}(100 - P_1)$.

Take, for instance, a concrete mix (designed to contain 4·5 per cent of entrained air) which has the following batch weights per m³.

	Weight (kg)	Absolute Volume (m³)
Air	0	0·0450
Cement	295	0·0936
Water	168	0·1683
Aggregate	1792	0·6931
Total	2255	1·0000

The bulk density of concrete containing 4·5 per cent of air is

$$W_1 = \frac{2255}{1·0} = 2255 \text{ kg/m}^3.$$

If the bulk density of fresh concrete by test were 2255 kg/m³, then the design percentage of air would be entrained. If it were

2290 kg/m³ (W_2)

then $P_2 = 100 - \dfrac{2290}{2255}(100 - 4·5) = 3$ per cent.

Therefore, the concrete actually contains 3·0 per cent of air, instead of 4·5 per cent for which the mix was designed.

It is worth noting that the theoretical bulk density of the concrete, on an air-free basis (W_0), is

$$\frac{2255}{1·0000 - 0·0450} \text{ or } 2361 \text{ kg/m}^3.$$

If its bulk density by test were found to be 2255 kg/m³, then it would contain

$$100 - \frac{2255}{2361}(100 - 0) \text{ or } 4·5 \text{ per cent of air.}$$

Other methods of determination of entrained air are given in Section 16.3, and procedures for the determination of bulk density of fresh and hardened concretes are given in References 23 and 48, and in Appendix I.

Tests for the analysis of freshly mixed, plain and air-entrained concretes are contained in AS 1012 Part 5 and BS 1881, and in the References of Section 14.1, while the air content of freshly mixed masonry-cement mortar can be calculated by the method described in AS 1316.

CHAPTER 6

WORKABILITY

6.1 GENERAL INFORMATION

Workability is that physical property of concrete which determines the amount of useful internal work per unit mass that is necessary to produce full compaction.[17] It is affected by the proportions of water and cement in a mix, the grading, shape, porosity and surface texture of the aggregate, and the presence or absence of surface-active agents, such as workability agents. The workability required for satisfactory placing and compaction is related not only to the properties of the concrete, but also to such external factors as the size, shape and surface roughness of formwork or moulds, the quantity and spacing of reinforcement, and the methods of conveyance and compaction employed (see Sections 7.1.2 and 7.1.3; and Glossary, Appendix VI).[38, 54-57]

6.2 MEASUREMENT

The workability of concrete may be determined approximately by the slump test (AS 1012 Part 3, BS 1881, ASTM C143, AASHTO T119*) or, more precisely, by the compacting-factor test (BS 1881). The latter is particularly suitable for testing concrete mixes of very low workability or zero slump. Alternative types of apparatus for determining workability are the flow table (ASTM C124 and C230), spread table, the "Kelly" ball penetrometer (ASTM C360), the "Vebe" (BS 1881) and "Wigmore" consistometers,[18, 19, 32] the "Thaulow" drop table,[32] the "Plastograph",[20] the "Tel-A-Slump",[21] the "Nasser-Ma",[61] the "Probe",[65] and "Simulated Placement",[73] meters. The consistence of mortar and grout is considered in Sections 24.3.5 and 27.2.2(b), and in Appendix I.

---

*AASHTO signifies American Association of State Highway and Transportation Officials.

154

6.2.1 SLUMP TEST

A hollow frustum of a cone (300 mm high and smooth inside) is made from 1·5 mm galvanised-iron sheet. The open base and top of this mould are of 200 mm and 100 mm internal diameter, respectively (Fig. 6.1). It is made from a segment circumscribed by arcs of 305 mm and 610 mm radii, and radial sides separated at the base by a 600 mm chord, an additional 25 mm at one end being provided for lap. Fresh concrete (containing aggregate not larger in size than 53 mm) should be sampled by the appropriate method specified in AS 1012 Part 1, BS 1881 or ASTM C172 (see Section 7.1.5).

The mould, after being cleaned and moistened internally, is placed on a level metal plate. It is held firmly in place and filled symmetrically in three layers, which are of approximately equal volume. Each layer is rodded with twenty-five strokes of a round-ended rod of 15 mm diameter and 600 mm length. The strokes are distributed in a uniform manner over the cross-section of the mould and they penetrate into the underlying layer. The concrete is struck off at the top of the mould, which is then withdrawn vertically. The average subsidence is measured to the nearest 5 mm, or 10 mm if over 80 mm, and recorded (with related data) as the slump (Plate 6).

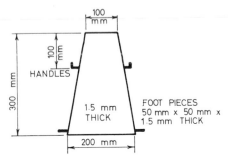

Fig. 6.1 Slump-test mould

6.2.2 SPREAD TABLE

A spread table, 700 mm square, consists of a galvanised-steel plate (3 mm thick) mounted on a timber frame. The surface is inscribed concentrically with a circle of 200 mm diameter and with two principal axes. One side of the table (which rests on an even, level, hard surface) can be raised by a handle to a height of approximately 40 mm. A stop bracket, as shown in Figure 6.2, limits the height of this movement.

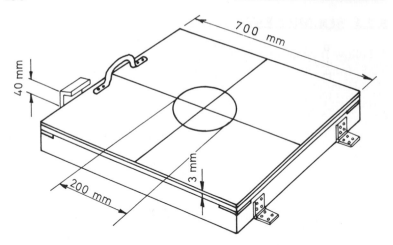

Fig. 6.2   Spread table

A hollow frustum of a cone is used as a mould and placed concentrically on the table, the cone being 200 mm high, with 200 mm diameter at the base and with 125 mm diameter at the top. The mould is filled in two layers in a way that is similar to that used in the slump test. After cleaning the surface of the table and withdrawing the mould, the table is raised gently by its handle to the stop level and allowed to fall, this operation being repeated fifteen times.

The average of the largest and smallest diameters of the spread specimen is determined and recorded as the spread index. A concrete mix that is designed for manual compaction should form a coherent layer without showing signs of segregation, bleeding or excessive laitance.

6.2.3 BALL PENETROMETER

The ball-penetrometer test (ASTM C360) is a convenient routine method of checking consistence, simply by measuring the depth to which a metal hemisphere of 150 mm diameter, weighted to $13 \cdot 60 \pm 0 \cdot 05$ kg, sinks into fresh concrete. The sample should be at least 200 mm deep and 450 mm wide, so as to avoid the effects of boundary conditions. Concrete may be tested in a wheelbarrow or formwork, provided that it is undisturbed by vibration and the weight is allowed to settle freely without impact. The reported result should be an average of at least three readings. As the results cannot be translated quantitatively into terms of slump, the test should not be used as a basis for the acceptance or rejection of concrete.

6.2.4 COMPACTING-FACTOR TEST

A compacting-factor apparatus measures the degree of compaction that is achieved by a given amount of work on a specimen of fresh concrete. The apparatus consists of two rigid, conical hoppers, which are mounted axially above a cylindrical mould. Each hopper is made of brass, bronze or steel and is polished internally. The lower ends of the hoppers are closed with tightly fitting, hinged trap-doors (which are 3 mm thick), provided with quick release catches (Fig. 6.3). The maximum size of aggregate in the concrete should not exceed 37·5 mm unless the apparatus is modified,[40] and standard sampling should be carried out for field tests (see Section 7.1.5).

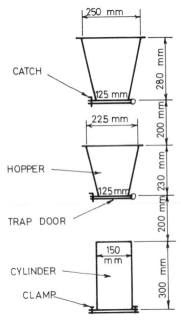

Fig. 6.3 Compacting-factor apparatus

After the upper hopper has been filled flush with concrete, by means of a hand scoop, the trap-door is opened so that the concrete falls into the lower hopper. During this operation, the cylinder is temporarily covered by two steel floats. As differences in workability arise from initial processes in the hydration of cement, the concrete is released from the upper hopper 10 minutes after the completion of mixing. At this age, the workability of the mix is near its value at the time of placing in formwork. In factory production, where moulding is quickly done, the time lag for comparable results may be reduced to 2 minutes.

The cylinder is uncovered and the trap-door of the lower hopper is opened for discharge into the cylinder. Mixes which stick in the hoppers can be helped through by pushing a bullet-nosed rod, of 15 mm diameter and 600 mm length, gently into the concrete from the top. The excess of concrete remaining above the level of the top of the cylinder is cut off by simultaneously sliding two steel floats across from the outside to the centre. The outside of the cylinder is wiped clean and the weight of partially compacted concrete in the cylinder is determined to the nearest 10 g.

The cylinder is refilled with concrete from the same sample in layers approximately 50 mm deep, the layers being heavily rammed or preferably vibrated so as to obtain full compaction. The top surface of the fully compacted concrete is struck off level with the top of the cylinder, which is then wiped clean. The weight of fully compacted concrete is determined by weighing to the nearest 10 g. Alternatively, the latter weight may be determined by calculation.

The *compacting-factor* is the ratio of the weight of partially compacted concrete to the weight of fully compacted concrete, expressed to the nearest second decimal place. Its value, incidentally, is not closely comparable with slump values.

The principle of the test is that the concrete is first brought to a standard state in the lower hopper, then a standard amount of work is done on it during its fall into the cylinder. The test shows that a mix with a compacting factor less than about 0·75 cannot be compacted satisfactorily by hand, while one with a value less than about 0·60 cannot always be compacted by vibration unless pressure is applied as well.

The workability of very dry mixes may be determined, alternatively, by a vibrated, compacting-factor test, whereby the standard cylinder of the normal, compacting-factor apparatus is fitted with a detachable, cylindrical hopper, which serves to extend the height of the cylinder by 50 mm. The cylinder and hopper, on being clamped to a small electromagnetic vibrating table, are filled loosely with concrete to the level of the top of the hopper. The table is set in vibration at 100 Hz at 1·5 g and, after 15 seconds, the specimen is topped up to the level of the hopper. Vibration is continued for 1 minute; additional topping up is done after 0·5 minute if necessary.

After 1 minute, the hopper is removed and the concrete is struck off with two floats to the level of the top of the cylinder. The cylinder is weighed and the ratio of the weight of concrete to the theoretical, compacted weight is taken as the vibrated, compacting factor. The principle of this test is similar to that of the normal, compacting-factor test, but with an increase in the amount of applied work. It is useful for the comparison of mixes of extremely low workability.[18]

Workability may be controlled in the field by balancing the compacting-factor cylinder of concrete, which is of requisite average workability, against a weight. For given materials, adjustments in the amount of mixing water which would be required to offset variations in workability could be registered directly on a dial to the nearest litre. The compacting factors for lightweight and ordinary concretes are comparable within each type of concrete, but are unrelated to each other.[41] With air-entrained concrete, the work done in the test is reduced by restraint, as cohesive concrete falls through the hoppers.

For compacting-factor applications on site, the partially compacted concrete in the 300 mm high cylinder, after being levelled off, is fully compacted by vibration. By measuring the resulting surface subsidence in millimetres, the compacting factor is readily determined directly as the difference between unity and the subsidence/300 ratio.

6.2.5 VEBE CONSISTOMETER

The Vebe apparatus, as developed by V. Bahrner, consists of a cylindrical container in which is formed a slump cone of concrete. A horizontal glass plate, fitted with a rod that moves in a guide, is placed on the concrete cone and a vibrating base is set in motion (Figure 6.4). The glass plate descends as the cone is converted into

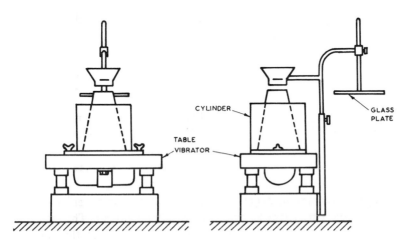

Fig. 6.4 Vebe consistometer

a cylinder, and the vibrator is switched off as soon as all of the underside of the plate is coated with cement paste. The end point is determined from an automatic recorder and the period of vibration in seconds (Vebe degrees or "VB°") is taken as a measure of the workability.

For purposes of quality control, the test is useful with fibre-reinforced or stiff, harsh mixes that require vibration (10–40 VB°).[49] A modified Vebe test, with less intense vibration for increased timing accuracy, can be used on air-entrained concrete of high workability.[69]

6.3 BASIC FACTORS

The compacting factor of concrete can be calculated with a reasonable degree of accuracy from such factors as the water/cement ratio, characteristics of the aggregate, and the mix proportions.[27]

6.3.1 WATER CONTENT

The compacting factor is related to the water content that is available for the lubrication of a mix. Experiments with neat portland cement paste[22] have shown that its consistence, as measured by a penetration of the plunger of a "Vicat" apparatus, increases suddenly when the water/cement ratio is slightly increased gravimetrically above 25 per cent. The compacting factor, therefore, is related to the water/cement ratio in excess of 0·25 by weight. This ratio is calculated from the water content of a mix in excess of that required to bring the aggregate to a saturated, surface-dry condition. When mixing test specimens, therefore, if the aggregate has been dried to constant weight, it is first brought to the saturated, surface-dry condition with the requisite amount of water (see Section 3.3.4(j)) and allowed to stand for 0·5 hour without evaporation.

6.3.2 AGGREGATE CHARACTERISTICS

The effect of particle size on workability may be estimated on the basis of a surface index, f_s, empirical values of which are given in Tables 6.1 and 6.3. The surface index of an aggregate is calculated by multiplying the individual values of particles within certain sieve sizes by the percentage of particles within each respective size and adding up the total. An example is given in tabulated form for Grading No. 1 in Figure 4.7, the resulting figure being reduced to a

TABLE 6.1   **SURFACE INDEX**

Sieve Size (mm) Range of Particles	Surface Index (f_s)	Calculation for Grading No. 1	
		Percentage of Particles	f_s
75 -37·5	0·5	—	—
37·5 -19·0	1	50	50
19·0 - 9·5	2	14	28
9·5 - 4·75	4	12	48
4·75- 2·36	8	6	48
2·36- 1·18	12	6	72
1·18- 0·60	15	5	75
0·60- 0·30	12	4	48
0·30- 0·15	10	3	30
Passing 0·15	1	—	—
		Total	399
		Total × 10⁻³	0·399

convenient size by multiplying it by 10^{-3}. Experiments have shown that mixes tend to segregate when the surface index of aggregate falls much below 0·5, due to inadequate fines to fill the voids in the coarse aggregate.

In addition, the effect of shape and angularity on workability may be estimated on the basis of an angularity index, f_A. Its value is determined from the expression

$$\frac{3f_H}{20} + 1 \cdot 0$$

where f_H is the angularity number; that is, the percentage of voids in aggregate in excess of 33 (BS 812). Analysis has shown that the angularity index, which is typified in Table 6.2, can be used to modify the surface index by simply multiplying them together ($f_s \times f_A$). The angularity index of a composite aggregate is calculated proportionately to the percentage of each type of material in a mix. For instance, the angularity index of an aggregate containing 70 per cent of rounded coarse aggregate and 30 per cent of fine aggregate is $1 \cdot 15 \times 0 \cdot 7 + 1 \cdot 8 \times 0 \cdot 3$ or $1 \cdot 34$.[22, 23, 55]

TABLE 6.2   **ANGULARITY INDEX**

Type of Aggregate	Angularity Index (f_A)
Rounded Stone	
Coarse	1·15
Fine	1·80
Irregular Gravel and Sand	1·90
Crushed Basalt or Granite	
Coarse and Fine	2·2–2·5
Crushed Limestone	
Coarse	2·05

6.3.3 MIX PROPORTIONS

Workability is dependent on mix proportions by volume. Where there is a significant variation in the specific gravity of the different sizes of particles, the surface index, for accuracy, should be determined on a basis of absolute volume. Furthermore, the grading and angularity of aggregate have been found to have negligible effect on workability when the aggregate/cement ratio (A_v) by absolute volume is reduced to 2. A modified, surface-angularity index, therefore, is

$$f_s \, f_A \, (A_v - 2)$$

where $A_v = \dfrac{3 \cdot 15 \, A_w}{G_A \, C_w}$, A_w = total weight of dry aggregate, C_w =

weight of cement, and G_A = average specific gravity of the aggregate.

(In the prevailing derivations, the specific gravities of cement and aggregate are taken as $3 \cdot 15$ and $2 \cdot 6$, respectively.)

From the foregoing factors, it is concluded that the compacting factor is a function of

$$\frac{W/C - 0 \cdot 25}{f_s \, f_A \, (A_v - 2)}$$

where W/C = water/cement ratio.

Figure 6.5 shows, for mixes at an age of 10 minutes, the relationship between compacting-factor, water/cement ratio, mix proportions, and the surface and angularity indices. The equation for compacting factor (cf) thus represented may be shown to be

$$cf = 0 \cdot 74 \; \frac{10 \, (W/C - 0 \cdot 25) + 0 \cdot 67}{f_s \, f_A \, (A_v - 2)}$$

This equation tends to overestimate cf values that are over $0 \cdot 95$, due to segregation. All values thus derived which are over $0 \cdot 95$ may be regarded collectively as $0 \cdot 95 +$. Comparisons between the values of calculated and measured compacting factor, using stable mixes and a wide variety of gradings, have shown agreement to within $\pm \, 0 \cdot 02$ in 75 per cent of the investigations and $\pm \, 0 \cdot 04$ in 97 per cent of such studies.

The foregoing factors, which contribute to workability, may be used in the design of a concrete mix. Take, for example, a mix with a water/cement ratio of $0 \cdot 5$ by weight for strength requirement, and a compacting factor of $0 \cdot 85$ for low workability. Anticipated

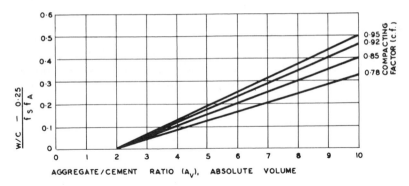

Fig. 6.5 Compacting and related factors

values of $f_s \pm 0.05$, for uniformly graded aggregate, are indicated in Table 6.3.

If 20 mm nominal-size, graded aggregate is used, the proportions of sand ($f_s = 1.1$) and 20 mm to 5 mm coarse aggregate that are required to give an f_s value of 0.6 are 1:1.67, respectively. With these proportions and the fine and coarse sands in Zone 2 of BS 882 (Table 6.3), the range of combined surface index is 0.67–0.53. For a mix with 20 mm nominal-size, graded aggregate, these values are within acceptable limits (i.e. about 0.5–0.8). The lower limit of this range provides against the risk of segregation, while the higher limit avoids a drop in strength (which would result from using an unusually sandy mix with ordinary methods of compaction).

TABLE 6.3   **SURFACE INDEX**

Aggregate	Surface Index (f_s)
40 mm with Sand	0.55
20 mm with Sand	0.60
20 mm–5 mm	0.30
10 mm with Sand	0.70
Sand*	0.91–1.29

* Zone 2 of BS 882.

The angularity index f_A, as a result of the examination or testing of samples may be taken as 1.9. The product, $f_s f_A$, thus varies from 0.67×1.9 to 0.53×1.9 or 1.27 to 1.01, the average value being 1.14. For the latter figure

$$\frac{W/C - 0.25}{f_s f_A} = \frac{0.5 - 0.25}{1.14} = 0.22$$

The corresponding value of A_v for a cf value of 0·85 in Figure 6.5 is 6·4.

For purposes of investigation, an estimate may be made now of the influence of a variation in the grading of the sand within Zone 2. This is done, firstly, on the assumption that the water/cement ratio is maintained at 0·5. In this instance

$$\frac{W/C - 0·25}{f_s f_A} \text{ ranges from } \frac{0·25 \text{ to } 0·25}{1·27 \quad 1·01}$$

or 0·2 to 0·25. With A_v equal to 6·4, Figure 6.5 indicates a range in cf of 0·82 to 0·88.

Secondly, the estimate is made on the assumption that the workability is maintained at 0·85 and the water/cement ratio is varied. This relationship applies where constant workability is an important requirement during placing and compaction. Here

$$\frac{W/C - 0·25}{f_s f_A} \text{ must remain constant at } 0·22$$

so that $W/C - 0·25 = 0·22\, f_s f_A$ and ranges from $0·22 \times 1·27$ to $0·22 \times 1·01$, or 0·28 to 0·22. The value of W/C ranges, therefore, from $0·25 + 0·28$ to $0·25 + 0·22$, or 0·53 to 0·47 by weight. These variations are probably as large as are desirable for good-quality concrete. If the range, either in workability or strength, were unduly large for suitable mix design, a less variable sand would have to be used. Finally, the aggregate/cement ratio is:

$$\frac{A_v G_A}{3·15} = \frac{6·4 \times 2·6}{3·15} \text{ or 5·3 by weight.}$$

Dividing the proportion of total aggregate into fine and coarse, in the predetermined ratio of 1 : 1·67, the mix proportions are then 1 part cement : 2 parts sand : 3·3 parts stone, with a water/cement ratio of 0·5 by weight. [51, 54-57]

6.4 WORKABILITY AGENTS OR PLASTICISERS

The workability of mixes and the quality of concrete may be improved by adjustments in the grading and inherent characteristics of the aggregate and by the use of air-entraining, wetting and cement-dispersing (water-reducing) agents. The former procedure should be given precedence over the latter for good-quality concrete. Chemical admixtures for concrete, including water-reducing agents, are dealt with in Appendix I. Air-entraining and wetting agents are primarily surface-active agents of anionic, cationic or nonionic type. Cement-dispersing (water-reducing) agents are typically sodium or calcium lignosulphonate, with or without a triethanolamine or calcium chloride accelerator (see Sections 10.1, 10.2, 22.3, 31.1.1, 32.1.1 and 33.2.3(a)).[37]

The calcium lignosulphonate is derived from the digestion (under pressure) of softwood chips in a solution of sulphuric acid and calcium sulphite, the content of wood sugar (e.g. xylose, with lesser amounts of glucose and mannose) in the finished product being limited to 2 per cent. A 33 per cent solution of lignosulphonate is used at the rate of 0·4–0·8 per cent by weight of cement, the latter proportion being used for set-retardation and ready-mixed concrete. With the foregoing additives, the quantity of mixing water must be reduced (usually by 5–15 per cent), if a given degree of workability is to be maintained (see Section 10.2).

In air-entrained concrete, minute and unconnected stable bubbles of air are distributed throughout the concrete to the extent of 3–6 per cent of its volume. An increase of entrained air from a value of 1 per cent to 6 per cent in a 1 : 6 mix, with a water/cement ratio of 0·50 by weight, increases the compacting factor by about 0·03. In a 1 : 9 mix with a water/cement ratio of 0·67, the corresponding increase is 0·07,[9] which is a substantial improvement (see Tables 4.6 and 4.16; and Section 6.2.4).

The amount of entrained air may be determined gravimetrically or (more usually) by an air meter, the latter being obtainable from the suppliers of air-entraining agents. The composition of air-entraining agents is dealt with further in Chapter 16. The sand content in air-entrained concrete should be reduced slightly, so as to keep the cement factor constant. The adjustment is equal to the air content minus the reduction of water, or about 60–75 kg/m³ of mix.

A wetting agent is characterised by its ability to collect at an interface between a liquid and a solid or gas. It increases the workability' of concrete by increasing the wetting efficiency of the mixing water and by helping to bring the cement and water into intimate contact. The improvement is less marked in low-slump mixes than in plastic mixes. Typical wetting agents are sulphated methyl and ethyl oleates, alkyl phenol ethylene oxide condensate, polyglycol monoricinoleate, and sodium sulphonate of dodecyl benzene. The quantity required depends upon the available proportion of active constituent and the amount of fine aggregate in the mix. It varies from 5 g to 15 g per 40-kg bag of cement, when the agent contains 80 per cent of active constituent. For practical use, a dilute stock solution is prepared and added by a 500 ml measure to the mix. Bituminous and paraffin wax emulsions can serve as permeability-reducing and workability agents (see Section 11.1.2(c)).

Cement-dispersing (water-reducing) agents, if suitably formulated and used, separate cement particles that are clumped together and favourably influence their associated setting and hardening properties. They reduce interparticle attraction and tend to entrain 1–2 per cent of air in a dense mix.

The three types of agents are added to concrete in requisite amount in the mixing water (as determined by tests for optimum results), so as to suit their performance characteristics. If more than one additive is to be used, they should (if chemically incompatible) be incorporated separately in a mix during normal batching. Accelerating agents, if used, can give rise to substantial increases in drying shrinkage and creep, even when incorporated at very low dosage rates (see Section 10.1).

When air-entraining and cement-dispersing agents are used together, the former is reduced in amount by one-half to two-thirds for mixing purposes. A drop in strength due to air-entrainment is usually more than compensated by cement-dispersion and a deliberate reduction in the water/cement ratio. Cement dispersion considerably increases the strength of pozzolanic cement concrete, but its power of deflocculation in a coarsely ground cement (e.g. high-alumina cement) may be only about one-quarter of its effectiveness in ordinary portland cement. Mixes must be properly designed for best results, an allowance being made for the water in additives that are in liquid form. An admixture of finely divided diatomite or fly-ash (see Sections 17.1–17.2) or hydrated lime (Section 18.1) improves the workability of portland cement mixes, but with differing effects on durability.

Workability agents do not turn bad mixes into good ones, but they are a means of modifying or improving the properties that are already inherent in high-quality concrete. Effective compaction and a reduced water/cement ratio, for a given degree of workability, lead to a reduction of permeability, drying shrinkage, possible efflorescence and coefficient of variation in compressive-strength test results. Air-entrainment, with an adjusted mix, is particularly effective in improving the watertightness and the resistance of concrete to freeze–thaw and seawater hazards.

Workability agents may be used to advantage in ordinary, pigmented, pumped, grouted fibre-reinforced and rolled types of concrete, also in lightweight-aggregate concrete, cement manufacture and cement mortar generally. The workability agents should be free from calcium chloride (for accelerated setting) when used in concrete that will be in contact with high-tensile steel (e.g. prestressed concrete), or subjected to marine or intense hydrothermal conditions (e.g. heated concrete floors), or when selectively used with special cement (see Section 3.1.2). Surface characteristics are impaired by bleeding, segregation, excessive steel trowelling, high concentration of admixture, and inferior curing (see Section 15.2.2(a)).

Workability agents should be used with care because, if used in excessive amount, they may cause substantial air-entrainment. Surplus air may be removed, however, by introducing an air-detraining compound, such as polymethylsiloxane emulsion or tributyl phosphate. Suspending, lubricating and thickening agents with thixotropic properties (e.g. polyethylene oxide condensate or hydroxypropyl methylcellulose, bentonite or paraffin-wax emulsion) minimise bleeding in wet mixes and cement grout (see Sections 15.2.2(d)(vi), 24.3.1 and 33.2.3(c)(iv); and Chemical Admixtures, Appendix I; also AS MP20, Part 2).[39, 60] Set-controlling, permeability-reducing, air-entraining and expanding agents are described in Sections 10.1–10.2, 11.1.2(c), 16.1 and 25.1–25.4, respectively.

REFERENCES

1. FERET, R. "Sur La Compacité des Mortiers Hydrauliques". *Annales des Ponts et Chaussées,* Paris, 4, 21 (1892), pp. 1-164.
2. FULLER, W. B. and THOMPSON, S. E. "The Laws of Proportioning Concrete". *ASCE Transactions,* 59 (December 1907), pp. 67-172.
3. BOLOMEY, J. "Determination of the Compressive Strength of Mortar and Concrete". *Schweizeriche Bauzeitung,* 88, 2 (July 1926), pp. 41-55.
4. BAKER, C. A. "A Rapid Field Method for Design of Concrete Mixes". *IE Aust. Journal,* 28, 4-5 (April-May 1956), pp. 119-27; 28, 9 (September 1956), pp. 250-51.
5. (i) DSIR, "Design of Concrete Mixes". *Road Note 4,* 1967.
   (ii) DoE, *Design of Normal Concrete Mixes.* HMSO, London, 1975.
6. McINTOSH, J. D. *Concrete Mix Design* and *Concrete Mix Design Data,* C & CA, London, 1966.
7. RAJU, N. K. *Design of Concrete Mixes.* Sefgal, Faridabad, India, 1974.
8. ACI COMMITTEE 211, 613. "Recommended Practice for Selecting Proportions for Normal and Heavyweight Concrete". *ACI Journal,* 26, 1 (September 1954), pp. 49-64; 66, 8 (August 1969), pp. 612-28; 67, 2 (February 1970), pp. 192-94; 70, 4 (April 1973), pp. 253-55; *ACI Manual of Concrete Practice,* 2; *ACI Publication,* SP-46; *ACI Journal,* 71, 11 (November 1974), pp. 577-78; 72, 2 (February 1975), pp. 46-49; 3 (March 1975), pp. 111-13.
9. WRIGHT, P. J. F. "Entrained Air in Concrete". *ICE Proceedings,* 1, 2, 3 (May 1953), pp. 337-58.
10. ERNTROY, H. C. and SHACKLOCK, B. W. "Design of High-strength Concrete Mixes". *Proceedings, Symposium on Mix Design and Quality Control of Concrete,* C & CA, Britain, 1954.
11. COLLINS, A. R. "The Principles of Making High Strength Concrete". *The Builder,* 177, 5575 (December 1949), pp. 832-35.
12. WALSH, H. N. *How to Make Good Concrete.* Concrete Publications, London, 1939.
13. FULTON, F. S. "Graphical Determination of the Grading of Combined Aggregates". *Concrete and Constructional Engineering,* 52, 5 (May 1957), pp. 187-90.
14. DRISCOLL, G. F. "How to Blend Aggregates to Meet Specifications". *Engineering News-Record,* 144, 1 (January 1950), pp. 45-48; 144, 15 (April 1950), p. 46.
15. HIGGINSON, E. C., WALLACE, G. B. and ORE, E. L., "Effect of Maximum Size Aggregate on Compressive Strength of Mass Concrete". *ACI Publication,* SP-6, 1963, pp. 219-55.
16. WALKER, S., BLOEM, D. L. and GAYNOR, R. D. "Relationship of Concrete Strength to Maximum Size of Aggregate", *Joint Research Laboratory Publication,* 7, National Sand and Gravel Association, Washington, 1959.
17. GLANVILLE, W. H., COLLINS, A. R. and MATTHEWS, D. D. "The Grading of Aggregates and Workability of Concrete". *DSIR Road Research, Technical Paper,* 5, HMSO, London, 1947.
18. CUSENS, A. R. "Measurement of the Workability of Dry Concrete Mixes". *Magazine of Concrete Research,* 8, 22 (March 1956), pp. 23-30.
19. WIGMORE, V. S. "Consistometer". *The Engineer,* 187, 4850 (January 1949), p. 29.
20. POLATTY, J. M. "New Type of Consistency Meter Tested at Allatoona Dam". *ACI Journal,* 21, 2 (October 1949), pp. 129-36.
21. TRAUFFER, W. E. *et al. Concrete Industries Year Book.* Pit and Quarry Publications, Chicago, 1976.
22. "The Workability of Concrete". *Supplement to Wimpey News,* 217 (George Wimpey, London; January 1960).
23. JENNINGS, R. D. "Design of Concrete Mixes". *Civil Engineering,* 45 (March 1975), pp. 40-47.
24. NEWMAN, K. "The Design of Concrete Mixes with High Alumina Cement". *The Reinforced Concrete Review,* 5, 5 (March 1960), pp. 269-301.

25. WALKER, S. and BLOEM, D. L. "Effect of Aggregate Size on Properties of Concrete". *ACI Journal,* 32, 3 (September 1960), pp. 283-98; 32, 9 (March 1961), pp. 1201-58.
26. WELCH, G. B. "High Strength Concrete Mix Design". *Constructional Review,* 34, 10 (October 1961), pp. 28-37; 35, 8 (August 1962), pp. 27-30.
27. MURDOCK, L. J. "The Workability of Concrete". *Magazine of Concrete Research,* 12, 36 (November 1960), pp. 135-44; 13, 38 (July 1961), pp. 79-92.
28. BROWNE, F. L. "Computation of Air in Air-entraining Concrete by the Gravimetric Method". *ACI Journal,* 23, 9 (May 1952), pp. 785-86.
29. HUGHES, B. P. "Rational Concrete Mix Design". *ICE Proceedings,* 17 (November 1960), pp. 315-32; 21 (April 1962), pp. 927-52.
30. *Design, Control and Characteristics of Concrete; Basic Guide to Concrete Construction.* C & CA Aust., Sydney, 1975.
31. COOKE, A. M. "A Guide to the Design of Concrete Mixes". C & CA Aust., *Technical Report,* TR36, 1974.
32. ACI COMMITTEE 211. "Selecting Proportions for No-Slump Concrete". *ACI Journal,* 71, 4 (April 1974), pp. 153-70.
33. SHACKLOCK, B. W. "Concrete Constituents and Mix Proportions". *Technical Report TR 11.004,* C & CA, London, 1974.
34. "Proportioning Concrete Mixes". *ACI Publication* SP-46; *ACI Journal,* 72, 3 (March 1975), pp. 111-13; 73, 3 (March 1976), pp. 173-74.
35. *High Strength Concrete.* National Crushed Stone Association, Washington DC, 1975.
36. WRIGHT, P. J. F. "The Flexural Strength of Plain Concrete. Its Measurement and Use in Designing Concrete Mixes". *DSIR Road Research Technical Paper,* 67, HMSO, London, 1964.
37. HOBBS, C. "Concrete Additives from the Viewpoint of the Large Contractor". *Chemistry and Industry,* 13 (March 28 1964), pp. 526-35.
38. HEATON, B. S. "Strength, Durability, and Shrinkage of Incompletely Compacted Concrete". *ACI Journal,* 65, 10 (October 1968), pp. 846-50.
39. ACI COMMITTEE 212. "Admixtures for Concrete". *ACI Journal,* 60, 11 (November 1963), pp. 1481-1523; 61, 6, 2 (June 1964), pp. 2053-58.
40. SMITH, M. R. "Test for Consistency of 6 in. Aggregate Concrete". *ACI Journal,* 63, 6 (June 1966), pp. 701-706.
41. MATHER, B. "Partially Compacted Weight of Concrete as a Measure of Workability". *ACI Journal,* 63, 4 (April 1966), pp. 441-50; 12 (December 1966), pp. 1467-68.
42. SMITH, I. A. "Design of Fly-ash Concretes". *ICE Proceedings,* 36 (April 1967), pp. 769-90; 39 (March 1968), pp. 489-503.
43. STORK, J. "Proportioning Concrete Mixes Using Digital Computers". *ACI Special Publication,* SP-16, 1967, pp. 41-75.
44. TYNES, W. O. "Comparison of Properties of Mass Concrete Containing 3- and 6-in. Maximum Size, Crushed Limestone Coarse Aggregate". *Technical Report 6-748,* United States Army Corps of Engineers, Vicksburg, 1966.
45. LI, S.-T. "Proposed Synthesis of Gap-Graded Shrinkage-Compensating Concrete". *ACI Journal,* 64, 10 (October 1967), pp. 654-61; 65, 4 (April 1968), pp. 341-45; 72, 3 (March 1975), p. 112.
46. FROST, R. J. "Rationalization of the Trial Mix Approach to Concrete Mix Proportioning and Concrete Control Therefrom". *ACI Journal,* 64, 8 (August 1967), pp. 499-509; 65, 2 (February 1968), pp. 158-60.
47. HOLLAND, H. "Design of Concrete Mixes for Estimating Purposes". *Civil Engineering and Public Works Review,* 62, 736 (November 1967), pp. 1251-58.
48. POPOVICS, S. "Estimating Proportions for Structural Concrete Mixtures". *ACI Journal,* 65, 2 (February 1968), pp. 143-50; 8 (August 1968), pp. 681-84.
49. HUGHES, B. P. and BAHRAMIAM, B. "Workability of Concrete: A Comparison of Existing Tests". *Journal of Materials,* 2, 3 (September 1967), pp. 519-36.
50. CAMERON, R. "Mix Design Simplified". *Concrete* (London), 2, 2 (February 1968), pp. 85-87.

51. HUGHES, B. P. "Rational Design of High-Quality Concrete Mixes". *Concrete* (London), 2, 5 (May 1968), pp. 212-22; 2, 7 (July 1968), p. 285.
52. WELCH, G. B. "Concrete Specifications (A.S. CA2, Amendment 5, 1968)". *Constructional Review*, 41, 8 (August 1968), pp. 22-26.
53. CANNON, R. W. "Proportioning Fly Ash Concrete Mixes for Strength and Economy". *ACI Journal*, 65, 11 (November 1968), pp. 969-79; 66, 5 (May 1969), p. 440.
54. POPOVICS, S. "Analysis of the Influence of Water Content on Consistency". *Highway Research Record 218*, United States Highway Research Board, 1968, pp. 22-33.
55. ALEXANDERSON, J. "Influence of Mix Composition on the Consistence of Fresh Concrete and Cement Mortar". *Proceedings 39*, Swedish Cement and Concrete Research Institute, Stockholm, 1968.
56. MURDOCK, L. F. and BLACKLEDGE, G. F. *Concrete Materials and Practice*, Arnold, London, 1968.
57. POWERS, T. C. *Properties of Fresh Concrete*, Wiley, New York, 1969.
58. PARROTT, L. J. "Production and Properties of High-strength Concrete". *Concrete* (London), 3, 11 (November 1969), pp. 443-48.
59. HERSEY, A. T. "Simplification of Mix Designs". *Concrete Products*, 72, 12 (December 1969), pp. 48-50.
60. *Chemistry of Cement*. Fifth International Symposium, Cement Association of Japan, Proceedings I-IV, Tokyo, 1969.
61. NASSER, K. W. and MA, C. M. "New Workability and Compacting Apparatus for Concrete". *ACI Journal*, 66, 9 (September 1969), pp. 720-24.
62. HARRIS, A. J. "High-strength Concrete: Manufacture and Properties". *Structural Engineer*, 47, 11 (November 1969), pp. 441-46; 48, 5 (May 1970), pp. 195-201.
63. STAMENKOVIC, H. "High Strength and Water Impermeability of Concrete as a Function of Surface Area of Aggregate". *Materials and Structures*, 3, 14 (March/April 1970), pp. 91-98.
64. MACINNIS, C. and THOMSON, D. V. "Special Techniques for Producing High Strength Concrete". *ACI Journal*, 67, 12 (December 1970), pp. 996-1002.
65. NASSER, K. W. and REZK, N. M. "New Probe for Testing Workability and Compaction of Fresh Concrete". *ACI Journal*, 69, 5 (May 1972), pp. 270-75; 11 (November 1972), pp. 707-708.
66. CANNON, J. P. and MURTI, G. R. K. "Concrete Optimized Mix Proportioning (COMP)". *Cement and Concrete Research*, 1, 4 (July 1971), pp. 353-66.
67. ALEXANDERSON, J. "Design of Concrete Mixes". *Materials and Structures, Research and Testing* (Paris), 4, 22 (July/August 1971), pp. 203-12.
68. MATTISON, E. N. and BERESFORD, F. D. "Production and Properties of High-Strength Concrete". *IE Aust Symposium on Concrete Research and Development, 1970-1973*, IE Aust, Sydney, 1973.
69. HUGHES, B. P. "Rational Design of Air Entrained Concrete". *ICE Proceedings*, 55, 2 (December 1973), pp. 841-53; 57, 2 (September 1974), pp. 545-46.
70. "C & CA Mix Series 1 to 4" on Medium Strength, Pumpable, Air-entrained and High-strength Concretes, C & CA, Britain, 1974.
71. "Why Concrete Code (BS CP110) Calls For More Cement". *ICE New Civil Engineer* (21 February 1974), pp. 28, 32.
72. PERENCHIO, W. F. "Very High Strength Concrete". *Bulletin RD 014*, United States PCA R & DL, 1973.
73. ANGLES, J. G. "Measuring Workability". *Concrete* (London), 8, 12 (December 1974), p. 26.
74. POMEROY, C. D. "Influence of Aggregate on the Properties of Concretes of Specified Cube Crushing Strengths". C & CA, London, TR42.504, 1975.

PART 3

PRODUCTION PROCEDURES AND CONTROL

CHAPTER 7

HANDLING OPERATIONS

7.1 MIXING

7.1.1 EQUIPMENT

Mixing equipment should be of adequate capacity for placing operations and capable of handling concrete with the largest aggregate and lowest slump that may be required on a project. Drum and paddle mixers should ensure an end-to-end exchange of materials by a cascading, folding or spreading movement of the materials. Drum mixers are of tilting, nontilting, split and reversing types. The tendency of fine, slightly moist mixes to ball is minimised in efficient paddle and pan mixers. Ranges in the speed rating of batch mixers of large to small size are indicated in Table 7.1. It is largely the time element of mixing or number of revolutions at rated speed, or the possible use of high-and-low frequency vibratory mixing[152] that influences the strength and quality of concrete.

TABLE 7.1   **SPEED OF MIXERS**

Type	Rate of Rotation	
	Container (rev/min)	Blades (rev/min)
Drum	14-20	14-20
Paddle	0	20-30
Pan	12-60*	18-100†
Pan	0	48* upon 36*

* Clockwise.
† Counterclockwise.

A rating plate on each mixer should designate its uses, rated capacity and speed of rotation. Batch sizes should not exceed the rated capacity of a stationary mixer by more than 10 per cent. They should not exceed 60 per cent and 80 per cent of the drum capacity of truck mixers and agitators, respectively. For efficient operation, plant and truck mixers should be equipped with batch-timing and reset-counting devices, respectively.

Standards for batch-type mixers are BS 1305 and 3963; those for ready-mixed concrete are AS 1379, BS 1926 and ASTM C94. Concrete that is discharged from a mixer or conveyance of any kind should be of such uniformity that, when samples are taken (at intervals of up to 20 minutes) at the one-sixth and five-sixths points of a batch, the difference between their slump-test values should not exceed one-half of their average slump. Further tests on mixer performance are given in BS 3963 and Reference 1. Weigh-batchers and metering devices should be regularly checked and maintained. Mixing equipment is shown in Plates 5 and 7[45, 170] and tests for rapidly analysing fresh concrete are given in Appendixes I and IV, 4(b).[11, 185, 188]

The inside surfaces of machines should be kept free from hardened concrete or mortar and worn blades. During the midday break, a mixer should be kept running with some water and a shovelful of stone screenings in the drum. At the end of the day's operation, it should be cleaned thoroughly by hosing and churning in the foregoing manner for about 10 minutes. Working parts should be kept well greased and, after being cleaned, a machine should be rubbed over with an oily rag so as to facilitate the subsequent removal of caked cement. Monitored equipment is being increasingly used at central mixing plants for the recycling of cementitious waste and washings for environmental protective purposes.

7.1.2 CHARGING OPERATION AND CONVEYANCE

Thorough mixing is essential to consistent workability, watertight-ness, control of quality, and strength. The first batch that is put into a clean, dry mixer should contain extra cement, sand and water to coat the interior surfaces. With stationary and mobile drum mixers, the batch is normally charged into the drum so that some water (5–10 per cent) precedes the entry of the aggregate and cement, and all of the water is admitted within the first quarter of the mixing period (Reference 1 and AS 1379). The solid materials are so arranged in the charging hopper that proportional amounts of each will be in all parts of the supply stream as it flows into the mixer, precautions being taken to prevent a loss of cement during the charging operation.[45, 59]

Concrete that is fully plant mixed (see Sections 3.1.3, 3.2.3, 3.3.5 and 3.3.6) may be transported afterwards in a truck agitator or mixer which is operating at "agitating" speed. When concrete is mixed in a paddle or pan mixer, best results are obtained by mixing the sand and cement together, adding the water and then introducing the coarse aggregate. For convenience in the charging of a small, revolving-drum mixer (usually with poor job control), the water, cement, sand and coarse aggregate may be added successively, if desired. Continuous mixing of cement-treated crushed rock and modified procedures for lightweight concrete are dealt with in Sections 26.2 and 33.2.2(d), while for incidental work, an extra 10 per cent of cement should be used when mixing is done by hand.

The mixing time for mixers with a capacity of 1 m^3 or less should be not less than 1 minute. A minimum time of 1·5 minutes is sometimes advantageous. For mixers of larger capacity, this minimum time should be increased by 0·25 minute for each m^3 (or part thereof) additional capacity. The mixing time is measured from the time that all of the aggregate and cement are in the mixer to the time when the mixer is half-emptied. In a dual-drum paver, this does not include the time taken to transfer a batch from one compartment to the other (about 9 seconds). Minimum mixing times may be more or less halved by using an efficient pan, triple-action paddle, vibratory or wobble-plate mixer.

With combined plant and truck mixing, the ingredients are intermingled in the plant mixer for about 0·5 minute and subsequently truck-mixed by 55–100 revolutions of the drum paddles at "mixing" speed. When concrete is to be completely mixed in a truck mixer, its mixing operation (55–100 revolutions at "mixing" speed) is started within 30 minutes of mingling the cement with the aggregate, all revolutions in excess of 100 being made at "agitating" speed. The efficiency of mixing steel-fibre reinforced concrete is increased by mechanically dispensing up to 2–3 per cent of fibres by volume in batches not greater than 75 per cent of the rated capacity of a truck mixer. Fibre-balling is thereby inhibited during the mixing operation.

In connection with composite batching, sieved fibres can be blown progressively into a reversing drum mixer, where they are dispersed uniformly over the other mix components and successively incorporated by the mixing action. The batch is discharged within a few minutes to prevent balling by prolonged agitation. Alternatively, the fibres are blended with coarse aggregate while being charged into the mixer, otherwise they can be mechanically sieved into a pan mixer equipped with a paddle that rotates in the same direction as the pan (contrary to usual practice, see Table 7.1).

Fibres with a low aspect ratio (Glossary, Appendix VI) are a component that can be readily introduced into a concrete matrix, particularly when it contains a workability agent and only a low proportion of aggregate particles larger than 10 mm in nominal size. These factors govern certain practical aspects of fibre mix design for uniform products that meet prescribed terms of performance. Composite spray, mandrel and vacuum-dewatering methods are appropriately used in making modular shell and conduit units (see Sections 3.4.2 and 23.4.2).

The grinding action of prolonged mixing increases the amount of fine material in a normal mix. This increases the surface area and absorptive capacity of the aggregate and (coupled with evaporation) causes a change in the slump, drying shrinkage and potential strength.[3] Plant-mixed concrete may be agitated for 7 hours in a temperate climate without loss of 28-day strength. Agitation for 3–4 hours may more than halve the slump and increase compressive strength by 10–20 per cent (see Section 7.2.2).[3]

In hot weather or under quick-stiffening conditions, the time of discharge may have to be reduced, but it is unlikely (except under the most unusual circumstances) that the period need be less than 1 hour. The temperature of concrete at the time of delivery should be between 15°C and 30°C when the outdoor temperature is below 5°C, and between 10°C and 30°C when the outdoor temperature is 5°C or over (AS 1379, BS 1926 and ASTM C94; Section 32.1.2).

Concrete which is transported in a truck mixer or agitator should normally be discharged within 1·5 hours of the introduction of the water to the cement and aggregate. Under specially favourable conditions, periods of up to 3–4 hours may be allowed. For control purposes, each delivery docket should be marked with the time at which the water is added to the cement and aggregate, and truck drivers should be trained to adhere to their schedule of duties for best results in the delivery of ready-mixed concrete. On slow-moving pumping jobs, the turning of tilting mixers may be reversed periodically, or the beginning and end discharges of two truck mixers may be blended for improved uniformity and continuity of the concrete in a full hopper (see Section 24.1.2).[59, 145, 146]

The bodies of nonagitating conveyances should be of smooth,

watertight, metal construction. They should be equipped with gates for a free-flow discharge and provided with watertight covers for protection (when necessary) against the weather. Discharge should be completed within 45 minutes after the introduction of the mixing water to the cement and aggregate. Well-proportioned, plastic mixes (lightly air-entrained), short hauls and smooth roads are desirable factors for the delivery of uniform concrete by this means. When transporting foamed concrete in truck mixers, its bubble structure is retained by slowly rotating the mix at 1–1·5 rev/min during transit and by rotating it at 8–10 rev/min for 30 seconds at the work site.

Dry batching (ASTM C387) may be used where there is a lengthy journey to the point of placement. Dry-batch cans or rubber bags are of the same capacity as the mixer served and they are emptied by inversion. The cement and aggregate should be kept separate, otherwise extra cement should be provided. If they are not kept separate, site mixing and placing should be done within 6 hours (unless the aggregate is particularly dry). Containers must be completely empty or clean before being refilled.[31, 128]

7.1.3 WORKABILITY OF TRANSPORTED CONCRETE

Precautions should be taken to preserve the workability of ready-mixed concrete by expediting its delivery and placement. In hot weather, recourse should be had to cold mixing water or ice, shading and sprinkling the aggregate, painting the equipment white and the use of a setting retarder if necessary. Late additions of powdered calcium lignosulphonate are preferable to using an extra amount of water in a mix, to offset a loss of slump during transit. Alternatively, centrally dry-batched materials should be mixed in a work-site mixer (see Section 32.2). High-alumina cement concrete must be handled as described in Sections 3.1.2(b) and 20.1.5(g).

The following measures should be taken to ensure reasonable uniformity of workability, from batch to batch, in transit-mixed concrete as it arrives at the point of placement.

Handle the aggregate so as to minimise variations in grading and moisture content.
Verify that mixers do not contain variable amounts of water before batching, and maintain close control of the mixing water and other ingredients at all stages of production.
Regulate the number of revolutions at mixing and agitating speeds.

Coordinate supply and demand by two-way radio between the trucks and batching plant, discharge delays being limited to 0·5 hour.

Effectively supervise the entire organisation.

Ordinary concrete is generally regarded as having the correct consistence when it is adequately plastic and workable, readily placed and compacted with appropriate equipment, and does not suffer objectionable segregation during transportation and handling. The slump of a batch of concrete (at the time of discharge) is expressed as an average of two tests, one on a sample which is obtained at the one-quarter point of the batch volume and the other on a sample that is taken at the three-quarters point. The average should be within 15 mm of a specified slump when the slump does not exceed 80 mm, and within 30 mm when the slump exceeds 80 mm (AS 1379). Concrete sampled for strength tests can be used for making slump tests.

Useful aids in maintaining uniform workability are the appearance of the concrete during mixing, a recording consistence meter (not of the torque type) on 2 m^3 or larger stationary mixers, and testing appliances for quickly assessing the consistence of batched concrete (see Chapter 6). Batch-to-batch uniformity may require, also, that the air content should be not more than 1·5 per cent (or 2 per cent for structural lightweight and blastfurnace slag aggregates) above a predetermined average value. The temperature of concrete, as mixed, should not vary much more than 3°C above or below an average value. Concrete with less than 10–20 mm slump is difficult to discharge from truck mixers (Appendix II, 3).

7.1.4 RETEMPERING OF PREPARED MIXES

If prepared mixes are remixed with additional water, the strength of the retempered concrete or mortar bears the same relationship to total water/cement ratio as that of concrete or mortar which is made originally with that water/cement ratio. A small increment of water may sometimes be added at stationary mixers, under careful supervision, to improve the workability of delayed batches. This is permissible if followed by additional mixing (equal to half of the requisite minimum period) and if the maximum allowable water/cement ratio is not exceeded.[4]

With ready-mixed concrete, additional water may be added to compensate for losses of up to 20 mm in slump during transit, provided that the strength is not reduced below the permissible minimum.[5] When some of the mixing water is withheld until the

Continuous mixer
(Concrete output: 90 m^3/h
Cement-treated crushed rock: 120 m^3/h)

Monorail transporter

Slip forming of wheat silos at Geelong

Slip forming
of reservoir tower

Concrete extrusion
at work site

Multiple-gate discharge
onto vibratory
casting table

mixer arrives at a job site, this water should be incorporated into the mass by 20–30 revolutions of the drum at mixing speed. An additional slump of 20 mm caused by adding water to an ordinary concrete mix lowers its 28-day compressive strength by 1·5 MPa and increases its permeability and drying shrinkage (see Tables 4.4 and 13.3; Figures 4.3 and 4.4; Sections 11.1.2 and 31.2.1).

7.1.5 SAMPLING FRESH CONCRETE

A representative sample of a batch for acceptance tests, such as are described in Chapter 14 and referred to in Appendix I, should be at least 30 l in volume. A smaller sample is permissible for routine slump tests (see Sections 6.2.1 and 7.1.1). For acceptance tests, a sample should be obtained by taking three or more, approximately equal, regularly spaced portions during the discharge of a batch from central and work-site mixers or truck mixers and agitators. Sampling from the latter two conveyances should be done by successively passing a receptacle through the discharge stream or by diverting it so as to discharge into a container. The rate of discharge of the batch for sampling purposes may be changed by regulating the rate of revolution of the drum paddles, but not the size of the gate opening. Portions taken exclude the first and last 0·2 m^3 of the batch. From mixers of 0·5 m^3 capacity, a single sample taken from the middle of the discharge suffices. A composite sample may consist of separate portions, each approximately 5 litres in volume, which are dug from different places in concrete that has been deposited in readiness for casting. These are mixed together on a nonabsorbent surface and protected from the weather, so as to prevent a loss or gain of moisture (AS 1012 Part 1, 1379 and 1480; BS 1881, CP110 Part 1; and ASTM C172).

Information to be recorded should include the date, time, and sampling technique; the specified strength and workability; the cement content, composition and fineness; the details of batch materials and proportions, admixtures, slump values, air content, ambient conditions; and the final location of the sampled batch (see Section 14.4.1 and Appendix II).

The results obtained from test specimens are representative only of the lot of concrete from which the samples were taken. Compressive strength may be taken as the average of three test specimens which are sampled from a batch. Where the adequacy of a structure as a whole is concerned, random samples should be taken in continuous work at the rate of at least three each day or six from each 150 m^3 of concrete that is placed within 2 days.

Fewer samples and specimens may be permitted on small or regular projects (see Sections 14.4 and 34.1.4(a)(ii); also Appendix II, 3).[78, 106, 134]

7.2 PLACING

7.2.1 PROCEDURE

Handling and placing operations should not cause a separation of concrete into coarse aggregate and mortar, the tendency to which increases greatly with increased slump and increased maximum size and amounts of coarse aggregate in a mix. Segregation does not correct itself in the course of other operations and, unless precautions are taken, it is particularly likely to take place at points of discharge. Should clusters of segregated coarse aggregate appear, they should be scattered and formed into homogeneous concrete, and steps should be taken to prevent a recurrence.

In all classes of work, concrete should not be caused to flow horizontally or on slopes in formwork, and inclined layers should be avoided at construction joints. The concrete should be placed in horizontal layers, not more than 600 mm deep, each layer being shallow enough to ensure that the previous layer is still soft and that the two layers can be integrated during compaction. In massive structures with control joints, lifts from 1·5–2·5 m thick are placed in layers by means of cranes or cableways with bottom-discharge buckets of 3–10 m^3 capacity.

Each layer of concrete should be thoroughly consolidated with vibrators or puddling sticks and spades. Manually compacted concrete should be spaded adjacent to the formwork so as to close voids in the surface. Handling and compacting equipment should be of adequate capacity and its manner and frequency of use should meet the scheduled rate of progress of good-quality construction. A monorail concreting skip (Plate 7) is very suitable for concreting long trenches or strip footings, which should be free from pools of water at the time of placement. Powered barrows are expeditious and economical on small work, and pumping or special equipment is used on large projects (see Illingworth, Appendix IV, 4(a) and Section 24.1).[205]

For direct placement, ready-mixed concrete may be discharged over an extended range by a folding, pivoted assembly, which is mounted at the rear of a truck mixer or agitator. The equipment comprises an adjustable belt conveyor and three-piece chute. By this means, if the consistence of the concrete is such that it can be carried by the conveyor when inclined at 30° and flows down the chute when this is inclined at 25°, the radius for placement can be varied up to about 7·5 m. The belt conveyor may be used for lifting concrete to first-floor level or into hoppers, and the contents of a 3 m³ truck can thus be discharged in about 10 minutes.[1, 5, 7-9, 15]

When concrete is discharged into a hopper, bucket or formwork, the equipment should be so arranged or operated that the concrete will drop vertically into the centre of the receptacle that is to receive it. If segregation is to be avoided, it is essential that the final portion of the drop should be vertical without interference. Concrete from a chute or buggy should not strike formwork at an angle and ricochet onto reinforcement or form faces. A cohesive mix tends to produce uniformity.

In narrow formwork, it is advisable that the concrete be discharged into a light hopper which feeds into a light, flexible, drop duct. The duct is suspended between adjacent layers of reinforcement to within 0·6 m of placed concrete. In confined situations, alternatively, the concrete may be fed through external ducts. Each duct has a pocket at the lower end, from which the concrete spills through ports (up to 1·2 m apart) and drops, with the aid of vibration, within the formwork.

With deep, narrow formwork, a water/cement ratio gradient may be used, the water content being successively reduced as each quarter depth of formwork is filled. The quality of the concrete thus placed tends to become equalised by water gain.[1, 5] Concrete can be dropped (e.g. down a shaft) through a pipe of 150 mm or 200 mm diameter and any length, and into a discharge hopper of sufficient size to give a remix effect.[6, 189] Vibrating pipes with intermediate and end-damping devices may be used to reduce the velocity of flow.[93] When discharging concrete, care should be taken to avoid the continuous formation of a conical heap. Some correct methods of handling concrete are illustrated in Figures 7.1 and 7.2, Plate 8, and Reference 30 of Part 2.[5]

7.2.2 DELAYED PLACING

A limit to the permissible delay in placing concrete that is not continuously agitated after being mixed is governed by, among other considerations, the setting times of the cement used. Concrete that is made with normal portland cement (having initial and final setting times of about 2 hours and 4 hours, respectively) has been found to have a maximum compressive strength when the preparation of test specimens is delayed 2–2·5 hours after the concrete is mixed.

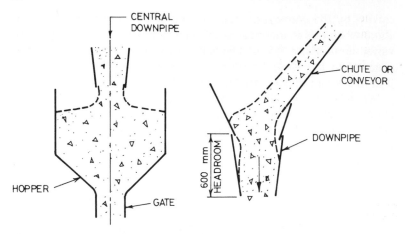

Fig. 7.1   Methods of handling concrete

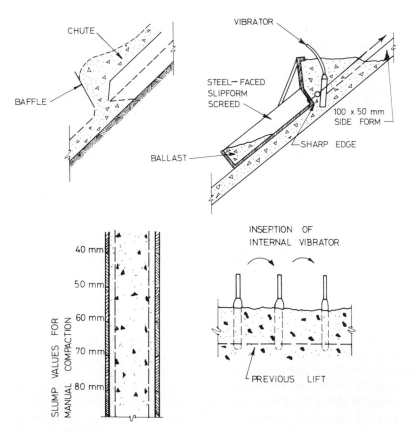

Fig. 7.2   Methods of handling concrete

The increase in strength due to this delay, as indicated in Figure 7.3, may be ascribed to a lowering of the water/cement ratio through losses of water by evaporation and by absorption in the aggregate. The beneficial effect of this factor offsets the disadvantages of partial hydration and setting of the cement during the first 2–2·5 hours. Beyond this period, the partial setting of cement causes the curve of strength development to fall, with the result that after a delay of about 4 hours the strength of the concrete at 7 and 28 days is again the same as it would be if placed when freshly mixed. Beyond 4 hours, the resulting strength falls rapidly in value.[10]

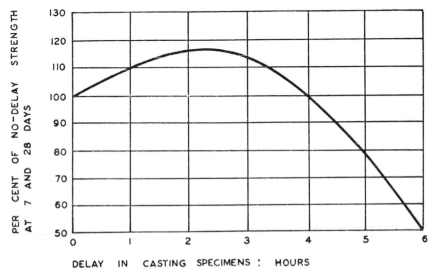

Fig. 7.3 Delayed placing

Concrete immediately after being mixed has been found to have a water content in its mortar component (i.e. cement and aggregate below 5 mm nominal size gauge) about 14 per cent lower than the theoretical value as determined from the formula

$$\frac{\text{Weight of water}}{\text{Weight of cement and fine aggregate}} \times 100 \text{ per cent.}$$

This loss may be attributed to both absorption and adsorption by the coarse aggregate and evaporation during the mixing period of 2–3 minutes. Further losses of water from the mortar occur at the rate of about 5 per cent per hour of delay, due to evaporation, adsorption by the aggregate and the fixation of water by hydration of the cement.

The effective lowering of the water/cement ratio may be taken as about 4-5 per cent per hour of delay while the concrete is in a plastic state. Thus, a delay of 2-2·5 hours would lower the water/ cement ratio by about 10 per cent (0·6 to 0·54 by weight) and cause a corresponding increase in strength of about 20 per cent. This result has been borne out experimentally in strength data obtained.

An inevitable effect of the lowering of the water/cement ratio is a reduction in the degree of workability of the mix. This must be taken into account when placing the concrete. The effect will depend upon climatic conditions, and the absorptive and adsorptive properties of the aggregate. It necessitates an increased expenditure of energy for the achievement of full compaction (see Section 23.1) without adding extra water.

When moulds are well-filled with ordinary concrete, delayed screeding, wood floating, and steel trowelling at an age of 3 hours can increase the average strength of concrete specimens about 20 per cent. The delayed placement of mortar and concrete causes reduced shrinkage.

7.2.3 COMPACTION

Concrete should be thoroughly compacted during the operation of placing, and worked around reinforcement or embedded fixtures and into corners of the formwork. When vibrators are used to aid compaction, experienced operators, reduced water/cement ratios, and strong formwork are required. If concrete placed in a wall or column is to be made monolithic with a slab above, the former should be allowed to settle, and be revibrated near the time of initial set of the cement, before the slab is placed and compacted by vibration (see Sections 23.1 and 23.2).[1, 14]

If concrete is inadequately compacted by rodding, tamping, vibration, rolling or pressure, air voids will remain in the hardened concrete. These voids are in excess of those due to water which is not needed for hydration. Figure 7.4 shows that the strength of concrete falls rapidly as the percentage of air voids increases. Even a 1 per cent reduction in compaction below maximum density lowers the strength by about 6 per cent. If concrete contains 10 per cent of air voids, its strength is less than half the strength it could have if it were fully compacted (see Section 23.1).

High-quality concrete requires effective compaction and extra work during placement. Inadequate compaction of a poorly graded mix may offset any gains in strength which are due to the use of a

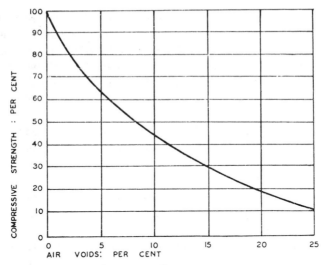

Fig. 7.4 Effect of air voids on strength

reduced water/cement ratio. Badly compacted concrete is likely to have unsightly patches of honeycombing, or highly permeable areas that may lead to the corrosion of reinforcement and spalling of faces. Every care must be taken in handling, placing, and compacting concrete to minimise segregation and differential settlement of the mass.[11] These factors affect, among other characteristics, the watertightness and durability of concrete, as dealt with in Chapters 11 and 15.[17, 190, 216]

7.2.4 CONSTRUCTION JOINTS

Concreting should be carried out continuously up to construction joints, which should subsequently develop the requisite strength. Construction joints in beams and slabs should be located normal to the line of thrust and where the shearing force is least. Joints carrying shearing force should be strengthened with steel dowels and diagonal stirrups. Alternatively, they should be serrated in profile or provided with joggles, suitable stop boards or formwork being used for the purpose. Joints in gravity dams are arranged horizontally and vertically so that any plane through the concrete cuts through a practical minimum of construction joints (see Section 30.1).

In vertical members, horizontal construction joints should be kept straight by filling formwork to a little above the bottom of 50 mm × 25 mm guide-battens. Bulges and offsets are avoided when the formwork is secured tightly against the concrete and tie-rods are close to each joint. Horizontal grooves (where permissible) may improve the appearance of construction joints in a wall surface. Vee-shaped or bevelled battens may be used for the purpose. Weakened-plane joints in pavements on the ground may be similarly formed underneath. At horizontal construction joints in walls, formwork can be rigidly anchored by resetting it tightly with only 25 mm of sheathing bearing on the concrete below the joint[1] (see Fig. 7.5). On horizontal joints in mass concrete, the working of surface concrete by job traffic should be kept to a minimum for subsequent effective jointing.

Good-quality concrete is easier to join onto than inferior concrete, which usually has a high water-gain and layer of laitance on the surface. Surplus mortar should be removed and the coarse aggregate exposed by a green-cut clean-up of the partly hardened surface. Stiff brooms may be used for this purpose on small areas. A strong jet of air and water, at approximately 700 kPa may be used on large areas.[1] The operation should neither ravel the surface below the desired depth nor form cloudy pools of water, which would leave a film on the surface when they dry. The treated surface should be kept continuously moist with wet hessian or preferably a 25 mm layer of saturated sand and, before concreting is resumed, the surface should be thoroughly washed and cleaned.

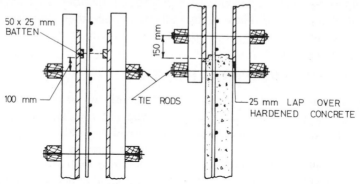

Fig. 7.5 Horizontal construction joint

Hardened surfaces that have not been green-cut should be thoroughly hacked, scabbled, roughened, or abraded and cleaned, so that all laitance, coatings on coarse aggregate, weak and loose materials are removed. Cutting and cleaning operations are carried out by the following means: power-driven picks and rotary wire brooms, pavement scarifier machines with revolving cutting-heads,[18] wet sandblasting by a cement gun, bushhammering, pneumatic scabbling, high-pressure water jetting and air jetting, wire brushing, sweeping, washing and industrial-vacuum treatment. Strong fine aggregate, passing a 4·75-mm sieve and retained on a 2·36-mm sieve, is suitable for sandblasting purposes, 1 m^3 being used per 50-100 m^2 (see Reference 74 of Part 6).

The prepared and cleaned surface is moistened with water, brushed with neat cement slurry and coated 3-10 mm thick with freshly mixed cement mortar. The cement slurry should be scrubbed into the surface with a fine wire or stiff broom. The mortar should have the same proportions of cement and sand as those in the fresh concrete and be spread to no greater depth than is required to form a sound joint. New concrete is placed and compacted while the bonding mortar is fresh.

A 5 per cent solution of calcium chloride (washed over maturely hardened portland cement concrete, just before applying portland cement slurry) and an admixture of calcium chloride (e.g. 2 per cent by weight of cement; see Section 10.1) in the bonding mortar are a very good aid to bond.[19] When applying the chemical, care should be taken to keep it off steel reinforcement, conduits and fixtures. Its effectiveness with ordinary portland cement, such as at the interface of two-course concrete floors (see Section 20.1.5(c)), is due to

A marked increase in the concentration of calcium compounds (which are present in the cement paste of fresh mortar or concrete) at the surface of the joint.

An improved surface penetrating property imparted to these compounds and a transitory development of needle-like crystals of calcium oxychloride at the interface (see Reference 9 of Part 1).[169]

The accelerated setting characteristic imparted to cement that is in intimate contact with the joint.

In column or deep narrow formwork where bonding is difficult, it is advisable to follow the mortar with 75 mm of concrete that can be readily placed. The concrete contains mortar in excess of that in the usual mix and a reduced size of coarse aggregate.[5] In mass concrete construction, the first 50 mm of new concrete should contain only one-half the proportion of coarse aggregate that is used in the regular mix. Further information on bonding is given in Sections 10.1, 20.1.5(c), 20.2 and 21.1.6. Movement joints in industrial concrete floors and liquid-containing structures are dealt with in Section 20.1.5(f) and BS 5337, respectively. Preliminary hardening at a joint face may be avoided sometimes by the use of a retarder (see Section 10.2).

CHAPTER **8**

FORMWORK

8.1 GENERAL INFORMATION

Formwork should be strong and rigid, and accessible for placing, working and vibrating concrete. It should be convenient to erect and strip without damage to hardened concrete, economical of material, capable of reuse where possible and conducive to a good surface finish. It should be true to shape, line and dimension; that is, properly supported and braced or tied to maintain its position and shape both during and after the placing of concrete. It should also be sufficiently tight; for example, with polyurethane-foam strips (AS K165 and K166, BS 3667 and 4021) at joints to prevent the leakage of mortar or grout (AS 1509, 1510 and 1082, BS CP110 Part 1, and Section 36.8).

Additional strength and tightness are required to withstand the increased pressure of concrete that is made plastic by vibration and (if necessary) the jarring action of external vibrators. It is desirable to give formwork an upward camber of 1 in 500 (see Section 33.3.2(b)), so that main beams and slabs will not have a sag when they have hardened and taken up their deflection under dead and live loads. The satisfactory design of formwork is an important factor in the economy and efficiency of reinforced concrete construction. It is illustrated in AS 1509, in References 1, 20-30, 84, 109, 110, 122 and 161, also in Figure 8.1. Finishes are featured in AS 1510, Section 8.1.1 and Chapter 28.

The surfaces of formwork which are to be in contact with concrete should be cleaned and, depending upon type, either thoroughly wetted or treated with a suitable release agent. A form oil protects steel moulds from deterioration but it should be kept off reinforcement. Steel forms should never be sandblasted or abraded to bright

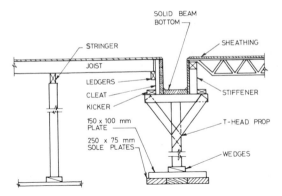

Fig. 8.1 Formwork and falsework for beams and slabs

metal. Temporary openings should be provided at the bases of column and wall formwork for the removal of debris. On completion of a casting programme, the dimensions of stripped members should be within specified limits of tolerance (see Section 3.4). Unsightly bulges and offsets should be avoided, by resetting and securely bolting formwork close to horizontal joints (Fig. 7.5) and stainless steel slipformers may be used to promote a smooth finish.[192, 196]

The holes for form ties should be of minimum size for assembly and, if found to leak, they should be plugged. The ends of form ties should be suitably designed and subsequently removed, so that a hole of minimum size is left and adjacent concrete will not be spalled. The filling of these holes, if necessary, should be done in a sound, inconspicuous manner. Sliding forms (Plate 8), on being raised, must be pulled tight against the concrete already cast. Liners and special coatings for forms and moulds are dealt with in Section 27.1.3.

The trend in formwork design is towards composite steel deck (Section 3.4.1(b)), large, structurally stable elements with 24 mm fibrous-cement sheets or 27 mm resin-faced plywood, built-in supports and striking devices, and mechanisation for economical repetitive use. Surface density and colour are improved by using liners with uniform absorbency and dampening them with a fog spray just before uniformly placing and compacting concrete in thin layers (see Section 23.1.3).[109, 143, 184, 186, 190-193, 201-203]

8.1.1 OFF-FORM, BOARD-MARKED FINISH

Board-marked concrete, with limited carbonation or hydration staining, is obtainable with the following operations.

Coarse-grained seasoned boards, specially selected with uniform absorbency or porosity, are used in tight, rigid formwork. The lower the absorbency, the lighter is the colour of the hardened concrete.

Form leakage is prevented with foamed polyurethane or polystyrene jointing strips and vee-shaped battens or fillets.

Soft grain (e.g. in back-cut Oregon or rip-sawn Baltic Pine) is first removed by even weathering, wire brushing, flame singeing or light sandblast treatment. Board grains are set in parallel to a regular pattern.

A stain-suppressing release agent (see Section 8.3), formulated and tested for off-form finish, is thinly and evenly applied.

Interfacial tension and possible surface pitting are reduced by fog-spray dampening of forming surfaces just prior to placing concrete with a low water/cement ratio.

Cohesive concrete, which must be consistently made, is placed in thin layers and continuously compacted by vibration. To minimise surface blemishes, internal vibrators should not touch the formwork and consolidation may be improved by carefully working the interfacial concrete with a blade and by employing revibration (see Sections 7.2.1, 8.1, 23.1.3 and 23.2).

Form boards, possessing uniform properties and tightening facilities, are kept in close contact with the concrete to prevent early differential efflorescence. Stripping must be regularly done at a tentative age of 24 hours. Groups of boards that are lacquer-primed and kept clean may be reused up to twelve times. Holes left by the plastic cones of formwork hardware are filled with matching precast plugs and suitably located construction joints are made as described in Section 7.2.4.

Exposed surfaces are immediately membrane cured, protected against strong draughts and rainfall, and kept uniformly moist for at least 1 week.

For uniform finishes, surfaces may be fluted and forming components employed should all receive the same treatment (see AS 1510).[143, 186, 193]

8.1.2 ROPE FINISH

Form liners to be used for casting a ribbed rope finish consist of grooved plywood boards, on which are mounted lengths of 24 mm diameter hemp rope located at 64 mm centres. The lengths of rope are bound at the ends and gun-pinned, on 250 mm centres, into a series of single loops on the liners. Serrated battens are tightly fitted where recessed construction joints are required at floor levels.

Structural backing forms are tie-bolted to the liners, these being initially treated with a release agent, and the ropes are wetted before placing the concrete. Designated grades of concrete (Appendix VIII, 5) for a thirty-storey, cast-in-place office project, for example, would be reduced from 40 MPa to 25 MPa as the construction increases in height.

After early-stage hardening of the placed concrete, the backing forms are removed and the lining boards are subsequently eased off. At a concrete age of 48 hours, four-loop groups of rope are successively winched away from the concrete and tie-bolt holes are filled with preformed plugs prior to curing the concrete.[206] Precast rope-finish panels are cast face-downwards with repetitive design details for economical moulding, erection, site connection, securement and jointing (see Sections 27.1 and 28.12, also Reference 156 of Part 5).

8.2 PRESSURE OF CONCRETE

8.2.1 VERTICAL FORMWORK

The lateral pressure exerted by ordinary concrete is affected by the degree of arching of the aggregate, the rate of hardening and setting shrinkage of the cement, the method of compaction, and the rigidity of the formwork. It can be measured under site conditions by an instrument known as a formwork pressure balance (Kinnear, Appendix IV, 4(b)). The following empirical relationships for concrete pressure[26, 30, 163] are indicated in Figure 8.2.

(a) Manually compacted concrete.
The maximum pressure, p_{max}, in kPa and the depth of concrete from the surface to the position of maximum pressure, H, in m are given by

$$p_{max} = 28 \cdot 5 \ R^{\frac{1}{2}} \text{ and } H = \frac{p_{max}}{17 \cdot 3}$$

where R = rate of placing in m/hour.

(b) Internally vibrated concrete.
For internally vibrated concrete, where the vibrators do not penetrate more than 600-900 mm, the maximum pressure occurs at

the same depth as before, but full hydrostatic pressure is considered to be developed, hence

$$p_{max} = 38 \cdot 8 \; R^{\frac{1}{4}} \text{ and } H = \frac{p_{max}}{23 \cdot 6}.$$

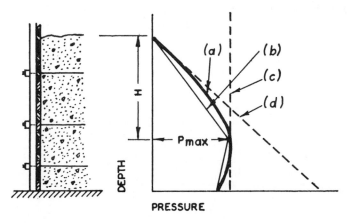

PRESSURE

Fig. 8.2 Distribution of pressure. (a) Actual pressure distribution for manual compaction and internal vibration. (b) Shape assumed for design. (c) Formwork is designed usually for this pressure. (d) Design curve for external vibration.

Values of the maximum pressure and the depth below the surface at which maximum pressure occurs for various rates of placing are given in Table 8.1 for an air temperature of 23 °C. At low rates of placing, the hardening of the cement has a more important effect than the arching action of the aggregate and conversely at

TABLE 8.1   **MAXIMUM PRESSURE**

| Rate of Placing (m/hour) | Maximum Pressure | | Depth to Position of Maximum Pressure (m) |
	Manually Compacted (kPa)	Internally Vibrated (kPa)	
0·6	24·0	32·5	1·4
0·9	27·5	37·5	1·6
1·2	30·5	41·0	1·8
1·5	32·5	44·5	1·9
1·8	34·5	47·5	2·0
2·1	36·5	49·5	2·1
2·4	38·0	52·0	2·2
2·7	39·5	54·0	2·3
3·0	41·0	56·0	2·4
6·0	52·0	70·5	3·0

high rates of placing. The time required to reach p_{max} ranges from about 2 hours for rates of 0·6-1·0 m/hour to less than 30 minutes for rates of more than 6·0 m/hour.

The maximum pressure is multiplied by a modifying factor, which is given in Table 8.2, to allow for the effect of air temperature at the time of placing. Lightweight concrete pressure is dealt with in AS 1509.

TABLE 8.2 **MODIFYING FACTOR**

Temperature (°C)	5	10	15	20	21	25
Modifying factor	1·53	1·30	1·13	1·02	1·00	0·93

On the assumption that the pressure varies linearly from zero at the surface to p_{max} kPa at a depth of H m below the surface, the pressure p kPa at any depth of h m (less than H m) is given by

$$p = p_{max} \frac{h}{H}.$$

If the depth of concrete to be placed is less than H m, the formwork is designed for a pressure p kPa. Although the upper part of vertical formwork may be designed for smaller pressures than the lower part, this refinement is seldom warranted, as uniform sizes and spacings of members are convenient to use and they simplify construction.

(c) Externally vibrated concrete.
The material is regarded as a fluid exerting the same hydrostatic pressure as a liquid weighing 2400 kg/m³ for the full depth of the vibrated formwork. Consequently, the pressure developed at a depth h m below the surface is given by $p = 0·24 \times 98·1 h = 23·6 h$ kPá (Appendix VIII).

8.2.2 HORIZONTAL FORMWORK

The vertical loading for design purposes is the dead load of the concrete and formwork plus the live load associated with construction, which may come on before the formwork is stripped. A live load of 2 kN/m² (2 kPa) is adequate for most construction, but sometimes it may be necessary to design for live loads up to 4 kN/m² (4 kPa) on restricted areas (AS 1170, Part 1).

8.2.3 TIMBER FORMWORK

Working stresses for the design of timber formwork are given in Table 8.3; also in AS 1684 and 1720.[21, 161, 172-174] Most hardwoods and softwoods, in a moderately seasoned condition, may be regarded as belonging to groups F5, F7 and F8. A glossary of formwork terms is given in AS 1082 and BS 4340.

TABLE 8.3   WORKING STRESSES

Stress Grade (MPa)	Basic Working Stress (MPa)				
	Extreme Fibre Stress ($F'b$)	Tension Parallel to Grain ($F't$)	Shear in Beams ($F's$)	Compression Parallel to Grain ($F'c$)	Modulus of Elasticity (E)
F2	2·8	2·2	0·36	2·1	4 500
F3	3·4	2·8	0·43	2·6	5 200
F4	4·3	3·4	0·52	3·3	6 100
F5	5·5	4·3	0·62	4·1	6 900
F7	6·9	5·5	0·72	5·2	7 900
F8	8·6	6·9	0·86	6·6	9 100
F11	11·0	8·6	1·05	8·3	10 500
F14	14·0	11·0	1·25	10·5	12 500
F17	17·0	14·0	1·45	13·0	14 000
F22	22·0	17·0	1·70	16·5	16 000
F27	27·5	22·0	2·05	20·5	18 500
F34	34·5	27·5	2·45	26·0	21 500

8.3 RELEASE AGENTS

An ideal release agent, when properly used, should produce a clean stripping action with a minimum of surface defects on the hardened concrete, such as stains, voids and impaired qualities for interfacial bond. A release agent should be selected carefully, experimented with where necessary and applied thinly and uniformly, usually before each placement of concrete. An insecticide sprayer or other pneumatic appliance may be used for the purpose, as an excessive application of some release agents is likely to cause surface pittings or blow holes. A membrane-curing compound with parting properties can be used as a bond-breaking medium between successive placements of concrete, such as in lift-slab and tilt-up construction. Depending on the nature of the formwork and the concrete, release agents can contribute to pleasing, off-form, textured finishes and the suppression of efflorescence (see Section 36.4).

The following typical release agents may be selectively used at the rate of 10-30 m²/litre on impervious formwork and moulds.

Fatty-acid compounds that react on formwork with calcium hydroxide (liberated in setting portland cement) to form a water-insoluble soap.

Acrylic stearate or silicone wax, or petroleum jelly mixed with 10 per cent of aluminium oleate or 25 per cent of silicone grease.[144]

A mixture of light mineral oil (125 seconds at 40°C, Saybolt) and neat's-foot oil (17-20 per cent of free fatty acid) in the proportion of 1·5 : 1, with or without a defoament.

Animal-fat emulsion, or wax and wool grease dispersed in mineral oil.

Saponaceous or soluble cutting oil (1 part mixed into 3 parts of water).

Moisture-cured polyurethane or a set-retarding compound (see Section 10.2).

Membrane-curing compounds (e.g. "P. J. Emulsion" on formwork, and microcrystalline wax or resinous emulsion where a break of bond is required between slabs of concrete; see Section 9.3).

Limewash (whitewash) with a nominal formulation of 5 kg of hydrated lime in 10 litres of water containing 1 kg of common salt. The suspension is covered, stirred occasionally, brought to suitable consistence with water the next day and mixed with alum at the rate of 6 g/litre for improved quality.

8.4 STRIPPING

Formwork should not be removed until the concrete is strong enough to support its own weight, together with any construction loads which may be placed on it. Table 8.4 sets out the minimum times which should elapse with different types of cement. The figures given are intended as a guide only and may need modification under special circumstances. For example, in very low ambient temperatures, where spans are long or where there is reason to suspect the concrete may not have gained strength as rapidly as usual, formwork should be left in place for longer periods.

Where formwork must be stripped early so that it can be reused, or in order that finishing operations can be carried out, the concrete should have reached a compressive strength of twice the stress to which it may be subjected at the time of stripping (BS CP110 Part 1 and CP114). Its strength may be determined from control

TABLE 8.4   MINIMUM STRIPPING TIME IN DAYS

Parts of Formwork Removed	Ordinary Weather (16-23°C)			Cold Weather (2-4°C)		
	OPC	HESPC	HAC	OPC	HESPC	HAC
Beam sides, walls and unloaded columns	2	1	1	5	4	1
Slabs, with props left underneath	5	3	1	10	8	1
Beam soffits, with props left underneath	8	4	1	14	9	1
Removal of props under slabs	7-14	5	1	21	11	2
Removal of props under beams	14-18	8	1	28	21	3

Note: OPC, ordinary portland cement. HESPC, high-early-strength portland cement. HAC, high-alumina cement.

specimens or by means of the nondestructive hardness-test apparatus described in Section 14.9. Formwork may also be removed before the concrete has achieved sufficient strength to withstand erection loads, provided the member is adequately shored to distribute the load to other members (e.g. lower floors) having the requisite load-carrying capacity.[165, 209]

Props should be kept in place until the concrete has reached at least 70 per cent of its specified 28-day strength (Appendix IV, 2; VIII, 5), care being taken to prevent supporting members from being overloaded. Depending on the weather, early stripping of concrete elements, some 12-48 hours after casting, may facilitate finishing operations while care is being taken to avoid profile damage. Bulkheads at construction joints should be left in place for at least 15 hours after concrete placement.

Unremoved forms assist initially in curing the concrete. On thick sections, the thermal insulating property of timber forms should be taken into account when considering the effect of adiabatic conditions and dimensional change due to heat of hydration of the cement (see Sections 3.1.1 and 30.1; also Reference 113 of Part 6). Without inducing thermal shock, form-stripping should be followed with appropriate moist curing. On large projects, laser beams can serve as a precise survey-control deflection detector and guidance system for alignment purposes, such as that required for the setting or resetting of formwork, slipforms and automated equipment.

In high-rise building construction, elastic-plastic load/deformation props can be used to optimise loadings on successive floors, so that each carries a load commensurate with its age-dependent, load-carrying capacity (Wheen, Appendix IV, 1).

CHAPTER 9

CURING

9.1 GENERAL INFORMATION

Curing is a procedure that is adopted to promote the hardening of concrete under conditions of humidity and temperature which are conducive to the progressive and proper setting of the constituent cement. The cementitious properties of portland cement concrete are largely due to the formation of silicate gels as reaction products of hydration. These develop at and within the surfaces of the cement particles and in the spaces between them (see Chapters 3 and 31).

The chemical process of hardening continues at a diminishing rate for an indefinite period, as long as moisture is present and the temperature is favourable. Early, progressive, effective curing for a practicable period is an essential operation for the adequate and uniform development of a cementitious binding matrix. It not only increases the compressive strength of concrete, but improves the requisite qualities of durability, impermeability, abrasion-resistance and general dimensional stability.

Methods of curing concrete include one or more of the following procedures, the effectiveness of which may be estimated from strength tests, relative humidity or colour gauges, or the water lost by concrete in a given time.[57]

Retaining preformed linings or the formwork closely against concrete for at least 5 days and keeping timber forms damp by wetting (see Section 8.4).
Retarding evaporation of newly laid slabs by using aliphatic (e.g. cetyl) alcohol, shading, wind shields and fog sprays (see Sections 31.2.4 and 32.2).
Ponding flat plates or slabs with water to a depth of 50 mm after initial retardation of evaporation, or temporary membrane curing

199

in hot weather, and the cement has reached its time of final set (Table 3.6). A cooling effect is induced by subsequent evaporation.

Covering surfaces with a wet layer of moisture-retaining material (e.g. hessian, underfelt, cotton mats, sacking, burlap, straw or sand) which must be kept continuously wet (Plate 9). White, vinyl-coated burlap requires moistening only at the time of application.

Immersing precast products in water or spraying exposed concrete surfaces regularly with water, thereby keeping them continuously wet. The water should be applied at an appropriate temperature, without introducing cycles of wetting and drying, so that surface crazing will not be caused by excessive thermal and dimensional changes. Soaker hoses and rotary rainwave sprays are useful facilities.

Laying a vapourproof membrane (e.g. polyethylene sheet or impermeable building paper) with lapped or sealed joins and held-down edges. It protects concrete against early damage by rain, creates a near-saturation surface condition and offsets early drying and carbonation shrinkage effects on large exposed surfaces. The regular wetting of a sand-covered membrane has a cooling effect on concrete slabs that are cast in hot weather (see Sections 9.3, 20.1.2(b), 20.1.5(e), 31.2 and 32.2).

Coating exposed surfaces with a membrane-curing compound near the time of initial set of the cement (see Sections 9.3 and 11.1.2(e)).

Keeping concrete moist, while raising its temperature by steam or hydrothermal curing (see Sections 9.4 and 9.5), or adiabatic conditions of hydration in stacked and covered moulds for accelerated hardening.

Cooling massively placed concrete by methods that will keep its temperature resulting from the heat of hydration of cement below 32°C (see Chapters 3 and 30).

9.2 MOIST CURING

Moist curing should be carried out under supervision, with a suitable allocation of labour and in such a way as not to impair the surface of the concrete. In order to meet demands for quality and economy, ordinary portland cement concrete should be moist cured for a minimum period of 7 days, preferably 10 days, at ambient temperatures above about 15°C. This period may be reduced to not less than 3 days, preferably 4 days, for high-early-strength portland cement concrete. It should be increased for slow-hardening cement concrete and when concreting is done in cold weather (see Section 32.1). Increased curing periods are desirable for high-quality concrete and concrete floors or pavements.

The resulting maturity and strength development of portland cement concrete are governed by the product of the average atmospheric temperature (expressed in °C plus 10°C) during the curing period and the age of the concrete in hours (see Section 9.4.4). The requisite maturity factors for the minimum curing recommendations are 4200°C hours, preferably 6000°C hours, for ordinary portland cement concrete and 1800°C hours, preferably 2400°C hours, for high-early-strength portland cement concrete. Some general relationships between curing and strength are illustrated in Figures 9.1 to 9.4. Figure 9.1 shows that the curing of portland cement concrete that has dried prematurely can be resumed where practicable by resaturating it.

The data in Figure 9.1 show the strength development of specimens that are cured at $23 \pm 2°C$ and tested wet (the specimens which are air cured are saturated prior to testing). The ambient temperatures used for testing purposes in different climatic regions are indicated in Section 14.4.1. For comparative purposes, the dry-tested strength of continuously moist-cured, dense concrete, at an age of 1 month or more, is 20-25 per cent greater than its equivalent wet-tested strength.

The durability of concrete is improved by moist curing, through a decrease in permeability and absorption. Nevertheless, fresh concrete that is exposed to the weather must, when necessary, be protected against the possibility of frost action. The retarded rate of evaporation of moisture from air-entrained concrete is conducive to incipient curing, although effective curing in general virtually ceases at a relative humidity below 85 per cent. Underground works of portland cement concrete are moist cured protractedly when located in still moist air of over 90 per cent relative humidity. Test specimens require lime-saturated water or fog-spray curing at almost 100 per cent relative humidity, in order to ensure thorough hydration and to prevent moisture loss by the heat of hydration of the cement (see Section 14.4).

The rate of hydration is related to temperature at very early ages and, because of this, precautions may be required to control the temperature of the concrete (see Sections 32.1 and 32.2). Mixing, placing and curing at high early temperatures (e.g. in tropical areas) may result in reduced potential strength. In Figure 9.2, the curing

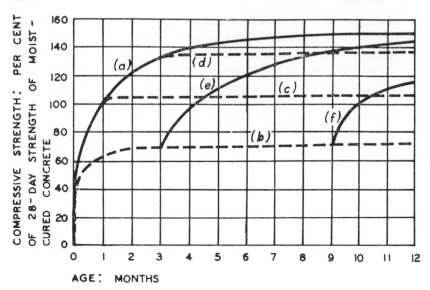

Fig. 9.1 Curing and strength relationship for portland cement concrete. (a) Continuously moist cured. (b) Continuously air cured. (c) Moist cured 1 month, then air cured. (d) Moist cured 3 months, then air cured. (e) Air cured 3 months, then saturated. (f) Air cured 9 months, then saturated.

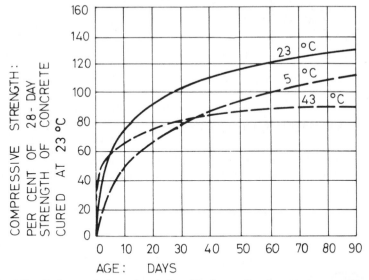

Fig. 9.2 Curing and strength relationship for portland cement concrete.

and strength relationship is for specimens cured in sealed containers, at the temperature indicated, and tested at 23°C.[32, 41, 42, 60, 61, 149]

Moist curing at an early age is particularly important for concrete that hardens rapidly and attains requisite strength with a reduced curing period. This necessity applies to concrete that contains a setting accelerator or is made with either high-early-strength portland cement or high-alumina cement. With the latter cement, which differs radically from portland cement, the side forms of concrete are removed within 6 hours and wetting is done with cold-water sprays, which are applied sparingly at first and then copiously for 24 hours. Figure 9.3 illustrates the adverse effect of warm moist curing and inappropriate ambient conditions on the compressive strength of high-alumina cement concrete (see Section 3.1.2(b)).

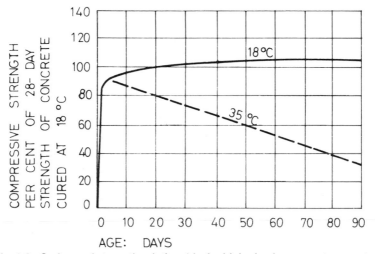

Fig. 9.3 Curing and strength relationship for high-alumina cement concrete.

9.3 MEMBRANE CURING

A curing membrane is a physical barrier to the evaporation of moisture from concrete. A curing liquid should dry within 4 hours to a continuous, coherent, adhesive film that is nontacky, nontoxic, nonslip, free from pinholes and harmless to the concrete. A clear liquid may contain a fugitive dye or white pigment to facilitate even coverage and to reflect sunlight. A curing compound should be used selectively or removed before surface treatments are applied, as in Sections 20.2, 28.2, 28.3 and 28.4; and it should be kept off waterstops when applied to the vertical-joint surfaces of dams.

A first-class membrane, if properly used, gives the equivalent (at 28 days) of 2 weeks moist curing. A plastic or vapourproof sheet can be a very efficient curing aid. Membrane curing is particularly useful on concrete pavements or where water is scarce. It should be applied to freshly laid concrete, either immediately after the finishing operation near the time of initial set of the cement, or just after the disappearance of free moisture within 1 hour after placement.

Excess bleed water, if present on slabs, should be skimmed off prior to correct finishing and curing. An immediate spraying of aliphatic (cetyl) alcohol diluted with 9 parts of water over freshly screeded or reworked surfaces, when used as a prelude to membrane curing, retards the initial evaporation of moisture during hot-weather concreting. Incipient rain damage, surface warping and plastic shrinkage cracking can be controlled by an effective choice and use of these means (see Sections 31.2.4 and 32.2).

The efficiency of membrane curing is approximately halved if applied to concrete that has been air cured overnight. Membrane curing, commencing the day of placement, is both inexpensive and highly beneficial for strength, surface soundness, abrasion resistance and impermeability development. It may be used before or after a period of moist curing. Concrete that has been stripped of formwork should be thoroughly wetted before being membrane cured.

The rates of application given here are for smooth surfaces and compounds with a high rating in curing efficiency. Rough surfaces require heavier coats, application being by means of a brush, broom, fibrous roller or spray appliance (see Sections 9.1, 12.1, 20.1.5(e) and 31.2.4).

Where suitable for use, typical media for membrane curing are

Plastic sheet (0·1-0·25 mm thick) of clear or white polyethylene, white burlap-polyethylene (ASTM C171), polyvinyl chloride (BS 1763 and 2739) or nylon. The regular wetting of a sand covering, when placed over this type of membrane, has a cooling effect on concrete slabs that are laid in hot weather.

Waterproof building paper (BS 1521 and ASTM C171), either plain or with a plastic sandwich (e.g. "Sisalthene").

Polyethylene emulsion, or a microcrystalline wax emulsion containing over 40 per cent by weight of wax (mp 70°C) in a naphtha vehicle, applied by spray, roller or brush at the rate of 3 m²/litre.

Bitumen emulsion of stabilised or mixing-grade type containing over 50 per cent of bitumen. Coverage is at the rate of 2 m²/litre or 1·5 m²/litre, where the membrane is to serve as a subsequent tack coat for bituminous surfacing.

Hydrophilic chlorinated rubber solution or epoxy resin compound at a rate of 5·5 m²/litre. It gives early rainfall protection and several coats may serve as finished surfacing (see Sections 20.2 and 21.1.6). They should be applied with care, under well-ventilated conditions.

Paraffin wax and oil emulsion (e.g. "P. J. Emulsion"), coverage being at the rate of 3 m²/litre. A typical formulation is shown in Table 9.1.

TABLE 9.1   **CURING COMPOUND**

Ingredients	Parts by Weight
Paraffin wax (mp 52-55°C)	3·75
Boiled linseed oil	5·65
Triethanolamine	1·00
Stearic acid (single-pressed grade)	1·25
Water	18·35
Total	30·00

All the ingredients, except triethanolamine, are heated together and stirred until the solid constituents are melted. The triethanolamine is added and stirring is continued at about 20 rev/min until the mixture cools. With hard water, the quantities of triethanolamine and stearic acid should be adjusted to ensure stability of the emulsion. If slight separation occurs, the emulsion should be reheated and stirred.

When a concrete slab is being cured with a vapourproof membrane, it should be laid closely against the finished surface and exposed edges of the slab, the sheets thus used being suitably lapped and held in place. In warm weather, if water is available for curing, the membrane is raised the next day and relaid after the concrete surface has been thoroughly wetted. Otherwise, a curing compound may be spray-applied in two coats at right-angles to each other or covered for a few days with a 20 mm layer of vermiculite granules, these being dampened in hot, windy weather.

A curing oil that is likely to inhibit the hardening of a concrete surface should be restricted in use on concrete pavements. For purposes of adhesion, a concrete surface should be as free from laitance as possible, prior to being coated with a bituminous or resinous compound in liquid form. When a bituminous coating is used in strong sunlight, it should be whitewashed so as to minimise a possible rise in temperature of the concrete. A warming bath at 40°C assists spraying in cold weather.

A membrane-curing compound does not usually blister after being applied to a concrete surface, because moisture can pass slowly through it by osmosis. Membrane curing can prevent efflorescence by stopping the free movement of water and soluble salts to the surface of the concrete. Tests for curing efficiency and water-retention efficiency are given in Section 14.5 and ASTM C156, respectively, while specifications on various membrane-curing materials are contained in ASTM C171, C309 and E96.[33]

9.4 STEAM CURING

9.4.1 TYPES

Low-pressure or atmospheric steam curing at 40-55°C for 10-12 hours. High-temperature steam curing at 65-95°C for 10-16 hours, the end temperature being 40-55°C. High-pressure steam curing at 180-188°C or 900-1100 kPa gauge pressure for 7-8 hours, the period being varied to suit the size and composition of the product.[180]

In each instance, freshly cast concrete should be maintained at 10-30°C for a few hours before being steam cured. A delay of 2-6 hours, depending on the temperature of curing, makes it possible to achieve 24-hour strengths 15-40 per cent higher than when steam curing is started immediately. The presetting period may be more than halved, however, for clinker concrete (see Sections 9.2, 32.1 and 32.2).

9.4.2 LOW-PRESSURE STEAM CURING

Low-pressure steam curing overnight enables concrete products to be demoulded or handled the next morning with a minimum of breakage. Moulds are thus released for daily reuse. This type of curing is particularly useful in cold weather, but it must be followed by moist curing for 7 days or more for requisite strength development at an age of 28 days. Alternatively, it can be followed to marked advantage by high-pressure steam curing.

9.4.3 HIGH-TEMPERATURE STEAM CURING

The objective of high-temperature curing is the attainment of high-early strength and the economical production of uniform quality products throughout the year. To this end, the seasonal curing cycle and the cement factor must be adjusted to suit the outdoor temperature conditions, the brand of cement and type of aggregate used, and the size and type of units being produced. The concrete should be sufficiently stiff initially to withstand safely the expansion of incidental air and basic components. More extensive steam treatment is required in winter than in summer, and care should be taken to prevent a too rapid rise in temperature of freshly moulded products, which could cause porosity and fine cracking (Plate 9).

Prestressed precast elements are steam cured in a chamber or in steam-enveloped moulds under cover on their casting bed. Large units are covered with double polyethylene-nylon or synthetic-rubber sheets supported on tubular-aluminium frames, and they are insulated where necessary with fibreglass pads. A typical sequence of operations for a double-line system with lightweight-aggregate concrete is as follows.

At 08.00 hours, open the moulds, remove hardened members and prepare equipment for the succeeding cycle.

At 11.00-13.00 hours, place and immediately cover the concrete so that, during a 4-hour preset period, there is no shrinkage cracking by evaporation and a temperature of 25-30°C is attained by the heat of hydration.

At 18.00 hours, apply steam so that the temperature rises at the rate of 11°C per hour to 72-75°C.

At 22.00 hours, keep the maximum temperature constant for 6 hours.

At 04.00 hours, close the steam supply and allow cooling to proceed at the rate of 6°C per hour.

At 07.00 hours, disc-cut, lance-cut or jack-release the tendons and remove the covers.

In alternative steam-curing cycles, the maximum rates of temperature rise and fall are limited to 24°C per hour. Tepid moist-curing conditions should exist during the cooling period and be continued for at least 24 hours. The maximum temperature may be reduced by 5°C for high-early-strength portland cement or increased by 5°C for dense-aggregate mixes. For early strength and crack control, the concrete temperature should fall within a range of 38-54°C at the time of transfer of the prestressing force. Moist curing for 3 days is followed by drying for 6 days before despatch. Specific conditions for curing prestressed concrete at elevated temperatures are given in AS 1481.[63]

Concrete building blocks, if compacted thoroughly by vibropressure, may be steam cured and dried in a 24-hour production cycle as follows.

A holding period of 2 hours is allowed after the curing chamber is filled. In cold weather, the chamber is preheated to between 25°C and 30°C and kept warm by bleeding in a little steam.

Steam is admitted so that the temperature rises at a rate of not more than 33°C per hour.

Maximum temperature is 85°C for lightweight and dense-aggregate blocks, as their rises in temperature are similar.

Live steam is required for 2-4 hours (with up to 1 hour at ceiling temperature), the period being governed by uniformity of maximum temperature, seasonal conditions and insulation of the chamber.

A soaking period of 8 hours or longer is allowed with the steam turned off.

Evacuation of steam and a 3-hour circulation of air are carried out at moderate velocity and with a temperature below 100°C.

A cooling period of 2 hours is provided before removal of the units to covered storage (Plate 9), where they are brought to equilibrium dryness (possibly with artificial carbonation) for dimensional stability.

An alternative procedure is to pass blocks on stillages through a tunnel 120 m long in a period of 24 hours. Over a length of 20 m, near the middle of the tunnel, there are six sprays through which saturated air is forced at a temperature of 95°C. The temperature of this air is lowered as it extends to the ends of the tunnel, where it is between 50°C and 55°C. The blocks are thus heated at an increasing rate to the hottest part of the tunnel and cooled as they near the exit.

A 24-hour curing and drying cycle, because of its sacrificial effect on potential 28-day strength, requires the use of densely compacted, strong mixes. With many products, the drying operation may be postponed and preceded by a period of moist curing in addition to the steam curing. The development of strength, watertightness and uniform quality is particularly desirable in certain types of concrete (e.g. thick sections of prestressed concrete or ordinary masonry units) and moist or membrane curing at a favourable temperature is essential for the development of these requirements.[133]

Steam is supplied in a dry, saturated condition, the relative humidity throughout the curing period being maintained at almost 100 per cent. The steam heats the mass in the chamber by giving off latent heat, while condensing to moisture on surfaces that are at a lower temperature. Moisture is added to the mass as long as it remains at or below the dew point temperature of the vapour envelope. When the temperature of the mass (inclusive of that due to heat of hydration of the cement) equals that of the vapour envelope, the mass starts to lose water if the steam supply is left on.

Should the equilibrium temperature be exceeded, the vapour pressure of the mass would be greater than that of the chamber atmosphere. This condition would cause a partial or differential dehydration of the products and a reduction in the efficiency of curing. Remedial measures consist in turning off the steam supply at the proper time, temporarily using overhead, hot-water fog sprays, or curing special units in closed moulds.

The equilibrium temperature in steam curing should be determined for each plant, class of product and type of aggregate. For these data, one or more concrete specimens are suspended from a scale balance by a chain, which passes through a small hole in the roof to a point midway between the ceiling and the floor. Readings of temperature and weight are taken at 15-minute intervals, until such time as the units show no further increase in weight. For hollow-cored building blocks, the equilibrium temperature is in the region of 70-80°C.

For large and small masses, 1-2 hours of extra holding time at the beginning of the curing cycle is, in general, more valuable than the same time spent in the soaking period.[35, 40, 49] An adequate holding period and adequate evaporation-free curing are prerequisites of minimum efflorescence. The longer the soaking period is extended, the better is the finished product. In a well-constructed steam chamber, the drop in temperature during this period is approximately 2-4°C per hour, this being more than halved with extra insulation.[171]

The steam is evacuated in 0·5-0·75 hour by an exhaust fan, the door of the chamber being opened some 200 mm at the same time. Units that are to be artificially dried by a current of hot air should be stacked apart or with their cores horizontal. Effective drying requires that moisture should move from the interior of units as fast as evaporation occurs from the surface. If the air stream causes too rapid drying (by velocity, temperature and variable action), case hardening and strains develop and the drying process becomes inefficient.[36] Units that are open stacked in stockpiles (on an impervious base and under cover) may be dried by a mobile heater and blower.

For minimum drying shrinkage, building blocks should have the average moisture condition that they would attain in service. Before being built into external walls, they must be dried to suit the average annual relative humidity of the locality in which they are to be used. General requirements on equilibrium moisture content or dimensional stability are given in Section 27.2 and Chapter 33. A rapid dryness test for masonry is described in Section 14.6.[37]

Unrestrained drying shrinkage is reduced at an early age by steam curing, and is reduced further in finished products by carbonation (see Section 9.6). The permeability of concrete is increased by steam curing, and is reduced subsequently by moist curing (Chapter 11).[133] It should be noted that if building blocks are prematurely or rapidly dehydrated in the steam chamber, subsequent wetting and rapid drying could restart hydration, bring calcium hydroxide to the surface and, through atmospheric carbonation, cause efflorescence. Durability, shrinkage, creep and bond effects are cited in Sections 19.1.2(d), 31.2.1, 31.3.3, 33.3.2(a) and 34.1.5.

9.4.4 MATURITY FACTOR

The strength development of a given concrete, as indicated in Section 9.2, is related to the product of curing temperature above a specified datum (°C), multiplied by time (hours). This product is known as the maturity factor. The relationship holds if conditions are similar or favourable and sufficient moisture is present at all stages for hydration of the cement. [34, 35, 134, 136, 148, 156] Typical relationships for a particular mix and brand of portland cement, but different aggregate and curing conditions, are shown in Figure 9.4. A transition in shape of the moist-curing relationship to that of high-temperature steam curing starts at about 45°C.[50, 62] In all curing cycles, the optimum temperature during the initial setting period is $23 \pm 2°C$, which could be accompanied by a change of 2500°C hours in maturity factor for a 28-day period of moist curing within this range of temperature (see Sections 9.2, 9.4.1, 32.1 and 32.2).[179]

The maturity factor for a particular curing cycle is the area of its temperature-time diagram reckoned from a datum of $-10°C$, this being the approximate temperature at which the hydration of cement ceases, for all practical purposes. Under comparable circumstances, the relationship may be used for determining the necessary maturity for a particular strength. With the high-temperature steam curing of prestressed concrete, for instance, a maturity factor of 1200-1500°C hours would be required to develop an adequate

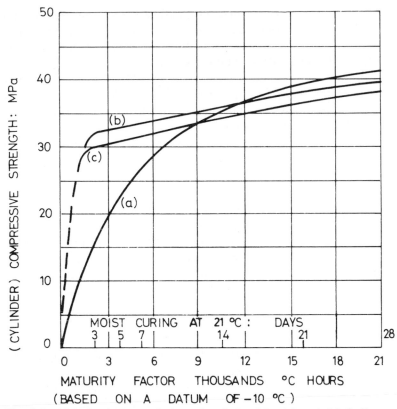

Fig. 9.4 Maturity factor and strength relationship. (a) and (b) Ordinary concrete. (c) Lightweight-aggregate concrete. (a) Moist curing. (b) and (c) High-temperature steam curing followed by moist curing. Mix: 1 : 1·5 : 3 concrete, water/cement ratio 0·4 by weight. Compaction by vibration and revibration.

compressive strength for the subsequent transfer of pretensioned load (BS CP110 Part 1 and CP116 Part 2, also Sections 4.3.4 and 34.1.4). In nonfreezing localities, a comparable maturity factor may be reckoned from a datum of 0°C (AS 1481).

A very large number of test results would be required if the effects of all variables in the relationship were to be studied. They would include relative humidity and whether precast units were mould-encased or demoulded during high-temperature steam curing. For highest strength results, consideration should be given to the mix proportions, type and gradation of aggregate, and to the characteristics of the concrete, in addition to the maturity factor. Some cements react better to steam curing than others, and the same applies to certain admixtures. Early strength is greatly increased by the use of high-early-strength portland cement and controlled steam curing (see Sections 3.1.2, 4.3.4 and 9.4.3).[168]

9.4.5 CURING CHAMBERS

Many factors must be considered in the design and construction of steam chambers. They include the plant layout, the shape, size and proximity of the chambers to a block machine, and the materials and methods of construction; the size of racks, the number of units per rack and per chamber, and the production rate of the block machine; the capacities of a fork-lift truck and a steam generator, the flow and distribution of the steam for curing; and the care of the health of operators.[39] The size of steam chambers is based upon the number of racks and their holding capacity, and the need for charging them within 2 hours. These factors are primarily related to the size of the unit to be cured (see Section 27.2). Racks with a capacity of seventy-two 200 mm units are 1625 mm long and 840 mm wide, and steam chambers are usually designed to carry 1000-1500 units.

Provision is made for a ceiling clearance of less than 300 mm and a 100 mm clearance around each rack, in addition to a 100 mm wall-protection kerb on each side. Adjacent to radiation coils, the clearance should be 150 mm. The chambers are 12-18 m long and high enough to take a fork-lift truck. They are of light construction and both vapourtight and thoroughly insulated, so as to maintain a saturated atmosphere at high temperature during the curing cycle (Plate 9).

A typical cross-section is shown in Figure 9.5, the footings being carried to a depth where the subgrade is unaffected by seasonal variation. Masonry cavity walls are provided with weepholes in the lowest horizontal joint and kept free from mortar droppings. Lightweight concrete is rendered to reduce moisture absorption and the outer face of the inner wall is sealed with a coat of high-melting-point bitumen or tar. Joint reinforcement is incorporated, and control joints at 4·5-6 m centres are placed in walls, roof and floor, the joints being sealed with high-temperature mastic.

The roof is insulated with rendered foamed concrete or vermiculite concrete, or with mats of mineral wool or foamed glass. The upper surface is sealed (when dry) with built-up bituminous felt and the lower surface with horizontal-retort tar (see Sections 33.4.2 and 36.5). The floor may be laid on rock-wool or other insulation. It is sloped 1 in 200 to a grating-covered drain extending across the front of the chambers, or to a water-sealed drain.

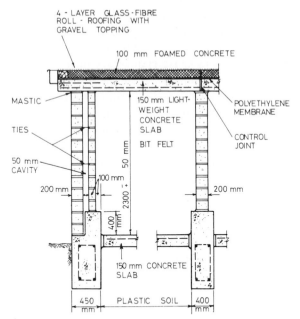

Fig. 9.5   Steam-curing chamber.

Ceilings that are crowned or sloped enable condensate to drain to the sides. Openings must be efficiently sealed and insulated with fabricated aluminium doors of swing-up type. Door jambs of concrete are faced with an anchored metal section that does not corrode. The steam supply line should be short, direct, well insulated, trapped to drain off condensate, and of ample diameter to ensure adequate pressure and capacity. A thermostatically controlled automatic valve in the steam line entering each chamber holds steam at a desired temperature under conditions of varying pressure.

Several methods are used for introducing steam into the chambers and maintaining suitable conditions of temperature and relative humidity. For instance, dry saturated steam may be injected at ceiling height from one end of a chamber. Its velocity should be from 1·2 to 1·7 m/second per linear m, so as to mix thoroughly with the air and distribute heat and moisture uniformly. A good practical length of chamber for this system is 17 m. Steam may be injected through a series of small orifices of increasing diameter in a 40 mm diameter steam pipe at or near the ceiling or floor level. In another system, a steam pipe is placed in a water trough.[51]

Steam injected at ceiling height may set up a vertical temperature gradient, but it has the advantage that condensate from it tends to settle on the units being cured. Steam injected near the floor may have a reduced temperature gradient, but it is likely to have a high content of condensate, which would reduce the amount that is deposited on the units. Overhead fog nozzles, supplied with hot water at over 200 kPa, are a supplementary means of counteracting a possible withdrawal of moisture from the heated units.

Heat and moisture may be separately controlled by injecting steam close to the ceiling and at the bottom of a set of heat radiation coils, which are fitted to the lower portion of the chamber walls. In the latter system, steam from 3 mm petcocks at 750 mm centres is forced upwards past six 50 mm diameter radiation tubes located at 150 mm centres and equipped with a steam-trap. The steam jets and coils are shielded by a 1 m deep baffle plate.[45]

If a drying period is to follow the curing cycle, the radiation coils only are supplied with steam. Each chamber is first evacuated by a centrally located exhaust fan, which operates through a 600 mm square, noncorrosive duct with a damper. The duct is placed at the opposite end of a single-door chamber, or in the centre of the roof of a two-door chamber. Temperature indicating or recording instruments are essential for efficient curing operations and control.

The capacity of the automatic steam generator should be in excess of plant requirements, as overall efficiency is improved when the generator is not operated at maximum load. For purposes of estimation, 0·75 kW boiler power is required per thirty 200 mm × 200 mm × 400 mm hollow-cored blocks cured simultaneously, exclusive of sundry plant demands for steam and hot water.[96] An extra 10 per cent of power capacity is required for uninsulated steam chambers, and another 10 per cent is required if heavyweight concrete is to be brought up to the maximum temperature that was indicated earlier for lightweight concrete.[171]

Information on a constant temperature and humidity fog-chamber for curing concrete specimens is contained in References 41 and 42, and design factors are dealt with in Reference 96.

9.4.6 HIGH-PRESSURE STEAM CURING

Concrete units of the highest quality can be produced by high-pressure steam curing in cylindrical steel autoclaves. The cementitious matrix of the mix is usually modified, as one-third or more of the cement content can be replaced to advantage by silica flour or fly ash. Where these siliceous components are mass produced for economical use, it is possible for a saving in the cost of materials to meet the greater part of term payments for plant installation (see Sections 17.1, 17.2 and 33.2.3(a)).[43, 134]

Concrete products thus produced have minimum shrinkage and volume change, maximum strength in 24 hours, an attractive light colour, and surfaces that will not effloresce or leach. These characteristics are largely due to the development of special hydration products, such as hydrogarnets and monocalcium silicate hydrate. The most suitable combination of all factors concerned is governed by the type of product being produced and its particular usage.[180]

After a suitable presetting period, precast units (such as hollow-cored blocks; stacked on their sides or open stacked) are wheeled on racks into the autoclave. Either a fork-lift truck with composite rail and rubber-tyred wheels or an automatic loading and unloading system is used for this purpose. Air is vented efficiently through thermodynamic traps, while saturated steam is brought up to gauge pressure or temperature, which is usually 1000 kPa or 185°C, in a 3-hour period. Thick units are steam heated and cured for a longer period than thin units.[47]

After a soaking period of 5-7 hours at maximum pressure, the steam is expelled quickly through a blowdown line and muffler, so that the pressure drops to atmospheric pressure in 10-15 minutes. The moisture content of the units is thereby reduced to a low figure. Vacuum treatment in an autoclave, which is strengthened with external rings, may be used for the same purpose. The autoclaved products are ready for use on being cooled.

Quick-opening doors on autoclaves are of ring-lock or wedge-lock type, and supports, except one which is fixed, are of roller or rocker type.[44] An autoclave resting on two supports has ring girders at each of them. Condensate is removed through a steam-trap, so as to prevent undesirable arching effects due to a temperature gradient. Heat is saved by insulating steam pipes and the autoclaves and by passing the condensate through a heat exchanger.

The capacity of a large autoclave, in terms of 200 mm × 200 mm × 400 mm blocks, is shown in Table 9.2. The daily output per autoclave is proportional to the number of curing cycles per day.

TABLE 9.2   **CAPACITY OF AUTOCLAVE**

Size		Racks		Total Number of Blocks
Diameter m	Length m	Blocks	Number	
2·6	26·2	72	25	1800
2·6	36·5	72	35	2520
2·75	36·5	102	35	3570
3·05	36·5	120	35	4200

Automatic equipment is employed for operating steam valves, recording pressures and temperatures and operating doors. The corrosion of steelwork is inhibited by using sulphur-free aggregate, venting off air, quickly expelling steam for dry results, and either coating steel surfaces yearly with an epoxy-based paint or by injecting about 500 ml of engine oil per cycle into the steam supply. The oil, which is automatically introduced at the rate of six drops a minute, does not stain the products. Alternatively, alloyed aluminium may be used effectively and economically in racks and pallets for high-pressure steam curing.[150] Table 9.3 shows the properties of dry saturated steam which apply if the autoclave is initially purged of air.

TABLE 9.3   **DRY SATURATED STEAM**

Temperature (°C)	Absolute Pressure (kPa)
100	101
120	199
140	361
160	618
180	1003
185	1124
190	1255
195	1398
200	1554

The boiler power per autoclave required by an automatic steam generator is indicated in Figure 9.6. The data pertain to 200 mm × 200 mm × 400 mm, hollow-cored, heavyweight blocks, and a 3-hour heat-up period in an insulated vessel to 185°C or 1000 kPa gauge pressure.[45] Out-of-doors insulation may consist of fibreglass enclosed in a 0·70 mm aluminium sheath.

Steam may be generated, alternatively, in a boilerless autoclave system. The system consists of autoclaves with hot oil coils immersed in water to a depth of 50 mm (as indicated by a level control probe), a condensate recovery-storage tank for each autoclave, a fired heater, an expansion tank and ancillary equipment.[94, 129] A presetting room (73°C and 100 per cent relative humidity) is heated by an oil pipe immersed in water in a metal trough between the tracks of mobile racks.[13]

The heating capacity of the system should be sufficient to bring steam in each autoclave to 1000 kPa gauge pressure in 2 hours. Hot oil from the fired heater, operating at 260-290°C, is circulated by a centrifugal pump. The hot water in an autoclave, after a

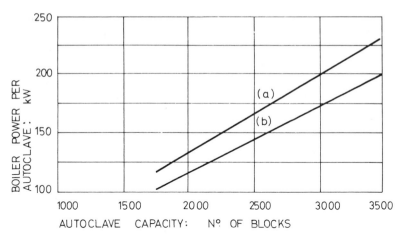

AUTOCLAVE CAPACITY: N° OF BLOCKS

Fig. 9.6   Steam generation. (a) Feed-water temperature = 10°C (75 kW = 1290 kg of steam per hour). (b) Feed-water temperature = 93°C (75 kW = 1480 kg of steam per hour).

5-hour holding period at full pressure, is transferred by steam pressure and pump to a condensate recovery-storage tank, where the charge is held at about 950 kPa until needed for the next curing cycle. The system can be totally automated and operated from a control panel without the services of a fireman.

High-pressure steam curing, where justified by market potential, opens up a new vista in the concrete products industry. It provides the building industry with an increased variety of superior products, such as high-grade masonry, asbestos cement, and lightweight calcium silicate hydrate (CSH) in plank or block form. It quickens output and increases consumer acceptance at a reduced cost, but it is inapplicable to concrete that is made with limestone aggregate. High-pressure steam curing is dealt with further in Sections 31.1.2 and 33.2.3.

9.5 HYDROTHERMAL CURING

9.5.1 HEATED MOULDS

With long precast or prestressed concrete units, heated moulds are a convenient means of obtaining high-early strength. Consideration must be given to presetting requirements, as indicated in Section 9.4. Hot oil is more commonly used than hot water for heating the moulds.

The oil is forced through pipe lines (30-50 mm diameter) installed close to the sides of hollow steel moulds, provision being made in the pipe lines for lineal movement during cycles of thermal change. Oil enters at 150-180°C and leaves at over 90°C. The concrete is brought to about 65°C in 4 hours; at the same time it is covered with a plastic membrane, tarpaulin or burlap sprayed with hot water so as to retain moisture.

Alternatively, hot air may be circulated through ducts, of 200 mm diameter, for example. Steam with a trace of oil may be circulated through perforated pipes under the casting beds, or through interfacial ducts in steel-lined concrete moulds. Structural members can be site-cast with enclosed-cavity forms made of polyester resin reinforced with fibreglass, these being evenly air-heated internally by means of electric elements that are controlled by a thermostat.

A high-strength mix, after the presetting and heat-up periods, can be brought to a cylinder compressive strength of 22·0 MPa (cube strength, 27·5 MPa) in 1 day (see Section 3.2.1, also Plate 9). Electric polyurethane covers or aluminium-foil heating elements, used with insulation, promote the strength-gain of concrete in cold weather (see Section 32.1.4).

9.5.2 INFRARED CURING

Concrete in steel moulds is kept moist for 1·5-2 hours and, to minimise subsequent drying, exposed surfaces are either coated with sodium silicate or covered tightly with oxidised steel plate. Heating is done by infrared rays for 2-4 hours, the temperature being brought to 90°C and kept there for a short period only, so that the reduction of water content is as little as possible.

The procedure is one of early hardening for handling or stripping purposes, and it should be followed by adequate moist curing for normal strength development. For an insulated container with a small air space, the power requirement is about 130 kW.h/m³ of concrete.[53, 166, 210]

9.5.3 ELECTRIC CURING

In very cold climates, alternating electric current is used for heating and hardening concrete, so that formwork may be stripped in the course of 2·5-4 days. In a 36-hour treatment, heating is limited to about 5-6°C per hour and a maximum temperature of 50°C during the first 24 hours. The temperature is then raised at twice the above rate to 70°C, where it is held until the termination of the heating period. The period of subsequent cooling should not be less than 24 hours. With a maturity factor of 2500-2800°C hours, portland cement concrete can thus attain about two-thirds of a 28-day target strength in 3 days.[55]

Electrodes are of galvanised-steel strips or steel rods for surface or internal use, respectively. The transformer used should be adjustable in steps from 10-20 volts to several hundred volts. Placing and curing operations should be so arranged that there will be a uniform temperature rise in every part of the concrete.[56] In mass concrete, the heat of hydration of the cement augments and extends the heating process.

Electric curing is expensive and costs about 25-30 per cent of the price of concrete. The consumption of current varies usually from 80-160 kW.h/m³ of treated concrete. Short-circuiting should be avoided in reinforced concrete members and, if demoulded precast units are to be electrically cured, they should be treated in a saturated atmosphere. The process can be used advantageously at off-peak tariff in the mass-production of long-line prestressed concrete elements.[52, 155, 166]

9.6 CARBONATION TREATMENT

Carbonation is the chemical combination of the hydration products of portland cement with carbon dioxide gas. When the gas is applied warm to cured concrete products under drying conditions, early-age drying shrinkage is reduced and strength and durability are improved after treatment.[58, 157, 158] Preliminary hydration of the cement matrix is essential for best results with combustion-gas curing.

Factors affecting the results are the concentration and pressure of the gas, the period and temperature of treatment, and the receptiveness or moisture content of the concrete. The artificial carbonation of concrete progressively produces water which, if allowed to accumulate, can form a moisture condition that will inhibit further reaction. Other aspects are dealt with in Section 31.4.

A 200 mm, hollow, concrete masonry unit requires approximately 0·5 kg of carbon dioxide for complete carbonation. This treatment would be expensive, therefore, if the gas were derived from commercially available 23 kg cylinders. Waste flue gas, containing 10-20 per cent of carbon dioxide, is an alternative source of supply for partial carbonation. The exhaust gas from lime kilns, steam-raising plant, or a controlled gas or oil burner and fan system may be used for this purpose.

The gas, if very hot, can be partly cooled and humidified by means of water sprays. With a gas or oil burner discharging into a combustion chamber, which is located in the rear wall of a conditioning chamber, fine-mist sprays can be directed where necessary against the hot face of a mild-steel, short up-draught duct. The carbonating gas thus produced is circulated vertically and longitudinally, by means of an overhead fan jutting out from the rear wall of the enclosure.

In a composite process, vibropressure compacted blocks, when preliminarily steam cured and partly dried, are carbonated immediately in a storage building. The flue gas is blown into the building and uniformly distributed through louvres in the roof or by medium of a perforated baffle. The blocks are carbonated for 18 hours or more, brought to equilibrium dryness and kept dry until used. A full treatment of concrete products for sulphate-resistance would require 6-8 hours at 200-300 kPa gauge pressure or 4-7 days at atmospheric pressure (Glossary, Appendix VI).

The following paragraphs summarise the subject.

Carbonation treatment is suitable for cured products in an unsaturated condition, suitable products being treated immediately after atmospheric steam curing.

The temperature should lie between 65°C and 100°C, there being little improvement at 40°C owing to limited receptiveness. The early-age drying shrinkage of concrete masonry units can be reduced some 40-50 per cent, or 50-60 per cent with extended treatment.

Fired gas or oil carbonation stops possible drip efflorescence on concrete masonry units in curing chambers. Carbonation treatment generally toughens surfaces and reduces ordinary efflorescence on concrete masonry, particularly when it has a medium pore structure or bulk density.

An excessive and differential carbonation effect in concrete units causes differential shrinkage, fine surface cracks, brittleness, and a reduced resistance to shock and weathering. Rapid surface drying and atmospheric carbonation of high-temperature steam-cured units have a somewhat similar effect.

Safety precautions include adequate ventilation during handling operations, as 100 parts per million of carbon monoxide in flue gas may prove lethal.

Porous concrete is carbonated more deeply and rapidly than dense concrete. Cast-stone products are vastly improved by artificial carbonation, while pozzolanic cement products have a reduced capacity. Autoclaved products, in contrast to ordinary products, do not develop significant shrinkage as a result of atmospheric carbonation.[45]

The carbon dioxide content of flue gas can be measured by an "Orsat" apparatus, which is obtainable from local suppliers of scientific equipment (BS 4587).

The resistance of concrete to sulphates is considerably increased by adequate treatment.

In automated masonry manufacture, accelerated burner curing is continuously applicable in an octagonal kiln provided with a circular track. Controlled atmospheres in each segment slowly heat the blocks to a temperature of 77-80°C, for a 2-hour soak treatment at this stage prior to cooling.

Fired natural gas or oil, or flue-gas carbonation treatment is economical to install and operate.[32, 39] Natural-gas burner curing is highly effective with minimal air pollution or atmospheric nuisance in built-up areas. Natural and bottled propane gases, being free from sulphur, minimise the corrosion of steel curing racks.

CHAPTER **10**

SET-CONTROLLING AGENTS

10.1 ACCELERATORS

The early strength of portland cement concrete can be increased appreciably by several means (see Sections 23.4 and 32.1), including the use of an accelerator (see Section 33.2.3(a) and Table 33.4) such as calcium chloride in the mix. Increased early strength during cold weather expedites demoulding or stripping and finishing operations, permits early loading of anchor devices, and affords resistance to damage by frost action or freezing temperatures at an early age.

The amount of calcium chloride that is used should not exceed 2 per cent by weight of cement in a mix. Under temperate conditions, 1 per cent should suffice, whereas if less than 0·5 per cent were used, the set could be retarded for a short period. The requisite amount for a given effect can be reduced in the presence of a lignosulphonate plasticiser, but the chemical has no hardening effect upon lime or pozzolan in a mix. Although calcium chloride reacts differently with different cements, the data represented in Figures 3.2 and 10.1 may be useful in estimating the strength development at different temperatures and ages.[1, 60, 169]

Calcium chloride (meeting AS 1478, 1479, or ASTM C494 and D98) is inexpensive and readily available. Ordinary flaky calcium chloride should contain at least 77 per cent of calcium chloride, and not more than 2 per cent of sodium chloride, 0·5 per cent of magnesium chloride and 1 per cent of other impurities, excluding water. After being formed into a solution, it is introduced into the mixing water in the requisite amount. For gauging purposes, a level litre container holds approximately 0·8 kg of flaky calcium chloride, which will dissolve in 1 litre of water.

The solution is used at the rate of 0·5-1 litre per bag of cement, and is measured either manually or by mechanical dispensing equipment as referred to in Section 3.2.4. In handling the solution, precautions should be taken to avoid the use of containers and valves with dissimilar metals, as these would corrode by galvanic action and a compound solution of calcium chloride and air-entraining agent would give a precipitate of calcium resinate and "gum-up" valves, interfere with the operation of meters and reduce the calibrated capacity of containers. Separate solutions of admixtures should be introduced separately into the mixer.

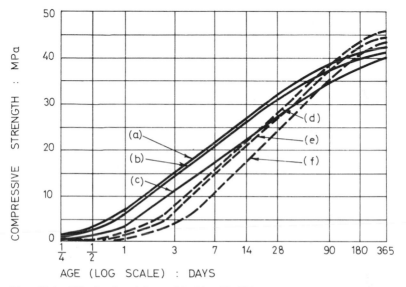

Fig. 10.1 Effect of calcium chloride ($CaCl_2$) on compressive strength. (a) 2 per cent $CaCl_2$. (b) 1 per cent $CaCl_2$. (c) No $CaCl_2$. (d) 2 per cent $CaCl_2$. (e) 1 per cent $CaCl_2$. (f) No $CaCl_2$. Mix: 1 : 2·5 : 4 concrete, water/cement ratio 0·53 by weight. ————, specimens made and cured at 23°C. – – –, specimens made and cured at 5°C for 37 days and cured at 23°C thereafter.

When calcium chloride is used with an air-entraining agent, the amount of agent may be reduced for a particular air content or degree of workability. An admixture of calcium chloride increases the workability of concrete, the slump being increased by up to about 25 mm. This feature may be used to get increased compaction and strength. Mixes which contain the chemical have an accelerated rate of heat development and stiffening. Delays in handling and placing should be avoided, and a proposed addition of calcium chloride to ready-mixed concrete should be deferred during warm weather or for long hauls.

Table 10.1 shows the upgrading effect of calcium chloride on the compressive strength of concrete which is mixed and cured under temperate to warm conditions. It is apparent that the emphasis of strength improvement by calcium chloride shifts from 1 day to 28 days as the initial temperature increases.[61]

Concrete mixes are usually designed from data derived from the behaviour of given mixes at a moderate ambient temperature. When such mixes with a similar cement are mixed and placed in a tropical climate, their resulting 28-day strength may be up to about 20 per cent below their anticipated target strength for temperate conditions. This effect is indicated in Figure 9.2. For tropical concreting, therefore, an admixture of calcium chloride can more than offset this discrepancy in 28-day strength, by virtue of the increment shown in Table 10.1.

TABLE 10.1   **STRENGTH INDEX**

Temperature (°C)		Increase in Compressive Strength with 2 per cent Calcium Chloride	
Mixing	Curing	1 day	28 days
20	20	1·80	1·20
30	20	1·62	1·25
20	30	1·23	1·19
30	30	1·01	1·28

This result is postulated as being due firstly to the influence of temperature on the setting time of the tricalcium aluminate in portland cement, and secondly to the accelerated rate of heat gain during the setting of the admixed cement. Strength advantages can therefore be imparted to concrete by calcium chloride throughout a range of mixing temperature. Steam-cured plain concrete is increased in strength at all ages when this accelerator is used (see Section 15.2.2(a)).[64] With concrete in service, a strong solution of calcium chloride can cause disruption by the formation of calcium chloroaluminate (see Section 19.1.1(b)).

An admixture of calcium chloride reduces the resistance of concrete to sulphate attack. Where sulphate conditions are encountered and concreting must be continued during cold weather, an additional bag of cement per m^3 of concrete is approximately equivalent, in early-strength effect, to 1 per cent of calcium chloride by weight of cement. The chemical increases the expansion caused by alkali-aggregate reaction and should be avoided with special cements (see Section 3.1.2).

Calcium chloride increases by 100 per cent the creep of 6-day cured concrete (of designated strength and slump) that is stored at 35 per cent relative humidity and loaded to 30 per cent of its compressive strength at 7 days. The effect is a little more than halved by effective moist curing or by proportionate loading at 28 days.

An accelerator of triethanolamine increases creep some 30 per cent at 7-day loading, but insignificantly at 28-day loading. It reduces potential ultimate strength partly because of air-entrainment.

A formate-based, nonionic accelerator, which is noncorrosive for special applications, is actively effective at low temperatures with minimal side effects.[65, 97]

Calcium chloride increases shrinkage from 10 per cent to possibly 70 per cent, a significant amount occurring within 24 hours of placing the concrete (see Section 31.2). Early moisture loss and initial settlement are reduced, and freezing-and-thawing durability is increased at early ages, but is subsequently reduced by its use. Marine structures excepted and stray electrical currents being absent, calcium chloride in a limited amount does not corrode steel reinforcement in dense concrete with adequate cover; its use in excessive amounts causes shrinkage cracks and reduced strength.[141]

In steam-cured prestressed concrete, the calcium chloride causes electrolytic corrosion of pretensioned, high-tensile and hard-drawn steel wires. The corrosion is less severe when products of this kind are cured in hot water, and it does not occur when steam curing is used on its own.[66, 126, 142] There is little risk of corrosion in post-tensioned work where calcium chloride is kept away from the steel. In the presence of moisture the chemical corrodes crimped copper strips at control joints (see Section 19.1.1), also iron granules in floor surfacing (see Section 22.1) and galvanised-steel sheet, steel elements or fixtures. Waterstops of rubber, PVC, or polypropylene are unaffected by it. Admixtures containing calcium chloride should not be used in the vicinity of dissimilar metals or in heated concrete floors, unless embedded metal tubes are protected with a closely bonded polyethylene or polyvinyl chloride plastic envelope (see Section 22.3.2).

A special use for an accelerator is in quick-setting mortar for plugging undercut leaks, using a pressure-relief tube for temporarily discharging flow water. The mortar consists typically of 1 part portland cement : 0·5-1 part pozzolanic silica flour and fine aggregate, by weight, mixed with a 30-40 per cent solution of sodium carbonate. It is applied in layers over a damp substrate and stiff cement-slurry primer that is mixed with the same set-accelerating solution.

Calcium carbonate is precipitated in interstices and capillary tracts by an interaction, under damp conditions, of the sodium carbonate with calcium hydroxide present in set portland cement. The absorption of air-borne carbon dioxide, by liberated sodium hydroxide, further generates sodium carbonate. The consequent growth of calcium carbonate, under moist cementitious conditions, promotes impermeability and the closure of fissures up to 0·2 mm wide.

In related applications, the setting rate can be reduced, if desired, by incorporating an alpha-hydroxy acid salt (e.g. sodium lactate) to the extent of about one-quarter of the activating accelerator (see Tables 5.2 and 33.4; also Sections 10.2 and 11.1.2(c), (f)). Leak stoppage can be effected also with composite cementitious mixes that are briefed in Section 3.1.2(b), and other accelerators are listed in Table 33.4 (Section 33.2.3(a)), the nonionic varieties being calcium formate and triethanolamine.

Care should be taken to avoid using any type of accelerator where it would have adverse secondary effects on shrinkage, strength and durability. Control in the time of set, or the rate and extent of strength development, may be obtained with other cements or mixes or with special cementitious nuclei, which have alternative hydrating and strength characteristics (see Section 3.1.2(e)(iii)).[147] A concentration of accelerator with deliquescent properties in mortar joints may cause dampness and mould growth on a plastered and papered surface. It delays drying in concrete floors.

When economically justified, stannous chloride may be used in lieu of a nonionic accelerator where reinforced concrete is to be used under marine, wet–dry cycle, steam-curing or hydrothermal conditions. It accelerates the development of strength and at the same time inhibits corrosion of steel reinforcement, whereas calcium chloride acts catalytically (Glossary, Appendix VI) in accelerating corrosion of the steel under identical conditions (see Sections 15.2.2(a) and 22.3; also Reference 60 of Part 4).

Concrete containing calcium chloride should have an adequate cement content, a low water/cement ratio and a high degree of compaction for external use. While additional cement may be a more expensive admixture than calcium chloride for cold-weather concreting work (see Section 32.1.1), it serves substantially to protect steel reinforcement or moulds against corrosion in lieu of introducing harmful migratory ions in solution for accelerated hardening.

While a method for determining the chloride content of a liquid is described in ASTM D512, a colorimetric test may be used for estimating the presence of a chloride admixture in hardened cement concrete or mortar. To this end, one or two drops of a 1 per cent solution of silver nitrate are placed on a wire-brushed surface of the concrete or mortar; and, after 1-2 hours, one or two drops of a concentrated solution of potassium dichromate are placed on the same spot. If in bright daylight the spot soon becomes bluish purple in colour rather than red, the indication is that a chloride compound is present. Where sodium chloride (e.g. in seawater) has not been introduced with the mixing water, the detected admixture may be assumed to be calcium chloride until investigated afterwards by chemical analysis (see References 144, 155, 172, 173 of Part 4).

10.2 RETARDERS

Certain admixtures may be used to retard the set of portland cement and the rate of stiffening of concrete.[1] They are dealt with in AS 1478, 1479 and MP 20, BS 5075 and ASTM C494. These admixtures are used for such purposes as to

Remove the tendency of some cements to exhibit false set.[67]

Counter the accelerating effect of high temperature and improve workability in hot weather.

Transport ready-mixed concrete for long distances.

Keep multiple lifts plastic for predetermined periods and facilitate placing without hardened joints.

Promote plastic deformation and suppress plastic cracking as a result of restraint and ineffective curing at early ages (see Sections 3.1.1, 31.2.1 and 31.2.4).

Eliminate early cracking due to formwork movement or beam deflections in continuous or composite bridge decks.

Ensure uniform stiffening of concrete for special work, on a 24-hour basis.

Compact and densify concrete by revibration and delayed or overnight finishing.

Expose decorative aggregate, or obtain a strong bond for rendering or plastering, by the treatment of formwork and removal of mortar with inhibited set.

Pump slurry long distances in cementation work and make oil-well cement and set-retarded, ready-mixed mortar (subsequently cured).

Reduce the maximum temperature rise in massive concrete structures by extending the heat-dissipation period.

The behaviour of retarders varies tremendously, not only with their composition but with different portland cements, mix proportions and ambient temperatures, and the mixing sequence.[151] Some retarders function as water-reducing or workability agents, with a beneficial effect upon 28-day strength. Their capacity for retardation is reduced sometimes by an accelerator. The requisite amount and effect of any particular retarder in a mix should be ascertained from technical advice, preliminary tests and experience. Deliberate water reduction with a set-retarding, cement-dispersing agent is indicated in Section 6.4.[120]

Retarders usually fall within the categories of cellulose products (e.g. calcium lignosulphonate); hydroxy carboxylic acids (e.g. adipic, citric, mucic, tartaric acids) or their salts; alpha-hydroxy organic acids (e.g. lactic, glycolic, gluconic acids) or their salts; carbohydrate derivatives (e.g. starch, sodium gluconate); gums; proteins; phosphates; borates; and sugars (sucrose, glucose, saccharides).[68] With a cellulose type of retarder, the addition of 0·2-0·5 per cent of its active constituent (by weight of cement) trebles or quadruples the period required for concrete to stiffen to a stage where it can no longer be vibrated.[69] Superphosphate and traces of lime-soluble lead or zinc compounds act also as set retarders (see Reference 45 of Part 1).

A 0·05 per cent admixture of sucrose (i.e. ordinary table sugar) retards the initial setting of ordinary portland cement by approximately 4 hours, improves workability and increases the 7-day and 28-day compressive strengths of concrete by about 8 per cent. A 0·2 per cent admixture of sucrose delays the initial hardening or strength development for several days.[70] Acids of the hydroxyl group have the characteristic of not causing air-entrainment.

Calcium and alkali lignosulphonates, containing a limited amount of wood sugar, are commonly used as workability agents, with or without an accelerator such as triethanolamine or calcium chloride. When the ambient temperature varies from 23°C, the amount of retarder required is increased or reduced by about one-fifth for a rise or fall of 10°C, respectively. Some setting times at 23°C of portland cement paste that contains retarders are illustrated in Table 10.2, the water/cement ratio of the paste being 0·35 by weight.[154]

For a prolonged set retardation, a retarder should be dissolved in a small proportion (e.g. 10 per cent) of the mixing water of a batch of concrete and incorporated after the batch has been mixed for at

TABLE 10.2   **SET RETARDATION**

Retarder	Quantity (per cent by weight of cement)	Setting Time (hours)	
		Initial	Final
None	—	4	6·5
Calcium lignosulphonate	0·10	4·5	7·5
	0·25	6·5	10
	0·50	12	17
Citric acid	0·10	10	14
	0·25	19	44
	0·50	36	130
Sucrose	0·10	14	24
	0·25	144	360

least 0·5 minute. This procedure gives the gypsum in portland cement a preliminary period in which to dissolve and react superficially with the particles of tricalcium aluminate, thereby coating them with calcium sulphoaluminate which reduces their rate of hydration. When the retarder is added, instead of reacting immediately with the tricalcium aluminate, it is now adsorbed preferentially by the silicate compounds of the cement, with a consequent marked reduction in their rate of hydration (see Section 3.1.1).[151]

The hardening characteristics of mortar which is wet-screened from concrete are measured by resistance of the mortar to

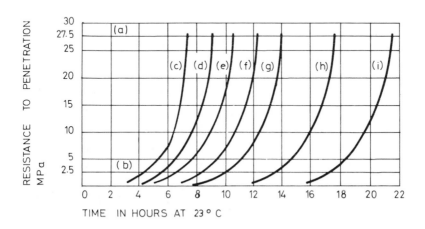

Fig. 10.2 Hardening of concrete. (a) Hardened concrete (28 MPa). (b) Vibration limit (3·5 MPa). (c) 1 per cent calcium chloride. (d) Plain concrete. (e) (f) (g) (h) and (i) Retarded concrete.

penetration by a Proctor needle, as described in Section 14.11. The resistance should not exceed 0·35 MPa within 3 hours of mixing, or 3·5 MPa at the time of revibration or resumption of placing in monolithic construction. The resistance, at temperatures above 18°C, should be at least 28·0 MPa within 3-4 hours of the time of attainment of 3·5 MPa. These requirements are illustrated in Figure 10.2. A value of 28·0 MPa is equivalent to a (cylinder) compressive strength of approximately 0·7 MPa (ASTM C403).[192]

Figure 10.3 illustrates how, with a retarder, revibration for 15 seconds may be extended to advantage over a period of time.[71] A retarder should not reduce the normal 48-hour and ultimate strengths and the potential durability of concrete. The normal effect of delayed placing is shown in Section 7.2.2.

Note: Workability, permeability-reducing, air-entraining, pozzolanic and expanding agents are described in Sections 6.4, 11.1.2(c), 16.1, 17.1-17.2 and 25.1-25.4.

Fig. 10.3 Revibration of concrete. (a) and (b) Retarded concrete. (c) Plain concrete. Mix: 1 : 2 : 3·5 concrete, water/cement ratio 0·55 by weight. Optimum period for revibration lies within the area bounded by the dotted lines.

Moist curing of spillway (Eppalock Dam, Victoria)

High-temperature steam curing on heated casting tables

High-temperature steam curing of concrete blocks

Waterstop
in concrete wall

Triple-cell outfall sewer, reinforced with cold-worked
deformed bars, located under a proposed runway
at Kingsford Smith Airport, Sydney

Sewage detention tanks
7.0 Ml capacity, at a
pumping station (Kew,
Victoria), built with
sulphate-resisting
portland cement
concrete

CHAPTER 11

PERMEABILITY AND ABSORPTION

11.1 PERMEABILITY

11.1.1 CHARACTERISTICS

Permeability is that property of a material which permits the passage of a fluid through its internal structure. Concrete is inherently porous or pervious to water, because not all the space between the aggregate particles becomes filled with solid cementitious material. Workable mixes require much more water than is necessary for the hydration of cement, while during mixing, some air is always entrapped in the mass, also chemical combination during the hardening process reduces the absolute volume of the cement and water and develops minute, interconnected voids.

Although high permeability and, to a less extent, absorption permit disintegrating agencies to enter and damage concrete, procedures are available for making it sufficiently watertight and durable for all practical purposes. The basic requirements are materials of good quality in proper proportions, and effective mixing, compaction and curing.

Studies of concrete mixes have shown that, during the setting period, the settlement of solid particles in placed concrete causes excess water to rise and form many channels or capillaries. Some of the water is trapped below the aggregate particles and horizontal reinforcement and some fills the fine interstices between cement particles. Hydration of the cement produces a gel which decreases the size of these incipient voids and increases the watertightness of the concrete, but the voids are never eliminated completely.

231

Thorough curing is necessary to secure watertight concrete. In concrete used under damp conditions (e.g. underground pressure-pipe lines), the cement becomes effectively hydrated and any early weeps "take up" as autogenous healing progresses (see Sections 31.1 and 31.3).

The watertightness of concrete may be of greater significance than compressive strength, particularly in water-retaining structures. This is not due to any serious loss of water through percolation, but because of a need to prevent disintegration resulting from frost action or the freezing and thawing of saturated porous concrete, also to prevent slow weakening through the dissolving out of slowly soluble components, and to stop the formation of unsightly deposits of efflorescence (i.e. alkaline salt and calcium carbonate) on the surface.

Permeability tests on concrete, as described in Section 14.10.1, are of value in determining

> The rate of leakage through walls of a concrete structure or pipe.
> The effect of variations in the cement, aggregate and procedures of mixing, placing and curing.
> The probable life of concrete as affected by the corrosive action of aggressive percolating waters.
> Basic information on the internal pore structure of concrete, which is related to such items as age, absorption, capillarity, dimensional change, induced tension and ambient conditions.
> The comparative effectiveness of impermeability-aid materials.

Note: A weatherproofness test is described in Section 14.7.5.

11.1.2 FACTORS AFFECTING WATERTIGHTNESS

The amount of water that percolates through concrete is a function of pressure and factors governing the condition of the concrete. The following factors, among others (see Sections 31.1-31.3) affect concrete.

(a) Water and cement.
For plastic workable mixes, the permeability increases with the water/cement ratio, there being a fourfold increase in porosity as the ratio increases from 0·6 to 0·7 (Fig. 11.1). A water/cement ratio of not more than 21 litres of water per bag of cement should be used for thin sections and not more than 25 litres per bag for massive structures. As dry mixes do not consolidate very readily,

more water is required for minimum permeability than for maximum strength. For hand-rodded concrete, permeability increases when the amount of water is reduced below that amount which will produce a slump of about 50 mm. Permeability decreases as the cement/voids ratio increases, and this relationship appears to be more definite than that between permeability and water/cement ratio. The mix for thin, reinforced concrete sections should not be leaner than 1 part cement : 4·5 parts aggregate, unless very well-graded combinations of aggregate are used and the concrete is thoroughly consolidated. Cements with a high fineness index and heat of hydration are highly crack-prone with inefficient curing.

With well-cured dense concrete and an optimum amount of mixing water, an increase of cement content above 335 kg/m³ (1 part cement : 6 parts aggregate) does not materially affect permeability. However, wet consistencies for placement in thin sections require a richer mix, and inefficient curing requires a greater cement content for equal watertightness. Increased fineness of cement improves the cohesiveness of mixes and the watertightness of concrete, provided

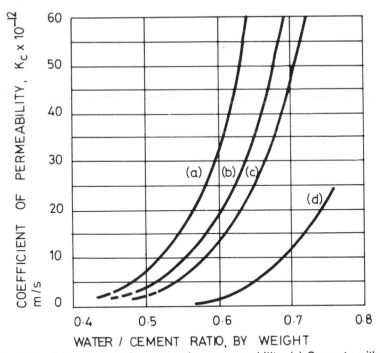

Fig. 11.1 Effect of water/cement ratio on permeability. (a) Concrete with aggregate of 125 mm maximum size. (b) Concrete with aggregate of 75 mm maximum size. (c) Concrete with aggregate of 37·5 mm maximum size. (d) Mortar with aggregate of 4·75 mm maximum size.

that possible shrinkage cracking is controlled. Cement with a slow hydrating and hardening characteristic tends to increase the permeability of concrete that is made with a high slump. Concrete can be made impervious to water under pressure (up to 5 MPa) by the use of an expansive or stressing type cement (see Section 3.1.2(a)).

(b) Aggregate.

As shown in Figure 11.1, the greater the maximum size of aggregate for a given water/cement ratio, the greater the permeability, probably because of the relatively large water voids developed on the underside of the coarser aggregate particles. Aggregate should be sound and of low porosity. Rounded and well-graded aggregate gives the lowest permeability. Aggregate that is poorly shaped should be avoided. It is very important that sufficient fine material be used, particularly that passing a 0·60-mm sieve and retained on a 0·30-mm sieve. An overall grading approximating that of Curve 3 in Figure 4.6 should prove satisfactory for the purpose.

(c) Admixtures.

Permeability-reducing admixtures are sometimes used to improve the watertightness of concrete. Many of them produce lime soaps, which form hydrophobic or water-repellent linings in the pores of the concrete. Water-repellent compounds consist of butyl, aluminium, calcium or ammonium soaps of the fatty-acid type (e.g. stearates and oleates), the amount of active constituent used being about 0·4 per cent (e.g. butyl stearate) by weight of cement.

Bituminous and paraffin wax emulsions and colloidal aluminium silicate are sometimes used in these compounds. Permeability-reducing admixtures, because of their plasticising effect or water content, should be used with a reduced quantity of mixing water. The water repellence shown by some chemicals in this category can become appreciably reduced in concrete at an early age (AS MP20, Part 1).

Where a small amount of air is entrained by these compounds, the mix should be adjusted as described in Section 16.1. They tend to lower strength and reduce the bond of subsequently applied cementitious or composite coatings, particularly where bleed water causes a weak layer of chemically dosed laitance to form on the surface. For integrally waterproofed mortar renderings, therefore, a waterproofed background concrete or mortar should be roughened, scratched or greencut to provide a strong mechanical key with the surface coat, which should be applied and cured as early as practicable for a monolithic result.

The best use of water-repellent compounds appears to be in walls, slabs and sections that are not subjected to tension or hydrostatic pressure, and where vapourproofness is not required. In water-retaining structures, the use of extra cement is generally more effective than the use of a permeability-reducing agent and the extra cost for equal effect is usually less.[99, 141] Water-repellent compounds are ineffective in lime-based mortar, because they inhibit the carbonation of hydrated lime.

Air-entraining and cement-dispersing agents, used singly or in combination, greatly contribute to watertightness by reducing settlement, presetting cracks, bleeding and capillarity of placed concrete. When the mix has been adjusted as described in Sections 4.3.3 and 16.1, it has a high degree of cohesiveness. Air-entrainment is a very effective means of reducing the movement of water in capillaries (Appendix III).

Alternatively, the capillary tracts of clean, damp, exposed concrete substrates can be closed against water migration (but without stopping vapour movement) by effectively using osmotic pressure (see Glossary) and a water-insoluble precipitate, e.g. calcium carbonate. This ingredient is produced from moist cement and mortar plugs and coatings, that are compounded with sodium carbonate and sometimes a set-controller, e.g. sodium lactate. Each coating, which is brushed well into the surface, is applied at the rate of 0·8 to 1·2 kg/m² and kept damp for several days (see Sections 10.1 and 11.1.2(f)).

With ordinary mixes, particularly those deficient in fines, a finely divided admixture of bentonite (see Section 24.3), pozzolan or hydrated lime has a useful effect in lowering permeability. If fly ash is used for this purpose, it should be of suitable quality, and hydrated lime (5 per cent by weight of cement) should not be used in untreated or unautoclaved concrete that has to resist aggressive waters.

The impermeability of concrete admixed with fly ash increases in time much faster than that of concrete made with the pure cement. The same degree of impermeability of two nominal-mix concretes, one gauged with a 20-30 per cent replacement of the cement by good fly ash, can be obtained in about a year. Air-entrainment, cement dispersion and a pozzolan may be used together to marked advantage in portland cement concrete (see Sections 6.4, 17.1 and 17.2).

Cement mortar may be damp-proofed by a 5 per cent admixture of mineral oil residuum; this proportion by weight of cement is doubled for concrete. The oil is mixed with freshly prepared mortar before coarse aggregate is added to the mix. The setting time is delayed some 50 per cent and the compressive strength and bond strength to steel reinforcement are reduced, the latter being large unless deformed rods are used. Bonding to roughened hard surfaces is improved by a preliminary application of plain cement grout and mortar.

Mineral oil residuum for damp-proofing should have a specific gravity of 0·93-0·94 at 25°C, and a 99·9 per cent solubility in carbon disulphide at air temperature. It should contain 1·5-2·5 per cent of bitumen insoluble in paraffin naphtha with a boiling point at 85°C. It should yield 2·5-4·0 per cent of residual coke and be of 40-45° Engler viscosity at 50°C, also it should show not more than 2 per cent loss in weight when a 20-g sample is heated for 5 hours at 160°C.

(d) Uniformity of concrete.

Minor defects or nonhomogeneous conditions in concrete, that would have no appreciable effect on compressive strength, influence initial leakage through concrete to a marked degree. The majority of leaks in concrete structures are probably due to such defects as cracks in the structure and void spaces in the concrete due to segregation, honeycombing, or differential settlement of the newly placed mass, rather than to inherent porosity of the cement paste or aggregate. To minimise these defects, the mix should be workable and both homogeneous and well compacted in place prior to being moist-cured.

Unless special care is taken in the design of mixes and fabrication of sections, the placed concrete pulls away from the underside of horizontal reinforcing rods or other obstructions to settlement and causes cracks to form over them. Should cracks occur, their width can be measured with an optical micrometer, a microscope, a plastic scale and magnifying glass, a set of standard machinist's gauges, or a radiographic or photostress technique. The plastic scale is marked with a series of black lines with thicknesses varying from 0·05 mm to 5 mm (see Sections 15.2.2(b) and 34.1.1). Cracks are revealed by a surface application of volatile liquid (e.g. methylated spirits) and they can be simply and effectively sealed by pressure-grouting or intrusion with plastic (see Reference 140 of Part 4 and Reference 134 of Part 8).[11] Where water-retaining members are in almost continuous contact with liquid, tolerable crack widths are up to 0·15 mm (see Reference 141 of Part 6).

Reinforcement should have adequate cover and the concrete should be air-entrained and, where practicable, reconsolidated by reworking of the surface and revibration at about the time of initial set of the cement. The effect of increased compaction by proficient vibration and revibration is to close fine seepage paths, caused by bleeding and upward water gain, and reduce the permeability of hardened concrete by over one-third (see Section 23.2).

Free water, carrying admixtures and potential laitance material in suspension should be skimmed off the surface of newly placed concrete, in order to secure a good bond with successive lifts. Construction joints should be greencut or roughened, cleaned and dampened prior to coating with cement slurry, mortar and concrete (see Section 7.2.4). Resinous bonding procedures are dealt with in Section 21.1.6; and movement joints must be suitably designed.

Joint accessories include internal and external waterstops (e.g. rubber, polyvinyl chloride or polypropylene); crimped copper strips (see Sections 10.1 and 19.1.1; also BS 1878 and ASTM B248); bonded, preformed, elastomeric or bitumastic strips; compressed cork; polychloroprene (e.g. "Neoprene") compression seals; pliable fillers; and joint sealing compounds to suit the width and type of joint (see Sections 20.1.5(f), 26.1, 28.12 and 31.2.1; also Appendix I). Typical jointing illustrations are given in Figure 30.3 and Plate 10.[187] Also, for estimating the proper placement of internal waterstops and sound concrete at joints, a strike-hammer test can be used as indicated in Section 20.1.1.

Cryogenic means of joining tunnel caissons in wet terrain include the freezing of ground gaps with piped liquid nitrogen (AS 1894). After placing polyethylene or polyurethane foam insulation (50 mm thick) against the frozen background and erecting temporary form-liners, each joint is pump-filled with high-early-strength portland cement concrete. This bears against a bondbreak strip of foam rubber to facilitate movement at impermeable joints (see Reference 128 of Part 6).

(e) Curing.
Continued hydration of the cement results in gel development which reduces the size of voids and increases the watertightness of the concrete. Figure 11.2 shows a great increase in impermeability with moist curing and age, the increase being even greater than that of strength with curing. Although, through different curing conditions, identical concrete samples may show widely differing initial permeability, these differences decrease if the concrete continues to be subjected to hydraulic conditions. Freshly placed concrete must be protected from drying conditions and kept damp for a minimum period of 3 days, preferably much longer.

Typical curing methods, in order of decreasing efficiency, are ponding, covering with wet sand and membrane curing of moist concrete (see Section 9.1).

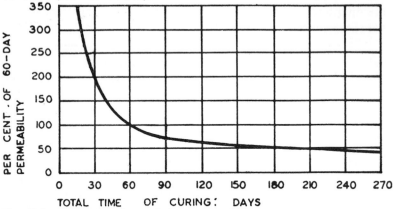

Fig. 11.2   Effect of curing period on permeability.

(f) Surface treatments.

Certain mortar formulations, which are typically portland cement, pozzolanic silica flour and fine aggregate, sodium carbonate and alpha-hydroxy acid salt, can be used in the presence of moisture for repairing leakages and partly closing capillary tracts with calcium carbonate. This by-product, which is water insoluble, is slowly soluble in waters containing free carbon dioxide or acid (see Sections 10.1 and 11.1.2(c)).

When permeability tests are being conducted on concrete pipes, porous patches may be sealed within 1 hour by using sodium silicate solution instead of water as the pressure medium. Internal surface treatments or linings are effective in reducing or preventing leakage from water-retaining structures. Liquid sealants should be applied in several uniform, continuous coats. Various membranes, with prescribed primers and adhesives, are as follows:

Bentonite backing boards, where water has low salinity.
Bituminous or polyisobutylene lining; ethylene propylene terpolymer (ASTM 3020) or rubber-bitumen polyethylene membrane; butyl, vinyl, or vulcanised natural rubber sheeting.
Cement slurry containing bentonite and calcium chloride, bagged on and cured. Pneumatic mortar for extensive repairs and lining.
Chemical sealants of calcium chloride and sodium silicate solutions in successive applications.
Epoxy resin coatings; resinous laminates with filler; chopped fibreglass or fibreglass cloth, nylon fabric being used for crack-

resistant seals over construction joints (Perkins, Appendix IV, 1). Mastic asphalt; clay-stabilised bitumen-emulsion seals with nylon scrim; and bituminous protective mortar and cove (see Section 20.2).

Paraffin wax emulsified in water or dissolved in volatile solvents.

Siliconate rubber-latex; polymer mortar; aluminium inner shell.

Note: Some of these treatments may be used for the membrane curing of high-quality concrete. They should be used as early as practicable and at the end of any bleeding of a concrete pavement. Natural rubber sheets, 5 mm thick, may be used for lining extensively repaired concrete reservoirs. They are kept flexible to cope with possible cracking, supported on metal strips over joints, secured *in situ* with noncorrodible bolts with sealed cap-nuts, and spark-tested at sealed-cover joins.[164] Materials and methods of waterproofing concrete bridge decks are examined in Reference 214.

(g) Design.

Design and constructional details for water-storage, water-excluding and sanitary engineering structures are given in AS 1481, BS CP102 and BS 5337; also the ACI Committee 350 Report (see References 65 and 171 of Part 4).[82, 114, 131, 212] Prestressed slabs can be compressed sufficiently to be waterproof (Chapter 34 and Reference 34 of Part 8). Jointing materials are indicated in Sections 11.1.2(d), 20.1.5(f), 26.1 and 31.2.1; and good workmanship throughout is essential.[187] Soft-water corrosion is studied in Section 19.2, and polymer concrete possessing low permeability is dealt with in Section 21.4.

11.2 ABSORPTION

Absorption is a physical process by which concrete draws water into its pores or capillaries (Chapter 16). In wetting and drying tests, the drying operation withdraws not only the free water that is mechanically held in concrete, but some of the colloidal water that is more tenaciously held in the cement gel. The absorptions and apparent porosities indicated by the tests are larger, therefore, than those associated with the ordinary temperature and humidity environment of concrete in service.

The repeated wetting and drying of concrete containing calcium or sodium chloride may cause fretting of the surface. This failure is due to the increased solubility of calcium hydroxide in chloride solution or the formation of chloroaluminates in the concrete (see Section 15.2.3(g)).

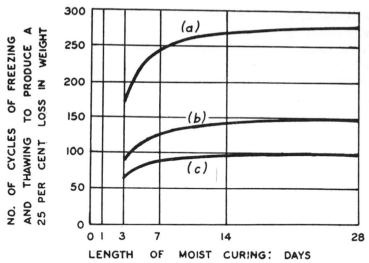

Fig. 11.3 Curing and durability relationship. (a) Water/cement ratio 0·45. (b) Water/cement ratio 0·62. (c) Water/cement ratio 0·80.

Water absorption is related to the resistance of concrete to frost action or cyclic freezing and thawing of water beneath the surface, the relationship holding more with the rate of absorption than with the total amount. Concrete should be dense, sound and adequately cured to minimise absorption, which is greater in hardened, lean mixes than in rich mixes. With aggregate of low porosity, Figure 11.3 shows that precast concrete panels should be cured for at least 10 days, if maximum resistance to disintegration by frost is to be obtained. Pore structure and water-absorption tests on concrete and aggregate are dealt with in Sections 14.10.2, 31.1.1, 33.3.2(e) and 33.3.3(e), also in Appendix I and Appendix IV, 4(b) (Hughes).

11.3 AIR-PERMEABILITY

The rate of passage of air through sections of sound, dense concrete appears to be inversely proportional to the thickness of concrete and directly proportional to the air pressure. With air pressure of up to 7·5 kPa and for thicknesses of 100-225 mm, the rate of seepage (litres/m^2. hour) is approximately 7·8 times the ratio of the pressure (kPa) to the thickness of the section (mm). The seepage is greater and more variable for poor-quality concrete than for sound material, and it may be increased several times by constructional blemishes and hair-cracks.

Factors that contribute to watertightness (see Chapter 11 and Appendix III) are, for example, increased cement content, pozzolanic admixture, fly ash (see Sections 17.1 and 17.2), thorough moist curing, prestressing and void discontinuity or impregnation. These factors are conducive to airtight concrete sections. A particular concrete that is moderately permeable to air may be substantially impermeable to some other gases. Concrete corrosion due to an aggressive, gaseous environment is dealt with in Section 15.2.3(d).

CHAPTER **12**

ABRASION AND
EROSION
OF CONCRETE

12.1 TYPES OF ABRASION

On different surfaces, wear is brought about in various ways, as in

Concrete floors and heavily trafficked footways (BS CP2006), where wear is caused by a rubbing action of foot traffic, light trucking and the skidding or sliding of objects over abrasive particles.

Concrete road surfaces, where the exposure factor is a rubbing and impact-cutting action of heavy trucking and automobiles (with or without chains) and accelerated by the presence of abrasive particles.

Airport runways, because of the impact and abrasion of high-pressure tyres during landing operations.

Jet-engine warm-up aprons and rocket-launching platforms, where disintegration takes place by blast and heat (velocities up to 1100 m/s and temperatures from about 700-1800°C for a few minutes.

Hydraulic structures, where impact abrasion or erosion occur through cavitation.

Underwater construction and conduit-inverts, where a cutting action is caused by abrasive materials carried by flowing water.

Concrete bunkers and chutes in heavy industry, where intense grinding, shearing and impact forces are set up by the movement of raw material.

242

The lining of rotary kilns, where rubbing and impact-cutting by clinker or expanded aggregate take place at elevated temperature.

Each of these examples is a study in itself, which is beyond the scope of these brief particulars. An abrasion test (see Section 14.2) must take into account the characteristics of the concrete being tested (e.g. composition, moisture content, state of stress and age) and the presence or otherwise of surface-hardened material. It must simulate also the nature and magnitude of the abrasive forces in service.

In general, the abrasion resistance of ordinary concrete without special finishing is a direct function of its compressive strength at an age of 1-3 months. For best results, it is essential to make good concrete (with low bleeding and shrinkage characteristics) and to cure it properly. The need for high-quality material and workmanship is indicated by the fact that the wear resistance of 40 MPa concrete is three times that of 20 MPa concrete (see Sections 3.2.2, 4.3.1, 5.1, 20.1.2(b)(iv), 22.1 and 26.3). A 24-hour delay in curing can cause a 60 per cent drop in potential wear resistance of a concrete topping, by comparison with one immediately cured with a plastic-sheet membrane followed by moist curing.

The type of aggregate used has a marked bearing on the later stages of wear of medium-strength and low-strength concretes, but its effect on wear resistance is small with concrete above 40 MPa. Under erosive conditions, abrasive forces may cut the surface and the underlying mortar matrix, release the aggregate from the mass and thus cause destruction of the concrete. Wear data are given in References 39 and 115 of Part 4 and the text relating strength to hardness (see Sections 5.1, 14.9.2 and 15.2.5).

As a guide to the design of an integral granolithic topping or surface-course mix for pavements with hard-wearing, nondusting qualities, abrasion tests have shown that a 1 : 1·5 : 2 cement, sand and gravel concrete has nearly three times the wear resistance of a 1 : 3 cement and sand mortar. Depending on the severity of abrasive exposure, the minimum 28-day (cylinder) compressive strength recommended for granolithic concrete is 30 MPa for moderate-duty industrial floors and 35-40 MPa for those expected to resist severely abrasive conditions.

A concrete pavement that is cured immediately and continuously for at least 2 weeks (see Sections 9.2, 9.3 and 20.1.5(e)) can be about twice as wear resistant as one cured for 3 days, following a 1-day delay after laying. Concreting in extreme weather requires special care to prevent impaired results (see Sections 9.2, 31.2.4, 32.1 and 32.2).[197]

12.2 EROSION

The causes of erosion (see Section 15.2.5) and some means of dealing with it are described.

12.2.1 CAVITATION

The pressure in fast-flowing water should be kept above its vapour pressure. If vapour bubbles form they will, on entering a zone of higher pressure, collapse with great impact. Their repeated collapse near the surface of concrete causes severe pitting. Damage from cavitation is not common in open conduits with velocities below 12 m/s. Concrete in closed conduits can become pitted by cavitation at velocities as low as 7·5 m/s, where the air pressure is reduced by the sweep of the flowing water. At higher velocities, the forces of cavitation, through·abrupt changes in direction and velocity, may reach an estimated value of up to 700 MPa. These forces are sufficient to erode away large quantities of high-quality concrete and penetrate through thick steel plates. Forces of cavitation in a closed conduit can be materially reduced by introducing air (at the area of low pressure) into the flowing water.

Design and construction must provide smooth, well-aligned surfaces, free of abrupt changes in slope or curvature, so as to ensure smooth uniform flow that does not pull away from the concrete surface. Transverse irregularities are usually limited to within 5 mm to prevent pitting downstream. Dense, strong concrete is required, with an impermeable surface that withstands without damage the impact of a 30 m/s jet of water.[83, 105] An absorbent form liner reduces the water/cement ratio to advantage at the surface of newly placed concrete. As cavitation tends to remove large stone particles, the near-surface aggregate should not exceed 20 mm nominal size. A cavitation test is described in Section 14.2.7.

12.2.2 ABRASION

Erosion by stone detritus can be as severe as that caused by cavitation, depending upon the quantity, shape, size and hardness of the particles being transported, the velocity of the water and the quality of the concrete. Because of this, diversion tunnels may be badly worn when dams are being built. Best-quality concrete is required, as abrasion-resistance increases with the strength of the concrete. At velocities below 12 m/s, surface smoothness and alignment are less important with abrasion than with cavitation. The diameter of particles transported (d mm) is related to the bottom velocity (V m/s) by the following formulae.

Below 6·7 mm, $V = 0.197 \, d^{4/9}$.
Above 6·7 mm, $V = 0.155 \, d^{1/2}$.

12.2.3 PROCEDURES

The denseness and strength of flat surfaces are increased by vacuum dewatering. Steel trowelling must be kept to a minimum and applied firmly after the concrete has stiffened. Early, continuous moist curing, usually for 14 days, is essential. Water spray or (white) membrane curing of unformed concrete should start just after its initial hardening, particularly in hot dry weather when crazing and cracking readily occur. The temperature of deposited concrete should not exceed 32°C, as its strength and durability decrease as the placing temperature rises. Protection is required against frost action and freezing.

Dentated sills and baffles in stilling basins, designed to dissipate energy, should be made of the best possible concrete. Where erosion occurs, periodical replacement of concrete will be required unless the cause is removed. A square-edge for repair work is readily obtained with a concrete saw. Cleaning is done by wet sandblasting and air–water-jet washing, the holes being kept wet for 12 hours or more and allowed to become saturated surface dry before new concrete is placed.

Where velocities exceed 12 m/s, grinding to ensure smoothness may be required. A tough-skin facing of fibre-reinforced concrete is distinctly beneficial where surfaces may be subjected to water-borne cavitation, erosion and impact (see Section 3.4.2 and Reference 145 of Part 1). Repairs may be made alternatively with the aid of epoxy-resin materials, mortar and concrete, as indicated in Sections 20.4.3 and 21.1.

CHAPTER 13

STATISTICAL CONTROL

13.1 VARIABILITY OF CONCRETE

For the controlled execution of a design, the strength of concrete should be such that only a small or nominal proportion of test results (e.g. 1 per cent or possibly 10 per cent) falls below a specified minimum strength. In prestressed-concrete work, control is required to ensure that the strength of the weakest part of the member to be stressed is as close as is economically possible to the mean strength. If only medium high strength is required, the cost of control can be balanced against the cost of allowing a greater margin of safety in the mix design. When very high strength is needed, however, it can be attained only be eliminating every possibility of occasional low strengths through a high degree of control, even if this is expensive. Specification references to these ends are given in Section 3.4.[74]

Test specimens are made in groups of three (or two) from different batches to minimise errors in testing. Provided that the specimens are properly made, stored and tested, the test results are a measure of the uniformity or quality of concrete as it leaves the mixer. They are indicative of strength in a structure under satisfactory conditions of placing, compaction and curing (AS 1480 and MP28, Section C4, Commentary on AS 1480).

A pattern of the test results on a major project is indicated by a histogram (Figure 13.1; see Reference 11 of Part 2). This type of diagram is drawn by splitting 100 or more test results into groups, which fall within progressive strength ranges, and plotting the number of results which fall in each group against the average strength of that group. For a given set of conditions and degree of

246

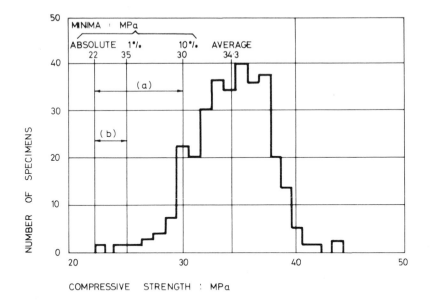

Fig. 13.1 Histogram. (a) 10 per cent of all specimens. (b) 1 per cent of all specimens. Total number of specimens: 315.

control, a large number of test results (about 1000) would approximate to a normal distribution curve. This curve represents the distribution of results which would be obtained if their variation were due only to chance or random effects.

The probable distribution of test results, under two conditions of controlled production, is shown in Table 13.1.

TABLE 13.1   DISTRIBUTION OF TEST RESULTS

Control of Production and Testing	Percentage of tests falling below:			
	Average of all	90 per cent of average	80 per cent of average	70 per cent of average
Excellent	50	10-15	1-4	1
Good	50	15-25	5-10	1

13.2 STATISTICAL ANALYSIS

13.2.1 CLASSICAL METHOD

The aim of concrete control is to achieve uniformity and economy in construction, while complying with predetermined requirements of strength and durability. No matter how well a job is controlled, there will be some variation in the strength, and the number of values above and below the mean will fall in some pattern similar to that shown below.[1, 48, 72-81, 95, 135-137, 200]

Where there is good control, the values are bunched close to the mean and the curve is steep; but where there is poor control, the values are spread out laterally and the curve is relatively flat. The distribution of the points tends to follow a normal probability curve, which can be used conveniently for measuring the degree of control that is being obtained or exercised on a job, and for predicting what mean must be obtained to have all strengths fall above a selected figure (AS MP28, Section C4, Commentary on AS 1480).

The arithmetic mean or sample mean X of control tests is the average of all results in a set. The deviation of each result is its difference from the arithmetic mean, and the variance of all the results is the sum of the squares of their deviations divided by their number less unity. The square root of the variance is the *standard deviation, S.*

$$S = \sqrt{\frac{\Sigma d^2}{n-1}} \text{ MPa}$$

where d = deviation from the mean, n = number of tests taken.

While at least twenty-four test results are needed for a determination of standard deviation, standard specifications (e.g. AS 1480) require at least thirty results. Since standard deviation is a measure of the spread of the results, its value is small for a very uniform product. The standard deviation, if expressed as a percentage of the arithmetic mean strength, is called the *coefficient of variation.* It is given by

$$V = \frac{100S}{X} \text{ per cent}$$

and is a measure of relative variability. A direct application of these principles to an adequate quantity of reliable test data, as illustrated in Table 13.2, is expedited by the use of a calculating-machine or computer.

TABLE 13.2   **STANDARD DEVIATION**

Date	Compressive Strength (MPa)	Deviation (MPa)	Squared Deviation
16.10.73	24·3	1·7	2·89
	22·1	0·5	0·25
	24·8	2·2	4·84
	20·5	2·1	4·41
	23·8	1·2	1·44
	21·2	1·4	1·96
19.10.73	20·4	2·2	4·84
	23·2	0·6	0·36
	23·2	0·6	0·36
	23·4	0·8	0·64
	21·7	0·9	0·81
	22·5	0·1	0·01
25.10.73	21·3	1·3	1·69
	21·7	0·9	0·81
	22·8	0·2	0·04
	21·4	1·2	1·44
	22·3	0·3	0·09
	20·8	1·8	3·24
29.10.73	24·1	1·5	2·25
	21·8	0·8	0·64
	23·2	0·6	0·36
31.10.73	25·6	3·0	9·0
	24·3	1·7	2·89
	25·2	2·6	6·76
5.11.73	22·4	0·2	0·04
	19·4	3·2	10·24
	23·1	0·5	0·25
	Total = 610·5		Total = 62·55

Example.

Mean of 27 tests $\overline{X} = 22 \cdot 6$.

Standard deviation $S = \sqrt{\dfrac{62 \cdot 55}{(27 - 1)}}$

$\qquad\qquad = \sqrt{2 \cdot 4058}$

$\qquad\qquad = 1 \cdot 551$ MPa.

Coefficient of variation $V = \dfrac{1 \cdot 551}{22 \cdot 6}$

$\qquad\qquad\qquad = 6 \cdot 9$ per cent.

The tabulated data are now plotted, according to their variations from the mean, in a distribution curve of compressive strength (Fig. 13.2).

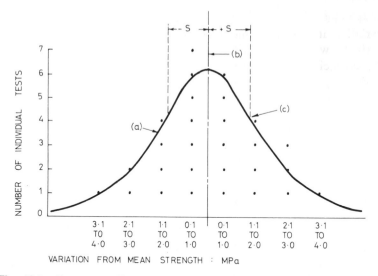

Fig. 13.2 Frequency distribution of compressive strength. (a) Theoretical distribution curve. (b) Mean = 22·6 MPa. (c) Inflection point. Standard deviation S = 1·551 MPa.

Each dot in Figure 13.2 represents a single test result lying within the strength range indicated. The curve, symmetrical about the vertical line indicating the mean strength, represents the theoretical distribution of an infinite number of tests. The inflection points of the curve occur at values equal to plus and minus the value of the standard deviation and, within this range, 68 per cent of all observations may be expected to fall. Theoretically, 94 per cent of the observations fall within a range equal to plus and minus twice the standard deviation and 99 per cent (100 per cent for a small number of tests) within a range equal to plus and minus three times the standard deviation.

The coefficient of variation is used not only as a measure of the variability or lack of uniformity in the results, but also to compare groups with different means. A low coefficient indicates good control and a high one indicates poor control (see Table 4.4). A coefficient of about 10 per cent is about as low as can be expected for field concrete work, and coefficients as high as 25 per cent are not uncommon. Should increased variability occur on a project, the cause should be investigated. It would be obvious that either the site control is showing some relaxation, or other factors (such as increased variation in the quality of the cement, decreased mixing efficiency and poor specimen making) are having an adverse effect.[34, 167, 170, 194, 207]

A point of interest is that if the standard deviation or actual variation in results remains sensibly constant, the coefficient of variation will decrease with increased mean strengths. Therefore, if the coefficient of variation is taken as the criterion, improved uniformity will be obtained from high-strength concrete with no improvement in control. In practice, high-strength concrete is usually manufactured with greater care and with better equipment than are used for producing ordinary concrete. This causes the results from high-strength concrete to show up still better than those obtained from ordinary concrete (see Reference 30 of Part 2).

TABLE 13.3 **STATISTICAL CONTROL**

Test Strengths Above Control Strength (per cent)	Coefficient of Variation (per cent)	Mean Strength Required (MPa)	Estimated Cement Content* (kg/m³)	Estimated Cost of Cement per m³ of Concrete ($A)†
70	5	20·5	257	11·24
	10	21·1	264	11·55
	15	21·7	269	11·77
	20	22·4	276	12·08
	25	23·1	283	12·38
80	5	20·9	261	11·42
	10	21·8	270	11·81
	15	22·9	281	12·29
	20	24·0	292	12·78
	25	25·3	305	13·34
90	5	21·4	266	11·64
	10	22·9	281	12·29
	15	24·8	300	13·13
	20	26·9	323	14·13
	25	29·4	350	15·31
99	5	22·6	278	12·16
	10	26·1	314	13·74
	15	30·7	365	15·97
	20	37·5	464	20·30
	25	47·9	715	31·28

* 40 mm nominal-size aggregate.
† Cost based on $1·75 per 40 kg bag of cement.

In Table 13.3, statistical methods have been used to show what mean strengths are required in order to have 70, 80, 90 and 99 per cent of the strength values fall above a selected value, such as 20 MPa.[60]

TABLE 13.4   **DEGREES OF CONTROL**

Coefficient of Variation (per cent)	Degree of Control	Standard Deviation (MPa for average compressive strength 40 MPa)
5·0	Well-controlled laboratory test	2
10·0	Excellent	4
12·5	Very Good	5
15·0	Good	6
17·5	Fair	7
20·0	Poor	8
25·0	Bad	10

It should be noted that under good control, with a coefficient of variation of 10 per cent, a mean strength of only 22·9 MPa is required to ensure that 90 per cent of the strength values fall above 20 MPa, but a mean strength of 29·4 MPa is required for a coefficient of variation of 25 per cent. This example illustrates the economy that could accompany good control on large constructional projects, based upon million cubic metre quantities of concrete with cement at $A 1.75 per 40-kg bag at the site. As work

TABLE 13.5   **FACTORS AFFECTING STRENGTH**

Factor	Probable Maximum Variation in Strength (per cent)
Cement from one source	25
Cement from different sources	50
Grading of aggregate	20
Bulking of fine aggregate	25
Batching	
By weight	8
By volume	
Good	15
Normal	30
Bad	50
Poor compaction	50
Handling, mixing and transporting	Unknown, but may be eliminated by attention to detail.
Temperature	Unimportant after 28 days, provided temperature is above freezing.
Making and testing specimens	30

control is improved and the coefficient of variation is reduced, a variation in cement strength becomes increasingly important. A coefficient of variation of 5 per cent in the strength of cement from one plant is normal, with double this figure when two or more plants are considered. Tables 13.4 and 13.5 show coefficients of variation for different degrees of control and some important factors that affect average compressive strength (see Table 4.4).

Statistical methods of control require a large number of results over a significant period, for the close assessment of data on variation. The data thus compiled can be used to

Gauge the success of a particular procedure of control in safeguarding the uniformity and quality of concrete for a structure.

Estimate the correctness of assumptions that were made in design calculations.

Indicate, for use in mix design, the extent of variation that may be expected under a particular set of conditions (Appendix II, 3).[185]

Carry out computer regression analyses of accelerated strength tests, accompanied with allied variables, as a periodical early check of cement and 28-day strength prediction factors in automatic batching (see Sections 3.1.4 and 3.3.6; also Reference 30 of Part 1).[98]

13.2.2 GROUPED-FREQUENCY METHOD

The statistical analysis of a series of compressive-strength results is simplified arithmetically by the use of grouped-frequency tables.[138] Histograms are plotted from tabulated tally marks, the test data are translated into small whole numbers and an arithmetical check is applied to the calculations. The procedure is best illustrated by an example, such as the analysis of the concrete programme that is represented in Table 13.6, where the following symbolic interpretations have been used.

X_0 = assumed average compressive strength at 28 days.

f = frequency of occurrence of the results within each class boundary.

d = deviation of each class from the assumed average, the measurements being in successive numerals of the class interval C (which is taken as 1 MPa) and given negative or positive signs in relation to the magnitude of X_0.

The application of this technique is described.

Step 1. Enter the test results as successive strokes in tally-mark columns where they cross within appropriate class boundaries. For example, a value of 32·4 MPa is entered with a mark within the class boundaries of 32-33 MPa. (Normally four columns are provided, with ten divisions each.)

Step 2. Form a histogram by blocking in the areas that are occupied by tally marks for all the results under review, and enter the numbers for each class in the frequency column headed f.

Step 3. Visually select on the histogram an approximate average compressive strength X_0 and draw a horizontal pencil line through the nearest class midmark division. This step can be expedited by an inspectional balancing of areas about the 90° sides of a symmetrically placed set-square.

Step 4. Enter successive numbers (starting from 0 at X_0) in the deviation column headed d, those that are upwards from X_0 being marked negative and those downwards being positive. (Blue and red inks may be used to clarify the distinction.)

Step 5. Complete the columns headed d^2, fd, fd^2, $f(d + 1)^2$, it being noted that the product fd is negative above X_0 and positive below it, and that $(d + 1)^2$ at any entry is d^2 in the one underneath. (Three cardboard masks as illustrated below the table may be used to isolate readily the numbers that are to be multiplied.)

Step 6. Enter the summations (Σ) of f, fd, fd^2 and $f(d + 1)^2$ at the foot of the table, the value of fd being signified as negative or positive.

Step 7. Apply Charlier's check on arithmetical accuracy.
$\Sigma f(d + 1)^2 = \Sigma fd^2 + 2 \Sigma fd + \Sigma f.$
$11131 = 10982 - 28 + 177 = 11131$, which checks.

Step 8. Determine statistical values from the following formulae.

Arithmetic average, $\overline{X} = X_0 + C \dfrac{\Sigma fd}{\Sigma f} = 49 \cdot 5 + \dfrac{-14}{177}$
$= 49 \cdot 42$ MPa (C being 1).

Standard deviation, $S = C \sqrt{\dfrac{\Sigma fd^2}{\Sigma f} - \left[\dfrac{\Sigma fd}{\Sigma f}\right]^2}$
$= \sqrt{\dfrac{10982}{177} - \left[\dfrac{-14}{177}\right]^2} = 7 \cdot 88$ MPa.

Coefficient of variation, $V = \dfrac{100\,S}{\overline{X}}$ per cent $= \dfrac{7 \cdot 88 \times 100}{49 \cdot 42}$
$= 15 \cdot 94$ per cent.

TABLE 13.6   **GROUPED-FREQUENCY TABLE**

Class Boundaries (MPa)	Class Midmark (MPa)	d^2	Deviation (d) $\oplus$	Frequency (f)	fd	fd^2	$f(d+1)^2$	Tally Marks
30								
	30·5							
31								
	31·5							
32								
	32·5	289	− 17	1	− 17	289	256	32·4
33								
	33·5	256	16	1	16	256	225	
34								
	34·5	225	15	2	30	450	392	
35								
	35·5	196	14	0	0	0	0	
36								
	36·5	169	13	1	13	169	144	
37								
	37·5	144	12	2	24	288	242	
38								
	38·5	121	11	7	77	847	700	
39								
	39·5	100	10	5	50	500	405	
40								
	40·5	81	9	10	90	810	640	
41								
	41·5	64	8	7	56	448	343	
42								
	42·5	49	7	3	21	147	108	
43								
	43·5	36	6	8	48	288	200	
44								
	44·5	25	5	8	40	200	128	
45								
	45·5	16	4	5	20	80	45	
46								
	46·5	9	3	8	24	72	32	
47								
	47·5	4	2	10	20	40	10	
48								
	48·5	1	− 1	2	− 2	2	0	
49								$\bar{X} =$
	49·5	0	0	13	0	0	13	49·42
50								
	50·5	1	+ 1	22	+ 22	22	88	
51								
	51·5	4	2	8	16	32	72	

$X_0 = 49·5$

Class Boundaries (MPa)	Class Midmark (MPa)	d^2	Deviation (d) $\oplus$	Frequency (f)	fd	fd^2	$f(d+1)^2$	Tally Marks
52								
	52·5	9	3	6	18	54	96	
53								
	53·5	16	4	6	24	96	150	
54								
$X_0 =$ 55	54·5	25	5	4	20	100	144	
	55·5	36	6	2	12	72	98	
56								
	56·5	49	7	5	35	245	320	
57								
	57·5	64	8	5	40	320	405	
58								
	58·5	81	9	2	18	162	200	
59								
	59·5	100	10	9	90	900	1089	
60								
	60·5	121	11	1	11	121	144	
61								
	61·5	144	12	1	12	144	169	
62								
	62·5	169	13	4	52	676	784	
63								
	63·5	196	14	2	28	392	450	
64								
	64·5	225	15	1	15	225	256	
65								
	65·5	256	16	1	16	256	289	
66								
	66·5	289	17	1	17	289	324	
67								
	67·5	324	18	1	18	324	361	
68								
	68·5	361	19	0	0	0	0	
69								
	69·5	400	20	0	0	0	0	
70								
	70·5	441	21	2	42	882	968	
71								
	71·5	484	22	0	0	0	0	
72								
	72·5	529	23	0	0	0	0	
73								
	73·5	576	24	0	0	0	0	

Class Boundaries (MPa)	Class Mid-mark (MPa)	d^2	Deviation (d) ⊕	Frequency (f)	fd	fd^2	$f(d+1)^2$	Tally Marks
74								
	74·5	625	25	0	0	0	0	
75								
	75·5	676	26	0	0	0	0	
76								
	76·5	729	27	0	0	0	0	
77								
	77·5	784	+28	1	+28	784	841	77·8
78								
	78·5				−548			
79								
	79·5				+534			
80								
$X_0 =$	49·5		⊕	177	−14	10982	11131	
$C = 1$				Σf	Σfd	Σfd^2	$\Sigma f(d+1)^2$	

$d \times f = fd$ — MASK 1

$d^2 \quad \times f \quad = fd^2$ — MASK 2

$(d+1)^2 \quad \times f \quad = f(d+1)^2$ — MASK 3

A final report on a project should include the total number of tests, the arithmetic average, the minimum and maximum results, the range between them, the percentage of the total number of tests that fall below a specified minimum value, and the ratio of the average strength to the specified minimum strength.

13.3 STANDARDS

Standard specifications for statistical control are listed.

Acceptance of evidence based on the results of probability sampling (ASTM E141).

Application of statistical methods to industrial standardisation and quality control (BS 600).

Choice of sample size to estimate the average quality of a lot or process (ASTM E122).

Code for the recording of industrial-accident statistics (AS CZ6).

Data-processing flow-chart symbols, rules and conventions (BS 4058; Appendix IV, 2(a)(ii)).

Dealing with outlying observations (ASTM E178).

Designating significant places in specified limiting values (ASTM E29).

Guide to inspection procedures (BS PD6452).

Guide to statistical interpretation of data (BS 2846).

Probability sampling of materials (ASTM E105).

Quality-control charts (BS 1313, 2564).

Quality-control systems (AS 1821-1823, (E)Z 501-503).

Quality control using small samples (BS 2635).

Terms used in quality control (AS 1057).

Use of the terms *precision* and *accuracy* as applied to measurement of a property of a material (ASTM E177).

CHAPTER 14

TESTS

14.1 GENERAL INFORMATION

Methods of sampling fresh concrete are described in Section 7.1.5, and terms relating to methods of mechanical testing are defined in ASTM Standard E6. Laboratory testing is necessary to establish whether materials meet the requirements of specifications; and universities and technological institutes will usually assist with regularly calibrated equipment.

Referee testing laboratories are registered in different countries (e.g. the National Association of Testing Authorities, Australia). A certificate of compliance with a specification should be obtainable from a supplier of material. Usually the cost of testing concrete is part of a contract, otherwise, the cost is borne by the vendor if the material does not comply with specification, and by the purchaser if it does. Duplicate tests are made in cases of dispute.

For structural concrete, unless otherwise specified, not more than one standard strength test in twenty should have an average compressive strength less than 90 per cent of the specified value. Odd consignments of mixed and placed concrete up to 15 per cent below the specified minimum 28-day strength may be accepted at a proportionate penalty rate of payment. An additional discrepancy of up to 10 per cent, where structurally permissible, may be accepted at a higher penalty rate of payment.[80, 213]

Control factors and coefficients of variation for concrete manufacture are given in Table 4.5 (see Section 4.3.1). Careful investigations, proof-engineered designs and tests, performance monitoring and progressive evaluation against collapse are basic prerequisites for major projects located in hazardous environments (see Section 15.2.1). Core tests, nondestructive tests or load tests are required on structures containing concrete below the specified strength (AS 1012 and 1480, BS 1881, CP110 Part 1 and CP114, ASTM C42, C174 and C513, also ACI Committee 318, Appendix IV, 2(a)(i)). Certain testing-laboratory temperatures, which are related to a range of operational climatic conditions, are indicated in Section 14.4.1.

In general, local cements of recent manufacture may be taken at the manufacturer's guarantee that they will meet the appropriate specifications (they will usually exceed them), but imported cements should be tested. Standards on cement and aggregate are listed in Tables 3.1 and 3.8. Tests on aggregate are described in Section 3.3.4. Specifications are itemised on sampling, air and cement contents, bleeding, compressive specimen and core strengths, drying shrinkage and moisture movement, flexural and tensile strengths, slump consistence, *bulk density* and the *rapid analysis of fresh concrete*. They are given in Appendixes I and IV, 4(b); and in References 11, 183, 185, 188, 195 and 199.

Bulking of moist sand is dealt with in Section 3.3.6(b), and the free surface moisture of aggregate is determined as described in Section 3.2.3. The air content of air-entrained concrete should be measured by an air meter or by gravimetric procedure, and controlled by adjusting the amount of air-entraining agent. Air meters are obtainable from the suppliers of air-entraining agents and their use is described in AS 1012 Part 4, BS 1881 and ASTM C231. A variety of testing applications and appliances are featured in Plates 11-14, in Reference 92 of Part 1, and in Appendixes I and IV, 4(b).[1]

Test guidelines for assessing and comparing the properties of fibre-reinforced (Section 3.4.2), polymer (Section 21.4) and cement concretes are suggested; those marked with an asterisk are basic to factual determinations. Measurements are made at 28 days, where appropriate. Standard tests may have to be modified or devised to deal with some new varieties of isotropic and anisotropic materials.

Abrasion resistance; absorption of water; and acoustic qualities (Appendixes I and IV, 5).
Deformation characteristics, Poisson's ratio and energy coefficients, based on the complete stress–strain curve to 1 per cent strain in compression (Section 14.9.4 and Appendix I).
Dimensional change of 28-day wet-cured specimens under standard and weather-exposure conditions (Section 31.2.1 and Appendix I).
Durability under sulphate, acidic and freeze–thaw conditions (Section 14.7).

Elastic modulus, including static* and dynamic values (Section 14.9.4 and Appendix I).

Fracture toughness* at the onset of cracking and its change with crack growth (Reference 134 of Part 1).

Flammability resistance and noxious fumes (Sections 15.2.4, 21.1.4 and Appendix I).

Porosity and permeability to water (Section 14.10).

Specific creep* at stresses of 5 MPa and 20 per cent of the compressive strength, applied to 28-day, wet-cured specimens under standard and highly humid conditions (Sections 21.4, 31.3.2 and Appendix I).

Specific gravity and specific heat (Appendix I).

Strength: compressive*, wet and dry, fatigue, flexural, impact and tensile splitting* (Sections 14.4 and 14.8).

Thermal conductivity and expansion coefficient* (Appendixes I, IV, 4(b) and IV, 5).

Work of fracture by tests on notched beams.[202, 203]

14.2 ABRASION AND EROSION TESTS

Abrasion and erosion are dealt with in Chapter 12 and Appendix I. Methods of test in corresponding sequence for comparative purposes are described here.

14.2.1 LOS ANGELES AND DISC ABRADERS

The abrasion resistance of concrete is a function of the wear resistance of the aggregate, as cement paste may have little resistance to abrasive conditions. The wear resistance of an aggregate is determined by a Los Angeles abrasion machine, as described in Section 3.3.4(f)(i), or a machine containing a loaded sample on an abrasive disc in accordance with BS 812.

14.2.2 REAMER

A reamer of 40 mm diameter is ground to a blunt, eccentric chisel face and cross-fluted with three 4-mm indentations. The tool is rotated counterclockwise at 134 rev/min in a modified drill press, under a load of 1·35 kN, the load being checked by a spring balance. The test specimen, 150 mm × 150 mm × 60 mm, is fastened to a turntable which rotates clockwise at 23 rev/min and with an eccentricity of 40 mm in relation to the reamer. Debris is blown off by compressed air.

The depth of the groove which is produced during successive runs of 1 minute is measured with a depth micrometer at four points. The rate of wear is represented by the average increase in depth per minute taken over three runs following the first. The test simulates the type of wear that exists in ore bunkers and chutes. It is applicable to concrete and wood for comparable, practical purposes.

14.2.3 RECIPROCATING ABRADER

This machine is used by Imperial Chemical Industries (ICI) in Britain. A sample of tile is clamped on to a steel plate, which is caused to travel backwards and forwards below a cast-iron hopper freely resting upon it. The hopper is filled with a closely graded sand, which flows out of holes in its base between the tile and the bottom of the hopper. The tile is thus exposed to the abrasive action of fine sand, which is continually being renewed, while the surface of the tile never becomes blinded with material abraded from it. The loss of material from the tile is measured by weighing. When the rate of wear reaches a steady value, a series of weighings are made which show the time (in minutes) required to abrade away a 0·025 mm thickness. This figure represents the "abrasion resistance". The higher the figure, the higher the resistance to wear. The relative life of a material can thus be calculated on the basis of 3 mm thickness. The results obtained are claimed to be in line with service behaviour.

14.2.4 ROLLING STEEL BALLS

Apparatus of the "Davis" (United States) or "Ebener" (Germany) type cause abrasion by rolling steel balls under pressure over a test surface. A concrete surface is prepared for testing by wood floating and light steel trowelling.[100, 101] In the Davis-type apparatus, a 4·5 kN load is applied to a 300 mm diameter, rubber-covered head rotating at 60 rev/min. The head bears on forty-one, 25 mm, steel grinding balls in annular formation over the test surface. The load is applied for 5 minutes and the loss in weight of a dry specimen is determined.

In the Ebener-type apparatus, a loaded grinding head and a turntable rotate in opposite directions. The depth of wear of 28-day specimens is determined from the weight lost in each of five test cycles.

Both types of test simulate practical wear conditions under the rolling, sliding and impact action of steel-wheeled dollies.

14.2.5 ROTARY CUTTER

An abrading cutter consists of a modified drill press in which a load of 80 N is applied to thirty-two, rotating dressing wheels that bear on a dry test specimen. Platform scales are used to calibrate the load (which is applied by a weighted lever) and the positions of individual dressers on the shaft are varied periodically to equalise wear. The driving head is rotated for 5000 revolutions at approximately 200 rev/min.[100, 102] Silicon carbide passing a 0.30-mm sieve is applied to the test surface at the rate of 3 g per minute and the loss in weight per unit area of abraded surface is determined. The average of three tests represents the wear that may be caused by heavy foot or wheeled traffic in service.

14.2.6 SHOTBLAST

In a "Ruemelin" shotblast apparatus, broken steel shot or zirconium oxide (2000 g passing a 1·18-mm sieve and retained on a 0·30-mm sieve) is ejected by compressed air at 620 kPa from a 6·5 mm diameter nozzle. The jet is directed against the surface of a 300 mm × 300 mm × 50 mm concrete test specimen, located 100 mm from the end of the nozzle. The shot is discharged at the rate of 500 g per minute. Eight tests are made at different locations on the specimen and, by weighing it dry before and after the operation, the loss in weight per test is determined (ASTM C418).[100, 102]

The test may be modified in procedure. It simulates the abrasive effect of solids in flowing water and it may be used for testing the abrasion resistance of the mortar matrix of concrete or of a protective film over the surface. Its effectiveness is reduced on a highly resilient surface film, which causes shot to rebound without cutting the surface.

14.2.7 CAVITATION APPLIANCE

Vapour bubbles are formed in water flowing through a slot-shaped Venturi throat and are carried downstream, where an increase in pressure causes them to collapse. Upstream and downstream pressures are adjusted to cause the centre of the area of collapse to coincide with the exposed area (75 mm × 270 mm) of a concrete specimen. Velocity in the throat is about 30 m/s and the period of exposure is 3 hours. From the measured volume of erosion, the erosion resistance is expressed as the number of hours required to erode 2·5 cm^3/cm^2 of exposed surface.[38, 83, 105]

14.3 CEMENT CONTENT OF HARDENED CONCRETE

Methods for determining the cement content of hardened portland cement concrete or mortar are described in published literature, including AS 1012 Part 15, BS 1881 and ASTM C85.[11, 12, 117-119] In a typical analysis the bulk density of an 8-kg or larger concrete sample is determined from its weight in a saturated, surface-dry condition, its apparent loss of weight in water at $23 \pm 2°C$ for volume evaluation and, finally, its weight after drying at $105 \pm 5°C$.

The sample, on being broken into 25 mm pieces, is quartered to at least 1 kg size and disintegrated by heating at 550°C. The process is repeated with very strong concrete, after water-saturating the cooled material. Dismembered particles of aggregate (retained on 4·75-mm and 2·36-mm sieves) are freed from matrix by physical and chemical means. The cement content of 50 g of the matrix, after grinding to pass a 0·075-mm sieve, is estimated by dissolving the cement constituent in hydrochloric acid and measuring the component amounts of calcium oxide and soluble silica.

Due allowance is made in the analysis for the composition of the particular portland cement used (e.g. 64·5 per cent calcium oxide and 21·5 per cent soluble silica; see Section 3.1.1) and a possible portion of these constituents that may be dissolved from aggregate during the testing programme. The estimated cement component, which may vary by up to more than 10 per cent from the actual value, is likely to be more accurate for a rich mix than a lean one.

14.4 COMPRESSIVE, TENSILE, FLEXURAL AND BOND TESTS

14.4.1 COMPRESSION

Specimens of properly mixed and sampled concrete should be fully compacted, transported, capped, cured and strength tested in accordance with AS 1012 (Parts 1 and 8-11), 1379 and 1480; BS 1881, 1926 and CP110 Part 1; ASTM C31, C39, C42, C92, C116, C172, C192, C330, C495 and C617.[1, 153] Slump tests (see Section 6.2.1) are made simultaneously on concrete sampled for strength tests (see Sections 7.1.3 and 7.1.5). Standard concrete cylinders (150 mm diameter × 300 mm height for aggregate up to 40 mm size) are fully consolidated, by evenly distributed rodding, in each of three equal uniform layers.

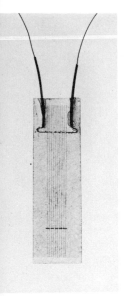

Electrical-resistance
wire strain gauge

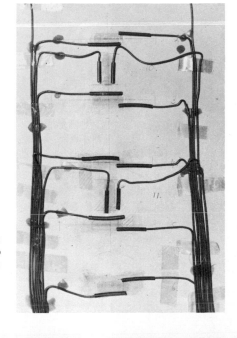

Similar gauges
attached to
test specimen

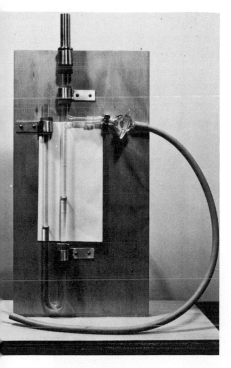

"Blaine"
air-permeability apparatus

Covermeter

Compaction
of specimen
with vibrating table

Compaction
of specimen
with tamping rod

Compaction
of specimen
with internal vibrator

Testing machine (1000 kN)
with concrete core and compressometer

For slumps of 5-25 mm, 30-50 mm, 55-75 mm and over 75 mm, the minimum number of strokes per layer (using a 600 mm × 15 mm diameter, round-ended rod) is typically forty, thirty-five, thirty, and twenty-five for each successive consistence. Holes remaining in the surface of each layer are closed by lightly tapping the sides of moulds. No-slump concrete can be vibratory compacted, in each of two layers, without causing segregation or excessive laitance.

The vibrational frequencies of internal and external vibrators are at least 100 Hz and 50 Hz, respectively, the diameter of the former being from 15 mm to one-fifth the least dimension of the moulds. Cube specimens (100 mm or 150 mm size for aggregate up to 20 mm and 40 mm, respectively) are fully compacted in each of two layers, either by vibration or manually with a 375 mm, 1·8-kg steel bar having a ramming face 25 mm square. The minimum number of strokes per layer, which varies with the type of concrete, is twenty-five for 100 mm cubes and thirty-five for 150 mm cubes, or otherwise as required to achieve full compaction (Plate 12).

Tag-identified specimens are stored initially in cover-plated moulds for 18-72 hours. In a tropical zone (up to 25° latitude), the testing-laboratory temperature is kept 4°C higher than 23 ± 2°C (which is applicable to a temperate zone). In temperate to cooler regions, the ambient temperature used for testing purposes is typically 20 ± 2°C. Specimens made in the field for 28-day strength tests are stored initially at 23 ± 10°C in a temperate zone and 27 ± 8°C in a tropical zone, the temperature tolerances for 7-day tests being halved. Specimen-making records should include items on the zone, compaction procedure, time on site and the measured maximum and minimum temperatures.

Moist curing at standard temperature is done in lime-saturated water or a fog room held at almost 100 per cent relative humidity. This is checked periodically by a sling psychrometer, and the method and duration of standard moist curing are carefully recorded. The moulded and capped (or ground) bearing surfaces of compression specimens should be plane to within 0·05 mm and squarely formed to within 0·5°.

For compressive strengths up to 50 MPa, the specimens are thinly capped (on a manually prepared test end) at least 1 hour before being tested, using a 3·5 ± 0·25 sulphur/1·0 silica (0.075 mm) ratio mix by weight at 165 ± 10°C. Special high-strength gypsum plaster paste, if used, should have a 2-hour compressive strength in excess of 30 MPa. Alternative caps are high-alumina or portland cement paste or mortar (1:1 mix), used after a 2-3 hour delay to minimise shrinkage. These (taken in sequence) are moist cured for 18-24 hours and 2-3 days, respectively.

For compressive strengths of 50-80 MPa, the specimens are sulphur capped at least 2 hours before being tested, the cap having a 50 mm cube strength in excess of 35 MPa. Portland cement mortar for the purpose is moist cured for 6 days. For strengths in excess of 80 MPa, the cube strength of the mortar must be in excess of 50 MPa. Laboratory cylinders, 2-4 hours after moulding, may be capped with cement mortar.

Cardboard moulds do not fulfil standard test requirements, and tend to give results 6-12 per cent lower than those obtained with steel moulds. This discrepancy is due to a possible movement of cardboard moulds containing unhardened concrete, and to variations in compaction and the conditions of early moist curing (ASTM C470).[85] Compressive-test results can be affected by specimen imperfections and nonstandard states of the test heads, and by friction between the loading surfaces of specimens. Capping effects are reported in References 86, 87 and 132.

The compressive strength of concrete in a structure, calculated to the nearest 0·5 MPa, is considered generally to be more truly represented by the apparent strength established by standard cylinders than by cubes. Nondestructive tests are dealt with in Section 14.9, and *in situ,* push-out cylinders (plastic inserts) or drilled cores (Appendix I) can be used for check-testing at the standard loading rate of 20 ± 2 MPa/minute.[2, 36, 75] Standards on Young's Modulus of Elasticity and bulk density determinations are listed in Appendix I. Items for sampling, moulding and test records and reports are scheduled in standard specifications, and specimen conversion data are illustrated in Section 4.4.

Test cylinders of no-fines concrete are consolidated in three layers, each being tamped eight times with a no-fines special tamper, as illustrated in CSA Standard A23.4. This appliance consists of a 2·5-kg rammer head and shaft, falling 305 mm within a vented, 52 mm diameter guide tube and striking a 149 mm diameter bearing plate, 5 mm thick. The total weight of the unit is 4·8 kg. Concrete block masonry is dealt with in Section 27·2.

14.4.2 TENSION, FLEXURE AND BOND

Tensile strength is commonly determined by a split test; that is, by loading cylinders (150 mm diameter × 300 mm length) on their sides along their full length. Unused, tempered-hardboard pads (330 mm × 25 mm × 5 mm) are placed on the bearing surfaces of specimens, which are loaded symmetrically thereafter at the rate of 100 ± 10 kN/minute or $1 \cdot 5 \pm 0 \cdot 15$ MPa/minute. The tensile splitting strength (MPa) is the breaking load (kN) divided by $22 \cdot 5\pi$ (Appendix I).[88-91, 113, 140] It is usually about one-tenth of the compressive strength and two-thirds of the flexural strength of a 150 mm $\times$ 150 mm beam specimen at least 500 mm long.

Direct tests are given in References 108 and 182. Means of verifying testing machines are specified in AS B128, BS 1610, ASTM E4, E74 and E83. Tests for tensile and flexural strengths, modulus of rupture and bond strength (AS 1012 and 1480, BS 1881 and CP114, ASTM C78, C234 and C293) are given in Appendix I and References 112 and 134. The split-cube tensile strength (MPa, 150 mm cube), if tested diagonally, is 0·2234 times the breaking load (kN).[92, 130] Concrete-ring tensile tests are documented by Malhotra in Appendix IV, 4(b). Machines for cylinder splitting tests must be designed for the purpose.[88]

For ordinary and lightweight concretes, the bond strength of plain steel reinforcement in flexural members may be one-third to one-half less than the equivalent strength derived from pull-out tests. This result is due to a very shallow cavitation that may form beneath horizontal steel reinforcement, through bleeding and settlement of the concrete. Cavitation can be reduced by attention to certain considerations given in Section 15.2.2(d)(vi) and by effective site control (ASTM C232 and C243, also Appendix II).[1] Further details are given in Section 33.3 and Appendix IV, 4(b).

14.5 CURING EFFICIENCY TEST

The efficiency index of membrane curing media (see Section 9.3) may be determined from flexure tests on beams (150 mm $\times$ 150 mm $\times$ 700 mm), using the expression

$$\frac{M - A}{W - A} \times 100 \text{ per cent}$$

and the average modulus of rupture of sets of specimens which have been cured under the following conditions: membrane M, water W and air A. The top edges of the specimens for conditions M and A are sealed to the mould with a beading of resinous material (ASTM C156 and C309), Faraday wax (AS 1757 and 1759) or resinous-wax mixture (see Section 3.3.4(e)(i)). Other tests are given in Reference 46.

Membrane-cured and air-cured specimens are stored in their moulds in a hot room ($50 \pm 2°C$ and 17 per cent relative humidity) for 26 days. They are then demoulded, immersed in water at $23 \pm 2°C$ and tested at an age of 28 days, the original top surface being tested in tension. Water-cured specimens are covered with damp hessian (which is kept moist by capillary action) immediately after initial set of the cement. They are demoulded after 18 hours, placed in water in the hot room until aged 26 days, and then immersed in water at $23 \pm 2°C$ until tested at an age of 28 days (AS 1012 Part 2 and Part 11, BS 1881; and Section 9.3).

The efficiency index of curing media may be graded as follows.

Grade	Index (per cent)
Excellent	60 or more
Good	40-60
Fair	30-40
Poor	Below 30

Other tests include ASTM C156 on water-retention efficiency (Appendix I) and the procedures described in Reference 46.

14.6 DRYNESS TEST ON MASONRY

An apparatus is available for measuring the moisture condition of a block in terms of relative humidity.[37] A representative sample block is broken into halves, one of which is broken by a hammer and cold chisel into dust-free lumps, 19·0-mm to 53-mm sieve size. The lumps are promptly placed in an expanded metal pail, which is sealed in a metal container with a quick-closing lid that is provided with blower, motor and hygrometer attachments.

The blower is operated until an approximate balance is obtained between the moisture in the concrete and the relative humidity of the air in the container. Direct hygrometer readings are corrected by means of a calibration curve. This curve is based on periodic tests of the hygrometer against known relative humidities, which are readily established within the container by five saturated-salt solutions. From a series of corrected test readings at intervals, a humidity-time curve is drawn to show the trend towards a final value of relative humidity. This value, which is obtained within 0·5 hour, indicates the moisture condition of the block and the average relative humidity to which the group that it represents may be exposed satisfactorily in service.

The test determines when various lots of blocks are dry enough, or what additional drying is required, to suit a specific use with dimensional stability. It indicates when blocks are at a suitable moisture condition for the application of glazes or other laminates. If a sample block is to be tested at a laboratory, it can be stored temporarily in a closed metal container to ensure constant moisture content (see Sections 27.2.1, 27.2.2, 33.3.2(b) and 33.3.3(b)).

14.7 DURABILITY TESTS

14.7.1 ALKALI–AGGREGATE REACTION— RAPID TEST

Mortar or concrete made with the aggregate under test and a high-alkali cement is sealed in three 0·5-litre fruit-preserving jars, which are stored in an inverted position after their contents have hardened. Before the test, the jars are coated with chlorinated rubber lacquer to prevent reaction between cement and the glass. Porous coarse aggregate is saturated with water prior to mixing. For 1 : 2 mortar, a water/cement ratio of 0·5 by weight is suitable. With concrete mixes, the water should be sufficient to make a plastic mix. The bottom of each jar is struck several times with the palm of the hand to settle the contents. At an age of 20-24 hours, about 25 ml of water is added to each jar and the cap is replaced firmly. If the jars are stored at an elevated temperature (e.g. 50°C), they should be placed in this storage before their contents have hardened. The jars are examined daily for about a month or until multiple cracking occurs.

The test is useful for the rapid determination in the field of the susceptibility of aggregate to alkali and the compatibility of given combinations of cement and aggregate. The development in 28 days of nothing more than a simple crack indicates very small expansion, which may be regarded as immaterial. Multiple cracking indicates an aggregate of high susceptibility to reaction with the alkali in cement (see Sections 3.3.4(e)(i) and 15.2.3(a)).

14.7.2 CHEMICAL TESTS

Specimens of cement mortar or concrete are subjected to fully or partially immersed storage or flow-over-the-surface tests for certain periods. At intervals between tests, the surfaces are wire-brushed to remove loosened or softened material. These tests should be programmed so as to represent, under intensified conditions, the particular or anticipated type of hazard that must be resisted in service.

The test results are gauged better from the average depth of scaling of specimens than from their recorded loss in weight as a function of time, because of the different rates of change in their dimension and volume through progressive deterioration under given aggressive conditions. In long-term (e.g. 10 years) sulphate-resistance tests, the time required to develop an expansion of 0·1 per cent correlates closely with an ultimate resistance performance of the concrete. Sample specimens are immersed in fresh water for carrying out comparative crushing, flexure and nondestructive tests, as well as measurements of dimensional change. The potential expansion of portland cement mortars exposed to sulphate is determinable by the test method of ASTM C452 (see Section 3.1.2(e)(viii)).

Tests for chemical resistance may be carried out as indicated in ASTM C267. Aggressive liquids for the purpose may consist of reagent solutions of 5 per cent concentration (maintained to within ±0·25 per cent) and held at the approximate average service temperature to within ±6°C. The chemicals used are magnesium or ammonium sulphate, ammonium chloride and sulphuric, lactic and acetic acids. Specimens of different size and shape may be used, provided that the volume of reagent is at least five times their volume (Plate 14).

In an accelerated sulphate test (carried out by the United States Bureau of Reclamation), specimens of 75 mm diameter × 150 mm length are alternately soaked in a 2·1 per cent solution of sodium sulphate at about 23°C for 16 hours and then dried in air at 55°C for 8 hours. The criterion of failure, by internal disruption, is a loss of 15 per cent in dynamic modulus of elasticity or an expansion of 0·2 per cent. In each of these tests, complete failure is indicated by a 40 per cent reduction in dynamic modulus or an expansion of 0·5 per cent. If dilute hydrochloric acid is added to a filtered water extract of corrosion products and then barium chloride solution is added, the formation of a white precipitate would indicate sulphate was present.

When testing asbestos cement products, specimens (each about 100 cm² total area) are immersed vertically for 24 hours in 270 ml of 5 per cent acetic acid solution at about 23°C. A separate solution and container are used for each test piece. The acid is titrated (before and after the test) against a 0·5N sodium hydroxide solution, using thymol blue as an indicator. The amount of acetic acid neutralised per cm² of specimen is calculated as described in AS A41. While the maximum allowable amount is 0·100 g/cm², the production figure for autoclaved asbestos cement pipe is usually no more than one-half this amount.

In the absence of long-term exposure tests, accelerated tests that suitably represent the nature of attack give a significant indication of the degree of corrosion and the degree of resistance that may be expected. There are certain exposures, however, to which concrete is inherently unresistant, and other materials of proven resistance should be used in these circumstances (see Section 15.2.3(b), (d)).

14.7.3 FREEZE–THAW TESTS

In Europe and North America, the temperature during winter months may stay below freezing point for long periods, but in the Snowy Mountains (Australia) the effect of frost action upon exposed concrete surfaces can be more severe. Here, it is usual to experience severe frosts at night and temperatures above freezing point during most days. Under these conditions, durability is more important than strength in the design of concrete mixes (see Section 15.2.3(c)).

Consequently, the Snowy Mountains Hydro-Electric Authority subjects cylinders (of 75 mm diameter and 150 mm length) to freeze-thaw cycles, in order to determine the durability of concrete under conditions somewhat related to those in service. Specimens made with aggregate passing 19·0-mm mesh are cured for 28 days at 23°C and 100 per cent relative humidity. They are then subjected to temperature cycles from $-12°C$ to $+23°C$ and back to $-12°C$ three times in each 24 hours. Ordinary concrete, without air-entrainment, disintegrates or loses 25 per cent of its original weight after 60-120 cycles. Well-designed, air-entrained concrete does not disintegrate, but frets at the surface. Durability is measured by the number of cycles that cause a 25 per cent reduction in weight, and it may be conveniently classified into the categories shown in Table 14.1.

TABLE 14.1 DURABILITY

Number of Cycles	Classification
Less than 100	Poor
100-300	Fair
300-500	Satisfactory
500-1000	Good
Greater than 1000	Excellent

Rapid freezing tests in water or air and thawing in water are described in ASTM Standards C290 and C291.

14.7.4 WEATHERING TESTS

Weatherometer and wetting and drying tests are used to accelerate mechanically the deterioration of surfacing materials and concrete. Typical procedures are described here.

(a) Weatherometer.

Specimens are rotated in an atmosphere heated to 60°C by arc lamps or a combination of infrared and ultraviolet light. Cold water is sprayed on the specimens (surface temperature up to 90°C) for 9 minutes in a 1-hourly cycle or 18 minutes in a 2-hourly cycle. Artificial weathering for 500-5000 cycles, varying from 10 to 95 per cent relative humidity, is commonly used in these tests.

(b) Wetting and drying.

A daily cycle of operation consists of drying 75 mm × 75 mm × 375 mm specimens in a cabinet for 8 hours at 60 ± 2°C, for example, under forced draught. The air is changed at frequent intervals and is directed to each specimen by means of ducts and baffles. The specimens are subsequently soaked in water for 16 hours.

Water at 23 ± 2°C enters the cabinet and is subsequently brought back to this temperature by the circulation of cool air over the surface. The soaking water is reused for a period of 30 days in order to minimise the effect of leaching. A testing cycle usually falls within this period. The test not only enables a good estimate to be made of the effects of weathering, but it accelerates the reaction (if any) between alkalies present in portland cement and reactive aggregate (see Section 15.2.3(g)).

14.7.5 WEATHERPROOFNESS

For weatherproofness or weather-penetration investigations, a large specimen-panel of walling (e.g. up to 6 m high and 5 m wide) is subjected to simulated storm conditions, which represent high-speed wind gusts and heavy rain. The slip-stream from the airscrew of an obsolete fighter aircraft capable of developing about 2 MW is kept steady at speeds of from 30 m/s to 45 m/s over areas of 9 m² or so for several minutes. Watersprays at the rear of the tail, which is ahead of the test specimen, simulate rain at up to 500 mm/hour. A dynamic weather-testing rig of this type is established at the Australian Experimental Building Station, Chatswood, New South Wales.

Alternative equipment for weatherproof testing, known as the "Sirowet" exposure-test rig, has been devised by the CSIRO Division of Building Research, Highett, Victoria. It is made up of reinforced polyester modules (each of 1 m²) which are bolted together to form a box about 7 m wide and 5 m high. On tightly clamping the box to a jointed building-facade mock-up, the surface under test is drenched by heavy water sprays that are fitted at the centre of each module.

Alternate pressure and suction conditions up to 3 kPa are produced in order to simulate wind gusts of up to 250 km/hour. These conditions are created by connecting the box through a flexible coupling to a 45 cm diameter fan driven by a 7 kW motor. Air leakage and water penetration tests, in addition to deflection deformation measurements, are instrumented to meet the requirements of ASTM Standards E283 and E331, and Loading Codes covering wind force referred to in Table 3.25. Other test procedures are presented in BS 4315 and Reference 127.[54]

14.8 FATIGUE AND IMPACT RESISTANCE

14.8.1 FATIGUE

Typical testing machines are the "Krouse-Purdue" and "Amsler" fatigue machines, which apply a pulsating load to cylinders (of 75 mm diameter and 150 mm height) by a hydraulic system acting on a piston without impact. They have capacities of 270 kN and 490 kN and can operate at 1000 and 500 cycles/minute, respectively. Ordinary and high-strength concrete specimens can endure more than 10^7 cycles at a 40 per cent stress level, some specimens can endure up to 5×10^3 cycles at an 80 per cent stress level, without failure.[123, 177, 208]

14.8.2 IMPACT

The impact resistance of concrete may be measured by the energy that is absorbed by 100 mm concrete cubes, when they are subjected to repeated blows from a ballistic pendulum. When impact is applied on a small, localised area, the impact resistance increases with the angularity and roughness of the surface of aggregate in the concrete. The increase is probably due to increased bond between coarse aggregate and the mortar matrix. Under impact loading air-cured concrete cubes crack more readily than water-cured cubes at an age of 28 days.[162]

The compressive strength, elastic modulus and ability of concrete to absorb strain energy under impact loading are significantly increased as the duration of impact is decreased. Work done at the United States National Bureau of Standards shows that the ratio of dynamic to static compressive strength for low-strength concrete (17 MPa static strength) ranges from 1·09 for a 0·9-second impact to 1·84 for a 0·00025-second impact. The ratio for high-strength concrete (45 MPa static strength) ranges from 1·13 for the former impact to 1·85 for a 0·00043-second impact. The ratio of dynamic to static modulus of elasticity increases, with decreasing duration of impact, to 1·47 for low-strength and 1·33 for high-strength concrete. The linear portions of the stress–strain curves become longer and steeper as the duration of the impact is decreased. The ratio of strain energy absorbed under dynamic loading to that absorbed under static loading reaches a maximum of about 2·2 for both types of concrete.

Such data are pertinent to the design of concrete pavements and structures that may be subjected to falling heavy objects, earthquakes, bomb-blasts or similar forms of impact loading.

14.9 NONDESTRUCTIVE TESTS

14.9.1 GAMMA-RAY AND NEUTRON TESTS

The density of concrete can be related to its absorption of gamma rays, such as those from cobalt 60. No unique relationship appears to exist between the absorption of gamma rays and compressive strength; other than that as the strength of concrete can be related to its density, so can its strength be related to its gamma-ray absorption.[178] This method may be used to determine the presence of water lenses or air voids (down to 5 mm diameter at depths up to 450 mm) in grouted ducts, the compaction of concrete and soils, and the soundness of welds in steelwork (BS 4408 Part 3 and Section 23.1.3).

The water content of concrete, sand and soil may be determined by means of neutrons. Fast or high-energy neutrons travelling in a material are converted into slow neutrons on impact with molecules of water. The proportion thus converted is a measure of moisture content.[134] Detected changes in moisture content can be used for estimating the carbonated depth of concrete. Moisture-measuring apparatus is referred to in Sections 3.2.3 and 20.2 (BS 4408).

14.9.2 HARDNESS TEST

Hardness testers (e.g. "Schmidt" spring hammer[115] and "Windsor" explosive penetration probe[176]) are much simpler and cheaper than equipment used in other forms of nondestructive tests for estimating the compressive strength of concrete.

For the spring hammer, a flat surface 125-150 mm square is smoothed with a carborundum stone. Readings are taken at six to ten points, about 25 mm apart, over the selected area. Strength results are estimated on the basis that the quality of concrete depends upon the hardness of the mortar matrix at the surface. They are affected, however, by the type and quantity of aggregate, smoothness of surface and method of use of the instrument. A coefficient of variation of 20 per cent is common (BS 1881 and 4408 Part 4).

A "Schmidt" spring hammer (Plate 13) is a quick, subsidiary means for determining the uniformity of concrete and its strength for removal of formwork or the release of stress in the tendons of prestressed concrete. Strength results are related to a rebound number on a calibration chart.[115, 198]

14.9.3 LOCATION OF REINFORCEMENT

The position of reinforcement in concrete can be determined by radar, radiography or low-cost electromagnetic instruments designated as "Covermeter" and "Pachometer" (Plate 11). The operation of the electromagnetic equipment is based upon the effect of steel, at depths of up to about 70 mm, on the field of an electromagnet (BS 1881 and 4408 Part 1).[116, 134, 135]

14.9.4 SONIC AND ULTRASONIC

(a) Resonance methods.

Sonic measurements are made of the fundamental or resonant frequency of vibration of a test specimen which has a simple geometric shape. A vibration exciter, which is fed from a variable-frequency oscillator, is used to drive the supported specimen. The resultant vibrations, either transverse or longitudinal (depending on the procedure adopted), are observed by means of a vibration pick-up and measuring amplifier. The frequency of the oscillator is varied until the amplitude of the vibrations reaches a maximum. The fundamental or resonant frequency is the lowest frequency at which this takes place. The dynamic Young's Modulus (E_d) of the concrete is calculated from this frequency, the dimensions of the specimen and the density of the concrete (BS 1881 Part 5 and ASTM C215).

The longitudinal resonant frequency of a specimen is many times higher than its transverse frequency. In somewhat fresh concrete, longitudinal vibrations are difficult to produce because of considerable damping. By similar procedures, the shear modulus *(G)* of concrete can be determined from the torsional resonant frequency of a specimen. Poisson's ratio (σ), which lies between one-third for early-age and one-sixth for mature concrete (Appendix I; Glossary, Appendix VI),[125, 181] may then be found from the formula

$$\sigma = \frac{E_d}{2G} - 1$$

(b) Pulse methods.

The velocity of propagation of sound through concrete is measured. A pulse of vibrations of ultrasonic frequency is sent from a piezoelectric crystal transducer held in contact with one face of the concrete being tested. The pulse is received after passing through the concrete by a crystal pick-up, which converts the mechanical pulse into a corresponding electrical signal. This signal is amplified and the time of transmission of the pulse is ascertained electronically. This can be measured to one-millionth of a second. A cathode-ray oscilloscope (with provision for the direct measurement of transit time) displays the transmitted and received signals. Knowing the distance traversed, the velocity of propagation of the pulse may be calculated. This is of the order of 3000-5000 m/s (BS 4408 Part 5, ASTM C597 and E317).[16, 103, 104, 134, 198]

Dynamic Young's Modulus (E_d) is calculated from the velocity of the pulse V by means of the formula

$$E_d = V^2 \rho \, \frac{(1 + \sigma) \, (1 - 2\sigma)}{(1 - \sigma)}$$

where ρ = density, and σ = Poisson's ratio.

From calculated values of E_d (kN/mm^2), the static modulus of elasticity *(E_s)* can be estimated from $1\cdot25 \, E_d - 19$ for normal aggregate concrete, or from $1\cdot04 \, E_d - 4$ for structural lightweight-aggregate concrete. The results thus obtained are correct generally to within $\pm \, 4 \times 10^3$ MPa (see Sections 31.3.2, 33.3.2(a) and 34.1.4(c); also BS CP110, Part 1 and Reference 148 of Part 7). The value of Poisson's ratio can be calculated from the resonant frequency of a specimen for one type of vibration and the velocity of propagation of another type, or from the velocity of propagation of two types of vibration. At 28 days, the static Poisson's ratio of lightweight concrete for design purposes may be taken as one-fifth.

Sonic and ultrasonic measurements can be made with an accuracy of approximately 1 per cent when tests are carried out under controlled conditions. The accuracy of the pulse method depends, however, upon the received signal being similar to the zero signal. If the concrete is of poor quality, this may not be the case. When an electrically vibrated mechanical hammer is used to produce the pulse, the force of the blow has an effect on the velocity of propagation. The surfaces require special preparation.

A formula for the relationship between E_d and compressive strength is of the form

$$E_d = K^n \sqrt{R}$$

where R = compressive strength, and K and n are constants.

In relating the measured pulse velocity to compressive strength, it should be noted that concretes which have the same compressive strength do not always propagate an ultrasonic pulse with the same velocity. The type of aggregate and the ratio of aggregate to cement, for example, affect the velocity. When using these methods, the assumption that concrete is isotropic is not necessarily true. Elastic modulus and maturity relationships are derived in Reference 159, safety requirements for electronic apparatus are given in BS 415, and relevant illustrations are given in Plate 13.

(c) Scope of these techniques.
The resonance method is being used extensively in many laboratories and has already taken its place in standard specifications for determining the dynamic modulus of elasticity of concrete. Its main advantages over the pulse method are

Simpler and lower-cost equipment.

It can be used on low-quality concrete, where the results from the pulse method become inaccurate due to large loss of energy within the material.

An average value is obtained, which is usually preferable to the more detailed information given by the pulse method.

The main disadvantage of the resonance method is that it is only practicable to use it for testing fairly small, simply shaped specimens. Young's Modulus of elasticity, derived from resonance tests, is higher than that obtained from conventional stress–strain measurements with progressive loading. The moduli thus obtained are therefore referred to as dynamic and static, respectively.

For a *particular* concrete (i.e. one of the same mix proportions and with the same type of aggregate), there is a direct relationship between compressive strength and longitudinal wave velocity, the higher the velocity the higher the strength. Variation in the quality of concrete in a structure can thus be ascertained. This is an important development for controlling the quality of concrete.

Where arrangements are made to obtain a relationship between strength and wave velocity, the second method can be used to ascertain the strength of concrete as it exists in a structure. This technique is the best yet devised for this purpose and enables the structural designer to think of concrete, not solely in terms of the compressive strength of cylinders or cubes (the significance of which is in doubt), but also in terms of ultrasonic wave velocity or Young's Modulus. For instance, velocities of 5000 m/s indicate good-quality concrete, whereas velocities below 3000 m/s reveal an inferior grade.[16, 215]

Measurements of velocity made over a large surface (such as a floor or wall slab, or column) show the presence of poor areas in relation to those of good quality. As a structure dries, there is a decrease in pulse velocity relative to comparable specimens cured under water. A small correction is usually made, therefore, before interpreting results in terms of strength.[204]

The wave-velocity technique can be used to indicate when formwork may be stripped safely, whether a precast structural member can be hoisted into place, or whether a member may be stressed in prestressed construction work. The test in each case is carried out on the actual member and not on a supposedly representative sample. By using pulse-echo technique on a slab, its thickness may be deduced to within 5 per cent.[139, 160] It may be used to investigate whether concrete has been well compacted and to determine the effects of attack by aggressive waters and by fire.

The presence of cracks in concrete can be detected by the pulse method. Cracks change the transmission time and increase the attenuation of the transmitted signal. Not only the size of cracks, but also their change with time can be measured.[119] The presence of moderate steel reinforcement does not prevent the successful application of the technique, which can be used for testing not only cement concrete, but other materials such as soils, ceramics, stone and bituminous concrete. It can be used for detecting defective areas of rock in a quarry. The nature of the rock formation encountered during tunnelling operations can be investigated, the pulse for this purpose being produced by seismic means.

"Schmidt" spring hammer test

Dynamic fracture of high-strength
calcium silicate hydrate

Ultrasonic test

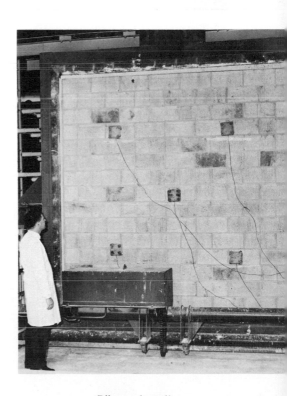

Pilot-scale wall
for fire resistance test

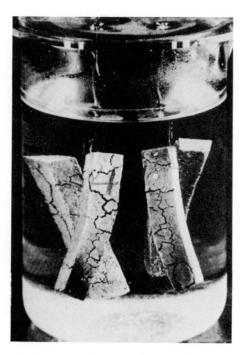

Test results of 1:3 mortar specimens, after being immersed in 10 per cent sodium sulphate solution for six months

Ordinary portland cement

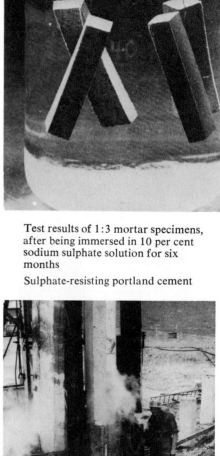

Test results of 1:3 mortar specimens, after being immersed in 10 per cent sodium sulphate solution for six months

Sulphate-resisting portland cement

Splicing concrete piles with thermally cured encased epoxy resin

(Dry process)

14.10 PERMEABILITY AND ABSORPTION TESTS

14.10.1 PERMEABILITY

Permeability tests are described in AS 1757 and 1759, or BS 473 and 550 for concrete roofing tiles, and in AS 1500 (Japanese Industrial Standard A5406) for hollow concrete blocks (see Section 27.2.1). The permeability of concrete is measured by the amount of water that passes through specimens, by virtue of percolation under pressure, also by diffusion as a vapour under differential pressure, and by capillarity and evaporation (see References 117, 140 and 185 of Part 4).[124]

The pressure type of permeability test is most commonly used for concrete, which should be cast and tested in such a way as to simulate the respective directional conditions in a project. A cylindrical specimen is sealed at its curved surface inside a suitable metal container, so that water under pressure may be applied to the top flat surface only. A hand-operated pump or a vertical water-pipe reservoir with a long water column, air-bleeder valve and pressure gauge is used for this purpose.[124] The water column is connected to the top of the container and compressed air is used to maintain a constant pressure on the column.[134] Air is evacuated initially from the system, and the percolating water which leaks through the bottom surface of the specimen is protected against variations in humidity of the air. A rain gauge may be used to collect the effluent from the free surface and prevent evaporation losses.

The outflow is less than the inflow during the early part of any test, because of absorption, but eventually they will agree. As water under compressed air readily absorbs some air, which would be released under reduced pressure inside a specimen and retard the flow, provision should be made to replace the column of water at intervals. This is done when the air content of water above the specimen reaches about 0·2 per cent, so as to prevent the release of air within the concrete.

A coefficient of permeability K_c can be determined from the formula

$$K_c = \frac{Q}{A} \times \frac{L}{H}$$

where $\frac{Q}{A}$ = rate of flow in m³/s per m² of area of cross-section under pressure and standard temperature conditions (20°C), and $\frac{L}{H} =$ ratio of percolation length to head of fluid.

In a water-permeability test rig for masonry blocks (AS 1500 and JIS A5406), a rigid metal cover is fitted with a transparent, 60 mm bore, graduated cylinder. It is clamped with six locking bolts and a sponge-rubber seal (10 mm thick) onto one face of a test specimen, which has been immersed in water at 23 ± 2°C for a period of 2 hours.

In a draught-free atmosphere, an initial test head is established 1 minute after water is first applied to the test face. The residual head is read, timed and recorded at intervals of up to 2 hours or until the water head falls below 100 mm. An initial head of 200 mm should not fall below 100 mm in 2 hours. Quicker control results are obtainable, for graphical presentation, by measuring a 100 mm loss of head from an initial head of 300 or 400 mm.

14.10.2 ABSORPTION

Absorption tests and related considerations are described in Sections 3.3.4(j), 27.2.1, 33.3.2(e) and 33.3.3(e); also in various standards, such as AS 1141, 1342, 1346, 1467 and 1500, BS 1881, ASTM C642 and those listed in Table 14.2 and Appendix I.

TABLE 14.2   1-DAY ABSORPTION TESTS

Product	Australian Standard Specification	Value (per cent)	British Standard Specification	Value (per cent)
Asbestos-cement decking	—	—	3717	28
Asbestos-cement pipes (pressure, sewerage and drainage)	1711	20	486; 3656	20
Asbestos-cement flue pipes	—	—	567	30
Eaves gutters	—	—	2908	7
Flags	—	—	368	6·5
Pipes and culverts	1342; 1392 1597	6·5-8	556; 4625	6·5
Tiles (roofing)	1757; 1759	10	473; 550	—

In the testing of precast concrete kerbs, channels and edgings, five units are taken from every thousand, and groups comprising three test pieces are taken from each of them. The specimens are approximately 100 mm cubes for kerbs and channels, and 100 mm

× 100 mm × 50 mm for edgings. Specimens have two moulded faces of 100 mm × 100 mm and four faces cut by a hammer and chisel (AS 1226 and A175, BS 340).

The specimens are dried in a ventilated oven at 105 ± 5°C for 72 hours, cooled in a desiccator for 24 hours and weighed. After being immersed in water at about 23°C for 10 minutes ± 10 seconds, they are removed, shaken and wiped to remove free water, and reweighed. For resistance to frost action, the average increase in weight of each group should not exceed 2·5 per cent at an age of 28 days.[121]

The 10-minute absorption of concrete kerbs that are pressed hydraulically at not less than 7·0 MPa is generally below 2·0 per cent. For durability, also, the aggregate crushing value of coarse aggregate and the sum of its flakiness and elongation indexes should comply with the recommendations given in Section 3.3.4(g). Should two or more groups fail in a test, the consignment that they represent should be rejected. If one fails, a further five should be tested, and if all groups then meet the requirement, the consignment is accepted.

Absorption tests, where required on concrete pipes, are supplementary to ultimate-load and hydrostatic-pressure tests. They are conducted on cut specimens, $1·0 \times 10^4$-$1·5 \times 10^4$ mm^2 in size. These are dried at 105 ± 5°C for 4 or more days to constant weight, boiled for 5 hours in water, cooled while immersed to 20 ± 5°C, drained for up to 1 minute and then wiped free of superficial water. The percentage absorption of the concrete (allowing for steel reinforcement present) is then determined gravimetrically.

14.11 PROCTOR HARDENING TEST

For estimating the hardening characteristics of concrete, mortar is sieved from the concrete through 4·75 mm mesh (see Table 3.21) and cast into two specimens (150 mm diameter and 150 mm height). These are vibrated for 5 seconds, covered and stored at a temperature which simulates that of the place where the concrete would be placed (see Section 14.4.1; AS 1012 Part 18 and ASTM C403).

Penetration tests are made with a standard Proctor needle at 0·5–2-hour intervals. On removing bleeding water, the mortar is penetrated at a uniform rate to a depth of 25 mm in 10 seconds, and the resistance to penetration is measured. The specimens are used alternately for successive penetrations, which are distributed symmetrically at 50 mm centres over the surface.

Proctor penetration needles are of 28, 20, 14, 9, 6·5 and 4·5 mm diameter size. The largest needle is used initially until the penetration resistance reaches the maximum pressure of the spring stock of the apparatus. Smaller needles are used as the mortar continues to stiffen and a record is kept of the time, size of needle and applied pressure for each penetration. For convenience in testing, a lever and a rack-and-pinion assembly may be used in conjunction with a platform scale.[69]

In conjunction with these tests, the compressive strength of similarly stored specimens of concrete is determined. Firstly, when the penetration resistance of the mortar reaches 28·0 MPa (Fig. 10.2). Secondly, at an age of 48 hours. The behaviour of a sample of concrete, 15 litres or more in volume, is observed at the same time. A pocket penetrometer is available for estimating the state of hardness of placed concrete, beyond which it should not be disturbed (see Section 10.2).[111]

REFERENCES

1. UNITED STATES BUREAU OF RECLAMATION. *Concrete Manual,* United States Government Printing Office, Washington, 1975.
2. BLOEM, D. L. "Concrete Strength in Structures". *ACI Journal,* 65, 3 (March 1968), pp. 176-87; *Concrete Products,* 72, 1 (January 1969), pp. 46-50.
3. MATTISON, E. N. "Effect of Extended Mixing and Agitating Time on Some Properties of Concrete". *Constructional Review,* 40, 11 (November 1967), pp. 20-28.
4. ACI COMMITTEE 613. "Recommended Practice for Selecting Proportions for Concrete". *ACI Journal,* 26, 1 (September 1954), pp. 49-64; 66, 8 (August 1969), pp. 612-28; 67, 2 (February 1970), pp. 192-94.
5. ACI COMMITTEE 304. "Measuring, Mixing, Transporting and Placing Concrete". *ACI Journal,* 69, 7 (July 1972), pp. 374-414; 70, 1 (January 1973), pp. 55-60.
6. CHOKSHI, C. K. "The Transport and Placement of Concrete". *Indian Concrete Journal,* 33, 9 (September 1959), pp. 316-21, 325.
7. DAFFY, J. "Concrete Placing". *Contracting and Construction Engineer,* 27, 6 (June 1973), pp. 4-33.
8. CIRIA, "Placing Concrete". *Report Profile 15 (Forwards),* Construction Industry Research and Information Association, London, 1969.
9. ILLINGWORTH, J. R. "Concrete Conveyors". *Concrete* (London), 6, 3 (March 1972), pp. 20-22; *ACI Journal,* 69, 4 (April 1972), pp. 201-207.
10. RAY, K., KARMAKAR, J. K. and DATTA, S. K. "Variation of Concrete Strength due to Delay in Placing". *Indian Concrete Journal,* 30, 2 (February 1956), p. 42.
11. FIGG, J. W. and BOWDEN, S. R. *The Analysis of Concretes.* HMSO, London, 1971.
12. MANDER, J. E. and MACEVER, D. Y. "Determination of Cement Content in Hardened Concrete by X-ray Spectrograph". *Pitt and Quarry,* 58, 4 (October 1965), pp. 127-33.
13. "Modern Automatic Block-making Equipment in Belgium". *Concrete Building and Concrete Products,* 41, 1 (January 1966), pp. 21-27.
14. HEATON, B. S. "Relationship in Concrete between Strength, Compaction and Slump". *Constructional Review,* 39, 2 (February 1966), pp. 16-22.
15. ACI COMMITTEE 207. "Symposium on Mass Concrete". *ACI Publication,* SP-6, 1963.
16. "Testing of Concrete by the Ultrasonic Pulse Method". *RILEM Recommendations NDTI,* Paris, 1972.
17. ACI COMMITTEE 309. "Consolidation of Concrete". *ACI Journal,* 68, 12 (December 1971), pp. 893-932.
18. WESTALL, W. G. "Bonding Thin Concrete to Old Pavements". *Civil Engineering* (United States), 28, 6 (June 1958), pp. 34-37.
19. SERKIN, W. "Plastic Surgery for Concrete". *Technical Journal,* Rocla Concrete Pipes Ltd, 38 (November 1967), p. 1.
20. WYNN, A. E. and MANNING, G. P. *Design and Construction of Formwork for Concrete Structures,* C & CA, London, 1974.
21. PEURIFOY, R. L. *Formwork for Concrete Structures.* McGraw-Hill, New York, 1964.
22. HUNTER, L. E. "The Design of Timber and Steel Formwork for Concrete". *Civil Engineering and Public Works Review,* 46, 538 (April 1951), pp. 246-47; 543 (September 1951), pp. 686-90; 546 (December 1951), pp. 936-37; 47, 547 (January 1952), pp. 57-59; 549 (March 1952), pp. 222-24; 550 (April 1952), pp. 313-14.
23. GORDON, C. and HOUGH, R. D. "Preconstruction Consideration for a Building of Exposed Concrete". *ACI Journal,* 70, 9 (September 1973), pp. 601-605.
24. SIMONE, V. J. De and CAMELLERIE, J. F. "Novel Structural Frame Combined with Slip-form Construction Results in Record Breaking Construction Time". *ACI Journal,* 62, 10 (October 1965), pp. 1225-35.
25. ACI COMMITTEE 347. "Precast Concrete Units Used as Forms for

283

Cast-in-Place Concrete". *ACI Journal*, 66, 10 (October 1969), pp. 798-813.

26. SAA, *Rules for Design and Construction of Formwork*, AS 1509; *Surface Finish*, AS 1510.

27. Cement and Concrete Association of Australia. *Symposium on Formwork*, Melbourne, 1971.

28. RICHARDSON, J. T. *Practical Formwork and Mould Construction*. Applied Science, Britain, 1976.

29. STEIN, J. and DONALDSON, P. K. "Techniques and Formwork for Continuous Vertical Construction". *Technical Report* TRCS 1, The Concrete Society, 1967.

30. ACI COMMITTEE 347. "Formwork for Concrete". *ACI Publication* SP-4 (3rd edn, 1973); *ACI Journal*, 71, 11 (November 1974), pp. 559-69; 72, 8 (August 1975), p. N25.

31. WILLIS, W. E. "Dry-cast Concrete". *Concrete Construction*, 14, 12 (December 1969), pp. 457-58.

32. PAPINEAU, D. "Block Curing and Drying With Carbon Dioxide". *Modern Concrete*, 32, 6 (October 1968), pp. 38-43; 7 (November 1968), pp. 36-39, 66; 10 (February 1969), pp. 56-62.

33. HOUSTON, B. J. "Use of Membrane-forming Curing Compounds on Concrete Surfaces that are to be Painted". *Technical Report* C-68-1, United States Army Corps of Engineers, Vicksburg, 1968.

34. ACI COMMITTEE 348. "Avoiding Gross Errors in Concrete Construction". *ACI Journal*, 72, 11 (November 1975), pp. 638-46.

35. HANSON, J. A. "Optimum Steam Curing Procedure in Precasting Plants". *ACI Journal*, 60, 1 (January 1963), pp. 75-100; 9 (September 1963), pp. 1287-99.

36. LEWIS, R. K. "Compressive Strength of Concrete Cores". *CSIRO Report*, L.18, CSIRO, Division of Building Research, 1972.

37. MENZEL, C. A. "Menzel Relative Humidity Method". *Concrete*, 64, 10 (October 1956), pp. 36-38.

38. PRICE, W. H. and WALLACE, G. B. "Resistance of Concrete and Protective Coatings to Forces of Cavitation". *ACI Journal*, 46 (1950), pp. 109-120.

39. COUPER, R. R. "Use of Combustion Products for the Accelerated Curing of Concrete Masonry". *Constructional Review*, 42, 4 (November 1969), pp. 65-69; 44, 2 (May 1971), pp. 63-68.

40. HANSON, J. A. "Optimum Steam Curing Procedure for Structural Lightweight Concrete". *ACI Journal, Proceedings*, 62, 6 (June 1965), pp. 661-72; 12 (December 1965), pp. 1671-73.

41. HONDROS, G. "A Constant Temperature and Humidity Fog-chamber for Curing Concrete Specimens". *Civil Engineering and Public Works Review*, 53, 627 (September 1958), pp. 1017-20.

42. KING, J. W. H. and TIMUSK, J. "Design and Construction of a Chamber with Controlled Temperature and Humidity". *Magazine of Concrete Research*, 17, 51 (June 1965), pp. 101-2.

43. COPELAND, R. E. "Economics of High Pressure Steam Curing". *Concrete Products*, 69, 3 (March 1966), pp. 70-76.

44. MAHER, J. B. "Considerations in Determining Size, Layout, and Details of Autoclaves". *Concrete*, 66, 5 (May 1958), pp. 28-31.

45. *Concrete Industries Yearbook*. Pit and Quarry Publications, Chicago, 1976.

46. CARRIER, R. E. and CADY, P. D. "Evaluating Effectiveness of Concrete Curing Compounds". *Journal of Materials*, 5, 2 (June 1970), pp. 294-302.

47. MENZEL, C. *et al.* "Menzel Symposium on High Pressure Steam Curing". *ACI Special Publication*, SP-32, 1972.

48. METCALF, J. B. "The Specification of Concrete Strength". British Road Research Laboratory, *Reports* LR299, LR300 and LR301, 1970.

49. ROSS, A. D. "Heat Flow in the Steam Curing of Concrete Products". *ICE Proceedings*, 1, 5, 6 (November 1956), pp. 695-702.

50. PLOWMAN, J. M. "Maturity and the Strength of Concrete". *Magazine of Concrete Research*, 8, 22 (March 1956), pp. 13-22; 24 (November 1956), pp. 169-83; 10, 28 (March 1958), p. 45.

51. "Curing Pipe with Supersaturated Steam". *Concrete,* 61, 8 (August 1953), pp. 3-5, 14.
52. ANDERSON, B. W. "Electrical Curing". *Concrete Building and Concrete Products,* 43, 3 (March 1968), pp. 34-39.
53. SPEKTOR, B. V. "Infra-red Heating in the Production of Concrete". British C & CA, *Translation,* No. 79, 1965.
54. BROWN, N. G. and BALLANTYNE, E. R. "The Sirowet Rig for Testing Weatherproofness of Building Facades". *CSIRO Report,* CSIRO, Division of Building Research, 1975.
55. BILLIG, K. "Electro-concrete". *ICE Journal,* 24, 8 (1944-45), pp. 368-78.
56. ITAKURA, C. "Electrical Curing of Concrete". *Concrete and Constructional Engineering,* 50, 5 (May 1955), pp. 211-14.
57. CARRIER, R. E. "Concrete Curing: A New Approach". *Concrete Construction,* 16, 7 (July 1971), pp. 279-81.
58. FREEDMAN, S. "Carbonation Treatment of Concrete Masonry Units". *Modern Concrete,* 33, 5 (September 1969), pp. 33-41.
59. BLOEM, D. L. and GAYNOR, R. D. "Factors Affecting the Homogeneity of Ready-Mixed Concrete". *ACI Journal,* 68, 7 (July 1971), pp. 521-25.
60. *Statistics: Tables and Summary.* Royal Melbourne Institute of Technology Press, Melbourne, 1974.
61. RIDLEY, T. "An Investigation into the Manufacture of High-strength Concrete in a Tropical Climate". *ICE Proceedings,* 13 (May 1959), pp. 23-34; 15 (February 1960), pp. 170-81.
62. BERGSTROM, S. G. "Curing Temperature, Age and Strength of Concrete". *Magazine of Concrete Research,* 5, 14 (December 1953), pp. 61-66.
63. NAVARATNARAJAH, V. "An Analysis of Stresses During Steam Curing of Pretensioned Concrete". *Constructional Review,* 40, 12 (December 1967), pp. 18-25.
64. DICKINSON, W. E. "Benefits of Using Calcium Chloride in Concrete Products". *Pit and Quarry,* 48, 8 (February 1956), pp. 190-92.
65. ERLIN, B. and HIME, W. G. "The Role of Calcium Chloride (and Nonchloride Accelerators) in Concrete". *Concrete Construction,* 21, 2 (February 1976), pp. 57-61.
66. EVANS, R. H. "Use of Calcium Chloride in Prestressed Concrete". *Proceedings, World Conference on Prestressed Concrete,* University of California, San Francisco, 1957.
67. ACI COMMITTEE 212. "Admixtures for Concrete". *ACI Journal,* 68, 9 (September 1971), pp. 646-76; 69, 3 (March 1972), pp. 189-90.
68. KENNERLEY, R. A., WILLIAMS, A. L. and ST JOHN, D. A. "Water-reducing Retarders for Concrete". *DSIR* (NZ), *Report* DL 2026, 1960.
69. TUTHILL, L. H. and CORDON, W. A. "Properties and Uses of Initially Retarded Concrete". *ACI Journal,* 27, 3 (November 1955), pp. 273-86; 2, 28, 6 (December 1956), pp. 1187-1200.
70. ASHWORTH, R. "Some Investigations into the Use of Sugar as an Admixture to Concrete". *ICE Proceedings,* 31 (June 1965), pp. 129-45; 33 (March 1966), pp. 518-27.
71. SCHUTZ, R. J. "Setting Time of Concrete Controlled by the Use of Admixtures". *ACI Journal,* 30, 7 (January 1959), pp. 769-81.
72. AMERICAN CONCRETE INSTITUTE. "Computer Applications in Concrete Design and Technology". *ACI Publication* SP-16, 1967.
73. EWAN, W. D. "When and How to Use Cu-Sum Charts". *Technometrics,* 5, 1 (February 1963), pp. 1-22.
74. HADDAD, G. J. and FREEDMAN, S. "Statistical Product Control". *Modern Concrete,* 33, 11 (March 1970), pp. 34-40; 12 (April 1970), pp. 34-36; 34, 1 (May 1970), pp. 30-33; 2 (June 1970), pp. 33-37.
75. ACI COMMITTEE 214. "Recommended Practice for Evaluation of Strength Test Results of Concrete". *ACI Journal,* 73, 5 (May 1976), pp. 265-78; 11 (November 1976), pp. 645-648.
76. WALSH, P. F. "Quality Control of Concrete by Compressive Strength Testing". *CSIRO Report* 15, CSIRO, Division of Building Research, 1974.

77. DAY, K. W. "Economical Control of Concrete Quality". *Australian Civil Engineering,* 10, 5 (May 1969), pp. 34-39; *Constructional Review,* 34, 7 (July 1961), pp. 44-56.
78. RILEY, O. and COOPER, S. B. "Concrete Control on a Major Project". *ACI Journal,* 68, 2 (February 1971), pp. 107-114.
79. CORDON, W. A. "Minimum Strength Specifications Can Be Practical". *ACI Journal,* 66, 7 (July 1969), pp. 539-45; 67, 1 (January 1970), pp. 63-69.
80. BURGESS, A. J. *et al.* "High Strength Concrete for the Willows Bridge". *ACI Journal,* 67, 8 (August 1970), pp. 611-19.
81. TEYCHENNE, D. C. "Treatment of the Variations of Concrete Strength in Codes of Practice". D of E BRE, *Current Paper* CP 6/74, 1974.
82. CREASY, L. R. and WHITE, L. S. "Water-excluding Structures". *ICE Proceedings,* 14 (September 1959), pp. 31-42.
83. ACI COMMITTEE 210. "Erosion Resistance of Concrete in Hydraulic Structures". *ACI Journal,* 27, 3 (November 1955), pp. 259-71.
84. GILL, H. R. *Concrete Formwork Designer's Handbook.* Concrete Publications, London, 1960.
85. BURMEISTER, R. A. "Tests of Paper Moulds for Concrete Cylinders". *ACI Journal,* 47, 1 (September 1950), pp. 17-24.
86. HENNING, N. E. "Concrete Test Moulds and Concrete Capping Materials". *ACI Journal,* 32, 7 (January 1961), pp. 851-54.
87. WERNER, G. "The Effect of Type of Capping Material on the Compressive Strength of Concrete Cylinders". *ASTM Proceedings,* 58 (1958), pp. 1166-86.
88. SHERRIFF, T. "Cylinder Splitting". *Concrete* (London), 9, 2 (February 1975), pp. 34-37; 5 (May 1975), pp. 23, 32.
89. DAVIES, J. D. and BOSE, D. K. "Stress Distribution in Splitting Tests". *ACI Journal,* 65, 8 (August 1968), pp. 662-69; 66, 2 (February 1969), pp. 157-59.
90. CHAPMAN, G. P. "The Cylinder Splitting Test". *Concrete* (London), 2, 2 (February 1968), pp. 77-85; 3, 5 (May 1969), pp. 173-74.
91. PINCUS, G. and GESUND, H. "Evaluating the Tensile Strength of Concrete". *Materials Research and Standards,* 5, 9 (September 1965), pp. 454-58.
92. HANNANT, D. J. "Tensile Strength of Concrete". *Structural Engineer,* 50, 7 (July 1972), pp. 253-58.
93. SLESAREW, YU. M. "Vibrating Pipes for Feed of the Concrete Mix". *Library Communication,* 887, DSIR, Building Research Station, 1959.
94. COOKE, J. "Boilerless Autoclaves". *Concrete,* 68, 11 (November 1960), pp. 16-19.
95. HALSTEAD, P. E. "Significance of Concrete Cube Tests". C & CA Britain. *Technical Report,* TRA421, 1969.
96. SHORE, W. J. "Block Kiln Arrangement—Good Design Proves Profitable". *Modern Concrete,* 24, 2 (June 1960), pp. 52-55.
97. HOPE, B. B. and MANNING, D. G. "Creep of Concrete Influenced by Accelerators". *ACI Journal,* 68, 5 (May 1971), pp. 361-64; 11 (November 1971), pp. 877-78.
98. WILLIAMS, E. J. *Regression Analysis.* Wiley, New York, 1959.
99. TROXELL, G. E., DAVIS, H. E. and KELLY, J. W. *Composition and Properties of Concrete.* McGraw-Hill, New York, 1968.
100. SMITH, F. L. "Effect of Aggregate Quality on Resistance of Concrete to Abrasion". *ASTM Special Technical Publication* No. 205, 1958, pp. 91-106.
100. FENTRESS, B. "Slab Construction Practice Compared by Wear Tests". *ACI Journal,* 70, 7 (July 1973), pp. 486-91.
102. UNITED STATES ARMY CORPS OF ENGINEERS. *Handbook for Concrete and Cement.* Central Concrete Laboratory, New York, 1949 (plus Supplements).
103. PHILLEO, R. E. "Comparison of Results of Three Methods for Determining Young's Modulus of Elasticity of Concrete". *ACI Journal,* 26, 5 (January 1955), pp. 461-69; 2, 27, 4 (December 1955), pp. 1-3, 472.

104. GALAN, A. "Estimate of Concrete Strength by Ultrasonic Pulse Velocity and Damping Constant". *ACI Journal,* 64, 10 (October 1967), pp. 678-84.
105. WATKINS, R. D. and SAMARIN, A. "Cavitation Erodibility of Concrete". *National Conference Publication,* 75/6, IE Aust. Symposium on Serviceability of Concrete, Melbourne, 1975.
106. WADDELL, J. J. *Practical Quality Control for Concrete.* McGraw-Hill, New York, 1962.
107. FORBES, W. S. *Ready-mixed Concrete.* Contractors Record, London, 1958.
108. EVANS, R. H. and MARATHE, M. S. "Microcracking and Stress-Strain Curves for Concrete in Tension". *Materials and Structures,* 1, 1 (January/February 1968), pp. 61-64.
109. LINTILL, W. H. *et al.* "Formwork Technology, and Film-faced Plywood, in Concrete Construction". *Civil Engineering and Public Works Review,* 60, 713 (December 1965), pp. 1795-1803; 62, 728 (March 1967), pp. 339-42; 67, 791 (June 1972), pp. 581-92.
110. O'BRIEN, J. "Principles and Practice of Slipform". *C & CA Aust. Report,* TR33, 1973.
111. *Pocket Concrete Penetrometer.* Soiltest, Chicago.
112. ACI COMMITTEE 408. "A Guide for Determination of Bond Strength in Beam Specimens". *ACI Journal,* 61, 2 (February 1964), pp. 129-35.
113. KOMLOS, K. "Determination of the Tensile Strength of Concrete". *Indian Concrete Journal,* 41, 11 (November 1967), pp. 429-36; 42, 2 (February 1968), pp. 68-76; 11 (November 1968), pp. 473-78, 82; 43, 2 (February 1969), pp. 42-49, 54.
114. "Slabs for Liquid-containing Structures. Design in Accordance with BS CP2007 (1960)". *Concrete and Constructional Engineering,* 56, 6 (June 1961), pp. 219-26.
115. KOLEK, J. "An Appreciation of the Schmidt Rebound Hammer". *Constructional Review,* 32, 1 (January 1959), pp. 22-30.
116. "Downward-looking Radar Test". Highway Research Board, Washington, *Highway Research News,* 53 (1973), pp. 36-39.
117. DEWAN, R. L. "Rapid Estimation of Cement Content in Mortar and Concrete". *Indian Concrete Journal,* 33, 4 (April 1959), pp. 132-33.
118. SENGUPTA, D. P. "Estimation of Cement Content in Hardened Concrete and Mortar". *Indian Concrete Journal,* 32, 11 (November 1958), pp. 389-90.
119. AMERICAN CONCRETE INSTITUTE. "Evaluation of Concrete Properties from Sonic Tests". *ACI Monograph,* 2, 1966.
120. HANSEN, W. C. *et al.* "Symposium on Effect of Water-reducing Admixtures and Set-retarding Admixtures on Properties of Concrete. *ASTM Special Publication* No. 266, 1960.
121. WRIGHT, P. J. F. "Durability of Concrete Kerbs". *The Engineer,* 211, 5496 (May 1961), pp. 855-59.
122. NEWMAN, W. M. "Moving Forms for Multi-storey Buildings". *Constructional Review,* 35, 6 (June 1962), pp. 22-27.
123. GRAY, W. H., McLAUGHLIN, J. F. and ANTRIM, J. D. "Fatigue Properties of Lightweight Aggregate Concrete". *ACI Journal,* 58, 2 (August 1961), pp. 149-62.
124. TYLER, I. L. and ERLIN, B. "A Proposed Simple Test Method for Determining the Permeability of Concrete". *PCA R & DL Journal,* 3, 3 (September 1961), pp. 2-7.
125. JORDAN, I. J. and ILLSTON, J. M. "Creep of Sealed Concrete Under Multiaxial Compressive Stresses". *Magazine of Concrete Research,* 21, 69 (December 1969), pp. 195-204.
126. LEGGET, R. F. "Failure of Prestressed Concrete Pipe at Regina. Saskatchewan". *ICE Proceedings,* 22 (May 1962), pp. 11-20; 25 (June 1963), pp. 201-6.
127. BORTZ, S. A. and LITVIN, A. "Investigation of Continuous Wire Reinforcement as a Replacement for Brick Ties in Masonry Walls". *ACI Journal,* 59, 5 (May 1962), pp. 673-86.
128. "Prepackaged Concrete Mix Scores on First Big Job Test". *Engineering News-Record,* 168, 10 (March 8, 1962), pp. 26-27.

129. "Hot Oil Used For Steam Generation in New Autoclaves at Utah Plant". *Modern Concrete,* 25, 10 (February 1962), pp. 46-48.
130. MALHOTRA, V. M. "Effect of Specimen Size on Tensile Strength of Concrete". *ACI Journal,* 67, 6 (June 1970), pp. 467-69.
131. CREASY, L. R. "Watertight Concrete Basements", *RCA Journal,* 1, 3 (May/June 1962), pp. 109-34.
132. SAUCIER, K. L. "Effect of Method of Preparation of Ends of Concrete Cylinders for Testing". United States Waterways Experiment Station, *Miscellaneous Paper,* C-72-12, 1972.
133. HIGGINSON, E. C. "Effect of Steam Curing on the Important Properties of Concrete". *ACI Journal,* 58, 3 (September 1961), pp. 281-98.
134. ORCHARD, D. F. *Concrete Technology.* I and II, Applied Science, London, 1973.
135. SOROKA, I. "Parameters for the Quality Control of Concrete". *Concrete* (London), 8, 12 (December 1974), pp. 41-43.
136. FULTON, F. S. *Concrete Technology.* Portland Cement Institute, Johannesburg, 1969.
137. McINTOSH, J. D. *Concrete and Statistics.* Spon, London, 1968.
138. MORONEY, M. J. *Facts from Figures.* Penguin Books, Mitcham, Vic., 1956.
139. GOLIS, M. J. "Pavement Thickness Measurement Using Ultrasonic Pulses". United States Highway Research Board, *Highway Research Record,* 218, 1968, pp. 40-48.
140. NARROW, I. and ULLBERG, E. "Correlation Between Tensile Splitting Strength and Flexural Strength of Concrete". *ACI Journal,* 60, 1 (January 1963), pp. 27-38.
141. SHACKLOCK, B. W. "The Use of Admixtures in Concrete". *Concrete and Constructional Engineering,* 57, 9 (September 1962), pp. 355-60.
142. MONFORE, G. E. and VERBECK, G. J. "Corrosion of Prestressed Wires in Concrete". *ACI Journal,* 32, 5 (November 1960), pp. 491-515; 12 (June 1961), pp. 1639-48.
143. MILLER, D. "Design of Timber Forms for Off-form Concrete Finishes". *Constructional Review,* 37, 12 (December 1964), pp. 29-42.
144. ARBER, M. G., ROBERTS, J. S. and VIVIAN, H. E. "Concrete Form Treatments". *Constructional Review,* 34, 6 (June 1961), pp. 27-33.
145. "What it Takes to be a Good Ready Mix Truck Driver". *Concrete Products,* 66, 3 (March 1963), pp. 38-50.
146. NICHOLSON, J. A. *Ready Mixed Concrete. Practical Aids for Operation as Suggested by a Producer.* Maclean-Hunter Publishing Corporation, Chicago, 1963.
147. SHIRAYAMA, K. "Germ of Crystallization as Admixture for Accelerating Hardening of Cement". *Occasional Report* No. 6, Building Research Institute, Ministry of Construction, Japan, 1961.
148. ACI COMMITTEE 517. "Atmospheric Pressure Steam Curing of Concrete". *ACI Journal,* 66, 8 (August 1969), pp. 629-46.
149. ACI COMMITTEE 308. "Curing Concrete". *ACI Journal,* 68, 4 (April 1971), pp. 233-43.
150. McGEARY, F. L. and FLUCKER, R. L. "Aluminium Racks and Pallets for Autoclave and Kiln Curing of Concrete Block". *Modern Concrete,* 27, 4 (August 1963), pp. 42-45, 64.
151. DODSON, V. H. and FARKAS, E. "Delayed Addition of Set Retarding Admixtures to Portland Cement Concrete". *ASTM Proceedings,* 64 (1964), pp. 816-29.
152. McMASTER, R. C. "Sonic Mixer Makes Concrete Up to Three Times Stronger". *Concrete Products,* 73, 2 (February 1970), pp. 54-55.
153. WAGNER, W. K. "Effect of Sampling and Job Curing Procedures on Compressive Strength of Concrete". *Materials Research and Standards,* 3, 8 (August 1963), pp. 629-34.
154. BRUERE, G. M. "Functions and Uses of Water-reducing, Set-retarding and Set-accelerating Admixtures in Concrete". *Constructional Review,* 37, 2 (February 1964), pp. 16-21.
155. MARTINET, C. "Electrical Process of Curing Prestressed Precast Concrete". *Concrete Building and Concrete Products,* 39, 1 (January 1964), pp. 29-31.

156. LEWIS, R. K. "A Summary of Investigations of Steam Curing of Concrete Related to Cement Characteristics". *Constructional Review,* 41, 3 (March 1968), pp. 18-25.

157. POWERS, T. C. "A Hypothesis on Carbonation Shrinkage". *PCA R & DL Journal,* 4, 2 (May 1962), pp. 40-50.

158. TOENNIES, H. T. and SHIDELER, J. J. "Plant Drying and Carbonation of Concrete Block. NCMA-PCA Co-operative Program". *ACI Journal,* 60, 5 (May 1963), pp. 617-33; 2, 61, 6 (June 1964), p. vii.

159. ELVERY, R. H. and EVANS, E. P. "Effect of Curing Conditions on the Physical Properties of Concrete". *Magazine of Concrete Research,* 16, 46 (March 1964), pp. 11-20.

160. BRADFIELD, G. and GATFIELD, E. N. "Determining the Thickness of Concrete Pavements by Mechanical Waves: Directed-beam Method". *Magazine of Concrete Research,* 16, 46 (March 1964), pp. 49-53.

161. Cement & Concrete Association of Aust. "Formwork Design and Detailing". *Contracting and Construction Engineer,* 26, 3 (March 1972), pp. 6-37.

162. GREEN, H. "Impact Strength of Concrete". *ICE Proceedings,* 28 (July 1964), pp. 383-96; 31 (July 1965), pp. 315-18.

163. OLSEN, R. H. *et al.* "Lateral Pressures of Concrete on Formwork". *ACI Journal,* 71, 7 (July 1974), pp. 358-61.

164. "Natural Rubber Saves a Reservoir". *Contracting and Construction Equipment,* 17, 10 (June 1964), pp. 90-91.

165. BLAKEY, F. A. and BERESFORD, F. D. "Stripping of Formwork for Concrete in Buildings in Relation to Structural Design". *IE Aust. Civil Engineering Transactions,* CE7, 2 (October 1965), pp. 92-96.

166. COOK, G. P. "Accelerated Hardening of Concrete in the Manufacture of Precast Concrete Units". *Constructional Review,* 37, 11 (November 1964), pp. 21-28.

167. ACI COMMITTEE 311. *ACI Manual of Concrete Inspection: Publication SP-2,* 1975.

168. ELVERY, R. H. and EVANS, E. P. "Effect of Curing Conditions on the Physical Properties of Concrete". *Prestress,* 14 (September 1964), pp. 5-18.

169. ROSENBERG, A. M. "Study of the Mechanism Through Which Calcium Chloride Accelerates the Set of Portland Cement". *ACI Journal,* 61, 10 (October 1964), pp. 1261-69.

170. BRAY, L. S. and KEIFER, Jr, O. "Check List for Batch Plant Inspection". *ACI Journal,* 61, 6 (June 1964), pp. 625-42; 12 (December 1964), pp. 1647-49.

171. SHORE, W. J. "Ideal Steam Curing System for Concrete Block". *Concrete Products,* 68, 4 (April 1964), pp. 54-55.

172. HURD, M. K. "Form Design Aids Based on New Lumber Sizes". *ACI Journal,* 70, 7 (July 1973), pp. 480-85; *ACI Special Publication,* SP-4, 1973.

173. PEARSON, R. G., KLOOT, N. H. and BOYD, J. D. *Timber Engineering Design Handbook.* Jacaranda Press, Melbourne, 1970.

174. Standards Association of Australia. *Construction in Light Timber Framing,* AS 1684.

175. CAMPBELL, R. H. and TOBIN, R. E. "Core and Cylinder Strengths of Natural and Lightweight Concrete". *ACI Journal,* 64, 4 (April 1967), pp. 190-95; 10 (October 1967), pp. 692-94.

176. NEVILLE, A. M. "Penetration Resistance Test". *Concrete* (London), 6, 3 (March 1972), p. 28.

177. MURDOCK, J. W. "A Critical Review of Research on Fatigue of Plain Concrete". *Bulletin* 475, University of Illinois, Engineering Experiment Station, 1965.

178. HARLAND, D. G. "A Radio-active Method for Measuring Variations in Density in Concrete Cores, Cubes and Beams". *Magazine of Concrete Research,* 18, 55 (June 1966), pp. 95-101; 19, 58 (March 1967), pp. 60-63.

179. LEWIS, R. K. "Steam Curing of Lightweight Concrete After Initial Moist Curing". *Constructional Review,* 38, 10 (October 1965), pp. 13-17; 11 (November 1965), p. 32.

180. ACI COMMITTEES 516 and 517. "High Pressure, Atmospheric and Low Pressure Steam Curing". *ACI Manual of Concrete Practice*, 3, Products and Processes, 1976.
181. ANSON, M. and NEWMAN, K. "Effect of Mix Proportions and Method of Testing on Poisson's Ratio for Mortars and Concretes". *Magazine of Concrete Research*, 18, 56 (September 1966), pp. 115-30; 19, 59 (June 1967), p. 118.
182. JOHNSTON, C. D. and SIDWELL, E. H. "Testing Concrete in Tension and in Compression". *Magazine of Concrete Research*, 20, 65 (December 1968), pp. 221-28.
183. EDWARDS, A. C. and GOODSALL, G. D. "Field Investigation of a Method for the Rapid Analysis of Fresh Concrete". *Transport and Road Research Laboratory Report*, LR560, Britain, 1973.
184. SAMUELSSON, P. "Voids in Concrete Surfaces". *ACI Journal*, 67, 11 (November 1970), pp. 868-74.
185. GHOSH, R. K. and CHATTERJEE, M. R. "Direct Adjustment of Cement Concrete Mixes from Field Strength Data". *Australian Road Research Board Journal*, 3, 6 (June 1968), pp. 42-52.
186. MORRISH, P. "Site Control of Concrete Surface Finishes". *Building in Concrete*, Building Science Forum of Australia, 15th Conference, Sydney, 1970.
187. ACI COMMITTEE 504. "Guide to Joint Sealants for Concrete Structures". *ACI Journal*, 67, 7 (July 1970), pp. 489-536; *ACI Manual of Concrete Practice*, 3, Products and Processes, 1976.
188. BAVELJA, R. "Analysing Fresh Concrete". *Concrete* (London), 4, 9 (September 1970), pp. 351-53; 6, 3 (March 1972), pp. 23-27.
189. TURNER, C. D. "Unconfined Free-fall of Concrete". *ACI Journal*, 67, 12 (December 1970), pp. 975-76.
190. READING, T. J. "The Bughole Problem". *ACI Journal*, 69, 3 (March 1972), pp. 165-71.
191. SHILSTONE, J. M. *et al*. "Placing and Compacting Architectural Concrete". *Concrete Construction*, 17, 11 (November 1972), pp. 536-38.
192. FISHER, G. H. "Concrete Set Control for Vertical Slipforming". *ACI Journal*, 69, 9 (September 1972), pp. 556-61.
193. NORTH, B. H. "Appearance Matters". *Concrete*, 7, 5 (May 1973), pp. 18-23.
194. ACI COMMITTEE 311. "Inspection of Concrete". *ACI Journal*, 72, 6 (June 1975), pp. 269-90.
195. C & CA Britain. "Rapid Analysis of Fresh Concrete". *Precast Concrete*, 4, 4 (April 1973), pp. 201-2.
196. O'BRIEN, J. B. "Principles and Practice of Slipform". *Contracting and Construction Engineer*, 28, 3-5 (March/May 1974), pp. 45-57,
197. CHAPLIN, R. G. "Abrasion Resistance of Concrete Floors". C & CA Britain, *Technical Report* 42.471, 1972.
198. ELVERY, R. H. "Estimating Strength of Concrete in Structures". *Concrete* (London), 7, 11 (November 1973), pp. 49-51.
199. FORRESTER, J. A. and LEES, T. P. "Apparatus for Rapid Analysis of Fresh Concrete". C & CA Britain, *Technical Report* TR42.490, 1971; *Concrete* (London), 9, 3 (March 1975), pp. 45-47.
200. ACI COMMITTEE 704. "Concrete Quality". *ACI Enchiridion*, E 704-4, 1975.
201. Editorial. "Concrete Surfaces—Form Liners". *Concrete Construction*, 19, 2 (February 1974), pp. 55-57.
202. BROWN, J. H. "Failure of Glass Fibre Reinforced Notched Beams in Flexure". *Magazine of Concrete Research*, 25, 82 (March 1973), pp. 31-38.
203. WALSH, P. F. "Linear Fracture Mechanics Solutions for Zero and Right Angle Notches". *CSIRO Technical Paper (2nd Series)*, No. 2, CSIRO, Division of Building Research, 1974.
204. TOMSETT, H. N. "Non-destructive Testing of Floor Slabs". *Concrete* (London), 8, 3 (March 1974), pp. 41-42.
205. ILLINGWORTH, J. R. "Transporting Fresh Concrete". *Concrete* (London), 8, 3 (March 1974), pp. 49-52.

206. "Rope Finish". *Constructional Review,* 47, 2 (June 1974), pp. 46-49; *Concrete Construction,* 20, 1 (January 1975), pp. 9-10; 9 (September 1975), pp. 411-13.
207. ROHDE, W. *et al.* "Responsibility In Concrete Inspection". *ACI Journal,* 71, 4 (April 1974), pp. 201-18; 10 (October 1974), pp. 529-32.
208. ACI COMMITTEE 215. "Fatigue of Concrete". *ACI Special Publication,* SP-41; *ACI Journal,* 71, 4 (April 1974), pp. 219-23.
209. WEAVER, J. and SADGROVE, B. M. "Striking Times of Formwork —Tables of Curing Periods to Achieve Given Strengths". *CIRIA Report* 36, Construction Industry Research and Information Association, London, 1971.
210. ALEXANDERSON, J. "Strength Losses In Heat Cured Concrete". *Proceedings* 43, Swedish Cement and Concrete Research Institute, Stockholm, 1972.
211. MORGAN, D. R. "Effects of Chemical Admixtures on Creep in Concrete". *IE Aust. Transactions,* CE16, 1 (August 1974), pp. 7-11.
212. PEDUZZI, A. "Tunnel Sealing with PVC Sheet". *Tunnels and Tunnelling,* 6, 4 (July/August 1974), pp. 36-37.
213. ROSENBLUETH, E. *et al.* "Bonus and Penalty in Acceptance Criteria for Concrete". *ACI Journal,* 71, 9 (September 1974), pp. 466-72.
214. MACDONALD, M. D. "Waterproofing Concrete Bridge Decks: Materials and Methods". *Transport and Road Research Laboratory, Report 636,* Britain, 1974.
215. DAVIS, S. G. and MARTIN, S. J. "Quality of Concrete and Its Variation in Structures". C & CA Britain, *Report 42.487,* 1973.
216. FORSSBLAD, L. *Concrete Compaction in the Manufacture of Concrete Products and Prefabricated Building Units.* Swedish Cement Association, Malmo, Sweden, 1971.

PART 4

DURABILITY AND STABILITY

CHAPTER 15

DURABLE CONCRETE STRUCTURES AND PRODUCTS

15.1 INTRODUCTION

A durable concrete is one which will withstand in satisfactory degree the effects of the service conditions to which it is subjected. The principles of making good plain concrete were known in the middle of the nineteenth century, but they were often abandoned in reinforced concrete because of the difficulties of placement, compaction and curing. Inferior quality thus led to rusting of the reinforcement and spalling of the concrete cover.

Reinforced concrete structures that are properly built and used will, however, give upwards of 50 years of service with little or no special maintenance. As was the case in early building projects, should the requirements for durability be ignored during design and construction, deterioration and failure can soon occur under rigorous conditions of service. The quality or permeability of the concrete and the thickness of cover determine largely the extent to which aggressive solutions may enter and attack both the concrete and steel reinforcement.

Many materials and techniques have been devised for improving the physical characteristics of concrete. For best results they must be used selectively and effectively, with commonsense and competent supervision. Where remedial work must be done, both the deterioration and its cause must be rectified as a function of regular maintenance. If the various factors affecting durability are recognised and dealt with, the ratio of effective life of high-quality to low-quality reinforced concrete may be 20 to 1, or even higher, under adverse conditions of exposure or service.

295

15.2 FACTORS AFFECTING DURABILITY

The concept of durability ranges from the general requirements of adequate structural strength to special considerations of enduring design which incorporates structural resistance to environmental attack under prolonged service conditions. Within these terms of reference, concrete structures and products can fail by

Collapse due to bad design or construction, or by overloading.
Corrosion of reinforcement.
Disintegration of the concrete itself.
Fire.
Wear.

These causes, or their component factors, are considered separately in this chapter, where they are cross-referenced to sources of relevant information in the text.

15.2.1 COLLAPSE

Structural collapse of concrete structures, although comparatively rare, is prevented normally by the correct usage of proper premises, design criteria and proof-engineered designs (see Table 3.25 and Appendix IV, 2).[1-8, 37, 73] Some information on the instrumentation, performance monitoring and progressive strength evaluation of full-scale structures is contained in Reference 67 and Rice and Hoffman, Appendix IV, 2(a)(i) and 4(b). Other basic causes of deterioration in reinforced concrete appear here as headings in the classification used throughout this book, while the related paragraphs summarise the considerations which apply in each case.[34, 151, 153]

15.2.2 CORROSION OF REINFORCEMENT

(a) Corrosive aggregrate and deleterious substances.

Considerations:

Admixtures containing corrosion promoters (e.g. ammonium, sodium or calcium chloride) to be avoided in concrete that will be subjected to steam-curing, hydrothermal, electromotive or marine conditions, or placed in contact with tendons in prestressed concrete. (See Sections 10.1 and 22.3; also Chapter 25 and Section 34.1.)[17, 86, 108, 117, 144, 191]
Aggregates containing sulphides and sulphates to be avoided (e.g. certain cinder, coke-breeze and iron-ore mine aggregates).

(See Sections 3.3.4, 15.2.3(e), 19.1.1. and 33.2.2(c)(ii).)

Bore or ground water and doubtful or potentially reactive aggregate to be tested; cement matrix around tendons to be kept at pH 12-14.[117, 133, 140, 141, 177]

Cathodic protection of wound-wire tendons in prestressed concrete. (See Sections 15.2.2(c), 22.3.1, 23.3.2 and 34.2.2)[17-19, 28, 38, 75, 117, 176, 177, 185, 186]

High-alumina cement with a sulphide constituent to be excluded generally from prestressed concrete. (See Sections 3.1.2(b) and 34.1.4(b).)

Hydrogen sulphide, atomic hydrogen, acidic ground waters and interfacial voids to be kept off prestressing steel in concrete water-retaining structures. (See Sections 22.3.1 and 34.1.4(b); also Glossary, Appendix VI.) [17-19, 117, 140, 141]

Low mixing-water salinity (see Section 3.2.1); durability in aggressive atmospheres.

Low rate of liquid absorption (see Section 15.2.3(b), (c), and low susceptibility of processed steel to stress corrosion or strain ageing after galvanising (see Table 22.1).

Protective coatings on steel reinforcement, provision being made for bond by the quality, roughness and shape of the surface. (See Sections 3.4, 21.1.6 and 22.3.)[17, 18, 117, 128, 140, 141, 176]

Strength accelerator (if any) under the foregoing conditions to be a corrosion inhibitor in dense reinforced concrete (e.g. stannous chloride; see Section 22.3.2); or steel to be coated with cement slurry and an inhibitor.

Surface impregnation of dry slender concrete piles in aggressive sulphate ground with coal-tar epoxy resin compound (e.g. three coats), the first coat being brushed on. (See Sections 20.2 and 21.1.)

Tests for stress-corrosion susceptibility of steel prestressing wire.[17, 117, 177]

(b) Cracking of concrete.

(i) Cracking due to dimensional change caused by drying, moisture and temperature. Considerations:

Aliphatic alcohol-evaporation retardant (see Section 9.1); supervision (Appendix II).

Chamfered re-entrant angles; and limitations of sectional change, rotational distortion and obstruction to settlement within forms.[4-8, 73]

Concrete pipes to be placed underground.

Continuous slabs, laid on well-compacted subgrades, to have 0·6 per cent cross-sectional shrinkage and temperature steel (mainly near the top) with strengthened remote control joints (see Section 26.3).[24]

Defects in construction.[196]

Dense concrete with a low water/cement ratio and carefully graded aggregate, with limited flakiness and fines (Chapters 3, 4, 7 and 23).

Differential drying shrinkage and movements across sections, due to high heat of hydration, thermal, carbonation and low-humidity conditions (see Sections 31.1-31.4).[73]

Drying and carbonation treatments of precast products after curing (see Sections 9.6 and 31.4); induced compression in continuous composite decks by flexion.

Early, progressive and effective moist or membrane curing (see Sections 9.1 and 31.2.4).

Fibre reinforced concrete and cement-impregnated reinforcing tapes (see Sections 3.4.2 and 15.2.4).

Finely ground portland cement, if used, to contain 2·5-3·5 per cent sulphur trioxide by weight, or an optimal amount for soundness with tricalcium aluminate.

Hard, impermeable aggregate with a moderate to low coefficient of thermal expansion (e.g. limestone; see Section 31.2.1).

Joint movement in panel cladding.[197]

Lean mixes and artificial cooling of large-mass work for controlled heat of hydration of cement and accompanying expansion, contraction and possible shrinkage cracking of the concrete (see Section 30.1).

Limited density gradient and warping, water content, segregation, bleeding, cavitation, and settlement over obstructions (e.g. horizontal reinforcement in deep beams; see Sections 11.1.2(d), 15.2.2(d)(vi), 33.2.2(d) and 33.3.2(a)).

Low-heat cement or portland blastfurnace cement, and cement that is not too finely ground or does not harden rapidly and develop a high heat of hydration (see Sections 3.1.1 and 31.2.1).

Minimum of segregation, differential settlement, laitance and admixtures which promote shrinkage (e.g. calcium chloride; see Chapter 7 and Sections 10.1, 31.2.1 and 31.2.3).

Nonshrink cement concrete, mortar and grout (see Section 3.1.2(a) and Chapter 25).

Preformed, fibrous-cement sheathing of reinforced concrete members for improved serviceability (see Section 3.4.2).

Prepacked, no-fines (protected reinforcement) and air-entrained concretes (see Sections 16.1, 24.4 and 33.3).

Pressure grouting of ducts in construction joints to compensate for drying shrinkage (see Section 24.3).

Proof-engineered, prestressed, noncracking or limited-cracking structural design (see Section 34.1.1).

Protection against vibration of hardened concrete, thermal shock and frost action at an early age (see Sections 8.4, 31.2.4 and 32.1).

Release of forms before precast units are steam cured (see References 136 and 142 of Part 6).

Removal of formwork to be delayed in cold weather (see Section 8.4).[73]

Repairs with epoxy or polysulphide epoxy resin, impregnated grout or nonshrinkage mixes (see Chapter 21, Section 24.3 and Chapter 25).[73]

Revibration of water-gained concrete and reworking of slab surfaces, using suitable vibrators and wood floats, in order to close fine seepage paths and relieve surface tension near the time of initial set (see Sections 20.1.5(d), 23.2 and 31.2.4).

Setting retarder monitored in the decks of continuous or composite bridges and the casting of deep sections (see Section 10.2).

Shrinkage-crack widths at the surface of integral slabs to be limited to 0·1 mm by means of shrinkage-control reinforcement (see Sections 22.3.2, 34.1.1, References 192-198 of Part 4 and Reference 165 of Part 5).[24]

Shrinkage and temperature reinforcement well distributed in adequate amount (e.g. 0·35-0·70 per cent) and low restraint by control and sliding joints (see Sections 20.1, 26.3, 27.2, 31.2 and 33.4).

Shrinking aggregate to be used with air-entrainment, kept in a moist condition, or avoided (see Reference 136 of Part 3).[79]

Surface wall-feature grooves, e.g. 20 per cent deep at 5 m centres or at waterstop locations.

Vapour barriers, ventilation and insulation where air moisture and temperature vary within wide limits.

Vibrovacuum dewatering of newly laid concrete slabs (see Section 23.4).

White, alkali-resistant and weather-reflective paint (see Sections 28.2 and 28.3).

(ii) Cracking due to excessive tensile stress and deflection.
Considerations:

Appropriate joints to relieve stress caused by end restraints.

Concrete masonry partition walls to be structurally reinforced in horizontal joints or grouted cores.

Deflection control by means of limited span/depth ratios, such as follows.[24]

Flexure	Beams		Slabs	
	Under Brick	Others	Under Brick	Others
Simply supported	13	17	20	25
Two-directional	—	—	23	28
Continuous	17	20	23	28
Cantilever	6·5	8	8	10

Deflection-control reinforcement, placed in the compression zone of flexural members, to be up to one-half of the positive moment tensile steel (see Reference 136 of Part 6, and Park and Paulay of Appendix IV, 2(a)(i)).[24]

Deformed bars, anchorages, spacers, stirrups, splices and distributing reinforcement (see Sections 3.4, 15.2.3(g) and 34.1.1).[4-8]

Design assumptions to be correlated with field conditions, and actual quantities of concrete used to be correlated with design estimates.

Designs to offset differential support or camber settlements and member shortenings due to low elastic modulus, soft or unsound aggregate and oversanded concretes under load.

Elements with a high degree of elasticity or flexibility for adaptation to intense local or shock stress (e.g. fender piles and moving ground: see Sections 33.4, 34.1, 35.1 and 36.1).

Limited elongation, restraint or overstressing.[4-8, 73]

Limited shrinkage stress at points of contraflexure in long continuous beams (see Section 31.2).

Maximum tensile design stress for cold-worked bars and welded-wire fabric to be limited to 170 MPa or, in the case of cantilevers, 110 MPa.

Oscillation and temperature gradient effects to be included in the stress and strain analyses of tall slender structures (see Appendix IV, 2).

Plastic distribution of stress in special structures to be estimated from models, consideration being given to the moisture content, type and age of the concrete (see Appendix IV, 2).

Positive and negative moment coefficients for structural flat slabs to be increased, where appropriate, by some 25 per cent.

Poststressing accessories for reducing tensile strain or cracks (see Section 34.1).

Preformed fibrous-cement sheathing of reinforced concrete members for improved serviceability (see Section 3.4.2).

Precast concrete beams to be kept moist until an *in situ* slab is cast in composite construction.

Prestressed, gap-graded concrete of high strength and elastic modulus, with low shrinkage, creep and permeability properties (see Section 23.1.5).

Prolonged moist condition applied to promote the autogenous healing of uncarbonated fissures (see Sections 11.1, 31.1 and 31.3).

Realignment of deflected elements by jacking (see Section 34.2.7).

Serviceability, vibration and reinforcing criteria to be satisfactorily met in conjunction with strength criteria in structural design (see Table 3.25 and Chapter 31; Reference 136 of Part 6; Glossary, Appendix VI; and Appendix IV, 2(a)(i)).

Shoring to be carefully removed from thin sections; deformations and reversal of stress to be limited during construction; falsework and scaffolding to be sufficiently strong and rigid and not overloaded (see Section 8.4).

Statically determinate and umbrella supports, and adequately strong skeletal elements.

Studies of foundation conditions, daily and seasonal variations, and long-term effects in mining and earthquake areas (see Section 36.1, AS 1289, BS 1377, ASTM Standards, Part 19).[9, 10]

Sudden large changes in amount and direction of reinforcement or built-in conduits to be avoided.

Welding at joints with high concentration of reinforcement (see Appendix I).

(c) Electrolysis.

Considerations:

Calcium chloride, fly ash or carbon black in ordinary reinforced concrete that is likely to be damp to be strictly limited within 500 m of electric transport tracks or where stray direct current in the steel reinforcement exceeds about 5 volts (see Section 23.5).[11, 12] Stray currents to be excluded from tendons.

Cement-dispersion (e.g. 0·25 per cent calcium lignosulphonate admixture) in ordinary reinforced concrete containing 1 per cent of calcium chloride or 30 per cent fly ash, by weight of cement, where the direct current carried is up to 20 volts.[12]

Direct current shunt on metallic pipes with a potential difference of over 8 volts in ordinary reinforced concrete structures.

Dense dry concrete and proper concreting practice.

Effective insulation of direct current power circuits and the joints of metallic pipes that may carry direct current.

Magnesium oxychloride to be kept dry and 50 mm away from steel reinforcement. Saline waters greatly increase electrolysis.

Potential gradient in wet plain concrete to be kept below 200 V/m.[11]

Precast products or cured, well-compacted concrete containing at least ten 40-kg bags of cement per m³. Waterproof coating or encasing in wet ground.

Reinforcement in foundation to be given at least 50 mm cover (see Section 3.4).

Reinforcement to be prevented from serving as an earth or forming either an anode or cathode in moist concrete, the former causing rusting and the latter destruction of bond.

Soil resistivity and chemical analysis of ground water (see Sections 19.1 and 34.1).

Synthetic rubber ("Neoprene") load-bearing pads, vibration absorbers and electrical insulators, particularly in wet saline conditions.

(d) Permeability of concrete due to various causes.

(i) Defective mix design. Considerations:
Air-entrainment, pozzolan or bentonite additive.[108]
Batching by weight (see Sections 3.1.3, 3.3.6 and 7.1.1).
Carefully graded fine aggregate in applied finishes.
Increased cement content in normal sections (e.g. 9-10·5 (40-kg) bags/m³), with concreting procedures to minimise voids and shrinkage effects (see Sections 11.1.2, 19.1.2, 19.1.3 and 31.2).
Mix to be designed correctly with low water content and cement with a limited rate of hardening (see Section 3.1 and Chapters 4, 6 and 23).

(ii) Faulty construction joints. Considerations:
Calcium chloride treatment and mortar bonding coat (see Section 7.2.4).
Concrete to have increased cement content and workability against prepared surface (see Section 7.2.4).

Continuous placement and unit-hardening of reinforced concrete.
Integral waterproofers omitted or used with precautions to obtain
adequate bond in two-course work (see Chapters 11 and 20).
Nonshrinkage mixes, mortar or grout in confined or closure work
(see Chapter 25).
Proper placement, treatment and sealing of joints with green-cut
or indented surfaces (see Sections 7.2, 9.1, 20.1, 26.1, 27.2, 31.2,
33.2 and 34.2.6).
Revibration of current and preceding unhardened lifts of concrete
(see Section 23.2).

(iii) Inadequate compaction and honeycombing. Considerations:
Mixes to be suitable for vibration (see Section 23.1.5).
Pneumatic mortar to be applied properly (see Section 24.2).
Pressure moulding and roller compaction (see Sections 23.1 and
23.3).
Revibration or recompaction near initial set (see Section 23.2).
Vacuum dewatering (see Section 23.4).

(iv) Inadequate cover. Considerations:
Alloy steel rods (containing manganese and silicon) in special
circumstances.
Dense concrete cover of adequate thickness to suit the class of
construction (see Sections 3.4, 19.1.3 and 22.3.1).[4-8]
Passivation of steel reinforcement with a rust inhibitor (e.g.
sodium benzoate, nitrite or chromate, or stannous chloride; see
Section 22.3).
Protective coating of steel without a reduction of bond strength
(see Sections 21.1.6 and 22.3).
Steel reinforcement to be secured adequately in position.

(v) Inadequate curing. Considerations:
Early evaporation and drying to be prevented (see Sections 9.1,
20.1.5 and 31.2.4).
Moist curing above 10°C for at least the first week (see Sections
9.2, 20.1.5 and 31.2.4).
No abnormal carbonation within 24 hours of placement (see
Sections 9.6 and 31.4).

(vi) Segregation, bleeding, cavitation, settlement and laitance.
Considerations:
Adjusted coarse-aggregate/fine-aggregate ratio and grading of
fine aggregate (see Sections 3.3.4(d) and 11.1.2(b)).
Air-entrainment (nonionic) or cement dispersion (see Sections
6.4, 11.1.2 and 16.1).

Bentonite additive (passing 0·075-mm sieve), 0·3-2 per cent by weight of water (see AS MP20, Part 2).

Careful manufacture, transportation, placement and compaction (see Sections 7.1, 7.2 and Chapter 23; also Sections 33.2.2(d) and 33.3.2(a)).

Colloidal grout made with high-early-strength portland cement. Long chutes and drops to be avoided.

Low water/cement ratio; absorptive form linings.

Paraffin wax emulsion (0·3 solids, 0·5-1·5 per cent by weight of cement; see AS MP20, Part 2).

Prolonged mixing or delay in placing concrete (see Sections 7.1 and 7.2).

Reduced water content of mixes toward the top of deep sections.

Removal of laitance and concrete with a high water-gain.

Reworking or revibration of exposed surfaces in the correct way (see Sections 20.1.5(d) and 23.2).

Suspending, thickening, water-retention, thixotropic or sedimentation-control agent (e.g. polyethylene oxide condensate or hydroxypropyl methylcellulose), 0·03-0·3 per cent by weight of water (see AS MP20, Part 2).

Tight formwork or nonleak lining and joins (e.g. polyethylene or polyvinyl chloride; see Sections 8.1 and 27.1.3) and effective site control (see Appendix II).

Vacuum and vibrovacuum dewatering (see Section 23.4).

15.2.3 DISINTEGRATION OF CONCRETE

(a) Alkali-aggregate reaction.

Considerations:
Admixtures in cement to nullify or quickly dissipate possible reaction (e.g. air-entrainment, see Section 16.1), and 20 parts of finely divided opaline or reactive silica per part of sodium oxide plus 0·658 potassium oxide in excess of 0·5 per cent by weight (see Section 3.3.3(a)(vii), and ASTM C114 and C441).[14, 32, 39]

Cement with an alkali content (sodium oxide equivalent) less than 0·6 per cent.

Expanded, acid-type, volcanic aggregate to be tested with high-alkali cement before use in heated concrete floors (see Table 33.1).

Impermeable concrete (see Sections 11.1 and 15.2.3(g)).

Low moisture content in matured concrete.

Mixing water to be of neutral composition or with a low content of sodium and potassium compounds (see Sections 3.2.1 and 20.1.2).

Portland blastfurnace-slag cement concrete (see Sections 3.1.1 and 3.1.2(e)(ii)).

Rejection of glass and reactive aggregate (see Table 3.8; and Sections 3.3.4(e) and 14.7.1).

Suitable mineralogy and petrography of the quarry to be used (see Section 3.3.3).[99]

(b) Attack by aggressive solutions.

(See also Section 15.2.3(d) and (f).) Considerations:

Aggregate, cement and concrete to be of adequate long-term strength and serviceability.

Aggressive solutions or waters to be kept out of contact with concrete by means of protective surface courses or treatments, or by impregnating dry concrete products with bituminous or coal-tar epoxy resin compounds by vacuum-pressure methods (see Sections 19.1.2, 19.2, 20.2, 21.1 and 21.3).[15]

Back-filling to be sulphate-free and densely compacted.

Bore samples, ground and drainage water, and industrial waste solutions to be analysed before construction starts. An historical investigation to be made of the site (BS 1377, 1924, CP2001; ASTM D420, D1452, D1586, D1587, D2113).[9-11, 73]

Calcium chloride to be omitted in concrete or admixtures (see Section 10.1).

Cement dispersion with or without air-entrainment (see Sections 6.4 and 16.1).

Continuous placement, stress-distributing reinforcement and properly made joints; prestressed concrete (see Sections 11.1.2(a), 19.1.2(d) and 34.1).

Data on the physicochemical properties of materials to be verified initially by accelerated and exposure tests (see Chapters 3 and 14).

Dissolution of calcium hydroxide, calcium carbonate and hydrated calcium silicates in set cement by free carbon dioxide in percolating aggressive ground waters or aquifers (see Sections 19.1, 19.2 and 31.4).[108, 140]

Excellent workmanship, control tests and competent inspection (see Appendix II and Chapter 14). Minimal bleeding and weak regions in concrete matrix.

High-alumina cement concrete to be placed, cured and used only under suitable temperature and humidity conditions (see Sections 3.1 and 20.1.5).

High-pressure steam curing of concrete with partial replacement of portland cement by silica flour or a pozzolan (see Sections 9.4, 17.1 and 33.2.3).

Increased cement content (e.g. 10·5-14 (40-kg) bags/m³) in thin and heavily reinforced sections (i.e. cement-factor mix design; see Sections 11.1, 19.1 and 20.1).[73]

Increased thickness to provide for possible loss in effective section (see Section 19.1).

Low permeability and absorption (see Sections 11.1, 14.10, 15.2.3(c), (f) and (g)).

Mature, dry, carbonated, dense concrete that has been effectively cured for increased resistance to sulphate attack, but not necessarily to acidic solutions (see Sections 9.1-9.6, 19.1.2 and 20.1; also Chapter 23 and Section 31.4; and Appendix III).

Polymer concrete (see Sections 15.2.4 and 21.4).

Potential expansion of portland cement concrete exposed to sulphate (see Sections 3.1.2(e)(viii) and 14.7.2; also ASTM C452).[73]

Pozzolan component with portland cement (e.g. 15-30 per cent by weight; see Chapter 17).

Pressure-moulding or pressure on fresh concrete while being cured (e.g. up to 5 MPa by pneumatic cells in joints; see Sections 3.2.2, 4.3.4, 23.1 and 23.2).

Preventive maintenance, including the regular cleaning of industrial floors (see Section 20.4).

Protection of reinforcement (e.g. adequate cover) particularly in the wetting and drying zone of marine structures (see Sections 3.4, 19.1.3, 21.1 and 22.3).[4-8, 15, 16]

Site and catchment surveys to include the effects of seasonal variation and likely flow of ground water.[10, 73, 151]

Special and polymer cement concretes (see Sections 3.1 and 19.1.2; also see Glossary, Appendix VI).

Underdrainage where sulphate salts in solution exceed 0·1 per cent.

Upgrade, decontaminate and drain locally polluted work sites.

Vitrified clay pipes with rubber ring joints, or plastic pipes (see Appendix I).

(c) Frost action, and repeated freezing and thawing.

Considerations:

Air-entrainment with an adjusted mix (see Sections 6.4 and 16.1).

Autoclaved concrete with a lime-silica base (see Sections 9.4 and 33.2.3).

Close control of grading (see Sections 3.3.4, 4.3.1 and Appendix II).

Considerable slope on copings, sills or projections on which water might collect,[16] and suitable bases under concrete roads (see Reference 37 of Part 5).

Early-strength cement, nonsaturation of exposed concrete in freezing weather, and protection against frost action (see Sections 3.1, 31.4 and 32.1).

Electric and moist curing (see Section 9.5); nonfreeze or dry sealed slabs (see Sections 20.1.4(c) and 20.2).

Freeze-thaw and water-absorption test studies (see Sections 14.7 and 14.10).[39]

Limited aggregate crushing value, and flakiness and elongation indexes (see Section 3.3.4(f) and (g)); and strong aggregate-cement bond.

Revibration, reworked surface and pressure moulding of fresh concrete (see Section 23.2).

Shrinkage stress to be kept low in mass concrete (see Chapters 30 and 31).

Uniform, high-quality concrete with a water/cement ratio not exceeding 0·45-0·55 by weight (see Table 4.5; and Sections 11.1.2 and 11.2).

Water absorption of dry 100 mm concrete cubes at an age of 28 days not to exceed 2·5 per cent in 10 minutes (see Section 14.10).

(d) Gaseous and bacterial corrosion.

Considerations:

Acidic effluents to be neutralised, diluted and cooled before entering concrete sewers, the flow being promoted with flushing.

Acid-resistant liners, glazes and joints; polyvinyl chloride liners; epoxy and furane resins or other durable treatment (see Sections 12.1, 15.2.5, 19.1.2, 20.2, 21.1 and 21.3).

Additional segmental thickness of main outfall sewers, at and above the daily high level of flow and at the crown (see Section 19.1 and Appendix III).

Admixture of copper powder passing 0·053-mm sieve (e.g. 1 per cent by weight of cement) for bacteriolytic action.

Autoclaved concrete with a lime-silica base (see Sections 9.4 and 33.2.3).

Bacterial production of sulphides to be retarded in sewers above waterline by such measures as chlorination (6 ppm/km of outfall) or partial purification; nonturbulent flow; periodic cleaning to ensure design velocity of flow; and precipitation of stale sewage with metallic salts or lime (pH raised to 8·5-9·0 by 150 ppm). (See Bacterial corrosion in the Glossary, Appendix VI.)

Calcareous aggregate (e.g. limestone of suitable quality; see Sections 19.1.1 and 19.1.2).

Dense, smooth, hard, crack-free surfaces.[146]

Dissolution of calcium hydroxide, calcium carbonate and hydrated calcium silicates in set cement by free carbon dioxide in percolating aggressive ground waters or aquifers (see Sections 19.1, 19.2 and 31.4).[108, 140, 175]

Drainage and dry structures, their surface moisture being below 20 per cent or 85 per cent relative humidity.

Hydraulic tractive wall-shear stress (e.g. 3·5 Pa) to be sufficient to inhibit the deposition of biological slime that causes a generation of hydrogen sulphide in trunk sewers.[134]

Impregnation of dry concrete with bitumen or bituminous epoxy resin compound and hardener by vacuum-pressure methods (see Section 21.1).[15]

Lead lining (0·25 mm) by metal-spray gun in two coats (BS CP3003).

Mature or carbonated concrete that has been effectively cured (see Sections 9.6, 19.1 and 31.4).

Measures and materials described in Section 15.2.3(b).

Pneumatic mortar to be made with high-alumina cement where structures are liable to sulphate attack (see Section 24.2).

Polymer concrete (see Sections 15.2.4 and 21.4).

Pressure and vibration processes of concrete pipe manufacture (see Section 23.3).

Rust inhibitor around steel reinforcement (see Section 22.3).

Sacrificial cover in concrete outfall sewers and tall chimneys, and the retention of concrete quality to adequately carry design stresses.

Sewer walls to be kept dry by forced ventilation, and force mains to be designed for continuous air injection (e.g. 15 litres per second per metre diameter).

Silicon tetrafluoride gas treatment under pressure to a depth of over 5 mm, although this serves only to palliate eventual collapse (see Section 22.2).

Smoke baffles and ducts in railway structures.

Special cement of requisite quality and high-quality concrete with increased cement content (see Sections 3.1 and 19.1.2).

Ventilated air space and independently supported corrosion-resistant lining in tall concrete chimneys (see Appendix IV, 2).

(e) Incompatibility of materials.

Considerations:

Corrosion promoters (e.g. calcium chloride) to be avoided under certain conditions (see Sections 10.1, 15.2.2, 19.1.1, 19.2 and 22.3; also Chapter 25 and Section 34.1).

Different brands of portland cement are inadvisable in two-layer work with steel fabric at the interface, and in prestressed concrete pipes that will be subjected to abnormal temperature and humidity or steam-curing conditions (see Sections 22.3 and 34.2.2).

Dissimilar nonferrous metals to be kept apart or insulated against galvanic corrosion (see Section 22.3.2).

Dry conditions or a moisture barrier (e.g. epoxy resin) to be placed between large abutting areas of damp portland cement concrete or structural steel elements in high relative humidity, and damp interactive material (e.g. gypsum plaster; see Sections 19.1.1(b) and 21.1.6).

Lime with magnesium oxide to be fully hydrated or admixed with calcium chloride (see Sections 18.2.1 and 33.2.3(a)).

Low differential in thermal diffusivity and coefficient of expansion.[11]

Magnesium oxychloride or Sorel cement surfacing to be laid properly on reinforced concrete. This means: ground floors must have an impermeable membrane underneath them; nearby steel reinforcement and metal conduits with less than 50 mm concrete cover must be protectively coated; cracks and sulphurous impurity are to be avoided; magnesium chloride solution is to be used to a minimum extent; a 5 per cent admixture of copper powder is to be used where humid conditions promote sweating; service conditions are to be kept dry; and surfaces are to be maintained at 3-6-month intervals with pale, boiled, linseed oil (BS 776, CP204; ASTM C275, C276; and see Reference 12 of Part 1).

Protection of ordinary or nonalkali-resistant fibreglass and various metals that are likely to be attacked by moist portland cement (see Sections 3.4.2 and 22.3.2; also References 12 and 65 of Part 1).[108]

Shrinkage and fluxing effects of plasticised vinyl tiles laid on bitumen.

Volumetrically stable or nonreactive aggregate that is free from decomposed stone and harmful materials (see Sections 3.3.1 and 15.2.3(a)).

(f) Leaching.

Considerations:

Aggressive water to be reduced in content of free carbon dioxide (see Section 19.2).

Air-entrainment and pozzolanic additive (see Chapters 16 and 17).[39]

Autoclaved concrete with a lime-silica base (see Sections 9.4.6 and 33.2.3).

Drainage of structures and freedom from permeable defects and cracks.

Fresh concrete to be protected from heavy rainfall.

High-alumina and portland blastfurnace-slag cement concretes (see Section 3.1.2).

High-pressure steam curing of a suitable mix. For example, for autoclaved, asbestos-cement pressure pipe: 3 parts portland cement, 2 parts ground sand passing 0·075-mm sieve, and 1 part teased asbestos fibre consisting of 2 parts chrysotile: 1 part crocidolite and amosite (AS 1711; BS 486, 567, 3656; see Sections 3.4.2, 9.4.6 and 33.2.3; also Reference 65 of Part 1).[82]

Impermeable or dense, cement-rich concrete with a low rate of absorption (see Sections 11.1, 11.2 and 19.2); dewatering action during manufacture (see Sections 23.3 and 23.4).

Lime-saturated water or fog-spray curing; carbonated concrete.

Protective surface treatment (see Sections 19.1.2, 20.2, 21.1 and 21.3).

Silicon tetrafluoride gas treatment under pressure (see Section 22.2).

Temporary hardness to be imparted to soft water (see Section 19.2).

(g) Repeated wetting and drying.

Considerations:

Aggregate to have a low volume change with changing moisture content, and to be free from adverse ageing effects, reactive iron pyrites and other unstable material (see Section 3.3).

Air-entrainment or pozzolan (see Chapters 16 and 17).

Calcium chloride or sodium chloride to be omitted in concrete or admixture (see Sections 10.1, 19.1 and 22.3).

Cement concrete to have high tensile strength, elastic properties, low shrinkage and increased richness under rigorous conditions (see Sections 31.1, 31.2 and 31.3).[11]

Cohesive mix, sufficiently plastic to continuously envelop steel reinforcement with compacting equipment.

Cracks to be less than 0·25 mm wide, due to restrained shrinkage, deflection and working stress in steel reinforcement (see Sections 22.3.2 and 34.1.1; and ASTM C76).

Dense, smooth, crack-free surface with a hardened cement skin left intact.

Impervious coatings or facing slabs to be on drained and ventilated sections that will be free from efflorescence (see Chapters 28 and 36).

Impregnation of cracks with epoxy resin and hardener, or cement grout under pressure (see Sections 21.1.6 and 24.3).

Increased impermeability (see Chapter 11 and Section 15.2.2(d)).

Mature or carbonated concrete (see Sections 9.6 and 31.4).

Precast concrete products to be dry at the time of use (see Sections 27.1, 27.2 and 33.3).

Prestressed concrete (see Chapter 34).

Strong bond between mortar matrix, aggregate and steel (see Sections 3.4 and 21.1.6).

Surface drainage, watertight joints and projections that throw rainwater clear of walls.

Uniform concrete with low shrinkage stress, water absorption and dimensional change (see Appendix III).

Vibration (adjusted mix) and high-pressure compaction or revibration of heavily reinforced sections (see Chapter 23).

Water content of mixes to be reduced as the height of placed concrete is increased (see Appendix III).

15.2.4 FIRE

Considerations:

Adequate cover and design of good-quality concrete (see Sections 3.4 and 34.1.3).[4-7, 117, 126, 160]

Antistatic precautions in inflammable atmospheres. For instance, conductive flooring, special footwear, nonspark wire brushes and beryllium-copper maintenance tools (see Sections 20.2(i) and 36.7.4).

Fire, smoke, oxygen-index, heat-release and safety ratings of plastic, rubber and textile furnishings.[170, 178]

Lightweight concrete (see Sections 33.3.2(f) and 33.3.3(f)).

Low coefficient of thermal expansion and content of free silica in aggregate (e.g. basic igneous rock, burnt clay or limestone; see Sections 3.3 and 31.2.1).

Main reinforcement to be bound with hard-drawn steel wire.[4-7]

Polymer concrete and fibre reinforcements to be evaluated, for particular applications, on the basis of fire resistance (see Sections 3.4.2 and 21.4).

Polymer insulation (e.g. polyurethane foam) to be noncombustibly lined for toxic-smoke control on ignition, or be tested and approved before use.

Protective board, plaster or sprayed insulation (see Sections 33.3.2(f) and 34.1.3(b)).

Refractory concrete or mortar and relief joints for high-temperature service (see Section 29.1).

Repair of fire-damaged concrete structures.[37, 148]

Safety measures against spontaneous combustion and fire hazards generally (see Appendix I): such as air-handling systems, AS 1668 Part 1; welding, AS 1674; glazing, BS CP153 Part 4; heating appliances, BS 1945.

Sprinkler system (AS CA16 and BS CP402.201) with duplicate water service; automatic fire alarm installation (AS 1670, BS 3116 and CP1019), laserbeam fire detector, or fire-barrier ceilings and doors for a reduced fire period and concrete cover (see Sections 3.4.1(b) and 34.1.3(b)).

Statutory requirements to be observed for the occupancy and size of buildings (e.g. uniform building regulations or codes; fire underwriters and national fire protection associations; see AS 1481 and BS CP3 Chapter IV; see Appendixes IV, 2(a)(i) and 4(b)).[154]

Structural elements to be sectionally designed, or thermally tested as prototypes, to provide requisite fire-resistance rating or stipulated fire endurance at a 90 per cent confidence level of prediction (AS 1480, 1481, 1530, 1668 Part 1 and CA57; also BS EN2 on fire classification and BS CP110 Part 1, 4547; see Plate 13).[4-7, 20-23, 126]

Thermal conditions to be limited to 250°C in portland cement concrete and 400°C in steel reinforcement.

15.2.5 WEAR DUE TO ABRASION AND EROSION

Considerations:

Admixture of cement-dispersing or water-reducing agent, air-entrainment being limited to 3 per cent (see Section 6.4 and Chapter 16).

Autoclaved concrete with a lime-silica base (see Sections 9.4.6 and 33.2.3).

Bleed water to be uncumulative beneath finished surfaces of freshly laid concrete slabs.

Coarse aggregate to be interlocked and close to the surface, as in prepacked concrete and cement-bound macadam (see Sections 24.4 and 26.1).

Dry, high-strength concrete with high impermeability and a minimum compressive strength of 30 MPa, preferably 40 MPa, at 28 days (see Sections 4.3.4, 11.1, 12.1, 20.1 and 23.1; also Appendix III).[39]

Early and continuous moist curing for 2 weeks (see Sections 9.2, 9.3, 12.1 and 20.1.5(e)).

Fired or ceramic bond to be developed in refractory concrete (see Section 29.1).

Gas-hardening treatment (e.g. silicon tetrafluoride under pressure; see Section 22.2).

Granular metal surfacing, steel grids, stainless-steel or cast-iron tiles, and vitrified clay bricks or heavy-duty floor tiles (see Sections 20.2(i) and 22.1; see illustration of Plate 15).

Hard, tough, well-graded aggregate with medium-rough surface, rounded edges and bulky shape, well bonded into the concrete at the surface (see Section 3.3 and Chapter 12).

Hydraulic-model studies for effective design.

Increased thickness of section to provide for loss by wear.

Metal-strip joints in industrial floors with a plane surface finish (see Section 20.1.5).

Resilient, abrasion-resistant coatings. For example: asphalt or polyvinyl chloride tiles; cold-laid or hot-laid mastic ahphalt; chlorinated rubber lacquer and grit; epoxy resin compound and polyamide hardener (either mixed with filler and applied 5 mm thick over a primer, or painted on in three coats and blinded with grit); magnesium oxychloride; and polyvinyl acetate cement slurry or mortar over dry concrete (with a vapourproof membrane), the service conditions being kept dry (see Sections 20.2(c), (e), (g), (h), (i), 21.1 and 21.2).

Rubber coating to cushion impact and reduce wear by abrasive materials in hydraulic structures (see Chapter 12 and Section 14.2.7).

Smooth surface and hydraulic flow without turbulence or cavitation (see Chapter 12).

Surface behaviour to be estimated from abrasion and erosion tests (see Chapter 12 and Section 14.2).

Surface-hardening liquids (e.g. metallic silicofluoride and sodium silicate). The latter, when applied to stone or carbonated surfaces, is followed by a solution of magnesium silicofluoride, calcium chloride or 1·5 per cent hydrofluoric acid (see Sections 20.2(a) and 20.2(b)).[11]

Thorough compaction without laitance or "drier" coats on concrete pavements (see Sections 20.1 and 20.2, also Chapters 23 and 26).

Vibrovacuum dewatering or absorptive form lining (see Section 23.4).

CHAPTER 16

AIR-ENTRAINMENT

16.1 CONCRETE

16.1.1 GENERAL INFORMATION

Freshly mixed concrete contains 4-5 per cent of air that is naturally entrapped in perceptible cavities which, by effective compaction, can be reduced collectively to below 1 per cent (see Section 7.2.3 and Chapter 23). The residual air is scattered at random throughout the mass and, in concrete of ordinary quality, is linked by a capillary system to exterior surfaces. The latter form of porosity is caused by excess mixing water which, on escaping by bleeding, leaves channels, runlets or capillaries within the concrete (see Chapter 11). The use of excess mixing water leads also to segregation, settlement and to the formation of laitance and efflorescence.

It is evident, then, that concrete can have built into it an easy means of ingress of liquids which may reduce its durability. These characteristics can be brought under control by introducing into ordinary or lean mixes an air-entraining agent, which subdivides the naturally entrapped air into minute disconnected bubbles. These with an optimum diameter of about 0·05 mm, are distributed about 0·13 mm apart throughout the mass.

The air is thus deliberately entrained in the form of 500-800 billion minute bubbles per m³ of concrete. This entrained air constitutes an additional fine aggregate, which lubricates the mix and enables the mixing water to be reduced by up to about 15 per cent for a given degree of workability. The possible reduction of mixing water, due to improved plasticity from air-entrainment, is much greater for lean than for rich mixes.

315

The residual water is still far more than that required initially for the hydration of cement. The excess water thus remaining is used largely in forming the skin of the innumerable, minute bubbles throughout the mass. It thereby becomes integrated temporarily in the mix, with a corresponding increase in cohesiveness and stability at reduced water/cement ratios.

Deliberate air-entrainment not only causes a marked reduction in the capillary structure of ordinary to lean concrete, but it also reduces appreciably the capillary forces which cause water to be drawn into the concrete. The temporary fixation of the residual excess water in the freshly placed concrete limits the rate of evaporation of moisture required for curing. The result, in exposed concrete, is a less severe drying-shrinkage gradient and loss of potential strength at an early age. These characteristics, coupled with high cohesiveness, are benefical in reducing plastic cracking and in improving the surface finish of large areas of newly laid concrete (see Section 31.2.4).

An air-entraining agent should produce stable, minute bubbles and comply with the requirements of AS 1478 and 1479, ASTM C226, C233, C260 and C494. It may be introduced in the form of an air-entraining portland cement (see Section 3.1.2(e)), or separately as a surface-active agent of anionic, cationic or nonionic type. The separate agent procedure is advantageous in controlling the amount of air that is incorporated in concrete under various conditions. Typical agents consist of saponified "Vinsol" resin or stabilised wood resin (see Section 33.2.3(b)); primary alkylolamide plus alkylaryl sulphonate; sodium abietate or secondary alkyl sulphonate; saponin or keratin compound; and a triethanolamine salt of a sulphonated hydrocarbon or fatty-acid glyceride. Their composition should prevent gumming in mechanical dispensing equipment (see Plate 5).[26, 129]

The proportion required in order to entrain 3-6 per cent of air varies from 0·01 to 0·05 per cent (on a basis of active constituent) by weight of cement. To facilitate measurement at the mixer, air-entraining agents are introduced as a solution at the rate of 30-110 ml per 40-kg bag of cement, depending on the composition and concentration of active constituent in the solution and the amount of air to be entrained under given conditions. An air-entraining agent may be used in conjunction with another admixture for cement dispersion, set control or water repellency, provided that it is incorporated separately and the amount of agent is reduced when the other admixture tends to entrain air (see Section 6.4, Chapter 10 and Section 11.1.2(c)).

The air content of concrete should be measured regularly at the site, and, as part of the requirements of mix modification, the quantity of the particular agent used should be adjusted promptly for uniform results. General procedures and air meters for measuring air content are briefly described in Section 16.3. Various factors affecting the content of entrained air are discussed here, these being considered in conjunction with the characteristics of the particular agent that is used.

(i) Cement. Different brands of cement may cause a large variation in air content. The requisite amount of agent, for a particular air content, varies not only directly with the specific surface area of cement, but also inversely with its alkalinity. The improvement in workability, with a given amount of agent, is much greater for a lean mix than a rich one.

(ii) Sand. The amount of air entrained increases notably with increasing proportions of sand and fractional-size particles between the 0·60-mm and 0·15-mm sieve sizes. Particles between the 0·60-mm and 0·30-mm sieve sizes actively entrain air and those between the 0·30-mm and 0·15-mm sieve sizes strongly retain air, but the presence of still finer particles inhibits air-entrainment (see Section 3.3.4(h)). Pumped concrete, which contains an increased fine-aggregate/coarse-aggregate ratio, should be gauged with strictly limited or zero amount of agent that entrains air (see Sections 6.4, 16.2 and 24.1).

(iii) Coarse aggregate. Mixes containing crushed stone entrain more air than those containing gravel, and the air content goes down as the amount and particle size of coarse aggregate go up.

(iv) Size of batch. The amount of air entrained increases as the size of batch decreases.

(v) Type of mix. Lean mixes entrain air more readily than rich mixes, while concrete containing fly ash may require several times as much agent as ordinary concrete for a given content of air. The air content increases with increasing slump, but drops rapidly beyond a value of about 180 mm.

(vi) Type of mixer. Different amounts of air are entrained by different mixers, which should operate to a standard mixing-time schedule.

(vii) Length of mixing time. The air content increases generally with lengthened mixing time, but it is reduced by prolonged mixing.

(viii) Temperature of concrete. An additional amount of agent is required with rising temperature, for the entrainment of a constant amount of air in the concrete.

The specified content of entrained air applies more specifically to placed concrete than to freshly mixed concrete and, with the exception of form vibration, all types of vibratory compaction may be used with air-entrained concrete. Vibrational effects, which are described in Section 23.1.6, are governed by the characteristics of the mix, also by the type, intensity and duration of vibration. The air content is usually reduced by not more than one-sixth to one-third by internal vibration.

Air-entrained concrete, if designed on a strength basis, should be proportioned initially with an increased target strength to compensate for an estimated reduction of, say, 10 per cent in compressive strength and 6 per cent in flexural strength (see Section 4.3.3). Alternatively, a nominal mix may be adjusted: firstly, by reducing its water content so as to maintain a given slump and, secondly, by decreasing the fine-sand content so that each cubic metre of hardened concrete will contain the specified quantity of cement. The adjustment in fine-sand content (corrected for bulking in volumetric batching) is equal to the air content minus the reduction of water. A subsequent analysis, if desired, is described in BS 1881.

Controlled air-entrainment in ordinary to lean concrete, coupled with mix adjustment, causes a marked improvement in homogeneity, impermeability, freeze-thaw durability and surface texture. For effective results, the air content should lie between a minimum value of 3 per cent and, for strength control purposes, a maximum value of 6 per cent $\pm$ 1 per cent. The cohesion and workability of an air-entrained mix, with a reduced water content, minimise the ill-effects of segregation, settlement and capillarity in hardened concrete. A secondary effect is a reduction of surface pitting through an accumulation of water globules against formwork (see Section 23.1.6).

Decreased settlement (ASTM C232 and C243) leads to improved bond strength in reinforced concrete, while reduced capillarity not only helps to minimise efflorescence and leaching, but also leads to increased resistance to attack by brine, seawater and sulphate solution. Under freezing conditions, air-entrainment accommodates an increase in volume of absorbed water without causing spalling. Under moderate conditions of alkali-aggregate reaction, air-entrainment relieves distress in concrete and prevents or reduces significantly its possible deterioration.[39]

The enhanced durability of air-entrained concrete is inadequate to resist rigorous acidic or bacterial agencies (see Chapter 19 and Glossary, Appendix VI). Air-entrainment reduces the wear resistance and elastic modulus of concrete, the latter being reduced approximately 700 MPa for each 1 per cent of air.[26] It has no effect generally on shrinkage, other than a beneficial effect on the drying-shrinkage gradient from the surface.

Controlled air-entrainment in ordinary to lean mixes has a justifiable use under certain conditions of conveyance, placement and exposure. Under no circumstances should it be used as a means for saving cement or as a substitute for correct concrete practice. Where air-entrainment would not produce a perceptible improvement in durability, such as in rich mixes, the air content need be no more than is necessary to improve the workability (see Sections 14.7.3, 15.2.2(d)(vi), 15.2.3(a) and (c)). Hand finishing without drag is facilitated by using a magnesium or aluminium float.

A highly air-entrained concrete, admixed with a small amount of polypropylene fibre, has thixotropic sculptural properties that facilitate manual ornamental moulding and decorative cladding formation (see Section 3.4.2; also John Laing, "Faircrete", Britain). This composite has high impact resistance, moderate thermal properties and a compressive strength of 20 MPa at 1300 kg/m^3 (see Reference 124 of Part 1).

Note: Workability, set-controlling, permeability-reducing, suspending, pozzolanic, expanding and foaming agents are described in Sections 6.4, 10.1-10.2, 11.1.2(c), 15.2.2(d)(vi), 17.1-17.2, 25.1-25.4 and 33.2.3(b), respectively.

16.1.2 CONVEYANCE OF AIR-ENTRAINED CONCRETE

As varying results can arise through the use of different air-entraining agents in central and agitating-truck mixed concrete, their characteristics should be known and their quantity controlled carefully during use. Agents which bring about an increase in volume during transit cause spillage from agitating trucks and an undesirable loss in strength of the concrete. The effect is influenced by, among other things, the type of agitating truck (e.g. the number of blades) and the grading of aggregate (e.g. gap or continuous grading). The type of sand used may have a marked effect on the amount of air entrained with a given amount of agent, which should be formulated so as not to "gum-up" automatic dispensers.

With nonagitating trucks, the transit of air-entrained concrete may be accompanied by a reduction of slump and air content, due possibly to compaction. The loss of slump and air is not a function of time or distance of transit above certain minima (e.g. about 5 minutes and 3 kilometres, respectively). The procedure may be used for transit times up to about 45 minutes, regardless of distance. The percentage of air entrained at the mixer should be near the upper limit allowed by specifications, as the compressive strength increases with the trucking. There is no evidence of segregation and there is little effect on the durability of the concrete. Moisture losses in transit should be minimised by the use of tarpaulin covers and by keeping dump trucks in good condition.

16.2 MORTAR

In the first known book on architecture, written in the year 16 BC by Vitruvius Pollio, reference is made to the practice of the ancient Greeks and the Romans of increasing the workability of mortar by beating it with sticks for several hours. Since 1936, the workability of concrete and mortar mixes has been increased by adding an air-entraining agent at the mixer. This additive reduces the surface tension of the mixing water, thereby increasing its wetting action, and causes frictionless aggregate to be incorporated in the mix in the form of microscopic air cells. The amount of air that is thus entrained in mortar is about three times that entrained in concrete and lies between about 10 and 20 per cent (ASTM C185). Tests on 1 : 3 cement-sand mortar, with a constant water/cement ratio of 0·55 by weight, have shown a 3 per cent strength reduction (from 28 MPa at 28 days) for each 1 per cent of entrained air. Air-entrainment is not used, however, in lime mortar.

A reduced quantity of mixing water, for a given degree of workability of cement concrete and mortar, largely offsets a reduction of strength caused by the entrained air. For instance, a water/cement ratio of 1·4 (by weight) for a composite 1 : 1 : 6 cement-lime-sand mortar (by volume) can be reduced to about 0·7 for a 1 : 6 cement-sand, air-entrained mortar with the same consistence. Both mortars possess good adhesive, wetting and cohesive properties, although their subsequent characteristics of plastic deformation under load are somewhat different. Sands to be used in mortar should be clean, graded and coarse to medium gauge; a coarse-gauge sand is particularly desirable for mixes that are to be used for spatterdash, undercoating and rendering work. An improvement in workability by air-entrainment is most noticeable in cement-based mixes with coarse, sharp sands. Standard specifications for mortars, including some relevant materials and their uses,[77, 78] are listed in Table 16.1.

With resin-based plasticisers, a maximum content of air is entrained by mechanical mixing for 3·5-4 minutes, further mixing producing a slight drop in air content. With detergent-type plasticisers, the percentage of air entrained is usually increased by mixing for longer periods. The quantity used in mortar varies from 30-150 ml per 40-kg bag of cement and, with some air-entraining agents, there may be a slight set-retardant effect on the cement.

In masonry work, the strength of jointing mortar should be weaker than the units joined with it. In addition to moderate strength, a suitable degree of plasticity and low dimensional change in the hardened state are requirements for satisfactory performance. These factors help to offset the ill effects of drying shrinkage of inadequately dried concrete building blocks, the moisture and thermal movements of wall components generally, and the expansion of burnt clay bricks caused by protracted but progressively diminishing hydration.

The unrestrained expansion of clay brick walls, particularly those made with cream or lightly burnt bricks, can reach more than 1·5 mm per m (length and height) within the first five years.[25] It is practically halved by using only well-burnt bricks and storing the bricks for about 6 months before use. Vertical expansion joints 25 mm wide, containing a highly compressible filler and sealant, are located at 15-m intervals in walls or up to 10-m intervals in parapets built of 1 mm/m potentially expansive bricks. Horizontal joints 15 mm wide are located beneath shelf angles or haunches at floor levels in reinforced concrete frames, this provision being reduced in steel-framed, load-bearing, clay-brick buildings (AS 1640, BS CP111, CP121 and CP122).[25]

A high-lime or low-cement composite mortar with nominal 1 : 3 proportions takes up, by plastic yielding, most of the expansion of clay bricks with age. Its moisture movement is only one-quarter to one-half that of an air-entrained cement-sand mortar (see Table 27.2; also Section 31.2.1). A low-stress control-joint composite can be formed with split-strip, closed-cell polyethylene foam. The back-up foam component is cleanly sealed with a highly durable, butyl rubber elastomer, after the surface strip is removed at the final stage of building construction (see Section 27.2 and Reference 72 of Part 5).

TABLE 16.1   **STANDARDS ON MORTARS**

Title	Standard Specification		
	Australian	**British**	**ASTM**
Aggregates for masonry mortar and grout	—	—	C144; C404
Air content of hydraulic cement mortar	—	—	C185
Bleeding of cement pastes and mortars	1012.6	—	C232; C243
Bond strength of mortar to masonry units	—	—	E149
Building sands from natural sources	—	1198-1200	—
Calcium sulphate in hydrated portland cement mortar	—	—	C265
Chemical-resistant resin-type mortars	—	—	(See Section 21.1.6(a))
Coarse and fine aggregates	1141; 1465; 1466	812; 882	C33
Compressive strength of hydraulic cement mortars	—	—	C109; C349
Drying shrinkage, moisture movement and volume change measurements	1012.13; 1346; 1500	1180; 1217; 1881; 2028; 2908	C157; C341; C426; C490; C596
External rendered finishes	—	CP221	—
False set of portland cement (mortar method)	—	—	C359
Flexural strength of hydraulic cement mortars	—	—	C348
Grout for masonry	1475; 1640	—	C476
Hydrated lime for masonry purposes	1672	890	C207
Internal plastering	CA27	CP211	—
Masonry cement	1316	—	C91
Masonry mortar	1640; A123	CP121; PD6472	C270
Mechanical mixing of hydraulic cement mortars of plastic consistence	—	—	C305
Mortar-making properties of fine aggregate	—	—	C87
Packaged dry combined materials for mortar and concrete	—	—	C387
Sands for external renderings, internal plastering with lime and portland cement and floor screeds	—	1199	—
Sands for mortar for plain and reinforced brickwork, block-walling and masonry	—	1200	—
Tensile strength of hydraulic cement mortars	—	—	C190
Use of hydraulic cement mortars in chemical-resistant masonry	—	—	C398
Water-retentivity of mortar	1012.6; 1316; 1640 Commentary	—	C91; C110; C270

An air-entrained cement mortar has an adequate degree of workability and strength in the hardened state for practical uses, but it is rigid when hard and less likely (in comparison with high-lime composite mortar) to minimise cracking in long masonry walls. Lime in a mortar mix is beneficial in prolonging its water retention on contact with a dry surface (ASTM C91, C110 and C270; also Sections 27.2.2(b) and 33.4), and its moderate effect generally on the compressive strength of brickwork made with composite mortar is illustrated in Table 16.2.[84]

These results were derived from tests on short, square columns made with medium-strength bricks (20 MPa) and loaded axially at an age of 3 months. They indicate that 1 part binder : 3 parts sand gives a mortar with optimum strength under these conditions and that a 50 per cent replacement of portland cement by hydrated lime (which reduces the strength of mortar by over 40 per cent) lowers the strength of brickwork by only 4 per cent. While compressive strength is a useful index of performance, consideration must be given to bond and shear strengths for wind-load resistance, and a probable variation of 25-35 per cent through differences in quality of workmanship. With low-strength bricks (10 MPa), 1 part cement : 2 parts lime : 9 parts sands gives a mortar that may be used, except at low temperatures, for optimum brickwork strength.[93, 150]

TABLE 16.2   EFFECT OF MORTAR MIX ON STRENGTH
OF BRICKWORK

Proportions of Cement and Lime in Binder by Volume		Ratio of Strength*				
Cement	Lime	1 : 1†	1 : 1·5†	1 : 2†	1 : 3†	1 : 4†
100	0	—	—	0·96	1·00	—
50	50	0·72	0·87	0·94	0·96	0·92
40	60	0·70	0·84	0·90	0·92	0·87
30	70	0·66	0·77	0·84	0·87	0·81
20	80	0·58	0·68	0·74	0·79	0·71
10	90	0·47	0·56	0·60	0·65	0·59
0	100	—	—	—	0·48	—

* Strength of brickwork built with composite mortar : strength of brickwork built with 1 : 3 cement mortar.
† Proportion of binder (cement and lime) to sand by volume.

Brickwork in ordinary cement mortar often shows more efflorescence than it would do in a high-lime composite mortar. A possible reason for this is that a dense joint offers a high resistance to the passage of moisture. In consequence, more of the water present evaporates from the face of the bricks and deposits there any salts that may be present. Clayey fine sand promotes this effect and minor alkalies in the cement may contribute to it. Sulphate compounds, on being leached from clay bricks, cause a marked expansion of cement mortar joints (see Sections 19.1.1(b) and 36.4).

Mortar is dealt with further in Table 5.2, Appendix I, Chapter 25 and Sections 27.2.2(b) and 33.4.3. Spatterdash is described in the Glossary (Appendix VI), strength development is discussed in Reference 85, and masonry cement is dealt with in Table 3.6 and Sections 3.1.2(c), 27.2.2(b), (k) and 33.4.3. It is particularly desirable to use a clean, graded, coarse-gauge to medium-gauge sand in masonry cement mortar. The bulk density of air-free mortar can be determined by the method appended to AS 1379.

The bond of rendering or plastering to *in situ* concrete can be increased by using a retarder on the formwork and wire brushing the surface immediately on stripping the concrete (see Section 10.2). The cohesion of sprayed and grouted mortars is increased by activation in a high-speed mixer, such as a "Colcrete" machine (see Reference 134 of Part 5).[102, 103]

16.3 METHODS OF MEASURING ENTRAINED AIR

16.3.1 THEORETICAL WEIGHT METHOD

A known volume (e.g. 10 or 20 litres) of air-entrained concrete is weighed and the bulk density determined. This is compared with the theoretical bulk density of similar concrete without entrained air. Unless frequent checks of the specific gravities and moisture content are made on the job and calculations are revised, variations are sufficient in amount to affect the apparent air content by as much as 2 per cent.

16.3.2 GRAVIMETRIC METHOD

The bulk density of air-entrained concrete is compared with the bulk density of similar concrete without entrained air, each being determined by weighing known volumes of plastic concrete. Air determinations by this method may deviate by as much as 1·5 per cent from the true value, owing to variations in mix proportion, water/cement ratio and consistence. A determination by gravimetric calculation is illustrated in Section 5.3 and ASTM C138.

16.3.3 PYCNOMETER METHOD

A known volume of plastic concrete is stirred with a known volume of water, thereby diluting the concrete to such an extent that the entrained air is released and escapes to the surface. Measurement of the resulting reduction in combined volume is made by a hook gauge, and the volume of air released is determined. Results by this method have sometimes been erratic, because of incomplete release of the entrained air.

16.3.4 PRESSURE METHOD

Air pressure is applied to a column of water enclosed above a known volume of concrete, and the resulting change in the height of water in the column is noted (see Fig. 3.11). Since air is the only compressible material in the system, the original volume of the air can be established by applying the pressure–volume relationship for gas at constant temperature. Modifications of the original Klein-Walker meter are currently being tried, applied air pressures varying from 30 to 200 kPa.

The effect of particle interference is less at low pressure than at high, and is completely eliminated in the United States Engineer Corps air meter, in which vacuum is used instead of pressure. A correction is applied to air meter results to compensate for the air content (if any) of aggregate. The pressure method is the most promising of the methods currently available for use in the field and the laboratory.[27] Standard specifications for determining the air content of freshly mixed concrete and mortar are listed in Table 16.3, the air content of hardened concrete being measured when required with microscopical and high-pressure apparatuses (ASTM C457).[98]

16.3.5 VOLUMETRIC METHOD

An air meter for this method consists of a bowl and top section, each of which is not less than 5 litres capacity. The top is fitted with a transparent graduated neck and screw cap at the top. The bowl is filled with freshly mixed concrete in three layers of equal depth. Each layer is rodded twenty-five to forty times (according to concrete consistences of over 80 mm slump to below 30 mm slump) and the sides of the bowl are tapped ten to fifteen times after each rodding. The concrete is struck off flush with the top of the bowl.

TABLE 16.3   **STANDARDS ON AIR CONTENT**

Method	Standard Specification		
	Australian	British	ASTM
Gravimetric	—	—	C138
Micrometric (hardened concrete)	—	—	C457
Pressure	1012 Pt 4	1881 Pt 2	C231
Volumetric	1012 Pt 4	—	C173; C185

. The top section is clamped on and water is added, without disturbance, through a temporary funnel. The bottom of the meniscus is brought level with the zero or top mark of the neck, with the aid of a rubber syringe. After inserting the screw cap, the unit is inverted and agitated until the concrete settles free from the base. With the neck elevated, the unit is then rolled, rocked, rested and jarred until no further drop in the water column takes place through removal of air from the concrete.

A foamy mass on the surface of the water is dispelled with an appropriate amount of isopropyl alcohol, which is added with the syringe in 1·0 per cent increments by volume of the bowl. The percentage air content of the concrete is equal to the reading at the bottom of the meniscus of the liquid in the neck, plus the amount of alcohol that was added. The method can be used with lightweight-aggregate concrete (see Section 33.2.2(d); Table 16.3).

CHAPTER 17

POZZOLANS

17.1 PORTLAND-POZZOLAN CEMENT CONCRETE

17.1.1 GENERAL INFORMATION

The term "pozzolan" is used to describe naturally occurring and artificial siliceous or siliceous aluminous materials, which in themselves possess little or no cementitious value, but will, in finely divided form and in the presence of moisture, chemically react with calcium hydroxide at ordinary temperatures to form compounds possessing cementitious properties. The word is derived from Pozzuoli, where the Romans found a volcanic ash from Mount Vesuvius that would form a cement, with hydraulic properties, when mixed with lime putty.

Materials with these characteristics are: amorphous silica (opal, diatomite, colloidal silica); volcanic ash, scoria, pumice and obsidian; burnt clay and shale (e.g. burnt at 600-900°C for 2 hours), brick-dust, calcined bauxite, black-coal fly ash and well-burnt furnace clinker; granulated blastfurnace slag; powdered glass (95 per cent passing a 0·053-mm sieve (see Sections 3.1.2(e)(ii) and 3.3.3); and bleached ultramarine pigment which potently prevents efflorescence and increases 7-day compressive strengths by up to 70 per cent).

A portland-pozzolan cement is manufactured by blending 10-30 per cent by weight of pozzolanic material with portland cement, either by simple mixing or by intergrinding with cement clinker. The calcium hydroxide liberated during the process of hydration of the cement combines slowly with the pozzolan to give it cementitious properties, thereby contributing to watertightness and long,

327

continued gain in strength of the concrete. All types of portland cement may be mixed with a pozzolan, the quantity of which is lessened if it has a high water requirement.

The pozzolanic action of such material varies widely, and should be determined by trial mixes and the standard physical tests for cement. Also it can be determined by accelerated activity tests, physicochemical analyses, and long-term strength and durability tests of moist-cured and adiabatic mass-cured specimens. The strength of portland-pozzolan cement concrete is at first lower than that of a straight portland cement concrete that is made with the same proportions of mix. The former gradually increases in strength with age and becomes stronger than the latter.

The hardening of mixes containing pozzolans can be accelerated, if necessary, by an admixture of 0·2-0·3 per cent of caustic alkali or 0·1 per cent of alkali silicofluoride (by weight of cement), or by activation by a 10-minute exposure to concentrated (up to 10N) hydrochloric acid. The latter treatment is ineffective with fly ash, as it appreciably increases the water required for mixing and approximately halves the early and 28-day strengths. Strength is increased by several months of moist curing, the rate of increase being accelerated by curing at a temperature above 23°C. Permanent efflorescence of concrete building blocks can be controlled by the use of a pozzolan and effective steam curing (see Section 36.4).

Pozzolans containing amorphous silica are more active than those containing crystalline silica, the rate of reaction depending upon the chemical composition, pore structure and fineness of the material. Pumicites and diatomites are among the most active pozzolans and, while opaline silica is favourable as an ingredient in a finely divided pozzolanic admixture, it is potentially reactive if incorporated as a constituent of ordinary aggregate (see Sections 3.3.3, 3.3.4 and 15.2.3(a)). Partial calcination sometimes improves pozzolanic activity. Some clays and shales when thus treated are increased in chemical reactivity and, when finely ground, are highly satisfactory as pozzolans. Fly ash is the pulverised fuel ash that is caught in the electric precipitators of steam power plants. It is suitable for use if it is of extreme fineness and suitable composition. Fly ash and clinker (if used) must be free from sulphides and sulphates.

Superfine diatomite[31] has a specific surface area ten times that of portland cement. Experiments made at the University of California have shown that a 6 per cent replacement of portland cement by this material, when it is interground with an air-entraining agent, can produce (for a given consistence) the following results.

The water requirement of fresh concrete and the drying shrinkage of hardened concrete made with the blended cement are decreased.

Increased concrete strength at all ages and much reduced bleeding.

The durability of the concrete is much increased.

A possible reduction in cement content of concrete, up to as much as two bags per cubic metre without sacrificing final strength.

Investigations at the United States Bureau of Reclamation, on the activity of portland-pozzolan cement, have shown the following.

The compressive strength of a lime–pozzolan sand mortar can be correlated fairly well with that of mortar which is made with a corresponding portland-pozzolan cement.

In a general way, the strength contribution of a pozzolan at late ages can be predicted from the compressive strength of a lime–pozzolan sand mortar at an early age, when curing is done at a temperature above 23°C.

A preliminary method of testing the activity of a pozzolan, therefore, is to determine the compressive strength of cylinders (of 50 mm diameter and 100 mm height) of mortar at 7 and 28 days. The mortar is made of 1 part (by weight) of hydrated lime, 2 parts of pozzolan and 9 parts of sand with a fineness modulus of about 3·0 (e.g. Leighton Buzzard or Ottawa sand). The specimens (in sealed metal containers) are cured successively at 23°C for 24 hours and 55°C until 4 hours before the time of testing, the final stage being at 23°C. The mix has a flow consistence of 110 ± 5 per cent, and the thermal tolerance is up to ± 2°C.

The average strength of at least three of the lime–pozzolan sand mortar specimens should be not less than 5·5 MPa at 7 days. Preliminary tests (see Table 17.1) must be supplemented with trial tests on actual concrete mixes, incorporating the pozzolan, under standard and simulated field conditions.

The more important effects accompanying the use of a portland-pozzolan cement in mass concrete construction are summarised.

Improved workability.

Reduction of segregation, bleeding and possible efflorescence.

No increase in drying shrinkage when air-entrainment is used.

Protracted impermeability to pure and soft waters.

Improved resistance to dilute sulphate solutions and weakly acidic waters, an increased sulphate-resistance being obtained by the use of sulphate-resisting portland cement, a limited content of pozzolan and an increased cement content (see Sections 3.1.2 and 19.1.2).

Relatively low rise in temperature while hardening with a consequent reduction in volume change and the amount of artificial cooling required.

Increased plastic flow or creep and stress-adjusting characteristics (see Section 31.3).

Reduced expansion by alkali–aggregate reaction (if any). For this purpose, pozzolans with a high content of opal (such as diatomite and opaline chert or shale) are much more effective as reaction inhibitors than those which have a high proportion of glass (such as pumicite, tuff and fly ash; ASTM C441, see Table 17.1).

Long, continued gain in compressive and tensile strengths after slow development of same at early ages. This feature is more marked in lean mixes than in rich ones.

Appreciable saving in cost over concrete made entirely with portland cement when the pozzolan is readily available.

Portland-pozzolan cement, with or without air-entrainment, is particularly suitable for use in mass concrete structures (such as dams and bridge piers), where low heat of hydration is desired; hydraulic structures of all kinds where watertightness is important; structures subject to attack from ground water, seawater or dilute industrial wastes; and underwater construction where the concrete is deposited by tremie or bucket. A pozzolan may be used as a partial replacement of the fines of sand, without a reduction of cement content, where high-early strength is required. A small partial replacement of both is practicable (see Section 19.1.2(d) and Reference 34 of Part 2).[112, 113]

17.2 PORTLAND FLY ASH CEMENT

17.2.1 GENERAL INFORMATION

Fly ash, a residue from the combustion of pulverised coal, is collected by electrostatic separators from the flue gas of power plants. Its composition and properties can vary widely (AS K152 or BS 1016 Part 14).

The following specification indicates the properties of a black-coal fly ash that is suitable for use as a pozzolan.

Specific surface area between 400 m^2/kg and 700 m^2/kg as determined by the air-permeability method. More than 85 per cent of the material, when wet sieved, should pass a 0·053-mm sieve (Plate 11).

Loss on ignition, less than 10 per cent (to ensure low carbon content).

Magnesia (as MgO), 4 per cent maximum.

Sulphate (as SO_3), 2·5 per cent maximum.

Silica, over 40 per cent.

On mixing 1 : 4 with portland cement, concrete of the same consistence as that made with cement alone should be produced by adding less than 3 per cent additional water.

A fly ash meeting these details, or the standards listed in Table 17·1 might be produced in any steam plant using a variety of black coal and employing proper pulverised-fuel equipment and conditions of combustion. The glassy, siliceous component of fly ash combines slowly over a long period with the lime that is liberated during the hydration of portland cement. The rate of reaction is increased by the following means.

Fine pulverisation of black coal prior to firing.

A naturally formed composition of amorphous silica and hydraulic lime (e.g. $4SiO_2 \cdot 2Al_2O_3 \cdot 1CaO$).

Subdivision of fly ash in a ball mill, although the contribution of fine spherical particles to the workability of a mix is offset by crushing them.

Effective moist curing near 40°C, or autoclave curing in saturated steam at approximately 1000 kPa pressure or 185°C.

TABLE 17.1   STANDARDS ON POZZOLANS

Title	AS	BS	ASTM
Blended hydraulic cements	1317	—	C595
Sampling and testing fly ash for use as an admixture in portland cement concrete	—	—	C311
Fly ash and raw or calcined natural pozzolans for use in portland cement concrete	1129; 1130	3892	C618
Specification for fly ash and other pozzolans for use with lime	—	—	C593
Test for effectiveness of mineral admixtures in preventing excessive expansion of concrete due to the alkali-aggregate reaction	—	—	C441

The early strength of portland fly ash cement mass concrete is normally lowered by replacing some of the cement with fly ash. With age, however, the strength gradually increases and eventually exceeds that of an equivalent normal mix. The rate of increase of strength after 28 days is greater than that of straight portland cement concrete or mortar.

For the achievement of a prescribed 28-day strength, fly ash (with low carbon content and 400 m^2/kg fineness) can be utilised in mixes as a 15-20 per cent replacement of cement or a partial replacement of cement and sand, the sand reduction being about 10 per cent by weight. A cement-dispersing agent improves the 7-day depressed strength and may facilitate hot-weather concreting by set retardation (see Sections 6.4 and 10.2).[113] These mixes are suitable for pumped concrete, workable bleed-resistant concrete and thick sections requiring a reduced rate of evolution of heat of hydration. In massive construction with a low rate of strength gain, the cement replacement may range from 20 to 30 per cent.

Portland fly ash cement concrete, with a small replacement of cement by fly ash, quickly attains strength equal to that of straight portland cement concrete, provided that both mixes have the same proportions. Drying shrinkage is usually greater for portland-pozzolan cement than for portland cement alone, but it is generally less with fly ash than with most of the natural pozzolans. As indicated in Section 3.1.2(e)(vii), a high-magnesia cement can be volume-stabilised with a siliceous pozzolanic admixture, such as fly ash (see Reference 10 of Part 1). An admixture of fly ash improves not only the durability of concrete but its resistance to penetration by gases or moisture (see Section 11.1.2(c)).

Investigations of the effect of percentage of carbon in fly ash on the strength and behaviour of concrete have shown the following.

A high-carbon fly ash decreases the efficiency of air-entraining agents and requires more water than a low-carbon fly ash.

There is no consistent relationship between carbon content of fly ash and strength.

Under proper conditions of curing, an improved concrete will result (particularly as regards weathering resistance) from about 20 per cent replacement of cement by fly ash with a reasonably high carbon content (up to 12 per cent).

The effect on drying shrinkage is slight and, in general, is the same as for straight portland cement concrete.

With few exceptions, the finer the fly ash, the stronger the concrete.

The appearance of concrete is unaltered by the introduction of fly ash into the mix, and all types of portland fly ash cement may be used (i.e. high-early-strength, low-heat, ordinary and sulphate-resisting portland cements blended with fly ash). Various sources of relavent information are contained in Table 17.1 and References 29-32. For purposes of quality control, preliminary tests must be followed regularly by control tests as concreting proceeds. Excellent results can be obtained by incorporating a pozzolan in the form of a slurry that has been activated by compressed air or high-speed mixing.

The design of fly ash concrete is illustrated in References 42 and 53 of Part 2. In all projects, thorough moist curing is essential for strength development and nondusting surfaces.

CHAPTER **18**

LIME

18.1 CEMENT–LIME CONCRETE

Various types of lime are described in the Glossary (Appendix VI) and, for best results, it is very desirable that prepared lime should be used in a fresh or an uncarbonated condition. Hydrated lime is commonly used in composite mortar, but it should not be added to concrete without considering the conditions to which it will be exposed. When a mix is difficult to work, it is better to vary the aggregate gradation than to improve workability by the addition of lime.

A large number of tests by Duff Abrams[88] has shown that the addition of lime beyond a certain percentage slightly reduces the strength of concrete, but this is not serious if the proportions in Table 18.1 are not exceeded. Normally, about one-half of these tabulated quantities are sufficient to notably improve the workability of an otherwise harsh mix.

TABLE 18.1   **LIMITING ADMIXTURES OF LIME**

Mix by Volume			Amount of Hydrate per Bag of Cement (kg/40 kg)
Cement	Sand	Stone	
1	1·5	3	3
1	2	4	4
1	2·5	5	5

The richer a concrete mix, the less does the addition of lime increase its workability. An admixture of lime for improved workability is, therefore, more effective in a lean mix than in a rich one. Dry hydrated lime, which is obtainable in nonstandard containers from various sources, may differ in purity, fineness, hydration maturity or bulk density. On the basis that high-calcium lime (specific gravity 2·35) is supplied in three-ply or five-ply paper bags of 25 kg and 0·035 m³ typical nominal capacity, the bagged bulk density of the hydrate may be taken for gauging purposes as 720 kg/m³.

The bulk density of loose hydrated lime, however, depends largely on its content of entrained air and may range from about 400 to 550 kg/m³. Where composite mixes are proportioned by volume, therefore, the lime should be gauged bag-wise or batched, either gravimetrically or volumetrically, in such a way that the correct nominal quantities of lime will be used on a specified basis. Corresponding bulk densities for portland cement (specific gravity 3·15) are 1500 and about 1200-1280 kg/m³, respectively and, as described in Section 3.3.6(b), corrections should be made in volumetric proportions for the bulking effect of damp sand.

When lime is prepared and used as a lime putty, care should be taken to incorporate the correct amount of lime and allow for the water content in mixing operations. To this end, a given volume of lime putty is considered to contain approximately an equal volume of hydrated lime when the bulk density of the putty is 1280 kg/m³. Lime putty may be prepared from hydrated lime or quicklime. Putty made from the former should be allowed to mature from milk of lime for not less than 16 hours, while putty made from the latter should be allowed to mature from milk of lime (after it has been run through a 0·60-mm sieve) for at least 2 weeks for best results, excess water then being removed and the putty protected from drying out (AS 1672 and BS 890).

Improved workability is caused by the lime acting as a lubricant and allowing the particles of the mix to slide more easily over each other at the time of compaction. At the same time, an admixture of hydrated lime decreases the rate of hardening of the concrete, which may have some significance for work in cold weather. Investigations have shown that the compressive strength of ordinary cement-lime concrete at 28 days must usually be estimated from the cement and water content of the mix. Indications are, therefore, that ordinary hydrated lime acts as a filler, which is beneficial in reducing voids by making good a possible deficiency in grading of fine aggregate.

On the other hand, hydrated, eminently hydraulic lime (when available) may be used to advantage in appreciable proportions with portland cement, or by itself as a binder for making concrete, where quick hardening is not required or high-early strength is unnecessary. The protection given to reinforcement by concrete is mainly due to its alkaline character, which can be increased by an admixture of hydrated lime. Standard specifications on lime are listed in Table 18.2 and composite mortars are dealt with in Sections 16.2, 27.2.2(b), (k), 33.4.2 and 33.4.3.

TABLE 18.2   **STANDARDS ON LIME**

Title	Standard Specification		
	Australian	**British**	**ASTM**
Building limes	1672	890	C5; C6
Chemical analysis of limestone, quicklime and hydrated lime	1672	890	C25
Hydrated lime for masonry purposes	1672	890	C207
Hydraulic hydrated lime for structural purposes	—	—	C141
Quicklime and hydrated lime for calcium silicate brick	1672	890	C49; C415
Sampling and testing of lime and limestone products	1672	890	C50; C110
Special finishing hydrated lime	—	—	C206

18.2 SOUNDNESS OF LIME

18.2.1 GENERAL INFORMATION

Unsoundness of lime is caused by unhydrated calcium and magnesium oxides, particularly the latter (if in excess of 3 per cent by weight of quicklime). A simple way of overcoming the delayed expansion of limes due to these components is to add about 2 per cent of calcium chloride (by weight of hydrated lime) to the gauging water. In lime containing 10 per cent or more magnesium oxide, an additional 1 per cent of calcium chloride is added per 5 per cent of magnesium oxide in excess of 10 per cent. The admixture of calcium chloride quickly reacts catalytically with these oxides to form both calcium and magnesium oxychlorides. These components, being in small concentration, are immediately hydrolysed and decomposed into calcium hydroxide, magnesium hydroxide and calcium chloride. Possible problems associated with the delayed hydration of calcium oxide and magnesium oxide in hardened products are thus readily prevented.

If it is not known whether hydrated lime that is to be used in mortar is sound or not, 2 per cent of calcium chloride should be added to the gauging water. No ill effect is caused by the addition of calcium chloride should the hydrated lime used be sound and not need such treatment. An admixture of calcium chloride accelerates the hardening and increases the workability of portland cement mixes. Other possible side effects on concrete and reinforcement should be considered (see Sections 10.1 and 15.2.2(a)).

Alternatively, where full soundness is required, lime containing magnesium oxide or overburnt calcium oxide should be slaked and soaked for at least 2 weeks. Other methods include hydration in a high-pressure hydrator at 850 kPa for 1 hour or more and (if the lime is hydrated and bagged) steam treatment in an autoclave at 200-300 kPa for 3-6 hours. The treatment given depends on the composition of the lime and the degree to which it is burnt and initially hydrated. For best results, steps should be taken to minimise or limit carbonation by the atmosphere of hydrated lime before it is used for making lime mortar, lightweight calcium silicate hydrate products and sand–lime bricks.[94, 95]

18.2.2 TESTS FOR UNSOUNDNESS

Test 1. A pressed bar of mortar (100 mm × 25 mm × 6 mm), containing 1 part by weight of lime mixed with 9 parts of sand, is subjected to the action of steam in an autoclave at a pressure of 1000 kPa for 5 hours. Following the treatment, a sound lime must show no signs of "cracking, crumbling, checking or alteration of form".

Test 2. A lime–sand mortar* containing 15 per cent by weight of lime, in a Le Chatelier mould, is autoclaved at a temperature of 220°C (i.e. 2000 kPa) for 3 hours. Under these conditions, the hydration of magnesium oxide is complete. An expansion of more than 3 mm is indicative of unsoundness and one of 3 mm or less of full soundness.

18.3 LIME–SOIL STABILISATION

When soils of high plasticity are treated with an admixture of hydrated lime, or either ground quicklime or lime slurry, their plasticity index is reduced significantly. There is a consequent improvement in shrinkage and drainage characteristics and the mixture, on being cured, shows increased compressive strength and durability with age. A wide range of soils are suitable for treatment, although the increase in strength obtainable may be very much more marked with some soils than with others. Where soils are of relatively low plasticity, they can be stabilised with lime and fly ash, these ingredients being admixed in requisite quantity (in the

---

*Sand gradation: 70 per cent passing a 0·850-mm sieve and retained on a 0·600-mm sieve; 30 per cent passing a 0·150-mm sieve. Water/lime–sand ratio: 6·5 per cent.

proportion of 1 : 2, respectively, by weight) and cured under conditions that promote pozzolanic action.[35, 89, 90]

The stabilising effect of lime in clayey soil is brought about by an immediate flocculation of the clay particles, which is associated with a substitution of exchangeable cations by calcium ions (see Glossary, Appendix VI). This reaction enables clay to become friable and readily processed when wet. It is followed by a slower pozzolanic action in the presence of moisture, whereby cementitious bonds are formed through a combination of the lime with fine siliceous aluminous material that is present in the soil.

Lime–soil mixes should be evaluated in a laboratory before they are used in the field. Depending on the nature of the soil and the project, the requisite proportion of hydrated lime lies between 2 and 8 per cent by weight of the dry soil. With most soils stabilised to a consolidated depth of 150 mm, some $2 \cdot 7$ kg/m^2 must be applied to introduce a lime content of 1 per cent.

For the construction of semirigid road bases and subgrades or the improvement of clayey work-site areas, the following operations are necessary.

Scarify and partially pulverise the soil.
Evenly spread the lime.
Intimately mix the lime and water with the soil.
Compact the mixture to maximum practical density.
Properly cure the stabilised course before placing on it a wearing surface or a complete pavement base course.

Soils with a high content of clay, after being partially mixed with lime, should be lightly compacted with a multityred roller and allowed to moist cure for 24-48 hours. This step promotes disintegration of the clay and facilitates the final mixing operation with a rotary mixer or stabiliser machine. An optimum moisture content should be approximately maintained, so as to facilitate compaction of the lime–soil mixture to 100 per cent of the maximum density obtainable by a standard light compaction test (AS 1289, BS 1377 and 1924, ASTM D558, D698, D1241 and D1557).

Clayey soils may be expeditiously conditioned and stabilised with lime and portland cement, the equipment used being similar to that required for soil–cement (see Section 26.5). Lime–soil surfaces should be either moist cured for 5-7 days (by sprinkling and rolling or covering with wet, base-course material; see Section 26.2), or membrane cured with a few coatings of bituminous emulsion. Workers engaged in handling, spreading and mixing lime should be supplied with protective clothing and closely fitting goggles.

CHAPTER 19

CHEMICAL ATTACK

19.1 AGGRESSIVE SOLUTIONS

19.1.1 OCCURRENCE

(a) Acid.
Several chemical reagents attack the cement matrix and, to a lesser extent, the aggregate of concrete. These include surface, brackish and aquifer waters (Glossary, Appendix VI) containing free carbon dioxide. Carbon dioxide forms a mild to strong aggressive carbonic acid solution that has a high capacity for dissolving calcium hydroxide and calcium carbonate and for both decomposing and dissipating hydrated calcium silicates from set cement, particularly in poor-quality concrete. Usually, in the medium term, the resultant leaching does not detract significantly from the basic serviceability of structures made with good-quality concrete (see Sections 19.2, 31.1, 31.2, 31.4 and 33.3.3; also see Reference 12 of Part 1).[108, 140]

Brown-coal, peat-marsh and moor waters may have a pH value down to about 4·5, due to aggressive carbonic acid, humic acid or even sulphuric acid (arising from pyrites or marcasite in the soil and a polluted industrial atmosphere). When the ground water is lowered by drainage, pyrites or marcasite in swampy ground and rock formations oxidise to form ferric sulphate and, in wet conditions, sulphuric acid (see Section 3.3.3(a)(viii)). The chemical composition of ground water can show marked variations due to changes in locative, seasonal, seepage, contaminating and drainage characteristics; while an incremental degradation of atmospheric moisture by sulphurous flue effluents can have an insidious effect on basic cementitious materials and corrodible metals.

Aggressive ground water can cause concrete sewer pipes to collapse at the invert by percolating along the bed, attacking and entering mortar joints, and corroding the invert internally as well as externally. Pipe joints should be made, therefore, with a rubber water-seal and high-alumina cement mortar or a sulphate-resisting cement mortar, instead of ordinary portland cement mortar. Weakly acidified solutions, including rain water, also attack copper waterstops and, on penetrating the concrete, promote the rusting of steel reinforcement.[37-39, 140, 151]

Acid attack above the daily high level of flow in outfall concrete sewers is due to sulphuric acid, the rate of deterioration being greater generally in hot than in cold climates. The action arises from hydrogen sulphide, which is produced from sulphur compounds in sewage due to the activity of metabolic organisms or sulphate-reducing bacteria. These organisms live in slimes found on sewer walls below the flow line.

High concentrations of sulphuric acid are formed on the moist walls above liquid level by the action of these sulphur-oxidising bacteria, thus leading to bacterial corrosion (see Glossary, Appendix VI). The progressive deterioration and expansion of ordinary portland cement also causes the development of a laminated structure in the zones of attack on the inside of the sewer. Aggressive industrial wastes and sewage with a high content of sulphate accentuate the problem, which results in a reduction of wall thickness in main concrete sewers of up to 6 mm per annum.[36, 146]

Lactic, citric and acetic acids, sugar solutions and vegetable oil in food processing plants are spilled onto concrete floors. A mild but persistent attack softens the surface and causes it to wear rapidly and unevenly. Acetic acid attack and gradual deterioration are caused by silage in concrete silos, while chemical attack in substructures and superstructures can arise from ground or industrial sources of sulphates (see Sections 14.7.2 and 15.2.3(b), (d) and (f)).[121]

Ammonium chloride and nitrate solutions, particularly the latter, cause strength reductions by leaching calcium hydroxide from the cement matrix of concrete. This attack can be controlled by the integral use of a pozzolan, which combines with calcium hydroxide during the curing period to form cementitious silicates.

(b) Sulphate.

Dry portland cement concrete in dry sulphate-bearing soil is not attacked. Sulphates in solution, however, cause severe attack by diffusing into the concrete through pores, causing softening or swelling by chemical action with the constituents of hydrated portland cement, and disruption by expansion as they crystallise. The rate of deterioration is governed mainly by the quality, capillarity and age of the concrete; the type of cement; the nature and concentration of the sulphate solution, its pressure gradient through the concrete; and the rate of evaporation of moisture from exposed surfaces. It is most severe with cyclic saturation and drying, and where a wicking action causes the solution to be drawn strongly along capillaries in the concrete (see Reference 111 of Part 1).[73, 108]

Attack may occur when the concrete is exposed to ground water that has become contaminated by percolating through gas industry sites, refuse tips, slag heaps, gypsum plaster spoil, terrain containing gypsum or pyrites, orchard soil dressed with sulphate-bearing fertiliser (then irrigated by flooding), and through industrial waste filling that contains sulphates and is placed behind retaining walls and under floors. It is worth noting that trenches in which pipes are laid often act as drains after the back-filling is in place; aggressive aquifer solutions may thus be brought into contact with the outside surface of the pipes far from the source of contamination of the ground water. In outfall sewers, attack is caused by hydrogen sulphide in sewage gas, the action being hastened when ventilation is poor and the temperature of the sewage is greater than 23°C. Other disruptive agencies include pyrites or marcasite and, in tall chimneys and railway tunnels, a condensate containing sulphur compounds from burnt fuel.[51, 68, 138, 171]

Seawater contains a little over 0·25 per cent of sulphate and, in marine structures, the deterioration of concrete is caused by alternate wetting and drying, wave and frost action, attack by magnesium sulphate and chloride, and corrosion of the steel reinforcement. The concrete just above highwater level is most vulnerable.

The following reactions may take place in portland cement concrete under moist sulphate conditions.

1. Conversion of hydrated calcium aluminate (and ferrite) to calcium sulphoaluminate or ettringite (and sulphoferrite).
2. Conversion of the calcium hydroxide to calcium sulphate.
3. Decomposition of the hydrated calcium silicates.

With calcium sulphate solution, reaction 1 occurs; with sodium sulphate solution, reactions 1 and 2 occur; with magnesium and ammonium sulphates, all three reactions can occur. The first three of these sulphates are represented respectively by gypsum, Glauber's salt and Epsom salt. In dense concrete, only partial conversion of calcium hydroxide to calcium sulphate occurs over long periods.

With strong calcium chloride solution, disruption is caused by the formation of calcium chloroaluminate. Solutions of sulphate and chloride salts, on penetrating concrete, promote the corrosion of steel reinforcement. Ground water containing 350 parts of chloride per million can severely attack steel; while hydrogen sulphide in very low concentration, either as a gas or dissolved in moisture, can cause the structural failure of steel that is under stress (this is called stress corrosion, see Glossary, Appendix VI).[17-19, 110, 117, 144-146, 155, 172-177, 187]

Increased resistance, or a reduced rate of attack by chemical reagents, can be achieved by several means, which should be applied both individually and collectively for effectiveness. They include suitable mixes, increased cement content and thickness of section, special cements, suitable aggregate, control of manufacture, maximum compaction, early and effective curing, special admixtures and surface treatments, and regular preventive or curative maintenance. Standards on site investigations are given in Reference 10 of Part 4; and methods of test for chloride and sulphate ions in water are given in AS 1289, BS 2690, ASTM D512 and D516.

19.1.2 PROLONGATION OF SERVICE

The best probable service performance of concrete, under aggressive conditions, may be estimated from chemical tests (see Section 14.7.2). It is related to the following factors and those listed in Sections 15.2.2(c), 15.2.3(b) and (d).

(a) Suitable mixes.
Suitable mixes for land structures and various sulphate conditions are indicated in Table 19.1. The special cements referred to are dealt with here, their prescribed minimum contents being increased for particular situations, such as thin sections of cast-in-place concrete. Mixes used in close proximity to acidic conditions should contain about 25-50 per cent more cement than is ordinarily used in concrete, the maximum water/cement ratio being 0·45 by weight (see Section 15.2.3(b) and BS CP110 Part 1).

As calcium sulphoaluminate forms most readily in concentrated solutions of calcium hydroxide, and since the disintegration of portland cement in sulphate solution is due chiefly to the formation of the first compound, the severity of attack may be reduced by adding active silica or a pozzolan to a concrete mix. This additive will combine with the calcium hydroxide, making much of it insoluble, and thus retard the formation of the sulphoaluminate (see Chapter 17 and Section 19.1.2(d)(iii)).

TABLE 19.1   CONCRETE EXPOSED TO SULPHATE ATTACK

Class	Ground Water Sulphate* (ppm)	Soil Sulphate* Total (per cent)	Soil Sulphate* 2:1 Water/Soil extract (g/litre)	Type of cement	Minimum cement content (kg/m³)† Aggregate maximum size: 40 mm	20 mm	10 mm	Maximum water/cement ratio for lower sulphate* limits by weight††
1	Less than 300	Less than 0·2	—	Ordinary portland or portland-blastfurnace	240	280	330	0·55
2	300–1200	0·2–0·5	—	Ordinary portland or portland-blastfurnace	290	330	380	0·50
				Sulphate-resisting portland or supersulphated	240	280	330	0·55
3	1200–2500	0·5–1·0	1·9–3·1	Sulphate-resisting portland or supersulphated	270	310	360	0·50
4	2500–5000	1·0–2·0	3·1–5·6	Sulphate-resisting portland or supersulphated	290	330	380	0·50
5	Over 5000	Over 2·0	Over 5·6	Sulphate-resisting portland or supersulphated with adequate protective coatings.‡	330	370	420	0·45

* Sulphates expressed as sulphur trioxide.

† Requirements for dense, fully compacted, properly cured concrete.

‡ See Section 19.1.2(e).

Note: These data apply only to dense concrete that is made with approved dense natural or furnace slag aggregate (see Table 3.8) and is normally placed in near neutral groundwaters (pH 6-9). These contain naturally occurring sulphates without other aggressive contaminants: such as mineral acids, the high content of free carbon dioxide in aquifers, and ammonium salts. Minimum cement contents apply generally to the lower classified limits of sulphate. Permissible class limits may be about halved for severe conditions (e.g. sulphate solutions flowing under hydrostatic pressure or migrating continually by capillary action through the concrete).

Pertinent molecular weights are as follows: Calcium sulphate ($CaSO_4$): 136. Sodium sulphate (Na_2SO_4): 142. Magnesium sulphate ($MgSO_4$): 120. Sulphur trioxide (SO_3): 80. Ammonium sulphate ((NH_4)$_2SO_4$): 132. Sulphate (SO_4): 96.

(b) Special cements.
The lower the content of tricalcium aluminate in portland cement of suitable composition (see Section 3.1.2(e)(viii), the better able is concrete made with it to resist sulphate attack. When this component is less than about 5 per cent, the cement is not only sulphate-resistant but an extra bag per m^3 will increase the sulphate resistance of concrete over 50 per cent. While the concrete can thus be made to resist calcium and sodium sulphate solutions (and attack by magnesium sulphate can be mitigated), it is still vulnerable to attack by ammonium sulphate (e.g. at gas works).

Special cements include supersulphate and high-alumina cements, and portland cements of sulphate-resisting, low-heat, pozzolanic and blastfurnace type. Those indicated in Table 19.1 conform with the recommendations given in BRE Digest 174 (1975) for dense, fully compacted concrete, which is to come in contact with naturally occurring sulphates (see Section 3.1.2(b), (e) and (g)).[73]

Supersulphate cement has given an acceptable service in dense concrete, which has been made with a water/cement ratio of 0·40 or less and brought into contact with mineral acids down to pH 3·5. Early and adequate moist curing (see Sections 9.1 and 9.2) is essential to stop evaporation and ensure the proper hardening of this concrete. High-alumina cement concrete may be used with discretion down to pH 4·0, provided that particular care is taken to ensure that it is effectively designed, made, placed and cured, and is used under only the most exacting and favourable conditions (see Sections 3.1.2(b) and 9.2, also Reference 12 of Part 1 and Reference 24 of Part 2).

Sulphate-resisting portland cement (preferably with a tricalcium aluminate content below 3·5 per cent) is slightly more resistant than ordinary portland cement on exposure to an acidity limit of pH 6. With more aggressive environments, long-term protective or tanking measures are a prerequisite for satisfactory serviceability. In each instance, the concreting procedure adopted should suit the particular characteristics of the special cement employed.[201]

Dense rich mixes, made with a water/cement ratio below 0·45 by weight, are much more resistant to chemical attack than ordinary mixes. For suitable performance in restrained sections, the specific surface area of the cement should be somewhat below 400 m^2/kg (see References 12, 26 and 111 of Part 1, see also Sections 3.4, 15.2.2(b) and 31.2).[145] Precautions pertaining to seawater and waters containing aggressive carbon dioxide are given in Sections 19.1.3 and 19.2; while testing and a typical use of sulphate-resisting portland cement are exemplified in Sections 3.1.2(e)(viii) and 14.7.2, also in ASTM C452 and in Plates 10 and 14.

(c) Aggregate.
Aggregate should be sound, impermeable and in general agreement with the requirements set out in Sections 3.3.1 and 3.3.4; AS 1141 and 1465, BS 882 and ASTM C33.

Contrary to expectation, studies indicate that the corrosion-resistance of main sewers can be increased considerably by substituting an acid-soluble aggregate (e.g. limestone) for an acid-insoluble one. A tentative explanation of this result is that, with siliceous or other insoluble aggregate in concrete, acid progressively attacks the cementitious matrix only, the aggregate particles at the surface dropping away as the matrix is destroyed. On the other hand, concrete containing calcareous aggregate can be attacked by acid over the whole of the exposed surface, the acid being neutralised to an increased extent and the cementitious matrix destroyed to a less extent in the process.

This type of concrete is attacked at a reduced rate also by sulphate under acid conditions. Suitable limestone is more effective than dolomite as aggregate for the purpose. The service-life of concrete sewers could be increased five-fold by this technique, provided that sacrificial concrete is included in design (e.g. 90 mm cover at the crown; see Section 15.2.3(d)).[40, 66]

(d) Manufacture of concrete.
Essential physical requirements of concrete are high quality, low permeability and low porosity. Contributing factors to these characteristics in placed and consolidated concrete are the following.

Mixes that are designed for maximum density (i.e. with low void content and surface area of aggregate) and have an increased content of cement (e.g. at least nine 40-kg bags/m³; see Table 5.1), that must meet standard specifications at the time of use.
Correct gauging and thorough mixing of cement with satisfactory aggregate and the minimum quantity of water for workability.
Placing without segregation, followed by thorough and uniform compaction.
Reconsolidation by high-frequency revibration and refloating operations near the time of initial set of the cement (e.g. between 2 and 3 hours after mixing).
Watertight construction joints.
Early and adequate, continuous moist curing for at least 14 days at normal temperatures (or 7 days after high-temperature, atmospheric steam curing) to produce low permeability.

In addition to the foregoing, consideration should be given to the following factors (where applicable) for increased durability.

1. Air-entrainment (3-6 per cent by volume) with or without cement dispersion, which are brought about usually by the addition of surface-active agents at the mixer. Taken individually or collectively these aids to durability improve the workability, uniformity and surface finish of concrete, and decrease its permeability. Their use should be accompanied by a reduction in the quantity of mixing water and a partial redesign of the mix. These adjustments are required in order to maintain a given degree of workability and content of cement per m^3 of concrete. Control factors required are indicated in Section 6.4 and Chapter 16. The durability of the concrete is greatly increased towards marine and freeze–thaw conditions, but not towards acidic solutions.

2. A pozzolan, of cement fineness, in portland cement concrete. In mass concrete, it is used as a 10-30 per cent (by weight) replacement of cement or cement and sand. In precast concrete, about half this amount is used as an additive or a partial replacement of fine aggregate.

Note: Precast concrete products must have a high cement content to start with, not only for purposes of durability but also for high-early strength and ease in handling. A pozzolan does not, of itself, make concrete immune to chemical or sulphate attack, but it can markedly assist in prolonging the life of concrete where pH values are above 5·0. Under marine conditions, a 30 per cent replacement of cement by a pozzolan is likely to decrease permanently the compressive strength of precast concrete. Steam curing promotes pozzolanic action and durability.

3. A combination of items 1 and 2, with a simply adjusted mix. A lessening of requisite mixing water by item 1 more than offsets any increase that may be caused by item 2 for a given degree of workability. The net effect in general is a reduction of permeability and drying shrinkage, and greatly increased durability of the concrete under many rigorous conditions.

4. High-frequency vibration of suitably proportioned low-slump or no-slump concrete, with or without high pressure, vacuum or revibration procedures for thorough compaction.

5. Prestressing by pretensioning or post-tensioning methods. When prestressed concrete is made with impervious aggregate

and adequate cover, a high degree of durability is associated with a dense, high-quality concrete that is free from cracks.[100]

6. Early and thorough moist curing, followed by drying and carbonation treatment or several weeks storage. Effective carbonation (see Sections 9.6 and 31.4) or steam curing (Section 9.4) at 55-70°C increases the resistance of concrete to sulphate attack, but not to acidic solutions.[139]

7. High-pressure steam curing or autoclave treatment of precast concrete products, where immediate and marked durability are required. Curing is done in saturated steam at 1000 kPa for periods of 8-12 hours. An admixture of pozzolan or ground siliceous sand can lead to an effective and economical use of portland cement as indicated in factor 2 described previously.

8. Permanent formwork of precast shells or units of high-alumina cement concrete, behind which is placed sulphate-resisting portland cement concrete.

9. Silicon tetrafluoride gas treatment under pressure (see Section 22.2).

(e) Surface treatments.

While surface treatments may tend to prevent the entry of destructive agents into concrete, and thus serve a useful purpose, their service life may be limited. The essential requirement for lasting satisfactory performance, therefore, lies in the inherent qualities of the concrete itself. Coatings are frequently used on portland cement concrete, but they are satisfactory only if correctly applied to first-class concrete and maintained. Special linings are mastic asphalt and stress-relieved stainless steel, also surface-primed, impermeable tankings (such as a double-seal coat of coal-tar epoxy resin or clay-stabilised bituminous emulsion, reinforced with fibreglass or nylon fabric membranes (see Section 11.1.2(f)).

Where deterioration is likely to occur through a prolonged exposure to chemical reagents, the dimensions of structural sections should be increased, so as to allow for some wastage.[66] In strongly acidic ground, subsoil and well-point drainage should be of acid-resistant construction, and lime-stabilised bedding and backfilling should be used in foundation trenches.

In areas that are inaccessible for surface renewal, a protective membrane may be placed under the concrete. This may consist of two or three layers of bituminous asbestos felt, which are lapped, sealed at joins, heavily coated with hot bitumen and covered with a single layer of stone screenings; this last layer is then pressed in. On placing the concrete, a permanent lining or protective membrane is thus formed. Alternatives are fibreglass polyethylene sheet (0·25 mm thick) or polyvinyl chloride sheet (BS 1763 and 2739) with lapped and sealed joins (see Section 21.3).

Surface treatments are intended primarily to supplement other methods of providing resistance to attack by destructive agents. They should be selected or designed and applied to suit particular projects (see Sections 11.1.2(f), 20.2 and 21.1).

19.1.3 APPLICATION TO MARINE WORK

For marine structures, portland cement concrete should be of modified or air-entrained type and well compacted. Sulphate-resisting, low-heat or blastfurnace portland cement, where economically available, should be used in preference to ordinary portland cement (see Section 3.1.2(e)). Some 25-30 per cent more cementing material should be used in precast units and the exposed faces of mass construction than is customary for ordinary concrete. The aggregate must be unaffected by seawater and construction joints should be avoided from 0·6 metre below low water to at least an equal distance above high water level. Methods for making dense, durable concrete should be followed and, where practicable, moving forms should be used to obviate joints in cast *in situ* concrete. Seawater should not be able to enter at any point.

The minimum clear cover over steel reinforcement at the intertidal zone should be 65 mm for precast units and 75-100 mm for cast *in situ* concrete. Great durability and a reduction of cover can be obtained by impregnating dry precast units with bitumen by vacuum-pressure methods. Underwater and coal-tar epoxy resins, and cadmium-coated steel[128] are very suitable in marine environments.

Reinforced concrete near seawater should be watertight and free from a calcium chloride admixture.[15, 38, 41, 108, 109] Underwater concreting and prestressed concrete are dealt with in Section 30.2 and Reference 74 of Part 6.[100, 117] Pretensioned-type, air-entrained concrete piles are extremely durable when made with an impermeable cover that is very little impaired with hair-line cracks. A tough-skin facing of fibre-reinforced concrete is an effective means of improving the performance of armoured marine breakwater units (see Section 3.4.2 and Reference 145 of Part 1).

19.2 PURE AND SOFT WATER CORROSION

Portland cement concrete is corroded in certain structures, such as pipe lines, where large volumes of pure or soft water flow for long periods. The severity of attack is related to the quality both of the water and the concrete, and it may be of a low order with best quality concrete and autoclaved asbestos-cement pipes. Soft water absorbs carbon dioxide from decomposed vegetation and the air (see Section 19.1.1). Also, snow and desalinated waters, in pure form, are often particularly aggressive (see Section 15.2.3(f)). As a first step toward disintegration, calcium hydroxide is removed in solution, this material being more soluble in cold water than in warm. If the water contains free carbon dioxide, its aggressiveness is increased and both calcium hydroxide and calcium carbonate go into solution as calcium bicarbonate, $Ca(HCO_3)_2$.[175]

In addition, the hydrated calcium silicates in set portland cement are stable only in contact with an alkaline solution; so that if calcium hydroxide present in set cement is removed, the silicates are decomposed with a liberation of calcium hydroxide, which in turn is removed. In this way, reactions can continue until most of the hydrated silicates are decomposed and the leaching action can be advanced to a stage where no cementitious material remains to bind the aggregate together (see Reference 12 of Part 1).[108, 140]

The pH value of the water, as a measure of its aggressiveness, must be considered in relation to the hardness of the water and the content of free carbon dioxide. Thus, soft water with a pH value of over 7 can be aggressive if its temporary hardness (which is removable by boiling) is negligible; that is, below 10 ppm of calcium and magnesium carbonate (present as bicarbonates). On the other hand, if the temporary hardness is above 50 ppm, serious corrosion is unlikely to take place if the pH value is as low as 5. If the temporary hardness is below 10 ppm, corrosion can be serious, even if the content of free carbon dioxide is negligible. At a higher state of temporary hardness, free carbon dioxide can be tolerated probably up to 10 ppm.[117, 169, 181]

The severity of attack in water-supply pipes of concrete (see AS 1342, 1392, BS 556 and 4625) or asbestos-cement (see AS 1711 and BS 486), and attack in mortar-lined pipes (see AS 1281 and 1516) may be reduced by treating water containing carbon dioxide with quicklime, or by running it through beds of crushed limestone to increase its temporary hardness. Fair to good results can be obtained by protecting the pipes with bituminous coatings.

Improved results can be obtained with an admixture of active natural pozzolan (e.g. opaline silica) in the region of 10 per cent by weight of cement, the cement factor of the mix being kept constant. Two coats of coal-tar epoxy resin or hot-applied, coal-tar pitch $(0.5-1.2 \text{ kg/m}^2)$ are an effective sealant of primed surfaces.

Superior durability is obtainable with autoclaved asbestos-cement pipes and bore casings, constituted with silica flour passing a 0.053-mm sieve and high-pressure steam cured at 800-1000 kPa for 8 hours (see Section 33.2.3). Consequently, in highly corrosive aquifer waters, with a pH value not less than 6 and free carbon dioxide content not exceeding 50 ppm, a long-term service life can be expected. However, in exceedingly aggressive waters (with a pH value of 5-5.5 and 250 ppm free carbon dioxide), exposure tests have shown up to 5 mm depth deterioration in 17 years. In bore casings, coupling joints can be sealed with rubber vee-rings and be made shear resistant by sliding three-strand Terylene rope, via tangential holes, into interfacial, matching grooves for locking purposes.[82, 180, 187]

Further means of minimising leaching (e.g. in aqueducts and tunnel linings) are special cement concretes made with polymer (see Section 21.4), aluminous, portland-blastfurnace, portland-pozzolan and low-heat portland cement (dicalcium silicate in excess of tricalcium silicate). Very dense concrete should be used with increased cement content, construction joints being made watertight and located at intervals close enough to prevent cracking. Moss control in water-races is dealt with in Section 28.7 and general standards on asbestos-cement pipes and concrete conduits are contained in Appendix I.[97]

High-temperature butyl and ethylene-propylene-terpolymer (EPT) rubbers are suitable for use in distillation plants containing brine and desalinated water at temperatures up to 175°C. Plastic linings for general protection are described in Section 21.3.

CHAPTER 20

CONCRETE FLOORS

20.1 INDUSTRIAL CONCRETE FLOORS

20.1.1 INTRODUCTION

Ordinary single-course concrete floors seldom satisfy the stringent requirements of industrial use, and it is usually necessary to apply a suitable surface course or treatment that is specially adapted to withstand the conditions under which the floor is likely to be used. The main considerations are that the floor should resist corrosion, impact, wear and variations in load and that it should be free from dust, taint and physical hazards (slipperiness, sparking and fire). Also it must be easy and cheap to maintain.

The protection of concrete floors against chemical corrosion is an important consideration in factories, particularly those in which food processing is carried out and products of fermentation or deterioration of foodstuffs may form. Portland cement concrete is subject to chemical attack by acids and other industrial compounds, which penetrate through pores and cracks and cause deterioration of the constituents of the hydrated cement. The extent to which this happens is governed, among other things, by the porosity of the concrete, the type and quantity of cement in the mix, the concentration of the acid or compound causing the trouble, the provision for surface cleansing and drainage, and the extent to which the surface has been abraded by wear.

Of the various types of floor finish that may be used in an industrial building, a granolithic concrete surface is the most common because of the moderate initial cost and its resistance, within reasonable limits, to both wear and corrosion. Special surfacing materials (such as acid-resistant tiles and metal grids)

351

are sometimes used to minimise local effects of corrosion and wear, but a well-laid concrete surfacing has a good resistance to abrasion, is easily laid and cleaned, and can give satisfactory service.

A hard and durable concrete floor rings like an anvil when struck with a hammer. To attain this degree of durability, it is necessary to give careful attention to the details of design, the selection of materials, increased cement content, the manufacture and compaction of the concrete to give maximum density (see Section 23.1), and effective curing and surface treatments (see Section 20.2). Floors expected to resist substantial service conditions require special consideration in the design and laying of the concrete, and in providing skilled supervision for the execution of the work (see Appendix II).

Other flooring details are described in Sections 22.1 (granular metal surfacing), 36.3 and 36.4 (residential and terrazzo concrete floors). See also Reference 186 of Part 5 and miscellaneous references on concrete slabs and *in situ* finishes (see BS CP204).
33, 65, 82, 115, 118, 149, 158, 159, 182, 183, 184, 188-190

20.1.2 CEMENT CONCRETE

(a) Ingredients.

(i) Cements. Five types of cement may be used: portland (three varieties), high-alumina (see Section 20.1.5(g)), supersulphate, pozzolanic and blastfurnace slag cements (see Sections 3.1.2 and 3.1.3). The three varieties of portland cement are ordinary, sulphate-resisting, and low-heat, of which the second and third are a little less vulnerable to chemical attack than the first. High-alumina and supersulphate cements are very resistant to chemical attack, but are not immune to it. Cements that must resist chemical attack should not be mixed indiscriminately with another cement, hydrated lime or calcium chloride. They must be properly stored and at the time of use should comply with the requirements of standard specifications.

(ii) Pozzolan. (See portland-pozzolan cement concrete in Section 17.1.)

(iii) Aggregate. The aggregate should be tough, hard and dense, and reasonably uniform in grading and moisture content. It should comply with specified requirements and tests for aggregate (see Section 3.3.4 and AS 1465; BS 812, 882 and 1198-1201, or ASTM C33, C40, C123, C144 and C294. The particles should be free from

decomposed rock, harmful materials and an excessive amount of fines. Storage should be under cover, and samples should be taken and tested for control purposes (see Sections 3.3.1-3.3.5).

The fine aggregate (natural sand) should be fairly free from organic matter, as determined by the colorimetric test. It should be free from clustered particles, washed and graded as follows.

Passing 4·75-mm sieve: 90-100 per cent.
Passing 2·36-mm sieve: 60-100 per cent.
Passing 1·18-mm sieve: 30-90 per cent.
Passing 0·60-mm sieve: 15-60 per cent.
Passing 0·30-mm sieve: 5-30 per cent.
Passing 0·15-mm sieve: 0-10 per cent.

The fractions for surface-course sand should be near the low limits of the foregoing grading. The coarse aggregate should consist of crushed stone (such as basalt, granite, quartz or diorite), washed pea gravel, or air-cooled slag. It must be free from dust and be graded as shown in Table 20.1. ASTM Specification C33, however, permits a higher proportion of fines in each nominal size. All aggregate must be free from nodules of clay or weak material, which would tend to collect at the top of laid concrete (see Sections 3.3.4(e) and 4.5; refer also to Curve 1 in Figures 4.5-4.7 for mechanical compaction).

(iv) Water. Fairly pure water for mixing purposes should comply with Appendix I, Standards (e.g. BS 3148). The quantity of mixing water in a batch of concrete includes the free surface moisture of its component materials (see Sections 3.2.1, 3.2.3 and 3.2.4).

TABLE 20.1 GRADING OF COARSE AGGREGATE

Sieve Size (mm)	Per cent Stone (Nominal Size 40 mm) Passing	Per cent Screenings Passing		Per cent Toppings (Nominal Size 10 mm) Passing
		Nominal Size 20 mm	Nominal Size 14 mm	
37·5	90-100			
19·0	0-20	90-100	100	
13·2			90-100	100
9·5	0-5	0-20	0-45	85-100
4·75	0-5	0-5	0-10	0-20
2·36	0-2	0-5	0-5	0-5

(b) Manufacture.
(i) Proportioning. Skilled concreting is needed for industrial floors (see Sections 4.3.1 and 4.3.4). Excess mixing water (see Sections

3.2.2, 31.2 and 31.4) leads to segregation, bleeding, blistering, laitance-formation, dusting, crazing and cracking. In conjunction with ineffectual curing, it causes surface warping, drumminess, cleavage, scaling, porosity and wear. A granolithic topping mix should be as dry and harsh as can be compacted by mechanical means.

(ii) Workability. Workability agents, preferably nonionic (e.g. "Teric N8"), are described in Section 6.4, air-entrainment (if any) being limited in surface-course concrete (see Section 15.2.5). The slump is not more than 80 mm (50 mm with vibration; see Section 4.3) for the base and 30 mm (5 VB°; see Section 6.2.5) for the surface course. When finishing with a power float, no-slump or 10 VB° concrete (see Section 6.2) may be used, the mix being made so stiff that when a sample is squeezed in the hand, only a slight amount of moisture is brought to the surface.

(iii) Mixing. (For details, see Sections 3.2.4, 3.3.6(a) and (b), 7.1.1, 7.1.2 and 7.1.3.) Low-slump, granolithic concrete for the surface course is preferably made in a paddle or pan-mixer, using direct batching or dry prebatching procedures at the work site (see Section 7.1.2).

(iv) Testing. Ordinary testing procedures on concrete are indicated in Sections 14.1 and 14.4.

Shear tests on drilled core specimens should show a bond strength between adjacent courses in a two-course concrete floor of at least 1·5 MPa at 28 days. The characteristic (cylinder) compressive strength of concrete in the base and surface courses should be not less than 25 MPa and 45 MPa, respectively, at 28 days (see Sections 12.1, 15.2.3(b) and 15.2.5, also see Appendix VIII, 5).

20.1.3 PREPARATION OF SUBGRADE

A well-compacted and well-drained subgrade is necessary to support a concrete floor on the ground. Earth filling is spread and well compacted in thin layers. The consolidation of the levelled subgrade should be uniform, giving everywhere the same bearing value; its condition can be judged approximately by walking on it. Light vibrating rollers, vibratory plates and power tampers are effective in compacting the subgrade. Service ducts, as well as stormwater and subsoil drains, should be installed where necessary. A clayey or unstable subgrade should be covered with a compacted layer of coarse granular fill (hardcore bed), about 100 mm thick. Materials for this purpose are 14-40 mm nominal size gravel, broken brick, crushed stone, weak rock or salamander, ashes, sand or crushed slag. They should be free from pyrites and gypsum or other sulphates.

A separator course should be laid on the prepared subgrade or bed, in order to improve the curing, density and impermeability of the underneath surface of a concrete slab. Integral waterproofers, while offering some resistance to liquid water penetration, do not materially reduce the passage of water vapour through concrete. An enduring water-vapourproof membrane is highly desirable where the surface course of a concrete floor is likely to be sealed with substances that are affected by sustained moisture. This matter is dealt with in published literature (BS CP102 and CP121.101). [42, 43]

An economical membrane consists of polyethylene or polypropylene film (0·125 mm thick) or impregnated fabric, polyvinyl chloride sheet (see Appendix I, Standards), or reinforced building paper (BS 1521) containing a plastic film. Plastic sheets should be lapped 150 mm or, where the subgrade is damp, sealed with water-resistant adhesive or pressure-sensitive plastic tape (BS J10). They should extend up the perimeter of the floor and, without being exposed to sunlight, connect with the damp-proof course in walls. [44]

The membrane, without being damaged, is laid in advance progressively. To this end, a hardcore bed should be blinded with fine material and formed smooth, work planks should be laid temporarily prior to concreting, internal vibrators should not penetrate the membrane, and plastic sheets should be 0·25 mm thick where careless workmanship may be encountered. If a membrane is not used, the trimmed and compacted subgrade or bed should be wetted prior to placing the concrete.

20.1.4 THE BASE COURSE

(a) Requirements.
When a floor is to be laid on unstable soil, it should be designed to carry the loads that will be imposed on it and to minimise the effects of subsequent movements of the soil. [47] The lower side, for instance, may be constructed as a waffle-shaped grid, with doubly reinforced ribs at not more than 4·5 m centres each way. Slabs with thickened edges or ribs have advantages in strength, but those of uniform thickness are easier to construct. In some circumstances, a suspended floor on piers and beams may be more effective than a concrete floor laid directly on the ground. Slabs supported around the periphery may be of uniform thickness or of waffle-form design.

Polystyrene-foam pans (or inverted boxes made of waxed fibreboard, polypropylene or asbestos-cement sheet) may be used as permanent formwork in grid floor construction. Where heavy loading from vehicular traffic or machinery is expected, the base course should be thicker and more effectively reinforced than would be necessary under ordinary conditions of loading.

The base course of a floor on the ground should be cast in simply shaped sections, the sizes of which are determined by such factors as convenience in laying, effectiveness of bond with a concrete surface course, and freedom from random cracking between contraction joints (Plate 16). Excessive bleed water (if any) is better skimmed off by dragging the surface with a 25 mm compressor hose than by mopping it up with sprinklings of cement and stone dust, or by allowing it to remain and accentuate laitance formation. Adequate provision for initial shrinkage will suffice for subsequent movement caused by changes in moisture and temperature. Control joints may be laid out in diamond formation across heavy traffic aisles and at rugged fixtures, so as to minimise possible damage to their edges in a surface course.

If a floor is to be exposed to the weather, it is advisable to limit the dimensions of sections to 4·5 m, but 9 m by 4·5 m sections may be used if the floor is fully enclosed. With light steel-fabric reinforcement (2-4 kg/m²), the sections may be extended in size to 9 m by 4·5 m or to 12 m by 6 m for outdoor or indoor conditions, respectively. If larger sections are laid in one operation, additional reinforcement (see Section 3.4), technical skills (see Appendix III) and weakened plane or dummy joints must be used to control or predetermine the lines along which cracking is most likely to take place.

Sections containing heating elements, set at a depth 40-50 mm from the top, should be doubly reinforced and limited in dimensions to 4·5 m and 6 m, respectively. The mains of these elements, in grid or coil formation, should be looped and loose-sleeved across joints; and the foregoing limits should be reduced 20-50 per cent with unbonded courses or ineffectual concreting methods (see Section 20.1.5(a)).

Weakened plane joints are formed by fillets of metal (greased on one side), bituminous fibre board, closed-cell polyethylene or foamed polystyrene to a depth of one-quarter to one-third of the thickness of the base course. They are embedded at the bottom or inserted in slots that have been cut by a vibrating blade. Control joints with rubber-based mastic, 5 mm thick, are required alongside walls and around floor sections, columns, manholes, stairways, sumps and machine bases. Slabs with weakened plane joints require main control joints, 10-20 mm wide, at 40-m intervals (see Section 26.1).

Precautions must be taken to prevent relative movement of the slabs and resultant surface lipping. To this end, either bevelled tongue-and-groove or chevron-shaped joints are used, and mesh reinforcement is allowed to cross weakened plane joints. Alternatively, bitumen-coated or partly sheathed steel dowels are cast into abutting slabs at joints. The dowels are of 20 mm diameter steel rod, 500 mm long, spaced about 600 mm apart (see Section 31.2.1).

In rigorous, cold climates and in the construction of heated concrete floors, durable insulation should be placed around the sides of the floor and given a 450 mm return underneath. This practice reduces loss of warmth around the perimeter, limits chilling of the surface and minimises possible condensation on enclosed areas. Materials for this purpose are 25 mm cork board, cellular vulcanised rubber, fibreglass board, pads of resin-bonded rock or slag wool, foamed polyvinyl chloride, polyurethane, polyethylene or polystyrene, or 75 mm precast lightweight concrete ($650-800 \text{ kg/m}^3$, dry). A modified polyurethane, two-part polysulphide or bituminous coating is applied to insulating materials to exclude water (see Section 36.3.2(c)).

(b) Reinforcement.
Fabric reinforcement is used in large floors or where the bearing capacity and uniformity of the subgrade are uncertain or unsatisfactory. Apart from any structural advantage, this distributes the cracking of the concrete and minimises the magnitude of individual cracks (see Section 36.3.2(a)).

Steel-fabric reinforcement should lie in a plane above the centre of the base course, but not closer than 25 mm to the upper surface. It should be securely fixed in position (supported on pipes which are moved along as concreting proceeds) or be spread over a portion of the concrete that has been laid and compacted (see Section 23.1.2(a) and (c)). The outside main wires of adjoining sheets should be fastened together at 900-mm intervals with 1·25 mm annealed wire. Top steel elements strengthen the corners of slabs.

(c) Ingredients.
It is rarely necessary to use a mix that is richer than about 1 part cement to 2 parts sand to 4 parts coarse aggregate (20 mm nominal size) by volume. This mix would be suitable for a 125 mm

base course, which should be regarded as the minimum thickness for an industrial floor. A suitable mix for a course that is 150 mm or more thick would be 1 part cement, some 2 parts sand and 4·25 parts coarse aggregate (28 mm nominal size) by volume, with 20-23 litres of water per 40-kg bag of cement.[115]

For floors in freezing chambers, the first course is a layer of no-fines concrete (100 mm or more thick), smoothed on top with screeded cement mortar to provide a bearing surface for an enduring membrane and insulation course. The mix proportions are 8 parts of 20 mm aggregate to 1 part of cement, with just enough water to coat each stone with a cement slurry. Air-entrained concrete is used for subsequent concreting and a soil-heating system may be installed to minimise frost heave (see Sections 33.4.2 and 33.4.3).[96]

(d) Laying.

Sections are formed by laying the concrete in alternate bays against side forms or stock boards that have been cleaned and oiled (see Section 23.1.2(c) and Plate 16). The forms or boards are removed when the concrete has set and the face of the set concrete is smeared with a mastic material before the intervening bays are laid. Tongue-and-groove joints are readily made, using suitably shaped forms, and asbestos cement or insulating board may be used peripherally as permanent forms.

As the base course must be bonded to and made monolithic with the surface course, the concrete should be screeded and compacted to correct profile and left with a roughened surface that will key readily (see Section 23.1.2(c)). To ensure a sound and strong surface finish, the concrete should be laid as dry as practicable. If laitance or weak mortar forms on the surface, through bleeding or overmanipulation of the concrete, it should be removed and the coarse aggregate exposed by a green-cut clean-up of the partly hardened surface. Stiff-bristle brooms are used in one direction for the green-cut and subsequent cleaning operations.

If a base course is allowed to harden and mature before the surface course is laid, effective roughening of the surface, moist curing (at least 1 week) and cleaning are prerequisites for a monolithic result. The excessive use of vibrating appliances with wet mixes should be avoided. An integral waterproofer should be omitted in the base course, as it may act as a repellent towards a cement slurry or mortar used for bonding purposes.

20.1.5 THE SURFACE COURSE

(a) Requirements.

Essential physical requirements of granolithic concrete for the surface course are high quality, low permeability and low porosity. These characteristics are provided by the following factors.

1. A mix that is relatively rich in good-quality cement, proportioned for maximum density with sound aggregate, carefully gauged and thoroughly mixed with the minimum amount of water for workability.

2. Thorough and uniform consolidation, preferably with the aid of a vibrating screed, or hand rollers and power floats.

3. Early and adequate curing to ensure maximum hydration of the cement and prevent defects arising through early drying shrinkage.

4. Watertight control and weakened plane joints.

The finished surface course must not be less than 20 mm thick if applied to a freshly placed concrete base, but must be 25 mm thick if laid the next day, and 40 mm thick if laid thereafter. A course that is used for resurfacing an old concrete floor should be at least 50 mm thick. One laid on an impermeable membrane or containing radiant-heating elements must be 75 mm thick, light reinforcement being used in large bays.[70, 84, 111, 115, 116, 130] As it is better that the surface course should be of uniform thickness, it is recommended that the surface of the base course be sloped to provide the required drainage fall. The surface course may be reduced in thickness if bonded *in situ* with epoxy resin compound and hardener applied at the rate of 0·5 kg/m² (see Section 21.1.6(a)). Thick courses are compacted in two layers.

A membrane should be incorporated wherever moisture must be prevented from moving through a floor, such as in a freezing chamber or in a multistorey building where an upper floor is exposed to wet conditions. The membrane usually consists of bituminous felt, with or without a layer of acid-resistant asphalt, laid with broken joints in two coats which finish 10 mm thick. It is bonded with bitumen to the body of the floor and is designed as a liquid-tight receptacle that will shed seepage moisture into drains.

The surface course is finished with a drainage slope which must be no flatter than 1 in 80 for shedding water or mildly corrosive liquid, and 1 in 40 for the most corrosive liquors. As a safety precaution, the slope of the floor is made transverse to the general direction of traffic. Waste liquors and accidental spillage should discharge quickly to a drain so that the floor will act as a drainage catchment as little as possible. As described elsewhere,[45] much care is necessary with details such as coves, drainage lines and cleansing facilities, liquid-tight joints with walls, drains and service ducts, freedom from stagnant pockets or shallow depressions, and the camber of suspended-floor members to offset deflection and creep effects under load.

There must be a clear understanding in advance as to the correct specification, creative procedure, equipment, materials and specialised skills required in laying an industrial concrete floor. If necessary, a trial bay for investigation and acceptance reference should be laid before a major project is started. The prospective surface soundness and wear-resistance of single-course work can be enhanced by vibrovacuum processing prior to power floating (see Sections 12.1 and 23.4.2). For service satisfaction, it is advisable to employ competent, experienced and reliable paviours of repute, also to investigate concreting principles and techniques, to supervise laying responsibly, to conduct tests, to keep records, and progressively to proof-check essential operations for proficient performance (see Sections 20.1.2(a), (b), 20.1.4(a), (d) and 20.1.5(a)-(f)).

(b) Ingredients.
When the concrete is to be mechanically compacted the mix is nominally 1 part of cement, 1 part of sand and 2 parts of 10 mm coarse aggregate (stone toppings or pea gravel) by volume. It may comprise 1 part of cement, 1·5 parts of sand and 1·5 parts of coarse aggregate for a manually compacted course. A compromise between the two proportions is permissible, and a leaner mix (e.g. the aggregate content increased by one-third) may be used for ordinary floors in very light industry. The maximum size of aggregate may be increased in courses 50 mm or more thick (see Section 20.1.2(a)(i) and (iii)), or in medium-strength concrete inserted under a stronger mix.

The water content, for a water/cement ratio of 0·3-0·4 by weight, is 12-16 litres per 40-kg bag of cement (see Section 20.1.2(b)). Where mild acid conditions are to be resisted by ordinary portland cement concrete, a suitable pozzolan may be used to replace 5-10 per cent of the sand by weight. A low slump (within a range of 0-30 mm) is essential.

(c) Laying.

The surface course may be laid mainly by hand methods, immediately after the base course is laid. An air-entrained mix with low water content for the base course minimises the risk of defects being formed by excess water bleeding to the surface. Before new concrete is laid over freshly placed concrete, any water that has bled to the surface should be removed with a hair broom. The new concrete is spread, raked, screeded and compacted by hand-tamping, vibrovacuum processing or surface vibrating appliances (see Sections 23.1.2(c) and 23.4); the first method is described here. When brought to correct profile, the surface is manually skipfloated and finished from crossbeams.

An early-bonded floor finish may be laid as soon as the base course has become firm enough to resist foot pressure without marking. In this instance, light hand rolling and power floating can be used to marked advantage. A convenient procedure is to lay some of the base course in checkerboard fashion in an afternoon and the corresponding surface course on the following morning. This allows ample time for finishing operations. The sections of surface course thus laid should be at least 3 days old before intervening sections are laid. The surface of the base course must be roughened by brushing, as indicated in Section 20.1.4(d), and then be cleaned by sweeping and air-jetting.

Soft mortar (consisting of 1 part of cement to 1 part of sand with 18 litres of water per 40-kg bag of cement) is brushed onto the partly hardened and lightly wetted base course just before spreading the surface course. This mortar is applied on the faces of side forms, on screed battens that are placed temporarily on mortar pads, and on joint fillets or abutting slabs (with a bond-breaking seal), so that the sides of finished sections will be free from honeycombing.

A well-hardened base course that has not been green-cut by brushing out excess mortar should be scarified by a machine with a revolving cutting head,[46] or scabbled by power-operated sharp picks or bush hammers. Both operations may be required on large sections. At least three-quarters of the surface and all laitance should be removed to expose a sound surface. Drastic chipping or hacking in several directions must be done at close centres around the perimeter of individual sections (i.e. in the vicinity of contraction joints and at the corners of slabs).[125]

The surface, prepared well ahead of concreting, is cleaned by brushing with stiff brooms and air-jetting. It is watered in advance, allowed to partly dry without leaving pools, and optionally sprayed or dampened with a 5 per cent solution of calcium chloride (see Section 7.2.4). Before the surface dries, a 1 part cement to 1 part sand slurry (paint consistence) is scrubbed into the damp surface and the surplus is brushed off. The freshly treated surface is then coated with a stiff 1 part cement to 1 part sand mortar to a nominal depth of about 3 mm, just prior to laying the surface course, so as to fill voids on the lower side of newly placed concrete. This mortar may be admixed with 2 per cent calcium chloride by weight of cement (see Section 7.2.4), and it should be spread up coves or vertical surfaces at joints.

An alternative interfacial treatment, for improving both the bonding and impermeability properties of joints, consists of cement and mortar coatings that are suitably formulated with a sodium carbonate admixture (see Sections 10.1 and 11.1.2(c)). If the base course contains an integral water repellent, the bonding slurry may be mixed with up to 0·05 per cent of wetting agent, by weight of cement, for ease in spreading (see Section 6.4).

A surface course that is laid on a partly or fully hardened base is spread, raked, and screeded in one or two layers (depending upon depth) to give the requisite thickness after compaction. Consolidation of the concrete before initial set of the cement is done with 100-kg–200-kg rollers (Plate 15). Three traverses are made in each of two directions, and additional compaction by tamping is done at the corners of slabs and alongside control joints as they are being formed. The surface should be checked with a template and adjustments made before the final stages of finishing.

If the surface course is very thick in local areas, it should be lightly reinforced over a supporting beam or wall and spread in several layers, each of which is rolled separately. If pigment is used in the top layer, it should comply with the requirements of AS K54 or BS 1014 and be incorporated on a weight basis.

(d) Finishing.

If the surface course is placed before the base course has stiffened or hardened, it is finished off manually by several wood floatings (over a period of up to 3 hours) and finally by steel trowelling. Otherwise, it is finished off to plane and grade by rotary-compactor power floating (see Reference 5 of Part 5) in two or three operations, followed by power and hand trowelling (Plate 15). The rotating disc of each power float must be taken over the joints in order to obtain a compact and flush surface finish. The concrete should have hardened sufficiently at the time of trowelling to prevent moisture and fine material being drawn to the surface. Hand steel trowels should give a ringing sound when drawn over the surface and, to avoid staining coloured concrete, celluloid, magnesium, aluminium or stainless steel trowels should be used (see Section 28.6).

As much as possible of the coarse aggregate should be near the surface, so as to resist abrasion and enhance the adhesion of any future surface coatings. The hydrated calcium aluminate and lime that are formed during the setting of portland cement cause a top coating of mortar to be vulnerable to chemical attack. To lessen its formation, and to control surface shrinkage and warping, it is necessary to avoid too much steel trowelling, the addition of water during finishing and the dusting of portland cement concrete with dry cement and stonedust or fine sand.

At a later date, fine material is sometimes removed from the surface by light buffing or burnishing, the floor being kept wet during the process. This is done by a power-operated machine with rapid-cutting abrasive stones or a wire-brush attachment. The ground surface is washed clean and residual pittings are subsequently filled with cement grout. Film is removed by rebuffing with fine-grained abrasive stones after 3 days of moist curing. Uniform colour in finishing is considered in Section 20.4.3.

(e) Curing.

Early and adequate curing is essential for best results. Shortly after the final trowelling of a day's placement of portland cement concrete, polyethylene sheet or other impervious membrane is laid with lapped joins the same day, and is held in place to prevent the evaporation of moisture from the surface. This sheet is lifted the next day, the concrete is thoroughly wetted with water and the sheet is relaid. It may be used on the next area of newly laid surface course (sheltered or exposed), provided that other means are utilised to cure the concrete continuously for 2 weeks (see Sections 9.1-9.3).

Moist curing is done with wet hessian blankets, underfelt or sand (25 mm thick), sprinkler hoses, ponded water, fog sprays, or an airtight, vapourproof membrane. Delayed or intermittent curing may prove to be unsatisfactory (see Section 31.2.4). Where building operations are to be continued after a floor has been laid, curing may be done with a scuff-proof or sand-covered membrane that is wetted for cooling effect in hot weather (see Section 32.2).

Pozzolanic concrete, in particular, is greatly improved by prolonged curing. The early and effective hydration of cement by proper curing substantially increases the strength, abrasion-resistance and durability of concrete. It helps to prevent curling and drumminess developing at the corners of slabs or bays through premature drying and shrinkage of the surface. A period of drying after curing improves the resistance of concrete to chemical attack. A portland cement concrete floor should be not less than 3 weeks old when put into industrial service.

(f) Jointing and caulking.

Joints in the surface course, which are made to coincide with those in the base course (see Section 20.1.4(a)), are either formed with twin-stripped polyethylene foam during the finishing operation or sawn to one-half depth at an age of 12-24 hours. Their effective width is about 8 mm. Joints formed by tooling should not have their edges raised above the plane of the floor. Those that are to be used under wet conditions should have sharp-cornered or saw-cut edges for the regular adhesion of a jointing compound without feather-edge fretting at the exposed surface.

When the concrete has cured and dried (and after the top portion of a split foam strip is removed), the joint gaps are thoroughly cleaned, primed and filled with elastomeric or trafficable grades of sealant (e.g. polyurethane). The compound selected should be firm, adhesive, ductile, of low potential shrinkage, and be resistant to liquids that are likely to be spilt on the floor (see Appendix I and Table 20.3). The properties of joint-sealing compounds should be assessed beforehand (see BS 2499; also Sections 11.1.2(d), 26.1, 28.12 and 31.2.1).

A fleximer mixture for ordinary jointing work may consist of 7 parts rubber latex, 1·5 parts water and 12 parts cement-based filler by weight. Where both acid and fat must be resisted (e.g. in dairies), the top 10 mm depth of joints must be sealed with highly durable, two-part polysulphide synthetic rubber, applied over an epoxy resin primer by means of an extrusion gun (see Sections 20.2 and 21.1; also AS 1527 and BS 4254).

Strong surface joints can be made of lead strip, inserted with 3 mm protrusion, that will be trafficked down flush. Otherwise, they may be made 12 mm wide and divided centrally with a square-edged, mild-steel flat, 5 mm thick by 25 mm or more deep. These divider strips are antirust-primed, caulked on each side and located with their top edge in the finished plane of the floor. They are held (on one side) to the base course either by splayed 10 mm diameter rods (150-250 mm long) welded to the strips at 1-m intervals or by M6 mm diameter bolts (40 mm long) which are screwed through lugs at 1-m intervals into expansion shield anchorages in the concrete.

(g) High-alumina cement concrete.

High-alumina cement concrete develops high compressive strength very quickly. It may be used, therefore, when a surface course must be put into service in a day or two. Several precautions must be taken, however, if the strength and durability of this concrete are to be maintained. For instance, the environment in which the concrete is to be used should not be continuously warm and moist, alkaline solutions should be kept off the surface, and the concrete should be specially made, laid and cured with cold water.

Concrete for the surface course, when made with high-alumina cement, should be of leaner composition than that recommended for portland cement. Increasing the proportion of high-alumina cement in the mix to above 1 part of cement to 1·5 parts of sand to 2·25 parts of clean stone toppings may not increase the strength of the concrete, because the amount of water that is required for complete hydration of the extra cement introduces defects in the texture of the concrete. The amount of water required is approximately 17-19 litres/40 kg of cement, and the concrete should be placed soon after it is mixed. The cement must not be mixed with stonedust or set-changing contaminants, and equipment that has been used with portland cement concrete must be thoroughly cleaned before being used with high-alumina cement concrete.

For reasons of economy, the base course of an industrial concrete floor is usually made with portland cement. It is prepared for subsequent bonding with the surface course and moist cured for 1 week (see Section 20.1.4(c) and (d)). The surface is coated with a slurry of 1 part high-alumina cement to 1 part sand and stiff mortar just before the surface course is laid (see Section 20.1.5(c), omitting calcium chloride). In laying high-alumina cement concrete, it is essential that temperature conditions are controlled and that curing with cold water follows immediately after early hardening (see Section 3.1.2).

20.2 SURFACE TREATMENTS

Surface treatments are used for such purposes as hardening the upper matrix of portland cement concrete, reducing the suction of liquids into surface pores, and minimising or preventing possible chemical attack. Consideration may have to be given to surface texture as a safety factor. The success of any particular treatment depends upon local conditions, types and concentrations of solutions present, temperature, load factors, wear, vibration, quality of substrate and workmanship, and usually a prerequisite curing and maturing period of 1 month.

As described elsewhere,[45] an industrial concrete floor must be properly laid and finished in the first instance if it is to give the best service. Surface treatments will not compensate for poor-quality concrete and their adhesion may be affected by underlying damp conditions. Floor insulation (particularly around the perimeter) is advantageous in preventing surface condensation which can lead to slipperiness in cold climates (see Section 20.1.4(a)).

Most sealants require a clean, hard, dry surface for adequate adhesion, and a weak surface skin should be removed to expose coarse aggregate. Surface scuffing may be done with a mechanical scarifier,[46] pneumatic scabbler, chipping hammer, bush hammer, wet rotary grinder, machine sander or steel-wool pad, wire brooms, high-pressure water jet, or a sand blast and vacuum appliance. Final cleaning can be done with compressed air, or a wet or dry industrial vacuum unit. Alternatively, or additionally, the surface is double-etched with hydrochloric acid and washed (see Section 20.4.2), care being taken to avoid saturation of the underlying concrete with acid.

A hardened concrete floor must be thoroughly dry before being coated with materials that would be affected by moisture from below. Drying from one face only, a slab of concrete may take about 4 months, if 100 mm thick, and about 8 months, if 150 mm thick, to meet the requirements of a dryness test (BS CP203).[43] The relative humidity of a pocket of air in equilibrium with a concrete surface (as determined overnight, i.e. 16 hours, by a surface hygrometer) should not exceed 70 per cent if the surface is to be covered with a water-sensitive coating. Hygrometer and electrical resistance dryness tests are specified in AS CA37 for resilient flooring. Alternatively, a small electronic "Aquameter", fitted with insertive prongs and classificatory scales, may be used for the purpose. A floor which is to be heated must be dried thoroughly before being sealed.

As many adhesive and surfacing materials are water-sensitive or poorly resistant to vapour pressure from below, a concrete floor on the ground should be constructed with an impermeable membrane (see Section 20.1.3). Where this membrane has been omitted, however, a vapourproof barrier (if required) may be built up with three coats of polyamide-cured epoxy resin to a thickness of at least 0·5 mm (0·5 kg/m^2). The viscosity of the resinous material is about 100 mPa.s at 23°C for priming purposes and 200 mPa.s for painting.

A primer is allowed to penetrate and harden for 12 hours, and the next coat should become touch dry before the final coat is applied. The potential characteristics of polyamide-cured epoxy resin are such that if one or more coats are applied to newly laid concrete, they will serve as a membrane-curing medium and develop a strong bond to a damp surface (see Section 21.1.6). The first coat of surface treatments should, in general, be strongly brushed with a stiff-bristle broom over and into a hardened concrete surface, and each successive coat should be oxide pigmented a different tone (0·025 by weight) and applied at right-angles to the preceding hardened one.

Dry sand (about 0·60-mm mesh sieve size) may be spread over an unhardened top coat to impart nonslip properties and a mechanical key to subsequent surface treatments. It would not be required if the solvents in subsequent sealants had a slight cutting effect on the hardened resinous surface. The sand is spread at the rate of 1·5 kg/ m². It is broad-cast or distributed evenly by means of a coarse sieve, fitted with a piece of strong paper punched with 5 mm diameter holes at 25-40 mm centres (see Section 20.2(c)).

Guidance on the use of various types of surface treatment is given in the following paragraphs. Rates of application are in m² per litre. These have been compiled from recorded and direct experience in connection with many problems relating to concrete floors in industry.[118] Increased quantities are needed on permeable areas.

(a) Metallic silicofluorides.
A surface-hardening mixture consists of 4 parts of magnesium silicofluoride and 1 or more parts of zinc silicofluoride. These constituents are supplemented with up to 1 part of lead or aluminium silicofluoride when the mixture is to be used for increasing the chemical resistance of concrete floors.

Three coats are applied and covered 24 hours later with two finishing coats of chlorinated rubber lacquer, boiled linseed oil, or a plastic or synthetic resin compound. A 30 per cent solution of metallic silicofluoride is diluted with two volumes of water for the first coat and one volume for the second, and is used neat for the third. The solutions are applied so that 1 litre will cover 2·5-3·5 m² in the first and second coats and 5·5 m² in the third. An interval of at least 4 hours is allowed between coats to ensure that the surface is sufficiently dry to allow the next coat to penetrate. When dry, the finished surface is washed thoroughly with water to remove encrusted water-soluble salts.

Silicofluorides react with the calcium hydroxide in portland cement and precipitate compounds in the pores, thereby densifying and hardening the concrete, but the treatment does not completely seal the pores (see Section 15.2.5). For this treatment and the next, the concrete should not contain an integral water-repellent agent nor be membrane-cured with a liquid compound. After moist curing, the concrete should be allowed to air dry until its age is at least 28 days. During application, silicofluoride requires good ventilation; also care must be taken as regards any flesh wounds.

(b) Low-alkali sodium silicate.
Sodium silicate is usually supplied as a solution containing 40 per cent of the material, and this is diluted with four volumes of water for the first coat, three volumes for the second and two volumes for the third. The solutions are applied until they cease to be absorbed, and each litre will cover about $3 \cdot 5$ m². Each coat is left for about 24 hours, washed with water and allowed to dry before the next coat is applied. Any unabsorbed material from the final application should be washed off (BS 3984).

This treatment is a common way of increasing surface hardness, but it will not withstand heavy wear (see Chapter 12 and Section 15.2.5). It gives a less durable result in the presence of aggressive solutions than treatment with silicofluorides, and should be supplemented with another form of protective treatment for increased resistance to corrosion. Freshly acid-etched decorative concrete should not be treated with sodium silicate.

(c) Chlorinated rubber lacquer.
A surface-penetrating grade of lacquer is used in preference to a thinned finishing grade. It consists of chlorinated rubber of low molecular weight, dissolved in a high aromatic solvent and suitably plasticised to improve the adhesive and lasting properties of the dried film. It is applied in three coats, with a drying period of 3-5 hours or longer between them; and 1 litre covers from $3 \cdot 5$-10 m² per coat.

An effective combination for factory floors is silicofluoride treatment, followed by two coats of penetrating grade, chlorinated rubber lacquer. The lacquer is applied to a sound, clean and dry surface, so that 1 litre covers $5 \cdot 5$ m² for the first coat and 10 m² for the second. Liquids must be kept off the treated surface for at least 24 hours. This treatment is highly resistant to attack by strong alkalies and strong oxidising acids of up to 5 per cent concentration, but not to aromatic solvents, animal fat, fatty acids and many oils.

Under severe conditions of corrosion, a priming treatment of chlorinated rubber is followed by two dressings of synthetic rubber solution applied at the rate of 1 litre to 3·5 m². Synthetic rubber solutions include such materials as polychloroprene (e.g. "Neoprene") and chlorosulphonated polyethylene (e.g. "Hypalon") mixed with a catalyst.

For a nonslip surface, a hard filler can be incorporated in finishing coats at the rate of 0·25 kg/litre. Fillers consist of silicon carbide, aluminium oxide, carborundum, fine sand and silica flour of 0·30-0·60-mm mesh sieve size (see Section 21.1.3). Filler increases the thickness of coating and, if a very coarse texture is desired, the surface may be given a stippled finish. Alternatively, a liquid coating may be covered with fine abrasive grit or dry sand, the partly embedded grains of which may (after the hardened surface has been swept) be further secured *in situ* with a light coating of the lacquer. Chlorinated rubber lacquer and chlorosulphonated polyethylene solution may be pigmented (e.g. light gray) for aesthetic effect.

(d) Boiled linseed oil.

The concrete surface should first be treated with a solution containing either 10 per cent ammonium carbonate, or 3 per cent phosphoric acid and 2 per cent zinc chloride. After drying for 24 hours, the surface is cleaned with wire brushes. This neutralising treatment may be omitted if the surface is well matured and dry, or has received silicofluoride or sodium silicate treatment.

Boiled linseed oil should have a specific gravity of 0·931-0·935 at 16°C and should contain not more than 1·5 per cent of unsaponifiable matter and 0·5 per cent total ash. It should also have a drying time at 20-25°C of not more than 24 hours (AS K4 and K5). Oils will not normally comply with these specifications unless suitable driers are included.

Three coats of the oil are applied, each being allowed to dry thoroughly. The first coat consists of a mixture of equal parts of oil and turpentine or other suitable thinner, applied at a temperature of 65-80°C in order to effect maximum penetration into the surface. Slightly diluted or undiluted oil at the same temperature is used for the succeeding coats. Only two coats (undiluted or slightly diluted) are necessary when they are applied after silicofluoride or sodium silicate treatment. To improve results during the first few years, another coat may be applied each year under dry conditions. Caustic alkali cleaning solutions will slowly saponify the oil.

(e) Plastic or synthetic resin compounds.

Various types of plastic reaction products (i.e. two-part mixes which harden *in situ* by chemical reaction) and plasticised synthetic resin are used for the protection of concrete floors. Those that are widely used include solventless liquid epoxy, epoxy or modified epoxy resin solution (see Table 20.4), furane, polyurethane, polyvinyl butyral resin, polyvinyl chloride and acetate copolymer, polyvinylidene chloride, and silicone acrylic copolymer compounds. Other suitable types of resin are cashew nutshell, coumarone, phenol furfural, and polyester. Some characteristics of the first group are given in Table 20.2. The tabulated resistance to sunlight applies to materials that are pigmented or formulated with light absorbers.

TABLE 20.2   **CHARACTERISTICS OF RESINS**

Type of Resin	Resistance to: Chemicals	Heat	Sunlight	Water	Wear
Epoxy	Good (A)	Good	Good	Good	Very Good
Furane	Excellent (A, B)	Good	Good	Good	Excellent
Polyurethane	Good (A)	Good	Good	Good	Excellent
Polyvinyl butyral resin	Good (C)	Excellent	Excellent	Excellent	Good
Polyvinyl chloride and acetate copolymer	Good (A, C)	Fair (E)	Fair	Good	Very Good (F)
Polyvinylidene chloride	Good (A, C)	Fair (E)	Fair	Good	Good (F)
Silicone acrylic copolymer	Fair (D)	Excellent	Excellent	Excellent	Good

Note: A. To acids, alkalies, animal fats, oils and solvents.
B. Strong oxidising acids excepted. All plastic coatings are damaged to a degree by these acids.
C. High resistance to fats and oils, but not to strong solvents (e.g. esters, ketones, and aromatic and chlorinated hydrocarbons).
D. Strong acids and solvents excepted.
E. Up to 70°C.
F. Hard grade.

It should be noted that unmodified epoxy resin is poorly resistant to the following hazards: acids of acetic, chromic, hypochlorous, nitric and sulphuric type; aniline, benzaldehyde, bleach liquors, bromine, calcium hypochlorite, carbon disulphide, chlorobenzene, furfural and hydrogen peroxide; gases of chlorine, fluorine and oxidising type; iodine, methylethyl ketone, nitrobenzene and sodium hypochlorite.[54] Modifying polymers for increased chemical resistance are indicated later in Table 20.4. Although polyester resin (see Table 20.5) is cheaper and cures more quickly than epoxy resin, it has less chemical resistance and a tendency to saponify in contact with fresh concrete, with a consequent reduction in bond strength.

Curing agents for epoxy compounds are amine and polyamide hardeners. For furane compounds, the hardener required is paratoluene mixed with sulphonic acid. For polyurethane compounds, the hardener required consists of both di-isocyanate and tri-isocyanate. The remaining resins in Table 20.2 harden by evaporation. The amine curing agent for epoxy resin imparts high chemical resistance, whereas the polyamide one develops desirable physical properties in the hardened compound. Where resin is hardened with paratoluenosulphonic acid, the concrete surface should be neutralised first with an application of metallic silicofluoride.

Interior and exterior grades are formulated for epoxy and polyurethane compounds. Priming and surfacing grades of resin solution and hardener are applied by firmly brushing with a circular motion, by using short-nap mohair or lambswool rollers, or a "squeegee", or (less preferably) by spraying. They all have strong adhesive qualities to sound, dust-free, dry surfaces. Three coats give good protection if used alone in sufficient thickness (0·50 mm), or two coats are sufficient if metallic silicofluoride treatment has been applied previously. As with lacquer treatment, 1 litre covers from 3·5-10 m² per coat.

Abrasion-resistant fillers are frequently incorporated to give nonslip surfacing and increased wear resistance over local areas, as indicated in Section 20.2(c) and the description of lacquer treatment. With polyurethane lacquer, filler is limited to 0·2 kg/litre. A trowellable grade of resinous mixture containing 1-2 kg of filler per litre has a coverage of 0·2 m²/litre for a 5 mm thickness. A hard-wearing, decorative surface treatment may consist of a solventless epoxy prime coat (applied to hold decorative paint flakes), and a polyurethane triple seal having a stippled or textured finish (see Section 20.2(c) and 21.1.6(c)). Surface treatment with methyl methacrylate is introduced experimentally in Reference 168 (see Section 21.4).

Pigments must have high opacity and resistance to alkali. Water must be kept off all but epoxy resin preparations until they are completely cured, a period of 3 days usually being sufficient. Appliances should be cleaned with suitable solvents immediately after use, and safetly precautions should be taken as indicated in Sections 21.1.6 and 21.1.7.[123]

(f) Bituminous seal.

A priming or tack coat is first applied to the concrete surface to ensure bond for the subsequent protective coats of bitumen. The primer should be a cut-back bitumen of such a consistence that it can be applied cold at the rate of 1 litre to 1·5-1·8 m² and allowed to dry. Aggregate should not be scattered over this primer, and even in warm weather the drying period should be at least 1 day. A film of bituminous emulsion is then uniformly applied at the rate of 1 litre to 0·9 m² and immediately covered with clean 3 mm grit or coarse washed sand before the emulsion turns black. The grit or sand is uniformly distributed at the rate of 1 m³ per 175 m² and is then lightly broomed and rolled.

Some hours later, the excess aggregate is carefully swept off and a second coat of emulsion is applied in the same way. This is covered with grit or sand but excess material is left on the surface for a day or two after rolling, to allow as much as possible to be worked into the bitumen by subsequent traffic. The excess material may then be swept off. A thin, uncovered continuous film of slow-breaking bituminous emulsion (e.g. 1 litre to 1·5 m²), applied just after the fresh concrete has hardened, may serve as a curing medium, as well as the subsequent priming coat in certain cases.

Bituminous seals do not resist nitric acid, fruit juices, oils, fats, blood, grease and other organic solvents, particularly at temperatures above 60°C. Bituminous surfacings are therefore not desirable for floors of dairies or abattoirs.

(g) Cold bituminous mastic mortar.

A layer of bituminous mastic mortar is applied to give a final depth of about 12 mm. The mortar is applied over a priming or tack coat made by diluting clay-stabilised bituminous emulsion with about 5 per cent cold water in order to reduce the emulsion to a thin brushing consistence, and at least 0·7 litre of this diluted emulsion must be applied per m². The priming coat must be well brushed into the surface, but it need not be allowed to dry before the mortar is laid. *Note:* Where a bituminous seal and bituminous mastic mortar are to be used contiguously on the same project, the priming coats thus indicated may be interchanged. An additional, light priming coat may be required over a bituminous membrane curing seal, and the mortar may be laid over an open-weave fibreglass fabric to prevent the possible occurrence of blistering (see Reference 214 of Part 3).

The mortar consists of a mixture of 1 part cement mixed dry with 4 parts clean, coarse to medium-sized sand by volume, to which a little water is added before mixing in 2 parts of clay-stabilised bituminous emulsion. The mix should be of such consistence that, when cut with a spade, it will remain firm and not slump. It should be used within 1 hour.

If trucking aisles only are to be surfaced with this material, the mix can be worked out to a feather edge. If it overlaps the priming coat, an inadequate bond may cause cleavage along the edges. High-alumina cement or silica flour should be used where the hazards are of acid type, but portland cement may be used in other cases. A mix containing cement hardens more quickly and gives a tougher surface than one containing silica flour and sand.

After being laid, the mortar is tamped, screeded and subsequently levelled with a wet wooden float. A few hours later, when the initial set has started, the surface should be refloated and steel trowelled (without excess) to dissipate stresses that tend to develop with loss of water content and cause slight shrinkage. Consolidation with a 150-kg hand roller is advantageous, but not essential and should be done as soon as the floor can be rolled without undue marking. The floor should be cured immediately afterwards with a suitable membrane or hessian kept moist for 24 hours. Early rapid drying of the surface must be avoided.

This type of flooring can be put into service several days after curing and, although slight indentations may be caused by early heavy trucking, this marking will disappear and traffic will definitely improve the surface. These mastic mortars are not recommended where organic solvents are used or where the floor will be continually wet and subjected to severe abrasion at the same time.

(h) Fleximer compounds.
These compounds are of two types, the more common consisting essentially of specially compounded rubber latex and either high-alumina or blended hydraulic cement with suitable fillers (BS CP204). It is characterised by strong bonding and surface-sealing qualities, and resists abrasion and corrosion by both dilute acids and alkalies.

In the second type, plasticised synthetic resin latex is used instead of rubber latex and confers resistance to abrasion and to corrosion by chemicals, oils, fats and solvents. This compound, if based upon polyvinyl acetate, must be modified for water resistance and used under dry conditions as much as possible. A vapourproof barrier is required in the floor.

A layer (about 5 mm thick) of fleximer compound over concrete with an even surface, may be put into service 3 or more days after it is laid. Both types of material are available as proprietary lines (see Section 21.2).

(i) Other surfaces.

Other surfacing materials or surface courses may be used where corrosive or abrasive conditions are severe. These include synthetic rubber solution as indicated in the description of lacquer treatment; mastic asphalt (BS CP204) (with or without embedded 50 mm gauge grids); vitrified bricks, and tiles of burnt clay (BS 3679, CP202 and ASTM C57, C112, C410), pressed concrete, polyvinyl chloride (AS 1884, 1889 and BS 3260),[48] stainless steel or cast iron. Tiles may incorporate carborundum to provide a nonslip texture and better resistance to wear.[49, 111] Where thin armour plating is needed, precut checker-plate tiles (e.g. 0·36 m^2) may be inverted and bonded with epoxy resin compound and mortar to prepared sound concrete (see Sections 20.2 and 21.1).

Burnt clay tiles (20 mm or more thick) may be used for general purpose factory floors, but should be from 30-50 mm thick if conditions are highly abrasive and acidic. To carry high impact and hot loads, vitrified bricks set on edge are required. Bricks and tiles may be set in any of the following materials which would give the best results: fleximer compound; supersulphate, sulphate-resisting, high-alumina or an acid-resistant cement mortar; solutions of bituminous rubber, tar rubber, synthetic rubber or synthetic resin (e.g. cashew nutshell and paraformaldehyde hardener; bitumen or coal-tar epoxy and polyamide hardener) with filler (see Sections 20.1.5(f) and 21.1.2). Herringbone construction, completely filled joints, adequate drainage falls and suitably located drainage channels are recommended.[45] Small mozaic tiles can be sheet-laid in epoxy resin.

A clay tile surface should have adequate provision for expansion (e.g. 5 mm wide fleximer joints at about 4 m centres in each direction). Expansion or control joints in a surface finish should coincide with those in the concrete floor. They should be provided where the flooring abuts fixtures and walls. A waterproof membrane should be placed between a portland cement concrete base and bricks or tiles with acid-resistant cement mortar bedding and joints,[42] but it may be omitted when a rubber or suitable resin-based compound is used instead of mortar. Metal tiles are set in bituminous mastic or rubber fleximer material (see Sections 20.2(g) and (h), respectively), which is laid in a semidry state about 40 mm thick (Plate 15). Chemical resistance of polyethylene and other plastics are specified in AS CA69 and BS CP3003.

TABLE 20.3 SURFACE TREATMENTS RECOMMENDED FOR PARTICULAR HAZARDS

Hazard*	Treatment**								
	(a)	(b)	(c)	(d)	(e)	(f)	(g)	(h)	(i)
Acids									
acetic, butyric, carbolic, carbonic, citric, formic, lactic, oxalic, phosphoric, tannic	x	x	x	x	x	x	x	x	x
hydrochloric, hydrofluoric, sulphuric, sulphurous	s	—	s	—	s	x	x	x	x
nitric	s	—	s	—	s	—	—	x	x
oleic, stearic	x	x	—	—	x	—	—	x	x
Ammonium fluoride	s	—	s	—	s	x	x	x	x
Beer	x	x	x	x	x	x	x	x	x
Carbonates†	x	x	x	x	x	x	x	x	x
Chlorides									
calcium†, potassium†, sodium‡	x	x	x	x	x	x	x	x	x
ammonium, copper, iron, magnesium, zinc	x	x	x	x	x	x	x	x	x
Creosote, cresol (lysol)	x	x	—	—	x	—	—	—	x
Formaldehyde, formalin	x	x	x	x	x	x	x	x	x
Fruit and vegetable juices	x	x	x	x	x	—	—	x	x
Glucose, sugars, molasses	x	x	x	x	x	x	x	x	x
Glycerine	x	x	x	x	x	x	x	x	x
Hydroxides†	x	x	x	x	x	x	x	x	x
Milk (lactic acid, sugar and fat)	x	x	—	x	x	—	—	—	x
Nitrates									
ammonium	s	—	s	—	s	x	x	x	x
potassium	x	x	x	x	x	x	x	x	x
Oils and fats	x	x	—	x	x	—	—	—	x
Phenol	x	x	—	—	x	—	—	—	x
Sauerkraut	x	x	x	x	x	—	—	x	x
Silage	x	x	x	x	x	—	x	x	x
Sulphates in solution									
ammonium	s	—	s	s	s	x	x	x	x
others	x	—	x	x	x	x	x	x	x
Sulphite liquor	x	x	x	x	x	x	x	x	x
Tanning liquor (acidic liquors only)	x	x	x	x	x	x	x	x	x
Urine	x	x	x	x	x	x	x	x	x
Vinegar	x	x	x	x	x	x	x	x	x
Water (acidified natural)	x	x	x	x	x	x	x	x	x
Whey	x	x	—	s	x	—	—	—	x
Xylol	x	x	—	—	x	—	—	—	x

* The following substances have no deleterious effect on concrete: bleaching powder, boracic (boric) acid, borax, calcium nitrate, fluorides (excluding ammonium fluoride), potassium permanganate, silicates.

** The special surface treatments shown are the minimum requirements on good-quality concrete. Poor-quality concrete requires more substantial protection. The recommendations given for treatment (h) refer to the

Note to Table 20.3 continued at foot of page 376.

Flexible sheet or tiles (AS K139, A180, 1884, 1889 and BS 3261, CP203) may be bonded to concrete with special adhesives. Rubber and vinyl sheets with an abrasive-filled surface stay nonslip under wet conditions.

Antistatic and conductive surfacing materials are used where electrostatic charges are likely to build up on a floor in an inflammable atmosphere. Relevant materials consist of earthed reinforcement, acetylene-black filler, special rubber, vinyl tiles with precision-ground joints, ceramic tiles, and fleximer-matrix terrazzo (see Sections 22.1 and 36.7). Standards (e.g. AS 1169, BS 2050, 3187, 3398 and ASTM C483) differ slightly in various countries. Antistatic footwear is specified in BS 2506, 3825 and means of controlling electrostatic charges are given in AS 1020.

20.3 SURFACE TREATMENTS FOR PARTICULAR HAZARDS

Table 20.3 shows special surface treatments, given in the preceding text, that are recommended for particular hazards.[50] With the exception of surface-hardening compounds, they should be applied to a surface that will ensure an adequate bond under working conditions. An increase in concentration of corrosive liquids can take place by evaporation.

The selection of one or more of these treatments will be governed by satisfactory performance under specific working conditions, and by cost. All will give useful service when properly applied, but the cheapest initially may not prove to be the most economical. Thin coatings are usually confined to light-traffic areas and exposure to cold dilute corrosive liquids, or incidental exposure to more severe conditions, as they are not permanent and need periodical renewal. Greater thickness of surface treatment and regular maintenance are necessary on traffic aisles and around machines. Increased cost, such as with the treatment (i), is often justified for certain difficult conditions.[118]

---

rubber-latex type of fleximer compound. s, two treatments are necessary, including (a). They are alternative to others shown for a particular hazard. For example, the treatment for hydrochloric acid hazard is (a) plus (c) or (e); (f); (g); (h) or (i) (see Section 20.2). x, alternative treatments. Selection of the treatment to be used will depend on economic considerations, the type and concentration of hazard involved, and the availability of materials.

† Special surface treatment is optional.

‡ Special surface treatment for sodium chloride solution is needed only when the floor is subject to alternate wetting and drying.

20.4 SAFETY, MAINTENANCE AND REPAIR

20.4.1 SAFETY

The following factors contribute to safety.

An even, nonslip surface on a floor designed for the purpose for which it is used (ASTM E445).

Floor insulation (particularly around the perimeter) to prevent surface condensation and consequent slipperiness in cold weather (see Section 20.1.4(a)).

Link mats for nonslip purposes. These consist of small rubber units held apart by stainless steel wire.

Safe manual handling of materials (AS 1339).

Safety colours (AS 1318) and warning signs (AS 1319), or a gradual transition in surface texture where appreciable differences occur in the coefficient of friction between adjacent surfacing materials.

Satisfactory working conditions. Environmental aids in winter include space or radiant heaters and, to some extent, red concrete.

Surface cleanliness and dryness.

Test for safety and protective footwear (BS 953).

Trafficways across slopes and curved at changes of direction, demarcation lines and handrails, and adequate lighting facilities provided where necessary (AS 1470).

Use of special or approved footwear (AS Z3 and BS 1870).

20.4.2 MAINTENANCE

Regular cleansing of a concrete floor is essential for maximum safety, hygiene and service performance. Grit and dirt should be removed to minimise the wearing effect of trucking. Any potentially corrosive liquid spilt on a concrete floor should be diluted with water and washed off soon afterwards. If corrosive liquids are not soon removed, their concentration and damaging effect could increase through progressive evaporation. Cleaning may be by brushing or mopping with warm soapy water or neutral detergent solution (low in sulphate content). Alkaline detergents and those containing sodium hypochlorite for bleaching purposes should not be used on high-alumina cement concrete.

Drip trays under machines and rubber mats or disused rubber tyres on unloading bays do much to prevent slipperiness and damage to the surface. Industrial trucks to be used on a concrete floor should be correctly designed with rubber or nylon tyres, and the operators should be properly trained to minimise the abrasive effect of tyre scrubbing due to skidding and quick starts and stops.

Concrete surfaces that have become dangerously smooth may be acid-etched to provide a nonskid texture or a suitable background for an appropriate surface treatment. The etching is done with commercial (33 per cent) hydrochloric acid, which is diluted with two to three (or more) times its volume of water in a plastic, 9-litre watering-can. The solution, which may be mixed with a dessert-spoonful of sulphonated detergent per 5 litres, is prepared to suit the nature of the concrete surface and is applied at the rate of 3 m² per litre.

The applied solution is allowed to react with the exposed cement for 5-10 minutes, or until the fizzing effect ceases, whereupon the surface is thoroughly scrubbed and rinsed copiously with water. The process may be repeated if necessary to produce a granular surface. Where drainage conditions are unfavourable, sawdust soaked with dilute acid solution may be spread 15-25 mm deep and, coupled with occasional raking to renew interfacial contact, left in place for about 1 hour. Drains or gullies, which might be damaged by acid, should be covered with hydrated lime beforehand.

The surface, after thorough washing, is allowed to dry. Acid-etching is most effective on good-quality concrete with wear-resistant properties. Where a surface is likely to wear smooth again within several months after etching, or it is in a satisfactory condition for priming and surfacing, a durable nonskid texture can be produced with an adequate application of filled chlorinated rubber lacquer or epoxy resin compound, the latter being mixed with a hardener (see Sections 20.2(c), 20.2(e) and 21.1.6(c)). Small defects should be repaired as they appear and the finishing coat of a thin protective treatment should be renewed at intervals.

20.4.3 REPAIR

Damaged areas may be scuffed, cleaned, primed and patched to feather-edge thickness with quick-hardening compounds such as filled epoxy resin compound and hardener, rubber-based fleximer, or bituminous mastic mortar. Cracks may be cut, cleaned and repaired with epoxy resin compound and hardener, with or without filler. For major repairs to concrete, the damaged material is cut out to a minimum depth of 40 mm, leaving holes with square or undercut edges. The holes are cleaned, dampened, coated with cement slurry and mortar, and filled with good-quality concrete. Epoxy resin compound and hardener (in the proportions of 3 : 1 by volume) may be used at the rate of 0·5 kg/m² as an alternative means of bonding new concrete (see Section 21.1.6).

Surfaces that are to be resealed or resurfaced must be prepared and cleaned. Solidified oil and grease should be scraped off the surface, and liquid oil should be soaked up with an absorbent material, such as hydrated lime, whiting, powdered diatomite, talc or portland cement. The affected areas are scrubbed clean with hot 10 per cent caustic soda solution and detergent solution.

A persistent stain on a garage floor can be removed with trisodium phosphate mixed with five to ten times its weight of water and enough whiting to form a paste. After applying it 10 mm thick for 24 hours, the surface is well cleaned with water. Badly oiled areas can be steam cleaned or soaked and scrubbed with degreasing oil (e.g. power kerosene containing a scouring agent, such as trisodium phosphate or cresylic acid), and then washed clean. Industrial wet-vacuum cleansing may be used on large equipment projects where splashing must be avoided. With volatile liquids, precautions must be taken against fire.

Residual stains are covered for some hours with a paste made of absorbent material and a solvent (e.g. benzol, white spirit, mineral turpentine, trichloroethylene or equal parts of acetone and amyl acetate). This paste is covered with a nonabsorbent sheet to retard evaporation of the solvent and promote fluxing of the oil. The solvent-fluxed oil is then withdrawn from pores in the concrete into the paste as it dries to a powder, which can then be brushed off. The surface may be scrubbed finally with hot detergent solution and rinsed clean. As trichloroethylene inhibits the cure of resinous adhesives, any traces of this chemical should be allowed to evaporate for several hours if the prepared surface is to be coated with an epoxy resin compound and hardener.[83]

Resurfacing with concrete is carried out as described in Section 20.1.5. For a uniform colour effect, care must be taken to ensure uniformity of compaction, water/cement ratio and curing in the top surface. To this end, the harder an area of topping is worked, the darker its colour becomes. This is because of a lowering of the water/cement ratio in the more densified zone and, particularly without proper curing, a relative reduction of hydration in the ferrite phase of the cement. The effect is accentuated by the formation of slight efflorescence on adjacent areas. Dry cement, if used with finish trowelling, causes dark stains that are integral with the surface.

CHAPTER 21

POLYMERS AND THEIR USES

21.1 EPOXY RESIN

21.1.1 GENERAL INFORMATION

Epoxy resin is a condensation product of bisphenol A and epichlorohydrin. This product, on being mixed with a chemically active reagent (curing agent or hardener), polymerises into a very strong and tough plastic (ASTM D1763). The pot life of the mixture varies with its formulation and temperature from about 15 minutes to several hours. At the time of use, it should be able to adhere strongly to a surface.

The resin cures with an exothermic reaction, and the higher the temperature of the reaction (such as in a large or heated mass), the shorter is the hardening period of the mixture and the greater is the possibility of heat distortion and cracking of the resin. At 23°C, epoxy resin and hardener become touch-dry within a few hours and the compound is completely cured within 1 week. Its application is facilitated if used in a shallow mass, which dissipates the heat of reaction and thereby prolongs the hardening time. The hardening time is decreased by high humidity and more than halved at temperatures above 40°C. The temperature of a concrete surface should be at least 16°C for surface-wetting and curing purposes. At temperatures below 13°C, curing is facilitated by the application of heat or the use of a special amine accelerator. Normally, light traffic can be allowed over a repaired area in 3-4 hours or a fully treated surface in 12-24 hours.

The properties of the cured resin may be varied over a wide range by the proper selection of resin, curing agent and curing procedure. They may be varied over an even wider range by incorporating modifying polymers, catalytic inhibitors, liquid diluents and inorganic fillers (see Section 21.1.3) into the uncured compound. By these means, improvements in handling characteristics and service performance may be obtained, possibly at reduced cost. The economics of epoxy resin compounds dictate that their use be supplementary to a material such as concrete, and not in place of it.[52, 131]

21.1.2 CATEGORIES

Epoxy resin compounds may be classified broadly as follows.

1. *Solventless liquid epoxy resin* with either an amine or a polyamide hardener. The quantity of polyamide hardener required is less critical than that for an amine hardener, and its curing rate is slower. Polyamide hardener (1-3 parts) may be blended with 1 part of the amine hardener, and suitable safety measures must be taken when an amine hardener is used extensively.

A solventless compound has negligible shrinkage and may therefore be applied in a heavy coat of up to 1 kg/m². Spreading rates are 1 litre to 6·5-7·5 m² for 0·125 mm thickness, 2·75-3·5 m² for 0·25 mm, and 1·5-1·8 m² for 0·50 mm thickness. If a filler is not used, the coating will harden with a high-gloss finish. Common modifications are indicated in Table 21.1, the modifying polymer being usually about one-third of the epoxy resin component. Modified resin should be stirred thoroughly before being mixed with hardener (See Appendix I).

TABLE 21.1   MODIFIED EPOXY RESIN

Modifying Polymer	Imparted Characteristic
Polysulphide synthetic rubber (e.g. "Thiokol")	High adhesion, chemical resistance and resilience (without undue sacrifice of strength)
Polyamide resin	High adhesion, resilience and resistance to impact
Bitumen (epoxidised-oil) or coal tar (horizontal retort)	High adhesion and water resistance and flexibility, good chemical resistance and adhesion, freedom from bleeding, and low cost (AS K172)

2. *Epoxy resin solution* with a curing agent of amine, amine adduct or polyamide type. The adduct is an amine which has been partly reacted with resin for improved curing effect. A cathodic metal primer contains 1 part by weight of polyamide-

cured epoxy resin (a solvent and suspending agent being extra) to 14 parts of superfine zinc dust (refer to Table 22.1, steel).

3. *Phenolic or other modified epoxy resin solution* with an amine adduct hardener. It has high chemical resistance and adequate adhesion to concrete, but on metal it requires a primer.

4. *Epoxy ester solution.* This compound is a one-pack product which hardens by the absorption of oxygen. It is softer than the other varieties of epoxy resin, and lower in chemical resistance and cost. It is strong in adhesion to sound surfaces generally and can be used as a good metal primer.

5. *Epoxy stoving compounds.* These materials are used only on metal and require heat curing.

6. *Epoxy resin foam* for sprayed or infill insulation with a unit weight of 36 kg/m^3 and a thermal conductivity value of 0·016 W/m.°C.

The foregoing resin compounds are designed for specific applications, as no single formulation will meet all engineering requirements. For instance, an epoxy adhesive (in liquid, paste and grout forms) is of polysulphide or other resin-modified type. A typical formula for an adhesive and binder compound is 10 parts by weight of epoxy resin, 4 parts of polysulphide polymer and 1 part of hardener, a variable amount of filler being used. Diluents (e.g. acetone, glycidyl ethers, certain alcohols and triphenyl phosphite) may be used in small quantities with epoxy compounds, so as to facilitate manipulation and surface penetration without detracting from strength (AS MA1.5 and BS CP3003).

21.1.3 FILLERS

Fillers are added to epoxy resin compounds to achieve selected improvements. These include the following: an extended pot life; increased hardness, wear resistance and load-carrying capacity; reduced coefficient of thermal expansion, shrinkage and cost; and possible use in a thick coat.

For premixed and surface-blinded coats, fine aggregate of the following material is used (where Mohs' scale of hardness is given in brackets): silicon carbide (9·5), aluminium oxide (9·0), emery and zircon (8·0), fine sand, fly ash or silica flour (7·0). Colloidal silica and modified bentonite impart thixotropic properties to mixes that are to be used on concrete walls, and fibreglass reinforcement can be used for additional strength (BS 3534 and ASTM C581, C582).

Fillers should be clean and dry and, depending upon the requisite surface texture of a coating, their particle size should fall within a 2·36 mm to 0·15 mm mesh sieve range. The voids in graded and single-size fillers vary from 30 to 50 per cent, respectively. For greatest strength development of a filled mix, the resin binder should more than fill the voids in the aggregate.

A mix that can be easily applied consists of 1 part resin with hardener and 1-3 parts of fine aggregate by volume. Its trowellability is improved by a small proportion of very fine particles in the filler. The flexibility or resilience of a cured resin mix decreases as the content of filler is increased. Mixes with a high content of filler (e.g. 85 per cent) are porous and rigid, and they should be sealed if used under exposed conditions. Much more filler can be mixed effectively with a solventless resin than with a resinous lacquer or solution. Pigmented grades of compound are used to resist the action of strong sunlight. Filled resinous compounds are suitably plasticised with cashew nutshell or pine oil. For aesthetic purposes, the filler of bitumen epoxy and coal-tar epoxy resins may be admixed with an oxide pigment (e.g. iron or chrome) at the rate of 1 kg/28 kg.[111, 127]

21.1.4 PROPERTIES

The hardened resins, taken collectively, have outstanding characteristics of adhesion, strength, toughness, resilence, resistance to impact, corrosion and, to a less extent, wear. Epoxy adhesives, being somewhat flexible or resilient when cured, are able to withstand the dimensional change effects of concrete (including drying shrinkage and creep), provided that the dimensional change of the adhesive is not expected to exceed 5 per cent.

An epoxy adhesive joint in structural concrete has a mean shear strength that is roughly equal to the split-cylinder tensile strength of the concrete. Its value, which is in excess of 2·75 MPa for short-term loading, may be more than halved in joints 2 mm thick by creep under sustained loading, or possibly doubled by sustained compression (e.g. 0·7 MPa) across joints. Superior results are further obtainable with epoxy mortar,[83, 136] while soft grades of the adhesive begin to yield at temperatures above 50°C (see Reference 80 of Part 8). General characteristics are given in Table 21.2 [53, 179] where a resilient grade of epoxy resin indicates one containing a polysulphide or polyamide-modifying polymer.

TABLE 21.2   **PHYSICAL CHARACTERISTICS**

Physical Characteristic	Grade of Resin		Comparative value of unfilled rigid grade of Polyester
	General Purpose	Resilient	
Strength (MPa)			
Compressive	140-280	21-210	70-250
Flexural	55-140	Up to 100	34-130
Tensile	14-83	0·7-28	14-70
Modular (MPa)			
Compressive	$2·8-12 \times 10^3$*	Up to $1·4 \times 10^3$	$2·1-4·1 \times 10$
Flexural	$2·8-10 \times 10^3$*	Up to $1·4 \times 10^3$	$2·1-5·9 \times 10$
Tensile	$2·1-7 \times 10^3$*	Up to $1·4 \times 10^3$	$1·0-4·1 \times 10$
Impact Load (Izod, kN.m/mm)	8·5-37	27-370	8-21
Hardness (Rockwell M)	75-110		60-115
Flammability (mm/min) (ASTM D635)	7·5 to self-extinguishing		40 to self-extinguishing†
Specific heat (kJ/kg.°C)	1·0-1·7		1·3-2·5
Thermal coefficient of linear expansion (per °C)	$3·1$*-9×10^{-5}	$5-8·1 \times 10^{-5}$	$7·2-11 \times 10^{-5}$
Thermal conductivity (W/m.°C)	0·17-0·87*	0·17	0·17
Heat distortion temperature (1·80 MPa; °C)	Up to 120	Not applicable	40-200
Maximum continuous service temperature (°C)	80‡	50	120-150
Specific gravity	1·115§-2·4*	1·0§-1·25	1·15-1·46
Water absorption on 24-hour immersion (per cent)	0·1-0·5	0·4	0·15-0·6

* Filled system. A two-third reduction in value is feasible with sustained loading.

† Chlorinated polyester has reduced flammability.

‡ 120°C if a liquid metaphenyline diamine hardener is used and 290°C if an anhydride hardener is used.

§ Unfilled system.

With few exceptions, the chemical resistance of epoxy resin is of a very high order.[54] Check tests should be made on selected materials, wherever possible, under actual or simulated conditions of service. The resin should be protected thermally in buildings for requisite fire-resistance, although epoxy-mortar bonded masonry is claimed to have passed a 4-hour fire test.[69] Data on resins are given also in References 122 and 137.

21.1.5 SURFACE CONDITION

Substrate surfaces must be strong, sound and clean; that is, free from friable material, laitance, contaminants and dust. Glazed surfaces should be lightly roughened with emery paper and steel surfaces should be thoroughly wirebrushed or sand-blasted (AS 1627). On concrete surfaces, bonding should be onto high-strength material or clean, exposed or partly protruding coarse aggregate (see Section 20.2). Bonding is normally onto dry surfaces (except where heavy resin is used below seawater), although concrete may be damp but not wet. In the latter instance, a solventless resin should contain a polyamide hardener and be brushed (but not sprayed) on to the surface for miscibility with moisture and subsequent bond.

The soundness of a concrete surface for bonding purposes may be estimated by inspection or by measuring the force required to pull off a 50 mm diameter pipe cap, which has been bonded to the surface with an epoxy resin adhesive. Curing of the resin may be accelerated by heating the pipe cap with a blow torch for about 1 minute. The bond strength of a coating to concrete can be determined by applying the test to a 65 mm diameter disc of the coating, which has been isolated by a hollow bit and drill. For satisfactory adhesion, the bond strength should exceed 0·7 MPa.[76]

21.1.6 TYPICAL USES

(a) Bonding.
Concrete, metal, timber, glass, stone, asphalt and plastic surfaces may be bonded together. A polyamide epoxy resin is required for strong adhesion to polyethylene film (see Table 21.1; also BS J10). Where technically suitable, concrete members can be joined together in about 15 minutes, using splice bars and a filled resin adhesive (i.e. 2 parts epoxy resin, 2 parts sand passing a 2·36 mm sieve and 1 part zircon flour by volume). A full monolithic bonding action between precast units is induced by applying a direct pressure during the hardening of thin joints.

Suitable surfaces must be provided for bonding purposes and, for fire resistance and maximum strength in the lift-slab system (see Section 27.3), steel plates welded to column reinforcement can be resin-bonded and then fillet-welded together. Concrete piles can be spliced quickly with thermally cured epoxy resin and hardener inside a galvanised-iron sleeve (Plate 14; see Reference 73 of Part 8). Strain gauges[72] and tension heads can be readily bonded to test specimens, and dowels or projecting reinforcement can be grouted into holes on one side of control or construction joints.[142]

Cast *in situ* concrete may be bonded to steel or hardened concrete surfaces. Clean steel reinforcement can be coated with epoxy resin and hardener and then either spattered with coarse sand or embedded with concrete cast around the unhardened resin, particularly when the resin becomes tacky at an age of about 0·5-1 hour. This process increases the bond strength by pull-out tests by amounts up to 500 per cent.[64, 91] A state of tackiness is required in order to prevent an emulsification of the liquid resinous compound by freshly placed concrete. The standing period for this purpose can be reduced to about 20 minutes by lightly sprinkling the resinous coating at this age with dry cement. The resin should have a high elastic modulus to minimise creep under sustained load (see Reference 168, Section 8, and Table 21.2).

For bonding fresh concrete to hardened concrete, the adhesive and hardener (immediately after mixing) are spread at the rate of 0·5 kg/m² shortly before the fresh concrete is placed, the concrete having a slump of less than 60 mm. Should a formulation contain solvent (for an increased pot life), the applied mixture should remain exposed for about 30 minutes to enable it to become sticky before being covered. Areas may be patched with a feathered edge, using coarse and fine concrete mixes over unset adhesive and hardener, the finished surface being membrane cured.

A glazed tile or ornamental veneer can be spot bonded or fully bonded to precast concrete panels, the panels themselves then being joined together if required. In highway work, continuously extruded kerbs are readily bonded to an existing roadway by means of two adhesive strips, each 25 mm wide. Sources of information on resinous mortars for chemical-resistant masonry (ASTM C279) are given in Table 21.3 and References 55 and 69; shear tests are given in ASTM E229.

(b) Sealing of cracks.
Small cracks can be lightly notched with a small grinding disc, thoroughly cleaned, primed with acetone and sealed with an injection of low-viscosity epoxy resin and hardener. A cake-icing gun or hypodermic syringe (with a 1·2-mm needle) or a caulking gun (with a disposable plastic cylinder and nozzle) may be used. Large cracks are primed with epoxy resin and hardener (in the proportions of 3 : 1 by volume), and filled with a trowellable epoxy mixture or paste. Deep fissures may be filled by pressure-grouting methods, using pressure nipples that are inset at 0·6-1·2 m intervals,

TABLE 21.3   STANDARDS ON CHEMICAL-RESISTANT
MORTARS

Title	ASTM
Absorption and apparent porosity of chemical-resistant mortars	C413
Bond strength of chemical-resistant mortars	C321
Chemical resistance of mortars	C267
Chemical-resistant resin mortars	C395
Compressive and tensile strengths of chemical-resistant resin mortars	C306, C307
Shrinkage and coefficient of thermal expansion of chemical-resistant mortars	C531
Use of chemical-resistant resin mortars	C399
Working and setting times of chemical-resistant resin mortars	C308

a vented surface seal and a lever-operated grease gun or a pneumatic extrusion gun (350-750 kPa). Sand, cement or pigment may be trowelled or brushed into the unhardened resin.[56, 132, 143]

Movement fissures can be sealed with a resilient epoxy resin compound, and split concrete can be repaired under suction with gently extruded thixotropic resin of about 1.5 Pa.s viscosity. In this operation, a 94 kPa vacuum is manipulated through a 100 mm^2 transparent pad, consisting of a 5 mm perspex sheet fitted with a foamed rubber gasket. Penetration of the resin is controlled by masking means or by manual regulation and inspection through the transparent pad. A hot-air gun may be used for drying cracks and equipment must be kept clean as indicated in 21.1.6(c).[200]

(c) Surface treatments.
Typical applications are as follows.

The antiskid surfacing of floors, roadways and stairway treads, using a mixed-in filler or a grit blinding that is tamped or rolled in. For example, a liquid compound that is spread on a clean roadway at the rate of 1 litre to 0·9-1·8 m^2 may be covered with 3 mm clean hard grit at the rate of 1 m^3 per 150-250 m^2.[71]

The priming of surfaces and joints prior to placing synthetic resin or rubber compounds, the primer being scrubbed or stiff-brushed into the surfaces at the rate of at least 1 litre per 3·5 m^2. This application may be followed by protective surface treatments which are required to withstand corrosive and abrasive conditions (see Section 20.2). Robust coatings over a primed surface should be 3-6 mm thick. Freedom from air bubbles or pin holes may be assessed by a nondestructive ionisation tester.

Tank linings (BS CP3003) should not be spark-tested, because of possible damage to them.

The impregnation of autoclaved asbestos-cement pipes for gas tightness, and the membrane curing of newly placed concrete (see Section 9.3). The end-protection of prestressed concrete members, the surface coating of aluminium casting moulds and the fabrication of resin moulds that are laminated with fibreglass. The provision of a moisture barrier under water-sensitive surfacing materials, two coats (each 0·25 mm thick) being applied to the prepared and liberally primed surface of a concrete floor. Further uses include the pigmented sealing of swimming pools, masonry walls, roofing tiles and slabs, the laying of thin terrazzo concrete, and the decorative treatment of surfaces with a mixture containing bronze, mica or fast-colour pigment.[123]

For these purposes, a resin compound is prepared for use by mixing the liquid components together at 21-32°C for 3 minutes, either manually or preferably with a portable electric drill fitted with a paint paddle. Filler in a limited amount is stirred in. In a larger amount, it is mixed with the liquid compound for 2-3 minutes on a clean steel plate, or in a mortar box or small pan concrete mixer, care being taken to minimise air-entrainment. For horizontal surface treatment, a filled mix should be distributed in a line-pile over an unset primer. It is screeded and wood-floated in a 300-900 mm width to requisite thickness and compacted. Final trowelling, at an age of 1-2 hours, is facilitated by wiping the steel trowels with lighting kerosene or an aliphatic solvent. Rebated bonded joints (10 mm wide) should be formed between new and hardened strip-applications, and a bitumen epoxy resin should not be fluxed or softened for a long period by solvent.

Terrazzo-type flooring 3-15 mm thick may be laid similarly, the brass divider strips normally required being spaced widely apart. The flooring is bonded with resin to a prepared concrete base course and ground smooth within 2 days. Alternatively, resin-bonded terrazzo tiles (225 mm square by 5 mm thick) can be laid with a rubberised adhesive and finished with an electric sanding-machine. Polyester resin may be used in lieu of epoxy resin as a pigmented matrix, the tiles in each instance being cut easily in a guillotine and laid with a close joint.

Automatic equipment can be used for measuring and dispensing purposes. Surface coatings can be applied by a multinozzle, airless spray gun, which can be set to cover a range of proportions of component materials. A glass-fibre cutter and blower attachment expedites the application of reinforced coatings.[53] Special equipment may be used for laying filled coatings over large areas or for applications under vacuum. A hot finishing screed will produce a high lustre.

Appliances must be cleaned progressively and immediately after use with resin solvents (e.g. acetone, toluol, xylol, methylene chloride, ketones or equal portions of alcohol and benzene). They may be cleaned otherwise (where practicable) by burning off the resin.

21.1.7 SAFETY PRECAUTIONS

Physiologically harmful volatile solvents, active diluents and amine hardeners (such as may be used in liquid epoxy resin), should be kept away from the skin. Safety measures include fire prevention, the use of a barrier cream, solvent respirators, cotton overalls and cap laundered daily, disposable plastic gloves (e.g. polyvinyl chloride), goggles, adequate ventilation, and soap and hot water for washing contacted skin. Full precautions should be taken by persons engaged in blending and mixing the materials in large amounts.

21.2 POLYVINYL ACETATE LATEX

21.2.1 GENERAL INFORMATION

Where prolonged contact is not made with water, an admixture of polyvinyl acetate (PVA) emulsion or latex in cement slurry or mortar may be used for special purposes. Its proper use can improve such properties as tensile and bond strengths, and resistance to impact and abrasion (see Section 20.2(h)). The following observations were made at the Massachusetts Institute of Technology and concern the use of a PVA latex in cement mortar. The latex contains 50 per cent of dispersed polymer and air-entrainment is minimised during mixing. Further data are given in References 119 and 147.

Maximum improvement in the tensile properties of latex-modified mortar is related to the following factors.

Intermediate-size dispersed particles in the range of 1-5 μm.
A latex/cement ratio of about 0·2 by weight, this ratio giving also the most workable mix. (Economically, this proportion is warranted only in special applications.)

A plasticiser (e.g. dibutyl phthalate or diethylene glycol) is not usually required in mortar, although it is essential in surfacing work.

Air curing only is required, although large areas should be covered at an early age, so as to prevent plastic cracking under strong drying conditions.

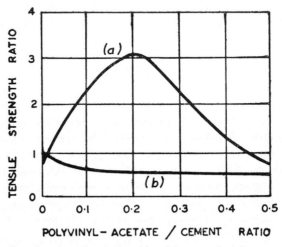

Fig. 21.1 Tensile strength relationship of latex and plain mortars. Curing: air and fog spray at 23°C. (a) 50 per cent relative humidity. (b) 100 per cent relative humidity. Age at test: 28 days.

Figure 21.1 indicates that the tensile strength of latex : 1 part cement : 3 parts sand mortar meeting the requirements is about three times that of moist-cured, plain cement mortar. The compressive strength of the air-cured latex mortar, with optimum admixture, is about 70 per cent of that of moist-cured, plain cement mortar at 28 days. Prolonged storage of the latex mortar in water or a highly humid atmosphere produces disruptive hydrolysis and acetic acid as a disintegrative, corrosive by-product.

Microscopic study of the latex mortar made with a water/cement ratio of 0·43 by weight shows a well-compacted matrix of cement gel with voids filled with polymer. In mortar with a 0·5 latex/cement ratio the cement hydration products are dispersed in a continuous phase of polymer. Using a red, oil-soluble dye, similar study of plasticised latex showed a layer of the plasticiser around the particles of polymer. The addition of plasticiser (e.g. 0·1 dibutyl phthalate) to the latex has the effect of reducing the degree of compaction of a dried mortar matrix and the tensile strength by about 30 per cent.

The modulus of elasticity of air-cured latex mortar, with a latex/cement ratio of 0·2, is about 0·2 times that of moist-cured plain cement mortar in compression and 0·3 in tension. A further increase in the ratio decreases the modulus further. The coefficient of thermal expansion of latex mortar is about 11×10^{-6} per °C, which is little different from that of plain cement mortar.

The resistance to abrasion is high and greatly increases with an increasing content of polymer, which acts as a very adhesive, resilient binder in preventing particles from being torn out by abrasive force. In surfacing work, the amount of plasticiser used (e.g. 0·1-0·15 latex) does not significantly affect this property.

The free shrinkage may be 100 per cent higher but this can be reduced by additional plasticiser. Under conditions of restraint, sudden loads and thermal stresses, the properties of adhesion, tensile strength, extensibility and energy absorption are sufficiently high to prevent or strongly resist cracking or rupture. Air-cured mortar with 0·2 latex/cement ratio has about four times the bond strength to old cementitious surfaces as moist-cured cement mortar.

Latexes with high water resistance are stabilised acrylic, styrene, butadiene and polyvinylidene (BS 4661), modified with 10 per cent of unplasticised PVA latex. Commercial grades have 40-50 per cent content of polymer. They may be applied to sound surfaces or used as an admixture. A delay in the setting time is significant with latex/cement ratios above 0·3. For functional compatibility, ammonia may be added to PVA latex to raise its pH value to 7·5. A few uses are given here.

21.2.2 FLOOR SURFACING

As the polymer is water-sensitive, concrete floors on the ground should have subgrade drainage and a vapourproof membrane below or an epoxy resin sealant on top (see Sections 20.1.3, 20.2 and 21.1.6).

The concrete is brought to a laitance-free, wood-float finish, or it is properly prepared and cleaned to ensure strong adhesion (see Sections 20.1.5(d), 20.2 and 20.4.2). Residual oil stains on an existing floor are coated with a slurry of latex and cement. This coat is scraped off after 24 hours and the process is repeated, if necessary, so as to produce an oil-free surface.

The surface is primed with finely dispersed plasticised polymer with adequate water-resistant properties. A composite acrylic or glyoxal formulation may be used for this purpose. A latex containing 50 per cent of polymer is diluted with an equal quantity of water, brushed liberally over the surface and allowed to dry.

Surfacing may be carried out with two coats of slurry, either sprayed on or spread with a soft broom at the rate of 1 litre per 3·5-4·5 m² with a drying period between coats. The slurry consists of 1 part of latex, 2 parts of water and 3 parts of fresh portland cement by volume. Alternatively, a mixture of 1 part of fresh cement and 3 parts of coarse-graded sand may be gauged into a stiff mortar with 1 part of the latex and 2 parts of water. It is screeded, wood-floated and steel-trowelled to a thickness of 6 mm, the same gauging liquid (not water) being used to ease the trowelling. A similar mix, but with only 1 part of water may be laid 3-6 mm thick and ground smooth, if desired, after 4 days.

The finished surface, in each instance, is air-cured for 1-2 days and opened to traffic. Under dry service conditions, the hardened surface is tough, antislip, oil-resistant and more than twice as abrasion resistant as regular concrete. The latex may be incorporated in the bonding slurry and the mixing water of magnesium oxychloride surfacing for an enhanced finish (see Section 15.2.3(e)). PVA emulsion bonding agents for internal use with gypsum building plasters are specified in BS 5270, and rubber latex is dealt with in Section 20.2(h).

21.2.3 RENDERING

A clean, unpainted surface is brushed liberally with a 1 : 1 mixture of latex and water, and allowed to dry. A 1 : 3 mixture of cement and sand is gauged with a 1 : 4 liquid mixture of latex and water. Levelling and finishing coats, each 6 mm thick, are applied and air cured. They are characterised by uniform surface tint, strong adhesion and low shrinkage values. A cement-dispersing agent may be incorporated in these materials (see Section 6.4) and, for increased resilience, the liquid may be gauged 1 : 3 by volume.

Wall and ceiling panels of asbestos cement, foamed plastic, gypsum plaster and heavy-grade building paper can be primed, sprayed, stippled or spatterdash coated (see Glossary, Appendix VI) with suitably formulated, similar materials. Suspending and dispersing agents may be incorporated to facilitate uniformity of thin applications.

21.3 PLASTIC LININGS FOR PIPES

Precast concrete pipes may be spun-lined with a thermosetting resin (e.g. epoxy, phenol furfural or polyester containing silica flour, diatomite, carbon, fibreglass, mineral wool or asbestos) as well as a hardener and thixotropic stabiliser. After surface-roughening internally and curing, the pipes are preheated to assist hardening of the plastic, which is sprayed from a revolving head and compacted by spinning. The thickness ranges from 5 mm to 10 mm for high resistance to chemical solutions, and the lining is extended over the ends of the pipes so as to ensure uniform protection. Fittings and couplings are similarly lined.

Concrete pipes from 750 mm to 3000 mm diameter are lined internally with plastic sheet, either wholly or (in sewers) above the anticipated minimum-flow level of fluid. Black, plasticised, filled-grade sheets of polyvinyl chloride (e.g. 2 mm thick) are anchored to the concrete by means of embedded tee-shaped ribs. The ribbed sheets are held by wires against an inner mould and, before the reinforcement is fixed in position, they are joined by strips of plastic, which are welded over butt joins with the aid of a hot-air blowpipe (AS K124, K139; BS 1763, 2739, 4203, CP3003; and ASTM D1593, D1755, D1927, D2123). [57, 66]

Ribbed plastic linings may be embedded in concrete pipes by vibration and pressure during manufacture, prior to being steam and moist cured. Alternatively, they may be monitored on a mandrel and pressure-grouted interfacially to precast units. Concrete manholes, standpipes, and structures subject to turbulence can be form-lined during *in situ* casting, and heat-welded liners are spark-tested at joins before pipelines are put into service.

Polyester resin, reinforced with fibreglass or nylon, may be used for wrapping and strengthening concrete pipes. Ribbed plastic linings facilitate the internal moist curing of large, cast *in situ* concrete conduits, these being otherwise prone to early evaporation and plastic shrinkage cracking under draughty or air-curing conditions.

Linings for asbestos-cement pipes are specified in ASTM C541; rigid polyvinyl chloride profiles for fitting sheet lining materials are given in BS 3835; and methods of testing plastics are set out in AS 1145; BS 2782, 4618; ASTM Standards, Parts 34-36. Polyethylene and polyvinyl chloride plastic pipes (e.g. up to 1000 mm diameter) and fibreglass reinforced sheets are referred to in Appendix I.[111, 120] Unplasticised polyvinyl chloride pressure pipes and fittings must be carefully handled, stacked, transported, laid on uniform-bearing bedding and properly jointed, so as to prevent subsequent fracture at locally induced stress points.

21.4 POLYMER CONCRETE

Polymer concrete is a composite prepared by impregnating hardened concrete with a monomer, or by mixing a monomer or polymer with aggregate or fresh cement concrete, the resin being subsequently polymerised, cured or hardened.[161-168] High-performance composites are not only expensive to produce but, with the possible exception of special, vibropressure-made proprietary composites, the impregnated grades are several times superior to the premixed variety (see Sections 4.5, 21.1, 21.2 and 23.3).

The impregnated type of composite is produced by thoroughly drying and evacuating the preformed concrete, then soaking it in a nonviscous monomer which, after draining off the excess, is integrally converted into a polymer. This is brought about by Cobalt-60 gamma radiation at ambient temperature or by thermal-catalytic initiation at 70-90°C, or by a combination of these techniques. The rate of penetration of the monomer into the pore structure of the concrete can be increased during the soaking cycle by applying pressure with nitrogen gas, this medium being used for safety reasons when handling an initiator-modified monomer.

The quality of the rigid matrix formed by polymerisation is strongly related to the degree of loading of the concrete (e.g. 3-8 per cent or more, depending upon porosity and degree of impregnation), the effectiveness of treatment and the type of monomer used. Selection of the monomer is normally based upon its cost, polymerisation rate and intending application. Research shows that methyl methacrylate results in a hard, glossy polymer with extremely high properties of structural performance. Methyl methacrylate can also be combined with 10 per cent by weight of trimethylolpropane trimethacrylate to achieve still better results. Another highly effective impregnant is "Curifax-8C", this having been formulated by the Atomic Energy Research Establishment at Harwell, Berkshire, Britain.

Depending on the particular monomer used, polymerisation may be inhibited by the presence of atmospheric oxygen and by integral stabilisers that militate against spontaneous decomposition. These restraints must be overcome with either high-energy radiation or thermal–catalytic intiation, and by using protective thermal water-bath or containment techniques. Typical thermal–catalytic initiators are azobisisobutyronitrile and benzoyl peroxide.

Irradiation, because of its ambient-temperature efficacy, causes a minimal loss of monomer by vaporisation. It produces a high degree of conversion, which can be marginally superior to that achieved by thermal catalysis. To minimise vaporisation, however, the activating temperature can be lowered by the use of promoters. Depending on the extent of specimen drying, dimensional changes can arise and internal stress developments can occur in partially dried concrete due to swelling of certain monomers such as methyl methacrylate.

The structural properties of polymethyl methacrylate concrete, rated by comparison with control specimens, are up to the data summarised seriatim below.

Compressive strength: four times.
Tensile strength: almost four times.
Modulus of elasticity: almost two times.
Modulus of rupture: almost four times.
Pull-out-bar bond strength: three times.
Flexural modulus of elasticity: almost 1·5 times.
Freezing and thawing: over five times.
Hardness-impact: more than 1·7 times.
Creep deformation: reduction by a factor of ten.
Water permeability: decrease to negligible values.
Water absorption: decrease by as much as twenty times.
Corrosion by distilled water and sulphate brines: reduction to negligible amounts.

The maximum compressive strength can be exceeded by extremely high-strength calcium silicate specimens (see Fig. 33.13) and, with age, can be approached by specially made, high-strength cement concrete. The stress-strain relationship of test-specimen composites is linear until cracking begins at 70-80 per cent of the short-term strength. Crack propagation is usually through the aggregate, rather than at its boundary, and the nominal failure pattern is explosive (see Section 14.1).

The specific creep of specimens at 1000 days (50 per cent atmospheric relative humidity) is similar to that of oven-dried, 28-day cured concrete that is loaded to 20 per cent of its immediate potential strength. By comparison with undried ordinary concrete, however, the specific creep is approximately one-fifth (see Section 31.3.2).

Potential usage applications of polymer concrete include corrosion-resistant desalination plants, sewer pipes, drain-tiles, precast tunnel liners, bridge decks, protective barricades, pressure hulls, marine and underwater structures, and prefabricated masonry panels. In-depth, safety-regulated investigations must be carried out wherever possible, in order to optimise production procedures and to derive data on the properties of composites for practical use (see Sections 14.1 and 15.2.4).[39]

CHAPTER 22

FURTHER PROCEDURES FOR DURABILITY

22.1 GRANULAR METAL SURFACING

The wear resistance of ordinary portland cement concrete floors can be increased with special granules (e.g. malleable cast iron, or nonferrous slag and corundum), which are embedded to a depth of 3 mm or more when the floors are being laid. The granules may be used with pigment for colouring purposes or with carbon-black for antistatic and conductive surfacing. The progressive carbonation of cement in a hardened surface allows iron granules to oxidise superficially and assume a brownish tint. The absorption of water into a floor is subsequently reduced.

Metallic granules are usually made from malleable cast-iron slivers, which are freed from cutting oil by caustic washing and heating to red heat in a rotary kiln. The product is milled, screened and graded suitably between 2·36-mm and 0·15-mm sieve sizes for purposes of density and trowellability. The granules are blended with a cement-dispersing agent (e.g. calcium lignosulphonate) so as to facilitate their embodiment with cement in the moist plastic surface of newly laid concrete.

The material is mixed with one-half its weight of portland cement (or 1 : 1 by volume) and applied uniformly as a dry shake in a double coat so as to form a monolithic finish. To ensure uniform quality and effective bonding, there should be sufficient moisture in the concrete to dampen the dry shake, without excess water being present or additional water being applied. When ambient conditions would cause rapid drying, the operation is facilitated by prompt workmanship on limited areas and by a set-retarding admixture (e.g. calcium lignosulphonate) in the concrete.

Each coat is floated immediately into the receptive surface of newly laid concrete and finished by refloating or delayed steel-trowelling (see Section 20.1.5(d)). Hard finish-trowelling is deferred until bleed water has ceased moving upwards into the stiffening topping or composite interface. The surface is then membrane cured effectively (see Section 9.3) or moist cured without contact with air (which could promote superficial rusting of iron granules at an early age). The quantity of metallic aggregate normally required varies from 3 to 6·5 kg/m^2 for moderate to heavy traffic conditions, the detailed procedure being adapted to suit heavy-duty conditions that may require a premixed topping, 10-20 mm thick (see· Section 20.1.5).

The concrete in the floor slab should be of sufficient potential strength to carry anticipated loading and resist indentation beneath the monolithic metallic surface. Single or two-course floors with minimal laitance may be used, the concrete being not leaner than a nine 40 kg-bags/m^3 mix (see Table 5.1) with a slump of not over 80 mm. For strong edges at joints, an adjacent shallow wedge of fresh concrete up to 15 mm deep may be replaced by metallic cement mortar containing particle suspending and dispersing agents.

Calcium chloride should be kept out of the concrete so as to prevent the metallic granules from popping out through excessive corrosion and expansion. Carborundum or emery grit may be used for nonslip surfaces or where service conditions are likely to be damp. Spark-resistant floors should not contain this grit or a swirl-floated, nonslip finish, and coloured floors should be straight-trowelled in one direction so as to harden with a uniform appearance. Where batching is done volumetrically, the bulk density (loose) of the metallic aggregate may be taken as 3000 kg/m^3 for gauging purposes (see Sections 3.3.6 and 5.1).

22.2 SILICON TETRAFLUORIDE GAS TREATMENT ("OCRATION")

The treatment of dry concrete products with silicon tetrafluoride gas is known as "Ocration", after Ocriet Fabriek of Baarn, Holland. The gas reacts with hydrated lime, which occurs as a by-product in the setting of portland cement, and gives rise to calcium fluoride and silica gel. The reaction is

$$2Ca\,(OH)_2 + SiF_4 = 2CaF_2 + Si\,(OH)_4.$$

The silica thus formed is deposited within the pores of the concrete.

The gas combines also with the hydrated calcium silicates and aluminates of hardened concrete, thereby increasing its density and durability.

Silicon tetrafluoride is made in a steam-jacketed and lead-lined generator, by stirring sodium silicofluoride and finely divided silica with an excess of sulphuric acid, while the temperature is raised to 60-70°C so as to accelerate the reaction. The gas generator is evacuated initially, the requisite quantity of sulphuric acid for reaction purposes is drawn in by the vacuum and the gas is formed immediately. It is passed into a gas storage vessel for use as required. The reaction is

$$2Na_2SiF_6 + SiO_2 + 2H_2SO_4 = 2Na_2SO_4 + 2H_2O + 3SiF_4.$$

Fluorspar may be used instead of sodium silicofluoride, but the latter is usually more economical.

The concrete products, after 2-3 weeks of moist curing, are heated and dried in a large autoclave (e.g. 3 m diameter and 18 m long) using superheated steam at an average temperature of 260°C. After blowdown of the steam, the hot products are allowed to dry further for a period of 10 hours and to cool for about 4 hours, so as to develop a degree of superficial porosity. The autoclave is then evacuated to form a vacuum of − 70 kPa, the silicon tetrafluoride gas is admitted and, after a 4-hour or 5-hour reaction period, the autoclave is evacuated of residual gas and unloaded. A complete cycle occupies about 1·5 days, and the depth of penetration of the treatment is governed by the intensity of drying, the concentration of gas and the density of the concrete.

The depth of the gas penetration is assessed as the minimum thickness of solid concrete which remains after a cut section is immersed for 15 minutes in a concentrated solution of hydrochloric acid at 23 ± 2°C. For concrete pipes in outfall sewers, the minimum depth on the inside may be taken as 5 mm. Ocrated concrete possesses palliative protective and bactericidal properties against aggressive agencies. Prestressed concrete products cannot be ocrated because of the effect of high-temperature drying on the high-tensile steel tendons.

22.3 CORROSION INHIBITION AND PROTECTION OF METALS IN CONTACT WITH CONCRETE

22.3.1 CORROSION POTENTIAL

Metallic corrosion is an alteration or oxidation process of chemical, electrolytic or electrochemical origin. It causes metallic ions to be discharged into an aqueous solution from anodic areas that occur at the bottom of surface craters or cracks and which function cathodically at the periphery. The action is promoted by such factors as a damp atmosphere and excessive cracking of the concrete (see Sections 15.2.2(a), (b), 31.2.1, 33.3.2(b), 33.3.3(b) and 34.1.1); hydrothermal, wetting-and-drying and aggressive electrolytic conditions; variations in pH value and electrostatic potential; internal stress and work-hardened characteristics of the metal; stray electric currents and the proximity of dissimilar metals under moist conditions, and concrete containing one brand of portland cement being placed, with interfacial steel reinforcement or conduits, against concrete containing calcium chloride or another brand of cement. [12, 13, 86, 117, 191]

A prime means of inhibiting metallic corrosion is to induce a state of electrolytic equilibrium, in which the tendency of ions to pass into solution is opposed by the tendency of inhibitive ions to stifle the process of dissolution. Corrosion inhibition is promoted in one or other of the following ways.

The formation of a passivation layer around the steel in reinforced concrete, giving lasting protection.

The chemical pretreatment of metal, so as to minimise variations of electrostatic potential in the surface.

The coating of clean metal with cadmium[128] or anticorrosive primers containing inhibitive pigments.

The sealing of surfaces with linear polymer resins which have a high resistance to water and aggressive solutions.

The cathodic protection of electrically continuous steel elements in reinforced and prestressed concrete structures that are installed in severely corrosive environments or subjected to stray-current interference. Appropriate design knowledge, testing criteria and monitoring surveillance of galvanic and impressed current systems are fundamental to their successful application (see Sections 15.2.2(a), (c), 23.3.2 and 34.2.2, also AS MA1.5). [17-19, 28, 38, 75, 117, 176, 177, 185, 186]

Forming weakened-plane joint

Rolling granolithic
concrete surface course

Power floating
after compaction

Power trowelling
before curing

Laying steel tiles

Vibrated Concrete
PLATE 16

Internal vibration of
reinforced concrete outlet
culvert at
Tullaroop Dam,
Victoria

Vibrating screed used
transversely on the lower
portion of a
concrete slab

Vibrating screed used longitudinally
on the upper portion of the same concrete slab

22.3.2 PROTECTIVE MEASURES

Under normal conditions of manufacture and exposure, steel reinforcement in damp portland cement concrete is passivated against corrosion by the alkalinity of calcium hydroxide (see Section 3.1.1) and its stabilising effect on natural oxide surface films. Prerequisites for this inhibitive property to be maintained are the following.

> Limited width of exposed cracks (e.g. 0·25 mm) and depth of penetration of the concrete cover by atmospheric carbonation (see Sections 15.2.2(b), 31.1-31.4 and 34.1.1).[192-198]

> Concrete of uniform high quality must be dense, impervious, durable and in close continuous contact with the reinforcement (see Section 7.2.3).[87]

> In locations with external exposure, steel reinforcement must be completely embedded to a depth (except in dense cement-rich concrete products) of at least 25 mm (see Sections 3.4 and 19.1.3).

Maximum crack widths should generally not exceed the following values.

> 0·1 mm for severe exposure (e.g. alternate wetting and drying).
> 0·15 mm for water-retaining members with almost continuous contact with liquid.
> 0·2 mm for concrete members exposed to the climate.
> 0·3 mm for protected concrete members.

In highly humid atmospheres, the pH value of protective concrete should not be reduced (mostly by carbonation) to below about 11·5.

Cement-rich mortar, mechanically applied over exposed, wound-wire tendons, should incorporate preceding and succeeding coats of cement paste. A subsequent protection with coal-tar epoxy resin prevents possible deterioration in aggressive ground waters (see Sections 15.2.3(f), 19.2, 21.1.6(c) and 34.2.2).

Saline mixing water (see Section 3.2.1(a)), chloride or sulphate-bearing aggregate (see Section 3.3.4(e)), calcium chloride or admixtures containing this accelerator should not be used when reinforced concrete is to be subjected to warmth and humidity, steam curing, hydrothermal, electromotive or wet, salty conditions (see Section 15.2.2(a)).[39, 58, 59, 80, 133]

Where the environmental conditions are seawater, salty ground, or a warm moist atmosphere containing chlorides or sulphur compounds, reinforced concrete must be properly manufactured with an adequate content of cement. Otherwise, even with a 75 mm concrete cover, corrosion cells are likely to form at close or remote intervals on the surface of the steel reinforcement (see Sections 19.1.1-19.1.3). The areas which function anodically cause progressive rusting of the steel, while those which function cathodically cause destruction of bond with the concrete. Subsequent cycles of wetting and drying tend to accelerate the corrosion and formation of hydrated iron oxide residue, the effects of which are visible in staining, cracking and spalling of the concrete.[101, 107]

In the fabrication of external members, chairs and fastening devices for reinforcement should not be amenable to corrosion at the surface. The consolidation of cast *in situ,* reinforced concrete should, for maximum density, include revibration near the time of initial set of the cement (see Sections 10.2 and 23.2). This operation eliminates the air spaces which develop with water-gain under steel reinforcement and coarse aggregate.

For increased resistance to corrosive conditions, an anodic inhibitor or oxygen-reducing agent may be added to the mixing water of reinforced concrete, in order to form a zone of strong passivation around the steel reinforcement. The efficiency of this zone in suppressing corrosion is assisted by an excess of the dispersed inhibitor in the surrounding concrete. It is very important that an adequate quantity of the inhibitor be used, not only to suppress corrosion generally, but also to prevent intensive corrosion developing at areas of attenuated protection. Two types of admixture may be used, namely, stannous chloride and sodium benzoate, nitrite or chromate. In Reference 157, studies are made of stainless-steel expansion-bolt anchors, designed for the noncorrosive fastening of fixtures in hydraulic concrete structures.

Stannous chloride may be used as a substitute for calcium chloride, under conditions where the latter would promote corrosion of steel reinforcement. When used in the proportion of about 1 per cent by weight of cement, this inhibitor accelerates the rate of early-strength gain of portland cement concrete and suppresses the corrosion of steel reinforcement, even under steam-curing conditions. It is essential that a solution of the material be used in a fresh condition and that the concrete be thoroughly compacted, so as to minimise oxidation of the material to stannic chloride. Its use is avoided with attenuated steel tendons.[60]

Sodium benzoate and nitrite are less efficient but safer to use than sodium chromate in limited amount. There is no change in the normal hardening characteristics of concrete as a result of their use. When the first two inhibitors are incorporated singly, as a 2 per cent

TABLE 22.1 **PROTECTION OF METAL***

Metal and Reaction	Remarks
Aluminium Vigorous in wet portland cement concrete mix, but only slight in dry concrete.	Galvanic currents severely corrode aluminium that is embedded in moist reinforced concrete containing chloride. While anodising gives no protection, high-alumina cement (alkali-free) has no effect. Surfaces should be degreased, cleaned, primed and painted with bitumen or, better still, a water-resistant resin containing corrosion-inhibiting pigments. Typical formulations are zinc phosphate or chromate in one of the following resins: epoxy, alkyd, polyvinyl chloroacetate or polyvinyl butyral (mixed with phosphoric acid). An etch primer improves the efficacy of bond between the metal surface and a subsequently applied protective, compatible, resinous coating (see Sections 20.2, 21.1 and 27.1.3).[124]
Copper No reaction.	Protection is required (as for aluminium) against magnesium oxychloride and "Keenes" quick-setting cement. Contact should not be made with ammonia or nitrogenous compounds (which may occur in foamed concrete) and calcium chloride in damp concrete or at wet joints.
Lead Moderately heavy in damp concrete.	Initial attack may be allowed on sections of adequate thickness because, as corrosion continues, a coating is formed which protects the metal from further attack. Protective treatment should be with bitumen for lasting adhesion and soundness.
Zinc As for lead.	The bond of galvanised-steel reinforcement in wet fresh concrete (made with white or low-chromate, under 65 ppm, cement) is impaired by release of hydrogen at breaks in the coating or where zinc-iron alloys extend to the surface.[80, 81, 104, 105] It should be kept away from solutions containing corrosion-promoting electrolytes. Bond is improved by weather oxidation, by chromate passivation treatment by the galvaniser, or by adding chromium trioxide to the mixing water (e.g. 100-200 ppm by weight). Damaged or uncoated contact areas should be zinc-rich painted (BS 4652).[17-19, 117, 176, 177] Strain-ageing embrittlement (see Glossary, Appendix VI) can occur in hot-dip galvanised, cold-worked, high-carbon or highly stressed steels. Precautions against this are given in AS 1214, 1650; BS 729 and ASTM A143. Aluminium or titanium stabilised steel and weldable oxygen-process steel are least susceptible (AS 1204; BS 4360 and ASTM A185).

TABLE 22.1   **PROTECTION OF METAL***
Continued

Metal and Reaction	Remarks
	Reinforcement bending at cherry-red heat (or stress relieving at 650°C and normalising of highly stressed parts at 850°C) should be done before galvanising. Severe cold-working or surface damage of susceptible steel should be avoided.

Steel
  No reaction if
  protection is
  adequate.
	Steel that will be partly or fully exposed to corrosive elements should be cleaned, primed and suitably coated.[152] Treatments include: zinc chromate (AS K211), phosphate or silicate; modified red lead, calcium plumbate (BS 3698) or lead chromate primer; a slurry of cement containing rubber latex and ammonium caseinate with or without potassium dichromate; pigmented resin compounds, such as those for aluminium; micaceous iron oxide in phenolic or alkyd resin; filled and pigmented, chlorinated rubber lacquer; zinc-rich or coal-tar epoxy resin. A film (0·150 mm or more thick) for adequate protection requires two to four applications. Consideration should be given to cadmium coatings,[128] hydrogen embrittlement,[17-19, 117, 133, 141, 176, 177] and bond strength (AS 1627, MA1.5 and BS CP2008).

* Dissimilar metals which may become moist should be kept apart [13] or insulated against galvanic or bimetallic corrosion. Tests for paint systems are featured in AS 1247, 1580; BS 1391, 3745 (metallic) and 3900. Relevant standards for the protection of steel are: AS 1627, K53, K108, K126, K127, K129, K143, K145, K172, K211, Int. 16A, MA1.5; BS 261, 1011, 1070, 2523, 2569, 3189, 3416, 3634, 3900, 4164, CP231, CP2008, DD24, PD420, PD539 and ASTM Parts 27 and 28. Abrasive blast cleaning is specified in AS 1627 Part 4 and BS 4232.

by weight addition to the mixing water, a passivation layer of ferric benzoate or nitrite is formed around steel reinforcement. Where the concrete is likely to be slightly porous, as in autoclaved lightweight concrete, the amount may be increased to 2 per cent by weight of cement. With the former concentration, a concrete mix should be designed with its target average strength increased by 15 per cent. About 75 per cent of the inhibitor has been recovered from ordinary concrete that has been weathered severely for 5 years. Cement slurry with an inhibitor can be effectively used as a coating on steel reinforcement before embedding it in concrete.[61, 106]

Metal conduits in a potentially corrosive environment should have a closely bonded envelope of polyethylene or polyvinyl chloride plastic, otherwise high-density plastic tubing may be used with means for accommodating strain or avoiding it at slab joints. Where high fire resistance is not required, clean steel reinforcement may be protected with two coats of polyamide-cured epoxy resin, the first coat being mixed with a corrosion-inhibiting pigment. For effective bond, the last coat is spattered with grit and allowed to harden or it is surrounded (when tacky at an age of about 0·5-1 hour) with newly placed concrete.[64, 199]

Reinforced concrete products can be protected substantially by drying and impregnating them with bituminous compounds by vacuum-pressure methods. Protective treatments on metal are indicated in Table 22.1. When epoxy resin is used as a structural bonding medium, consideration should be given to its low fire resistance and elastic modulus, the latter having an effect on creep under sustained load (see Sections 21.1.4 and 31.3).

REFERENCES

1. FELD, J. *Construction Failure*. Wiley, New York, 1968; *ASCE Civil Engineering*, 43, 6 (June 1973), pp. 89-92.
2. SZECHY, C. *Foundation Failures*. Concrete Publications, London, 1961.
3. REESE, R. C. "Safety Requirements in Structural Design and ACI 318-71". *ACI Journal*, 70, 10 (October 1973), pp. 669-79.
4. Standards Association of Australia. *SAA Concrete Structures Code, AS1480; Prestressed Concrete Code*, AS1481.
5. British Standards Institution. *Structural Use of: Concrete*, BS CP110; *Reinforced Concrete in Buildings*, BS CP114.
6. British Standards Institution. *Structural Use of: Precast Concrete, including Composite Construction*, BS CP116; *Prestressed Concrete*, BS CP115.
7. ACI COMMITTEE 318. "Building Code Requirements for Reinforced Concrete". See Appendix IV, 2 (a) (i).
8. ACI COMMITTEE 340. *Ultimate Strength Design and Design Handbook;* also C & CA Aust., *Australian Reinforced Concrete Design Handbook 1976*. See Appendix IV, 2 (a) (i).
9. Institution of Civil Engineers. "Foundations". *Civil Engineering* CP4, 1954.
10. *Site Investigations*. Standard Specifications AS1726, BS CP2001 and CP101, also ASTM D1452.
11. KLEINLOGEL, A. *Influences on Concrete*. Frederick Ungar, New York, 1950.
12. KONDO, Y., TAKEDA, A. and HIDESHIMA, S. "Effect of Admixtures on Electrolytic Corrosion of Steel Bars in Reinforced Concrete". *ACI Journal*, 31, 4 (October 1959), pp. 299-312.
13. ADDLESON, L. "Corrosion". *The Architect and Building News*, 231, 24 (14 June 1967), pp. 1043-47; 25 (21 June 1967), pp. 1097-1102; 232, 29 (19 July 1967), pp. 113-19.
14. PEPPER, L. and MATHER, B. "Effectiveness of Mineral Admixtures in Preventing Excessive Expansion of Concrete Due to Alkali–aggregate Reaction". *ASTM Proceedings*, 59 (1959), pp. 1178-1203.
15. WAKEMAN, C. M., DOCKWEILER, E. V., STOVER, H. E. and WHITENECK, L. L. "Use of Concrete in Marine Environments". *ACI Journal*, 29, 10 (1958), pp. 841-56; 2, 30, 6 (1958), pp. 1309-46.
16. LEA, F. M. and DAVEY, N. "The Deterioration of Concrete in Structures". *ICE Journal*, 32, 7 (1949), pp. 248-95.
17. SHREIR, L. L. *Corrosion, Vols I and II*. Newnes, London, 1965.
18. UHLIG, H. H. *Corrosion and Corrosion Control*. Wiley, New York, 1971.
19. LOGAN, H. L. *Stress Corrosion of Metals*. Wiley, New York, 1966.
20. *Fire Tests on Building Materials and Structures*. Standard Specifications AS1530 and BS 476.
21. ELSTNER, R. C. *et al.* "The Fire Endurance of Concrete". *Concrete Construction*, 19, 8 (August 1974), pp. 389-416.
22. American Society for Testing and Materials. *Fire Tests on Building Construction and Materials*. ASTM Standard E119.
23. SALSE, E. and LIN, T. D. "Structural Fire Resistance of Concrete". *ASCE Journal*, 102, ST1 (January 1976), pp. 51-63.
24. GUENTHER, N. "Design Requirements for Reinforced Concrete Structures to Reduce Cracking". *National Conference Publication*, 75/6, pp. 41-43. IE Aust. Symposium on Serviceability of Concrete, Melbourne, 1975.
25. "Design of Clay Brickwork Expansion Gaps". Brick Development Research Institute, *Techniques (2nd Series)*, 4, University of Melbourne; Aust. Experimental Building Station, *Notes on the Science of Building 134 and 135*, 1974.
26. IVEY, D. L. and TORRANS, P. H. "Air Void Systems in Ready-Mixed Concrete". *Journal of Materials*, 5, 2 (June 1970), pp. 492-522.
27. MENZEL, C. A. *et al.* "Symposium on Measurement of Entrained Air in Concrete". *ASTM Proceedings*, 47 (1947), pp. 832-913.

28. GOURLEY, J. T. "Cathodic Protection of Prestressed Concrete Pipe". *Concrete Institute of Australia, News,* vol. 2, 2 (June 1976), pp. 3-6, Sydney. *Rocla Technical Journal,* December, 1976.
29. "Symposium on Use of Pozzolanic Materials in Mortars and Concrete". *ASTM Special Technical Publication,* No. 99, 1949.
30. KENNERLEY, R. A. and CLELLAND, J. "An Investigation of New Zealand Pozzolans". *Bulletin,* No. 133, DSIR (NZ), 1959.
31. FABER, J. H. *et al.* "Fly Ash Utilization". *Information Circular 8348,* United States Department of Interior, Bureau of Mines, Washington, 1967.
32. GIPPS, R. de V. and BRITTON, A. H. "Local Pozzolana Used in Concrete for Koombooloomba Dam". *IE Aust. CE Transactions,* CE2, 2 (September 1960), pp. 65-77.
33. Australian Experimental Building Station. "In-situ Concrete Drives and Paths". *NSB 140,* 1975. See Appendix VII, Notes on the Science of Building.
34. "Sun, Salt and Sewers in the Middle East". *ICE New Civil Engineer,* 148 (19 June 1975), pp. 32-3; 181 (19 February 1976), pp. 20-3; 183 (4 March 1976), p. 43.
35. Highway Research Board, United States. "Preconditioning and Stabilizing Soils by Lime Admixtures". *Bulletin 262,* Washington, 1960.
36. POMEROY, R. "Protection of Concrete Sewers in the Presence of Hydrogen Sulphide". *Water and Sewage,* 107, 10 (October 1960), pp. 400-403.
37. CHAMPION, S. *Failure and Repair of Concrete Structures.* Contractors Record, London, 1961.
38. DIAMANT, R. M. E. *Prevention of Corrosion.* Business Books, London, 1971.
39. POWERS, T. C. *et al.* "Durability of Concrete". *ACI Journal,* 72, 4 (April 1975), pp. 167-71; *ACI Publication,* SP-47, 1975.
40. GREGOR, A. J., GILCHRIST, F. M. C. and SUSKIN, R. "Corrosion of Concrete Sewers". Council for Scientific and Industrial Research, South Africa. *Research Report 163,* Pretoria, 1959.
41. Cement and Concrete Association. *Proceedings of the FIP Symposium on Concrete Sea Structures,* Britain, 1973.
42. Department of Scientific and Industrial Research. "Damp-proof Treatments for Solid Floors". *Building Research Station Digest,* No. 86, HMSO, London.
43. WATERS, E. H. "Failures of Floor-surfacing Materials on Concrete Slabs—Their Causes and Prevention". *Building Study 2,* CSIRO, Division of Building Research, Highett, Victoria, 1960.
44. "Installing Continuous Membrane". *Architecture, Building, Engineering,* 37, 6 (June 1959), p. 29.
45. "Corrosion-resistant Floors in Industrial Buildings". *Building Research Station Digest 120,* 1959. HMSO, London.
46. WESTALL, W. G. "Bonding Thin Concrete to Old Pavements". *Civil Engineering* (United States), 28, 6 (1958), pp. 34-7; *Indian Concrete Journal,* 44, 9 (September 1970), pp. 377-78.
47. New Zealand Portland Cement Association. "Concrete Floor Design". *NZ Concrete Construction,* 6, 8 (August 1962), pp. 135-44; 10 (October 1962), pp. 174-83; 11 (November 1962), pp. 198-202.
48. Department of Scientific and Industrial Research. "Floor Finishes Based on Polyvinyl Chloride (PVC) and Polyvinyl Acetate (PVA)". *Building Research Station Digest No. 65,* HMSO, London.
49. SWEET'S CATALOG SERVICE. *Architectural Catalog File* (Section 13). F. W. Dodge Corporation, New York, 1976.
50. ROBERTS, A. G. "Organic Coatings, Properties, Selection, and Use". United States National Bureau of Standards, *Building Science Series 7,* 1968.
51. MATHER, B. "Sulphate Soundness, Sulphate Attack and Expansive Cement in Concrete". United States Army Corps of Engineers, *Miscellaneous Paper C-69-8,* 1969.
52. DIETZ, A. G. *Plastics for Architects and Builders.* Massachusetts Institute of Technology Press, Cambridge, Mass., 1969.

53. KUENNING, W. H. "Nature of Organic Adhesives and Their Use in a Concrete Laboratory". *PCA R & DL Journal*, 3, 1 (January 1961), pp. 57-69.

54. EVANS, V. *Plastics as Corrosion-Resistant Materials*. Pergamon Press, London, 1966.

55. BIKERMAN, J. J. *Science of Adhesive Joints*. Academic Press, New York, 1961.

56. MATTISON, E. N. "Air Gun Injects Epoxy Resins". *Building Materials*, 4, 4 (April/May 1963), pp. 40-1.

57. CARVAJAL, C. C. "The Protection of Concrete Pipes against Attack by Aggressive Soils and Waters by the Use of Plastic Coverings". *Proceedings, Third International Congress of the Precast Concrete Industry*, Stockholm, 1960.

58. SARAPIN, I. G. "Corrosion of Wire Reinforcement in Heavy Concrete with Added Calcium Chloride". *Translation No. 444, CSIRO*.

59. MIYOSHI, S., FOKODA, N., AWAYA, H. and TAMURA, Y. "Effect of Calcium Chloride on Corrosion of Steel in Reinforced Concrete". *Library Communication No. 1027*, DSIR, Building Research Station.

60. ARBER, M. G. and VIVIAN, H. E. "Inhibition of the Corrosion of Steel Embedded in Mortar". *Australian Journal of Applied Science*, 12, 3 (September 1961), pp. 339-47.

61. TREADAWAY, K. W. J. and RUSSELL, A. D. "Inhibition of the Corrosion of Steel in Concrete". *Current Paper CP 82/68*. Building Research Station, Britain, 1968.

62. DOBINSON, K. W. "Preventing Corrosion of Steel Structures Exposed to a Marine Atmosphere". *IE Aust. Journal*, 32, 9 (September 1960), pp. 183-88.

63. FANCUTT, F. and HUDSON, J. C. "The Choice of Protective Schemes for Structural Steelwork". *ICE Proceedings*, 17 (December 1960), pp. 405-30; 21 (March 1962), pp. 558-86.

64. ROBERTS, J. A. and VIVIAN, H. E. "Studies in Reinforcement—Concrete Bond". *Australian Journal of Applied Science*, 12, 1 (March 1961), pp. 104-39.

65. "Concrete In Residential Construction". *Concrete Construction*, 20, 11 (November 1975), pp. 489-525.

66. CARROLL, W. J. "Dual Protection for a Concrete Sewer System". *Civil Engineering* (United States), 32, 1 (January 1962), pp. 44-47.

67. HURST, H. T. and MASON JR, J. P. H. "Instrumentation Required for Structural Evaluation of Full-scale Buildings". *Agricultural Engineering*, 43, 5 (May 1962), pp. 280-4.

68. ZAR, M. "Concrete Chimneys". *ACI Journal*, 59, 3 (March 1962), N.L. pp. 15-23.

69. HUHTA, R. S. "The House That's Glued Together". *Concrete Products*, 64, 9 (September 1961), pp. 39-42.

70. ELECTRICAL DEVELOPMENT ASSOCIATION. *Construction and Finish of Floors That are to be Electrically Warmed*. EDA, London.

71. MACKENZIE, T. T. "Thin Epoxy Layer Restores Concrete Bridge Deck". *Civil Engineering*, 32, 3 (March 1962), pp. 42-43.

72. HONDROS, G and MOORE, G. J. "Epoxy Resin Protection of Electrical Resistance Strain Gauges for General Use in Concrete". *Civil Engineering and Public Works Review*, 57, 671 (June 1962), pp. 756-58.

73. "Concrete in Sulphate-bearing Soils and Groundwaters". *Building Research Establishment*, Britain, *Digest 174, 1975; Structural Use of Concrete*, BS CP110 Pt 1, 1972, plus Amendment No. 1, 1974, on restricted usage of H-A C (Section 3.1.2 (b)).

74. BALLARD, W. E. "Application and Inspection of Protective Coatings on Structural Steel". *ICE Proceedings*, 22 (June 1962), pp. 227-34; 24 (February 1963), pp. 291-95.

75. WRIGHT, H. J. "Cathodic Protection". *ICE Proceedings*, 21 (April 1962), pp. 811-24; 24 (February 1963), pp. 288-90.

76. ACI COMMITTEE 503. "Use of Epoxy Compounds With Concrete". *ACI Journal*, 70, 9 (September 1973), pp. 614-45; *ACI Manual of Concrete Practice*, Part 3, 1976.

77. DSIR, "Mortars for Jointing". *Building Research Station Digest No. 126,* HMSO, London.
78. ACI COMMITTEE 524. "Guide to Portland Cement Plastering". *ACI Journal,* 60, 7 (July 1963), pp. 817-34; 2, 61, 3 (March 1964), pp. 1923-36.
79. DSIR "Shrinkage of Natural Aggregates in Concrete". *Building Research Station Digest (2nd Series) 35,* HMSO, London.
80. EVERETT, L. H. and TREADAWAY, K. W. J. "Use of Galvanised Steel Reinforcement in Building". *Current Paper CP3/70,* UK Building Research Station, Britain, 1970.
81. NUTT, J. G. "Galvanized Reinforcement in Concrete". *Symposium on Zinc Coatings for the Protection of Steel,* Australian Zinc Development Association, Melbourne, 1968; *ACI Journal,* 69, 1 (January 1972), p. 80.
82. C & CA AUST. "Concrete Floors for Houses". *Technical Manual T16M,* 1975.
83. JOHNSON, R. P. "Glued Joints for Structural Concrete". *Structural Engineer,* 41, 10 (October 1963), pp. 313-21; 42, 4 (April 1964), pp. 135-41.
84. DSIR. *Principles of Modern Building, Vols I and II.* HMSO, London, 1959.
85. "Quality Control of Cement Rendering Mixes". *Supplement to Wimpey News,* 261 (September 1963), pp. 1-4. (George Wimpey, London); *The Builder,* 205, 6278 (September 13, 1963), pp. 529-30.
86. BROCARD, M. J. "Corrosion of Steels in Reinforced Concrete". *Library Communication 1193.* DSIR Building Research Station.
87. ABELES, P. W. and FILIPEK, S. "Corrosion of Steel in Finely-cracked Reinforced and Prestressed Concrete". *Concrete and Constructional Engineering,* 58, 11 (November 1963), pp. 429-34.
88. SEARLE, A. B. *Limestone and Its Products.* Benn, London, 1935.
89. STOCKER, P. T. "Diffusion and Diffuse Cementation in Lime and Cement Stabilised Clayey Soils". *Australian Road Research, 5,* 6 (December 1974), pp. 51-75.
90. RAYMOND, S. and SMITH, P. H. "Use of Stabilised Fly Ash in Road Construction". *Civil Engineering and Public Works Review,* 59, 690 (January 1964), pp. 70-72; 691 (February 1964), pp. 238-40; 692 (March 1964), pp. 361-63.
91. WELCH, G. B., CARMICHAEL, A. J. and HATTERSLEY, D. E. "Epoxy Resin Concrete". *Civil Engineering and Public Works Review,* 57, 671 (June 1962), pp. 759-62; 672 (July 1962), pp. 905-6.
92. FAIRWEATHER, V. "Salt Damaged Bridge Decks: Cathodic Helps". *ASCE Civil Engineering,* 45, 9 (September 1975), pp. 88-91.
93. COPELAND, R. E. and SAXER, E. L. "Tests of Structural Bond of Masonry Mortars to Concrete Block". *ACI Journal,* 61, 11 (November 1964), pp. 1411-52.
94. AZBE, V. J. "A Super Rotary Kiln". *Rock Products,* 67, 7 (July 1964), pp. 56-63; 8 (August 1964), pp. 96-100.
95. PARSONS, M. F. "A New Approach to the Vertical Shaft Kiln". *Pit and Quarry,* 57, 2 (August 1964), pp. 124-29.
96. BROWN, W. G., PENNER, E. *et al.* "Adfreezing and Frost Heaving of Foundations". *Canadian Buildings Digests CBD61* (revised, June 1965) and *CBD128,* 1970, National Research Council, Ottawa; National Research Council Canada, *DBR Papers 13674* and *13848,* 1973; *Canadian Geotechnical Journal,* 11, 3 (March 1974), pp. 323-38.
97. LOCHER, F. W. "Resistance of Concrete to Aggressive Carbonic Acid". *Beton (D),* 25, 7 (July 1975), pp. 241-45; *ICE Abstracts,* 9 (October 1975), 75/1483.
98. ERLIN, B. "Air Content of Hardened Concrete by a High-pressure Method". *PCA R & DL Journal,* 4, 3 (September 1962), pp. 24-29.
99. SWENSON, E. G. and GILLOTT, J. E. "Alkali Reactivity of Dolomitic Limestone Aggregate". *Magazine of Concrete Research,* 19, 59 (June 1967), pp. 95-104.
100. QUIRIN, E. J. "General Review of Prestressed Concrete in Marine Construction". *PCI Journal,* 9, 6 (December 1964), pp. 44-51.

101. MINISTRY OF TECHNOLOGY. "Protection Against Corrosion of Reinforcing Steel in Concrete", *Building Research Station Digest (2nd Series) 59,* HMSO, London.
102. MANOHAR, S. N. "Production Properties and Applications of Concrete". *Indian Concrete Journal,* 41, 7 (July 1967), pp. 262-75.
103. SOROKA, I. and KORIN, U. "Effect of Mechanical Activation on Properties of Mortar". *Materials Research and Standards,* 7, 1 (January 1967), pp. 11-17.
104. BIRD, C. E. "The Influence of Minor Constituents in Portland Cement on the Behaviour of Galvanized Steel in Concrete". *Corrosion Prevention and Control,* 11, 7 (July 1964), pp. 17-21.
105. CORNET, I. and BRESLER, B. "Corrosion of Steel and Galvanized Steel in Concrete". *Western Region Conference,* National Association of Corrosion Engineers, Phoenix, Arizona, 1964.
106. GOUDA, V. K. and MONFORE, G. E. "A Rapid Method for Studying Corrosion Inhibition of Steel in Concrete". *PCA R&DL Journal,* 7, 3 (September 1965), pp. 24-31.
107. MOZER, J. D., BIANCHINI, A. C. and KESLER, C. E. "Corrosion of Reinforcing Bars in Concrete". *ACI Journal,* 62, 8 (August 1965), pp. 909-31; 2, 63, 3 (March 1966), pp. 1723-30.
108. BICZOK, I. *Concrete Corrosion and Concrete Protection.* Hungarian Academy of Sciences, Budapest, 1972.
109. MORGAN, R. G. *et al.* "Concrete Sea Structures". *FIP Symposium,* USSR, 1972; *Civil Engineering and Public Works Review,* 67, 796 (November 1972), pp. 1161-62; *Constructional Review,* 39, 1 (January 1966), pp. 19-27.
110. LEA, F. M. "The Action of Ammonium Salts on Concrete". *Magazine of Concrete Research,* 17, 52 (September 1965), pp. 115-16.
111. *Specification.* The Architectural Press, London, 1976.
112. PRICE, W. H. "Pozzolans—A Review". *ACI Journal,* 72, 5 (May 1975), pp. 219-24; 11 (November 1975), pp. 648-49.
113. RYAN, W. G. J. and ASHBY, J. B. "Development and Use of Wangi Fly Ash in Ready Mixed Concrete". *IE Aust. Journal,* 38, 9 (September 1966), pp. 229-38.
114. MINISTRY OF TECHNOLOGY "Painting Metals in Buildings". *Building Research Station Digest (2nd Series) 70 and 71,* HMSO, London.
115. ACI COMMITTEE 302 "Recommended Practice for Concrete Floor and Slab Construction". *ACI Journal,* 65, 8 (August 1968), pp. 577-610; 66, 2 (February 1969), pp. 147-50; 8 (August 1969), p. 609.
116. BARTON, J. J. *Electric Floor Warming.* George Newnes, London, 1967.
117. PHILLIPS, E. "Survey of Corrosion of Prestressing Steel in Concrete Water-retaining Structures". *Technical Paper 9,* Department of the Environment and Conservation, Australian Water Resources Council, Canberra, 1975.
118. ACI COMMITTEE 515. "Guide for the Protection of Concrete Against Chemical Attack by Means of Coatings and Other Corrosion-Resistant Materials". *ACI Journal,* 63, 12 (December 1966), pp. 1305-92; or *ACI Manual of Concrete Practice,* 3, Products and Processes, 1976.
119. HOSEK, J. "Properties of Cement Mortars Modified by Polymer Emulsion". *ACI Journal,* 63, 12 (December 1966), pp. 1411-24; 2, 64, 6 (June 1967), pp. 1609-12.
120. "Sewerage Main Plastic Lined". *Australian Civil Engineering and Construction,* 8, 1 (January 1967), pp. 22-25.
121. United States Highway Research Board. "Symposium on Effects of Aggressive Fluids on Concrete". *HR Record 113,* 1966.
122. ALLEN, H. G. "Characteristics of Plastics and Their Use in Structures". *ICE Proceedings,* 36 (March 1967), pp. 533-55; 38 (September 1967), pp. 139-42.
123. STANDARDS ASSOCIATION OF AUSTRALIA. *Installation of Epoxy Resin Floor Toppings.* AS CA54.
124. WOODS, H. "Corrosion of Embedded Material other than Reinforcing Steel". *Bulletin 198,* PCA Research and Development Laboratories, 1966; *ACI Monograph 4,* 1968.

125. ACI COMMITTEE 325. "Design of Concrete Overlays for Pavements". *ACI Journal*, 64, 8 (August 1967), pp. 470-74.
126. ASHTON, L. A. "Fire and the Protection of Structures". *Structural Engineer*, 46, 1 (January 1968), pp. 5-12.
127. SUN, P.-F. *et al.* "Properties of Epoxy-Cement Concrete Systems". *ACI Journal*, 72, 11 (November 1975), pp. 608-13.
128. BIRD, C. E. and STRAUSS, F. J. "Metallic Coating for Reinforcing Steel". *Materials Protection*, 6, 7 (July 1967), pp. 48-52.
129. JOHNSON, K. L. "Improved Air-entraining Admixtures for Concrete". *ACI Journal*, 65, 5 (May 1968), pp. 402-11; 11 (November 1968), pp. 991-92.
130. COPPER DEVELOPMENT ASSOCIATION. *Copper for Radiant Heating.* CDA, London.
131. ACI COMMITTEE 503. "Epoxies with Concrete". *Special Publication SP-21,* American Concrete Institute, 1968.
132. TANDON, M. "Use of Epoxy Components for Crack-sealing Under Pressure". *Indian Concrete Journal*, 42, 8 (August 1968), pp. 502-5, 513.
133. SZILARD, R. "Corrosion and Corrosion Protection of Tendons in Prestressed Concrete Bridges". *ACI Journal*, 66, 1 (January 1969), pp. 42-59; 7 (July 1969), pp. 595-99.
134. ROBERTSON, W. J. "Controlling Hydrogen Sulphide Corrosion in Sewers". *Australian Civil Engineering*, 10, 6 (June 1969), pp. 41-43.
135. "Zinc-coated Reinforcement for Concrete". *Building Research Station Digest 109,* 1969. HMSO, London.
136. JOHNSON, R. P. "Properties of an Epoxy Mortar and Its Use for Structural Joints". *Structural Engineer*, 48, 6 (June 1970), pp. 227-33.
137. FOLIE, G. M. "Application of Glass Reinforced Plastics in Structural Engineering". *IE Aust. Journal*, 42, 9 (September 1970), pp. 107-111.
138. BOWDEN, S. R. "Analysis of Sulphate-bearing Soils". *Current Paper 3/68.* Building Research Station, Britain, 1968.
139. McGHEE, K. H. and BROWN, H. E. "Investigation of Durability of Steam Cured Concrete". *Report PB184946,* Virginia Highway Research Council, 1969.
140. JONES, A. P. "Protection of Prestressed Concrete Pipes by Cement Mortar Coat". *Technical Journal,* Rocla Concrete Pipes, 50-52 (February/September 1971), pp. 1-2.
141. MOORE, D. G. *et al.* "Protection of Steel in Prestressed Concrete Bridges". *NCHRP 90,* United States Highway Research Board, 1970.
142. BRITT, G. B. "Rapid Extension of Reinforced and Prestressed Concrete Piles". *Concrete* (London), 5, 1 (January 1971), pp. 9-12.
143. LETMAN, J. A. and HEWLETT, P. C. "Concrete Cracks. A Statement and Remedy". *Concrete* (London), 8, 1 (January 1974), pp. 30-34.
144. ANDREWS, D. A. "Determination of Chloride in Concrete". *Concrete* (London), 5, 11 (November 1971), p. 342.
145. SWENSON, E. G. *Performance of Concrete.* Toronto University Press, Ottawa, 1968.
146. ACI COMMITTEE 350. "Concrete Sanitary Engineering Structures". *ACI Journal*, 68, 8 (August 1971), pp. 560-77.
147. F.-YANNAS, S. A. and SHAH, S. P. "Polymer Latex Modified Mortar". *ACI Journal*, 69, 1 (January 1972), pp. 61-65.
148. JAMES, M. L. and TIMSON, D. J. "Fire-damaged Concrete Structures —Rebuild or Repair?" *Concrete* (London), 6, 6 (June 1972), pp. 22-27.
149. ACI COMMITTEE 302. "Concrete Floor Finishes". *ACI Journal*, 70, 6 (June 1973), pp. 416-29.
150. BRICK DEVELOPMENT RESEARCH INSTITUTE. "Design Charts of Permissible Compressive Forces in Walls and Piers". *BDRI Notes 15-17;* "Control Joints in Brickwork". *BMA (NSW) Note 7, 1973.*
151. SUBSURFACE INVESTIGATION FOR DESIGN AND CONSTRUCTION OF FOUNDATIONS OF BUILDINGS. *ASCE Proceedings,* 98, SM5 (May 1972), pp. 481-90; SM6 (June 1972), pp. 557-78; SM7 (July 1972), pp. 749-64; SM8 (August 1972), pp. 771-85.
152. "Painting: Iron and Steel". *Digest 70,* Building Research Establishment, Britain, 1973.
153. PUGSLEY, A. *Safety of Structures.* Edward Arnold, London, 1966.

154. BRANNIGAN, F. L. *Building Construction for the Fire Service*. National Fire Protection Association, Boston, 1971.
155. BROWNE, F. P. and BOLLING, N. B. "A New Technique for Analysis of Chlorides in Mortar". *Journal of Materials*, 6, 3 (September 1971), pp. 524-31.
156. O'BRIEN, J. "Design and Construction of Industrial Floors". *Technical Report 35*, C & CA Aust, 1973.
157. ARIDIS, E. M. "Jointing to Concrete Structures". *IE Aust. CE Transactions*, CE15, 1 and 2 (November 1973), pp. 63-68.
158. "Concrete Floor Construction". *Concrete Construction*, 18, 11 (November 1973), pp. 515-68; *NZ Concrete Construction*, 19, 1 (February 1975), pp. 8-16.
159. PANAK, J. J. and RAUHUT, J. B. "Behaviour and Design of Industrial Slabs on Grade". *ACI Journal*, 72, 5 (May 1975), pp. 219-24.
160. GUSTAFERRO, A. H. "Fire Resistance of Architectural Precast Concrete". *PCI Journal*, 19, 5 (September/October 1974), pp. 18-37.
161. SOLOMATOV, V. I. *Polymer-cement Concretes and Polymer Concretes*. United States National Bureau of Standards, Springfield, Virginia, 1972.
162. POMEROY, C. D. "Evaluating Modified Concretes". *Concrete* (London), 7, 5 (May 1973), pp. 34-36; 6 (June 1973), pp. 32-34.
163. CHANG, T.-Y. P. "Physical Properties of Premixed Polymer Concrete". *ASCE Proceedings*, 101, ST11 (November 1975), pp. 2293-2302.
164. ACI COMMITTEE 548. "Polymers in Concrete". *ACI Journal*, 70, 11 (November 1973), pp. 764-68; *ACI Publication SP-40*, 1973. *World Construction*, 28, 8 (August 1975), pp. 22-35; *Polymers in Concrete*, Construction Press, Lancaster, Britain, 1976.
165. LEVITT, M. *et al.* "Comparison of Concrete Polymer Composites Produced by High Energy Radiation". *PCI Journal*, 18, 3 (May/June 1973), pp. 35-41; 19, 1 (January/February 1974), pp. 109-13.
166. KUKACKA, L. E. *et al.* "Concrete-Polymer Composites". *ASCE Proceedings*, 97, ST9 (September 1971), pp. 2217-27.
167. FOWLER, D. W. and FRALEY, T. J. "Polymer-impregnated Brick Masonry". *ASCE Proceedings*, 100, ST1 (January 1974), pp. 1-10.
168. MEYER, A. H. "A Polymer as a Surface Treatment for Concrete". *ASCE Journal*, 100, ST6 (June 1974), pp. 1205-10.
169. NORDELL, E. S. K. *Water Treatment*. Reinhold, London, 1961.
170. AHRENS, H. W. and ZAHRADNIK, B. "Oxygen Index Rating of Plastics as a Guide to Their Behaviour". *Special Report BOU 29*, pp. 1-3, South African CSIR, NBRI, Pretoria, 1973.
171. ACI COMMITTEE 350. "Concrete Sanitary Engineering Structures". *ACI Manual of Concrete Practice*, 2, 1976.
172. BERMAN, H. A. "Determination of Chloride in Hardened Portland Cement Paste, Mortar and Concrete". *Journal of Materials*, 7, 3 (September 1972), pp. 330-35.
173. BEN-YAIR, M. "Effect of Chlorides on Concrete in Hot and Arid Regions". *Cement and Concrete Research*, 4, 3 (May 1974), pp. 405-16.
174. ATIMTAY, E. and FERGUSON, P. M. "Early Chloride Corrosion of Reinforced Concrete—A Test Report". *ACI Journal*, 70, 9 (September 1973), pp. 606-11; 71, 3 (March 1974), pp. 143-44.
175. SAUMAN, Z. "Effect of CO_2 on Porous Concrete". *Cement and Concrete Research*, 2, 5 (September 1972), pp. 541-49.
176. ACI COMMITTEE 222. "Corrosion of Metals in Concrete". *ACI Publication SP-49*, 1975; *ACI Journal*, 72, 7 (July 1975), pp. 365-68.
177. AILOR, W. H. *Handbook on Corrosion Testing and Evaluation*. Wiley, New York, 1971.
178. AUSTRALIAN POSTMASTER-GENERAL'S DEPARTMENT AND CSIRO. "Very Early Smoke Detection Apparatus", *VESDA Information Kit*, 1974.
179. SUN, P.-F. *et al.* "Properties of Epoxy-Cement Concrete Systems". *ACI Journal*, 72, 11 (November 1975), pp. 608-13; 73, 5 (May 1976), pp. 301-302.

180. "Dewatering Brown-coal Open Cuts". *IE Aust. Journal,* 45, 12 (December 1973), Proprietary Supplement on Autoclaved AC Pipes and Couplings; also Jas Hardie Corrosion-Evaluation Chart, Camellia, NSW 2142.
181. KJENNRUD, A. "Permeability-Corrosion of Mortar and Concrete Under Attack by Water". *Report 232,* Norges Byggforsknings Institutt, Norway, 1974.
182. WALSH, P. F. "Design of Residential Slabs-on-Ground". *Technical Paper (2nd Series) No. 5* (2nd edn), CSIRO DBR; "Residential Floors: Concrete Slab-on-Ground Construction for Victoria". *Pamphlet,* CSIRO DBR, 1975.
183. DEACON, R. C. "Concrete Ground Floors; Their Design, Construction and Finish". *Publication 48.034,* C & CA Britain, 1974.
184. BARNBROOK, G. "Concrete Ground Floor Construction for the Man On Site: Parts 1 and 2". *Publications 48.035, 48.036,* C & CA Britain, 1974.
185. SIEDSES, H. "Experience in the Durability of Rocla Wire Wound Prestressed Concrete Pipes". Concrete Institute of Australia, *CIA Concrete Conference Proceedings,* Melbourne, 1975.
186. KIMBER, H. R. R. "Durability, Performance and Problems of the 36" Diameter Prestressed Concrete Section of the Morgan-Whyalla No. 2 Pipeline". Concrete Institute of Australia, *CIA Concrete Conference Proceedings,* Melbourne, 1975.
187. PEYTON, J. J. "Victorian Arts Centre". *Construction Review,* 48, 3 (August 1975), pp. 28-33.
188. HOLLAND, J. E., WASHUSEN, J. and CAMERON, D. *Seminar on Residential Raft Slabs.* Swinburne College of Technology, Melbourne, 1975.
189. "Criteria for Selection and Design of Residential Slabs on Ground". *Publication 1571.* United States National Academy of Sciences.
190. WASHUSEN, J. "Behaviour of Experimental Raft Slabs on Expansive Clay Soils in Melbourne". M.E. Thesis, Swinburne College of Technology, Hawthorn, Vic., 1976.
191. REHM, G. and MOLL, H. "Corrosion of Steel in Concrete". Building Research Station, Britain, LC1054, 1961.
192. LENDA, R. "Corrosion of Reinforcement in Reinforced Concrete as Related to Crack Width and Cover"; and "Prediction of Shrinkage and Control of Cracking Due to Shrinkage in Concrete". *Technical Reports U24 and U25,* Sydney University, 1974.
193. DANILECKI, W. "An Investigation into the Effect of Crack Width on the Corrosion of Reinforcement in Reinforced Concrete". Building Research Station, Britain, LC1514, 1969.
194. DAKHIL, F. H. *et al.* "Cracking of Fresh Concrete as Related to Reinforcement". *ACI Journal,* 72, 8 (August 1975), pp. 421-28.
195. LUTZ, L. A. "Analysis of Stresses in Concrete Near a Reinforcing Bar Due to Bond and Transverse Cracking". *ACI Journal,* 67, 10 (October 1970), pp. 778-87.
196. CAMPBELL-ALLEN, D. and HELFFENSTEIN, H. L. "Cracking in Concrete—Attributed to Defects in Construction". *Research Report R279,* University of Sydney, 1976.
197. KARPATI, K. K. and SEREDA, P. J. "Joint Movement in Precast Concrete Panel Cladding". *ASTM Journal of Testing and Evaluation,* 4, 2 (March 1976), pp. 151-56.
198. ACI COMMITTEE 224. "Causes, Mechanism, and Control of Cracking in Concrete". *ACI Publication SP-20,* 1968.
199. MATHEY, R. G. and CLIFTON, J. R. "Bond of Coated Reinforcing Bars in Concrete". *ASCE Journal,* 102, ST1 (January 1976), pp. 215-29.
200. MATTISON, E. N. "Epoxy Resin Repair of Cracked Concrete". *Rebuild,* 1, 4 (August 1976), pp. 3-4. (Newsletter issued by CSIRO, Division of Building Research, Australia.)
201. BYERS, W. G. "Field Procedure for Evaluating Potential Sulphate Damage to Concrete". *ACI Journal,* 73, 8 (August 1976), pp. 443-44.

Part 5

FABRICATION PROCESSES AND FINISHES

CHAPTER **23**

MECHANICAL COMPACTION

23.1 VIBRATED CONCRETE

23.1.1 GENERAL INFORMATION

Compaction by vibration is most effective with stiff mixes that are specially designed for the purpose, such as those with a slump of 0-50 mm and a decreased fine-aggregate/coarse-aggregate ratio (e.g. 0·35-0·40 by weight), these mixes being unsuitable for manual means of compaction. The effect of high-frequency vibration is to

Impart a vibratory motion to the particles of a mix, bring cement particles into a state of resonance and considerably improve workability.

Enable gravitational forces, sometimes assisted by extra pressure, to pull and pack the coarse particles of aggregate together.

Drive out surplus air and water and minimise voids in the matrix and faces of the concrete.

Densify the mass and necessitate the filling of formwork or moulds with some 5-10 per cent of additional concrete.

Permit early attention to form-stripping, demoulding, surface-finishing and curing.

These characteristics lead to improvements in quality and economy because, firstly, the effective compaction of concrete with a reduced content of mixing water is conducive to increased strength, impermeability, wear-resistance, dimensional stability and durability and, secondly, a permissible increase in the proportion of aggregate and rate of placement are basic requirements for an

417

economic concrete in large quantities. If a concrete mix has sufficient workability for it to be compacted readily by hand puddling, such as with a highly mortared high-slump mix, there is no advantage in vibrating it. The vibration of this mix would cause segregation and the formation of a very wet layer of mortar on the surface, with very little improvement (if any) in quality and strength.

Vibration imparts qualities of mobility and cohesiveness to stiff mixes containing an increased proportion of coarse to fine aggregate. Vibration for 2-3 minutes causes such mixes closely to approach full compaction, while further vibration (inadvertently up to several hours) has no deleterious effect on the concrete. If a mix has not been properly designed for vibration, it is preferable to reduce its slump rather than minimise the amount of vibration given to it. For the most efficient compaction, care should be taken to ensure that stiff harsh mixes are not undervibrated (see Section 7.2.3), and lightweight-aggregate concrete (see Section 33.2.2(d)) or more workable mixes are not overvibrated with consequent segregation and streakiness.

For practical purposes, a vibrated mix may be considered to be compacted sufficiently when air bubbles cease to appear and only sufficient mortar appears, without excess, to close the top surface interstices. With lightweight-aggregate concrete and fairly dry fine mixes, it is sometimes desirable to restrain the vibrational effect with top pressure (e.g. 10-300 kPa). In this way, the quality, strength and surface finish of certain products can be improved considerably. As mixes of different consistence are best handled with different types of equipment, the properties of mixes and their means of compaction should be selected to secure an economic concrete with requisite quality and effective compaction (see References 17 and 216 of Part 3).

23.1.2 APPLIANCES

Vibrational efficiency is influenced by many factors including the type, size and performance of vibrators, the direction and duration of vibration, the superimposed weight of concrete and the characteristics of the mix. Vibrators are actuated by means of a rotary eccentric mass, an oscillatory percussion mass or an electromagnetic pulsator; and they are driven by electric power, compressed air or a petrol engine. For safety under damp conditions, the power to electric vibrators should be transformed to meet statutory requirements (e.g. about 65 volts in certain countries).

Vibrators are of three types: internal, external and surface. Their vibrational efficiency may be gauged by the time taken to compact a given mass of suitable concrete under a certain condition of test. In general, this value may be taken as being roughly proportional to the acceleration, or to the product of the amplitude and the square of the frequency of vibration.[1] Different types of vibrator may be used simultaneously, a marked advantage being obtained when the direction of vibration is parallel to the pull of gravity. To ensure effective compaction, it is necessary to have an amplitude in excess of 0·065 mm at frequencies greater than 100 Hz (Glossary, Appendix VI) and an acceleration in excess of 2·5 g for a sufficient duration inside the concrete.[2, 5]

While a frequency in the region of 150 Hz is required to produce a resonance of cement particles, there seems to be an optimum frequency of vibration for most mixes. As this frequency gives the fastest rate of compaction for a given value of acceleration, no benefit may be derived by a large increase in the frequency, unless the amplitude or power input is increased as well. In practice, as there is an optimum water/cement ratio for each duration of vibration with a given frequency and amplitude (and *vice versa*) the characteristics of the equipment available may be a governing factor in determining the workability of a mix that should be used for efficient compaction on the site. Mix workability under simulated vibratory and placement conditions may be ascertained by a tentative device illustrated in Reference 73 of Part 2.

The ideal type of vibrator is one in which frequency and amplitude can be varied as desired, ranging from relatively low frequency and large amplitude for initial compaction to high frequency and low amplitude for final compaction. The movements of particles are thus made proportionate to their diminishing interstices for progressive interlock and compaction. For purposes of investigation, several devices may be used for measuring vibration: a vibrograph, vibration meter, stroboscope, piezoelectric accelerometer and cathode-ray oscilloscope. With the exception of vibratory pendulum-type heads, the moving parts of vibrators that are subject to wear should be kept well greased.

(a) Internal vibrators.
Most internal vibrators for immersion use have spud heads varying from 25 mm to 100 mm, with a power input between 0·75 kW and 5·0 kW, common sizes for reinforced concrete being 40-65 mm. The spacing of reinforcement and the proposed size of head should be coordinated at the design stage of a project. Large

units, up to 150 mm diameter, are used in massive construction. Effective compaction is obtained usually with a frequency of 100-120 Hz at an acceleration of 4 g to 10 g (i.e. 4-10 times that due to gravity), the frequency being evaluated under working conditions. Vibrations with higher frequency are used for calibration under idling conditions and for dispersing a fine matrix around the interlocking coarse particles of stiff lean mixes (Plate 16).

Internal vibrators of spring-headed pendulum and turbo-pneumatic type have vibrational frequencies of 150-200 Hz and 300-330 Hz, respectively, these frequencies being made possible by the elimination of conventional bearings. When selecting equipment to suit a particular class of work, it can be expected that coarse aggregate and mass concrete will be affected mainly by suitable cycles of medium frequency, and fine aggregate and structural concrete by those of a higher order.

(b) External vibrators.

External vibrators are of form and table types, the former being used on formwork and the latter for precast work. They are more powerful than internal vibrators, in order to transmit vibrational energy through formwork or moulds to the concrete. Form vibrators operate in service with a frequency of 50-100 Hz at an acceleration of at least 4 g. Vibrators of turbopneumatic type have frequencies of up to 270 Hz. External vibrators are clamped firmly to formwork or moulds, which must be strong, rigid and airtight. They are spaced about 900 mm apart horizontally and located some 250-400 mm below the surface of placed concrete.

A common vibrational effect of multiple units that lack synchronisation is a fundamental vibration on which is superimposed another; this thereby induces a range of vibrational activity in the various particles of a mix. Vibrators with variable frequency and amplitude may be adjusted for synchronous action between the formwork and concrete, and thus minimise a possible pumping action which would form air and laitance defects in the concrete. A high frequency and small amplitude would be desirable, for instance, near the top of formwork where the weight of concrete is relatively small. Form vibrators are efficacious for thin or heavily reinforced sections, where the formwork is up to 750 mm apart, and for steel moulds that are mounted on rubber pads.

Vibrating tables with a rigidly built, steel deck mounted on flexible supports should be synchronised to ensure a uniform vertical motion, their frequency under full load being 50-100 Hz at an acceleration of 4 g to over 12 g. They are actuated by electromagnets or eccentric-mass vibrators, the former being unidirectional in action and the latter being geared to rotate similarly in opposite directions in pairs. Otherwise, a suitable suspension should be used to eliminate the horizontal component.

Vibrating tables vary in size up to about 1200 mm by 3000 mm and in load capacity up to about 2 tonnes. For operational adjustment, the frequency can be changed by varying the motor speed or pulley ratio and the amplitude can be altered by varying the magnetic field of electromagnets, the eccentric moment, or the characteristics of the vibrator mountings. A vibrating table in a concrete factory is installed with mould-handling facilities near a concrete mixer, two synchronised tables being used for the compaction of concrete in long moulds.

These appliances are an efficient means of compacting stiff, harsh mixes for precast structural products and test specimens, provided that the acceleration is adequate under full working load.[3] Large moulds should, for preference, be of light but rigid metal construction and be clamped onto tables for unison of action. Filling is done during vibration, either continuously or in uniform shallow lifts. Lightweight-aggregate concrete should be loaded lightly to facilitate uniform compaction and a smooth top surface. Table vibration is sufficient when the concrete has developed a smooth surface and a tough plastic behaviour under firm pressure with the open hand. This condition may be reached in 10-20 seconds.

Compaction by impact vibration is obtainable by allowing unrestrained moulds of robust construction to bounce freely on a vibrating table. Jolt or shock vibration, which is effective with stiff mixes, has 100-250 cadences per minute and an amplitude of up to 15 mm.

(c) Surface vibrators.

Surface vibrations of pan, hammer and screed type operate directly on laid concrete, so that their mass assists substantially in obtaining compaction. These units are actuated with a large amplitude and used on shallow sections for best results. Pan vibrators consist of a plate, up to about 0.8 m² in size, with up-turned edges, operating at 50-100 Hz. They are used for embedding coarse aggregate in internally vibrated mass concrete and for compacting layers of concrete up to 150 mm thick. Two or three passes are advisable on concrete more than 100 mm thick, the air liberated by internal vibrators in deep work being allowed to escape first. Electric or pneumatic vibrating hammers, operating at 8-16 Hz are fitted with an end-plate for compacting stiff concrete in small moulds (e.g. test specimens and cast stone) resting on a firm, resilient base. They may be used externally on the stiffening members of formwork and moulds and for applying a certain amount of revibration. Test specimens of 100 mm or 150 mm size are filled in two or three layers, respectively, each of which is compacted directly for 5-10 seconds, the completed specimens being kept away from transmitted vibration.

Vibrating screeds, which span the concrete between side forms, are used for striking-off and compacting the top surface of wide sections up to 150 mm thick. They consist primarily of a timber or steel beam (the latter for preference) on which is mounted a vibrator, a shock-cushioned engine and possibly a pair of rubber-padded end-rollers. Tandem units consist of twin straight screeds, which may encase four, tensioned, steel straps that impart a vigorous, vibratory, slapping action uniformly to the concrete. Designs are standardised in lengths from 1·2 to 4·5 m and from 4·8 to 9 m. A screed may have a bull-nosed front edge for surmounting and compacting concrete that has been laid with a small surcharge and levelled evenly (Plate 16).

The vibratory action, with a frequency of 50-75 Hz at an acceleration of 4 g to 6 g, can be stopped and started when the screeds are stationary. Their intensity of action and rate of travel in one to three passes, on concrete of suitable consistence (see Table 4.16), should be such as to ensure adequate compaction and a satisfactory finish. Two-layer compaction may be done cross-wise on the first and longitudinally on the second layer, temporary cross-battens being used as guides and steel-mesh reinforcement (if required) being laid during a brief interim.

Surface vibration can be supplemented with internal vibration, in slabs more than 200 mm thick and with broom or hessian finishing for a skid-resistant texture. The output of vibrating screeds varies from about 40-75 m³ of compacted concrete per day. The rate is increased up to 400 m³ per hour in wheel-mounted or crawler-mounted pavers fitted at the sides with internal vibrators (see Sections 26.3 and 26.4).[43] Vibrating rollers offer a speedy and economical means of compacting large areas of low-slump, harsh-mix mass concrete.

23.1.3 OPERATION

Internal vibrators are inserted vertically and systematically (at regular intervals not normally exceeding 450 mm) into concrete that has been spread in even layers 300-600 mm thick. In large sections, however, the depth of layers may be increased to about 900 mm. For uniform compaction, the radii of vibrational action (usually 300-600 mm) are made to overlap and the vibrational period is regulated from 5-15 seconds in a quickly responsive mix, to 2-3 minutes in a slowly responsive one, the period being increased where horizontal reinforcement is present. The compactive energy of these vibrators increases with their diameter and depth of operation. Their rapid deep penetration, up-and-down jiggle action and final quick withdrawal drives out air and compacts the activated mass.

The withdrawal rate for closure of head-formed holes is about 75 mm per second and eccentric-mass vibrators should not be allowed to idle unduly when withdrawn from the concrete. In deep mass work, the spud heads of vibrators with long flexible drives may be let down to the bottom of formwork at intervals of up to 900 mm. After each lift of concrete has been placed, they are started and allowed to rise by buoyancy as entrapped air escapes and compaction proceeds. In very massive sections, where spread concrete can be levelled with a bulldozer, compaction may be done with gang-mounted vibrators actuated vertically on a tractor (see Section 30.1).

Reserve vibrators should be provided while others are being serviced, to ensure that sufficient capacity is available for compaction at the proposed rate of mixing and placing (see Section 7.2). For high-quality structural concrete, the rate of placing stiff harsh mixes in 450 mm lifts may not exceed 2·5 m^3 per hour per vibrator with a 1-1·5 kW input. This rate may not exceed 0·8 m^3 per hour in prestressed concrete. For mixes with a moderate to ready vibrational response, an indication of vibrator output is given in Table 23.1, it being evident that many more vibrators would be required on a given site for the former mixes than for the latter type, in order to ensure adequate consolidation of all portions of the concrete.

TABLE 23.1   VIBRATOR OUTPUT

Type of Work	Type of Vibrator	Capacity (m^3 of concrete per hour)
Gravity dam	Heavy-duty, two-man internal and surface	10-30
Heavy open forms, such as large foundations and retaining walls	Medium-duty internal	7·5-15
Medium to light, moderately open to narrow forms	Light-duty internal or external	4·0-7·5

To ensure uniform, nonstreaky results, internal vibrators should not come within 1·2 m of a leading face, neither should they push concrete laterally nor come within 75 mm of formwork. Their heads should be immersed to the full depth of a layer and for a short distance into the preceding one. Any danger of surface pitting against formwork can be reduced by suitable fog-spray dampening just before placing concrete, by restricting the height of layers, by imparting acceleration to concrete in the direction of gravity, and by working the interface with a blade (see Section 8.1).

With form vibration, the concrete should be placed in layers only 50-70 mm deep, so as to ensure uniformity of build-up and freedom from entrapped air. The vibrators, on being moved up, should operate just below the surface of the placed concrete. The top 600 mm of walls and columns should be puddled manually or vibrated internally, in order to close air voids and interfacial gaps which could occur under a reduced weight of concrete. Air-entrained concrete is inadvisable with form vibration, as it could cause an accumulation of large air bubbles adjacent to the formwork (see Sections 7.2.3 and 16.1.1).

The requisite duration of vibration can be judged by a mortar-rich appearance at the top of the concrete and a change in "feel" or sound of the vibrators, the latter becoming a tone of constant pitch. On major works, the optimum spacing of vibrators and the degree of compaction obtained can be determined by the use of irradiation equipment or, more generally, standard test moulds. The former requires a source of radioactive isotopes in a probe, a scintillation counter and a shield (see Section 14.9.1). For making specimens, standard test moulds are vibrated under representative site conditions for subsequent determinations of compressive strength and bulk density (see Sections 14.4.1 and 23.1.2).

An empirical method of simulating internal vibrational effects is to place several detachable moulds in the base of a large container, their distance from a job vibrator varying from 150 to 450 mm. On filling the container with concrete that is to be vibrated, a heavy lid is used to represent an additional mass of concrete. The amount and rate of subsidence of the lid during vibration is recorded by a stylus on a rotating drum. This device can be used for checking the vibrational response of trial and site mixes, and for estimating suitable requirements of consistence and vibration for effective compaction. External vibrational effects can be determined by clamping the moulds onto formwork, midway between two vibrators, or onto a loaded vibrating table. The defective compaction of site concrete may be remedied by revibration within 3 hours of placement, this period being extended for set-retarded concrete (see Section 10.2).

Steel reinforcement should be wired together and supported properly in position to ensure its accurate location during vibration. The bond of new concrete to steel and to hardened concrete is improved by proper vibration and revibration. With improperly designed mixes, an accumulation of excess water underneath horizontal reinforcement will greatly reduce its bond strength (see Sections 14.4.2 and 33.3.2(a); also Appendix I).

Formwork with airtight joints must be designed to resist increased pressure (see Section 8.2) and, where external vibrators are used, an oscillatory or pumping action. Robust steel forms and moulds are more efficient than timber ones, which absorb vibration more readily. Timber battens and bracings should be increased at least 25 per cent above those commonly used, boarding should be tongued and grooved or lined, connections should be screwed and wedges secured in place. Absorptive linings, covered with hessian or muslin, minimise surface pittings under formwork that slopes toward the concrete. The vertical formwork of intensely vibrated concrete can be stripped in 1-2 hours under favourable conditions, thus permitting early surface-finishing with requisite curing.

23.1.4 DUAL VIBRATION AND VIBROPRESSURE

In block-making machines, where vibration is applied to the under-side of moulds, increased density in a short time is obtainable by dual vibration. This system imparts to the moulds a greater momentum in a downwards direction than in an upwards one: the effect of downwards momentum is to compact a mix, while that of the lesser, upwards momentum is to prevent loosening of it.

In a power-operated machine making three blocks at a time, for instance, two eccentric-mass vibrators are placed vertically at each end of the mould. The upper vibrator, which rotates at twice the speed of the lower one, is synchronised so that both masses are at the bottom of their respective cycles once in two revolutions of the upper shaft, which carries a smaller off-centre mass than the lower one. After one revolution of the upper shaft, the large mass is at the top of its stroke and the small one is again at the bottom, so that its centrifugal force is deductible from the greater one.

The vibrations are thus alternately additive and subtractive, and the vibrators at each end of the mould are geared so that both ends rise and fall together. The device promotes uniformity of surface texture and compaction, increases output, and improves compressive strength by about 6 per cent in comparison with single vibration.

Economical mixes with a compacting factor below about 0·6 are best compacted by high-frequency external vibration and pressure, the latter being applied by a heavy spring-loaded plate on large units or hydraulically on small units (see Section 3.2.2 and Fig. 33.3). A piece of wire gauze and a porous pad or lining in moulds promote the escape of excess water, and bond break in large moulds is facilitated by compressed air (see Section 27.1.3). Applied pressures of 10-300 kPa are likely to increase the early compressive strength by some 10-20 per cent. Freyssinet, by a combination of vibropressure and steam curing, obtained 1-day values of 140 MPa.

Still higher values have been obtained experimentally with calcium silicate hydrate by the CSIRO (see Fig. 33.13).

With vibrated lightweight-aggregate concrete, the tendency of the coarse particles to bounce to the surface can be restrained by applying top pressure or a wire-mesh tamper, or by vibrating for 4 seconds only at an. acceleration of 7·5 g and an amplitude of not less than 0·4 mm. Vibrating rollers with smooth or lattice-sheet wheels are very effective on concrete slabs, while spinning and rolling processes are eminently suitable for the manufacture of concrete pipes (see Section 23.3). High-acceleration vibration, assisted by compression, is particularly useful for purposes of immediate demoulding in automated production. Vibropressure improves the efficacy of stress transmission at the ends of prestressed concrete members.

23.1.5 VIBRATABLE MIX DESIGN

In mixes designed for manual compaction (see Section 4.3), a continuously graded coarse aggregate is accommodated in a mortar matrix which forms a high proportion of the mix. A mix of this type could be adjusted to suit moderate vibration by keeping constant in the matrix the relationship of cement, sand and water, and increasing progressively the proportion of coarse aggregate. The behaviour of several vibrated mixes would be observed and mix proportions chosen that would combine economy of materials with suitable workability for compaction under given vibrational conditions.

The combined-aggregate/fine-aggregate ratio may thus be increased some 20 per cent, which would be equivalent to a reduction in cement content per m³ of up to 15 per cent, the ultimate strength of the concrete in each instance being the same. Alternatively, by keeping the aggregate/cement ratio constant and reducing the water/cement ratio, the compressive strength could be increased by about 40 per cent (Fig. 4.3) and the unrestrained drying shrinkage could be reduced by about 30 per cent (see Fig. 30.1). An estimation method of high-strength mix design, using a coarse overall grading of aggregate, has already been described (see Section 4.3.4). With lightweight-aggregate concrete, a 20-30 per cent content of well-graded ordinary sand in the combined aggregate is effective in imparting shape stability to demoulded products and improving their strength (see Section 33.2.2).

With stiff, harsh mixes, a single-size aggregate (e.g. 20 mm nominal size) can become so closely packed and interlocked by intense vibration that the next size of particle for void-filling purposes need not exceed 5 mm nominal size. A similar characteristic applies to fine aggregate, the coarse particles of which are blended with an optimum proportion of particles between the 0·60-mm and 0·15-mm sieve sizes. These fines facilitate the dispersion of cement, promote workability, prevent bleeding and help to produce satisfactory finishes and strength (see Section 4.5).

For these mixes, a design procedure developed by Stewart[4] consists in choosing a water/cement ratio for requisite strength or durability and a suitable aggregate/cement ratio for purposes of workability, economy, minimum shrinkage or creep, and full compaction. The following equation (based on the absolute volumes of mix components) is then used to calculate the requisite bulk specific gravity of the combined aggregate.

$$S = \frac{Y}{X + \dfrac{Y}{G} + 0 \cdot 32}$$

where S = bulk specific gravity of the combined aggregate (see Section 3.3.4(b)), Y = combined aggregate/cement ratio by weight, X = water/cement ratio by weight and G = average specific gravity of the combined aggregate, the value of which is generally in the region of 2·7 (see Sections 3.2.3, 3.3.4(j) and 4.3.2).

To facilitate the use of this equation, Table 23.2 contains typical relationships between aggregate/cement and water/cement ratios for intensely vibrated mixes of medium workability. A suitable value of Y may thus be extrapolated for a given value of X and used when evaluating S.

TABLE 23.2   **MIX RATIOS**

Aggregate/Cement (Y)	Water/Cement (X)
2 : 1	0·33
3 : 1	0·36
4 : 1	0·40
5 : 1	0·44
6 : 1	0·49
7 : 1	0·54
8 : 1	0·60
9 : 1	0·67

As rich mixes must accommodate a larger volume of cement paste than lean mixes, the bulk specific gravities of the combined aggregates S will be found to vary from about 1·4 in the former to 2·2 in the latter. For a particular value of S, the proportions of sand and coarse aggregate can be estimated from their bulk specific gravities for a certain degree of compaction (see Section 3.3.4(b)). For this purpose, a standard container is filled with dry aggregate, while tapping the sides a number of times (e.g. twelve times) with a hammer, so as to obtain a degree of compaction which represents that obtained by stone in the concrete. After repeating the procedure several times, the bulk specific gravity of the material is determined by dividing its weight per unit volume (kg/m³) by 1000 (see Reference 136 of Part 3).

Table 23.3 indicates, for various mixes, the percentage of sand that is required in combined aggregate.[4]

TABLE 23.3   **SAND IN COMBINED AGGREGATE**

Mix Proportions	Grading (mm)	Percentage Sand in Combined Aggregate Containing:	
		Crushed Rock	River Gravel
1 : 4-1 : 9	37·5-4·75	20-25	23-28
1 : 4-1 : 7·5	19·0-4·75	24-32	25-30
1 : 4-1 : 6	9·5-4·75	28-38	27-33

The foregoing procedure for a trial mix can be illustrated with the following data. Target cylinder compressive strength required is 30 MPa. The specific gravities of stone and sand are 2·70 and 2·65, respectively, their bulk specific gravities being 1·45 and 1·55, respectively. From Figure 4.3, the requisite water/cement ratio is 0·5 and, from Table 23.2, the corresponding aggregate/cement ratio is 6·2 by weight. Assuming that 28 per cent of sand will be required in the combined aggregate, then

$$G = 0.72 \times 2.70 + 0.28 \times 2.65 = 2.72.$$

$$S = \frac{6.2}{0.5 + \dfrac{6.2}{2.72} + 0.32} = 2.0.$$

Amount sand required $= \dfrac{2.0 - 1.45}{2.0} \times 100 = 27.5$ per cent.

The mix proportions are therefore 1 part cement : 1·7 parts sand : 4·5 parts stone, with a water/cement ratio of 0·5 by weight.

Lean mixes with an increased maximum size of aggregate can be compacted by intense vibration, provided that the matrix possesses sufficient cohesion and capacity to promote interparticle adhesion and workability with stability. In establishing a suitable mix, a particular vibrator will indicate adjustments that should be made to suit its energy characteristics and intensity. Subsequent site control is essential for uniform results. It includes attention to the storage, grading and batching of components (see Sections 3.3.4, 3.3.5, 3.3.6, 6.2 and 7.1); adjustments in aggregate proportions to offset changes in consistence through variations in sand grading; and the proper spacing and duration of vibrational means to effect full compaction (see Section 23.1.3 and Appendix II).

23.1.6 VIBRATIONAL EFFECTS

Concrete containing an excess of wet mortar becomes, when vibrated, unduly plastic and difficult to consolidate. Mixes and gradings that are suitable for hand punning therefore often show vibrational instability in the form of direct segregation or rotational and lateral flow of the main components.

Segregation causes a mix to separate into horizontal or vertical strata, usually with a heavy local accumulation of cement slurry and water. The end result includes heterogeneity and laitance (Glossary, Appendix VI), water pockets under stone particles and horizontal reinforcement, differential shrinkage and strength properties, and reduced bond to horizontal steel reinforcement (see Sections 14.4.2 and 33.3.2(a)). Segregation can be prevented by reducing the original content of water and fine sand. For surface vibration, the mortar content should be only just sufficient to bind coarse particles together at the surface. A deficiency of suitable mortar for void-filling purposes in vibrated concrete causes unbound coarse particles to collect on the top. A fatty surface should preferably be removed while unhardened, for subsequent bonding purposes (see Section 7.2.4).

Rotational instability occurs (particularly in narrow sections) in mixes that contain an excess of sand with a high proportion of fines. It causes air to be sucked into and entrapped in the flowing mass, with a consequent reduction in potential strength. Several measures may be taken to prevent it. These include a change of grading and proportions, so as to reduce the fractional content of very fine sand and water. The rigidity of the mix may be improved by increasing the maximum size of coarse aggregate or by raising the frequency of vibration. Vibrating tables should be of ample capacity or acceleration and be free from rotational motion, if of eccentric-mass type.

The effect of vibration on air-entrained concrete is to reduce its original air content, usually by not more than 15-30 per cent under average site conditions. This limit can be increased to over 50 per cent with prolonged high-frequency vibration in narrow formwork or moulds, but it is usually very small in massive sections. Adequate vibration should not be sacrificed to secure a specified air content, provided that it is sufficient to ensure a satisfactory resistance to frost action.

The air content is reduced more by internal than by surface vibration, the period required to remove a given amount being longer for low-slump than for high-slump concrete. The rate of loss is highest during the first few seconds, particularly with high-frequency vibration, and it decreases as the duration of vibration increases. Form vibration, which causes small particles to migrate outwards, should not be used with air-entrained concrete because of a tendency to increase aeration at the interface.

Increased air-entrainment by truck agitation and the possible unrestricted use of multiple admixtures should be considered in ready-mixed concrete. The amount of air entrained initially should be controlled to give that required finally, revibration (if used) accounting for a nominal reduction of 1 per cent. Measurements of entrained air should be made on samples of concrete that have received typical agitation during, firstly, conveyance and, secondly, vibratory compaction. Localised honeycombing is likely to be caused by leaky forms, undervibration and defective vibrator work in densifying the concrete (see Sections 7.1.5, 16.1.1, 16.1.2, 16.3, 23.1.3 and 23.2).

The effect of vibrating correctly designed concrete, under controlled conditions, is to improve substantially its degree and uniformity of compaction at small extra cost. The result is a marked improvement in density, quality and strength (see Section 7.2.3

Inserting cage
reinforcement
into steel mould

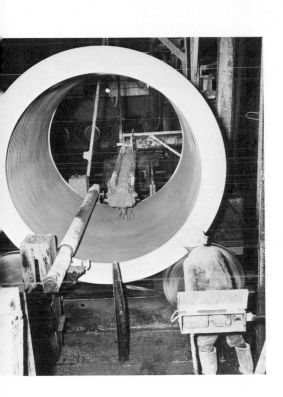

Spinning
and compaction
of placed concrete

Demoulding
after low-pressure
steam curing

Casting core-pipe

Spiral wrapping
with high-tensile
steel wire

Coating
with ejected
mortar

and Chapter 13). The bulk density of ordinary unreinforced concrete (e.g. 2400 kg/m^3) is brought into the region of 2500 kg/m^3 or more, and the elastic modulus is raised to approximately twice that of manually compacted concrete in order to reduce flexural deflection.[5]

23.2 REVIBRATION

Until the colloidal gel structure of hydrating cement reaches a state of considerable rigidity, a mass of concrete when revibrated will behave as a thixotropic, plastic material. The response to revibration falls into three categories.

In the first, the slowly stiffening mass is readily plasticised and freed from residual air.

In the second, which starts at about the time of initial set of cement, revibration causes the mass to disrupt slightly and then reconsolidate with possible expulsion of free water. The energy and time required for this operation are much more than are required for the initial vibratory compaction.

In the third, the concrete is so stiff that reconsolidation of the disrupted mass could not be effected with ordinary vibratory equipment.

Revibration may be done internally, externally or superficially. Internal revibration is effective as long as it is physically possible to insert an internal vibrator into the mass and cause it to become plastic and reconsolidate. Best results are obtained near the time of initial set, or about 2-2·5 hours after mixing under ordinary climatic conditions, additional concrete being placed initially to allow for subsequent compaction. Initial vibration of lightweight concrete (see Sections 23.1.1 and 33.2.2(d)) and high-slump mixes (if used) should be limited in order to minimise segregation.

Revibration can increase the compressive strength of concrete some 15-20 per cent, which is similar to the increase obtainable by delayed placing (see Section 7.2.2). The bond strength of plain reinforcing rods may be increased by revibration from 30 to 50 per cent. In comparison with unvibrated reinforced concrete, the bond strength at first slip is increased about 100 per cent. Reworking the surface to help overcome incipient plastic cracking is mentioned in Sections 7.2.2, 20.1.5(d) and 31.2.4.

The period over which revibration can be applied may be extended by the proper use of a set retarder to about 10 hours (see Section 10.2). A series of layers in a large deep section may thus be revibrated in a continuous operation, which may be facilitated by a network of buried internal vibrators. The beneficial effects of revibration are due to the removal of residual voids and water pockets, particularly from the top section of a placement, and the possible reduction of capillaries and surface pittings against formwork almost to nil (see Reference 190 of Part 3).[5]

Revibration is particularly useful where impermeability is required and where it is practicable to ensure that complete reconsolidation will take place. It is advantageous in the manufacture of precast and prestressed products, in the casting of thin reinforced sections, and in the further compaction of large volumes of concrete, when it can be made effective within a reasonable period of time. Beneficial results have been obtained in the construction of large water tanks, box-shaped or shaft-shaped bridge piers, and the hull shells and bulkheads of reinforced concrete ships which need a rich mix in addition to revibration.[6, 7]

23.3 CONCRETE PIPE MANUFACTURE

23.3.1 CENTRIFUGAL PROCESS

The centrifugal or vibrospin process, which was originated by W. R. Hume in 1910, has been brought to a high state of development in Australia and is used by concrete pipe manufacturers in many parts of the world. The detailed procedure is adjusted to minimise segregation, to keep cage reinforcement in the correct position and to suit the size of the product, which may range from 100 mm to 3000 mm diameter and 1·2 m to 6 m length.

The concrete in a large plant is mixed in an elevated pan mixer and fed evenly by an advancing endless belt to a rotating horizontal pipe mould. Moulds of different sizes are supported on discs in several machines and are handled by a travelling gantry crane, both to the machines and afterwards to steam-curing chambers. A typical mix contains eleven 40-kg bags of cement per m³, the water/cement ratio being from 0·27 to 0·32 by weight and the "Vebe" workability value (see Section 6.2) from 8° to 25°, depending on the size and class of pipe being made. The sand used has a fineness modulus of about 2·5, with at least 10 per cent passing a 0·30-mm sieve and not more than 5 per cent passing a 0·15-mm sieve. According to the wall thickness, the stone aggregate varies from 10 mm to 20 mm nominal size.

Each mould is rotated with a peripheral velocity of 4-5 m/s and, during filling, it is pulse vibrated through disc supports at frequencies of from 8 to 130 Hz. During rotation the filled mould is screeded and rolled by a sleeved internal shaft. The rate of spinning is then increased gradually, so as to compact the concrete finally under a centrifugal force equivalent to fifty times that of gravity. For pipes of small to large diameter, the peripheral velocity at this stage varies from 10 to 25 m/s for periods of 3-15 minutes or more (Plate 17).

The water exuded during spinning is removed with long-handled trowels, the final water/cement ratio being from 0·22 to 0·27 by weight. The pipes are partly cured in low-pressure steam (see Section 9.4) and demoulded on emerging at the far end of the steam chambers. The moulds are reassembled and returned to the production cycle, and pipes for special purposes are transferred to a fog-spray building or spraying area for further curing for 3-7 days. Pipe-stacking ramps, vacuum lifters, and gantry or mobile cranes facilitate handling operations. Special characteristics of the process and its products are referred to at the end of the next section.

23.3.2 ROLLER-SUSPENSION PROCESS

The roller-suspension process of vibropressure compaction, which was conceived and developed in Australia in 1943 (Rocla Pipes Ltd, Australian Patent 118079), has been adopted by over a hundred concrete pipe factories in more than twenty countries. In this process, a steel pipe mould is hung horizontally on a rotating steel roller with two end-bearings. The mould rotates because of friction between its end-rings and the roller, and concrete is fed evenly to the mould by an advancing endless belt.

For thorough compaction and maximum strength, the concrete should be stiff and cohesive, and only a light coating of damp cement should show on a steel rod when it is rubbed with a handful of fresh concrete. The water/cement ratio, according to pipe diameter and mix materials, is between 0·27 and 0·32 by weight. During sequential overfilling of the mould, and its forwards and reverse spinnings, the roller bears against the concrete with high pressure that is accompanied by vibration of the roller and centrifugal force. The pressure is determined by the weight of the mould and its contents, and the narrow width of contact between roller and concrete.

Cumulative, semiautomatic batching (see Section 3.3.6(a)(ii)) expedites the production of 3000 mm diameter by 5-7 m long pipes structured with shear-gridded, double-ring reinforcement. Precuring areas are of high relative humidity in a fully enclosed, draught-free factory. All pipes are cured at 50-65°C for at least 8 hours before demoulding and then cured 16 hours (see Sections 9.2 and 9.4.3). Moulded cylinders, compacted with simulated vibropressure, are compared with cored specimens. Compressive strengths are commonly 50 MPa and 70 MPa at 24-hour and 28-day ages, respectively (see Sections 3.2.2 and 23.1.2(b), (c); also Plate 18).

Each of the foregoing processes can be used for the lining of steel pipes and cylinders with mortar or concrete (AS 1281 and Reference 1 of Part 3). Concrete pipes produced by each process have high density, abrasion resistance and strength, low absorption and smooth finish. Special cement (e.g. sulphate-resisting portland cement; see Section 3.1.2(e)(viii)), surface treatments (see Section 19.1.3(e)), and plasticised, filled-grade, black polyvinyl-chloride anchored linings (see Section 21.3) are used for sulphate-bearing ground, corrosive effluents, and hydrogen sulphide sewer-gas resistance (see Section 19.1.1).

Standard specifications on concrete pipes and related conduits are listed in Section 19.2 and Appendix I. Rubber jointing-gaskets, which may be of skid-ring or rolling-ring type, are illustrated in Reference 45 of Part 3. Polymer concrete and prestressed concrete pipes are referred to in Sections 21.4 and 34.2.2, respectively. Cathodic protection of the latter, which is implemented with sacrificial anodes buried in aggressive soils, militates against possible corrosion of the stressed wire spiral that surrounds each pipe core (see Sections 15.2.2(a) and 22.3.1; Reference 28, Section 4).

23.4 VACUUM AND VIBROVACUUM DEWATERING OF CONCRETE

23.4.1 GENERAL INFORMATION

Since the formulation of the water/cement ratio law, which states that *the strength of concrete is in inverse ratio to the amount of water used in the mix, provided the concrete is fully compacted,* it has been recognised that all the desirable properties of properly compacted concrete are improved by using a minimum water/cement ratio. Mixes with low water/cement ratios are usually more difficult to place than wet mixes with water contents well above that necessary initially for hydration of the cement. The use of low water/cement ratios is made practicable by means of

Vibratory compaction, necessitating the use of strong formwork. Cement dispersion and, with certain qualifications, air-entrainment.

Increased cement content in proportion to the amount of water necessary to ensure workability of the mix for manual compaction. This results in increased cost and shrinkage of the concrete.

Removal of excess water by porous form linings sheathed with muslin, or by pressure treatment in varying intensity. This includes vacuum dewatering which, instead of reducing the water content of concrete at the time of mixing, removes unwanted mixing water from the concrete after it is placed in the formwork or moulds.

23.4.2 VACUUM DEWATERING

Vacuum dewatering is thus an extension of the water/cement ratio law, and is a means of effecting compaction and early hardening of freshly placed concrete by the utilisation of atmospheric pressure at 70-80 kPa. The procedure, briefly, is that concrete is mixed with sufficient water to produce a plastic mix, so that when placed it will fill the forms properly, after which the excess water not needed for the hydration of the cement is squeezed out by atmospheric pressure and removed by suction. The concrete is so compacted in this way that it can carry its own weight by direct bearing or strut action before setting.

The following benefits in concrete construction can be obtained by the process.

1. The 3-day compressive and indirect tensile strengths are notably increased. The 28-day strengths approximate those of well-compacted, nonprocessed concrete, which has the same initial water/cement ratio as that remaining in the vacuum-processed concrete before the time of initial set of the cement.

2. Concrete floors can be finished immediately after placement, particularly during cold or unsettled weather, and precast products can be handled at an age of 10 minutes.

3. Differential and lateral drying shrinkage movements of slabs are reduced proportionately to the elimination of "free water", prior to the setting and moist-curing stages, thereby minimising possible surface curling and increasing the requisite spacing of control joints.

4. Resistance to a cyclic test of freezing and thawing, as a measure of durability, is increased several times.

5. The withdrawal of surplus air and water from concrete, adjacent to the vacuum forms or mats, evens out batch variations in water content and results in a surface that is free from pit holes and laitance.

6. The resistance to wear of vacuum-processed concrete in pavements, and the spillways and aprons of dams, is 2·5 times that of ordinary concrete.

7. The absorption of moisture is reduced by up to 30 per cent, but the impermeability of the concrete may not be improved unless an air-entraining agent is used and the process supplemented with vibration.

8. Vacuum forms are reusable a great many times. They can be removed within 1 hour of placing concrete, or in about half the time normally required for concrete construction, depending on the strength of the concrete and the load to be carried. The filter pads, with proper care, can be used on about 10 000-15 000 m² of surface before replacement.

9. The quality and cost of monolithic *in situ* structures, built with vacuum forms, are highly competitive by comparison with those of structures built with precast products.

10. Successive placements are well bonded together due to the high rate of processing, which equals that of continuous placing. The bond strength between concrete and steel reinforcement is increased by up to two-thirds, compared with that of manually compacted concrete.

The vacuum process is generally applied directly to the surface of placed concrete, or to an area of slab immediately after it has been struck-off or compacted with a double-beam vibratory screed, while internal vacuum dewatering may be employed in special cases. In the surface process, that part of the form which is in contact with the concrete consists of vacuum mats. For floor or freeway work, the mats used are of pliable or rigid construction. Pliable mats comprise a series of filter pads, which are enclosed with an airtight top cover (up to 40 m² or double 30 m² in size) that can be rolled-up on two pipes for ease in handling by two workers.

Rigid mats, 1-2 m wide by 1-6 m long, are assembled to form a single or connected band across the end of placed concrete. They are basically an airtight backing of plywood, metal or fibreglass-reinforced plastic that is faced underneath with a filter screen. Behind the screen are a series of channels or passage-ways, formed typically of light, expanded or grooved metal, through which the extracted water flows to the suction outlet.

The filter screen consists of wire gauze (e.g. 0·30 mm mesh), felt (10 mm thick), fibreglass and/or nylon loosely woven cloth. Butter-muslin or canvas, if used, is treated with a copper solution so as to harden the fibres. The screen is stretched and cemented to the edges of each mat, which are then sealed with a band of sponge rubber for airtightness.

On the back of each mat is an outlet nipple for attaching a hose from the manifold of the vacuum system, while angle sections are sometimes provided (about 150 mm from the edges) for stiffness and as a means for lifting mats with cross-ropes. Long rectangular mats, such as are used vertically for walls, are provided with intermediate seals or cut-offs at 600-900 mm centres, enabling a reduction of hydrostatic pressure to be made on the forms by deaeration of the concrete as it is placed in successive lifts.

The forms are closely assembled in a way that will allow movement as the concrete is compressed, otherwise fissures will develop in the processed concrete. As the hydrostatic pressure on forms can be confined to that of a single lift, the internal formwork of concrete cavity-wall construction need consist only of bituminous paper supported by timber studs.

The depth to which the vacuum is effective in removing water depends on the length of time for which it is applied. For short periods, this depth appears to be about 300-400 mm from the face of the mat, horizontally or vertically. The effective radius of vacuum from the surface of internal-vacuum perforated tubes is from 150 to 200 mm.

Columns with a large cross-section may be processed simultaneously from the outside with a vacuum form and from the inside with a perforated, rubber-sheathed, pressure-and-vacuum tube. The water collecting in an internal vacuum tube is run off by gravity or sucked out through a central tube extending almost to the closed end of the vacuum tube.

In general, treatment under a vacuum of 70 kPa for 20 minutes will result in the withdrawal of up to one-third of the water in a mix originally containing 25 litres of water per 40-kg bag of cement.

Transparent inserts in the vacuum lines are used to provide observation points and a vapour barrier, when placed beneath a ground slab, prevents the possible withdrawal of subgrade air and water through the concrete.

Vacuum outfits are usually assembled to deal with up to two dozen or more mats. The main suction line is of 50 mm diameter reinforced vacuum hose connected to a 6 m (or longer) manifold of lightweight steel tubing provided with nipples for connection to, say, five to twenty mats with separate lengths of flexible 25 mm diameter vacuum hose. At the mat end of each hose line is a soft, rubber, suction cup, somewhat similar to that used by plumbers for clearing clogged drains. A stop-cock located near the cup is normally closed to maintain the vacuum in the line. The suction cups on the ends of the hose lines are pressed over the outlet nipples of the mats and the stop-cocks opened; the vacuum instantly seals the cups to the mats, which are thereby connected to the pump.

After 5-20 minutes (depending on the mix, thickness, consistence and vacuum), the stop-cock on the first mat is shut, releasing the cup so that the mat can be picked up and moved ahead for immediate reuse. No difficulty is experienced in keeping pace with the concreting of paving slabs, and the processing operators work from the mats at all times without having to walk on the concrete. About 170 m² of 150 mm slab can be processed by two mat workers per hour.

A truck-mounted pump driven by a power take-off from the truck transmission and capable of serving a line of sixty-five vacuum mats has connections for three hose lines to manifolds. The pump may be a "Nash Hytor" liquid-ring unit, or a "Dynavac" or "Fuller" oil-sealed rotary unit of 38-42 litre/s capacity at 70 kPa vacuum. Water drawn off from the concrete through a main hose line is trapped into 200-litre drums alongside the truck-mounted pump. The drums are so connected to the vacuum system that the flow of air and water can be switched without interruption to an empty collection tank, whenever a filled drum has to be disconnected. The overriding principle in setting out the vacuum line is that it should be as short and free from resistance losses as possible.

When placing concrete to be vacuum-processed, provision must be made for a reduction in depth or thickness by the removal of water and air, and the closing of voids by compaction of the concrete under a pressure of from 70 to 80 kPa over the area being treated. For example, a 175 mm slab of fresh concrete with a slump of 200 mm will contract in depth about 20 mm to no-slump consistence, when it is in an admirable state for further compaction by superficial mechanical vibration, prior to finishing and effective moist curing.

The amount of cement removed with the water is negligible, being about one-thousandth part only. The filter cloth and mats are kept clean and in good working condition by hosing after use. The fabric must be well wetted before the first treatment of concrete and, if there are signs of drying between uses, it should be wetted again.

Concrete is readily processed hard enough to permit the stripping of vertical forms within 1 hour after the last lift is placed, thereby enabling forms to be reused the same day. The strength of fresh concrete after vacuum treatment is about 100 kPa. Concrete walls therefore can have the forms struck for a height of 4·5-9 m, according to whether one or two faces have been treated. The supporting forms of floor, deck and roof slabs can be removed and reused after a period of 3-7 days.

Where a tessellated surface finish is required, heavy 65 mm by 65 mm wire mesh fastened to the underside of vacuum mats, or a coarse expanded metal with 30 mm by 65 mm openings at the back of the filter fabric instead of next to the concrete, will form indentations in a pavement surface for tyre grip or provide a good bond for subsequent surfacing.

The finishing of floor surfaces, which are hard enough after processing to be walked on without causing an imprint, can be commenced immediately with a power-driven rotary float. It is followed by power trowelling where a very smooth surface finish is specified. If a surface topping is applied, excess water contained within it is withdrawn by the relatively dry concrete to which it is applied, thus minimising shrinkage crazing. Thorough moist curing should be commenced immediately on completion of the finishing operation or within 1 hour after vacuum treatment.

A combination of the vacuum process with external vibration has greater potentialities for massive structures than for thin ones. If simultaneous vibration-vacuum treatment is applied to thin structures, the duration of vibration should be much shorter than that of suction. Vibration is effective in improving the contact between stiff concrete and the seal of the suction-chamber, and in driving excess water from the interior of plastic concrete to the surface for removal by suction. It reduces water absorption by breaking capillary paths and promoting compaction.

For the treatment of concrete pavements, a surface vacuum-vibrator has been developed. A shallow vacuum chamber (6 mm deep) of steel sheeting is permanently fixed to the bottom face of a surface vibrator, that is of spring-suspended platform type and is handled by two operatives. The bottom steel sheeting of the vacuum chamber is perforated with holes 3 mm in diameter and is covered outside by a filter cloth. A discharge socket is welded to the top of the chamber and linked to the vacuum pump by a reinforced hose. Stiffening webs are placed between the top and bottom sheets, extending radially from the suction outlet. A container is placed directly on the vibrating platform, or near to it, to collect the expelled waste water. Vibration for 15-30 seconds is followed by vacuum for 30-90 seconds.[8]

23.4.3 VACUUM LIFTER

In the mass production of concrete slabs, assembled reinforcement is placed in a mould fixed to a vibrating table, the concrete is filled from an overhead travelling hopper, and the vibrator is switched on for 10-15 seconds to produce a smooth surface. One or more suction pads, the full size of the slab, are lowered by a hand-operated crane and a vacuum of 80 kPa is applied and maintained for 40-60 seconds for a slab 50-75 mm thick.

The side portions of the mould are then lowered by a lever and, while the vacuum is maintained, the overhead crane removes the chamber from the mould with the newly placed concrete slab adhering by suction. Ribbed slabs or channels are manufactured in a similar way. The strength of concrete in slabs 100 mm thick, vacuum dewatered for 90-120 seconds, corresponds to that at an age of 8-10 hours.

23.4.4 REMARKS

The extraction of water is facilitated by limiting surplus amounts of aggregate fines in a mix, and by limiting the void content of air-entrained concrete to about 5 per cent. A 20-30 per cent increase in treatment time is required with air-entrainment, because of the air-void expansion in capillaries and its retarding effect on extraction of water. Freshly processed concrete contains less free water than does ordinary concrete for the gradual hydration of the cement. Therefore, it requires abundant watering or spraying during the first week for best results.

Vacuum dewatering presents a method of obtaining high-grade concrete with a low water/cement ratio, in a way that is easy and economical to apply. It can be combined with vibration and systems of prestressing, and is more effective in increasing load-carrying capacity when applied to the compression, rather than to the tension side of reinforced members in flexure. Modified fabrication and compaction techniques, using fibre-reinforced cement or mortar, include flexible spray suction sheet-moulding and rotating suction mandrel-moulding methods in the manufacture of flat, folded and tubular composites (see Section 3.4.2).

23.4.5 VACUUM MIXING

Concrete mixed in a vacuum gives up most of the air contained in the mix. In vacuum mixing, the ingredients are mixed dry, water is added and the vacuum produced in a special mixer. The mix is increased in bulk density by 6 per cent and the resulting concrete has greater strength (10-15 per cent), impermeability and abrasion resistance than ordinary concrete.

23.5 ELECTRO-OSMOSIS

A general definition is given in the Glossary, Appendix VI. The anode, which is supplied with direct current, consists of a round iron bar, while the cathode is a perforated steel tube. If the voltage is between 40 and 70 volts, water passes to the cathode and can be drawn off through the perforations. Before the concrete hardens, it should be vibrated to minimise pores from which water has been withdrawn. The temperature rise caused by the passage of the electric current should not be allowed to exceed 50°C, so as to prevent evaporation of water and consequent increase in porosity.

The potential difference required is about 1 volt/cm. Should it exceed 1·7 volts/cm, electrocataphoresis would occur, in which not only water but cement milk passes to the cathode, and causes a lowering of the cement factor per unit volume of the concrete. A further increase in the potential difference would cause electrolysis, in which water is decomposed into oxygen and hydrogen. A further increase would completely dry the concrete by electrothermal means. Sufficient water must be left in the concrete, however, for purposes of hydration. Each kind of concrete has its own strength limit, above which separation of water ceases to be economical.

Electro-osmosis by itself is not sufficient to increase the strength of concrete, but must be supplemented by closing the resultant pores by tamping, compression or vibration. Vibration is not only economical but, because of the partly colloidal nature of concrete, it enables a stiff, partly hardened mass to be converted into a plastic state and consequently become consolidated. Small currents are required, but the formwork should be covered with bituminous or tarred paper, or felt panels (5 mm thick) impregnated with bakelite, in order to insulate workers from them.

Electro-osmosis, coupled with vibration, is suitable for accelerating the drainage and increasing the strength of wet concrete mixes and soils. It is used, in conjunction with bentonite, to form damp-proof foundation courses and vertical tanking membranes in fine-grained soils. A bentonite suspension is introduced through perforated tubular anodes, driven into the soil or wall around the area, and it is caused to permeate towards a metal grille cathode during application of a low-voltage direct current.[122]

CHAPTER 24

PUMP AND PNEUMATIC PLACEMENT

24.1 PUMPED CONCRETE

24.1.1 GENERAL INFORMATION

Pumping of concrete is advantageous where other means of distribution are difficult, such as in underground work or structures where access is restricted (see References 1, 5 and 6 of Part 3). [5, 9, 80, 106, 139, 159, 166, 169, 188] It is economical and successful under the following conditions.

A mix of medium workability with aggregate not coarser than 53 mm.

Medium-fineness cement, with controlled aggregate-grading, mixing, placing, curing, and execution of the work.

Increased fine-aggregate/coarse-aggregate ratio, cement and water contents, shrinkage-distributing reinforcement, crack and creep potential.

Advantages include the following.

No other traffic is needed to transport concrete within the working radius of the pump.

Marked savings in manpower, mobile conveyances and time.

A constant and high rate of delivery (e.g. 20-150 m³/hour) according to the make and size of the pump.

Flexibility of operation through an easy switching of pipelines.

Simplified handling without segregation of the concrete.

24.1.2 PLANT AND PERFORMANCE

(a) Pumps.

Concrete pumps of piston, pneumatic and squeeze-pressure type are available in a range of sizes for 100-200 mm diameter pipelines.

(i) Piston pumps. Units of 20-150 m³/hour capacity are of one to four piston design, two pistons with reciprocal forwards and retraction strokes for steady discharge being commonly used. Quick-acting, rotating-plug valves provide a residual closure opening that can be adjusted to suit the maximum size of aggregate. This manipulation saves wear through the shear of coarse particles. Reverse-cycle pumping (i.e. from the line back into the hopper) is helpful when handling marginal mixes and stoppages or removing blockages.

A 0·2-2 m³ receiving hopper is equipped with remixing blades that help maintain the uniformity and consistence of mix. Oversized stone particles are excluded by finish screening of the coarse aggregate (see Section 3.3.5) or by passing the concrete through a suitable screen. Then both gravitational force and piston suction draw the concrete into the pump through an open inlet valve (Fig. 24.1). The pumping efficiency is about 70-80 per cent.

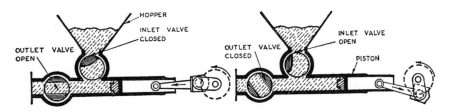

Fig. 24.1 Cross-section of pump

(ii) Pneumatic pumps. The concrete is delivered into a pressure hopper (that is then tightly sealed) and compressed air is admitted at the top from a tank reservoir. This pushes the concrete through a delivery pipe to an air-bleed, reblend, pressure-discharge vessel (0·5-1·0 m³ capacity) located at the end of the line. On large projects, several machines discharge successive slugs at near steady flow.

High air consumption necessitates that a 700 kPa compressor should have a free-air per minute rating some ten times the gross pump capacity. With care.to minimise segregation, displacement of reinforcement and damage of formwork, the equipment can be used for shotcreting and placing concrete behind tunnel forms (see Section 24.2.6).[10, 12, 106, 159, 188]

(iii) Squeeze-pressure pumps. Remixing paddles in an inlet hopper push concrete into a flexible hose, which skirts the inside periphery of an adjacent metal-drum or pumping chamber. Hydraulically powered rollers, monitored on a planetary drive, rotate on this hose and squeeze the concrete forwards into the delivery line. A high vacuum maintained in the pumping chamber returns the tube to its normal shape and induces a steady flow of concrete. Alternating reverse and forwards action during short delays can mobilise concrete in the pipeline and, by incorporating ancillary gear (such as a reducing cone, pneumatic nozzle and set-accelerator meter), the discharge can be modified for shotcreting purposes (see Section 24.2.6).

(b) Pipelines.
Pipelines are formed of multiple sections (3 m long) with special couplings for quick assembly and dismantling. Supplementary fittings include short sections, various bends, swivel joints, reflux and switch valves, clamps, ejector plugs (e.g. pump-maker's sponge rubber ball), blow-out fittings, pendant connectors for upfilling formwork, air vents for downhill pumping, supporting and handling accessories, and a mobile crane or mechanical distributing boom (e.g. 30 m long). Flexible-link pipes of reinforced rubber are used for connecting main lines successively to branches. Aluminium lines, if used instead of steel, can cause reactive expansion with high-alkali cement and a 20-30 per cent strength loss.

(c) Performance.
A piston pump with 120 mm delivery line, powered by a 60 kW diesel engine at 2500 rev/min, will deliver up to 35 m³/hour over a distance of 200 m horizontally or 60 m vertically, these distances being reduced by pipeline bends (see Table 24.1). A 90 kW at 2800 rev/min power unit will deliver up to 75 m³/hour over similar distances, larger units having up to twice this capacity.

Supergrade or stage-remix pumps have a working range of up to 400 m horizontally or 120 m vertically, the pipeline being provided with extra strong couplings, secured with a rubber seal and safety locking device. The rate of pumping is affected by the characteristics of the aggregate and cement, operational power and pressures, type and usage condition of equipment, cleanliness, length and diameter of the pipelines, ambient and exposure conditions (see Chapter 32), skilful organisation and operation, and the workability of the mix. Operational rules, equipment inspections and provisions against such hazards as hose whips, pipeline blowbacks and high-velocity ejections are safety precautions in pumping.

(d) Pipeline data.

The equivalent horizontal distances of 150 mm diameter pipelines, to allow for increased resistance due to lifts and bends, can be estimated from the pipeline equivalents in Table 24.1. In this procedure, the horizontal equivalents are substituted for the actual lengths of bends in each pipeline. The frictional resistance of rubber hose is over twice that of steel pipe.

TABLE 24.1   **PIPELINE EQUIVALENTS**

Pipe Feature	Horizontal Equivalent (m)
1 m vertical lift	8
90° bend	12
45° bend	6
22·5° bend	3

The content of concrete in pipelines must be based on their actual length; typical unit contents being as follows.

8·5 litres/m of 100 mm diameter pipe.
18·5 litres/m of 150 mm diameter pipe.
32·5 litres/m of 200 mm diameter pipe.

(e) Set-up plant.

Effective planning of pump locations, pipeline layout, placing sequence and the continuous supply of uniform concrete (at the fastest practicable rate) are prerequisites to efficient performance. Approach areas must be of suitable formation to carry a stream of trucks delivering premixed concrete. The pipelines are firmly supported with a minimum of bends and shortened as work recedes from the furthest points of distribution.

A preliminary pump-test run is made without concrete before truck deliveries begin, and concrete pumping is kept as continuous as feasible to expedite flow at a line velocity of, for example, 1-2 m/s.[169] In the event of a pipeline blockage, lifting gear expedites the clearance of affected sections and, for emptying and flushing purposes, adequate supplies of 700 kPa compressed air and water under pressure are essential.

On slow-moving projects, tilt-drum-type mixing may be reversed in direction periodically, or the beginning and end discharges of two truck mixers may be blended for improved uniformity and continuity of supply with a full hopper. With site batching, two mixers free-flow the concrete continuously into the pump hopper at the designated maximum delivery rate.

(f) Concrete.

For pumping efficiency, the concrete is proportioned so that its consistence is neither too wet nor too dry and its mortar matrix will serve as a pipeline lubricant. To meet requisite structural strength, service durability, pipeline lubrication and minimal segregation, the mix is normally not leaner than 1 part cement : 7 parts combined aggregate, coupled with a grading within the envelopes typified in Figure 24.2. Those shown for 19·0 mm and 26·5 mm maximum size can be extended to 37·5 mm and 53 mm maximum sizes. The slump ranges from 60 to 100 mm and the fineness modulus of the sand is from 2·0 to 3·0, the tolerance from average being 0·2.

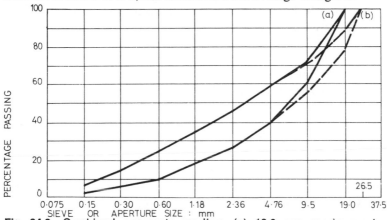

Fig. 24.2 Combined aggregate grading. (a) 19·0 mm maximum size. (b) 26·5 mm maximum size.

A medium sand grading, with optimum percentages passing the 0·15 mm and 0·30 mm sieves, is typified here.

Passing sieve size (mm)	0·15	0·30	0·60	1·18	2·36	4·75
Percentage by weight	7±3	22±8	46±14	71±14	88±10	96±2

The upper limits refer to fine sand and the lower to coarse sand (see Table 3.11), the effect of additional finer fractions being to increase the water requirements, void content and shrinkage of concrete.

Crushed stone passing the 9·5-mm sieve detracts from workability during pumping. Therefore the amount indicated on smooth combined-grading curves may be about halved for a starting mix. For estimating trial proportions of different size gradings of dry-rodded stone aggregate, which are to be blended with sands of fineness modulus ranging from 2·0 to 3·0, the likely volumes of stone aggregate per unit volume of concrete are also typified here.

Max. size (mm)	19·0	26·5	37·5	53·0
Sand FM2.4	0·59-0·66	0·63-0·71	0·67-0·75	0·70-0·78
Sand FM3.0	0·54-0·60	0·58-0·65	0·62-0·69	0·65-0·72

Extra bins facilitate the adjustment and control of continuous gradings of aggregate that are compatible with the least practicable amount of sand for pumpability. The maximum size of aggregate to be selected is limited to one-quarter of the diameter of the pipeline, and set-retardation can be used to offset some stiffening of the mix during pumping. This characteristic varies with the ambient exposure conditions, joint tightness and length of the pipeline.

A slightly lower percentage of fines passing the 0·30-mm sieve is advantageous in cement-rich and pozzolan-extended mixes. Up to 3 per cent of entrained air, by limited usage of an air-entraining agent, improves the workability and cohesion of mixes, and reduces blockages caused by bleeding during delays in placement. Higher air-entrainment may cause a pneumatic loss of discharge by reciprocating-pump action through a long pipeline. A fly ash extension of mixes containing a lignin-type, cement-dispersing or water-reducing agent gives beneficial pozzolanic and low air-entrainment properties to tunnel linings (see Sections 6.4, 16.1.1, 16.2 and 17.2).

With lightweight-aggregate concrete, pumping pressure forces much of the mixing water into the aggregate and greatly reduces the slump. While this water no longer forms part of the water/cement ratio and benefits moist curing, the resultant stiffening must be offset by prior vacuum processing or water saturation of the aggregate, and by mix modifications that have a minimal detriment on requisite properties of the hardened concrete.[159] Presoaking of lightweight aggregate is arranged to reach the average 24-hour absorption, and free water is allowed to drain for several hours before use in order to minimise variations in mix consistence. In all instances, a full-scale, field pumpability test must have been completed before a mix can be regarded as being acceptable for use.

(g) Operation.

Operating instructions must be observed in order to minimise blockages. Before concrete is pumped, the pipeline is wetted and primed by inserting an ejector plug, charging the hopper with 1 part cement : 2 parts sand grout, starting the pump and, after a second batch, admitting concrete to the hopper just before it empties. Air locks must be avoided by keeping the hopper charged and stopping the pump if the concrete falls to a low level in the hopper.

If pumping is to be stopped for more than a few minutes, the pipeline is kept free by running the pump for two or three strokes every few minutes, otherwise the pipeline should be emptied and cleaned. A signalling system (e.g. visual or bell type) is installed between operators at the pump and delivery points for coordinated work. Batching is stopped when the concrete in the pipeline is sufficient to complete the task in hand.

The pipeline is emptied by pumping until the hopper is empty, disconnecting the pipeline close to the pump, inserting an ejector plug and capping the opening with a blow-out fitting that enables compressed air to be admitted or exhausted through valves. As the plug and concrete are forced pneumatically along the pipeline, they are located manually by hammer-tapping. The rate of discharge is controlled by adjusting the two air valves, so that strong air pressure is not released suddenly at the discharge end of a pipeline, thereby preventing the possible collapse of narrow or confined formwork.

The emptied pipeline and pump should be cleaned promptly, the washing water being kept clear of placed concrete. An effective method is to force through the pipeline a 6 m cylinder of water contained between ejector plugs, the water being admitted through an attached 22·5° bend. During dismantling, each section of the pipeline is sloped and washed thoroughly, particularly at the spigot and socket ends, for effective reuse with tight joints. At the end of placement within a building, the pipeline may be cleaned in reverse with the aid of a dual air-and-water adapter.

In large circular tunnels, a 90° invert of lower slump concrete (e.g. 50-60 mm) is first placed by a conveyor belt, which is provided with a drop chute and laterally moving terminal. The concrete is vibrated as soon as it is dropped, slipformed to guide-rail alignment, and float finished at time intervals to suit the consistence of the concrete. Bleed water along the lowest part of the invert must be promptly removed by dipping from temporary recesses or by burlap blotting for a satisfactory performance in subsequent service.

24.2 SPRAYED OR PNEUMATIC MORTAR

24.2.1 GENERAL INFORMATION

Sprayed or pneumatically placed mortar is high-quality, dense-aggregate or lightweight-aggregate mortar, which is mixed and ejected into place under pressure. It is useful for repair work, shell construction, surface coatings generally and placement where access is difficult or formwork would be expensive. The initial mixing may be dry, wet or foamed; the first is dealt with in detail and the others are mentioned as a modification of dry mixing.[10-12, 188] When used for wet sand-blasting purposes, the equipment reduces the dust cloud of dry sand-blasting by 80 per cent.[189]

24.2.2 MORTAR

Best-quality materials should be selected for use (see Sections 3.1, 3.2 and 3.3). The fine aggregate should be coarse-screened sand which is washed and graded from coarse to fine, as specified in Sections 20.1.2, 24.1.2(f) and ASTM C33. Within close limits, its fineness modulus is from 2·5 to 3·3 (see Sections 3.3.4(c) and 3.3.5(a)) and, at the time of mixing with cement, its moisture content should be from 3 to 6 per cent by weight.

Normally the mortar consists of 1 part cement : 4 parts moist sand by volume (see Section 3.3.6(b)), but less sand may be used for high impermeability (e.g. horizontal surfaces and the coated prestressed concrete walls of cylindrical liquid-retaining structures and high-pressure pipes). For liquid-retaining structures, the mortar should have the following proportions after application: 40 kg of portland cement : 3·6 kg of hydrated lime : 0·085 m³ of fine aggregate passing a 2·36-mm sieve (AS 1481). The materials are mixed for 1·5 minutes and used within 0·75 hour. Water is added at the nozzle, in the requisite amount to ensure satisfactory hydration and placement, without sagging and excessive rebound. The water/cement ratio varies from 0·35 to 0·5 and the rebound of sand varies from 20 to 40 per cent by weight.

24.2.3 PLANT

The plant usually consists of: a concrete mixer; a vertical, double-chamber, pneumatic gun; delivery hoses for air, mixed material and water; an ejector nozzle; and an air compressor capable of delivering

Repair work
on the downstream face
of Ridgeway Dam,
Tasmania

Drilling
grout holes
from the crest

Grouting plant
in operation

Concrete paver in operation

Slip-form pavers moving in tandem

240 litres/s at 550 kPa. The upper chamber of the gun receives and pressurises the dry mix and delivers it to the lower chamber, where a feed wheel forces the material continuously and uniformly into a delivery hose. Recharging is done through the upper chamber, which has two conical doors for inlet and air-lock purposes.

Air pressure at the gun must be sufficient to cause material to be ejected at the nozzle at a uniform velocity of 120-150 m/s. For this purpose, a steady pressure of 350-400 kPa is required for a 30 m length of hose, or up to 500 kPa for high lifts or a hose up to 150 m long. For finishing work the pressure is reduced to 200-350 kPa.[10]

Water at the nozzle is introduced through a control valve and a perforated manifold. Its pressure is approximately 100 kPa higher than the highest air pressure required for placement. The nozzle lining should be replaced when badly worn and all equipment must be cleaned internally at least once each 8-hour shift.

24.2.4 APPLICATION

The surfaces of concrete and masonry must be prepared, where necessary, to a sound condition with bush hammers, wire brushes or sand blast, and thoroughly cleaned, wetted and scoured with an air-and-water blast and air blast. Mortar should be placed within 1 hour of surfaces being wetted, in order to facilitate hydration and bond.

In structural layers, steel mesh or fabric reinforcement is fastened in place with anchor dowels or tied to existing steel rods. The reinforcement should clear existing hard surfaces by at least 6 mm and the proposed surface by 25 mm (see Section 3.4). It should cause little interference with uniform placement and sand pockets at laps should be avoided by not tying fabric wires together in parallel.

Shooting is done in thin, rapidly repeated passes of the nozzle, which is held at right-angles to the working face and about 1 m from it. Lap marks are minimised by moving the nozzle continuously so as to trace an ellipse, about 0·2 m high and 0·6 m long, each revolution moving the trace forwards about 0·15 m on the long axis. Rebound occurs freely from a moist hard surface until a film of cement is formed and promotes bond. Rebound material should be blown or discharged clear of the work and not reused if uniform results are to be obtained.

The thickness applied at one time is limited to 25 mm on overhanging surfaces and 75 mm on vertical surfaces. A coat should reach initial set, but not final set before the next one is applied. It is broomed lightly and dampened to improve bond.

Construction joints are made with sloped surfaces which are cleaned, wetted and scoured before the resumption of the work.

Taut wires delineate the lines of finished surfaces. About 1 hour after placement, the surface may be trimmed with a sharp-edged screed or tool, which is worked upwards with a slicing motion. It is finished by lightly wood-floating or steel-trowelling, or applying a flash coat (about 2 mm thick) near the time of initial set of the main coat.

The work must be done by skilled operatives in fine, calm weather òr under cover. It should be inspected thoroughly (see Appendix II, 2(b)), hammer tested to detect unsound areas and repaired immediately where necessary. Adjacent fixtures or surfaces should be covered where cleanliness is required. As soon as the first signs of drying appear, the finished surface should be moistened with a fine spray of water. It is then membrane cured with two coats of sealing compound (see Section 9.3), or moist cured with wet hessian or burlap for 7 days (see Section 9.2).

24.2.5 CHARACTERISTICS

Mesh-moulded or specially cut cylinders,[10] moist cured at $23\pm2°C$, should develop a compressive strength of not less than 20 MPa at 7 days and 27·5 MPa at 28 days, the latter being usually above 35 MPa. Although a large proportion of fine voids in the mortar are not interconnected, an ordinary mix is sufficiently porous to allow moisture to penetrate. This feature is not necessarily detrimental, as evinced by the high resistance of the mortar both to wetting and drying, and to frost action.

Depending upon the components and the quality of the mortar, its drying shrinkage may range from 75 per cent to twice that of ordinary concrete. Should small cracks develop in large exposed surfaces after a period of some months, they should be protected from possible future weathering. In engineering structures, this should be done by sealing the cracks with synthetic rubber solution (e.g. "Neoprene") or the entire surface with boiled linseed oil (see Section 20.2 (d)).[11]

24.2.6 MODIFICATION

The foregoing procedure in modified form is used for the application of shotcrete or wet mixes generally. In one process, for instance, a paddle mixer discharges plasticised mortar or fine concrete into an adjacent hopper, which feeds the mix into an enclosed plunger conveyor system underneath. The conveyor consists of a horizontal driving chain, on which are fixed a series of discs that pass through a closely fitting rubber tube. Strong air-jets halfway along this tube cause the conveyed material to be extruded through a delivery hose and discharge nozzle. The material is ejected either directly or as a high-velocity spray propelled with additional air that is injected at the nozzle. Early hardening can be effected, where necessary, by metering into the flowing mix an air-suspended dry accelerator (see Section 24.1.2(a)).

The process can place wet-mixed material uniformly with no appreciable rebound loss and at a horizontal radius of up to 150 m. It can be used also with dry-mixed material and for sand-blasting purposes. The discharge nozzle is held about 450 mm from the working face with wet mixes and twice this distance with dry mixes.

Various units (ranging in capacity from 3 to 5·5 m³/hour) require a 120 litre/s air compressor operating at 700 kPa. A greater quantity of air is required for larger units (e.g. 9 m³/hour) or those driven by an air motor. The unit in each instance is flushed with water and air at the beginning and end of a working cycle. As the shrinkage potential of this type of material exceeds that of good-quality concrete, its practical application should include the use of mesh reinforcement, contraction joints, and both early and effective moist curing. Pneumatic pumps and their use for tunnel lining with shotcrete are indicated in References 10, 12, 106, 159, 188 and 189.

24.2.7 FIBROUS MIX

With fibrous mixes (see Section 3.4.2), the spray equipment used should be preferentially of the wet-batch type, this being characterised by a low-velocity delivery with minimal rebound loss. Steel-fibre mixes, subjected to evaluation trials, have manifested in the matrix the development of a two-dimensional loose matting. This formation facilitates the build-up of laminates possessing a considerably improved flexural strength, but the effect is not reflected in direct strength determinations. The dry-batch process with nozzle moistening, however, may prove superior on subterranean surfaces that are in a variably wet condition. A prickly texture remaining on the final profile is smoothed off with a fibreless coating.

Core specimens (at least 40 mm diameter by 120 mm long) are drilled selectively from site-applied laminates for the purpose of determining their physical and fibrous properties. Stringent safety precautions must be taken when spraying steel-fibre mixes, in order to avert any painful fibre penetrations of operatives. Protective apparel to be worn includes full face masks, leather gauntlets and full aprons or back-fastening boiler suits.

Sprayed fibrous mixes are suitable for use in large projects, such as tunnel and mine linings, slope and rock-face stabilisation, sea defence work, and refractory linings that require the incorporation of stainless steel fibres (see Reference 136 of Part 1).

24.2.8 REPAIR OF COLUMN

Figure 24.3 shows the repair of a column where concrete cover has flaked away. All loose concrete and dust are removed and any exposed reinforcement is freed from loose rust. Smooth surfaces are roughened and mesh reinforcement is fixed at least 6 mm clear of the old concrete. Battens are fixed to two sides, the concrete is well wetted, and new mortar is applied and screeded to a level surface at the time of initial set. When these two sides have hardened, the battens are removed and the remaining sides are coated, the mortar already placed acting as screed guides (see Reference 1 of Part 3 and References 1, 37 and 148 of Part 4).[10-12, 188] Ordinary concrete methods may be preferable, however, where the steel of reinforced concrete structures has been affected by marine corrosion or fire. Equipment and service, including protective respirators and clothing, are available from specialist consultants.

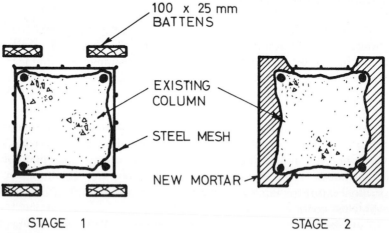

Fig. 24.3 Repair of column.

24.3 CEMENTATION OR PRESSURE GROUTING

24.3.1 GENERAL INFORMATION

Cementation or pressure grouting is the forceful injection of cement, fly ash cement or other grout into cracks, voids or fissures in structures or the ground. The cement should be cyclone separated so that 98 per cent will pass a 0·075-mm sieve. Bentonite (colloidal clay), bituminous and chemical grouts may be used where factors other than strength are important. High impermeability can accompany the use of bentonite, because it can absorb about three times its own weight of nonsaline water and can swell some seven times its dry bulk volume (see Reference 108 of Part 4).

Where voids and fissures to be filled are less than three times the diameter of the particles in a suspension grout, chemical grouting with special plant may be used for improved penetration. It consists of one of the following processes: successive injections of sodium silicate and calcium chloride; a single injection of chrome lignin; or a dual-pump single injection of proprietary chemicals and catalysts (e.g. "AM-9", "TDM" and "Cemex-A"). The materials in the third process cause instantaneous gelation, even in wet ground or below water tables, after a predetermined period. A rubbery gel thus formed is somewhat flexible in structure, moderately strong, highly impermeable and resistant to attack by fungi and dilute chemical solutions.

For the application of cement grouts in general, grout pumps of various size range from manually operated units to power-driven units with capacities varying up to 2 litres/s and up to 7000 kPa. Special pumps are equipped with by-pass assemblies, pressure gauges and operational accessories for safety and control. Fresh cement is used (with or without a set retarder) and the equipment is cleaned after each series of grouting operations or at 3-hour intervals.

When the acceptance rate of a soil is fairly large and constant, a single-line system may be used, but when the rate is variable a circulatory system is employed. In this system, a grout agitator is placed between the grout mixer and the pump, and a return duct from the injection point discharges grout into this vessel. Special appliances are used at the injection point of a circulatory system for grout flow and pressure control. Typical uses [13-17] are in grouting of ducts, in repair of structures, in consolidation of ground and in water cut-offs, these all being done by specialists.[14] Before making a main excavation, reinforced concrete enclosures can be formed by the bentonite-slurry, tremie-displacement trench method (see Section 30.2.3 and BS CP2004, Foundations).

24.3.2 GROUTING OF DUCTS

Ducts in prestressed concrete, rock anchorages, contraction joints and butted-assembly elements are pressure-filled with suitably formulated grout (see Sections 15.2.2(d)(vi), 25.1, 30.1, 34.1.10 and 34.2.3; also BS CP110 Part 1, and References 19, 117, 133, 141 of Part 4 and Reference 139 of Part 8).[18, 20, 164]

24.3.3 REPAIR OF STRUCTURES

Structures may tend to become unstable because of cracks or voids. Cracks may be caused by settlement or damage, while voids are often found in what appears to be solid brickwork or masonry of great thickness but which is, in fact, merely a thin skin containing dry rubble filling. In such cases, the filling of the cracks or voids with cement normally restores stability. This is done by drilling holes in the structure at carefully selected points and pumping in a cement grout, which is sufficiently thin to ensure complete penetration. Flaky aluminium powder may be incorporated in the grout, which is of colloidal, activated, cement-dispersed or plasticised type (see Section 25.1; also Plate 19).

24.3.4 CONSOLIDATION OF GROUND

Ground of fairly hard nature but loose texture (e.g. certain types of made-up ground) may be consolidated and increased in bearing capacity by cementation. The process is useful for stabilising embankments and foundation soils, arresting the settlement of structures, and jacking-up subsided concrete slabs. Pipes are driven into the ground and the cores within the pipes removed by means of an earth auger. In granular soil, the pipes may be jetted to the desired depth. Cement grout is then pumped into the ground through the injection pipes which are withdrawn as grouting progresses. Perforated bore casings may also be used to facilitate the penetration of grout into ground that is to be stabilised.

24.3.5 WATER CUT-OFFS

Water cut-offs are constructed in a similar manner to that used for ground consolidation, except that boreholes are usually much deeper and are normally drilled and lined. When grouting fissured rock, sections of a borehole can be isolated and grouted in stages with injection accessories called packers. On other occasions, when the ground is firm and noncaving, a short injection pipe may be fixed at the top of the hole or cemented in place. Water cut-offs are used for the following.

Below and around dams, using either cementation or slurry-trench concrete walling to prevent formation seepage.

Around basement, tunnel and deep excavations (e.g. mine shafts) to prevent ingress of water. With cementation, the ground may be excavated at a steep or even vertical slope without the use of timbering. In tunnel boring, pressurised bentonite slurry is used for face stabilisation and spoil removal by pump to the surface.

For drilling and sealing oil wells, using thixotropic stabilising fluids and special cement grout with a water/cement ratio of 0·4 by weight (see Section 10.2). Its consistence can be checked by a torque viscometer or flow cone (Appendix I).[19, 142, 164, 173]

24.4 PREPACKED CONCRETE

Prepacked or preplaced-aggregate concrete is made by injecting portland cement and sand grout under pressure into the bottom voids of a compacted mass of clean graded coarse aggregate of at least 28 mm nominal size.[19] The grout is admixed with a wetting agent (as an intrusion aid), a pozzolan (e.g. fly ash) and flaky aluminium powder (unpolished variety) to the extent of 0·005-0·02 per cent by weight of cement (up to 2 dessertspoonfuls per 40-kg bag). The pozzolan increases the flowability and subsequent strength of the grout and also minimises bleeding. The grout may be activated by high-speed mixing and the fineness modulus of the sand should be between 1·2 and 2·0 to ensure best pumping performance (see Reference 5 of Part 3).[134]

The aluminium powder gives a slight expansive effect to the grout, by the generation of hydrogen gas through chemical reaction with alkaline constituents of the unset cement. It is first blended with 50 parts of pozzolan or ground pumice stone by weight, the amount of blend used varying from 120 g per 40-kg bag of cement for concrete having a temperature of 23 °C, to 180 g or more for a temperature of 5 °C. As the aluminium powder has a tendency to float on water, the blend is incorporated in the mix before the water is added. If the expansion of the grout is restrained, the strength of the hardened grout is not appreciably weakened and may be slightly increased. At moderate temperatures, the reaction starts at the time of mixing, but after 45 minutes it becomes very weak. At temperatures above 30°C, the reaction may be completed in 30 minutes and subsidence may take place until the concrete takes its initial set. At 4°C, the reaction may not be completed for several hours. Nearly twice as much aluminium is required at 5°C as at 23°C to produce the same amount of expansion (see Reference 1 of Part 3).

In the raising of the Mundaring Weir, Western Australia, the grout used to percolate thoroughly between the stones consisted of three parts cement : 4 parts sand : 1 part fly ash (by weight), 25 g wetting agent (100 per cent active constituent), and 7·5 g of flaky aluminium powder per 40-kg bag of cement, with a water/cement plus fly ash ratio of 0·53.

The aggregate with about 45 per cent of voids when compacted is wetted or preferably inundated with water. As the grout is pumped into the mass, the water is displaced from the voids and a dense concrete with a high content of aggregate is created. Formwork should be watertight and freely vented at the top. Grout is introduced through 80 mm diameter GWI pipes spaced at not more than 1·5 m centres. Slits 10 mm wide by 150 mm long are cut on opposite sides of the pipes and spaced 450 mm apart. These allow the grout to emerge into the voids when pumped into the pipes through 25 mm diameter air hoses. The air hoses are pushed to the bottom of the pipes and gradually withdrawn as the grout level rises, so that about 900 mm or more of each hose remains immersed in the grout. The metal pipes may be thus withdrawn, also, if required.

The mixed grout is pumped continuously from a drum agitator by diaphragm-type sludge or mud pumps. Each pump keeps about four air hoses supplied with grout and is stopped every hour so that the system may be washed out with water and kept free from precipitated sand or setting grout. A break of 6 hours in grouting operations per day does not develop a cold joint during this period.

The level of the grout in situ is checked by a float on the end of a thin metal survey band, the float being weighted so that it passes through water, but floats on grout. After the voids are filled, a slight head should be maintained until the grout has taken its initial set. For this purpose, restraint near the top of the mass can be maintained by previously sealing it with about 150 mm of concrete or by applying a venting form that is suitably held down. A venting form consists of a timber frame that is sheeted successively with expanded metal, fly wire and muslin, which permits air and water to escape when grout is forced against it.

On a large project, the temperature of the grouted concrete may rise about 22°C in 2 days. Average values of diffusivity for concretes containing various coarse aggregates are given in References 12 and 113 of Part 6. This property is an index of the facility with which mass concrete will change in temperature (see Sections 3.1.1 and 30.1; also Reference 15 of Parts 1 and 6).

Prepacked concrete can be used as steel-pipe-column infill, or where concrete would be difficult to place and where tight, well-bonded repair or refacing work must be done in mass concrete. It develops strength properties similar to those of ordinary concrete and, with careful field practice, it is suitable for underwater construction (see Section 30.2.4).[20] Under ordinary conditions, following proper curing, the drying shrinkage of prepacked concrete is about one-half that of ordinary concrete. Early sandblasting or water-jetting, when applied to a wall, is a useful technique for producing a decorative exposed-aggregate finish (Naturbetong, Section 28.9.2).

CHAPTER **25**

NONSHRINKAGE
MORTAR OR GROUT

Low-shrinkage or nonshrinkage mortar, grout or fine concrete can be made with expanding agents of aluminium powder, malleable-iron granules or sulphoaluminate cement. The agents and mixes selectively used should be compatibly formulated for particular projects, uniformly manufactured, properly placed in confined spaces, and subsequently moist cured for effective results (AS MP20, Part 3). Special care is required to avoid any prospective interaction of these materials with highly stressed or high-tensile steel elements (see Sections 15.2.2(a), 34.1.4(b) and 34.1.10(ii)). The mixing of dry prebatched materials is effectively carried out in paddle, pan or vane-type mixers.[50]

Acceptance testing prerequisites for expansive agents or mixes and expanded specimens include their correct preparation and sampling, structural or chemical composition, and finished product uniformity. Physical characteristic determinations at $23 \pm 2°C$, or under other prescribed conditions, cover such factors as flow-cone or flow-table consistence, mortar slump (see Section 27.2.2(b)), gas generation, volume changes, bleeding and air content (see Appendix I), penetration-resistance setting times (see Section 14.11), and the restrained-expansion strengths of hardened mixes in compression and indirect tension (AS MP20, Part 4).

25.1 ALUMINIUM POWDER

A portland cement mortar or grout is made of 1 part of cement and 1·5-2 parts of fine or well-graded sand, the water/cement ratio being 0·5 by weight. Flaky aluminium powder, in the

460

proportion of 0·005 per cent by weight of cement (2 g per 40-kg bag), is incorporated in a blend with fine sand or a pozzolan (see Section 24.4). Wetting, cement-dispersing and suspending agents are used to minimise mixing water, drying shrinkage, settlement and possible bleeding in tall ducts (see Sections 6.4 and 15.2.2(d)(vi)).

Batches are of limited size, so that freshly prepared mixes can be placed immediately and expand in a confined space under conditions of restraint. As no expansion takes place after the mass has hardened, subsequent drying shrinkage may be regarded as a prospective characteristic. Ambient influences on the rate of generation of hydrogen gas and the resulting proportionate dosage of expanding agent required are indicated in Section 24.4. The mix not only increases the bond strength of horizontal steel reinforcement in deep beams, but also is useful for filling joints, tightening injected grout, bedding heavy objects and improving the homogeneity of grouted concrete.[21]

25.2 IRON GRANULES

Malleable cast-iron slivers are freed from cutting oil by caustic washing and heating to red heat in a rotary kiln. They are milled and graded into two size-grades: one for general use passing a 0·60-mm sieve and one for small clearances passing a 0·15-mm sieve. A corrosion promoter (e.g. sulphur, ammonium chloride or calcium chloride) and a cement-dispersing agent are incorporated, so as to promote limited expansion under moist oxidising conditions at an early age and to minimise the water required for a given degree of workability.

Typical mortar mixes consist of portland cement, metallic admixture and suitable sand in the proportion by weight of 1:1:1 or 1 : 0·5 : 1·5, respectively, with no more water than is necessary for effective placement. The proportions for a particular class of work are governed by experience, test data and the requirements of the project. They may be extended, for 100 mm gaps, with 1·5 parts of 10 mm coarse aggregate. Tight compaction of the material in a cavity is ensured by tamping it *in situ* and restraining subsequent movement, the adjacent surfaces being presoaked and primed ahead (in limited areas) with a 1 part cement : 1 part iron bonding slurry. Expansive materials should be thoroughly mixed, used within 15-30 minutes, finish compacted (see Section 27.2.2(k)) and damp cured immediately for 3 days.

A ram-in mortar, made with ordinary portland cement and a water/cement ratio of 0·40 by weight, can attain successive compressive strengths of up to about 35 and 50 MPa at 1 and 3 days. With high-early-strength portland cement, these strengths can reach up to 48 and 60 MPa, the components being equiproportioned, cast in 100 mm cubes, and damp cured at 23°C. Mobile grouts, made with these cements and a water/cement ratio of 0·50 by weight, have strengths 7-14 MPa below these values. Alternative brands of cement and techniques of fabrication can cause a similar effect at early ages.

The subsequent hydration and oxidation of the malleable-iron granules occur preferentially to a similar action on the surface of large metal fixtures that may be present, and the rate of corrosion decreases with progressive densification and impermeability of the mix. The admixture should not be used in contact with high-tensile steel or small-diameter reinforcement, such as in prestressed concrete.

Typical uses of these mixes are for tight bedding, caulking, underpinning, watertight joints, anchor bolts, repair work, pressure-injected grout and double-frustum, bar-splice, cast steel sleeves. Mixes with a low content of iron granules may be used for pneumatically applied mortar or for fine concrete for watertight applications. The corrosion of exposed iron particles will speckle a surface, but the effect may be reduced by the application of an appropriate surface treatment.

25.3 THIXOTROPIC GROUT

For bedding heavy-plant equipment, potentially expansive portland cement mortar (free from gas or corrosion promoters) is initially dry-batched with ettringite-forming and sedimentation-control ingredients (see Sections 3.1.2(a) and 15.2.2(d)(vi)), fine silica sand and rust-inhibited, malleable-iron granules. It is controllably mixed with water to form a self-levelling grout which, at equitable temperature, develops cohesive, bleed-free, thixotropic (see Glossary, Appendix VI), nonshrink and high-strength properties. The practical conditions to be catered for include repetitive dynamic loading, hydrothermal or thermal movements and stability of uniform bearing.

After levelling the bedplate of a mechanical appliance with temporary shims, the concrete base is water-saturated overnight and freed of excess water by pneumatic or vacuum means. Tight sideforms and air-bleed vents are fitted, then grout under low-head pressure (e.g. 200 mm) is poured into the interspace from a headbox on one side. Placement is continuous and the flow is activated by mild agitation or strap-drag work for up to 0·5 hour.

Early moisture-loss by evaporation is prevented by ponding or wet curing for 3 days and membrane curing is subsequently applied to exposed surfaces. Grout of 25-30 seconds flow-cone consistence (see ACE 79 and Appendixes I and VII) should be bleed-free in an 800-ml, 2-hour settlement test. Typical compressive strengths of 50 mm cubes (ASTM C109) exceed 20, 30 and 45 MPa at 3, 7 and 28 days, respectively, greater values being obtainable with more viscous grouts of the same dry-batch composition.

25.4 SULPHOALUMINATE CEMENT

Expansive hydraulic cement and some of its uses are described in Section 3.1.2(a). For effective results, the cement should be stored in vapourproof containers (e.g. polyethylene-lined bags), mixed with the minimum amount of water for requisite workability and compaction, and thoroughly moist cured. Typical nonshrinkage mixes consist of 1 part cement : 3 parts good-quality sand for cavities less than 25 mm in width; 1 part cement : 1·5 parts sand : 2·5-3 parts stone of 10-14 mm nominal size for wider cavities; and 1 part cement : 2 parts sand : 4 parts stone of 14 mm nominal size for cavities over 75 mm in width.

Contact surfaces, which should be structurally sound and clean, are first dampened and coated with 1 part cement : 1 part sand slurry or mortar. The expansive action, which accompanies the hydration of newly placed mixes, can be accommodated without causing disruption of bond within the material or in the surrounding mass. Expansive cement mixes have a variety of uses for infilling spaces under restraint in building construction and repair work. They include spaces in anchorages, under bedplates, between precast concrete elements, and around pipes passing through concrete.

CHAPTER 26

CONCRETE PAVEMENTS

26.1 CEMENT-BOUND MACADAM

Cement-bound macadam is known also as cement-penetration macadam and colloidal concrete. A concrete pavement in this category may be readily constructed by consolidating coarse stone aggregate to requisite thickness and profile, and completely filling the voids with cement and sand grout. The aggregate imparts a high degree of wear resistance to the surface, but it must be free from fine particles that would hinder penetration of the grout. The grading of aggregate for different types of pavement is given in Table 26.1.

TABLE 26.1 COARSE AGGREGATE

Type of Pavement	Minimum Thickness (mm)	Macadam, 100 Per Cent:	
		Passing sieve (mm)	Retained on sieve (mm)
Roadways and heavy tractor yards	150	63	37·5
Private drives and stock yards	100	53	26·5
Footpaths	75	37·5	19·0

A consolidated, granular subgrade or bed, 75 mm thick, is desirable to absorb excess water from the grout without removing cement. A sandy formation need only be shaped and well compacted by a power-operated vibrating roller (e.g. 4 tonne, 2500 cycles per minute). If it is only lightly compacted, hessian or flexible high-density polyethylene mesh may have to be laid before the aggregate is spread, so as to confine the sand to the subgrade and

464

ensure that subsequent grouting will be for the full depth of aggregate. A heavy soil or clayey formation should have mitre drains of crushed stone inserted at 15 m intervals. These are located in herring-bone fashion about the centre line and connected with a downwards slope to side drains. This subgrade, on being prepared and rolled, should be overlaid with the granular bed described.

In residential streets, service-pipe conduits are laid transversely below subgrade level for future house connections. Mitre drains are unnecessary where service-pipe conduits serve as subgrade drains. Seepage from the latter must be directed into underlying side drains connected into gully-pits. These lines of drainage are covered with hessian and sand so as to keep them free from grout later on.

For increased load-carrying capacity, pavements may be reinforced and thickened by 50 mm at the edges or increased in thickness generally. Slabs up to 6 m wide are aligned by deep kerbs, steel forms or 50 mm thick boards, securely held in place to line and level and backed with soil. A grout repellent (e.g. paraffin wax; see Section 9.3) may be applied to concrete channels to facilitate their final clean-up. Side forms are oiled and the coarse aggregate for roadways or reinforced slabs is laid and rolled in two layers. Rolling reduces the thickness of loosely spread coarse aggregate by about one-third, the maximum thickness (loose) of a single layer for initial rolling being about 150 mm.

The first layer, for a 150 mm pavement, is uniformly spread about 125 mm thick (crown depth). It is rolled to correct grade and camber with a 6–10-tonne traditional roller or a 1-tonne self-propelled vibrating roller (4000 cycles per minute), proceeding along the sides and working gradually towards the centre. Where hard-drawn, steel-wire fabric is to be included, 3.5 kg/m^2 oblong-mesh or the equivalent is laid, lapped 150 mm at side-joins, 300 mm at end-joins and securely tied together. A further 100 mm of coarse aggregate is uniformly spread and rolled to template shape with cross-falls of 1 in 35.

Note: Pavements only 100 mm thick may be adequately consolidated with a 330 kg hand-guided, self-propelled, vibrating roller (4000 cycles per minute). The compacting effect of an ordinary roller is increased tenfold or more by vibration. Before grouting in hot weather, the macadam may be cooled by sprinkling with water.

Main control joints 20 mm wide and of full depth are formed at the end of each half-day's work or at 45 m centres. They may be formed of bituminous fibreboard at a stop form, the board being topped with a clipped-on temporary fillet, so as to form a groove

20 mm wide by 40 mm deep for subsequent sealing. The depth of groove varies from one-sixth to one-quarter of the depth of the course and joint sealing may be effected with preformed synthetic rubber (e.g. "Neoprene") strip, installed without stretching or twisting. Steel dowel rods, 20 mm diameter by 600 mm long, are rigidly located at 300 mm centres. They are sheathed with 0·08 mm polyethylene film (close-fitting with a sealed end) for half their length and rubber-capped 25 mm at the sliding end. Cold-poured joint sealants for concrete pavements are guidelined in BS 5212. Mesh reinforcement does not cross the main joints. Secondary joints of weakened-plane or dummy type are formed along the crown of roadways and transversely at 15 m centres. Closer joints are used in cold-weather projects.[37]

Slots to accommodate sealing compound may be formed by vibrating the vertical leg of a 50 mm tee-section blade, about 3 m long, into the surface after it has been grouted and rolled. A temporary fillet (e.g. foamed polystyrene) is inserted and the adjacent profile is made good by the manipulation of material at the surface finishing stage. Alternatively, a flexible, tubular joint-former, 10 mm wide, may be set permanently in place. This type of fillet is of 0·30 mm galvanised sheet metal, either straight or parabolic in profile and in units up to 3·6 m long with interlocking ends.[22-24] After the concrete has hardened, the fillet is grooved with a wheel and subsequently caulked (see Reference 98 of Part 6).[182]

The grout may be made in ordinary mixers, but for large projects it is made in special proprietary mixers (e.g. "Sunderland" and "Colcrete"). The proportions of portland cement and sand may be varied from 1 : 1·5 to 1 : 2·5, respectively, by volume to suit the nature of the work. The cement must be fresh and the sand should be graded from 4·75-mm gauge to fine, so that its fineness modulus is from 2·2 to 2·5. Blending may be required to achieve this value. The water content of the grout should be the minimum for pouring consistence and not more than 30 litres/40-kg bag of cement. A cement-dispersing agent or wetting agent may be used as a water-reduction aid, but not for causing more than limited (if any) air-entrainment.

The grouting operation, which should proceed downhill, must leave a slight excess of grout on the surface. It is immediately followed by light longitudinal rolling with a 1-tonne or 2-tonne roller until all entrained air is expelled. Stone screenings, 14-20 mm nominal-size gauge, are evenly spread over the grout-flushed surface at the rate of 1 m³ per 70-90 m² and lightly rolled. Surface inequalities from template shape are corrected by using screenings and grout, and by heavy cross-tamping with a 300 mm × 100 mm steelshod screed board. Any rerolling required at this stage must be light and not cause damage to joint fillets.

Light transverse rolling, with a 225 mm diameter concrete-pipe roller, is applied to smooth the surface and remove free water by forcing it over the side forms. Additional smoothing, if necessary, is done with a 1·5 m × 150 mm × 5 mm flat, rubber "squeegee". A nonskid texture is produced by cross-belting with a heavy canvas belt, not less than 150 mm wide. The surface may be striated or smoothed further by light brooming or wood-floating near the time of initial set of the cement.

Effective moist curing is started as soon as practicable and continued for at least 10 days. Side forms are removed in 2 days and loose curing material is removed after 14 days. If required, concrete may be sawn at joints after it has reached a compressive strength of 7 MPa. After temporary joint-fillets have been removed or sawn out, or tubular joint-formers have been crimped down by a weighted wheel, the joint recesses thus formed are finally cleaned (when dry), primed and filled with hot bitumen and rubber compound.

In footpaths, the surface voids are filled with clean 10 mm stone toppings, and butt joints are placed at 2 m centres. Cement-grouted macadam, of 1 part cement : 2·5 parts sand : 10 parts stone composition, requires 0·5 of a 40-kg bag of cement (25 bags per tonne) per m², per 100 mm consolidated depth. Test cores of cement-grouted macadam have shown an average compressive strength at 28 days of 24 MPa, and a modulus of rupture which is 0·5-0·6 that of mixed and rolled concrete of similar composition.

26.2 CEMENT-TREATED CRUSHED ROCK

Cement-treated crushed rock, containing between 1 and 3 ± 0·3 per cent by weight of portland cement, is an economical and convenient medium for the construction of cement-stabilised pavement bases 150-200 mm or more thick over a prepared stable subgrade (see Sections 20.1.3 and 26.1; also AS 1289; BS 1377, 1924 and ASTM D422, D698, D1195, D1241, D1557, D1883). Increased thickness is needed for the construction of airfield pavements, the surface in each instance being protected with double bituminous seal coats or bituminous concrete some 50 mm or more thick. [130, 131]

Crushed rock used in this way must be of sound quality, with a Los Angeles abrasion-test value usually not above 30 per cent (see Section 3.3.4(f)). Alternative gradings for maximum density are given in Table 26.2, with a stipulation that the fraction passing a particular sieve should be between 60 and 80 per cent of that passing a sieve of twice the aperture.

TABLE 26.2   CRUSHED ROCK

Nominal Size (mm)	Per Cent Passing Sieve: mm							
	37·5	26·5	19·0	9·5	4·75	2·36	0·425	0·075
40	90-100	—	60-80	40-65	30-50	20-40	10-20	4-10
28	—	90-100	80-95	50-75	35-60	25-45	10-25	5-10
20	—	100	90-100	60-80	40-60	30-45	15-25	5-10

The rock fines passing a 0·425-mm sieve should have a lower liquid limit not exceeding 25, and a plasticity index not exceeding 6, while the sand equivalent of the fraction passing a 4·75-mm sieve should not be less than 50 (see Glossary, Appendix VI). When less than 2 per cent of cement is to be mixed with the crushed rock, the last two limits should be made 4 and 50, respectively. Cylindrical test specimens are prepared by mixing appropriate crushed rock with 2 per cent by weight of cement and compacting the mixture at optimum moisture content (e.g. 6-7 per cent by weight) to maximum density. The compressive strength thus obtained at 7 days should lie within a range of 1·5-3·5 MPa (BS 1924 and ASTM D558, D1632, D1633).

The material may be mixed *in situ* by single-pass equipment or preferably by a revolving blade, rotary pan or continuous mixer system at a plant which is capable of supplying at least 400 m³ per day. The crushed rock is separated into at least two sizes, including a division at the 4·75-mm sievé size with oversizes and undersizes limited to 15 per cent. These sizes are stored separately and combined to meet the grading and test requirements described (Plate 7).

The cement is gauged by weigh-batching on a separate scale, and the various sizes of aggregate in volumetric batching are measured in separate calibrated hoppers that are adjustable in size and completely filled. Weigh-batching is used for preference and the mixing time is at least 30 seconds (see Section 7.1.2). In continuous mixing, provision is made at the plant for interlocked feeders, revolution counters on drive shafts, bin-level indicators, wall vibrators on fine-aggregate bins, calibrating equipment, and automatic shutdown devices in the event of failure of a feeder or the depletion of a storage bin.

The mixture is transported under cover to prevent moisture loss and spread over a dampened subgrade by a self-propelled screeding and tamping spreader. The layer thus spread, if about 130 mm thick, would be reduced during compaction to 100 mm. The top consolidated layer of a pavement should be at least 75 mm thick. Laying operations are carried out continuously, so that the material is placed and compacted within 2 hours of mixing, and the delay between the spreading of adjacent lanes does not exceed 0·5 hour.

The material is compacted to 100 per cent Modified AASHO maximum density (see Glossary, Appendix VI) by one or two 10-tonne to 12-tonne smooth-wheel rollers, the rear wheels of which apply a pressure of 62·5 kN/m and face in the direction of travel of the spreader. Each layer receives no fewer than two coverages by the rear wheels per 25 mm of its compacted depth. Layers of over 100 mm and up to 150 mm consolidated depth may be compacted by one pass of this roller and four passes of a 4-tonne vibrating roller without causing segregation.

Each new lane is butted against the previous one after it is trimmed and lightly dampened, and the ends of runs are trimmed square and staggered in alignment. At the finishing stage, motor graders trim surplus material to waste and pneumatic-tyred rollers complete the compaction to a surface tolerance of 10 mm in 3 m. The compacted material is moistened with a fine water spray until the next layer is spread or a curing membrane is applied to the top layer. For the latter purpose, a bituminous emulsion is sprayed at the rate of 0·8 litre/m² within 8 hours of the final rolling, and is lightly covered with sand or grit at the rate of 1 m³ per 220 m². The surface is closed to traffic for 3 days prior to being protectively coated with appropriate bituminous material.

26.3 PREMIXED CONCRETE CONSTRUCTION

Information on the design and construction of traditional concrete pavement slabs, for different soil conditions and traffic intensities, is contained in References 22-25, 32-43, 147-149, 165, 183. Related factors are dealt with in Chapters 4, 5, 7, 9, 12, 15, 16, 20, 23 and 31. Heavy-duty road slabs, designed to carry 1500 to over 3000 vehicles per day 20 years after construction, are laid on a compacted granular or cementitious base course, 75-150 mm thick.

These slabs, ranging from 175 to 250 or 275 mm thick prior to surfacing (if any), are reinforced 60 mm from the top with oblong-mesh fabric, its minimum weight being either 3·5-5 kg/m² between transverse joints or 5·7 kg/m² for continuously reinforced pavements. Sealed control joints, 25 mm wide, are spaced up to 72 m apart or 50-55 m in cold weather, other movement joints being spaced some 18-24 m apart (see Sections 26.1 and 31.2.1).

Transverse joints are fitted with sliding dowels, 20-30 mm diameter by 500-700 mm long at 300 mm centres. Longitudinal joints up to 4·5 m apart are tied with either special fabric reinforcement or steel rods, 12 mm diameter by 1000 mm long at 600 mm centres. The concrete, which is often air-entrained, should have a minimum (cylinder) compressive strength of 30 MPa at 28 days (see Sections 4.3, 12.1, 15.2.3(c), 15.2.5 and 20.1.2).[37]

For nonslip riding qualities, a striated finish can be formed by a burlap drag or belt, or a broom of wire, stiff-fibre or soft-bristle type. Improved traction on roads in wet weather is also obtainable by cutting shallow, longitudinal grooves, 3 mm wide and 20 mm apart, into the surface. For aircraft-landing runways, the grooves are preferably transverse, and fibre-reinforced concrete is very durable when properly used in the pavement construction (see Section 3.4.2).[180]

Mechanical plant includes such items as special concrete pavers (see Section 23.1.2(c)), extrusion machines for kerbs and gutters,[121] under-the-road drills for jacked service pipes, and concrete saws and hollow drills. The design and application of concrete overlays, for reconditioning purposes, are described in References 33, 116-118, 158, and in Reference 18 of Part 3. Standard specifications on concrete kerbs and channels are AS A175 and BS 340 and, for pavement slabs, the permissible surface tolerance is 3 mm in 3 m.

Reconstructed aggregate for a base course can be made by crushing demolition concrete and asphaltic materials, while sub-pavement objects, substances and conditions can be made manifest by means of portable radar (Plates 8 and 20; see Reference 116 of Part 3). Reinforced concrete, road and bridge work specifications are included in Section 3.4.1(a), while shrinkage and cracking are studied in Sections 15.2.2(b), 22.3.2 and 31.2.[138, 187]

26.4 SLIP-FORM PROCESS

The slip-form paving process is an expeditious way of laying a concrete slab over a gravel road that has been shaped to profile and grade. Stakes and string lines are set to line and grade, so that shallow side tracks can be formed by a grader. After the compacted subgrade has been prepared, a crawler-mounted paver provided with electronic-sensor controls and side slip-forms then spreads, compacts, extrudes and finishes a concrete pavement at the rate of up to about 1·5 km per day (Plate 20).

Air-entrained concrete with a low slump is deposited from transit mixers or from the jib of a mobile batching plant. It is levelled to requisite loose depth by an adjustable strike-off screed (hydraulically controlled) that advances and retracts. Dowel rods at joints may be placed by hand. Two surface vibrators and tamping bars consolidate and puddle the mix, and push particles of coarse aggregate slightly below the surface. An adjustable extrusion plate (with a bullnosed front edge and which is 1000 mm deep) further compacts and trowels the surface under the 15-tonne weight of the machine. Tandem work enables reinforcing mats to be laid between lifts.

A 600 mm rubber belt with a lateral movement of 100-200 mm then dresses the surface of the pavement. Three slip-forms, each 5 m long, are added to the 7 m length of the machine and held in position by cross-braces. They support the concrete for a further 8-12 minutes during straight-edging and hand-floating operations. Finishing is done with a burlap drag or rubber flap and a curing compound is sprayed over the slab.[126, 148] Experimental concrete rail tracks have been similarly laid in Britain.

26.5 SOIL-CEMENT

Soil-cement is a highly compacted and hydrated mixture of soil and portland cement. It is an effective way of converting a wide range of soils into a durable, low-cost pavement base with a subsequently applied bituminous surface. For equal pavement deflection, the load-carrying capacity of soil-cement 150 mm thick is the same as that of 200-225 mm of compacted, first-class granular material. With well-equipped crews and modern machines, 1·5 km or more of roadway per day can be constructed in soil-cement. The process is economical where rapid progress is desirable and good granular materials are unobtainable locally at reasonable cost.

Sulphate-resisting portland cement should be used with soil containing sulphates and, in all soil-cement work, the mixture should be compacted to maximum density at optimum moisture content. The proportion of portland cement needed is established by standard laboratory tests and normally lies within a range of from 5 to 15 per cent by weight.[26-30, 127] Semiflexible pavements should have a minimum compressive strength (unconfined) of 1·75 MPa or a CBR value (wet) of 200-300 per cent at 7 days (see AS 1289 and ASTM D1883).[37]

The soil of a roadway is first scarified and pulverised (if necessary), then shaped, wetted and rolled to profile and grade. The cement is spread in requisite amount and the component materials are thoroughly mixed and spread *in situ* by a single-pass stabiliser machine. The moist soil-cement mixture is compacted immediately by rollers of tamping, vibratory, segmented or multiwheel type. It is shaped with a motor grader and rolled smooth.

The surface is then moistened and membrane cured with stabilised bitumen emulsion, which is applied at the rate of 1 litre per 0·75-1·5 m². Sand is applied to prevent pick-up by passing traffic. A bituminous wearing-course is applied as soon as practicable, the type and thickness being governed by expected traffic, availability of materials and local practices. Shrinkage cracks develop at an early age. They do not affect the performance of the base provided that it meets adequate standards of strength, uniformity and durability. For this purpose, simple field tests are used as a check on basic controls (see AS 1289, BS 1377, 1924 and ASTM D422, D558, D559, D698, D806, D915, D1195, D1241, D1556, D1557, D1632-35, D2167). Lime-soil stabilisation (with or without fly ash) is described in Section 18.3, and soil-cement slope protection for earthdams (e.g. using 2 m × 150 mm overlapping horizontal layers) is dealt with in Reference 31.

CHAPTER 27

PRECAST CONCRETE

27.1 CONCRETE PRODUCTS IN GENERAL

27.1.1 GENERAL INFORMATION

Precasting techniques are being used to an increasing extent in the repetitive production of modular, prefabricated and standardised building elements. These products are suitable for the quick assembly of structural framework, floors and roofs, load-carrying walls and cladding, and curvilinear components in dense, lightweight and prestressed varieties of concrete.[120]

Factors contributing to the growth of these all-in developments are

Modular standardisation in systems building (AS MP14 and BS 2900; see Module in the Glossary, Appendix VI).

High-strength concrete can be moulded accurately with an acceptable surface finish, and the resultant products can be tied together structurally by post-tensioning procedures.

Wide-span portal frames and floors can be constructed in precast prestressed concrete at reasonable cost (see Chapter 34).

Casting schedules can be carried out independently of climatic conditions with a high degree of certainty, efficiency, tolerance and skill.

Preliminary and full-scale load tests can be used to determine actual values of strength, dimensional stability, impermeability and deflection.

Fabrication can be done with a minimum number of closely integrated operations and with a high degree of completeness at each stage.

Savings accrue through improved operational programming with package-deal building, mechanised erection and elimination of formwork, internal plastering and external scaffolding (see Appendix IV, 3, 4).

473

The erection time of dry building construction can be brought down, to economic advantage, to about one-half of that required normally for conventional *in situ* concrete construction (Plates 21 and 22).

27.1.2 INDUSTRIALISATION

Stationary and conveyor methods of production are used for making precast concrete elements. In the stationary method, all the processes concerned are carried out at the place where the casting is done. These include initial low-pressure steam or hydrothermal curing and demoulding operations, before the products are removed for after-curing and any subsequent treatment that may be required. In the conveyor method, the processes of production have their respective assembly stations and the moulds and products are moved after each operation. Both of these methods require ample mould capacity, labour and space.[113-115, 152, 172, 176-178]

Increased productivity and economy can be achieved by utilising the thixotropic properties of concrete near the time of initial set of the cement. In a state of partial stiffness, the concrete can be activated or made workable by mechanical agitation or vibration (see Section 23.2 and Glossary, Appendix VI). This characteristic and the period in which it is most pronounced depends, among other things, on the mix proportions, type of cement and the use of plasticisers in the mix. It may be utilised in continuous casting processes where the concrete is extruded as a hollow or solid section, or where it is rolled into inverted ribbed panels, the elements thus formed being moved forward by a conveyor system and cured on the way (see Section 34.2.8).

The casting area of a site-production scheme, with a layout to suit the stationary use of moulds, should have a solid, well-drained, working surface, such as would be provided by a concrete slab or either level bearers or balks. The moulds are laid out with suitable gangways and end working spaces, so as to minimise the transport of concrete from a nearby elevated mixer (see Section 7.1). The concrete may be handled by a traversing endless belt, an electrically operated skip on a monorail, or a bottom-opening skip and transporter crane. Accessibility is provided for lifting and transporting operations, such as by fork-lift trucks, conveyor systems or a crane of suitable load capacity and working radius.[160]

A moist-curing area, which should hold at least 15 days' production, is arranged in bays with identifying casting dates. The products are stacked from the production side of the bays and removed from the despatch side, so that both operations may proceed simultaneously if necessary. Facilities for safe-handling purposes with a minimum of bending or shock include lifting inserts, supporting frames, and special lifting and transporting appliances (e.g. a mobile slewing gantry or tower crane, fork-lift or crane-mounted trucks, bogies or trailers).

Areas for steel reinforcement should provide for the storage and fabrication of stock, and the storage of at least 2 days' output of made-up cages. The equipment on large projects includes an electric bar-bender, a stirrup bender, reinforcing shears, a spot welder and possibly a butt welder, a hydraulic jack, and the means of handling high-tensile steel tendons in prestressing work.

27.1.3 DESIGN OF MOULDS

Moulds to be used for the manufacture of vibrated concrete should be sufficiently rigid or braced to prevent flutter and ensure dimensional accuracy in the finished product. For external vibration, the vibrators should be mounted on stiffening ribs that will ensure an efficient transmission of activity to the concrete. The resulting amplitude and compaction are improved if the moulds are supported on rubber mats or mountings of suitable hardness. For table vibration, the moulds should be stiffened mainly at their base for requisite strength or rigidity.

Various materials (including timber, metal, concrete and plastics) are used for mould-making purposes. Timber moulds are suitable for concrete that is to be compacted manually, and they are reasonably satisfactory at frequencies of 50-70 Hz. Because of their high damping capacity, however, the moulds tend to absorb vibrations of higher frequency. For satisfactory service, the timber must be seasoned and sufficiently thick or braced to have adequate stability and strength. Nominal thicknesses are 50 mm for pallets, 38 mm for sides and 12 mm plywood for panels that are sealed at the edges and backed with steel-angle frames.

About twice as many pallets as sets of sides and ends are needed, to allow early striking and reuse of the latter. Imported softwood rather than hardwood should be used in making these moulds, which should be given two or three coats of a suitable sealant before being put into service. Typical sealants are shellac lacquer, emulsions of styrene butadiene, silicone or polystyrene acrylic resin, or (better still) a compound of polyurethane resin and hardener (see Section 20.2(e)).

High-grade finishes for special work can be obtained with mould liners. These include a variety of materials such as moisture-resistant fibreboard, resin-faced and bitumen-laminated building paper, plastic sheeting and patterned rubber, the last-mentioned being used without a release agent (see Section 8.3). Plastic sheeting may consist of polyester, polyethylene, polyvinyl chloride, styrene butadiene or polystyrene resin. Tucks or sealed joins in a resinous sheathing prevent leakage at the corners of moulds or forms and the formation of cavitational areas in the concrete. Asbestos-cement sheeting, wood-wool (BS 3809), foamed plastic and cardboard tubes may be used as permanent concrete liners or core-formers.[110]

Metal moulds are made of mild steel or aluminium, the former being the more common. Steel moulds are ideal for high-frequency vibration, vibropressure compaction, high-temperature steam curing and fifty or more uses per item of product.[151] Sizable moulds can be fabricated with quick-acting cramps or a toggle-action release and pivotal sides, so as to simplify dismantling before extracting each product. Steel sheeting used in the panels of these moulds should not be less than 3 mm thick.

Aluminium moulds should be treated initially in a manner to prevent them sticking to the cast concrete. One or other of the following treatments may be used: boiling in tallow or heating in oil; anodising and applying two coats of chlorinated rubber or epoxy resin lacquer; or priming with a compound of zinc chromate, vinyl butyral resin and phosphoric acid hardener, then applying two coats of polyurethane resin and hardener (see Table 22.1).

Special moulds for repetitive use may be made either of concrete with a smooth-rubbed and shellac-coated or resin-coated surface, or plastic-bonded fibreglass (see Section 21.1.6).[44-46] Concrete moulds are suitable for internal vibration only. For ornamental work, special moulds may be made of plaster, gelatine or carved polystyrene foam.[110] A high-gloss effect on high-alumina cement products can be obtained primarily by casting a slurry or mortar containing this cement against sheet glass.

With few exceptions, a release agent (see Section 8.3) should be applied to the moulds before casting concrete so as to facilitate their separation. The withdrawal of large castings from moulds can be assisted by injecting compressed air or water between their horizontal surfaces. For instance, water pipes may be installed with small steel plates covering their orifices, the ordinary pressure of water supply (about 325 kPa) being sufficient to loosen a casting and float it free from the base of its mould. Hydraulic jacks and hammer vibration may be used to facilitate the initial release of large products before extracting them by mechanical means.

Where large numbers of identical reinforced concrete slabs are required, such as wall panels of room size, they can be precast economically in vertical battery moulds, either at a building site or at a central casting depot near several building projects. In this procedure, two master slabs for each battery are cast face downwards on a smooth plane surface. They are specially reinforced, matured to an age of at least 3 weeks, braced with walings and studs, and surface hardened before being used successively in battery moulds.

These master units are placed upright with their smooth faces separated by end spacers, so that a reinforced concrete slab can be cast, consolidated and cured in the intervening space. The resulting three slabs are moved apart for further casting purposes, temporary battens being used as end covers, and this process is repeated until, say, fifteen slabs have been similarly cast. The production and erection cycles, which are usually organised with a tower crane and several batteries in operation, are coordinated weekly to meet the one-storey requirements of a building programme.[146, 157] Tilt-up construction, prefabricated with miscellaneous surface finishes, is comprehensively described in References 62 and 150.

27.1.4 CASTING PROGRAMME

For a site-production scheme, the casting programme is linked with the number of precast elements required on a project during a specified period, the quantities of each type being scheduled on a time basis. From working drawings the requisite units are grouped according to common dimensions or shape, so that mould adaptation can be used to best advantage. For example, where precast beams of the same section are required in varying lengths, a mould is made to suit the longest one with a provision for stopping-off, so that those of shorter length can be made in the same mould.[1]

The total quantities of each type of element having been scheduled on a time basis, their day-by-day or week-by-week requirements are adjusted so as to level out the irregularities of anticipated demand. The delivery dates are brought forward 3 weeks or more so as to allow for curing, drying and contingent delays. The finally adjusted schedule is adopted as the casting programme of each component, for which the production cycle is determined on a daily basis, an alternate-day basis, or a time interval to suit the expected demand.

The operational objective is to produce about the same quantity of precast concrete per cycle of manufacture and make the task in each one similar. For this purpose, an estimate is made of the following items.

The requisite number of casting moulds and their component parts with different frequencies of use. A composite factory-on-wheels system for house-casting in 48 hours is cited in Reference 179.

The size of areas needed for production and storage purposes. The quantities of concrete and steel reinforcement required per cycle, which would govern the type and size of plant required; the progressive quantities of cement and aggregate; the labour requirements for the undertaking; and the storage capacity for the materials, including those covered and elevated for cement, fine aggregate and steel reinforcement.

The date by which the production area must be equipped in order to start operations.[181]

27.1.5 CURING

A watertight sheathing or protective cover over the newly cast concrete serves to promote early curing, which should be started as soon as practicable after initial set of the cement. Hydrothermal or steam-curing facilities, after an initial setting period, accelerate the early hardening of concrete, which may then be demoulded or handled the next morning in cold weather. Newly moulded concrete, separated by a membrane and stacked under impervious insulating covers, gains early strength adiabatically.

27.1.6 CONTROL OF QUALITY

On sizable projects, the operations are geared to produce on schedule the requisite elements for functional requirements, and mixing is usually done at the casting plant when specifications call for special grades of high-quality concrete. For a low coefficient of variation, a great deal of attention is usually paid to the control of mix quality, particularly with regard to regularity of grading and constant moisture content, and to the proper means of weigh-batching, mixing, placing, compacting, curing and testing of the concrete (see Chapters 3, 4, 6, 7, 9, 23 and Appendix II).

The testing laboratory for quality-control purposes should contain the following minimum accessories: sets of sieves and specimen moulds; air meters, moisture meters and balances; a portable concrete-testing machine (preferably with an electric pump attachment); sampling, measuring, capping and curing facilities; colorimetric-test and workability-test equipment; a demountable-type extensometer; micrometer or dial gauges and drying apparatus; a desiccator and relevant appliances as warranted by circumstances (see Chapter 14). Regular grading, moisture and workability tests enable mix adjustments of a minor nature to be made through the weigh-batching equipment.

Proper supervision and inspection should ensure that, among other things (see Appendix II), the following requirements are fulfilled.

Regular sampling and testing programmes are carried out.

Steel reinforcement, inserts, openings and clearances in the elements to be used are completely as prescribed.

Concrete of the proper mix is placed and compacted in accordance with specifications and acceptable standards of surface finish.

Units are not moved or handled until proper concrete strength has been attained.

The elements are dimensionally accurate within specified tolerances, which may be limited to plus or minus 2 mm in cross-section, a 5 mm deviation in alignment in 6·5 m and plus or minus 5 mm in length.

Erection and handling methods are technically correct.

Erected units are adequately shored, propped and braced for purposes of true alignment and safety.

Cast-in reference marks are used to prevent an accumulation of tolerances.

Exposed concrete finishes, bonding details, watertight fittings and jointing connections are carried out in accordance with plans and specifications (see Sections 8.1.1, 8.1.2 and 27.1.7).[110, 111]

27.1.7 ECONOMY AND DESIGN

It is a current theory that for quick service and economy, superstructures should not be built, but should be assembled from precast structural elements. These should be of the largest possible size in relation to functional purposes *in situ* and clearance required during transport, thereby enabling handling operations to be minimised by the use of mobile, slewing, tower or climbing cranes at the assembly sites (see Appendix IV, 4 and Table 27.1). For special purposes, a helicopter may be used to advantage.

In the precasting industry generally, the trend is toward high-quality, standardised products that are uniform in size and physical characteristics and as light as practicable in weight. The structural elements for dry building construction are made attractive in appearance and are able to take thin finishing coats of properly applied wallpaper or paint. Related trends are towards packaged compounds, extensive off-site preparatory work, and improved productivity through mechanisation, incentives and research, planned continuity of output and the engagement of trained operatives of requisite calibre and status (see Section 36.8).

At the design stage, attention should be given to crane reach and loading, and the possible economic use of two medium-sized cranes instead of a large one. While plant investment may not vary radically between precast and cast *in situ* building systems, a reduction in completion time can cause savings in plant expenditure and overhead site costs generally. The man-hours required for hoisting and placing various precast elements may be only one-fifth to one-quarter of those required to produce the equivalent elements by *in situ* means.

TABLE 27.1   CRANES AND SCAFFOLDING

Title	Standard Specification	
	Australian	British
Scaffolding code	1576	CP97, 3 Pts
Scaffolding equipment	1450; 1575	CP97; 1139
Scaffolding machines	B231	—
Scaffolding planks	1577; 1578	2482
Crane and hoist code	1418	—
Overhead travelling electric	1418	466
Power-driven derrick, jib, mobile and tower cranes	1418	CP3010; 327; 357; 2452; 1757; 2799
Power-driven mast hoist	1418	3125
Stresses in cranes	1418	2573

When estimating the site-production costs of precast concrete projects, consideration should be given to the following items.

An adequate volume of repetitive production, and the long-term or short-term capital provision for equipping and writing-off a suitably located casting yard.

The casting layout and schedule, which should show the location and erection sequence of all precast units required on a project. A substantial reduction in unit cost, over that of formwork for similar work, through the repetitive use of casting moulds, and by a possible elimination of internal plastering and external scaffolding.

Additional reinforcement that may be required for the satisfactory handling of large units.

Lifting and transporting equipment for the conveyance of precast products to the curing and drying area and to the site of operations.

Site-assembly equipment, safety measures, fixing connections and weathertight jointing (see Section 28.12).

Approximately equal man-hours of labour, on a reduced scale, for precasting and site work.

Requisite sampling and testing operations as an integral part of qualitative production.

Partition panels must be kept dry, installed efficiently and skim-finished effectively if they are to have an economic advantage over plastered block partitions. The overall effect in a large superstructure is a saving by precast work over *in situ* work of up to about 50 per cent in completion time and 10 per cent in total cost, apart from the capital benefit to be derived from an earlier use of the structure.

Design loads on buildings are specified in technical statutes (e.g. AS 1170 Parts 1 and 2, and BS 449 and CP3(v)). Design stresses in concrete are featured in Sections 3.4 and 34.1.4; see also AS 1480 and 1481; BS CP110, CP111, CP114-CP116 and CP2007; and refer to ACI Committees 317, 318, 324, 326, 333, 340, 435, 438, 512, 533 and 543 listed in Appendix IV, 2(a)(i).[161, 178]

Typical connections with precast concrete are dealt with in References 47-58, handling and erection guidelines being given in Reference 176. Decorative panels, with requisite details for fixing and jointing, are described in Sections 28.8-28.12 and References 57-68, 112, 128, 129, 135 and 163.

Standard specifications on precast units are listed in Table 14.2 and Appendix I. Cast stone is documented in AS A22, BS 1217 and 4357, and translucent concrete is introduced in Section 35.3 and the Glossary, Appendix VI. Operational economics are considered in Section 36.9 and Appendix IV, 3; other topics being dealt with in References 112, 140, 156, 167 and 171.

27.2 CONCRETE MASONRY

27.2.1 GENERAL INFORMATION

Concrete building blocks are classified according to their physical properties and usage as follows.

Grade 4 for lightly loaded internal walls.
Grade 12 for general structural and nonstructural uses.
Grades 15 and 20 for special highly stressed applications.

Standard specifications on masonry in general are contained in Table 27.2.

Blocks are made in sizes based nominally on a 100 mm module, including an allowance for a 10 mm joint. Typical sizes, in addition to those of half blocks and closers, are as follows.

Length: 400 mm.
Width: 100, 150, 200 and 300 mm.
Height: 100, 150 and 200 mm.

Hollow blocks, complying with these widths ($\pm$ 2 mm), are made with minimum shell thicknesses of 25, 25, 30 and 35 mm and minimum web thicknesses of 25, 25, 25 and 30 mm, respectively, unless requested otherwise by statutory authorities. Concrete bricks, by comparison, are nominally 100 mm square by 200, 300 or 400 mm long.

When sampling specimens for individual tests, they are selected at random from an identifiable lot and suitably marked for reference. The quantity of blocks taken is five for each of compressive strength and moisture content, and three for each of maximum potential drying shrinkage and water permeability tests. Up to twice these quantities are required when testing concrete bricks. Samples awaiting testing are stored on pallets in a dry place, but those taken for moisture content tests are weighed immediately or sealed separately in preweighed airtight containers or bags.

Standards in different countries (see Table 27.2) vary in the detailed procedures required for determining the physical properties of building blocks and bricks.

Compressive-strength specimens are initially surface rectified on bearing areas with 25 MPa gypsum plaster paste. They are then sulphur capped or bedded with fresh strips of 6 mm thick hardboard for reference and routine testing purposes.

The minimum characteristic compressive strength of blocks, when tested air-dry on net areas, is indicated in Table 27.3. In this

TABLE 27.2 STANDARDS ON MASONRY

Title	Standard Specification		
	Australian	British	ASTM
Aggregates for concrete masonry	1465-1467†	882; 1198-1200*	C144; C404
Blockwork Code Unreinforced blockwork	1475	CP111*; CP121†	—
Brickwork Code	1640	3921*; CP121†	—
Burnt clay and shale building bricks	1225; 1226; 1640*	3921; 4729; DD34	C62; C67; C126; C216
Calcium silicate (sand-lime) brick	1653; 1640	187	C73
Chemical-resistant masonry units	—	—	C279
Concrete building blocks	1500	2028*; 1364*	C90*; C129*; C140; C145
Concrete building bricks	1346	1180*	C55; C67
Hollow, loadbearing concrete masonry units	1500*	2028*; 1364*	C90
Hollow, nonloadbearing concrete masonry units	1500*	2028*; 1364*	C129
Hydrated lime for masonry purposes	1672	890	C207
Lightweight concrete aggregates	1467	3681; 3797	C331
Masonry cement	1316	—	C91
Mortar and grout for masonry construction (including fine aggregate and bond strength)	A123; 1475*; 1640*	2028*; CP121†; PD6472	C87; C270; C398; C476; E149
Precast concrete blocks	1346*	2028; 1364	—
Refractory and insulating firebrick (see Table 29.5)	—	—	—
Structural recommendations for loadbearing walls	1475*; 1640*	CP111	—
Test for drying shrinkage of concrete blocks and bricks	1500*; 1346*	2028*; 1364*; 1180*	C341; C426
Walling: Brick and block masonry	1475*; 1640*	CP121, Pt 1†	—

* Related information is included in this specification, which has a title different from that shown here (see Table 3.8).
† Inclusive of acoustic, thermal and fire resistance values.

TABLE 27.3 COMPRESSIVE STRENGTH

Gradation	Minimum Characteristic Strength (MPa)
4	4
12	12
15	15
20	20

classification, the characteristic compressive strength is the difference between the average of the individual compressive strengths and 0·43 times the five-specimen group range. The latter is the difference between the highest and lowest strengths in the group. When testing structural concrete bricks, the minimum characteristic transverse strength required is 1·4 MPa (AS 1346).

The moisture content, total absorption and density of specimens can be calculated from direct and water-displacement weighings of specimens, when they are cycled through initial moisture, 48-hour oven-dry and 24-hour saturation conditions. The average moisture content of concrete blocks and bricks, as delivered, should not exceed 80 kg/m³. A relative-humidity method of determination is described in Section 14.6 and ASTM C427.

Tests for determining maximum potential drying shrinkage are described in AS 1346, 1500, BS 2028 and ASTM C426. Its value in the first two standards is estimated from the drying shrinkage of three prisms. These are either full-brick specimens or pieces which have been cut from three untested sample blocks, so that each test piece obtained therefrom measures 1250 mm² in cross-sectional area and 200 mm in length. The prisms are subsequently dried to practically constant length from a 4-day water immersion state to an oven-dry condition at $50 \pm 1°C$ and approximately 17 per cent relative humidity (see Section 33.3.2(b)). For satisfactory performance, the average value obtained at $23 \pm 2°C$ should not exceed the following limits.

0·06 per cent for units to be used in external walling.
0·08 per cent for units to be used in internal walling.

A water-permeability test on hollow concrete blocks, which have been presoaked for 2 hours, consists of applying a head of water to one face of a specimen by using a gasket-sealed metal cover, that is clamped on and fitted with a graduated cylinder of 60 mm bore (JIS A5406 and AS 1500). Having established an initial test head in 1 minute, the residual head is read at intervals up to 2 hours or until it has fallen 100 mm. An initial head of 200 mm in reference tests may be extended to 400 mm for accelerated control tests with graphically illustrated results. When dealing with concrete bricks, this test is substituted by one for the initial rate of absorption of the bed face of oven-dry specimens in 1 minute. The indicated suction is a contributory factor in the design of an effective bonding mortar (AS 1640, Commentary).

The mix proportions of masonry units are related to their method of manufacture. For instance, those made in manually operated moulds may consist by volume of 1 part of cement and 5 parts of coarse-graded sand, or 1 part of cement, 2·5-3 parts of well-graded sand and 3-3·5 parts of 10-5 mm clean stone toppings (see Section 3.3.6(b)). The proportions for units made by power-operated machines consist of 1 part of cement and 10-14 parts of aggregate; a pan-type or paddle-type mixer being used for the preparation of relatively dry mortar or fine-concrete mixes.[144] For coloured blocks, 3-5 per cent of oxide pigment, by weight of cement, is adequate.

In order to obtain a fine surface texture, the ratio of fine aggregate to coarse aggregate is made greater than is customary in ordinary concrete mixes and the maximum size of aggregate is limited to 10 mm. An optimum proportion of fines passing the 0·15-mm sieve is needed in lean, no-slump, masonry concrete in order to improve its impermeability when made into masonry units (see Appendix III). Table 27.4 illustrates a grading of sand with a fineness modulus of 3·65 which would be suitable for moulding purposes in a power-driven, vibro-pressure machine.

TABLE 27.4   **FINE AGGREGATE**

Sieve (mm)	Cumulative Percentage Retained	Remarks
4·75	20	15 per cent retained on each of
2·36	40	1·18 mm, 0·60 mm and 0·30-mm
1·18	55	sieves.
0·60	70	
0·30	85	Fineness modulus equals $\frac{365}{100}$ or 3·65.
0·15	95	

For requisite structural strength and stability, it is important that the units should be cured properly (see Sections 9.4.3-6 and 9.6), dried to the region of an equilibrium moisture content (see Sections 33.2.2(d) and 33.2.3(d)(iv); also 33.3.2(b) and 33.3.3(b)), and employed in accordance with appropriate design and usage procedures, such as are indicated in the relevant specifications of Tables 3.25 and 27.2. Bond beams, joint reinforcement and control joints are used for the distribution or relief of various stresses and the control of cracking in concrete masonry.

Current trends in the production and uses of concrete masonry units incorporate a variety of industrial techniques. They include such items as high-speed automation wherever practicable; product conditioning to control shrinkage and efflorescence characteristics (see Sections 9.4.6, 9.6, 31.4 and 36.6); surface glazing and polymer impregnating procedures (see Section 21.4);[162, 174] mortarless lightweight and reinforced masonry fabrications (see Sections 3.4, 21.1.6, 33.1 and 33.4.3); [73-79] preassembled modular panel systems;[89] and structural applications of stressed floor and roof elements (see Section 34.1).[86, 123, 124] Further technical information is contained in statutory documents (see Table 27.2) and References 69-79, 185 and 186; while expansion-gap provisions for clay brickwork are incorporated in Section 16.2 and Reference 25 of Part 4.[186]

27.2.2 MASONRY BUILDING CONSTRUCTION

(a) Dry blocks.

Concrete blocks, adequately dried and delivered, should be protected from the weather until they are built into a wall. They should be stacked in narrow stock piles and spaced apart with the cores of hollow blocks vertical. The bottom courses should be raised on stillages or planks free from contact with the ground, and the stacked units should be covered with building paper, tarpaulins or other suitable means that will enable air to circulate freely through the stock piles (see Sections 33.2.2(d) and 33.2.3(d) (iv); also 33.3.2(b) and 33.3.3(b)).

For dimensional stability, the moisture content of masonry units at the time of use should approximate the equilibrium value that may be expected after they are built into a structure. Blocks are not normally wetted before or when laying them and, during cessation of work, the top of a wall should be covered to prevent the entry of rainwater or other moisture. Blocks should be tested for dryness before use (see Sections 14.6 and 27.2.1) and, if their moisture content is found to be unduly high, it may be necessary to introduce effective drying facilities or to provide for subsequent and satisfactory attainment of dimensional stability at the site.[86]

(b) Mortar and grout.

The bond strength of mortar to masonry units (ASTM E149) is affected by a number of factors. These include the surface texture of the bedding faces and rate of absorption of the units; the type and quantity of the cementing matrix; the workability and water-retentivity of the mortar; and the quality of workmanship in operative service. Mortars for walls exposed to ordinary ambient conditions are dealt with in Sections 16.2 and 33.4.3, also in the standard specifications represented in Tables 16.1 and 27.2. Where masonry walls are likely to be subjected to severe stresses, sea air or frost action, the strength and durability of the mortar should be made adequate for the structural service required.

A typical mix for use in both substructures and superstructures is 1 part of portland cement, 1 part of hydrated lime and 6 parts of sand, or 1 part of masonry cement and 3·5-4·5 parts of clean, graded, coarse to medium gauge sand, the proportions of which are based on dry nominal volumes and gauged as indicated in Sections 3.3.6(b) and 18.1. For increased plastic deformation or light loading conditions in a superstructure, or for use with lightweight concrete blocks, an alternative mix for use above the damp-proof course consists of 1 part of portland cement, 2 parts of hydrated lime and 9 parts of sand or 1 part of masonry cement and 4-5 parts of clean graded sand by volume.[96] Under cold winter conditions, however, the foregoing stronger mortar should be used.

The mortar, when tested for water retention (see Table 16.1), should have a flow after suction of not less than 70 per cent; but where masonry units are expected to have a high rate of moisture absorption, the flow value should be in the region of 85-90 per cent to avoid undue stiffening of the mortar. For reinforced, highly stressed applications, the mortar used may range from 1 : 0-0·25 : 3 to 1 : 0·5 : 4·5 parts by volume of portland cement, hydrated lime and fine aggregate. The foregoing mortars, when site mixed and moulded into several 50 mm by 25 mm square prisms, have typical 28-day compressive strengths of 3·0, 1·0, 11·0 and 5·5 MPa, respectively. Their laboratory-mixed counterparts are about 50 per cent stronger.

Grout used in reinforced block and brick masonry, for filling cavities over 50 mm wide, is typically 1 part by volume of portland cement, 0-0·1 part of hydrated lime, 2·5-3 parts of mortar sand and 1·5-2 parts of 10 mm aggregate. The workability of the grout (as measured by a 100 mm to 50 mm diameter by 150 mm high-slump cone) is in the region of 100-130 mm and the 28-day characteristic strength, when determined with 150 mm by 75 mm square prisms, is in the region of 15 MPa. For moulding field-test prisms, masonry units are placed on a nonabsorbent base and lined with 0·2 mm permeable paper or paper towel, prior to pouring and puddling triplicate specimens in two layers.[78]

Building blocks and mortar should be placed progressively near walling work for operational convenience. The mortar should be kept plastic and cohesive by being remixed frequently on a mortar board without the addition of water. Mortar that is to be used for filling cavities in masonry should have a cement/sand ratio not exceeding 1 : 3 by volume, a water/cement ratio not exceeding 0·6 by weight, and no visible separation of the components on standing.

Grout infill in the cores of hollow-unit walls, erected up to 8 m height, should be rodded, vibrated or puddled and reconsolidated in 1·2 m lifts placed at 0·5–1-hour intervals. Horizontal flow of grout is controlled at about 8 m centres, by placing mortar on cross-webs and blocking bond beam units with masonry bats set in mortar. The grouting of any section between control barriers should be completed in 1 day with a continuous series of lifts.[73-79] Rendering is dealt with in Section 27.2.2(k).

(c) Substructure.

Continuous reinforced concrete footings or foundation beams are required to distribute, as evenly as practicable, the vertical and lateral loads from a building to its foundation. They should be designed to suit the estimated loads and bearing capacity of the foundation material, and its expected behaviour due to possible variations in composition, depth, slope, settlement, seasonal conditions and drainage. Their depth should be to a zone where the moisture content of a clay soil is relatively constant, or where fine sand and silt are unaffected by freezing.

For one-storey and two-storey construction, where the walls do not carry laterally the wind load from roof structures, strip footings (on average uniform foundation material) may be proportioned so that their width is twice the overall thickness of wall, but not less than 250 mm, and their depth is 1·33 times the wall thickness. Reinforced concrete footings of strip, raft and pier-and-beam type are illustrated in References 81-85, the first-mentioned being reinforced top and bottom with two or three 12 mm diameter steel rods that are spaced by 6 mm diameter ligatures at 600 mm centres. Buildings on moving ground are dealt with in Section 36.1.

Where a floor level is within 1200 mm of a continuous reinforced concrete footing, the hollow concrete blocks of substructure walls should be capped with a bearer course of solid masonry, in order to provide for anchorages and the distribution of loads from timber beams and bearers. For this purpose, a strip of metal lath or wire gauze wide enough to cover core spaces can be placed in a mortar joint under the top course, and the cores are then filled with mortar or concrete, which is floated to a smooth surface. Where the substructure walls are 1200 mm or more high, they should be capped with a bond beam or its structural equivalent.

Erection of
lightweight-
aggregate concrete
panel, showing
safety rail and crane
radio control

Erection sequence
with bolted
joints and dowel
connectors which
pass through
floor slabs

Precast Concrete
PLATE 22

Erection of glazed
decorative wall unit

Erection of three-storey
wall unit

Chemistry Building,
Australian National University, Canberra

Ventilation is required under timber floors and in the cavities of hollow blocks and double-leafed walls. To this end, openings are spaced evenly around the exterior of buildings at a level immediately below that of the floor system and above any obstruction, such as damp-proof courses and bond beams across the cavity of walls. Ventilators should have open spaces equivalent to not less than 3000 mm^2 per linear metre of exterior wall. Vents should be provided at wall heads below bond beams or other cavity obstruction, and the openings in internal substructure walls should be at least four times as large as those provided externally.

(d) Damp-proof course.

The block course on which a damp-proof course is to be laid should be flushed up with mortar and formed into an even bed. The material used should comply with the requirements of either AS Int. 326 or 327, or BS 743 or CP102 as appropriate for bituminous damp-proof courses with a metal centre or fibre-felt base. Alternative materials for the purpose include black (0·25-0·5 mm thick) polyethylene or polypropylene film, ethylene propylene terpolymer (1·25-1·50 mm thick), siliconate rubber-latex cement mortar, and mastic asphalt (BS 1097 and 1418).

Where a damp-proof membrane crosses a hollow masonry wall 140 mm or more thick, the membrane may be pierced for purposes of internal drainage and ventilation with holes up to 40 mm diameter at the centre of each core in the blocks. Flashing material should be used above bond beams or lintels, below sills and parapet copings, through parapet walls just above roof level, adjacent to weepholes in cavity walls, and between concrete slab or timber-joist inserts and abutting walls.

(e) Cavity walls.

A cavity block wall, containing two leaves that share applied lateral loads, should meet the following requirements.

Each leaf, not under 100 mm nominal thickness, is separated by a cavity 50-75 mm wide.

The leaves are bonded with noncorrosive metal ties, 5 mm diameter, one tie being used for up to 0·4 m^2 of wall area or otherwise as designed specially to suit wider cavities.

The minimum nominal thickness of wall in single-storey construction is 250 mm, the inner leaf being thickened for structural efficiency in multistorey construction (AS 1475, 1640 and BS CP111, CP121 Part 1).[78]

Wall ties of appropriate shape and size (see AS Int. 324 and BS 1243, also References 78 and 190) are staggered in alternate courses, their insertions being at 400 mm intervals vertically and up to 1 m horizontally. Additional bonding ties are placed at intervals not exceeding 1 m around openings and within 300 mm of their boundaries. Lateral forces are counteracted by shear transfer facilities, including adequate wall bracing during construction, and wall ties are embedded at least 50 mm in the mortar joints of each leaf. In composite-coursing masonry, adjustable metal ties are installed between the leaves.

Weepholes, where required to drain away moisture, should be located in the first course above a damp-proof course, bond beam or other obstruction, the holes being approximately 50 mm high and located in each perpend. Mortar droppings in a cavity wall should be caught on a cavity-width batten, which is raised and cleaned progressively with laying, or they should be cleaned off wall ties and removed from temporary openings at the bottom of the wall. A similar procedure is required for composite structural columns and piers:

Rules governing special bonding patterns, such as the bonding of masonry walls with masonry header units, are established in statutory specifications (see Table 27.2). They include recommendations made, *inter alia,* by ACI Committee 531 on concrete masonry structures as set out in Reference 78. In masonry veneer construction, a 90 mm (actual thickness) leaf is attached across a 25-50 mm cavity to a timber load-bearing frame, using metal ties and 25 mm × 4 mm galvanised clouts at intervals not exceeding 400 mm vertically and 1·2 m horizontally.

(f) Bond beams.

Bond beams are of structural and nominal types of reinforced concrete element, and they are built integrally with masonry walls for the purpose of carrying flexural and tensile stresses in each respective instance. These beams should be of continuous construction around corners. Suitable provision should be made for the spacing of structural control or dowelled joints (see Figure 27.1) and for making the beams integral with the underlying structure.

This requirement should help to minimise any undesirable effect due to differential vertical movements (including those due to ambient conditions) in the two leaves of cavity walls, particularly when they exceed one storey in height. It can be effected by superimposed masonry not less than 2·4 m in height or by suitable anchorage to the masonry below. A procedure for the upper cavity walls of a two-storey building, incorporating a lightly loaded continuous external leaf, may consist of 12 mm diameter anchor rods at about 3 m intervals, the rods being secured into the perimeter of a concrete floor diaphragm and protected by a coating of antirust compound or concrete (see Section 22.3). In this instance, provision is made for taking up early vertical shrinkage of the masonry.

Structural bond beams and other diaphragms, including concrete floor and roof slabs, should be designed as structural elements transmitting lateral and vertical loading on the walls to other connecting structural elements. They are used at floor and eave levels, and at the heads of window and door openings. They may be omitted from any of these positions, however, when that position is within 600 mm of an equivalent structural element or within 1·2 m of a continuous reinforced concrete footing or foundation beam.

Nominal bond beams, which are usually U-shaped masonry units infilled with reinforced concrete, are a means of distributing minor loads and assisting in the control of cracking. They can be used instead of joint reinforcement immediately below wall openings, and as a substitute for structural bond beams at the top of one-storey or two-storey walls, where the effective length between lateral supports does not exceed 4 m and lateral forces from roof structures are not being carried by the beams. Their width should be not less than the width of the supporting wall or leaf thereof and their depth should be at least 100 mm (nominal).

The minimum reinforcement is one 12 mm diameter rod for 100 mm internal walls, or two similar rods near the base of U-shaped units in walls of greater width. At splices, the lapped length is 700 mm for plain and cold-twisted square bars, and 400 mm for deformed bars. Timber plates, embedded in mortar on concrete, are fastened where necessary at the top of walls by 10 mm diameter anchor bolts at 1·5 m centres. Design data on reinforced concrete masonry lintels, where required, are exemplified in Reference 90.

(g) Control joints.
Control joints, which serve primarily as contraction joints across walls, are installed, firstly, for the purpose of controlling the shrinkage movements of walls due to their initial drying to an equilibrium moisture content and, secondly, to control the effects of subsequent thermal, moisture and distortional movements of the

walls. They are incorporated at places where tensile stresses may be expected to concentrate and cause cracking, and they are designed for such purposes as weatherproofness and the accommodation of lateral forces on the walls. Articulated methods of masonry construction are well illustrated in Reference 186.

The maximum horizontal spacing of control joints in nonreinforced masonry (Table 27.5) is governed by the ambient conditions of usage, quality of units and mortar, and the joint reinforcing details. Suitable additional locations are at significant changes in wall section, sides of wall openings, adjacent inbuilt movement joints and at situations near wall intersections and angular configurations. The maximum spacing may be used where joint reinforcement is omitted and replaced by bond beams spaced at 1·2 m on centre.

TABLE 27.5   MAXIMUM SPACING OF CONTROL JOINTS IN
                          NONREINFORCED MASONRY

Maximum Panel	Vertical Spacing of Joint Reinforcement		
	None	600 mm	400 mm
Length/height	2	2·5	3
Length	9·6 m	12·0 m	15·6 m

The magnitude of anticipated long-term movements in control joints may be estimated as follows.

1. For concrete or calcium silicate brickwork: the maximum potential drying shrinkage of the units (e.g. 0·06 to 0·08 or 0·035 per cent, respectively), plus an allowance for thermal movement at the rate of 0·01 mm/m·°C. A clear weatherproofing treatment on external walls lessens their relative wetting and drying movements, which influence the effective spacing and design of control joints (see Section 27.2.2(k)).

2. For clay brickwork: the maximum potential long-term expansion of the units (e.g. 0·05-0·15 per cent), plus an allowance for thermal expansion at the rate of 0·008 mm/m·°C (see Section 16.2).

Vertical expansion joints generally should not close to a width of less than 10 mm when spaced at centres that limit their maximum movement to 15 mm. The corresponding limiting width of horizontal expansion joints is 3 mm, plus an allowance for the compressed thickness of a flexible joint filler. Additional allowances are to be made for the dimensional changes of skeletal frame structures (see Section 16.2, and AS 1475, 1640, BS CP111 and 121).[71-79, 87-89]

Continuous vertical joints are formed of full-length and half-length blocks that are separated by a closed-cell polyethylene sheet. Those exposed to the weather are primed on the sides but not on the back and sealed with a soft rubber, vinyl acrylic terpolymer or two-part polysulphide compound. The sealant is applied to a depth of 9-12 mm for widths over 12 mm and up to 12 mm, respectively. Typical details are illustrated in Figure 27.1 and References 87 and 88, also in proprietary literature.

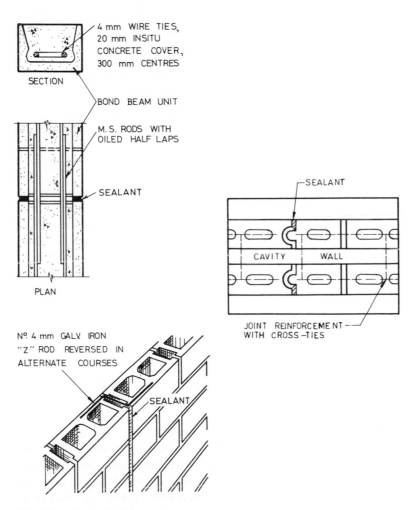

Fig. 27.1 Control joints in masonry

(h) Joint reinforcement.

The facing and backing leaves of masonry walls may be bonded with prefabricated joint reinforcement, consisting of at least two 4 mm diameter wires with one corrosion-resistant, 3 mm diameter cross-wire serving as a tie for each 0·2 m² of wall area. Joint reinforcement is spaced up to 600 mm on centre (see Table 27.5) and its minimum mortar cover is made 15 mm in weather-exposed walls. [86, 91]

In nonreinforced masonry, joint reinforcement is normally located in courses immediately above and below wall openings, at floor and roof levels, and near the tops of walls in order to control cracking. It should extend beyond openings for not less than 600 mm or to the ends of wall panels and, on delivery to the work site, it should be kept clean and flat (see Section 3.4).

Joint reinforcement is omitted in the vicinity of bond beams or similar structural restraints and terminated at control joints. It is lapped at least 300 mm or sufficiently at splices for the continuity of tensile strength. A cross-wire of welded-wire fabric is located at or within 75 mm of each end being lapped and special ties, splices or loops are formed at corners. Walls resting on flexural bases, which are subject to deflection under load, are strengthened against bridging effect by adequate low-level reinforcement.

In reinforced masonry walls, the minimum horizontal and vertical reinforcing areas are not less than 0·07 per cent of the corresponding wall sections; the total of the two reinforcing percentages is at least 0·2 per cent, and the product of their spacings in metres is not in excess of 3·0. Vertical reinforcement is provided at wall openings, intersections, ends and corners; while in solid-grouted walls, the spacing of horizontal reinforcement is kept within 1·2 m on centre.

(i) Stiffened walls.

Masonry walls may be laterally supported or stiffened by bond beams, floors, roofs, intersecting walls or buttresses provided with adequate bond or anchorage. The thickness of an intersecting wall or buttress should be not less than one-half that of the supported wall or 100 mm, the former element projecting not less than one-sixth the effective height of the supported wall or 600 mm, whichever is the greater in each instance.

Requisite bond or anchorage to intersecting walls or buttresses can be secured by one of the ensuing building techniques. Masonry bonding units are inserted, at every fourth course, into the supported wall to at least half its thickness or 100 mm, whichever is the lesser. Corrosion-resistant tie bars are installed not more than 1·2 m apart vertically. They are 6 mm thick by 32 mm wide and 700 mm long with 50 mm right-angle bends at each end, these being embedded in cores filled with mortar or grout supported by a metal lath.

Design rules for determining the slenderness ratio (see Glossary, Appendix VI) and stability of masonry walls and piers are set out for reference in standard building documents (e.g. AS 1475, 1640 and BS CP111).[153, 154] In buildings of not more than two storeys with an average floor live loading not exceeding 1·5 kPa or 1·5 kN/m², for instance, the effective dimension/thickness ratio of walls is not expected to exceed a range of 27-30. Where a section of wall is supported at both ends, an effective dimension for calculating this ratio is the length between centre lines of the supporting elements less four times its thickness.

Intersecting walls at corners must be bonded fully together. An intermediate wall that is nonloadbearing, however, may be tied to another wall with joint reinforcement or strips of metal lath at 400 mm centres. Wall columns or pilasters, with lateral ties in the mortar joints, may be formed by filling the cores of hollow blocks with reinforced concrete, the steelwork of which is fabricated with 6 mm diameter ligatures at every course. The reinforcement is held in place by temporary wire loops through the mortar joints and, either the first-course blocks are cut for cleanout purposes at the base, or the corefill is placed in lifts of about 1·0 m concurrently with the blockwork.

(j) Laying of blocks.
Hollow concrete blocks are laid with a shell bedding that is kept off the cross webs, while solid blocks are given a full bedding of mortar. Hollow blocks also are given a full bedding on concrete footings and at damp-proof courses. Before laying the first course, blocks should be aligned temporarily in position so as to check their selection for a project. A full bedding of mortar is then spread and furrowed with a trowel, so as to ensure plenty of mortar along the bottom edges of the blocks.

After laying the corner blocks accurately, the remainder of the first course of masonry is laid so that all units in it are aligned, levelled and plumbed correctly. The thicker shell-face of hollow blocks is placed uppermost, so as to provide a larger area on which to apply a mortar bedding; while the spreading of mortar on vertical shell-faces is facilitated by standing several of the units on end and preparing them in one operation. Each block is then placed over its final position and pushed downwards into the mortar bed and against the block that was laid previously. To ensure well-filled vertical joints, each of the adjacent faces of blocks that are to be butted together should be coated with mortar beforehand.

All corners at one lift are built accurately up to 900 mm higher than the central portion of a wall, and are stepped back in half-block lengths so as to present a symmetrical pattern to the work in progress. The blocks are checked and adjusted slightly as they are being laid, so as to ensure that they will be true to alignment, level and plumb. Right-angles are verified by their conformance to a three-four-five-sided triangle.

The height of each course is checked by a graduated gauge rod as the work rises, and a mason's level is used to make sure that the face of each block coincides with the plane of the wall. The horizontal spacing of blocks is checked progressively by holding against the wall a mason's level or straight-edge, for example, which is placed across the lower leading edges of the several stepped courses of blockwork.

Concrete masonry units which must be cut or chased should be shaped in a dry state with a masonry saw, guillotine or hammer and chisel, before being built into place with a low moisture content. Laying operations in general are expedited by the use of a tightline for each course of units and, when gripping a block, by tipping it backwards slightly so that the upper edge of the course below can be seen and used as a guide for placement. On rolling the block in hand into a vertical position, it is pushed and tapped into place (see Appendix IV, 4(a) for equipment).

The mortar should be spread ahead to form a cohesive bedding 10 mm thick, for only a small group of blocks, and these should be laid and adjusted while the mortar is soft and plastic. As each block is built into place, the extruded mortar is cut off with a trowel and is either reused on the vertical shell-faces of the block thus laid, or is returned to the mortar board and remixed with fresh mortar.

Care must be taken to keep wall surfaces clean and free from smears of mortar during construction. Any mortar droppings that stick to a wall should be allowed to dry before being removed with a trowel. The residual mortar, when hard, is removed by rubbing with a small piece of block and brushing clean. Various classes of masonry, such as concrete blocks or bricks, and clay or calcium silicate bricks, should not be used together indiscreetly in mixed types of construction, because of disruptive differences in their dimensional-change characteristics.

Where masonry walls are being erected on exposed sites, they should be built not higher than ten times their thickness, unless adequately braced or structurally supported. Foundation walls with back-filling should also be braced to resist horizontal pressure.

(k) Weatherproofing.

For weathertightness and clear uniform lines in masonry walls, the mortar joints (on becoming thumb-print hard) should be tooled either concave or vee-shape in profile, so as to compact the mortar tightly against the adjacent masonry. A jointing tool for horizontal joints is made from a 600 mm length of 15 mm diameter rod or 12 mm square bar, which is upturned at one end and provided with a handle. After tooling the horizontal joints, the vertical joints are tooled with a small double-bent jointer, any mortar burrs being trimmed off subsequently with a trowel or by rubbing with a burlap bag. Care should be taken to prevent the dislodgement of masonry which, at this stage, would break the mortar bond and allow water to penetrate the joints (see Section 27.2.2(b)).

Walls which have accommodated whatever drying shrinkage they had, or which have aged for at least 3 weeks, may be weather-proofed by surface treatment. Two coats of suitable paint (e.g. pigmented cement-based, chlorinated rubber or acrylic-based) are usually sufficient (see Sections 20.2(c), 28.2.7 and 28.3), although single-leaf masonry walls may require to be rendered (see Section 16.2).[92-97] For the latter purpose, the composition of rendering mortar is made similar to that of jointing mortar, which may be varied according to the season, degree of exposure, and density of the blocks. The backing and finishing coats of an external rendering (15 ± 3 mm thick) are usually of similar composition (see Sections 3.1.2(c), 16.2, 33.4.2 and 3). The mixes, which are improved by activation with high-speed mixing,[134] are given a wood-float finish to produce a textured surface that will breathe.

Large eaves and such fixtures as flashings, weepholes, splash shields, and drip edges on projections help considerably to deflect rainwater from the surface of walls and keep them dry in service. Surfaces which have become grimy should be cleaned thoroughly, by steam or pneumatic means, for example, before being surface-treated. Constructional details should be in accordance with the best building practice, such as recommended by modern manufacturers of concrete masonry units and illustrated in their technical catalogues. These include design manuals that are prepared for national or international usage with reinforced and nonreinforced concrete masonry systems (Plate 23). Precautions to be taken against possible surface efflorescence are delineated in Section 36.6.

498

CONCRETE TECHNOLOGY

27.3 LIFT-SLAB SYSTEM

The lift-slab system is an expeditious way of erecting repetitive flat-slab buildings and spatial structures of reinforced or prestressed concrete or fabricated steel. Lift slabs, after being cast in tiers on a concrete base, are raised on sliding collars and fastened to steel or precast concrete columns. The slabs are adaptable in structural design to incorporate such features as service conduits, asbestos-cement waffle boxes, lightweight aggregate, and either simple or propped cantilevers. The drainage of balconies is facilitated by a slight taper.

In multistorey buildings, column sections carrying lifting collars are erected on prepared footings and the concrete base is laid with a plane smooth finish. The base and subsequent lift slabs are membrane cured with a silicone-based compound, which serves as a separating agent. After the concrete base has hardened adequately, the lift slabs (with embedded collars) are cast consecutively and sprayed with the curing compound. Each slab serves as a soffit form for the next one, preparational work on which is started the following day. Before the top slab is lifted it should be membrane cured for 14 days or until its cylinder compressive strength reaches 17 MPa.

On each column section there is mounted a reciprocal, electrically controlled hydraulic jack, from which two threaded rods of high-tensile steel are connected symmetrically, through threaded bushes, to a collar in the slab to be lifted. More than one slab can be lifted simultaneously, and column splices are located so as to ensure adequate resistance to buckling during the lifting operations. Should a jack capacity of 700 kN be inadequate on a column (which is very seldom), two jacks may be used on it.

The jacks are connected to a mobile hydraulic generator which, with a crane for extending the columns, is carried on the top slab. The jacks lift the slabs in 10 mm strokes, which are repeated at the rate of up to six per minute. By means of a hand-held electric console, the levels of the slabs can be controlled to an accuracy of 1 mm. The slabs for the upper floors of a high-rise building are parked near the top of the partly erected columns, while lower slabs are being fixed thereto. The jacks are removed temporarily to enable columns to be extended and spliced (Plate 23).

Slab-to-column connections are made with shear pins or load wedges and welded shear plates.[98-104] Concrete columns are spliced with steel plates, epoxy resin and fillet welds (see Section 21.1.6(a)), the plates being recessed for fire resistance. Horizontal loads during lifting operations are resisted by the stiffening effect of completed connections, concrete service towers, shear walls or temporary bracing. After the first few slabs have been fastened, the initial storeys are completed for early occupation and, after the roof is in place, the jacks and hydraulic pumping equipment are lowered by hoist or crane.

Prefabricated lightweight walls and partitions are installed so as to allow for possible deflections, which are limited to 15 mm, unless they are counter-cambered by prestressing (see Section 34.1)[119, 125] The large amount of assembly work at ground or basement level endows the system with safety, speed and economy. At an average erection rate of one floor per week or less, a multistorey building can be erected in about two-thirds of the time required normally for cast *in situ* construction, with a resultant saving in structural cost of up to about 20 per cent. A notable example of the system is Kindersley House, Sydney, which was erected in 1960 to a height of eighteen storeys. Recent structures in Victoria utilising prestressed concrete lift slabs include the sixteen-storey Plaza Flats, St Kilda.

The system can be dramatically varied by lifting prefabricated floors and suspending them by hangers from heavy cantilevers, which are mounted on slip-formed concrete core towers.

27.4 JACKBLOCK CONSTRUCTION

The roof of a multistorey building is constructed at ground level and raised as a block by hydraulic jacks. Supporting walls of the roof and the top storey are constructed beneath the roof. The whole of this structure is jacked up to allow the same sequence of operations to follow. Lifting is carried out for each successive storey until the whole building is completely constructed. The process is speedy and economical to implement.[105, 133]

CHAPTER **28**

SURFACING PROCEDURES

28.1 SACK-RUBBED FINISH

Formed concrete surfaces are patched, cleaned, thoroughly wetted and allowed to reach a lightly damp condition. Mortar is then rubbed over the surface with clean burlap pads or sponge-rubber floats so as to fill the voids. The mortar consists of a plastic mix of 1 part cement : 1·5-2 parts sand (passing a 0·60-mm sieve) by volume. Some 30 per cent of the cement may be replaced by white portland cement or hydrated lime for a pale colour effect.

While the mortar in the voids is still plastic, the surface is sack-rubbed with a similar but dry mix without the addition of water. This operation stiffens the mortar in the voids, ensures that they are filled, and removes excess material from the surface. The finish mortar is moist cured (e.g. with wet hessian) for at least 3 days. Finishing operations should be done in the late afternoon, at night, on cloudy days or under cover, but never in direct sunlight, so as to promote early hydration of the cement. If the surface is not cured, dry loose material can be rubbed or brushed off during the following day. Some surface irregularities that occur in moderation may be regarded as a characteristic feature, rather than a blemish, of untreated concrete surfaces (see Section 16.2).

28.2 CEMENT-BASED PAINT

28.2.1 PREPARATION OF SURFACE

The surface, which has aged for at least 3 weeks, should be thoroughly cleaned and made free from efflorescence, form oil, loose material, dirt and dust. Acid-etching, wire brushing, sand-blasting, abrasive stones and washing are used for the purpose (see Sections 20.4.2 and 36.4).

Areas not to be sprayed should be properly masked before work is commenced and any overspray should be removed by dry-brushing 2-3 hours after application. The backing surface is thoroughly dampened but not soaked with water (e.g. several fine spray treatments) and painted just after the water sheen has gone off.

28.2.2 APPLICATION

On concrete masonry and external concrete surfaces, the first coat should be applied with a stiff-bristled scrubbing brush to ensure maximum adhesion and weathertightness. Immediately before applying this coat, the mortar joints are painted to ensure that they will be properly sealed. The entire wall surface is then painted and kept moist for at least 12 hours before the application of the second coat.

It is sometimes advantageous to "bag-up" the surface with a 1 part cement : 1 part sand grout prior to application of the first brush coat to a uniform thickness. A two-coat film about 0·4 mm thick normally has a 5-year life on smooth concrete, the rate of coverage of the paint being 2 m²/litre of two coats. About twice this thickness and quantity of paint are required on a rough-textured surface.

Weather conditions at the time of application should ideally be cool and dry. As the paint should be protected from drying out too quickly, it should be shielded from direct sunlight by a double thickness of hessian hung from scaffolding. On the other hand, not only will too wet a surface destroy bond, but rain on the surface within 12 hours after application may destroy the work. Excessively hot or near-freezing weather should be avoided.

28.2.3 CURING

Curing is essential to allow the mixture to hydrate under favourable conditions. A fine spray should be applied regularly after an interval of 6-12 hours, and for a period of 2-3 days after the second coat. Calcium chloride in the paint has a hygroscopic and set-accelerating effect, but if curing is not applied the dry paint may rub off. Membrane curing may be used (see Sections 9.3 and 28.1).

28.2.4 SPRAYING EQUIPMENT

Compressed air is supplied at a pressure of 50-70 kPa to a pressure pot where the mixture is agitated. From the pot, the mixture is forced along a hose up to 5 m long to the nozzle.

Air from a compressor is supplied at a maximum pressure of 250 kPa. Too high a pressure will cause fogging, rebound, waste of material and a dusty, porous, paint film; too low a pressure will result in inadequate bond. The choice of correct pressure for atomisation with the nozzle 250-300 mm from the work is largely a matter of experience. All equipment should be cleaned with clean water after use.

28.2.5 CONSISTENCE

The paint powder and water are mixed in approximately equal proportions, by first adding half the requisite quantity of water and stirring, allowing to stand for 10 minutes, then adding the remainder of the water and agitating. When strained through a 0·30-mm sieve, the mixture should have a paint-like consistence.

28.2.6 COMPOSITION

Proprietary paints are superior to job-mixed paints and storage should be in airtight containers. A typical composition is given in Table 28.1.

TABLE 28.1   **CEMENT-BASED PAINT**

Ingredient	Per Cent by Weight
1. White portland cement	80
2. Hydrated lime (and filler)	10
3. Titanium dioxide or zinc sulphide	3
4. Oxide pigment	3
5. Calcium or aluminium stearate	1
6. Calcium chloride	3

For fish ponds and swimming pools, portland cement is 90 per cent and lime is 0. In tints darker than ivory, titanium dioxide or zinc sulphide is 0 and oxide pigment is 6 per cent. Tinting colours must be alkali-resistant. For protected surfaces, paint powder may be mixed with up to half its volume of dry sand passing a 1·18-mm sieve (BS 4764).[107]

28.2.7 MODIFICATION

To form a tough "breathing"-type film, an emulsified polymer of stabilised silicone acrylic, styrene butadiene or polyvinyl type can be added in lieu of one-quarter of the mixing water; or a tinted latex in this category can be mixed with white cement, sand and water to suit the substrate. Other types of paint are described in Sections 20.2, 21.1, 21.2 and 28.4; also BS CP231 on Painting and References 107, 108 and 143.

28.3 RESIN-RUBBER PAINT

A solvent-type of copolymer of pigmented, acrylic-stabilised styrene butadiene may be used in two coats on exterior masonry or concrete surfaces. It is an excellent traffic-marking paint, as it forms tough films with exceptional resistance to extreme weather conditions and various chemical solutions. Maximum protection on concrete roofs is provided by using flaky aluminium filler and asbestos or fibreglass reinforcement in the binder (see Section 36.5). The substrate must be thoroughly clean, sound and dry.

28.4 SILICONES

Silicones are of organic solvent type, containing about 5 per cent of dissolved silicone resin, and dispersion-in-water type containing 3 per cent of methyl siliconate. The first variety is quick-drying and requires a clean dry surface, whereas the second one can be applied to a sound surface which is not quite dry. Silicone products, on penetrating concrete or mortar, deposit a water-repellent lining in the exterior pores or capillaries. As these passages remain unsealed, however, the treated surface will not exclude water applied under pressure, but it will allow water vapour to escape from inside a masonry wall for breathing purposes (BS 3826).

Surfaces with openings exceeding about 2 mm, after being cleaned by brushing, should be given a prior treatment with cement-based paint, mortar, or other suitable surface-binding material. As silicone solutions and dispersions are clear liquids, a fugitive dye is a useful aid in assuring complete surface coverage. They should be applied liberally with vigorous brushing or with the aid of a low-pressure spray (e.g. 70 kPa) so as to flood the surface and promote penetration to a depth of about 5 mm. The rate of application varies with the porosity of the surface and the solid content of the product, and ranges from 2-3·5 m²/litre (or heavier) for the first coat to 3·5-5·5 m²/litre for the second coat.

With a siliconate product, the second application must be made within 2-3 hours of the first for effective penetration. Appropriate safety and fire-prevention precautions include ample ventilation (see Section 21.1.7). A double application should give water-repellent protection for at least 5 years, but it may not arrest the slow weathering of stone masonry walls. Because of adhesion difficulties, water-mixed or water-solvented materials should not be applied to hardened silicone treatments.

Modifications for open-textured surfaces or increased water-repellency consist of silicone acrylic lacquer or siliconate acrylic emulsion, with or without a suitable filler and a cationic amine additive (e.g. 1 per cent) for improved adhesion in the presence of interfacial moisture. Since the resinous content of these compounds is at least 40 per cent, a siliconate acrylic emulsion should be diluted with an equal volume of water for flow-on priming purposes, but be used neat for subsequent surface treatment. Primed clean surfaces and seams can be sealed with silicone synthetic rubber, and damp-proof courses can be formed with a siliconate rubber-latex compound incorporated in a cement mortar joint.[175]

28.5 STAINING SOLUTIONS

The colouring of hardened concrete by a staining solution is effected by a chemical reaction between certain metallic salts and set cement. Materials suitable for this purpose are ferrous sulphate and ferrous chloride for fawns, browns and tints up to rusty red. Ferric chloride gives an orange colour, while a bluish green can be obtained with copper sulphate. The final colour is governed by the strength of the solution, the staining period and the quality and age of the concrete. As the strength of the solution will have to be determined to suit individual requirements, no hard and fast rules can be given on quantities, other than that they may be varied from 25-200 g of metallic salt per litre of water. On a concrete floor, which should be lightly dampened beforehand, 1 litre of solution will cover about 5·5 m² in one application. Sulphate-resisting portland cement will outlast ordinary portland cement after staining treatment.

For different shades of green, two separate solutions are prepared: one containing 100 g of copper sulphate per litre of water and the other 100 g of ferrous sulphate per litre. The latter is added to the former until the desired shade of green is obtained. As the ferrous sulphate solution is the stronger in colouring effect, it is added to the copper sulphate solution to a limited extent only. A note should be made of the amount added so that the depth of colour may be controlled. The effect of the solution is not immediate and a period of 3-5 days may have to elapse before the final colour is reached. It would seem preferable, therefore, to give a few applications of a weak solution, rather than one application of a strong solution, in order to get a uniform stain with the desired tint. Means for minimising efflorescence are included in Section 36.6.

Copper sulphate solution should be stored in a wooden, copper, glass or earthenware vessel, as electrochemical action takes place between copper sulphate and most metals. This will change the colour of the solution (e.g. in an iron vessel, the depth of green becomes uncontrollable).

Superior results can be obtained by using coloured portland cement or oxide pigment (AS K54 and BS 1014). Aluminium frames in new concrete or masonry walls should be cleaned and painted with clear, durable synthetic resin, so as to prevent rainwater stains below. Alternatively, Grade A anodising (e.g. bronze type) will substantially resist aggressive wash-down and give protection against the copper throw-off from overhead wires of nearby trains or trams. Anodic oxidation coatings on aluminium for architectural applications are specified in AS 1231.

28.6 STAIN REMOVAL

(a) Bronze and copper stains.
These stains can be removed by a mixture of 1 part by weight of powdered ammonium chloride and 4 parts of powdered talc or diatomite, which is made into a paste with ammonium hydroxide solution. A thin layer is applied to the affected areas and allowed to dry before it is removed, three applications being required to remove severe stains.

(b) Ink stain.
The affected areas can be bleached with commercial sodium hypochlorite solution containing 12·5 per cent available chlorine. This solution is applied in one of several ways: by flooding; by pressing soaked pieces of flannel against the surface and covering them with nonabsorbent material to prevent evaporation; and by spreading a prepared paste where necessary. A residual brown stain can be removed by the method applicable to iron stains, and a residual bluish stain can be eliminated by an additional bleaching with ammonia water.

(c) Iron stain.
A solution of 1 part of sodium citrate in 6 parts by weight of warm water is made into a paste with whiting. It is applied to the stain

with a trowel and scraped off when dry. Alternatively, the surface is thoroughly wetted by painting with the solution at 5-minute intervals, or by covering for 0·5 hour with cotton wool soaked with the solution. This treatment is followed by a pat of moist whiting that has sodium hydrosulphite crystals liberally sprinkled on the contact face. After 1 hour, the treated surface is washed with clean water.

(d) Oil stain.
Remove free oil (see Section 20.4.3), apply a paste made with 10 per cent trisodium phosphate solution and rinse after 24 hours.

(e) Smoke stain.
A paste of trichlorethylene and powdered talc is applied to the stain and covered with nonabsorbent material to prevent evaporation. The stain should be scrubbed before and after this treatment, and ventilation must be adequate to remove fumes. Alternatively, a cleansing and bleaching treatment may be employed.

(f) Wood stain.
A bleaching treatment is applied with strips of white cloth soaked with sodium hypochlorite solution, a prolonged period of moist contact being required.

28.7 MOSS CONTROL

Mosses, fungi and lichens on surfaces and roofs can be destroyed by an application of 3-5 per cent aqueous solution of either copper nitrate or sulphate or sodium pentachlorophenate. The copper compounds are very effective in that they give residual protection under exposed conditions for several years. Copper nitrate is less damaging to concrete than copper sulphate, but both cause light staining of concrete surfaces and electrochemical corrosion of galvanised iron, soldered joints and aluminium fixtures. Enamelled, plastic or glass receptacles should be used and roof spouting should be flushed with water while copper solutions are being applied.

An average-sized dwelling roof may require 2 kg of copper sulphate crystals dissolved in 45 litres of water, and treatment is best done after preliminary scraping and brushing during the summer. The treated residual growth withers and can be removed with a wire brush in about a week. As mosses and lichens require water for subsistence, prolonged damp conditions should be countered with good surface or subsoil drainage, and the possible use of a fungicide in surface-penetrating preparations.[109]

In concrete water-races at high altitude, where profuse moss growth may reduce carrying capacity by about 10 per cent, both types of solution can be effectively applied successively after the surface of the channel has been scraped and brushed clean. Copper pentachlorophenate is formed as a reaction product. A wetting agent (e.g. 0·01 per cent sodium pyridinium chloride) is required for a uniform spray treatment with copper-based solution. The surface, after being allowed to dry, is finally treated with commercial sodium silicate, diluted 1 : 3 and mixed with a potential coagulator (e.g. 1-2 per cent ammonium chloride).

New alpine water-races can be kept free of moss growth for many years by applying to the hardened concrete several coatings of low-alkali sodium silicate solution (without a coagulator; see Section 20.2(b)), or by using an admixture of 0·7 per cent of penta-chlorophenol by weight of cement. This problem, while arising in hydroelectric power schemes in alpine country, is seldom of importance in water-supply undertakings because of their relatively low altitude. Algal growth in large pipelines can be inhibited by a dosage of about 5 ppm of chlorine for a few hours at weekly intervals.

28.8 AGGREGATE FACING BY THE AGGREGATE-TRANSFER PROCESS

28.8.1 PROCEDURE

The aggregate-transfer method is a means of embedding selected aggregates in the face of concrete cast in place. Since only a single layer of aggregate is used for the surface, the method permits the use of aggregates which would be too expensive if used throughout the concrete.[110, 111]

Plywood panels 4 mm thick are cut to size, and strips of plywood coated with paraffin wax are pinned to the edges to form a tray. The depth of the tray is equal to the thickness of the largest aggregate to be used. In the tray, a mixture of an adhesive and filler is spread to a depth equal to half the nominal size of the aggregate. The aggregate is then sprinkled over the surface and pressed firmly into the matrix with a float faced with sponge rubber. When the matrix reaches the top of the tray, the tray is tilted to allow surplus aggregate to fall off. Any gaps are made good, and the prepared panel is left at least 4 days for a water-soluble cellulose adhesive to harden, or a shorter period if a stronger adhesive is used.

The prepared panels, which may be of any size of plywood (up to 2400 mm × 1200 mm), are placed within the formwork and fixed to it with staples. The sides of the trays are left in position as long as possible to prevent the aggregate from being dislodged. When these sides are removed, the joints between the panels are filled with strips of caulking compound. If the panels are pushed firmly against those already in place, a well-filled joint of about 1 mm will be formed.

The placed concrete should have a slump of about 80 mm. Compaction should be with an internal vibrator, care being taken to keep it from the panel. When the concrete has hardened, the formwork is removed, leaving the plywood bottoms of the trays in position. These are then removed and the aggregate remains embedded in the face of the concrete. Any adhesive covering the face is removed by washing and scrubbing.

28.8.2 ADHESIVES AND FILLERS

These may consist of a water-resistant adhesive, such as a mixture of nitrocellulose, dammer gum and acetate; an animal glue that just gels at 15°C; or a water-soluble cellulose adhesive. The adhesive must be strong enough to secure the aggregate in place while the concrete is being placed, but not so strong that the plywood cannot be removed without damage. Sand which passes a 1·18-mm mesh sieve has been found to be the most suitable filler. It does not shrink (as sawdust does if used) and, if the sand is chosen for its colour, the layer adhering to the concrete between the particles of aggregate adds to the final appearance.

28.8.3 AGGREGATE

Aggregate should be as nearly as possible of single size. Flat or elongated particles should not be used as they will not be properly embedded in the concrete (see Section 3.3.4(g)). Gradings of 9·5 mm to 4·75 mm; 13·2 mm to 6·7 mm; 9·5 mm to 8·0 mm; and 13·2 mm to 9·5 mm have all proved suitable. The texture of 9·5 mm to 6·7 mm and 19·0 mm to 13·2 mm gauge aggregate of one colour will be discernible at distances of 6 and 15 m, respectively.[111] Bulk storage of aggregate for a project or decorative glaze coatings[141] will ensure uniform colour.

Ribbed masonry, Concert Hall, Perth

Lift-slab system
in operation

Nuclear Power Plant,
United States of America

Placing special aggregate mix over newly cast
lightweight-aggregate concrete backing slab

Removing surplus stiffened mortar from tilted surface

28.9 AGGREGATE FACING BY REMOVAL OF MORTAR

28.9.1. SET-RETARDATION

Within 2-3 days after casting, the concrete surface matrix (which is set retarded) is brushed or washed off uniformly to expose the coarse aggregate. A set retarder is brushed uniformly over the interior of moulds or formwork, which should have watertight joints. Baffle sheets or drop chutes are used temporarily in vertical formwork and compaction tools or vibrators are kept away from coated surfaces so that retarder will not be displaced by concrete during placement.

Various types of retarder are indicated in Section 10.2. A typical retarder for the purpose consists of 1 part molasses to 3 parts water by volume. The solution is mixed with a blend of equal parts of whiting and fine sand, the quantity used being sufficient to obtain a stiff brushing consistence. A small quantity of glue-size is added to help bind the mixture. The retarder thus prepared may be used directly on vertical surfaces or in diluted form on the horizontal surfaces of formwork and concreting is done within a few hours. A set-retarding powder may be spread uniformly over a freshly placed concrete slab, just after the disappearance of free moisture. An application of 75 g/m^2 is effective to a depth of about 3 mm. Retarder-impregnated paper may be used as a form liner.

The depth of exposure finally obtained should not exceed half the size of the smallest particles of coarse aggregate. Residual cement may be removed, if desired, by washing the surface with dilute hydrochloric acid and water. Exposed-aggregate surfaces should be properly moist cured. They may be used for architectural purposes or as a good bonding base for subsequent concrete or rendering.

28.9.2 MECHANICAL MEANS

The hardened matrix of concrete is removed by bush hammers or a sand blast and vacuum closed-circuit appliance (e.g. "Vacu-Blast"). Dust-free operation is provided with surface-dry sand passing a 2·36-mm mesh sieve. Alternatively, the surface is eroded by an abrasive-in-water jet, the nozzle of which is held at an angle of 30-60° to the work and about 300 mm away from it. The aggregate is thus exposed to a uniform depth, patterns being formed by masking the surface with vinyl tape.

An abrasive jet-appliance (e.g. "Vaqua Processing Unit") consists of a hopper mounted on a venturi tube, which provides the suction necessary to induce the abrasive (AS 1627 Part 4) into a water jet. The hopper has upper and lower compartments, so that sandblasting and recharging operations can proceed concurrently. Compressed air is delivered to an injector at 700 kPa and at a rate of 120 litres of free air per second.

The rate of treatment is roughly 12 m²/hour when concrete is a few weeks old and 6 m²/hour when the concrete is 1 year old. The rate is increased on a saturated concrete surface. The procedure may be mechanised to minimise operating cost. In comparison with dryblasting, a much reduced quantity of silica dust is generated. Operators should wear a face mask in an atmosphere containing silica dust.[132]

A "Himat", high-speed, water-jetting appliance, operating fine jets of water at up to 100 MPa, can expose the aggregate of concrete in a few seconds. In industrialised application, the water is ejected at the rate of 30-35 litres/minute and recycled through a 5000-litre storage tank. Varying types of nozzle and pressure can be used for cutting, abrading and descaling work.[168]

"Naturbetong" exposed-aggregate walls and beams with a mural finish are made by vibratory-packing prewashed aggregate (passing a 37·5-mm sieve and retained on a 19·0-mm sieve) in successive layers within tight formwork. This is assembled with joint-seal strips of polyurethane foam. The voids are upwardly grouted with cement-dispersed, white or coloured cement equally proportioned with a pozzolan-blended sand (see Sections 6.4, 17.1, 17.2 and 24.4).

The grout is activated by high-speed mixing and injected into the prepacked aggregate by means of a screw-type pump connected by hoses to heightenable pipes. These, at 0·6 m centres, were located initially through the full depth of material before being down-drawn for the grouting operation. The formwork is externally vibrated and inspection holes are bunged as the grout level rises.

Form-stripping is carried out as soon as practicable within 24 hours, and the facade presenting a fairly uniform set characteristic is promptly sandblasted or water-jetted to specified depths, in two stages, to patterned template or sketched mural design. The exposed aggregate surface is blown clean, moist cured for 2 weeks with protection against direct-sun and drying-wind exposure, and finally silicone-treated for water repellency. The result is uniformly strong, impermeable and little affected by drying shrinkage (see Sections 28.4, 31.2 and 36.6).[170]

28.10 AGGREGATE EMBEDMENT

Panels are cast horizontally with low-slump concrete and given a green-cut surface (see Section 7.2.4). White cement mortar is spread within 2 hours of placing the concrete and decorative aggregate is tamped or rolled manually into the surface. Surplus mortar can be green-cut and removed by spraying with water near the time of initial set of the cement, the mould being tilted to promote surface drainage. A premix may be used similarly.

The concrete should contain 10 40-kg bags of cement per m^3 and have a minimum cylinder compressive strength of 30 MPa at 28 days. Limited table vibration is advisable. The panels should be moist or membrane cured until demoulded and cured subsequently on both sides for at least 1 week (see Sections 9.2, 9.3, 9.4 and 11.2.1). They are covered and stored for a further 3 weeks before use.

Panels vary in size from 0·5 m^2 × 50 mm thick for man-handling to 11 m^2 × 75 mm thick or 18 m^2 with increased thickness or reinforced ribs for mechanical handling. The edges may be thickened to facilitate jointing and oversizes are avoided to save time in site fixing. Light fabric reinforcement, for handling strength, should be placed within the backing concrete with a minimum cover normally of 25 mm on the weather side. Tying wire should not be exposed to weathering and if the cover is reduced to below 20 mm suitable measures to prevent corrosion should be used (see Sections 3.4 and 22.3.2). Decorative work and its applications are illustrated in References 57-68, 110, 111, 115 and 141, and in Plates 21 and 24.

28.11 THERMAL TEXTURING

Thermal texturing is a fast, flame-spalling process that can be used on concrete surfaces for skidproofing or bonding purposes, and on stone, particularly granite and quartzite, for decorative surfacing. Successive parallel sweeps are made with a 3000°C flame from an oxyacetylene descaling tip, the head of which is mounted on a traversing carriage and held about 25 mm from the surface to be roughened. The treatment causes minute chips to flake off and bring about controlled, superficial spalling. A thin film of water may be used to reduce retained heat, and safety measures include either a cowl over the flame or a face mask for the operator.

Thin veneer slabs down to 15 mm thick may be produced economically with a dramatic, sparkling effect on stone of a

crystalline nature. Differences in texture are produced by varying
the flame angle, speed, distance of flame above the surface and the
direction of traverse. Linear speeds vary from 0·6-3·0 m/min,
the greater the speed the smoother the treated surface. With a
180 mm flame width, a concrete slab can be textured at the rate of
23 m²/h, the gas consumption per hour being 4250 litres of
oxygen at 200 kPa and 2400 litres of acetylene at 70-100 kPa
pressure. Precast concrete and natural stone claddings on buildings
should be used as recommended in BS CP297 and CP298; also in
Section 28.12.[112, 155]

28.12 USE OF DECORATIVE FACING SLABS

Building stone and precast artificial stone of which wall slabs are
made are both microporous. Precautions must be taken when using
them to prevent efflorescence or staining by the migration of salt-
bearing moisture through the material (see Section 36.4). Fixing
and jointing details for precast concrete and natural stone claddings
are illustrated in BS CP297, CP298 and in References 60-68, 112,
115, 128, 129, 135, 146, 156, 163. The following items should be
considered when using facade panels.

> Self-draining vented air-spaces (with weepholes where necessary)
> behind facing slabs, at open-drained joints and around
> downpipes (see Fig. 30.3).
> Effective flashings, damp-proof courses, bituminous rubber
> coatings and vapourproof membranes in walls, parapets and
> floors.
> Jointing mortar (see Sections 16.2 and 27.2.2(b)) and sealing
> compounds that are nonstaining and of suitable composition for
> the project in hand (ASTM C509; see Sections 20.1.5(f) and
> 31.2.1).[136, 137]
> Noncorroding metal cramps, ties, trims, bolts, clips, dowels,
> brackets and supports; independent fixing, anchorage and support
> of slabs, steel reinforcement, and dry construction where
> practicable.[112]
> Movement joints to accommodate the dimensional changes of
> slabs due to drying shrinkage, ambient atmospheric conditions
> and the effects of internal heaters or physical instability.
> Drip edges on projections for shedding rain water and, in open-
> textured panels exposed to weathering, a water-repellent
> admixture for early-age self-cleaning and efflorescence-
> suppression purposes.

Facade drainage to minimise differential colour changes caused by variable rain-water washings on building surface profiles.[184]

Various joints can be used for panel cladding, namely, gap-filled, cover-strip, lapped, and open-drained varieties. Many attempts to improve the performance of joints have led to the open-drained variety, which incorporates a rain shield, an air chamber and a sealant at the rear for weather protection. In this way, the quality of a sealant is preserved by its being kept dry and shaded from solar radiation (see Fig. 30.3).

A vertical joint of this type, for efficient operation, should meet the following requirements.

1. A minimum width of 9 mm and a maximum cyclic movement of ± 20 per cent, so as to avoid blockage by debris, development of capillarity and possible fatigue of the sealant.

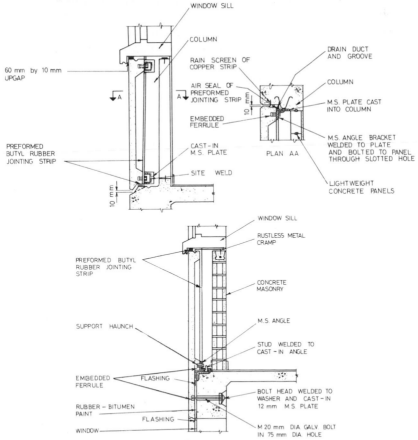

Fig. 28.1   Precast panel connections.

2. Dimensional characteristics that are sufficient to accommodate the tolerance limits and shrinkage of the panel.

3. An impediment to horizontal flow of water on the surface of the panels.

4. A vented rain screen or baffle and drainage duct system.

5. The outwards deflection of entrapped water by upstand flashings inserted at horizontal joints.

6. Equalised air-duct pressures that hinder water penetration past the rain screen and a readily installed airtight sealant barrier located at the rear.

In lightweight forms of construction, a horizontal rebated joint should be sealed at the back with a preformed jointing strip or with weak mortar that will prevent rain being blown up between the slabs. Alternatively, a horizontal joint containing an upsloped and upstepped bed may be sealed with jointing mortar and a jointing strip or mastic. In heavy forms of drained construction, the upstand of a rebated joint should be 100 mm high, left open at the front, flashed and effectively sealed at the rear. Its width should be not less than 9 mm so that the internal air seal can accommodate movement without capillary formation.

Some typical fastening procedures for precast concrete panels with open-drained joints are illustrated in Figure 28.1. They represent a partial application of the described principles to selected illustrations of panels published by the Cement and Concrete Association of Australia.[68] In practice, however, many designers may prefer to fill the face of joints with, perhaps, a metal trim or a mastic or plastic filler. Working details should be developed to suit each project through a study of appropriate literature, technical consideration and consultation, and weatherproofness tests on trial assemblies (see Sections 14.7.5, 27.1.7 and also BS CP297).

REFERENCES

1. "On-site Precasting for Multi-storey Office Building". *Concrete Building and Concrete Products.* 42, 3 (March 1967), pp. 149-53.
2. KRAFT, L. M. "Compaction of Concrete Slabs by Vibration". *ACI Journal,* 68, 6 (June 1971), pp. 462-67; 12 (December 1971), pp. 977-78.
3. CUSENS, A. R. "The Influence of Amplitude and Frequency in the Compaction of Concrete by Table Vibration". *Magazine of Concrete Research,* 10, 29 (August 1958), pp. 79-86; 11, 31 (March 1959), pp. 40-42.
4. STEWART, D. A. *The Design and Placing of High Quality Concrete* (2nd edn), E. and F. N. Spon, London, 1962.
5. ACI COMMITTEE 309 "Consolidation of Concrete". *ACI Journal,* 68, 12 (December 1971), pp. 893-932; 69, 6 (June 1972), pp. 344-46.
6. EVERARD, K. A. and BHAGAT, K. B. "Effects of Revibration of Concrete on Compressive Strength and Modulus of Elasticity". *Indian Concrete Journal,* 44, 8 (August 1970), pp. 346-51, 365.
7. VOLLICK, C. A. "Effects of Revibrating Concrete". *ACI Journal,* 29, 9 (March 1958), pp. 721-32; 2, 30, 3 (September 1958), pp. 1267-71; 7 (January 1959), pp. 769-81.
8. BILLIG, K. "Vacuum Concrete". *ICE Journal,* 29, 3 (January 1948), pp. 243-69.
9. EDITORIAL. "Concrete Pumping Techniques and Equipment". *International Construction,* 13, 8 (August 1974), pp. 62-77.
10. ACI COMMITTEE 506. "Recommended Practice for Shotcreting". *ACI Journal,* 63, 2 (February 1966), pp. 219-45; 7 (July 1966), p. 732; 9 (September 1966), pp. 1013-16.
11. RYAN, T. F. *Gunite: A Handbook for Engineers.* C & CA, Britain, 1973.
12. DAFFY, J. "Shotcrete in Engineering Construction". C & CA Aust. *Technical Report TR29,* 1971.
13. RAO, M. L. S. "Grouting of Structures in River Valley Projects". *Indian Concrete Journal,* 34, 5 (May 1960), pp. 171-74; 7 (July 1960), pp. 270-73; 10 (October 1960), pp. 398-400.
14. NORTHWOOD, Britain. "Grouting Design and Practice". *Consulting Engineer,* 33 (Suppl.), 10 (October 1969), 40 pp.
15. MOLLER, K. "Grouting Now". *Consulting Engineer,* 36, 8 (August 1972), pp. 44-49.
16. "Symposium on Grouting". *ASCE Proceedings,* 87, SM2 (April 1961), pp. 1-145.
17. ISCHY, E. and GLOSSOP, R. "An Introduction to Alluvial Grouting". *ICE Proceedings,* 21 (March 1962), pp. 449-74.
18. MAHAFFEY, P. I. "Grouting Prestressing Ducts". *Contracting and Construction Engineer,* 26, 4 (April 1972), pp. 6-25; C & CA Aust. *Advisory Note 34.012;* Aust. EBS *Researches and Facilities RF43.*
19. ACI COMMITTEE 304. "Preplaced Aggregate Concrete for Structural and Mass Concrete". *ACI Journal,* 66, 10 (October 1969), pp. 785-97; 67, 4 (April 1970), pp. 360-62.
20. AKATSUKA, Y. and MORIGUCHI, H. "Strengths of Prepacked Concrete and Reinforced Prepacked Concrete Beams". *ACI Journal,* 64, 4 (April 1967), pp. 204-12.
21. MENZEL, C. A. "Some Factors Influencing the Strength of Concrete Containing Admixtures of Powdered Aluminium". *ACI Journal,* 14, 3 (January 1943), pp. 165-84.
22. CRITCHELL, P. L. *Joints and Cracks in Concrete.* Spon, London, 1968.
23. KOSTER, W. *Expansion Joints in Bridges and Roads.* Spon, London, 1969.
24. ATKINSON, T. A. "Design Techniques and Current Standards for Heavy Duty Roads in Plain and Normally Reinforced Concrete". *Constructional Review,* 40, 6 (June 1967), pp. 22-30; 10 (October 1967), pp. 21-28.
25. PERSSON, B. O. E. "Continuously Reinforced Concrete Pavements with Elastic Joints—Elastically Articulated Pavements". *Nordisk Betong,* 13, 2 (1969), pp. 107-38.

26. Highway Research Board (United States). "Soil-cement". *Highway Research Record 255,* 1968. PCA Chicago, 1969.
27. *Soil-cement Construction Handbook.* Portland Cement Association, Chicago; Cement and Concrete Association, Australia and Britain, 1969.
28. "Soil-cement Inspector's Manual". Portland Cement Association, *Publication SC16.*
29. JOHNSTON, C. M. "Western Avenue, Lagos: The Design and Construction of a Soil-cement Pavement". *ICE Proceedings,* 20 (September 1961), pp. 107-40.
30. DEPARTMENT OF SCIENTIFIC AND INDUSTRIAL RESEARCH. "Determination of the Cement or Lime Content of Cement- or Lime-Stabilized Soil". *Road Note 28,* Road Research Laboratory, London, 1960.
31. "Soil-cement Slope Protection for Earth Dams: Field Inspection and Control". *Information Sheet CB13,* Portland Cement Association, 1967.
32. ACI COMMITTEE 325. "Design Procedure for Continuously Reinforced Concrete Pavements for Highways". *ACI Journal,* 69, 6 (June 1972), pp. 309-19; 12 (December 1972), pp. 782-84.
33. ACI COMMITTEE 325. "Design of Concrete Pavements and Overlays". *ACI Journal,* 30, 1 (July 1958), pp. 17-51; 3 (September 1958), pp. 315-20; 6 (December 1958), pp. 669-78; 8 (February 1959), pp. 829-38; 12 (June 1959), pp. 1413-20; 2, 31, 3 (September 1959), pp. 1493-97; 59, 11 (November 1962), pp. 1569-86; 64, 8 (August 1967), pp. 470-74; or *ACI Manual of Concrete Practice,* Part 1, 1976, Pavements and Slabs.
34. BRITISH STANDARDS INSTITUTION. *Code of Practice for Traffic Bearing Structures—Pavings.* BS CP2006.
35. BRITISH ROAD RESEARCH LABORATORY. "Methods of Texturing New Concrete Road Surfaces to provide Adequate Skidding Resistance". *Report LR290,* 1970.
36. ACI COMMITTEE 316. "Practice for Construction of Concrete Pavements and Concrete Bases". *ACI Journal,* 70, 8 (August 1973), pp. 545-70; 71, 2 (February 1974), pp. 81-82; 72, 4 (April 1975), pp. 171-74.
37. BRITISH ROAD RESEARCH LABORATORY. "A Guide to the Structural Design of Pavements for New Roads". *Road Note 29,* 3rd edn and *TRRL Report LR460,* HMSO, London, 1970-72.
38. WINTER, S. and JONES, G. "The Crack Control Criterion". *Civil Engineering and Public Works Review,* 65, 767 (June 1970), pp. 622-23.
39. O'GRADY, V. P. "Road Pavement Thickness Design Methods In Australia". *IE Aust Journal,* 31, 3 (March 1959), pp. 51-55.
40. MUNCE, B. R. *et al.* "Design, Instrumentation and Construction of a Continuously Reinforced Concrete Road Pavement". *IE Aust Symposium on Concrete Research and Development, 1970-1973,* Sydney, 1973.
41. EDITORIAL. "Residential Road Pavements in Plain Concrete". *Constructional Review,* 39, 8 (August 1966), pp. 13-18; 9 (September 1966), pp. 18-26.
42. CRONEY, D. "Design of Concrete Road Pavements". *Technical Paper PCS20,* The Concrete Society, London, 1967.
43. SHARP, D. R. *Housing Estate Roads.* Cement and Concrete Association.
44. HANSON, J. A. "Plastic Forms for Architectural Concrete". *ACI Journal,* 31, 11 (May 1960), pp. 1137-48.
45. MORGAN, D. A. "Glass Fibre Moulds". *Concrete Building and Concrete Products,* 35, 4 (April 1960), pp. 141-42; 5 (May 1960), pp. 185-86.
46. ROBERTS, J. A. and VIVIAN, H. E. "Form Liners for Concreting". *Constructional Review,* 34, 5 (May 1961), pp. 21-25.
47. WALKER, H. C. *et al. PCI Manual on Design of Connections for Precast Prestressed Concrete.* Prestressed Concrete Institute, Chicago. 1973; *PCI Journal,* 14, 6 (December 1969), pp. 14-58.
48. *Connection Details for Precast Prestressed Concrete.* C & CA Aust, Sydney, 1966.
49. ACI-ASCE COMMITTEE 512. "Recommended Practice for Manufactured Reinforced Concrete Floor and Roof Units". *ACI Journal,* 63. 6 (June 1966), pp. 625-36; 12 (December, 1966), pp. 1495-97, NL29: 64, 4 (April 1967), p. 185. "Precast Structural Concrete in Buildings".

ACI Journal, 71, 11 (November 1974), pp. 537-49; *ACI Manual of Concrete Practice,* Part 3, Products and Processes, 1976.

50. EDITORIAL. "Which Grout?" *Concrete Construction,* 19, 10 (October 1974) pp. 501-504; 20, 5 (May 1975), pp. 210-14.

51. ACI COMMITTEE 324. "Minimum Requirements for Thin-section Precast Concrete Construction". *ACI Journal,* 59, 6 (June 1962), pp. 745-55; 60, 2 (February 1963), p. 171.

52. "Structural Floors". *The Architect and Building News,* 231, 21 (24 May 1967), pp. 909-20; *Concrete Products,* 71, 11 (November 1968), pp. 37-67.

53. MORRIS, A. E. J. *Precast Concrete Cladding.* Fountain Press, London, 1966.

54. GLOVER, C. W. *Structural Precast Concrete.* Contractors Record, London, 1963.

55. LEWICKI, B. *Building with Large Prefabricates.* Elsevier, New York, 1966.

56. BIRKELAND, P. W. and BIRKELAND, H. W. "Connections in Precast Concrete Construction". *ACI Journal,* 63, 3 (March 1966), pp. 345-67; 9 (September 1966), pp. 1027-31.

57. BURHOUSE, P. "Connexions (Joints) in Structural Concrete". *Engineering Papers 46,* British Building Research Station, 1967.

58. *Precast Concrete Recommended Practice.* Concrete Institute of Australia, Sydney, 1974.

59. "Decorative Concrete". *Concrete Building and Concrete Products,* 42, 3 (March 1967), pp. 143-45; 4 (April 1967), pp. 195-207; 5 (May 1967), pp. 529-69.

60. HERBERT, M. R. M. "Open-Jointed Rain Screen Claddings". *Current Paper 89/74,* British BRE, 1974.

61. RITCHIE, T. "Facings to Framed Buildings". *Architecture and Building* 32, 7 (July 1957), pp. 277-78; 8 (August 1957), pp. 321-22.

62. EDITORIAL. "Tilt-up Concrete Panels". *Concrete Construction,* 16, 10 (October 1971), pp. 419-62.

63. ROSTRON, R. M. "Light Cladding: Erection". *The Architects' Journal,* 132, 3405 (July 1960), pp. 115-22.

64. RENSAA, E. M. "Joints in Precast Concrete Building Frames". *The Engineering Journal* (Montreal), 43, 8 (August 1960), pp. 64-67.

65. ROSTRON, R. M. "Heavy Cladding Panels". *Architectural Review,* 129, 771 (May 1961), pp. 355-64; 772 (June 1961), pp. 427-32.

66. BISHOP, D. "Weatherproof Joints Between Precast Concrete Panels". *The Builder,* 211, 6190 (5 January 1962), pp. 31-35.

67. C & CA Britain. "Precast Storey-height Panels Form the Outer Walls of Hide Tower, Westminster". *Concrete Quarterly,* 52 (January-March 1962), pp. 3-5.

68. C & CA Aust. "Wall Panels of Precast Concrete". *Constructional Review,* 36, 1 (January 1963), pp. 14-23; 42, 3 (August 1969), pp. 34-39.

69. ISAACS, H. "Ultimate Compressive Strength of Grouted Hollow Concrete Block Masonry". *Constructional Review,* 48, 2 (May 1975), pp. 36-47.

70. GAGE, M. "Concrete Blockwork". *Architects' Journal,* 151, 14 (April 1970), pp. 845-906.

71. *Concrete Block Construction* and *Masonry Simplifiea.* American Technical Society, Chicago, 1973.

72. SORENSON, C. P. and TASKER, H. E. "Cracking in Brick and Block Masonry". *Technical Study 43,* Commonwealth Experimental Building Station, Australia, 1965.

73. *Specification for the Design and Construction of Load-bearing Concrete Masonry.* NCMA, Arlington, Va, 1975.

74. *Design of Loadbearing Concrete Blockwork to BS CP111.* C & CA Britain, Data Sheets 2a and 2b, 1967.

75. AMRHEIN, J. E. *Masonry Design Manual* and *Reinforced Masonry Engineering Handbook.* Masonry Institute of America, Los Angeles; BDRI, Melbourne, 1973.

76. *Reinforced Concrete Masonry Design Manual—1973.* Concrete Masonry Association of Australia, Sydney, 1973.

77. *Reinforced Masonry*. Hazard Products, San Diego, Calif., 1970.
78. ACI COMMITTEE 531. "Concrete Masonry Structures—Design and Construction". *ACI Journal*, 67, 5 (May 1970), pp. 380-403; 6 (June 1970), pp. 442-60; 72, 11 (November 1975), pp. 614-27; 73, 5 (May 1976), pp. 296-300; *ACI Manual of Concrete Practice, Part 3,* Products and Processes, 1976.
79. ASCE–IABSE Technical Committee 27, *Planning and Design of Tall Buildings: Masonry Structures*. International Conference, Lehigh University, Bethlehem, 1972.
80. HERSEY, A. T., *et al.* "Avoiding Disasters in Concrete Pumping". *Concrete Construction*, 20, 6 (June 1975), pp. 233-40.
81. *Soils and Foundations:* 3. Building Research Station, Digest 67 (Second Series), 1966.
82. Aust. EBS. "Footings for Small Masonry Buildings". *Notes on the Science of Building, NSB 2, 6 and 9.*
83. BRITISH STANDARDS INSTITUTION. "Foundations and Substructures for Non-industrial Buildings of Not More than Four Storeys". BS CP101.
84. Aust. EBS. "Concrete Floors on the Ground in Domestic Construction". *Notes on the Science of Building, NSB 36; Building and Construction,* 44, 2229 (12 March 1968), pp. 3-4.
85. Aust. EBS. "Pier-and-beam Construction". *Notes on the Science of Building, NSB 41 and 74* and *Technical Study No. 42.*
86. *Concrete Industries Yearbook*. Pit and Quarry Publications, Chicago, 1976.
87. Aust. EBS. "Expansion and Control Joints". *Notes on the Science of Building, NSB 57 and 112.*
88. MANSFIELD, G. A., SIRRINE, C. A. and WILK, B. "Control Joints Regulate Effects of Volume Change in Concrete Masonry". *ACI Journal* 29, 1 (July 1957), pp. 59-70.
89. "Prefabricated Brick Masonry Specification". *Technical Note 40A*. Brick Institute of America, 1974.
90. C & CA Aust. "Reinforced Concrete Masonry Lintels for Concrete Masonry Buildings". *Constructional Review*, 33, 6 (June 1960), pp. 24-27.
91. STANDARDS ASSOCIATION OF AUSTRALIA. *Lightweight Reinforcement for Brickwork*. AS Int. 325.
92. SAA. *Internal Plastering on Solid Backgrounds*. AS CA27.
93. BSI. *Internal Plastering*. BS CP211.
94. BSI. *External Rendered Finishes*. BS CP221.
95. CEMENT AND CONCRETE ASSOCIATION OF AUSTRALIA. "Rendering Concrete". *Constructional Review*, 35, 1 (January 1962), pp. 32-37.
96. ACI COMMITTEE 524. "Portland Cement Plastering". *ACI Manual of Concrete Practice,* Part 3, Products and Processes, 1976.
97. DSIR. "External Rendered Finishes". *Building Research Station Digest 131,* 1960.
98. "Lift-slabs Connected to Precast Columns". *Engineering News-Record,* 158, 26 (27 June 1957), pp. 69-70.
99. SEFTON, W. "Multi-storey Lift-slab Construction". *ACI Journal*, 29, 7 (January 1958), pp. 579-89.
100. McCONNEL, K. "Lift-slab Technique on New Office Block in Sydney". *The Architects' Journal*, 130, 3373 (December 1959), pp. 675-82.
101. GURFINKEL, G. "Assembly Hall, University of Havana". *ACI Journal*, 65, 1 (January 1968), pp. 20-28.
102. BENSON, F. R. "Lift Slab Design and Construction". *Reinforced Concrete Review*, 5, 8 (December 1960), pp. 495-527.
103. "Lift-slab Method of Construction". *Concrete and Constructional Engineering,* 56, 3 (March 1961), pp. 120-29.
104. GREEN, N. B. "Factors in Design and Construction of Lift-slab Buildings". *ACI Journal*, 59, 4 (April 1962), pp. 527-50.
105. ADLER, F. "A Jacking Method for Multiple-storey Construction". *Concrete and Constructional Engineering*, 58, 3 (March 1963), pp. 137-40.
106. TUTHILL, L. H. "Tunnel Lining with Pumped Concrete". *ACI Journal,* 68, 4 (April 1971), pp. 252-62; 10 (October 1971), pp. 799-803.

107. ACI COMMITTEE 616. "Portland Cement Paint, and Guide for Painting Concrete". *ACI Journal,* 21, 1 (September 1949), pp. 1-16; 2, 22, 4 (December 1950), p. 16 (1-3); 28, 9 (March 1957), pp. 817-32; *ACI Manual of Concrete Practice,* Part 3, Products and Processes, 1976.
108. "Painting Clay Masonry". *Technical Note 6;* "Damp-proofing and Water-proofing Masonry Walls". *Technical Note 7.* Structural Clay Products Institute, Washington, 1961.
109. DSIR. "Control of Lichens, Moulds and Similar Growths on Building Materials". *Building Research Station Digest No. 47.*
110. GAGE, M. *Guide to Exposed Concrete Finishes.* C & CA, Britain, 1971.
111. WILSON, J. G. *Exposed Concrete Finishes: Vol. I, Finishes to In-situ Concrete; Vol. II, Finishes to Precast Concrete.* Contractors Record, London, 1965.
112. "Stone Veneer Cladding". Aust. EBS. *Technical Study 46,* 1969; and *Fourth BR Congress,* Sydney, 1970.
113. DIAMANT, R. M. E. *Industrialized Building,* Iliffe, London, Vol. I, 1964; Vol. II, 1965; Vol. III, 1968.
114. VAISSAC, A. "Mechanical Equipment for Precast Concrete Works". *Concrete Building & Concrete Products,* 41, 8 (August 1966), pp. 447-53.
115. ACI COMMITTEE 356. "Industrialization In Concrete Building Construction". *ACI Publication SP-48,* 1975; *ACI Journal,* 72, 5 (May 1975), pp. 246-48.
116. EDITORIAL. "Resurfacing Pavements With Thin Bonded Concrete". *NZ Concrete Construction,* 6, 9 (September 1962), pp. 160-63.
117. MELLINGER, F. M. "Structural Design of Concrete Overlays". *ACI Journal,* 60, 2 (February 1963), pp. 225-37.
118. REDDY, K. R. and RADHAKRISHNAN, S. "Upgrading an Airfield Pavement by Concrete Overlays". *Indian Concrete Journal,* 37, 3 (March 1963), pp. 90-94, 113.
119. OMSTED, H. "Notes on Post-tensioned Lift-slabs". *PCI Journal,* 10, 6 (December 1965), pp. 84-90.
120. ZIVERTS, G., *et al.* "System Building". *Concrete Products,* 69, 10 (October 1966), pp. 36-62; 71, 1 (January 1968), pp. 34-41.
121. CHENEY, A. "Kerb and Guttering Construction". *The Builder* (Adelaide), 40, 18 (May 1963), pp. 8-9.
122. HOLMES, W. J. "Electro-osmotic Damp-proofing—'Active' or 'Passive'?" *Architect and Building News,* 233, 10 (6 March 1968), pp. 378-79.
123. "Stressed Floors". *Constructional Review,* 47, 4 (December 1974), pp. 24-27.
124. KRONE, B. and GERO, J. S. "A Comparative Investigation of Prefabricated Flooring Systems". *Building Forum,* 1, 4 (December 1969), pp. 98-110.
125. OMSTED, H. "Creep of Prestressed Lightweight Concrete". *PCI Journal,* 11, 6 (December 1966), pp. 40-45.
126. Editorial. "Trends and Developments in Slipform Pavers". *International Construction,* 11, 5 (May 1972), pp. 50-61.
127. BHANDARI, R. K. M. "Drying Shrinkage of Cement Stabilised Mixtures". *ARRB Journal,* 5, 7 (May 1975), pp. 9-23.
128. KNIGHT, T. *Drained Joints in Precast Concrete Cladding.* National Building Agency, London; 1967.
129. ALVIN, J. E. "Making and Fixing Exposed-aggregate Slabs". *Concrete Building and Concrete Products,* 38, 11 (November 1963), pp. 501-3.
130. WILLIAMS, M. L. "Current Practice in the Use of Cement-treated Crushed Rock by the Country Roads Board of Victoria". *Constructional Review,* 36, 9 (September 1963), pp. 23-27.
131. WILLIAMS, H. C. and PURDAM, R. K. "Modification of Natural Gravels and Soils by the Addition of Small Amounts of Cement". *Constructional Review,* 37, 1 (January 1964), pp. 20-24.
132. SHUSTONE, J. M., *et al.* "Mechanical Finishing of Hardened Concrete". *Concrete Construction,* 17, 11 (November 1972), pp. 526-30, 549-51.
133. ADLER, F. "Jig for the Construction of a 17-storey Block of Flats at Barras Heath, Coventry". *ICE Proceedings,* 27 (March 1964), pp. 433-64; 32 (September 1965), pp. 76-96.

134. CHEEDEVILLE, M. J. "Prepacked Aggregate Concrete and Activated Mortars". *DSIR BRS Library Communication LC1190,* 1963.
135. "Fixing and Jointing Precast Concrete Wall Panels". *Concrete Building and Concrete Products,* 39, 12 (December 1964), pp. 705-13; 40, 1 (January 1965), pp. 43-49.
136. ROSTRON, M. "Jointing and Sealing". *The Architects' Journal,* 137, 9 (27 February 1963), pp. 453-56; 10 (6 March 1963), pp. 513-19; 17 (24 April 1963), pp. 885-90; 19 (8 May 1963), pp. 1009-14.
137. DSIR. "Jointing with Mastics and Gaskets". *Building Research Station Digest Nos 36 and 37* (Second Series), 1963.
138. NAGATAKI, S. "Shrinkage and Shrinkage Restraints in Concrete Pavements". *ASCE Journal,* 96, ST7 (July 1970), pp. 1333-58.
139. *Concrete Pumping.* Concrete Institute of Australia, Sydney, 1977; British BRE, HMSO, London, 1972.
140. ACI-ASCE COMMITTEE 512. "Suggested Design of Joints and Connections in Precast Structural Concrete". *ACI Journal,* 61, 8 (August 1964), pp. 921-37; 2, 62, 3 (March 1965), pp. 1697-1703; 2, 62, 6 (June 1965); p.v. *ACI Manual of Concrete Practice,* Part 3, Products and Processes, 1976.
141. TAUBER, E. and MURRAY, M. J. "Coloured Stone for Exposed Aggregate Panels". *Constructional Review,* 41, 7 (July 1968), pp. 20-23.
142. REINHARDT, P., *et al.* "Grouting of Post-tensioned Prestressed Concrete". *PCI Journal,* 17, 6 (November-December 1972), pp. 18-25.
143. DSIR. "Painting Asbestos Cement". *Building Research Station Digest No. 38* (Revised), 1963.
144. McINTOSH, J. D. and KOLEK J. "Concrete Mixes for Blocks". *Concrete Building and Concrete Products,* 40, 2 (February 1965), pp. 83-97.
145. COPELAND, R. E. "Procedures for Controlling Quality in Block Plants". *Concrete Products,* 68, 5 (May 1965), pp. 58-70.
146. EDITORIAL. "Battery Moulds Used for On-site Precast Housing". *International Construction,* 13, 12 (December 1974), pp. 40-44.
147. MOSS, J. K. "Continuously-reinforced Concrete Road Paving". *International Construction,* 14, 11 (November 1975), pp. 20-24.
148. PORTLAND CEMENT ASSOCIATION. *(i) Thickness Design for Concrete Pavements; (ii) Concrete for Light-Traffic Roads; (iii) Construction of Concrete Pavements with the Slip-form Paver.* PCA, Chicago, 1966.
149. *Optimum Size Coarse Aggregate for Portland Cement Concrete Design.* American Concrete Paving Association, Oak. Brook. Ill., 1972.
150. "Tilt-up Wall Construction". C & CA, Aust *Manual T20* and *Report TR30; Concrete Construction,* 16, 10 (October 1971), pp. 419-62; 19, 1 (January 1974), pp. 7-10; 12 (December 1974), pp. 601-602.
151. ZIVERTS, G. J. "Forms for Precast Concrete". *Concrete Products,* 67, 11 (November 1964), pp. 50-53; 12 (December 1964), pp. 36-40.
152. BROWN, W. P. "Industrialized Precast Concrete Building Construction, Melbourne, Australia". *Proceedings, FIP, VII Congress,* New York, 1974 (C & CA, Britain).
153. KRONE, R. H. and POLLITZ, R. N. "Spacing of Lateral Supports for Masonry Walls". *ACI Journal,* 62, 2 (February 1965), pp. 231-38; 63, 6 (June 1966), 2, p. iii.
154. MINISTRY OF TECHNOLOGY. "Strength of Brickwork, Blockwork and Concrete Walls". *Building Research Station Digest No. 61* (Second Series), HMSO, London.
155. McDANIEL, W. B. "Marble-faced Precast Panels". *PCI Journal,* 12, 4 (August 1967), pp. 29-37.
156. ACI COMMITTEE 533. "Precast Concrete Wall Panels". *ACI Journal,* 66, 4 (April 1969), pp. 270-75; 10 (October 1969), pp. 814-20; 67, 4 (April 1970), pp. 310-40; 68, 7 (July 1971), pp. 504-13; *ACI Manual of Concrete Practice,* Part 3, Products and Processes.
157. CRAIG, C. N. "Battery-cast Cladding Panels". *Building,* 213, 6497 (27 November 1967), pp. 171-74; 6499 (8 December 1967), pp. 143-44.
158. HIGGINS, G. E. and PETERS, C. H. "Repair of Spalled Concrete Surfaces with Thin Concrete Patches". *Report LR217,* Road Research Laboratory, 1968.

159. ACI COMMITTEE 304. "Placing Concrete by Pumping Methods". *ACI Journal,* 68, 5 (May 1971), pp. 327-45; 11 (November 1971), pp. 869-70.
160. WHITE, J. "Transportation and Erection of Precast Concrete Units". *Precast Concrete,* 1, 9 (September 1970), pp. 239-49.
161. *Australian Reinforced Concrete Design Handbook.* C & CA, Aust., Sydney, 1976.
162. TAUBER, E., CROOK, D. N. and MURRAY, M. J. "Glazing of Portland Cement Concrete". *Constructional Review,* 43, 1 (February 1970), pp. 58-65; and *Magazine of Concrete Research,* 22, 72 (September 1970), pp. 149-54.
163. PCI COMMITTEE. "Architectural Precast Concrete Joint Details". *PCI JOURNAL,* 18, 2 (March/April 1973), pp. 10-37; 6 (November/December 1973), pp. 117-20; and *Architectural Precast Concrete.* PCI, Chicago, 1973.
164. FREEDMAN, S. "Grouting and Protection of Post-tensioning Tendons". *Concrete Construction,* 18, 5 (May 1973), pp. 210-11, 254-57; 6 (June 1973), pp. 277-78, 306; 9 (September 1973), pp. 421-22, 440-44.
165. FINNEY, E. A. *Better Concrete Pavement Serviceability.* Iowa State University Press and ACI joint publication; *ACI Monograph 7,* 1973.
166. ILLINGWORTH, J. R. *Movement and Distribution of Concrete.* McGraw-Hill, London, 1972.
167. DAFFY, J. "Breakwater Units". *Contracting and Construction Engineer,* 28, 8 (June 1974), pp. 22-51.
168. BARRON, J. and NICHOLLS, E. C. "Working With Water". *Concrete* (London), 7, 10 (October 1973), pp. 22-24.
169. TOBIN, R. E. "Hydraulic Theory of Concrete Pumping". *ACI Journal,* 69, 8 (August 1972), pp. 505-10.
170. KARP, J. J. "Naturbetong". *ACI Journal,* 70, 8 (August 1973), pp. N10-N12.
171. IRELAND, V. "Standard Beam-Column Connection for Framed Precast Concrete Structures". *Contracting & Construction Engineer,* 28, 2 (February 1974), pp. 47-56.
172. RICHARDSON, J. G. *Precast Concrete Production.* C & CA Britain, Wexham Springs, Slough, 1973.
173. HEYNES, R. F. "Cement Grouting". *Concrete* (London), 8, 1 (January 1974), pp. 43-45.
174. FOWLER, D. W. and FRALEY, T. J. "Polymer-impregnated Brick Masonry". *ASCE Journal, Structural Division,* 100, ST1 (January 1974), pp. 1-10.
175. DAS, A. "Damp-proofing of Existing Buildings by Latex-Siliconate". *Building Digest 98,* India CBRI.
176. WADDELL, J. J. "Precast Concrete: Handling and Erection". *Monograph 8,* American Concrete Institute, 1974.
177. EDITORIAL. "Housing Modules". *Constructional Review,* 49, 1 (March 1976), pp. 40-47.
178. RATHS, C. H. "Design of Load Bearing Wall Panels". *PCI Journal,* 19, 1 (January/February 1974), pp. 14-61.
179. STICKLER, Jr., C. W. "Factory-on-Wheels Creates Poured Concrete Homes in Just 48 Hours". *Concrete Products,* 75, 10 (October 1972), pp. 50-51.
180. "Fibre-reinforced Concrete". United States Federal Highway Administration, *Highway Focus,* 4, 5 (October 1972), 98 pp.
181. RICHARDSON, A. *Precast Concrete Production.* Concrete Publications, London, 1973.
182. "Prefabricated Insert Eliminates Sawing of Pavement Joints". *Concrete Construction,* 19, 5 (May 1974), p. 238.
183. "AASHO Interim Guides for Design of Pavement Structures". *NCHRP Report 128,* United States Highway Research Board, Washington, 1972.
184. *Protection from Rain.* Department of the Environment, Britain, 1971.
185. *Concrete Block Construction,* and *Masonry Simplified.* American Technical Society, Chicago, 1973.
186. *A Guide to Masonry Construction.* CMA Aust. and South Australian Clay Brick Association, 1975.

187. ACI COMMITTEE 316. "Roadways and Airport Pavements". *ACI Publication SP-51; ACI Journal,* 72, 8 (August 1975), pp. 435-39.
188. GULLAN, G. T. "Shotcrete for Tunnel Lining". *Tunnels & Tunnelling,* 7, 5 (September/October 1975), pp. 37-47.
189. MATTHEWS, A. A., *et al.* "Use of Shotcrete for Underground Structural Support". *ACI Publication SP-45,* 1974.
190. GRIMM, C. T. "Metal Ties and Anchors for Brick Walls". *ASCE Journal,* 102, ST4 (April 1976), pp. 839-58.

PART 6

FABRICATION PROCESSES AND FUNDAMENTALS

CHAPTER 29

HEAT AND RADIATION RESISTANCE

29.1 REFRACTORY CONCRETE

29.1.1 MATERIALS AND MIXES

Refractory concrete is of two varieties, ordinary and superduty or very-high-temperature grades. They are made respectively with grey high-alumina cement (see Section 3.1.2(b)) and white calcium aluminate cement (e.g. "Secar 250"), refractory or high-temperature aggregate and water. The first-mentioned hydraulic cement is monocalcium aluminate ($CaO \cdot Al_2O_3$), while the second is tri-calcium penta-aluminate ($3CaO \cdot 5Al_2O_3$), which has an extra high alumina content with negligible amounts of iron oxides and silica. Each of these cements has a normal initial setting period of 2-4 hours and subsequently hardens with very great rapidity (see Sections 3.1.1 and 3.1.2).[104]

Types of concrete made with various cements and aggregates, and their service temperature limits under load, are given in Table 29.1 and Fig. 33.9. Ordinary portland cement concrete begins to disintegrate at 350-400°C due to the dissociation of hydrated lime into quicklime and water. In soaking heat, a 50 per cent factor of safety against spalling is required with siliceous aggregate, which undergoes a marked change in crystal structure and volume at 573°C. Siliceous sand mortar and ordinary concrete containing heat-resistant aggregate may be used to 300°C.[99]

High-temperature aggregate, in increasing order of refractoriness from 1500°C to 1800°C, consists of: molochite (i.e. fired kaolin; 1500°C); sillimanite (i.e. fired kyanite; 1550°C); calcined bauxite; high-alumina firebrick; chrome and chrome-magnesite (1650°C);

527

TABLE 29.1   **TYPES AND TEMPERATURES**

Cement	Aggregate	Type of Concrete	Service Temperature Limit
High-alumina or portland*	Siliceous sand or gravel	Structural	250°C
High-alumina	Fine-grained basic igneous rock (e.g. basalt and dolerite). Clay brick, scoria and expanded shale	Heat-resisting	1000°C
	Firebrick	Ordinary refractory	1350°C
	High-temperature	Superduty refractory	1350-1600°C
Calcium-aluminate	Firebrick	Superduty refractory	1450°C
	High-temperature	Superduty refractory	1500-1800°C

* Sulphate-resisting to resist attack by sulphur-bearing condensate.

carborundum (i.e. silicon carbide; 1700°C); and corundum (i.e. fused bauxite or fused alumina; brown: 1700°C, white: 1800°C).[48] Ordinary insulating refractory concrete consists of 1 part high-alumina cement and 4-6 parts by volume of vermiculite (hot-face limit, 1000°C), calcined diatomite (1000°C), porous fireclay (1350°C), or expanded clay or shale (hot-face limit, 1000-1200°C). Old firebrick (i.e. aluminosilicate) must be cleaned of slag and jointing material before being crushed and used as refractory aggregate.

The maximum size of aggregate used should not exceed one-fifth the dimension of the thinnest cross-section, with an upper limit of 40 mm to facilitate consolidation of the placed concrete. To enhance workability and assist in forming a fired or ceramic bond, the part of the combined aggregate that is capable of passing a 0·15-mm mesh sieve should be from 10 to 15 per cent by weight of the whole. Soaked half bricks may be used as "plums" in refractory concrete foundations. Ordinary mixes are given in Table 29.2.

Increased strength and reduced refractoriness accompany an increase in cement content. Mixes with up to twice the quantity of cement shown are advantageous for operating temperatures below 1000°C (e.g. kiln car tops and dampers). If the service temperature is near the upper limit given for the concrete, a reduction in cement content will raise the refractoriness of the mix. Prepared castable refractories are obtainable from proprietary sources.

TABLE 29.2   MIXES FOR ORDINARY REFRACTORY
CONCRETE

Mixes for Various Thicknesses	Quantities of Cement and Firebrick Aggregate per m³ of Concrete
Mix A (150 mm thick or over)	
Aluminous cement: 1 m³	0·29 m³ or 419 kg
Fine aggregate (3·35 mm down): 2 m³	0·58 m³ or 684 kg
Coarse aggregate (19·0-3·35 mm): 3 m³	0·87 m³ or 804 kg
Mix B (40-150 mm thick and for general purpose)	
Aluminous cement: 1 m³	0·32 m³ or 458 kg
Fine aggregate (3·35 mm down): 2 m³	0·63 m³ or 751 kg
Coarse aggregate (19·0-3·35 mm): 2 m³	0·63 m³ or 625 kg
Mix C (for mortar jointing, or work of small dimensions)	
Aluminous cement: 1 m³	0·39 m³ or 558 kg
Fine aggregate (3·35 mm down): 2·5 m³	0·97 m³ or 1176 kg

The workability of ordinary refractory mixes is greatly improved by an admixture of up to 10 per cent, by weight, of ground raw fireclay or bentonite to the cement, or by a wetting or air-entraining agent. A mix for stopping leaks is made of equal parts by volume of high-alumina cement, fine firebrick aggregate, ground raw fireclay and short asbestos fibre mixed with water.

For superduty refractory concrete, the mixes given in Table 29.3 are satisfactory for the majority of applications. The quantities of materials (dry) required to make 1 m³ of this concrete are given in Table 29.4.

TABLE 29.3   SUPERDUTY REFRACTORY CONCRETE

Calcium-aluminate Cement	Aggregate	
	Sieve Size	Volume
1	19·0 mm to dust	4-5
1	9·5 mm to dust	3-4
1	3·35 mm to dust	2-3

Steel reinforcement should be used only where really necessary and placed in positions where it will be relatively cool. A wide steel mesh is preferable to steel rods and the reinforcement may be fabricated with spring compensators for increased elastic deformation. The steel elements should be of sufficient diameter to prevent them from being oxidised away by hot gases reaching them through cracks in the concrete.

Reinforcement should be connected to steel exposed to the atmosphere and it should not be heated above 400°C, to obviate spalling and yielding effects.[1] It must be coated with paper, card-

TABLE 29.4   **QUANTITIES**

Aggregate	Mix Preparations		Mass in kg/m³ of Concrete	
	Volume	Mass	Cement	Aggregate
Molochite				
9·5 mm down	1 : 4	1 : 5	320	1600
3·35 mm down	1 : 3	1 : 3·8	400	1520
Sillimanite				
9·5 mm down	1 : 4	1 : 5·8	320	1840
3·35 mm down	1 : 3	1 : 4·4	400	1760
Corundum				
9·5 mm down	1 : 4	1 : 8	320	2560
3·35 mm down	1 : 3	1 : 6	400	2400

board or thick grease which (by burning off at an early age) allow differential thermal movement without rupturing the concrete. As there is no bond between the two materials, refractory concrete construction cannot be designed in the traditional way.

The reinforcement for pneumatically applied mortar linings to furnace stacks and temperature vessels consists of 75 mm × 75 mm × 3·15 mm welded mesh. It is fastened to hooked-end studs, blank nuts on edge, or 6 mm vertical rods crimped at 300 mm intervals, each of which is welded to the scaled and sand-blasted steel shell. The reinforcement must be pegged or dowelled firmly to an existing concrete surface, such as in an exhaust stack or a missile release structure, after it has been chipped and cleaned. Steel mesh should be covered by at least 20 mm of refractory mortar, preferably 25 mm and, to prevent segregation of an ejected mix, no two wires should touch each other in parallel at laps.

A 65 mm thick lining is customary for self-supporting stacks, but 40 mm is sufficient for guyed stacks and temperature vessels. The thicker lining is applied in two coats, the second before the underlayer has become hard, but is sufficiently firm to resist the gunning pressure. The thinner lining is built up to the full thickness at once, over areas of about 1 m². The usual mix is one volume of high-alumina cement to three volumes of aggregate passing a 9·5 mm screen and well graded, the fineness modulus being between 2·5 and 3·5. The final mix as placed would be about 1 : 2·5 due to loss of aggregate by rebound.

In rocket-launching structures, the lining has to withstand a high thrust for about 10 seconds, the temperature of the flame being 1800°C. A superduty refractory lining at least 75 mm thick, using white corundum aggregate, should therefore be used.[2]

29.1.2 MIXING, PLACING AND CURING

The mixer and tools should be clean and free from portland cement, lime or plaster. Permanent, insulating formwork, such as asbestos fibreboard, should not be bonded with free sodium silicate. These materials seriously upset the natural setting and hardening characteristics of high-alumina cement. Porous refractory aggregate (e.g. crushed firebrick) is first mixed with water for 3 minutes, or until initial suction is taken up. Excess water is drained off, the aluminous cement is added and the mix is brought to a plastic consistence with a further 2 minutes mixing. Aggregate that is not very absorbent is mixed dry with cement until the batch is of uniform colour. The amount of water added should be only sufficient to effect a plastic consistence and thorough consolidation when the concrete is placed. Mixing should be for at least 3 minutes.

Timber formwork is made watertight before use by a plastic or waterproof-paper lining or by plastic clay at the joints. Hardened concrete surfaces at bonded-type construction joints must be well wetted just prior to placing new concrete against them. The bond at joints can be broken, where necessary, by a coating of oil and temporary partitions or stop-ends in the formwork enable large structures to be cast in alternate bays. The consolidation of placed concrete is done manually or by vibratory means, the ramming of insulating concrete being very light, and the trowelling of screeded and wood-floated surfaces should be delayed until near the time of initial set of the cement. Pneumatic mortar should not be heavily trowelled, as any dragging action on the unset mortar may initiate cracking or cleavage. However, major protuberances may be gently sliced off with a steel trowel and a smooth surface obtained by lightly brushing with a soft brush occasionally dipped in water.

Placed concrete and mortar must be kept wet and cool for 24 hours, starting at the time of initial hardening of the concrete. Nonsupporting formwork should be stripped as soon as possible, usually within 4-6 hours of casting, and effective curing be started with cold water by continuous sprays or immersion. Steel forms need not be stripped early, but they should be kept cool with water played on the outside. Failure to cure the concrete by copious watering or spraying for 24 hours, starting a few hours after casting, can lead to a serious loss in quality. A pneumatic mortar lining in a chimney stack should be cured by fog sprays, while the top is covered to prevent through-draught (see Section 3.1.2(b) and BS 4207).

29.1.3 REMARKS

Refractory concrete has a high compressive strength by hydrated bond when first made and, when in service, by a physicochemical reaction on the fired face (between the cement and the aggregate) whereby a fired or ceramic bond is produced. On heating the concrete for the first time, its compressive strength at temperatures below those causing ceramic bond drops to about one-fifth of that possessed originally. With ordinary refractory concrete, the cohesion of the cement is low between about 700°C and 1000°C, the temperature being higher with superrefractories. The residual strength is adequate for the maintenance of physical form in ordinary industrial applications, but at temperatures in the region of 1100°C, the strength (subsequent to cooling) is increased by the formation of a ceramic bond.

The abrasion-resistance of concrete to be used below about 1000°C (e.g. in precast fireplaces) can be increased physically by the use of high-alumina aggregate, such as crushed saggars and fine-grained basic igneous rock, including stone grit that is low in silica content (see Table 29.1). Silica sand is not used; and pressure-vibration with a low water/cement ratio, followed by proper curing, is advisable. Vitrification at a reduced temperature can be brought about by an admixture of sodium silicofluoride. For special abrasion-resistant linings of, say, jet cells and rotary kilns, precast units may be prefered to a state of vitrification before use.

With superduty refractory concrete, a ceramic bond is obtained with crushed firebrick aggregate when the temperature reaches 1200°C and about 1400°C when corundum is used as the aggregate. The fusion points of superrefractories are very little higher than their maximum working temperatures. These temperatures are determined by refractoriness-under-load tests and correspond to the temperature at which the subsidence under a load of 193 kPa amounts to 10 per cent when the concrete is heated on all sides. Refractory concrete walls, however, are usually heated on one face only, the bulk of the concrete remaining at a lower temperature than that of the hot face. They easily withstand the loading which, in any case, is much higher than that ever reached in practice.

Working conditions may, in addition to high temperature, include such factors as chemical corrosion, thermal shock and abrasion or erosion, and they must be taken into account when selecting a refractory. In this regard, all refractory concrete is highly resistant to thermal shock and corrosion by aggressive condensate in flues. Calcium aluminate cement, because of its

purity, is highly resistant to slags, particularly when used with
white corundum.

Calcium aluminate corundum concrete may be used for casting
resistance-element holders in electric furnaces. It has a long life
also as a lining in the clinkering zone of rotary cement kilns. The
cement is not attacked by carbon monoxide and it has special
applications in the chemical, oil-refining and nuclear fission
industries. Superrefractories can be used as a hot-face lining to
ordinary refractory concrete, provided that the temperature at the
interface does not exceed 1300°C. Expansion grooves (about 3 mm
wide) may be made at intervals to relieve thermal stresses under
working conditions at the internal surface.

Thermal conductivity at 1000°C varies from 1·11 W/m.°C for
firebrick and molochite aggregate concrete, to 1·44 for sillimanite
and 2·31 for corundum aggregate concrete, their bulk densities
(fired) being 1920, 2160 and 2800 kg/m³, respectively. The
thermal conductivity at working temperature of insulating refractory
concrete (480-1280 kg/m³) varies from 0·108 to 0·505 W/m.°C.
A relationship between inside and outside wall temperatures, where
refractory concrete is equivalent to 1·33 standard brick, is shown in
Figure 29.1.

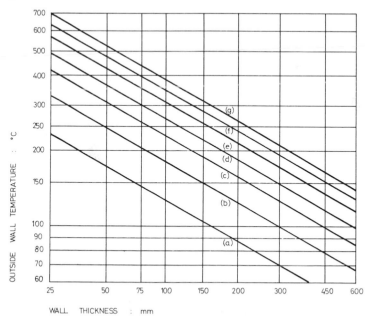

Fig. 29.1 Relationship between inside and outside wall temperatures. Inside
wall temperatures: (a) 300°C; (b) 500°C; (c) 700°C; (d) 900°C; (e) 1100°C;
(f) 1300°C; (g) 1500°C.

Refractory concrete construction (with firebrick aggregate) has the same dimensions at 1100°C as it had when cast, because the small contraction caused by the elimination of water on first firing practically neutralises the thermal expansion. Small refractory units or structures can therefore be cast monolithically. In large structures, joints can be confined to thin or butt joints every 1·2-1·5 m to allow for slight contraction on cooling. The joints close again when furnaces made with firebrick aggregate concrete are reheated. When corundum is used as aggregate, expansion joints are required, the coefficient of thermal expansion of the concrete being $2·5 \times 10^{-6}$-$5·0 \times 10^{-6}$ per °C (which is low). Whatever maximum temperature is reached on first heating corundum concrete, the resultant contraction after cooling is a little more than 0·3 per cent.

A high-density mix is used for concrete off-peak-power storage heaters (see AS 1206 and BS 3456, Section A11), and refractory concrete of all kinds is ready for use some 24 hours after casting. A period of air drying is desirable, but not essential. The first heating to 600°C should be done slowly, rising 50°C per hour, but subsequent heating (and cooling) may be done at any speed.[3] Standards on refractories in general are given in Table 29.5.

29.2 HEAT-RESISTANT PAINT

Pure silicone resin can resist temperatures up to 650°C, silicone or butyl titanate can serve up to 375°C, and silicone alkyd resin can go to 200°C, each paint being mixed with aluminium or zinc pigment. Continuous temperatures up to 180°C are resisted by an intumescent paint, "Flintkote 702" bituminous emulsion and silicone-type coal-tar epoxy resin paint, 0·25 mm thick (see AS 1530 and BS 476).

29.3 RADIATION SHIELDING

Radiation shielding should provide protection against X-rays, gamma rays and neutrons. These rays must be attenuated sufficiently to prevent permanent damage to persons exposed to them (BS 3385). The dosage of residual radiation that can be taken by a person working in it for 48 hours a week should not exceed the equivalent of 0·3 Röntgen. (The Röntgen is the international unit for quantity of X-ray and gamma radiation.)

Data are available for use in the design of concrete shields for protection against X-rays and gamma rays of energies up to

TABLE 29.5   **STANDARDS ON REFRACTORIES**

Title	AS	BS	ASTM
Air-setting refractory mortar for boiler or incinerator*	—	—	C178
Alumina-silica-base castable refractories for boiler furnaces	—	—	C213
Bulk density, porosity and water absorption	1774.5	—	C357; C493
Castable refractories	—	—	C401
Chemical analysis of refractories	R28	1902	C571-75
Chrome brick, chrome-magnesite brick, magnesite-chrome brick, and magnesite brick	—	—	C455; C544
Classification of silica refractory brick	—	—	C416
Crushing strength and modulus of rupture	1774.1	—	C133; C268
Disintegration of refractories in carbon monoxide	R31†	—	C288
Drying and firing shrinkage of castable refractories	R31*	1902	C179; C269
Fireclay and high-alumina refractory bricks	1617	—	C27
Fireclay refractories for boiler and incinerator services	—	—	C176
Fireclay refractories in the petroleum industry	—	1758	—
Insulating firebrick for linings of industrial furnaces	—	—	C434
Methods of testing refractory materials	R31†	1902	C38; C113
Mullite refractories	—	—	C210; C583
Panel spalling tests	R31†	—	C38; C107; C122; C180
Preferred sizes of refractories	1618; 1619	2496; 3056	—
Pyrometric cone equivalent (PCE) of refractory materials	1774.10	—	C24
Refractories for			
boiler and incinerator services	—	—	C106; C176
heavy and moderate duty stationary boiler services	—	—	C64; C153
malleable iron furnaces and annealing ovens	—	—	C63
molten glass	—	—	C621; C622
steel pouring pits	—	—	C435
Refractoriness under load	R31†	—	C16; C546
Refractory granular dolomite	—	—	C468; C492
Sampling and testing of insulating firebrick	R31†	2973	C93; C155
Silica refractory brick	—	5187	C416; C439
Single and double-screened ground refractory materials	—	—	C316
Size and bulk density of refractory brick	1618; 1619; R31†	1902	C20; C134*
Specific gravity	R31	1902	C135; C604
Test code for kilns for clay ware and refractories	—	1081	—
Warpage of refractory brick and tile	—	1902	C154
Zircon refractories	—	—	C545; C576

* Further tests are specified in ASTM Standards, Part 17.
† Methods 0-25.

16×10^{-12} J. The calculation of thickness of shielding is based upon the concept of half-thicknesses and the cumulative effect of succeeding layers of material, each of which halves the intensity of radiation penetrating it. Radiation may be divided into three types: direct, scattered and leakage. The first is dealt with in National Bureau of Standards *Handbooks 73 and 76*[1] on protection design, and in Reference 44. Scattered radiation is about 0·1 per cent of direct radiation and radiation leakage is indeterminable. Radiation is not restricted to the primary radiation from a radioactive material, but all bodies (including air) subjected to primary radiation become radioactive and subsequently emit secondary radiation, which may be injurious to health.

At a given distance, thicker protection is required against high-intensity radiation than against low. The same thicknesses of steel, ordinary dense concrete and calcareous stone weaken X-rays and gamma rays in proportion to their density (i.e. 7·8 : 2·3 : 2·7). As the intensity of radiation increases, the rate of increase of thickness required is much greater for lead than for concrete and steel. This feature consequently limits the use of lead to protection against radiation of low intensity. Heavy concrete is a practical means of attenuating X-rays and gamma rays, of which the latter are more penetrating than the former, and a concrete shield of high water content is very effective in slowing down and capturing neutrons.

Concrete made with limonite ore (containing from 8 to 12 per cent water of crystallisation) is as efficient as water for this purpose. Compared with ordinary dense concrete, a 900 mm wall of this concrete is superior by factors of 20 for gamma radiation and 280 for neutron shielding. (A concrete reactor shield is about 3 m thick for biological protection and a neutron shield should not become dry.) Shielding against neutrons is much more sensitive than that against gamma rays to changes in composition of the concrete, particularly in relation to water content and admixtures (if any) of certain elements such as lithium, boron and cadmium. For fast neutrons, this effect is less marked and the half-thicknesses required will increase fairly regularly with increasing neutron energy.[5-10, 13, 39, 72]

A concrete shield must fulfil the requirements of thorough compaction and homogeneity, requisite thickness, and the ability to operate at high temperatures for indefinite periods without drying out. The aggregates used include such materials as scrap iron, cast and pig iron, steel punchings, iron shot, ball bearings, ilmenite ($FeO \cdot TiO_2$), haematite (Fe_2O_3), magnetite (Fe_3O_4), goethite

($Fe_2O_3 \cdot H_2O$), limonite ($2Fe_2O_3 \cdot 3H_2O$) and barytes ($BaSO_4$), although not all of them will enable concrete to be made with bulk densities of 4500 kg/m^3 or more. For instance, concrete made with barytes aggregate containing 80 per cent barytes and 10 per cent iron oxide weighs 3500-3700 kg/m^3. Barytes aggregate is of a friable nature and decrepitates if heated beyond 350°C. It has a hardness of 3 (as compared with 7 for quartz) and a specific gravity of 4·2. Concrete made with it has a poor resistance to weathering if exposed externally. On the other hand, concrete containing well-graded, crushed cast-iron aggregate or a mixture of steel punchings and ilmenite sand can be made with a bulk density of over 4800 kg/m^3 (ASTM C637, C638).

Heavy concrete is fabricated by puddling or aggregate immersion, grout intrusion or prepacked construction, as well as by conventional mixing and placing methods. By the first-mentioned method, interlocking blocks of trapezoidal or complex shape are cast by feeding and compacting steel or iron punchings into cement mortar made with portland cement, iron-shot aggregate and a plasticising or wetting agent. With heavy aggregate and intense vibration, precasting minimises formwork pressure and both the cost and possible distortion of the formwork. Concrete with a bulk density of 6000 kg/m^3 can thus be made with ten 40-kg bags of cement per m^3, a water/cement ratio of not much over 0·4 by weight, a cylinder strength of over 40 MPa and a flexural strength of about 3·5 MPa at 28 days. A steel-trowelled finish can be easily applied and the blocks should be set with closely fitting and well-bonded joints.

In the second method, aggregate is placed in position and grout is pumped continuously under pressure into the voids. When processed barytes is used, the aggregate is graded in sizes of 20 mm, 28 mm, 40 mm, 75 mm and up to 125 mm, with everything less than 14 mm nominal size discarded, and it is packed *in situ* with 33-40 per cent of voids. The grout consists of 1 part of cement to 1·5 parts of crusher-run, heavy fine aggregate passing a 0·60-mm sieve, 0·06 per cent of wetting agent (based upon 100 per cent active constituent) and 0·01 per cent of flaky aluminium powder. It is prepared with a water/cement ratio of 0·5 by weight. Prepacked concrete, made with well-graded steel aggregate and magnetite, goethite or limonite grout has a bulk density of up to 6000 kg/m^3.*

---

*Magnetite is somewhat tougher than limonite and has a specific gravity of 4·8 to 5·2, as against 3·6 to 4·0 for limonite. Goethite is usually stronger, denser and more uniform in quality than limonite.

Higher bulk densities can be obtained by using fine steel shot (e.g. 30 μm $\times$ 10 nominal size) as aggregate in the grout (AS 1627, Part 4 and BS 2451).

Grouting hoses of 20 mm diameter are kept immersed in the grout in 50 mm diameter slotted sounding tubes that are used for checking the level of the grout. Vent pipes allow for the escape of air as the grout rises. Where obstructions to pouring concrete in the formwork are numerous, the prepacked method of placing has much to commend it for casting a shield with voidless, crack-free, homogeneous material.

With conventional mixing, the same grading, mix proportions (by volume) and types of compaction are used as for ordinary mass concrete. A mix of 1 part cement to, say, 8 parts (by weight) of barytes aggregate (40 mm nominal size), with a water/cement ratio of 0·5, plasticised with a wetting agent and mixed for at least 2 minutes, has a slump of 80 mm and a 7-day compressive strength of 20 MPa. A leaner mix may consist of 1 part cement : 4·5 parts barytes sand : 6·5 parts barytes coarse aggregate (40 mm to 7 mm nominal size), with a water/cement ratio of 0·66 by weight, and a slump of 100 mm. Formwork must be braced to withstand the extra pressure set up by heavy concrete. Pressure is reduced by continuously placing the concrete in slowly rising lifts. Internal vibration is usually supplemented with external, but both must be carefully done when heavy aggregates are of a friable nature or segregation is imminent.[115]

The densities of hardened prepacked concrete and conventionally mixed concrete, made with similar aggregates of 40 mm nominal size, are approximately the same. The strength of the former is one-half to two-thirds that of the latter and its shrinkage characteristic is also less, being only about one-half that of ordinary dense concrete.

Barytes concrete has relatively low drying shrinkage, but its coefficient of thermal expansion is approximately twice that of ordinary dense concrete. Values of specific heat, thermal conductivity and diffusivity are approximately one-half those obtained with natural aggregate concrete. Additions of iron oxide increase the thermal conductivity of concrete approximately 1 per cent for each 1 per cent of oxide added.

Standard specifications for aggregate in concrete (see Section 3.3.1) stipulate that they shall be free from certain deleterious substances, including sulphides (which occur in iron pyrites). Substances that are chemically active, when present in sufficient quantity, can adversely affect the physical properties of concrete through disruptive expansion. Caution must be exercised in using sulphur-bearing aggregate, because of its corrosive influence on steel, and in executing work during hot weather (see Section 32.2).

As the rusting of steel and iron requires a supply of water and oxygen in an environment with a pH value of about 8 or less, and the pH value of portland cement concrete mixes is between 9 and 10 (atmospheric carbonation seldom reduces it to below 8), iron-based aggregate in sheathed mass concrete is unlikely to undergo physico-chemical changes connected with rusting and cause expansion. Scrap punchings from steel fabricating shops are usually covered with a film of oil or grease, and it should be removed by steam cleaning or other means before the material is used. Scrap reinforcing rods, cut to order in short lengths, are more consistent in grading than random punchings and are likely to give superior results.

The physical properties and aggregate-making qualities of ore vary with the source, and they should be investigated carefully before such aggregate is procured and processed. It is advisable to test aggregate from ore for stability of performance under simulated working conditions. In order to withstand a working pressure of over 1·4 MPa and to limit the service temperature of portland cement concrete to below 300°C, the hot face of a normal shield is sheathed by a heavy steel lining with a carbon dioxide cooling system behind it. Crack distribution and attenuation (within 1 mm size) are effected by deformed steel reinforcement, some 0·2 per cent near both faces of the shield.

Portland cement concrete is the most economical material for shielding against X-rays and gamma rays of energies greater than 80×10^{-15} J. The added cost of using heavy aggregate is often justified by the resultant saving in weight and space of the barrier. Where heavyweight aggregate concrete is to be subjected to large blast loadings, its thickness may be governed by structural needs rather than shielding requirements. Prestressing techniques may be used in protective shields with an effective cooling system, heat-resistant insulation and surface sealant. The walls are checked before use for their effectiveness in attenuating radiation (BS 4975 and ASTM E185; see Plate 23).[34, 41, 42, 46, 58, 66, 67, 79]

Should high-density ferrophosphorous or ferrosilicon slag aggregate be used in a concrete shield, adequate ventilation is needed to remove an early emission of hydrogen. Its volume, which is dependent on conditions of free moisture and temperature, may exceed twenty-five times that of the concrete. Data on ionising-radiation shielding and lead bricks for radiation shielding are specified in BS 4094 and BS 4513 respectively. High-density, heavyweight concrete can be tailored to fit small volumes and serve as the counterweights for cranes. Certification recommendations for personnel engaged in nuclear-concrete inspection and testing are contained in Reference 8.

CHAPTER 30

MASS CONCRETE

30.1 MASSIVE CONSTRUCTION

Massive concrete construction, such as that used in dams, should have low thermal volume change, a high degree of impermeability and resistance to cracking, and enough strength to carry loads safely or without undue increase in sectional dimensions. Before concreting begins site foundations must be examined and prepared thoroughly, and fissured areas must be pressure grouted (see Section 24.3).

Of prime importance is the rise in temperature of freshly placed concrete, due to the exothermic reaction that accompanies the hydration of cement. In ordinary construction, the heat of hydration is soon dissipated and thermal expansion remains within acceptable limits. As the mass increases in size, the additional concrete serves as insulation in reducing the rate of dissipation of heat. By using prestressed rock anchorages near the upstream face of gravity dams, however, their design mass can be reduced some 30-40 per cent. The temperature produced by heat evolution in massive ordinary concrete with a low cement content may range from 40°C to 50°C, a high cement content causing a further increase of 10°C to 30°C (see Sections 3.1, 10.2, 24.4 and 34.2.3; see Plate 25).[11-15, 19, 61, 113]

The greatest change in volume is caused by the difference in temperature between the maximum reached during hardening and the final stable temperature of the mass. Unless this range of temperature is controlled, excessive tensile stress may arise from steep temperature gradients between the interior and exterior, and by restraint on volume change. Should the stress exceed the tensile strength of the concrete, the mass may fracture to the possible detriment of its monolithic nature, designed stress distribution and potential service life. Cracks thus created are located at boundaries at bedrock, inspection chambers, contraction joints and exterior surfaces (see Section 31.2.1 and Reference 105 of Part 1).[69, 73, 106]

Two basic methods of temperature control are employed: precooling and postcooling. In the first method, concrete components are precooled to as low as 4°C or cooled with chipped ice that serves as mixing water. Liquid nitrogen (− 184°C) may be used as a refrigerant to hold water in a lagged silo at 2°C (AS 1894). This medium enables coarse aggregate to be cooled to 10°C or less in 20-30 minutes, as a means of bringing the concrete temperature at placement to below 20°C in hot weather (see Fig. 9.2 and Section 32.2).[94, 118]

In the second method, cold water is circulated through a zigzag pipe system embedded in the concrete, the flow being reversed every 24 hours. Each lift is cooled for a sufficient length of time to ensure that the temperature of the mass is kept below 30°C at 5 days. Efficacious monitoring of this operation may be done by automatic temperature recording and metering means of the coolant.[128] By postcooling from 3 to 6 weeks, an arch dam can be brought to a stable temperature, or 3°C below it, prior to the requisite grouting of contraction joints (see Reference 15 of Part 1).[70, 107]

The following materials and measures are used in addition to direct coolant procedures for temperature control.

Low-heat portland cement, or in certain instances, other special cement with a low heat of hydration (see Sections 3.1.1 and 3.1.2).
Lowest practicable cement content in the interior of the mass. Partial replacement of the cement (e.g. 20 per cent) by a pozzolan (see Section 17.1).[130]
Set retardation by a lignosulphonate admixture (see Sections 6.4 and 10.2).
Favourable construction procedures for suitable temperature gradients and uniform quality, such as a limited thickness of lift and difference in height between adjacent monoliths some 15 m thick.[11-19, 24, 36]

In massive concrete structures, the minimum cement content is governed primarily by the requisite compressive strength and by such qualities as impermeability and weathering resistance at the exterior. The compressive strength of the concrete should be about 30 MPa at 90 days or four times the computed stress in the structure at an age of 1 year. With effective facilities, vibratory compaction and constructional techniques, it has been found possible to use as little as 3 (40-kg) bags of cement per m³ of concrete for the interior of a dam, provided that its exposed faces contain 5·5 bags per m³ to a depth of 1·5-2·5 m. The nominal size of coarse aggregate in these mixes is 150 mm and 75 mm, respectively. In less massive work with lower quality control, the cement content may be up to 5·5 and 7 bags per m³, respectively (see References 15, 34 and 44 of Part 2; also Section 23.1.3).

The production of satisfactory lean mass concrete requires the use of carefully graded aggregate up to 150 mm maximum size, a low water/cement ratio, air-entrainment (see Section 16.1), close control and special equipment for the fractionation and recombination of fine aggregate into optimum gradation. This type of equipment consists of rod-mills or rake classifiers and hydraulic sizers. Great technical and economic advantages arise from the use of pozzolanic concrete with water curing extended from 2 to 3 weeks.[61]

Lifts are usually 1500-2300 mm thick, and they are placed with an exposure period of 3-14 days between them. During winter months, surfaces should be insulated in situations where they are likely to be damaged by very low ambient temperatures or early frost action and cracking, while the interior is in a state of thermal expansion (see Section 32.1).[22] Joints can be made watertight with vertical waterstops (polyvinyl chloride, rubber or crimped copper; see Section 11.1.2(d)), with rubberised-bitumen cores that can be remelted *in situ* (e.g. at multiple buttresses) by circulated hot oil, by cooling and grouting operations,[12, 24] and by the use of proper green-cut, clean-up and placement procedures (see Sections 7.2 and 9.1). By means such as these, coupled with project-performance instrumentation (see Appendix IV, 4(b)), it is possible to deal satisfactorily with thermal volume change, which is otherwise a fundamental problem in massive concrete construction.[61, 74]

30.2 UNDERWATER CONCRETING

Underwater concrete may be placed on a prepared foundation, where necessary, by means of bags, skips, tremie tubes or by using grouted coarse aggregate. The concrete should be placed without falling through water, without puddling and with as little disturbance as possible. Plasticised or air-entrained concrete may be used to facilitate placement and 10 per cent additional cement should be used to compensate for possible loss of cement. In marine work, the cement content of the mix (with or without a pozzolan) may be increased to 10·5 (40-kg) bags per m³.

The surface of the concrete may be roughly levelled by heavy screeds worked off rails by two divers.* Slurry which has formed on concrete placed the previous day should be removed before concreting is resumed.

30.2.1 BAGGED CONCRETE

Hessian or canvas bags are partly filled (two-thirds to three-quarters) with concrete and placed soon afterwards in a manner similar to that used in protective sandbag walling. The bags are tied or preferably sewn up and lowered to a diver, who builds them in place and flattens each bag in the process. They are built in bond, with the mouth of each bag away from the outside surface. Courses of bags may be held together by driving steel spikes through them after they are placed in position.

Concrete for filling in small pockets is sometimes sent down in bags to the diver, who empties it into position. A variant of this method is the use of a canvas bag or "sausage", about 450 mm in diameter and 1·2 m in length, which is filled with concrete and lowered to the diver. While the bag is suspended, the diver releases a slip-knot at the bottom and allows the concrete to flow out. An "elephant's trunk" is a similar device, with the lower end of the bag turned up and secured by a chain. Bagwork in general is slow and laborious and therefore expensive, and should only be used where quantities do not justify more elaborate methods or where other considerations make it necessary.

30.2.2 SKIPS

Concrete is placed through the bottom doors of a self-tripping skip, which is closed at the top with a canvas cover or metal lid. After tripping, the skip is raised slowly until all the concrete has flowed out.

30.2.3 TREMIE TUBES

Large volumes of concrete, more than 900-1500 mm thick, are expeditiously placed by tremie tubes, which extend from slightly below the surface of freshly placed concrete to above the water surface. The tubes are in 1 m, 2 m and 4 m lengths, 150-250 mm diameter, and have quickly detachable joints. A 200 mm diameter

---

*Diving procedure is included in "Foundations", *Civil Engineering CP No. 4,* Institution of Civil Engineers, London.

tube is suitable for a 150 mm slump mix containing 40 mm nominal-size aggregate.

Each tremie tube is kept full continuously, by keeping the discharge end buried in newly placed concrete and by admitting sufficient concrete through the bottom door of a hopper. The capacity of the hopper is at least equal to that of the tube which it feeds. When a large area is to be concreted, tremie tubes are located at about 2·5-5 m centres. They are supported by staging or barges and can be moved vertically, but not laterally. The tubes are replenished successively from hoppers, so that the level of the placed concrete is gradually raised over the whole area.

Tremie tubes are charged by inserting at their top a bag filled with straw or similar material. As the bag is pushed down the tube by concrete, water is expelled from the bottom. The rate of flow of concrete in a tube may be varied by raising or lowering it, care being taken to keep the discharge end buried in concrete. When about 600 mm thickness of concrete has been placed at a tube, the hopper feeding it is transferred to the next one. By this means, concrete can be placed on a slope of up to about 20 per cent.

Retarded, air-entrained concrete may be used to advantage in mass tremie work, the initial set of the cement being retarded about 50 per cent. An appropriate admixture improves the flow and cohesive properties, promotes uniformity and reduces the rate of heat generation of the concrete (see Sections 10.2, 16.1 and 30.1). The tremie system can be effectively used for the construction of underground concrete diaphragm walls by the bentonite-slurry, tremie-displacement trench method (see Section 24.3.1).[74]

This technique enables reinforced concrete enclosure walls to be constructed by trenching or adjacent-boring methods, before making a main excavation in built-up areas. The bentonite-mud, which gives temporary ground support, is progressively reused; and cage-reinforced, tremie-placed concrete walls are successively placed and secured with prestressed anchors, as the main excavation is being carried out. Precast units may be used as a modified form of concrete placement.[132-135]

30.2.4 HYDRO VALVE

This equipment consists of a flexible nylon pouring pipe (e.g. 600 mm diameter), with a rigid cylindrical shield at the bottom. Concrete fed from a hopper moves downwards under gravity but, by having to force the pipe walls apart under increasing water pressure, the concrete can be slowly and gently placed on previously deposited concrete to a finish tolerance of 100 mm.

Single or multiple units are suspended from a crane, pontoon or gantry, and concrete placement to requisite level is progressively effected with an angle of repose of 1 in 5. The concrete is otherwise handled by traditional means (see Section 7.2.1), and the system facilitates the casting of reinforced concrete slabs underwater.[109]

30.2.5 PREPACKED CONCRETE

Coarse aggregate (28 mm or upwards in size) is placed, compacted and infilled with colloidal grout. The grout is pumped through pipes which reach to the bottom of the aggregate. It is made by intimately mixing cement, sand and water in a special mixer, or by adding cement-dispersing and plasticising agents to ordinary grout. Colloidal grout is highly plastic and can be used in underwater work without appreciably absorbing water. Further information is given in Section 24.4.

CHAPTER 31

FUNDAMENTAL CHARACTERISTICS

31.1 SETTING AND HARDENING*

31.1.1 MOIST CURING

Setting is the chemical action which begins when water is added to cement and causes a gradual disappearance of the plastic nature of the paste. Hardening is the development of strength that accompanies the chemical combination of cement and water.[17, 43, 47, 51, 92, 95] A given volume of cement paste which is free of air consists of solids and water. One of the reactions that takes place in the paste is represented by the equation

$$3CaO \cdot Al_2O_3 + 3CaSO_4 \cdot 2H_2O + 25H_2O = 3CaO \cdot Al_2O_3 \cdot 3CaSO_4 \cdot 31H_2O$$

Here, two solids in the original cement react with water to form a new solid phase, which occupies more space than that occupied by the original two solids, but less than that of the three components.

This is true for all of the reactions which occur between cement minerals and water (see Section 3.1.1). As hydration proceeds under air-curing and drying conditions, concrete shrinks in volume to a mass occupied by solids and voids incompletely filled with water. Two major causes of this deformation are a change in fundamental structure of the mass and the hydrostatic tension that results through a loss of free water. On the other hand, if concrete which has set under water is kept in a saturated state, it absorbs water slowly and swells.

---

*Factors concerning Setting and Hardening, Dimensional Change, Deformation and Carbonation (described in this Chapter) are complementary to those concerning Durability in Section 15.2.

Whatever the age of a specimen, an abrupt shrinkage occurs when specimens are taken out of water and placed in air. If excess water in the cured mortar or concrete is allowed to dry out very slowly, the stresses due to shrinkage, even if restrained, will be largely dissipated by creep. The moist curing of specimens prior to drying to a given equilibrium moisture content, however, has no appreciable influence on the final shrinkage of concrete, subject to all other factors being the same.

The newly formed solids consist of a lattice of very small particles of mineral gel, about 10 μm in size, in which are dispersed crystalline reaction products, soluble ions and grains of unhydrated cement. The fineness of the structure of these particles is a factor governing the stability and quality of the matrix, and depends upon the difficulties that oppose a rapid circulation of the ions as setting of the paste proceeds at atmospheric temperature. These difficulties are increased if the paste is compressed and the quantity of water is decreased.

The setting phenomenon can be accelerated by heating the mixture, but this treatment would increase the size of hydrate crystals and it causes a reduction in the strength of the concrete. A matrix with a very fine structure and high strength can be quickly produced if the paste is made with a low water/cement ratio and compressed while being heated to 100°C or more.

The setting process will become active again if the physical conditions are changed so that hydration products can grow by the movement, association and concentration of water and ions or saturated solutions of basic elements in the cement paste. When cement paste is allowed to set at atmospheric temperature, the cement gel particles do not grow in size as hydration proceeds, but new particles (similar in size and water content to those first formed) are produced (see Section 31.3.1).

Pores remaining in cement paste after the formation of hydrates are of two kinds: gel pores (or very small pores) between the gel particles and capillary pores (or much larger pores) between aggregates of gel particles. Heat of hydration is developed between the initial and final set; the amount of heat developed may be small and dissipated quickly, or be considerable and cause expansion exceeding the shrinkage characteristic. The maximum temperature of hydration usually occurs while the mixture is hardening and it gives rise to thermal volume change and thermo-osmosis (see Sections 3.1.1 and 30.1).[102, 117, 120]

The permeability of hardened cement paste rapidly increases with the water/cement ratio, the temperature of curing (when above 93°C) and (in steam curing) the rate of heating at an early age. The permeability of concrete is increased when pores are formed mainly in one direction. The watertightness of steam-cured products is markedly improved, however, if they are subsequently cured for at least 7 days in humid surroundings. The water in hardened cement paste consists of

Chemically combined or nonevaporable water in such compounds as $Ca(OH)_2$ and $3CaO \cdot Al_2O_3 \cdot 3CaSO_4 \cdot 31H_2O$.
Capillary or evaporable water in a state of triaxial tension in both the gel and capillary pores.
Water vapour in these pores.
Adsorbed water or water that is bound by surface forces.

The results of X-ray diffraction[35] and electron microscope studies have indicated that hydrated calcium silicates ($1 \cdot 5\ CaO \cdot SiO_2 \cdot 1 \cdot 0$-$2 \cdot 5H_2O$) in cement gel are extremely fine crystals with a layer structure. Such crystals are capable of holding water either between the layers or in interstices in their structure. The volume changes of ordinary cement products that accompany drying, carbonation and wetting are due largely to the loss or gain of interlayer water by these crystals (see Section 31.4).[37, 120] Suitable lignosulphonates (see Section 6.4) used in concrete modify the hydration of tricalcium aluminate, with a beneficial effect generally on mechanical strength, due to a formation of interlocking, acicular crystals or certain thin, hexagonal, platy or crumpled, foil-like products.[45, 47, 49, 88]

31.1.2 STEAM HARDENING

Low-pressure and high-pressure steam-cured products differ in the type of binder that is formed. That of the former is (in addition to other minerals) essentially a dicalcium silicate hydrate such as hillebrandite, while that of the latter (provided that silica is present in sufficient amount) is essentially a monocalcium silicate hydrate, designated mineralogically as tobermorite.[25, 26, 33, 51] The reactions of the formation of these solids are, in their simplest forms, respectively as follows.

Cement + Water = $2CaO \cdot SiO_2 \cdot nH_2O$ + $Ca(OH)_2$
Cement + Silica + Water = $CaO \cdot SiO_2 \cdot nH_2O$

Cement, if used as a raw ingredient of the binder in autoclaved products, contains about 62 per cent CaO and 21 per cent SiO_2. It is therefore too greatly deficient in silica to form monocalcium silicate hydrate. In practice, about 30-40 per cent of the cement should be replaced by silica flour or its equivalent in other forms of siliceous dust, such as fly ash, shale dust and aggregate dust. This adjustment is necessary for the proper development of a high-quality product, and it results in a saving of 10-20 per cent in the cost of materials (see Section 9.4.6 and Reference 180 of Part 3).

The drying shrinkage of these products appears to be due to movement of the interlayer water in the crystals of their binder. In this respect, the lattice of the binder of low-pressure cured products undergoes a larger volume change than does the lattice of tobermorite during movement of the interlayer water. This may result in the shrinkage of unrestrained masonry walls built with high-pressure, steam-cured blocks being only one-third to one-half that of walls built with low-pressure, steam-cured blocks and subjected to the same drying conditions. The dimensional stability of the latter blocks may be improved by supplementary carbonation treatment (see Section 9.6).

31.2 DIMENSIONAL CHANGE

31.2.1 ORDINARY CONCRETE

Concrete and mortar pass through three states: from mixing to the beginning of setting; setting and hardening; and drying at some stage after final set. Dimensional change is usually considered only in the hardening and drying states, on the ground that tensile stresses can arise only as a result of restraint of volume change of the hardened mass. The shrinkage starts with internal thermal movements due to heat of hydration at an age of 1-2 days (CEB-FIP and NAASRA, Appendix IV, 2(a)(i)).[14, 19, 27-30, 38, 53-57, 68, 85, 95, 96, 106, 122-124, 136, 141-143]

Factors affecting the shrinkage of concrete with time are: the quantity of mixing water; fineness index and heat of hydration of the cement; proportions of the mix; pore structure; size of specimens; carbonation; and humidity conditions surrounding the concrete (see Section 34.1.6). With given materials, shrinkage is closely related to the quantity of mixing water per unit volume of concrete. The water requirement of mixes with a given cement content is increased by using a high-slump, fine sand and small coarse aggregate. Crack control is briefed in Sections 15.2.2(b)(i) (ii), 22.3.2 and 34.1.1; also Appendix III.

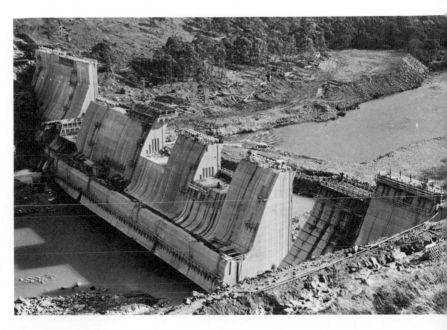

Anchor heads
after post-tensioning
the tendons to 2000 kN

Tendons about to be inserted
in ducts to bedrock

Anchor heads showing
at the crest of
Catagunya Dam, Tasmania

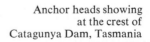

National Mutual Centre,
Melbourne

IBM Centre, Sydney

BP House, Melbourne

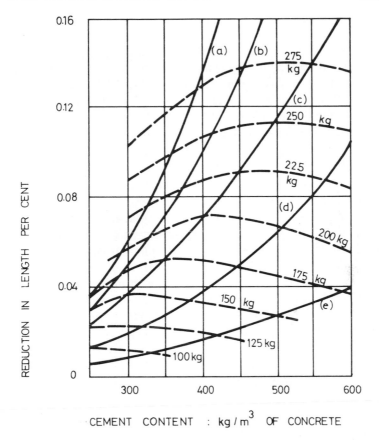

Fig. 30.1 Drying shrinkage of air-cured concrete at an age of 6 months. ————, water/cement ratio by weight. – – – –, water (kg/m³ of concrete). Water/cement ratios: (a) 0·7; (b) 0·6; (c) 0·5; (d) 0·4; (e) 0·3.

These items therefore cause increased shrinkage. It would be approximately constant over a wide range in cement content if the water content were kept constant per unit volume of concrete. Otherwise, it follows the pattern indicated in Figure 30.1, the results of which were obtained on 430 mm × 125 mm × 125 mm test prisms cured initially for 7 days in water and then in air at 23° ± 2°C and 50 per cent relative humidity.

Shrinkage depends to some extent upon the mineralogical composition of the cement, as indicated by the ratio

$$(a + b - 2c)/(a + d + c)$$

where a represents the tricalcium silicate, b the tricalcium aluminate, c the tetracalcium aluminoferrite, and d the dicalcium silicate, each unit being expressed in per cent. The effect of tricalcium aluminate is much greater than that of the other compounds in portland cement.[27] Wherever a low-cracking tendency is required, because of plastic deformation of a cementitious matrix, preference should be given to a slow-hardening cement (see Section 3.1.1).[54] Restraining components (i.e. aggregates, unhydrated grains of cement and stable microcrystalline products of hydration) serve from the beginning of drying to reduce shrinkage.

Aggregate affects shrinkage in several ways. It is low for aggregate of large maximum size, rounded shape or smooth surface texture, because of the reduced paste and water requirement for a given degree of workability. The shrinkage of a cementitious matrix is restrained more by large particles than by small, with a consequent reduction in overall shrinkage. Aggregate with a high modulus of elasticity induces a similar effect. Shrinkage increases with clay contamination, and with the amount of void space within and around the particles of aggregate. Certain shrinkage aggregate (see Section 15.2.2(b)(i)) and porous aggregate (such as sandstone and some varieties of lightweight aggregate) can allow normal shrinkage values to increase by up to 40 per cent or more (see Section 33.3.2(b)).[127]

For a given concrete, the magnitude of shrinkage can be estimated from the water requirements for different consistencies and sizes of aggregate. With regard to the former, each 30 mm increase in slump requires approximately an additional 3 per cent of mixing water, which increases shrinkage by about 6 per cent. For a given consistence, about 10 per cent more water is required with 20 mm aggregate than with 40 mm and, consequently, the shrinkage is increased by about 20 per cent. The reduction of restraint to shrinkage by the cementitious matrix would account for another 20 per cent, thereby bringing the total increase to about 40 per cent.

On the assumption that small specimens of concrete (which is made with a 100 mm slump and 40 mm nominal-size aggregate) have an ultimate shrinkage of 0·05 per cent, a series of estimated values are given in Table 30.1 for various consistencies and maximum sizes.

Both the shrinkage and expansion of concrete are associated with its pore structure which, in the case of hardened concrete in the dry state, is up to 15 per cent or more of the volume of the material. If the mixing water does not escape, no shrinkage occurs. This characteristic is indicated by the large difference in dimensional change between specimens that are left in the air and those that are maintained in a wet condition or enclosed in a waterproof envelope (see Fig. 30.2 and Appendix I).

TABLE 30.1   ULTIMATE SHRINKAGE

Slump (mm)	Nominal Size of Aggregate (mm)	Shrinkage (per cent)
50	20	0·063
100		0·071
150		0·079
50	40	0·044
100		0·050
150		0·056
50	50	0·037
100		0·041
150		0·045

The rate and amount of shrinkage are affected by the relative humidity of the air surrounding the concrete. For instance, if concrete is exposed to relative humidities of 25 and 75 per cent for 9 months, it will shrink about 10-12 per cent more and less, respectively, than if stored at 50 per cent for the same period and at 23°C. Under unfavourable climatic and building-site conditions, the rapid drying of large areas can cause plastic cracking in newly laid concrete (see Sections 11.1.2(c), (d) and 31.2.4).[108]

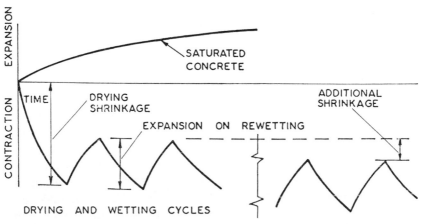

Fig. 30.2 Types of dimensional change

Workability agents (see Section 6.4) have little effect in themselves on drying shrinkage, provided that they do not contain a setting accelerator, such as calcium chloride or triethanolamine. The latter materials have an adverse effect on drying shrinkage. For instance, with 2 per cent of calcium chloride by weight of cement, the shrinkage of concrete is likely to be increased by 40-70 per cent.[27] Carbonation effects on shrinkage are discussed in Sections 9.6 and 31.4 and, when steam curing is used, its influence in reducing shrinkage may be taken as being similar to its effect on creep (see Sections 31.3.3 and 34.1.5; also Reference 156 of Part 3).[60, 63-65]

Thick sections have not only an internal heat-of-hydration expansion (see Section 30.1), but also an external drying shrinkage, because moisture in the underlying concrete is fed only slowly outwards and consequently delays and reduces the internal contraction. For large members that are exposed to the weather, the drying shrinkage may be only 0·1-0·4 times that of small laboratory specimens stored at 23°C and 50 per cent relative humidity. These effects and their causes should be considered when estimating, from laboratory specimens (see Fig. 30.1), the shrinkage of concrete structures (AS 1481; Fintel and Khan, Appendix IV, 2(a) (i)).

Test measurements on large specimens, when exposed to drying on all surfaces, are complicated by the stresses and cracking that arise through shrinkage being more rapid near the surface than at the interior. These disadvantages can be overcome by obtaining test results from specimens having linear measurements of not less than 150 mm. They are cast in special moulds with one end open and allowed to dry out from this exposed surface only.

The long-term drying shrinkage of ordinary hand-compacted concrete, when stored under laboratory conditions for a period of up to 5 years, has been observed to amount to about 0·053 per cent. That of vibrated concrete is 0·043 per cent and of vacuum concrete is 0·037 per cent. The shrinkage of a 1 : 3 mix is about twice that of a 1 : 6, and the expansion of concrete continually immersed in water is about one-fifth of the shrinkage that the same concrete would have if it were continually exposed to the air.

Other long-term tests carried out on moist-cured, diverse concrete specimens (100 mm diameter), stored in air at 21°C, have shown that the shrinkage at 50 per cent relative humidity is about 1·4 times that at 70 per cent relative humidity. Of the average shrinkage values obtained, in all test conditions over a 20-year period, some two-thirds of the total shrinkage occurred within the first 3 months, nearly 80 per cent within the first year, and the remaining 20 per cent in the ensuing 19 years. An additional 5 per cent occurred between 20 and 30 years.[106] Fibre reinforcement for controlling crack propagation in cement composites is entered for review in Section 3.4.2.

The coefficient of thermal expansion of concrete lies within a range of $7 \cdot 0 \times 10^{-6}$-$12 \cdot 5 \times 10^{-6}$ per °C. The coefficient is mainly influenced by the type of aggregate, and it is assumed to be constant for each type within the ordinary range of ambient temperature. Typical values are given in Table 30.2, where all factors except that of aggregate are kept the same. Comparative values are approximately $10 \cdot 8 \times 10^{-6}$ per °C for mild steel and $16 \cdot 2 \times 10^{-6} \pm 5 \cdot 4 \times 10^{-6}$ per °C for hardened cement paste, the latter being influenced by age, composition, degree of hydration and moisture content (Loubser, Appendix IV, 4(b)).[53, 87, 110, 111] Thermal data for plastics are given in Table 21.2.

TABLE 30.2   COEFFICIENT OF THERMAL EXPANSION

Type of Aggregate (Fine and Coarse)	Coefficient of Thermal Expansion $\times 10^{-6}$ per °C
Quartzite	12·8
Sandstone	11·7
Sand and gravel	10·8
Granite	9·5
Basalt	8·6
Expanded clay and shale	7·6
Limestone	6·8
Clinker	5·9

The value for limestone varies according to its composition, and may approach that for siliceous materials. For ordinary concrete and conditions, an average value of 10×10^{-6} per °C is used. On this basis, a variation in temperature of 50°C will cause a change in length of 15 mm in a concrete wall 30 m long. By comparison, a 50 per cent reduction of relative humidity from 100 per cent will cause a shrinkage of 19 mm.

These changes set up stresses that are likely to cause cracking around openings, where walls abut, and at changes in wall section, unless precautions are taken to prevent it. Shrinkage stresses in cylindrical tanks and in composite prestressed concrete construction are investigated in References 40 and 121 of this Part and of Part 8. Thermal shrinkage and creep at elevated temperatures are surveyed in Sections 29.1 and 29.3; also in References 84 and 97. Plastic deformation is reviewed in Sections 31.2.4 and 31.3.

Means to control the effects of dimensional change are needed in concrete floors (see Sections 20.1.4 and 20.1.5); deflections of reinforced masonry and concrete structures (see Reference 75 of Part 5),[129] lightweight concrete sections (see Sections 33.3 and 33.4); mass concrete (see Section 24.4 and Chapter 30); mortar jointing and rendering (see Section 16.2); placed concrete (see Section 7.2.4); precast elements and walls (see Sections 27.1-27.4, 28.8-28.12, 31.2.2, 33.3 and 33.4); prestressed concrete (see Chapter 34); refractory concrete (see Chapter 29); road pavements (see Sections 26.1-26.5); and *in situ* slabs used in conjunction with precast or prestressed beams (see References 9-17 and 79 of Part 8, also Reference 87 of Part 5 [Aust. EBS NSB 112]).[103]

Control joints in reinforced dense concrete buildings are placed about 30 m ± 6 m apart, depending on the type and location of construction and the dimensional differential to be accommodated. For instance, a heated concrete floor system may have control joints at half the spacing recommended for the exterior walls and roof of a reinforced concrete structure. A building frame with adequate temperature reinforcement in wall beams may have control joints at increased intervals where wall panels are fitted into keyways in the columns.[27, 56]

The use of double columns and beams should enable control joints to have a separating effect, in order to permit necessary movement. Exterior joints can be sealed or weatherproofed by the effective use of continuous, flexible waterstops (moulded polyvinyl chloride or rubber, or crimped copper strip; see Sections 10.1 and 19.1.1); weatherproof flashings and sliding cover plates; bonded, preformed, bitumastic and synthetic rubber gaskets; bitumen-impregnated polyurethane foam or compressed cork strips; highly compressible, closed-cell polyethylene sheet or rod; and high-performance elastomeric compounds. These include rubber-bitumen mastic, silicone latex, two-part polysulphide or polyurethane, vinyl acrylic terpolymer, polychloroprene and butyl rubber-based sealants. Flat concrete roofs require insulation (see Section 33.4) and suitable sliding freedom on walls, this being facilitated with 5 mm resilient rubber strips or bearing pads clad with expanded polyethylene or polystyrene (see Sections 11.1.2(d), 20.1.5(f), 26.1, 28.12 and 36.5; also Appendix I and Reference 187 of Part 3).[50, 74, 78, 98, 116, 123, 126, 140]

An open-drained joint consists of a vented rain screen or baffle, an internal air chamber and drainage duct, and a weather-protected air seal installed at the rear of the section (see Section 28.12). Should the rain screen consist of a high-performance sealant and closed-cell polyethylene rod back-up, the central cavity formed between it and the rear air seal would be vented (to preserve ambient pressure conditions) by inserting 9 mm polyvinyl chloride ducts at intervals through the sealant, these being outwardly sloped at 45°. The external surface is profiled to impede the passage of water which may tend to flow horizontally across the joint. A variety of control joints are illustrated in Figure 30.3, which includes a suggested alternative to filling the external face of a column joint with sealant and repairing it periodically. The external surface is shaped so as to impede the passage of water which may tend to flow horizontally across the joint.

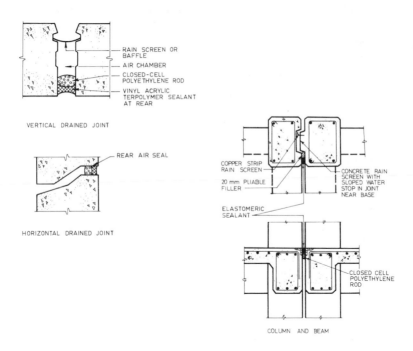

Fig. 30.3 Control joints in reinforced concrete

Centrally located, dumb-bell waterstops are used in full-movement wall joints of hydraulic structures. In the middle of a ground slab, concrete compaction is difficult around this type of waterstop, which may become folded or displaced during concreting. External flexible waterstops, located along the lower edge of ground joints, provide an effective membrane barrier to the passage of hydrostatic ground waters. The thin outer segment of a central duct in each unit is designed to rupture when the joint opens.[114, 119]

31.2.2 CONCRETE MASONRY

In concrete masonry, the tendency of all types of building block to crack is not proportional to their shrinkage, because of differences in extensibility or the strain that a block can withstand in tension before it cracks. This characteristic may be evaluated from measurements of the modulus of elasticity and tensile strength, and is expressed as a percentage of the drying-shrinkage strain of an unrestrained individual block. Once values for the extensibility of different types of concrete are established, the measurements of drying shrinkage become more useful for predicting the performance of building blocks in service. Concrete will not crack if its tensile strength at any age is in excess of the actual stress set up in it at that age. A drying test and building requirements are described in Sections 14.6, 27.2, 33.3 and 33.4 (see Hedstrom, Appendix IV, 4(b)).

31.2.3 SUMMARY ON SHRINKAGE

Factors which influence the shrinkage of concrete may be summed up as follows.

1. Water/cement ratio (see Fig. 30.1).

2. Nature, specific surface area and proportion of the cement.

3. Nature and granulometry of the aggregate, and the proportion of fine aggregate.

4. Rate and period of settlement of the aggregate in fresh concrete.

5. Proximity of reinforcement to the surface.

6. Setting admixtures and expansion-producing additives or aggregate.

7. Thermal and humidity conditions, carbonation, steam curing, pressure, and the time effect during placing, curing and storage (see Section 31.2.4).

8. Pore, mineral gel and crystal structure of the binding matrix.

9. Loss of interlayer water from the binding matrix, the former being minimised in precast units by drying them thoroughly before use.

10. Hardening, atmospheric, loading, inertia and plastic flow effects when the concrete is in service.

11. Dimensions of the test pieces and products (see Sections 31.2.1 and 34.1.6).

The cracking of concrete is governed by the extensibility and length of its section, as well as by its shrinkage characteristics. Standard specifications on drying shrinkage are given in Appendix I, while crack widths and their assessment in reinforced concrete design are referred to in Sections 11.1.2(d), 15.2.2(b), 22.3.2 and 34.1.1.

31.2.4 PLASTIC SHRINKAGE CRACKING

Field investigations indicate that the principal cause of early cracking of newly laid slabs is rapid drying at the surface while the concrete is still plastic. The rate of evaporation of moisture from concrete is influenced greatly by atmospheric conditions. With concrete and air at the same temperature, a change of conditions from 23°C and 70 per cent relative humidity can have the following effects: the rate is increased fourfold by a breeze at 4·5 m/s, tenfold by one at 9 m/s, and fiftyfold should these conditions become 32°C, 10 per cent relative humidity and 11 m/s.

A rapid-drying condition removes moisture from the surface faster than it can be replaced by capillary attraction. Even in cold weather, rapid drying is possible if the temperature of placed concrete is high compared with that of the air. The evaporation of water from a concrete surface tends to lower its temperature and accentuate the differential dimensional effect of drying shrinkage by one due to thermal change. The following procedures, which keep the evaporation rate below 10 litres/m²·h, will minimise plastic shrinkage cracking (see Section 31.2.1).

1. Dampen the subgrade, formwork and surfaces of adjacent concrete.

2. Dampen the aggregate if it is dry and absorptive.

3. Erect windbreaks to reduce wind velocity over concrete surfaces.

4. Provide protective awnings or reflective sunshades to control the surface temperature of the concrete.

5. Avoid excessive temperature developments in the concrete or temperature differences between the concrete, curing water and air.

6. Lower the temperature of concrete in hot weather.

7. Avoid overheating the concrete in cold weather.

8. Use tested air-entrained concrete, using less air-entrainer with a set retardant or other agent (see Sections 6.4, 10.2, 16.1, 16.3 and 32.2).

9. Minimise concrete bleeding and settlement, and firmly rework and compact the surface with wood-floats near the time of initial set of the cement, or as soon as plastic shrinkage cracking occurs in the preset stage (see Sections 20.1.5(d) and 23.2).[76]

10. Vacuum or vibrovacuum process newly laid concrete slabs prior to power-floating (see Section 23.4).

11. In the precuring period, retard the initial evaporation of moisture from freshly formed or reworked concrete surfaces by spraying them with a film of aliphatic (e.g. cetyl) alcohol diluted with 9 parts of water.

12. Protect newly placed concrete against early rapid drying, either with temporary coverings or by a fog spray, so as to promote plastic deformation of the matrix and a relief of stress by creep.[89]

A fog spray is an effective means of increasing humidity and preventing evaporation from the surface of exposed concrete at an early age. It is useful until other moist-curing or membrane-curing materials, or rotary rainwave sprays can be utilised (see Sections 9.1-9.3 and 32.2). Early moist curing should be continued until the concrete develops sufficient tensile strength to resist a limited amount of drying shrinkage (see Section 31.3.1).

If Figure 30.2 were inverted, it could suggest that warping and crazing of a concrete slab would result if while the underside remains damp, the surface of the fresh concrete were allowed to dry rapidly or were subjected to cycles of alternate drying and wetting. The latter effect is induced by intermittent hosing of the slab. To minimise warping and crazing, the driest mix that can be laid satisfactorily should be used, and curing should be as early, as uniform and as long as practicable. Other factors to be considered for minimising the cracking of concrete are indicated in Sections 11.1.2(d), 15.2.2(b), (d) and 20.1.5(e); and in this chapter generally.

31.3 DEFORMATION

31.3.1 GENERAL PRINCIPLES

Concrete may be likened in its behaviour to a complex pseudosolid to which the laws of thermodynamics may be applied. It can, within limits and with sufficient time, adjust itself to external conditions. For example, a local increase in compressive stress will cause ion-charged moisture in the stressed region to migrate in such a way as to relieve the stress. The redistribution of moisture results in shrinkage in regions losing moisture and swelling in regions gaining it. Also, an increase of mechanical pressure reduces the triaxial tension in capillary water and this could cause it to become more fully charged with ions from adjacent hydrates (see Glossary, Appendix VI). These changes manifest themselves externally as a gradual yielding to applied load that is known as creep. A reduction of internal stress in an overloaded zone by this means reduces the risk of failure.

The inherent properties of adaptability make concrete structures possible. In addition to moisture movement and migration of ions, each small community of particles undergoes considerable distortional deformation. For a given stress, the rate of deformation is quite marked if loads are applied while the concrete is drying. After the gel particles have acquired stable positions, the rate of and capacity for creep (for a given stress) are materially reduced (see Section 3.2.2).

Any deformation of concrete which is not infinitesimally small is made up of a series of sudden finite movements that are irreversible and cause minute fractures. These bring into play a phenomenon known as autogenous healing, which is an extension of the characteristics of the setting of cement paste. By this means, fractures will heal themselves if the distance between their surfaces is not excessive, and if there is sufficient moisture present and time to allow a further formation of hydrates through a movement of ions in newly formed, enlarged pores.

Cracking by drying shrinkage is minimised and autogenous healing is maximised by proper procedures in curing. In breakage fissures, the healing action is notably strengthened when adjacent surfaces are kept wet for 1-3 months and in close proximity (particularly with transverse pressure), and a bonding material is formed containing crystals of calcium carbonate and calcium hydroxide (see Section 3.1.2(e)(iii)). Piped water that tends to inhibit autogenous healing is outlined in Section 19.2.[20, 59, 121]

When considering the stability of concrete structures and members, within acceptable limits of stress and deformation, it is advisable to deal not only with external conditions for mechanical equilibrium, but also with conditions for structural equilibrium within the material.[52] With a quantitative knowledge of the latter, such phenomena as shrinkage, creep and plastic flow can be dealt with fundamentally and broadly, instead of solely on an empirical basis (see Sections 3.1.1, 3.2.2 and 14.4.2).[55, 62, 71, 77, 95, 100]

31.3.2 EFFECTIVE MODULUS DUE TO CREEP

Strain due to elasticity occurs as soon as a stress is imposed, and it remains constant. When the stress is sustained for any length of time, the concrete undergoes progressive creep (see Fig. 31.1). The rate of increase of strain caused by creep decreases with time and thus the total creep approaches an ultimate value (see Glossary, Appendix VI). The effects of creep can be taken care of in calculations, by reducing the ordinary elastic modulus to an effective modulus (AS 1480 and MP28 (Section C9), 1481, BS 1881 and ASTM C469).

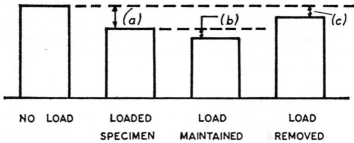

NO LOAD   LOADED   LOAD   LOAD

SPECIMEN   MAINTAINED   REMOVED

Fig. 31.1   Deformation with sustained load. (a) Elastic deformation.
(b) Creep. (c) Permanent deformation.

Table 31.1 shows in MPa the effect of mix proportions on the elastic modulus of dense-aggregate concrete. Elastic modulus values of structural grades of concrete (see Appendix VIII, 5), having a bulk density between 1250 and 2600 kg/m³ (Appendix I; normally about 2400 kg/m³), are formulated in Section 34.1.4(c) and graphically profiled in AS MP28, Section C9 of AS 1480. Other sources of information are BS CP110 Part 1, CP115 Part 2; ACI Committee 318, Appendix IV, 2(a)(i); and Reference 1 of Part 3, supplemented with References 54 and 148 of Part 7. Dynamic modulus and lightweight concrete values are included in Sections 14.9.4, 33.3.2(a) and 33.3.3(a).[13, 19, 81-84, 91, 93, 95, 125]

TABLE 31.1   **EFFECTIVE MODULUS**

Mix by Weight	Modulus of Elasticity, E_c (MPa)	Effective Modulus	
		At 12 Months	Ultimate Value
1 : 1 : 2	$36 \cdot 5 \times 10^3$	$11 \cdot 7 \times 10^3$	$9 \cdot 7 \times 10^3$
1 : 2 : 4	$23 \cdot 4 \times 10^3$	$5 \cdot 2 \times 10^3$	$4 \cdot 3 \times 10^3$
1 : 3 : 6	$20 \cdot 0 \times 10^3$	$3 \cdot 2 \times 10^3$	$2 \cdot 8 \times 10^3$

For the purpose of calculations involving time, it is desirable to separate the true elastic strain from the strain due to creep. For the 1 : 2 : 4 mix, the strains per unit stress are

$$\text{Elastic strain} = \frac{f_c}{E_c} = \frac{1}{23 \cdot 4} \times 10^{-3} = 42 \cdot 7 \times 10^{-6}.$$

$$\text{Combined strain at 12 months} = \frac{1}{5 \cdot 2} \times 10^{-3} = 192 \cdot 3 \times 10^{-6}.$$

$$\text{Combined ultimate strain} = \frac{1}{4 \cdot 3} \times 10^{-3} = 232 \cdot 6 \times 10^{-6}.$$

Therefore the strain due to creep alone at 12 months equals $149 \cdot 6 \times 10^{-6}$ and the ultimate value or ultimate specific creep equals $189 \cdot 9 \times 10^{-6}$, say, 190×10^{-6} per MPa. From this, an approximate curve may be sketched showing the strain due to creep against time. It may be observed that about 80 per cent of the ultimate creep is completed after 12 months and much of the long-term remainder takes place in the following year. Creep is affected by such factors as the following.

1. The type and content of cement, setting accelerator and pozzolan (see Sections 10.1, 17.1 and 17.2), the water/cement ratio and the characteristics of adsorbed water.

2. The type, grading, elastic modulus and fine-aggregate/coarse-aggregate ratio (see Sections 33.3.2(b) and 34.1.5).[32, 81]

3. The type of curing and the maturity factor (see Sections 9.4.4 and 31.1).

4. The elastic modulus and water content of the concrete.

5. The age or strength of the concrete when loaded and during the loading period (AS 1012 part 16 and ASTM C512 describe a test procedure).[106]

6. The intensity, nature and duration of loading, including multiaxial stress (see Section 34.1.5 and Reference 125 of Part 3).[77]

7. The incidence of cyclical wetting and drying, and the ambient thermal and humidity conditions in service.

8. The quantity of compressive and flexural reinforcement.[32, 101]

9. The size of the member (e.g. the creep of a large specimen may be only about one-half that of a small one at a certain age).[28]

10. The acceptable capacity of concrete structures for serviceability (AS 1480 and MP28, Section C10; BS CP 110, Part 1).[21, 27, 50, 53, 68, 80, 83, 85, 95-97, 100, 101, 105, 115, 129, 136-139]

The effect of various combinations of these factors may be difficult to interpret with precision from certain empirical data, which are available with specified parameters (see References of Section 31.3.3). Nevertheless, for estimating purposes, the curve shown in Figure 31.2 is indicative of average creep results for unrestrained ordinary concrete loaded at an age of 28 days (Fintel and Khan, Appendix IV, 2(a)(i)).[13] Creep estimation data for prestressed concrete are contained in Section 34.1.6(b).

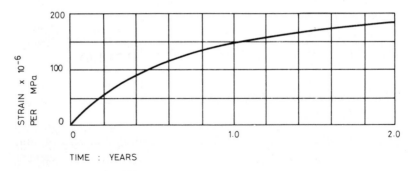

Fig. 31.2 Specific creep of 1 : 2 : 4 concrete. Ultimate value: 190×10^{-6} per MPa.

31.3.3 THEORIES OF CREEP IN CONCRETE

Creep occurs in the hardened cement matrix around strong sound aggregate. It may be attributed to: internal stresses caused by changes in the capillary structure of the cement paste; micro-cracking; slippage along planes within the crystal lattice; and gradual loss of water from the cement gel. The creep of neat cement can be as much as fifteen times that of concrete. It thus acts as an extremely viscous liquid. Several actions take place in creep and these include true creep (viscous in nature and leading to gradual transfer of load from the paste to the aggregate) and shrinkage, which are affected by variations in curing conditions, humidity and temperature (see Section 31.2).[13, 14, 21, 27, 32, 38, 53-57, 68, 71, 77, 80, 85, 90, 95-97, 99, 101, 105, 106, 112, 122, 136-139]

The creep of high-temperature, steam-cured concrete (produced with ordinary portland cement) is about 80 per cent of that of moist-cured similar concrete of comparable strength. The corresponding proportion for high-early-strength portland cement concrete is about 70 per cent, and this proportion can be reduced to below 30 per cent by using high-pressure steam curing and a suitably designed mix (see Section 9.4.1).[60, 63-65] With setting accelerators, the increased creep of ordinary concrete, loaded at 28 days, is up to one-third that obtained if loading is done at 7 days (see Section 10.1 and AS 1481).

Prestress losses in prestressed concrete, due to the effect of creep and a still greater effect due to shrinkage, are about three times as great in dry areas as in moist climates. Increased initial creep is induced in the concrete radiation shields of atomic reactors, due to increased thermal conditions, moisture mobility and activated ionic processes in the early stages of deformation. The functional performance of concrete generally is thus related, in different ways, to adaptive deformation and relieved stress concentrations through the influence of creep (see Sections 34.1.5 and 34.1.6).

31.4 CARBONATION EFFECTS

The hydrolysis of cement during the curing of concrete splits off calcium hydroxide from the calcium silicates. Where carbon dioxide contacts the calcium hydroxide, not only is crystalline calcium carbonate formed, but 2·4 kJ of heat are liberated per g of calcium oxide involved in the reaction. Also, for each molecule of carbon dioxide absorbed, one molecule of water is liberated.

If the water thus formed is removed and the concrete is kept partly dry, carbonation proceeds more rapidly than if the concrete were to remain wet. With diminishing humidity, the removal of water from surfaces and capillaries increases the access of carbon dioxide to the interior of the concrete. Under conditions of rapid drying, coupled with a steep moisture gradient within the material, shrinkage of the surface in conjunction with internal restraint may bring about cracking and warping.

Carbon dioxide not only combines principally with the calcium hydroxide of hydrated cement, but it also attacks and decomposes the major constituents into calcium carbonate and hydrated silica, alumina and ferric oxide. The rate of carbonation is increased by the concentration of carbon dioxide, the pressure of application, a drying atmosphere (relative humidity about 30 per cent or less), and a high degree of porosity in the concrete. The absorption of carbon dioxide is greatest when a cement-paste matrix contains about 10-15 per cent of its original free water. Under ordinary atmospheric conditions, the action is confined to the exterior of the mass and the surfaces of capillaries or cavities that are not filled with water (see Section 31.3.1 and Reference 175 of Part 4).

The depth of carbonation of concrete or mortar is a function of its density or impermeability, the ambient atmospheric conditions and the period of exposure. For instance, 85 per cent of the calcium oxide in portland cement could become carbonated, if specimens of ordinary cement concrete were stored for about 30 years in air possessing very favourable conditions for carbonation.

The beneficial effects of carbonation, after an initial period of hydration of portland cement, are an increase in strength and hardness, and a decrease in long-time volume change and permeability. A marked increase in strength accompanies the formation of a skeletal structure of calcium carbonate in set cement. In the carbonated compounds of hydrated cement, the separated silica (which is in an amorphous state) also contributes some cementing value, probably through a reduction of pore space and accompanying autogenous adhesion. The high strength of calcium carbonate in a substantial state of development is indicated by that of marble, namely 80-140 MPa.

Some effects of carbonation have been observed in 1 part cement : 3 parts sand mortar specimens, which were stored for 1 day in moulds and 6 days in water at 23°C. Control specimens were then stored in moist carbon dioxide free air. Others were placed successfully in a vacuum for 2 hours and in carbon dioxide at 350 kPa for 10 days, the pressure being increased to 1000 kPa for a further 15 days. The strength results obtained at an age of 32 days are given in Table 31.2. Volume-change tests were made on mortar bars which were cured and treated similarly, but with carbon dioxide treatment at 1000 kPa for 1 month. The volume change of the carbonated bars, under alternate wet and dry conditions, was only 30-40 per cent of that of uncarbonated specimens.

For best results by artificial carbonation, the cement gels in a matrix must first have reached a stable condition by hydration. If ordinary concrete or mortar were exposed to excessive carbon

TABLE 31.2   **STRENGTH RESULTS**

Characteristic (MPa)	Control Specimens	Carbonated Specimens
Compressive strength (cube)	34	58
Tensile strength (briquette)	3·2	5·2
Modulus of elasticity	26×10^3	38×10^3

dioxide during the first 24 hours of its life, a crazed or soft cretaceous surface may form. Normally, after a period of 24 hours, carbon dioxide not only ceases to be detrimental, but also enhances the quality of concrete or mortar. Should combustion heaters be used temporarily during winter construction, therefore, damage by premature carbonation must be prevented during the first 24 hours after placement. This is done by exhausting combustion gases to the outside atmosphere and providing ample ventilation, or by applying a membrane-curing compound or film as soon as possible. Methods of set control in cold weather are given in Chapter 32.

Waste combustion gases from lime kilns, if available, are an economical source of carbon dioxide. The relative strength of carbon dioxide, in waste gas and the ordinary atmosphere, can be estimated on the basis that the latter contains approximately 0·05 per cent of carbon dioxide by weight. Combustion gases for carbonation treatment should have a low sulphate content (see Section 9.6).

Concrete slabs on the ground are subject to drying and carbonation shrinkage of the top surface, while the underside remains close to its original dimension. Curling characteristics, which thus tend to be induced, can be minimised by thorough compaction, early membrane and moist curing, and jointing at reduced intervals (see Sections 9.2, 9.3, 20.1.4(a), 20.1.5(c), (e), (f) and 23.1.2(e); also Chapter 26). Warped floor slabs can be made flat again by saturation or ponding and sawing into panels of reduced size.

CHAPTER 32

CONCRETING IN EXTREME WEATHER

32.1 COLD WEATHER

32.1.1 GENERAL INFORMATION

The rate of hardening of all portland cements slows down in cold weather and is very low at a temperature slightly above 0°C. As the temperature rises, hardening proceeds at an increasing rate, providing the concrete has not been frozen. The early finishing of concrete floors and handling of precast products can be assisted physically, however, by dewatering them with vacuum processing (see Section 23.4).

With buildings, bridges and pavements, it is necessary to keep thin sections from freezing and enable the concrete to gain strength properly. With dams, piers and other forms of mass concrete, in which the heat from hydration of the cement is a problem in any season, it is necessary to minimise temperature gradients in the material and prevent local damage by frost to corners and exposed surfaces.[31, 131, 144]

Air-entrained concrete should be used when placing is done at low temperature, otherwise a much higher degree of protection is required. The concrete must not be placed against frozen surfaces. In conditions of frost at night, all surfaces with which the concrete is to come into contact should be covered at night or allowed to warm up before work begins.

Artificial heat should be used with special care with high-alumina cement, as this cement evolves considerable heat during setting and is damaged if the temperature rises above 29°C during its early life. Great differences in temperature in different parts of a structure may cause uneven shrinkage and cracking and should be avoided.

An admixture of calcium chloride is recommended under ordinary circumstances when the mean air temperature is below 4°C, so as to obtain increased strength at the end of the period of protection. It should be used with care and caution as indicated in Sections 10.1, 15.2.2(a) and 22.3. In mass concrete, calcium chloride is unnecessary except near surfaces. If triethanolamine is used as a setting accelerator (see Section 33.2.3(a)), it should be used sparingly. Its action is nonionic and most effective if used in warm water. An antifreeze admixture must be performance-tested before use.

Where a setting accelerator is not recommended, additional cement (up to 30 per cent), high-early-strength portland cement or a blend of high-early-strength and ordinary portland cements should be used instead. In the former instance, the sand content of the mix should be reduced, so as to avoid increased shrinkage through an excess of fines. Appropriate measures should be taken to provide proper curing conditions and to prevent or minimise early rapid drying of the newly placed concrete (see Section 31.2.4).

Recommended hydrothermal conditions (BS CP110 Part 1)[22] for concreting at low temperatures, fall into two classes.

1. Heating of the water and aggregate, or otherwise using electrically heated formwork,[31] to ensure an adequate concrete temperature.

2. Protection of the placed concrete by covering, insulation, or a warming enclosure.

32.1.2 TEMPERATURE OF PLACED CONCRETE

Provided that it is not permitted to freeze, concrete placed at low temperatures will develop greater strength and durability than concrete placed at high temperatures (see Section 9.2). Not only is a high temperature of freshly mixed concrete objectionable, but initial temperatures should not exceed the following recommended minima by more than 6°C or 11°C at the most (see Section 31.1.1).

Concrete mixed in the proportion of about 1 part cement to 6 parts aggregate (40 mm nominal size) and placed in thin sections should have a temperature after placement not less than 10°C for the first 3 days. To allow for a loss of heat between mixing and placing, the temperature of the concrete when mixed should be 13°C when the ambient temperature is above −1°C; 16°C when it is between −18°C and −1°C; and 18°C for temperatures below −18°C. These minima are increased 3°C for 1 : 5 mix concrete, with 20 mm nominal-size aggregate, placed in thin sections. They are reduced 3°C for mass concrete with 75 mm nominal-size aggregate and about 1 : 8 mix, and 6°C for a 1 : 12 mix of mass concrete with 150 mm nominal-size aggregate.

The temperature of freshly mixed concrete can be determined from the fraction

$$\frac{T_a\,W_a\,+\,T_c\,W_c\,+\,4{\cdot}5T_w\,W_w}{W_a\,+\,W_c\,+\,4{\cdot}5W_w}$$

where T = temperature (°C), W = weight, and subscripts a, c, w refer to aggregate, cement and water (including free water in the aggregate), respectively.

32.1.3 HEATING OF MATERIALS

Aggregate is difficult to heat uniformly to a predetermined temperature, but the temperature of mixing water can be readily adjusted by blending hot and cold water, so as to maintain the temperature of the concrete within a range of 6°C. The desired temperature of the concrete can usually be obtained by heating only the mixing water when the ambient temperature is above −1°C and the aggregate is free from ice. For temperatures below −1°C, the aggregate is usually heated also.

With mixing water at 60-70°C it is rarely necessary to heat the aggregate to temperatures higher than 16°C. When the stone aggregate is dry and free from ice, frost or frozen lumps, fresh concrete with an adequate temperature can be obtained by heating the mixing water to 60°C, and the sand up to about the same temperature. Cement should not come into contact with water at a temperature above 60°C, this being averted by mixing the aggregate and water before adding the cement.

The mixing water should be heated under controlled conditions and in sufficient quantity to minimise fluctuations in temperature from batch to batch. If the water and aggregate (when either is heated to above 38°C) are brought together in the mixer before the cement is added, water with a temperature up to boiling point may be used, provided that the aggregate is cold enough to bring the temperature of the water and aggregate mixture to a temperature normally not exceeding 16-27°C. An air-entraining agent should be incorporated in each batch, after partial mixing of the components at a nominal average temperature.

The heating of materials should be uniform, since wide variations in their temperature will noticeably vary the slump and water requirement of the concrete. Aggregate is heated by steam in pipes or, on small jobs, thawed by careful heating over large ducts in which fires are maintained.

When aggregate in stockpiles, hoppers, bins or trucks is thawed or heated by electric heating pokers or piped steam, the surface should be covered with tarpaulins to effect a uniform distribution of heat and to prevent a frozen crust being formed. If steam is confined in a pipe heating system, difficulties arising from variable moisture in the aggregate are avoided. When steam jets are liberated in aggregate, variable moisture thus formed can be troublesome. If they are used for thawing large quantities of aggregate under emergency conditions, it must be done well in advance of batching, so as to obtain a fair uniformity of moisture content and temperature.[31, 86] Mixers can be designed for steam-injection uses.

32.1.4 PROTECTION OF PLACED CONCRETE

As most of the heat of hydration of portland cement while hardening is developed during the first 3 days, heat may not have to be applied to maintain concrete at correct temperatures if the heat thus generated is effectively conserved. It is rarely practicable to do concrete work in weather which is so cold that the concrete cannot be kept at a suitable temperature by insulation. This may consist of airtight, closed-cell, polyethylene insulating blankets, layers of hessian or sacking under tarpaulins, or insulated forms, with or without the assistance of electric polyurethane covers or aluminium-foil heating elements.

Recommendations for protection are given in Table 32.1 for concrete walls and floor slabs above ground, and in Table 32.2 for concrete slabs laid on the ground at a temperature of 4°C, the concrete being placed at 10°C in each instance.[22] The requirements are calculated for blanket-type insulation with a conductivity of 0·036 W/m·°C for a thermal gradient of 1°C per 45 mm. The values given are for still air conditions and will not be realised where wind causes infiltration of air.

Closely packed straw used under canvas or sandwiched between waterproof paper fastened at the edges can be considered as a loose-fill type of insulation if draughts are kept out. The relative values of insulating materials that may be used are given in Table 32.3, there being little change in the insulating value of dead-air

TABLE 32.1   **INSULATION FOR WALLS AND SLABS ABOVE GROUND**

Wall Thickness (mm)	Minimum Air Temperature Allowable (°C) Thickness of Blanket Insulation:				
	10 mm	20 mm	30 mm	40 mm	50 mm
Cement content 180 kg/m³					
150	9	6	3	0	− 2
300	6	1	− 4	− 9	− 15
450	4	− 3	− 11	− 19	− 26
600	4	− 5	− 14	− 24	− 33
900	1	− 8	− 17	− 27	− 36
1200	1	− 9	− 19	− 30	− 39
1500	1	− 9	− 20	− 31	—
Cement content 360 kg/m³					
150	8	3	− 3	− 9	− 14
300	3	− 8	− 18	− 29	− 40
450	− 1	− 17	− 33	− 49	− 66
600	− 3	− 21	− 38	—	—
900	− 5	− 26	− 46	—	—
1200	− 6	− 28	− 49	—	—
1500	− 6	− 30	− 50	—	—

TABLE 32.2   **INSULATION FOR GROUND SLABS**

Slab Thickness (mm)	Minimum Air Temperature Allowable (°C) Thickness of Blanket Insulation:				
	10 mm	20 mm	30 mm	40 mm	50 mm
Cement content 180 kg/m³					
100	*	*	*	*	*
200	10	9	8	6	5
300	7	3	− 1	− 6	− 10
450	− 3	− 6	− 16	− 25	− 35
600	− 0	− 15	− 30	− 42	− 44
750	− 5	− 24	− 44	—	—
900	− 9	− 33	—	—	—
Cement content 360 kg/m³					
100	10	10	10	9	9
200	5	− 1	− 7	− 14	− 20
300	0	− 12	− 11	− 37	− 49
450	− 6	− 28	− 50	− 71	− 93
600	− 15	—	—	—	—
750	− 23	—	—	—	—
900	− 33	—	—	—	—

* Owing to the influence of a cold subgrade on these thin slabs, insulation alone will not maintain their temperature at the required 10°C in cold weather. In such cases, additional heat is necessary to maintain the required temperature in the concrete. It is provided by placing at higher temperatures, preheating the ground, using electric resistance wire under

Note for Table 32.2 continued on page 573

space with each 10 mm increase in width of the space. Frost damage to fresh concrete slabs can be prevented with a pumped blanket of urea formaldehyde foam, 25 mm thick. This layer is expendable or reusable as a subgrade separator course.

TABLE 32.3   RELATIVE VALUES OF INSULATING MATERIALS

Insulating Material	Equivalent Thickness (mm)
100 mm commercial blanket insulation	100·0
100 mm loose-fill, fibrous type insulation	100·0
100 mm insulating board	75·8
100 mm sawdust	61·0
100 mm timber	33·3
100 mm dead-air space	23·4
100 mm damp sand	2·3

The insulation should remain in place until the concrete reaches a (cylinder) compressive strength of 10 MPa. The time required to reach this strength (approximately) is tabulated in Table 32.4.

TABLE 32.4   REMOVAL OF INSULATION

28-Day Compressive Strength with Curing at 23°C (MPa)	Temperature of Concrete during Curing (°C)	Days to Reach 10 MPa with 2 Per Cent Calcium Chloride	
		Ordinary Portland Cement	High-early-strength Portland Cement
40	10	2	2
	5	4	3
	2	6	5
30	10	4	3
	5	6	4
	2	8	7
25	10	5	4
	5	7	5
	2	10	8
20	10	6	4
	5	9	6
	2	13	10

the insulation, or by other means, depending on the severity of the prevailing weather. If combustion heaters are used, premature carbonation of the fresh concrete should be avoided (see Section 31.4).

32.1.5 ENCLOSURES

Enclosures for heating may be made of wood, canvas, gypsum wall board, fibre insulation board, plywood, "Sisalkraft", tarred paper or other suitable material, provided that it is reasonably tight and safe for wind and snow loading. Fire is an ever-present hazard with temporary housing of this nature and should be taken into account when planning a form of procedure. Jets of exhaust steam are safest from this standpoint; the next safest being airplane heaters located outside the enclosure and blowing heated air into it. Open fires, braziers and salamanders do not provide circulation and should be avoided or rarely used.

When concrete slabs are being laid, tarpaulins or other readily movable coverings (supported on frames) should follow closely the placing of the concrete, so that less than 1 metre of finished slab is exposed to the outside air. Layers of insulating material placed directly on the concrete are effective also in preventing freezing. Housings and enclosures should be left in place for the requisite period of protection. Sections may be temporarily removed for short periods, provided that the uncovered concrete does not freeze, to enable additional formwork or concrete to be placed in position.

During the second 3-day period after placing, the temperature of the concrete should be kept above freezing point. The two 3-day periods may each be reduced to 2 days when high-early-strength portland cement is used. At the end of the period of protection, temperatures may be permitted to fall gradually by amounts not exceeding the following in 24 hours.

For relatively thin work, 28°C with 20 mm nominal-size aggregate and 22°C with 40 mm nominal-size aggregate.
For mass concrete, 17°C with 75 mm nominal-size aggregate and 11°C with 150 mm nominal-size aggregate.

Curing with water or exhaust steam during the period of protection may be followed by membrane curing with a sealing compound, applied during the first weather with temperatures above freezing after protection is removed.

32.1.6 REMOVAL OF FORMS

In warm weather, curing provided by formwork is usually inferior to that of moist curing, which should be applied properly at the earliest age practicable. In cold weather, unstripped formwork (other than steel) offers insulating protection to the concrete. With suitable insulation, formwork (including steel) will, in many instances, provide adequate protection without supplementary heating. In heated enclosures, formwork serves to distribute the warmth and prevent temperature gradients through local heating. Insulated stacked moulds induce adiabatic curing conditions.

An indication of the minimum time for stripping formwork at low temperatures is given in Section 8.4. The earliest permissible time for removal of props, shores and falsework should be determined from the strength of job-cured specimens of the concrete. The minimum strength required to support dead and imposed loads during the early life of the concrete should be complied with. Shoring may have to be reinstalled for a short period to support temporary live loads without exceeding permissible stresses in the concrete.

The curing period required in cold weather may be estimated by adding to the ordinary curing time the number of days on which the air temperature does not rise above 7°C. Notes on electric curing are given in Section 9.5.

32.1.7 ECONOMICAL PROTECTION

As most concrete structures are intended for a useful life of many years, adequate protection from low temperatures, proper curing and careful supervision are required not only to obtain satisfactory 28-day strength, but also to prevent the formation of frost-bitten corners, dehydrated areas and cracks resulting from overheating or the use of excessive amounts of cement or calcium chloride. The cost of adequate protection need not be excessive and should not be skimped with a consequent sacrifice of durability.

Concrete placed in cold weather may cost up to 10 per cent more than that placed under more clement conditions. An overall economy may be obtained through the correct use of extra portland cement, high-early-strength portland cement or calcium chloride, so as to increase the strength attained during a certain period of protection or to reduce the period of protection required to attain a certain strength.

32.1.8 RECORDS

Records should be kept (with stated dates and hours) of the ambient temperature and weather conditions, the temperatures at several points within the enclosure (if any) and on concrete surfaces, corners and edges (in sufficient number to show the highest and lowest temperatures of the concrete). The temperature of concrete may be taken by holding a thermometer against it, under a heavy cover of insulation, until a steady reading is obtained. Permanent records of a concrete project should include those of temperature and frost at night.

32.2 HOT WEATHER

32.2.1 GENERAL INFORMATION

Concreting in hot weather is characterised by evaporation losses of mixing and curing water, early stiffening and early-strength development through rapid hydration of the cement. It is accompanied by increased surface shrinkage, possibility of cracking and a loss in 28-day strength of up to 20 per cent (see Sections 9.2, 20.1.5(e) and 31.2.4).[23] These effects are minimised by

> Concreting under less rigorous ambient conditions or at night.
> Keeping cool the constituent elements or ingredients; avoiding excessive mixing and delays or midday placement; work-site mixing; and shading mixers, conveyances and pipelines, or coating them with white or reflective paint (see Sections 7.1.2-7.1.4 and 30.1).[94]
> Spraying formwork without leaving excess water internally prior to concreting. Dampening lowers the temperature of surfaces by evaporation.
> Using a tested set-retarding, cement-dispersing (water-reducing) or plasticising agent (e.g. a suitable lignosulphonate) which improves the workability and possible compaction, cohesiveness, crack suppression, or ultimate strength of hot-weather concrete (without an increase in water content), or promotes a given workability with requisite quality control and a reduced content of mixing water (see Sections 3.1.1, 6.4 and 10.2).[23]
> Proper protection, shading and curing as soon as practicable.

32.2.2 MATERIALS, MIXES AND PROCEDURES

The temperature of new concrete should be kept below 30°C in mass concrete (see Section 30.1). A change of 1°C in a concrete mix (8·5 40-kg bags of cement/m³; water/cement ratio, 0·6 by weight) will accompany any one of the following changes in temperature: 9·0°C for cement, 3·6°C for water or 1·6°C for aggregate. A mixture of equal parts of water and crushed ice used as mixing water will reduce the concrete temperature by about 10°C. Water lines should be buried, shaded, insulated or coated with reflective paint, and aggregate should be dampened regularly.

The slump of ordinary concrete may decrease by 2·5 mm per °C rise in temperature above 23°C. The 23°C consistence may be maintained with additional water. The amount required, with normal handling delays, would be 1·0 litre per m³ for 1°C rise in concrete temperature above 23°C. Additional water tends to cause increased shrinkage, laitance and reduced strength.

Contact surfaces should be dampened before the concrete is placed, compacted, finished and cured. These operations should be done as rapidly and effectively as possible with adequate skilled personnel. Placement should be continuous in batch volumes of manageable size, and early drying should be minimised by the use of wind-breaks, whitewashed hessian or aluminium-foil covers with closed sides on frames, and by means of fog sprays or a film of aliphatic (e.g. cetyl) alcohol. Incipient plastic shrinkage cracks may be removed by revibration or reworking of the surface near the time of initial set (see Sections 7.2.2, 20.1.5, 23.2 and 31.2.4).

Continuous water curing for 14 days or more should be started as soon as possible after finishing the placed concrete. The curing water should be at a temperature near that of the concrete, so as to avoid temperature-change stresses. Test results should be accompanied by field records of air temperature, relative humidity, unusual weather conditions, and the temperature of the concrete before and after placement (see Section 9.1). Early curing of slabs may be done with white plastic sheets, or a scuffproof or sand-covered vapourproof membrane that is kept cool by wetting (see Section 9.3).

REFERENCES

1. PROTZE, H. G. "Structural Refractory Concrete". *ACI Journal*, 28, 9 (March 1957), pp. 871-87; 2, 6 (December 1957), pp. 1355-57.
2. MONCKTON, B. R. "Development of Refractories for Use on a Dry Deflector for a Missile Launcher". *IE Aust. Journal*, 33, 3 (March 1961), pp. 105-12.
3. ROBSON, T. D. *High-alumina Cements and Concretes*. Contractors Record, London, 1962.
4. National Bureau of Standards (United States), *Handbook 73 and 76*, 1961.
5. ACI COMMITTEE 349. "Criteria for Reinforced Concrete Nuclear Power Containment Structures". *ACI Journal*, 69, 1 (January 1972), pp. 2-28; 10 (October 1972), pp. 650-56; 73, 6 (June 1973), p.v.
6. DAVIS, H. S. "Effects of High-temperature Exposure on Concrete". *Materials Research and Standards*, 7, 10 (October 1967), pp. 452-59.
7. KESLER, C. E. *et al.* "Concrete for Nuclear Reactors". *ACI Special Publication SP-34*, 1975. *ACI Journal*, 70, 1 (January 1973), þp. 43-54; 6 (June 1973), p. N30.
8. ACI-ASME COMMITTEE 359. "Code for Concrete Reactor Vessels and Containments". *ACI Journal*, 70, 5 (May 1973), pp. 323-27; 71, 6 (June 1974), pp. 306-12; 72, 7 (July 1975), pp. 338-46.
9. DAVIS, H. S. "Aggregates for Radiation Shielding Concrete". *Materials Research and Standards*, 7, 11 (November 1967), pp. 494-501.
10. GREENBORG, J. "Neutron Attenuation Mechanisms in Concrete Shielding". *Journal of Materials*, 4, 2 (June 1969), pp. 251-81.
11. CARLSON, R. W. "Temperatures and Stresses in Mass Concrete". *ACI Journal*, 34 (March/April 1938), pp. 497-515.
12. RAWHOUSER, C. "Cracking and Temperature Control of Mass Concrete". *ACI Journal*, 16, 4 (February 1945), pp. 305-46; 17, 2 (November 1945), pp. 348 (1-24).
13. MUKADDAM, M. "Creep Analysis of Concrete at Elevated Temperatures". *ACI Journal*, 71, 2 (February 1974), pp. 72-78.
14. ACI COMMITTEE 207. "Mass Concrete for Dams and Other Massive Structures". *ACI Journal*, 67, 4 (April 1970), pp. 273-309.
15. DUNSTAN, R, H. and MITCHELL, P. B. "Results of a Thermocouple Study in Mass Concrete in the Upper Tamar Dam". *ICE Proceedings*, 1, 60 (February 1976), pp. 27-52.
16. DE COURCY, J. W. "Movement in Concrete Structures". *Concrete* (London), 3, 6 (June 1969), pp. 241-46; 7 (July 1969), pp. 293-96; 8 (August 1969), pp. 335-39.
17. HRB COMMITTEE. "Symposium on the Structure of Portland Cement Paste and Concrete". *Special Report 90*, United States Highway Research Board, 1966.
18. ROPER, H. "Shrinkage, Tensile Creep and Cracking Tendency of Concretes". *Australian Road Research*, 5, 6 (December 1974), pp. 26-35.
19. ACI COMMITTEE 224. "Control of Cracking in Concrete Structures". *ACI Journal*, 69, 12 (December 1972), pp. 717-53; 70, 6 (June 1973), pp. 430-34; *ACI Publication SP-20*, 1968.
20. MENON, R. G. "Autogenous Healing of Concrete". *Indian Concrete Journal*, 33, 5 (May 1959), pp. 161-62, 178.
21. PENNY, R. K. and MARRIOTT, D. L. *Design for Creep*. McGraw-Hill, New York, 1971.
22. ACI STANDARD 306. "Recommended Practice for Cold Weather Concreting". *ACI Journal*, 62, 9 (September 1965), pp. 1009-34; 63, 3 (March 1966), pp. 305-6; 2, 3 (March 1966), pp. 1739-40; 2, 63, 6 (June 1966), p. iii; and *ACI Monograph No. 3*, 1966; ACI Convention Committee, Canada. "Behaviour of Concrete Under Temperature Extremes". *ACI Publication SP-39*, 1975.
23. ACI COMMITTEE 605. "Recommended Practice for Hot Weather Concreting". *ACI Journal*, 68, 7 (July 1971), pp. 489-503; 69, 1 (January 1972), pp. 70-73.

24. NICOL, T. B. "Warragamba Dam". *ICE Proceedings*, 27 (March 1964), pp. 491-546; 31 (August 1965), pp. 361-83; *IE Aust. Journal*, 36, 10-11 (October/November 1964), pp. 239-62; 37, 3 (March 1965), pp. 71-82; 38, 3 (March 1966), pp. 49-52.
25. COPELAND, L. E., KANTRO, D. L. and VERBECK, G. "Chemistry of Hydration of Portland Cement". *Proceedings, Fourth International Symposium on the Chemistry of Cement*, Washington, 1960.
26. BRUNAUER, S. "The Role of Tobermorite Gel in Concrete". *Journal of the Reinforced Concrete Association*, 1, 7 (January/February 1963), pp. 293-309.
27. KESLER, C. E. *et al.* "Creep, Shrinkage and Temperature Effects". *Planning and Design of Tall Buildings. ASCE-IABSE International Conference*, Lehigh University, Bethlehem, Pennsylvania. *Proceedings*, vol. III, pp. 757-64, 1972.
28. HANSEN, T. C. and MATTOCK, A. H. "Influence of Size and Shape of Member on the Shrinkage and Creep of Concrete". *ACI Journal*, 63, 2 (February 1966), pp. 267-90; 9 (September 1966), pp. 1017-22.
29. ILLSTON, J. M. and ENGLAND, L. "Creep and Shrinkage of Concrete and Their Influence on Structural Behaviour". *Structural Engineer*, 48, 7 (July 1970), pp. 283-92.
30. ELVERY, R. H. and SHAFI, M. "Analysis of Shrinkage Effects on Reinforced Concrete Structural Members". *ACI Journal*, 67, 1 (January 1970), pp. 45-52.
31. ARTOYD, T. N. W. *et al.* "Winter Concreting". *Concrete* (London), 1, 10 (October 1967), pp. 339-50; *Civil Engineering and Public Works Review*, 60, 710 (September 1965), pp. 1324-53; *Concrete Construction*, 15, 10 (October 1970), pp. 350-69; 11 (November 1970), pp. 395-96.
32. WOLHUTER, C. W. "Creep". *Prestress*, 18 (December 1968), pp. 10-20.
33. HELLER, L. and TAYLOR, H. F. W. *Crystallographic Data for the Calcium Silicates*. HMSO, London, 1956.
34. HORNBY, I. W., VERDON, G. F. and WONG, Y. C. "Elastic Tests on a Model of the Oldbury Nuclear Station Prestressed Concrete Pressure Vessel". *ICE Proceedings*, 34 (July 1966), pp. 347-67; 36 (April 1967), pp. 881-82.
35. KANTRO, D. L., COPELAND, L. E. and ANDERSON, E. R. "An X-ray Diffraction Investigation of Hydrated Portland Cement Pastes". *ASTM*, 60 (1960), pp. 1020-35.
36. CROWLEY, A. K. "Thermal Control of Concrete". *Sydney Water Board Journal*, 11, 4 (January 1962), pp. 122-27.
37. COPELAND, L. E. and SCHULZ, E. G. "Electron Optical Investigation of the Hydration Products of Calcium Silicates and Portland Cement". *Journal of the Portland Cement Association Research and Development Laboratories*, 4, 1 (January 1962), pp. 2-12.
38. MEYERS, B. L., BRANSON, D. E. and SCHUMANN, C. G. "Prediction of Creep and Shrinkage Behaviour for Design from Short Term Tests". *PCI Journal*, 17, 3 (May/June 1972), pp. 29-45.
39. WOLLENBERG, H. A. and SMITH, A. R. "Low-radioactivity Concrete". *Journal of Materials*, 3, 4 (December 1968), pp. 757-79.
40. DAVIES, J. D. "Stresses in Cylindrical Tanks Due to Shrinkage". *Concrete and Constructional Engineering*, 57, 5 (May 1962), pp. 193-96.
41. TAYLOR, R. S. and BURROW, R. E. D. "Basis of Design of Prestressed Concrete Pressure Vessels". *Structural Concrete*, 3, 5 (September/October 1966), pp. 225-60.
42. HARRIS, A. J. *et al.* "Prestressed Concrete Pressure Vessels for Nuclear Power Stations". *PCI Journal*, 10, 5 (October 1965), pp. 17-27.
43. HANSEN, W. C. "Porosity of Hardened Portland Cement Paste". *ACI Journal*, 60, 1 (January 1963), pp. 141-56.
44. FOSTER, B. E. "Attenuation of X-Rays and Gamma Rays in Concrete". *Materials Research and Standards*, 8, 3 (March 1968), pp. 19-24.
45. YOUNG, J. F. "Hydration of Tricalcium Aluminate with Lignosulphonate Additives". *Magazine of Concrete Research*, 14, 42 (November 1962), pp. 137-42; 16, 49 (December 1964), pp. 231-32.
46. "Prestressed Concrete Nuclear Plants for Electricite de France". *International Construction*, 10, 8 (August 1971), pp. 2-7.

47. GUPTA, P., CHATTERJI, S. and JEFFERY, J. W. "Effects of Various Additives on the Hydration Reaction of Tricalcium Aluminate". *Cement Technology*, 1, 2 (March/April 1970), pp. 59-66.
48. CHESTERS, J. H. *Refractories, Production and Properties*. Iron and Steel Institute, London, 1973.
49. CHATTERJI, S. "Electron-optical and X-ray Diffraction Investigation of the Effects of Lignosulphonates on the Hydration of C_3A". *Indian Concrete Journal*, 41, 4 (April 1967), pp. 151-60.
50. BUDGEN, W. E. J. et al. *Design for Movement in Buildings*. Symposium Proceedings, Concrete Society, London, 1970.
51. TAYLOR, H. F. W. *The Chemistry of Cements*, vols I and II. Academic Press, London, 1964.
52. HSU, T. T. C., SLATE, F. O., STURMAN, G. M. and WINTER, G. "Microcracking of Plain Concrete and the Shape of the Stress-strain Curve". *ACI Journal*, 60, 2 (February 1963), pp. 209-24; 12 (December 1963), pp. 1787-1824.
53. NEVILLE, A. M. *Creep of Concrete: Plain, Reinforced, and Prestressed*. North-Holland, 1970; *Properties of Concrete*. Pitman, 1968; "Hardened Concrete, Physical and Mechanical Aspects". *ACI Monograph No. 6*.
54. COUTINHO, A. De S. "Influence of the Type of Cement on its Cracking Tendency". *Technical Paper 216*, National Civil Engineering Laboratories, Lisbon, Portugal, 1964.
55. REICHARD, T. W. "Creep and Drying Shrinkage of Lightweight and Normal-Weight Concretes". *Monograph 74*. United States Department of Commerce, National Bureau of Standards, 1964.
56. COHN, E. B. and WALL, W. A. "Military Personnel Records Centre Built Without Expansion Joints". *ACI Journal*, 29, 12 (June 1958), pp. 1103-10.
57. HEIMAN, J. L. "Long-term Deformations in the Tower Building, Australia Square, Sydney". *ACI Journal*, 70, 4 (April 1973), pp. 279-84.
58. LEWIS, D. J. and IRVING, J. "Operational Stresses in Nuclear Prestressed Concrete Pressure Vessels". *Civil Engineering and Public Works Review*, 63, 743 (June 1968), pp. 673-76; 744 (July 1968).
59. DHIR, R. K. et al. "Strength and Deformation Properties of Autogenously Healed Mortars". *ACI Journal*, 70, 3 (March 1973), pp. 231-36.
60. HANSON, J. A. "Prestress Loss as Affected by Type of Curing". *PCI Journal*, 9, 2 (April 1964), pp. 69-93.
61. HETHERINGTON, J. W. "Construction Methods Used in Raising Tenterfield Dam". *IE Aust Transactions*, CE17, 1 (April 1975), pp. 37-39.
62. POPOVICS, S. "A Review of Stress-Strain Relationships for Concrete". *ACI Journal*, 67, 3 (March 1970), pp. 243-48; 9 (September 1970), pp. 752-56.
63. KLIEGER, P. "Some Aspects of Durability and Volume Change of Concrete for Prestressing". *PCA R & DL Journal*, 2, 3 (September 1960), pp. 2-12.
64. KEENE, P. W. "Concrete Cured in Steam at Atmospheric Pressure". *Prestress*, 13 (December 1963), pp. 4-28.
65. NEPPER-CHRISTENSEN, P. and SKOVGAARD, P. "Drying Shrinkage of Low-pressure Steam-cured Concrete Units". *RILEM International Conference on Problems of Accelerated Hardening of Concrete in Manufacturing Precast Reinforced Concrete Units*, Moscow, 1964.
66. GILL, S. *Structures for Nuclear Power*. Contractors Record, London, 1964.
67. CAMPBELL-ALLEN, D. "Prestressed Concrete Pressure Vessels". *Constructional Review*, 41, 1 (January 1968), pp. 22-28.
68. REESE, C. "Designing for Effects of Creep, Shrinkage, Temperature in Concrete". *ACI Publication SP-27*, 1971; *ACI Journal*, 69, 3 (March 1972), pp. 179-84.
69. RUDD, F. O. "Prediction and Control of Stresses in Concrete Block". *ACI Journal*, 62, 1 (January 1965), pp. 95-104.

70. ROBERTS, C. M., WILSON, E. B. and WILTSHIRE, J. G. "Design Aspects of the Strathfarrar and Kilmorack Hydroelectric Scheme". *ICE Proceedings,* 30 (March 1965), pp. 449-87.
71. WARD, M. A. and COOK, D. J. "Mechanism of Tensile Creep in Concrete". *Magazine of Concrete Research,* 21, 68 (September 1969), pp. 151-58.
72. CAMPBELL-ALLEN, D. and THORNE, C. P. "Thermal Conductivity of Concrete". *Magazine of Concrete Research,* 15, 43 (March 1963), pp. 39-48; 16, 49 (December 1964), pp. 233-34.
73. SIMS, F. W., RHODES, J. A. and CLOUGH, R. W. "Cracking in Norfork Dam". *ACI Journal,* 61, 3 (March 1964), pp. 265-85; 9 (September 1964), pp. 1213-18.
74. "Symposium on Concrete Construction in Aqueous Environments". *ACI Special Publication SP-8,* 1964.
75. KWEI, G. C. S. "Relation Between Creep and Increase in Strength and Hydration from Time of Load Application Onwards". *Indian Concrete Journal,* 39, 4 (April 1965), pp. 141-45.
76. RYELL, J. "An Unusual Case of Surface Deterioration on a Concrete Bridge Deck". *ACI Journal,* 62, 4 (April 1965), pp. 421-42.
77. ILLSTON, J. M. and JORDAAN, I. J. "Creep Prediction for Concrete Under Multiaxial Stress". *ACI Journal,* 69, 3 (March 1972), pp. 158-64.
78. SIEV, A. "Crack-proof Joint Between Reinforced Concrete Roof and Walls". *ACI Journal,* 62, 7 (July 1965), pp. 850-52.
79. MATHER, K. "High Strength, High Density Concrete". *ACI Journal,* 62, 8 (August 1965), pp. 951-60; 2, 63, 3 (March 1966), pp. 1731-32.
80. CORLEY, W. G. and SOZEN, M. A. "Time-dependent Deflections of Reinforced Concrete Beams". *ACI Journal,* 63, 3 (March 1966), pp. 373-86; 9 (September 1966), pp. 1033-39.
81. COUNTO, U. J. "Effect of the Elastic Modulus of the Aggregate on the Elastic Modulus, Creep and Creep Recovery of Concrete". *Magazine of Concrete Research,* 16, 48 (September 1964), pp. 129-38; 17, 52 (September 1965), pp. 142-51.
82. HANSEN, T. C. "Influence of Aggregate and Voids on Modulus of Elasticity of Concrete, Cement Mortar, and Cement Paste". *ACI Journal,* 62, 2 (February 1965), pp. 193-216; 7 (July 1965), p. NL18; 9 (September 1965), pp. 1181-84; 2, 63, 6 (June 1966), p. iii.
83. BAZANT, Z. P. "Prediction of Concrete Creep Effects Using Age-Adjusted Effective Modulus Method". *ACI Journal,* 69, 4 (April 1972), pp. 212-17; 70, 6 (June 1973), p. v.
84. CRUZ, C. R. "Elastic Properties of Concrete at High Temperatures". *Journal of the PCA, R & DL,* 8, 1 (January, 1966), pp. 37-45.
85. BATE, S. C. C. and LEWSLEY, C. S. "Environmental Changes, Temperature, Creep and Shrinkage in Concrete Structures". *Current Paper 7/70,* Building Research Station, Britain, 1970.
86. SIKTBERG, C. Y. "An Effective System for Control of Moisture in Concrete Aggregates". *Modern Concrete,* 30, 2 (June 1966), pp. 52-56.
87. BROWNE, R. D. "Thermal Movement of Concrete". *Concrete* (London), 6, 11 (November 1972), pp. 51-53.
88. VERBECK, G. "Cement Hydration Reactions at Early Ages". *Journal of the PCA, R & DL,* 7, 3 (September 1965), pp. 57-63.
89. CORDON, W. A. and THORPE, J. D. "Control of Rapid Drying of Fresh Concrete by Evaporation Control". *ACI Journal,* 62, 8 (August 1965), pp. 977-85; 2, 63, 3 (March 1966), pp. 1733-34.
90. CAMPBELL-ALLEN, D. and HOLFORD, J. G. "Stresses and Cracking in Concrete Due to Shrinkage". *IE Aust. CE Transactions,* CE12, 1 (April 1970), pp. 33-39.
91. PLOWMAN, J. M. "Young's Modulus and Poisson's Ratio of Concrete Cured at Various Humidities". *Magazine of Concrete Research,* 15, 44 (July 1963), pp. 77-82; 16, 49 (December 1964), p. 235.
92. BRUNAUER, S. and COPELAND, L. E. "The Chemistry of Concrete". *Scientific American,* 210, 4 (April 1964), pp. 80-92.
93. NEWMAN, K. "Properties of Concrete". *Structural Concrete,* 2, 11 (September/October 1965), pp. 451-82.

94. "Concrete is cooled by Liquid Nitrogen Injection System". *Construction Methods and Equipment,* 52, 12 (December 1970), pp. 36-37.
95. C & CA Conference Committee. *The Structure of Concrete and Its Behaviour under Load.* International Conference, London, 1965. Cement and Concrete Association, Britain, 1968.
96. MEYERS, B. L. and BRANSON, D. E. "Design Aid for Predicting Creep and Shrinkage Properties of Concrete". *ACI Journal,* 69, 9 (September 1972), pp. 551-55; 70, 6 (June 1973), p. v.
97. ZOLDNERS, N. G. et al. "Temperature and Concrete". *ACI Special Publication SP-25.* 1971. *ACI Journal,* 64, 2 (February 1967), pp. 97-103; 68, 4 (April 1971), pp. 276-81.
98. HRB SYMPOSIUM COMMITTEE. "Symposium on Joints and Sealants". *Highway Research Record No. 80, 1965; Special Publication of ACI Committee 504,* 1971.
99. HARMATHY, T. Z. "Thermal Properties of Concrete at Elevated Temperatures". *Journal of Materials,* 5, 1 (March 1970), pp. 47-74.
100. STEVENS, R. F. "Deflexions of Reinforced Concrete Beams". *ICE Proceedings,* 53, 2 (September 1972), pp. 207-24.
101. MILLER, C. A. and GURALNICK, S. A. "Creep Deflection of Reinforced Concrete Beams". *ASCE Proceedings,* 96, ST12 (December 1970), pp. 2625-38.
102. VERBECK, G. "Pore Structure". *Research Bulletin, 197,* PCA Research and Development Laboratories, 1966.
103. EVANS, E. P. and HUGHES, B. P. "Shrinkage and Thermal Cracking in a Reinforced Concrete Retaining Wall". *ICE Proceedings.* 39 (January 1968), pp. 111-25; 40 (August 1968), pp. 539-68.
104. PETZOLD, A. and ROHRS, M. *High Temperature Concrete.* Maclaren, London, 1970.
105. L'HERMITE, M. R. et al. "Physical and Chemical Causes of Creep and Shrinkage of Concrete". RILEM Symposium, Munich, 1968. *Materials and Structures,* 2, 8 (March/April 1969), pp. 103-62.
106. TROXELL, G. E., RAPHAEL, J. M. and DAVIS, R. E. "Long-time Creep and Shrinkage Tests of Plain and Reinforced Concrete". *ASTM Proceedings,* 58, (1958), pp. 1101-20.
107. TUTHILL, L. H. and ADAMS, R. F. "Cracking Controlled in Massive, Reinforced Structural Concrete by Application of Mass Concrete Practices". *ACI Journal,* 69, 8 (August 1972), pp. 481-91.
108. CADY, P. D., CLEAR, K. C. and MARSHALL, L. G. "Tensile Strength Reduction of Mortar and Concrete Due to Moisture Gradients". *ACI Journal,* 69, 11 (November 1972), pp. 700-705.
109. SCHOEWERT, L. C. and HILLEN, H. F. "Underwater Transporting of Concrete with the Hydro-valve". *ACI Journal,* 69, 9 (September 1972), pp. 584-88.
110. MAHER, D. R. H. "Effects of Differential Temperature on Continuous Prestressed Concrete Bridges". *IE Aust. CE Transactions,* CE12, 1 (April 1970), pp. 29-32.
111. EDITORIAL. "Thermal Properties of Precast Concrete". *PCI Journal,* 16, 3 (May/June 1971), pp. 33-43.
112. SHALON, R. and BERHANE, Z. "Shrinkage and Creep in Mortar and Concrete in Hot-humid Environment". *Symposium on Concrete and Reinforced Concrete in Hot Countries,* Haifa; *RILEM Proceedings,* BRS Israel, 1971, pp. 309-32.
113. ACI COMMITTEE 207. "Effect of Restraint, Volume Change, and Reinforcement on Cracking of Massive Concrete". *ACI Journal,* 70, 7 (July 1973), pp. 445-70.
114. HOFF, G. C. and HOUSTON, B. J. "Nonmetallic Waterstops". *ACI Journal,* 70, 1 (January 1973), pp. 7-13; 7 (July 1973), pp. 496-97.
115. ACI COMMITTEE 304. "High-density Concrete: Measuring, Mixing, Transporting and Placing". *ACI Journal,* 72, 8 (August 1975), pp. 407-14.
116. WATSON, S. C. "Compression Seals in Architectural/Industrial Uses". *ACI Journal,* 70, 10 (October 1973), pp. 699-700.

117. KARNAUKHOV, A. P. *et al. Pore Structure and Properties of Materials.* RILEM Proceedings of International Symposium, Prague, Academia, 1973.
118. Commonwealth Industrial Gases. "Liquid Nitrogen for Concrete Cooling". *Contracting and Construction Engineer,* 28, 3 (March 1974), p. 59.
119. PARMENTER, B. S. "Design and Construction of Joints in Concrete Pavements". Transport and Road Research Laboratory, Britain, *TRRL Report LR512,* 1973.
120. FAGERLUND, G. "Influence of Pore Structure on Shrinkage, Strength and Elastic Moduli". *DBM Report 44,* Lund Institute of Technology, Lund, Sweden, 1973.
121. MACPHERSON, J. D. "Cylindrical Reinforced Concrete Water Tanks With Thin Walls and High Hoop Stresses". *IE Aust. Journal,* 38, 4-5 (April/May 1966), pp. 87-94; 9 (September 1966), pp. 249-50.
122. CAMPBELL-ALLEN, D. "Prediction of Shrinkage for Australian Concrete". *IE Aust. CE Transactions,* CE15, 1 & 2 (November 1973), pp. 53-57, 62.
123. *Prevention of Cracks and High Stresses at Bearing Edges in Building Structures.* SK Bearings, Pampisford, Cambridge, Britain.
124. BECKER, N. K. and MACINNIS, C. "A Theoretical Method for Predicting the Shrinkage of Concrete". *ACI Journal,* 70, 9 (September 1973), pp. 652-57; 71, 3 (March 1974), pp. 145-48.
125. BALDWIN, R. and NORTH, M. A, "Stress-Strain Relationship for Concrete at High Temperatures". *Magazine of Concrete Research,* 25, 85 (December 1973), pp. 208-12.
126. SCHUTZ, R. J., *et al.* "Architectural Precast Concrete Joint Details". *PCI Journal,* 19, 2 (March/April 1973), pp. 10-37; PCI Reprint with 1974 Committee Revisions.
127. HOBBS, D. W. "Influence of Aggregate Restraint on the Shrinkage of Concrete". *ACI Journal,* 71, 9 (September 1974), pp. 445-50; 72, 3 (March 1975), pp. 114-16.
128. Editorial. "An Underground Railroad is Built on Site Above Ground". *Concrete Construction,* 19, 9 (September 1974), pp. 453-56.
129. ACI COMMITTEE 435. *Deflections of Concrete Structures. ACI Special Publication SP-43,* 1974.
130. MATHER, B. "Concrete of Low Portland Cement Content in Combination with Pozzolans". *ACI Journal,* 71, 12 (December 1974), pp. 589-99.
131. SADGROVE, B. M. "Freezing of Concrete At An Early Age". *C & CA Report 42.503,* 1974.
132. NASH, K. L. "Diaphragm Wall Construction Techniques". *ASCE Journal,* 100, CO4 (December 1974), pp. 605-20.
133. BOYES, R. G. H. "Uses of Bentonite in Civil Engineering". *ICE Proceedings,* 52, 1 (May 1972), pp. 25-37; 54, 1 (February 1973), pp. 169-71.
134. NORTHWOOD, Britain. "Structural Diaphragm Walls". *Consulting Engineer,* Suppl. to 33, 6 (June 1969), 20 pp.
135. Institution of Civil Engineers. *Ground Engineering* (1970) and *Diaphragm Walls And Anchorages* (1974). ICE Marketing Department, London.
136. "Serviceability of Concrete". *IE Aust. National Conference Publication 75/6,* Melbourne, 1975; AS 1480, MP28, Sect. C10 on Serviceability; and BS 5337.
137. RANGAN, B. V. "Prediction of Long-term Deflections of Flat Plates and Slabs". *UNICIV Report R-140,* University of New South Wales, 1975.
138. WARNER, R. F. "Axial Shortening in Reinforced Concrete Columns". *UNICIV Report R-143,* University of New South Wales, 1975.
139. ROSE, M. A. "Deflection Measurements". *Ultimate Strength Design,* C & CA Conference, Melbourne, 1975.
140. "Expansion Joints in Buildings". *BR Technical Report 65,* National Academy of Sciences, Washington, 1974.
141. HUGHES, B. P. "Controlling Shrinkage and Thermal Cracking", and "Early Thermal Movement and Cracking of Concrete". *Concrete* (London), 6, 5 (May 1972), pp. 39-42; 7, 5 (May 1973), pp. 43-44.

142. CAMPBELL-ALLEN, D. and HEFFENSTEIN, H. L. "Cracking in Concrete—Its Extent and Causes". *Research Report R253,* University of Sydney, 1974.
143. German Industrial Standard DIN 1045. *Concrete and Reinforced Concrete—Design and Construction.* (Standards Association Australia, 1976.)
144. BOYD, D. W. "Normal Freezing and Thawing Degree—days from Normal Monthly Temperatures". *Canadian Geotechnical Journal,* 13, 2 (May 1976), pp. 176-80.

PART 7

NEW MATERIALS
AND PRODUCTS

CHAPTER **33**

LIGHTWEIGHT CONCRETE

33.1 INTRODUCTION

Lightweight concrete was known in the early days of the Roman Empire, when pumice plums were embedded in mixes for the walls and domes of large temples. Further developments were deferred until early this century, when a growing impetus in the building industry brought about a progressive introduction of new and improved types of lightweight building material on a commercial basis. A variety of factors, in addition to research work and operational enterprise, have since led to the gradual acceptance of lightweight concrete as a commonplace building material.

For example, there is a growing need in built-up areas for reduction in the dead weight of tall buildings, wide-span concrete girders, flat plates, roofing construction and spatial structures generally. A saving of dead weight for a given live-load capacity is highly desirable for precast concrete elements, which must be transported some distance and manipulated into place either mechanically or manually. In some localities, it is advantageous to utilise industrial by-products or to manufacture lightweight aggregate, where natural aggregate of good quality is not readily available. As lightweight concrete can be made with a wide range of specific characteristics, there is ample scope for its selective use in modern building construction.

With few exceptions, however, lightweight concrete is a little more expensive than an equal volume of ordinary concrete at the ready-to-use stage. Justification of its use lies primarily in reduced weight, requisite strength and durability, improved thermal insulation and

589

good fireproofing qualities. Compensatory savings arise through ease of handling and working, speed of fabrication, reduced costs in formwork, transport and erection, and economies in the artificial heating or cooling of large buildings. There being no virtue in lightness of weight for its own sake, except for special purposes in building construction, it follows that lightweight concrete construction is warranted where savings in overall cost or functional advantages arise through its use.

Lightweight concrete falls into several categories according to variation in physical properties. In this regard, the bulk density (AS 1480 and ASTM C567) of the material falls within a range of 160-2000 kg/m³; the drying shrinkage varies from less than that of ordinary concrete to several times more; the insulating value runs from three to over twelve times that of ordinary concrete; and the compressive-strength/bulk-density ratio can have a wide range of values, with an upper limit under special circumstances in the region of 200. In order to embody a particular set of characteristics in a product for a specific purpose, lightweight concrete can be manufactured by one or other of the following procedures.

1. Leaving voids between coarse-aggregate particles which are bound together with cement, thereby creating a no-fines concrete.

2. Using various kinds of vesicular, cellular or expanded aggregate in a mix, so as to make a suitable grade of lightweight-aggregate concrete.

3. Entrapping small cells or interstices of air or gas in a cementing matrix which in itself may be light in weight, thus forming a foamed concrete of which a special variety is lightweight calcium silicate hydrate.

33.2 MATERIALS AND MANUFACTURE

33.2.1 NO-FINES CONCRETE

No-fines concrete had its origin in the Netherlands and, since 1923, it has been accepted widely in Britain and elsewhere for residential buildings. The concrete, which is a mixture of stone screenings and cement slurry, has a high proportion of continuous voids but practically no capillary paths.[1]

The mix is proportioned usually with 1 part by volume of cement and 8 parts of coarse aggregate, and with a water/cement ratio of 0·35-0·45 by weight (0·5-0·7 by volume). A richer mix, say 1 part of cement to 6 parts of coarse aggregate, may be used for the lower storeys of tall buildings and for structural-grade lightweight aggregate with a rough surface texture. The aggregate consists of washed gravel, crushed igneous rock or limestone, lightweight aggregate, crushed hard-burnt clay bricks or air-cooled blastfurnace slag, which is graded in size from 9·5 mm to 19·0 mm. The amount of mixing water required varies with the type of aggregate and its moisture content, only sufficient being used to form a cement slurry that will cover each particle of aggregate without running off and partly filling the voids. A fairly dry mix, therefore, must be used for satisfactory results.[124]

The aggregate is dampened during a short period of mixing before the cement and the remainder of the water are added. Adequate mixing of the concrete limits the output of a 0·5 m³ drum mixer to about 8·5 m³ per hour. The concrete is placed as soon as possible and only sufficient rodding is given to it to ensure that the formwork is evenly filled.

33.2.2 LIGHTWEIGHT-AGGREGATE CONCRETE

As with other specialised building materials, the production and marketing of lightweight aggregate are largely local in scope. As shown in Table 33.1, mineral lightweight aggregate may be classified into several categories: volcanic, sedimentary and industrial by-product types of material. With minor variations in procedure, concretes using these aggregates are mixed and placed in much the same way as ordinary concrete. Lightweight aggregate with a low bulk density, which is suitable for making insulating concrete, may consist alternatively of expanded polystyrene beads[147] or heat-expanded, carbonised particles that are made from dampened wheat kernels.

The economical use of lightweight volcanic aggregate is influenced by the distance of the source from built-up areas, while furnace clinker is diminishing as a source of lightweight aggregate because of a widespread conversion of power stations to fuel oil and pulverised coal. Alternative industrial by-products are available in large quantities, and in certain areas these are being utilised extensively for the production of foamed blastfurnace slag and sintered fly ash.

Lightweight aggregate for structural application consists of expanded clay and shale, foamed slag and sintered fly ash, while that for concrete with high thermal insulation includes exfoliated

TABLE 33.1   CLASSIFICATION AND PROPERTIES OF MINERAL LIGHTWEIGHT AGGREGATES

Type	Material	Composition	Texture in Natural State	Texture when Used	Bulk Density, loose expanded aggregate (kg/m³)
Volcanic		Rock Type			
	Pumice	Acid	Frothy, extremely vesicular	Frothy, extremely vesicular	480-880
	Scoria	Basic	Clinker-like, highly vesicular	Clinker-like, highly vesicular	640-1000
	Tuff	Acid	Porous	Porous	800-1200
	Obsidian	Acid and Intermediate	Nonporous	Cellular, frothy	400-800
	Perlite	Acid	Nonporous	Cellular, frothy	80-240
Sedimentary	Clays, shales, slates*	Hydrated aluminium silicates	Nonporous	Cellular	560-1000
	Vermiculite**	Hydrated aluminium silicates	Micaceous	Accordion-like	65-190
Industrial by-products	Clinker†	Sintered furnace residue; low in combustibles, sulphide and sulphates	Cellular	Cellular	720-1000
	Foamed blastfurnace slag††	Best slag is 35-38 per cent SiO_2, 44-47 per cent CaO + Al_2O_3, MgO, Fe_2O_3, low S	Glassy	Cellular	320-970
	Sintered fly ash‡	Over 40 per cent SiO_2, 10-20 per cent C + low MgO and S	Very fine	Cellular	640-970

See Notes on page 593

vermiculite and expanded perlite. The process of bloating clay and shale for aggregate was discovered at the time of World War I, and the demand for this type of aggregate has grown tremendously since World War II. This result has been brought about by an increased availability of high-quality expanded aggregate at reasonable cost, and the practical proof that it can be used in concrete to attain the same compressive strength as that possessed by ordinary structural concrete.

For batching purposes, lightweight aggregate with coated rounded particles is more effective than crushed sintered aggregate in producing a workable mix with a low water content. As the bulk density of different fractions of lightweight aggregate varies inversely with their particle size, it is necessary (with weigh-batching) to increase gravimetrically the proportion of fine components, in order to provide a desirable distribution of particle sizes. For structural concrete, the maximum bulk density of coarse and fine grades of lightweight aggregate is limited to 880 and 1100 kg/m^3, respectively. Data on the sampling, grading and testing of lightweight aggregate for concrete are given in AS 1467, BS 3681 and 3797, and ASTM C35, C330, C331, C332 and C495.

The first major application of lightweight-aggregate concrete in Australia took place in 1959 in connection with the ten-storey Bank of Adelaide building in Melbourne. Subsequent examples in this locality include such landmarks as the twenty-six-storey Consolidated Zinc building, the twenty-five-storey National Mutual Centre, the twenty-two-storey BP (Australia) and nineteen-storey Customs House buildings, the twenty-storey Domain Park and twenty to thirty-storey Victorian Housing Commission flats, the seventeen-storey Royal and Globe Insurance block, and the fifteen-storey Commonwealth Offices. Buildings in the Sydney area, to mention but a few, include the twenty-two-storey Pearl Assurance House and Reserve Bank office block, the twenty-two-storey IBM Centre, the twenty-storey NRMA headquarters, the three-storey Royal North Shore Hospital on a clay foundation, the Westfield Development at Dee Why that incorporated six 1100 m² lift slabs, and the fifty to fifty-two-storey Australia Square and Park Regis projects (Plates 26 and 27).[134]

Notes for Table 33.1

* Ferruginous types are generally suitable.
** Formed by alteration of micas in several types of rock.
† Best residues are from high-temperature combustion of coal and coke.
†† Zinc smelter slag also may be satisfactory.
‡ Residue from high-temperature combustion of pulverised black coal.

Of the innumerable applications of lightweight-aggregate concrete abroad, some notable illustrations are the seventy-storey Lake Point Tower, the 182 m high Twin Towers, 145 m high Brunswick Building, and forty-storey Jupiter Centre, all in Chicago; the fifty-storey American Hotel, 143·5 m high CBS Building, thirty-five-storey Carol Towers and 91·5 m high Tower East apartments in New York City; the thirty-seven-storey Hopkinson House in Philadelphia, and a 122 m diameter dome at the University of Illinois, the twenty-two-storey Hilton Hotel in Pittsburgh, and the twenty-two-storey flats for the Battersea Borough Council in Britain.

The tremendous possibilities of prestressed lightweight-aggregate concrete in structural work are indicated in its growing applications to precast mullions, ribbed and hollow elements for wide-span floors and roofs, lift slabs, waffle-type floors and flat plates, curvilinear and troughed roofing systems, and concrete road-bridge construction. A notable exemplification of its structural efficiency is shown in the extension of a hangar at Heathrow Airport, London, where a main beam of 61 m span with secondary beams of 18·3 m span were fabricated in prestressed concrete containing sintered fly ash.

(a) Volcanic aggregates.

(*i*) *Pumice* is a cemented volcanic ash or honeycombed lava containing elongated parallel cells, which were formed by steam when escaping from the molten rock. For lightweight aggregate, the material should have a minimum hardness of 4·0 on the Mohr scale. It is crushed, washed to remove dust and clay, and graded into suitable size (see Section 3.3.3(b)(ii)).[2, 127]

(*ii*) *Scoria* is a vesicular volcanic rock of basic composition and concrete made with it as aggregate has a lower water absorption and drying shrinkage than pumice concrete of about the same strength.

(*iii*) *Obsidian*. The raw material, which is a glassy volcanic rock with a small quantity of entrapped moisture, is granulated and fed into a rotary kiln with a small percentage of ground quartz or quartzite (see Section 3.3.3(b)(ii)). As the obsidian fuses, each particle becomes coated with unfused silica flour which prevents the mass from sticking. Coated spherical pellets with a vesicular structure are thus produced.

(*iv*) *Perlite*. The ore, which is a variety of obsidian containing 2-5 per cent of entrapped moisture, is ground and graded so that the particle size is between 2·36-mm and 0·30-mm sieves for concrete aggregate and 1·18-mm and 0·30-mm sieves for plaster aggregate (see Section 3.3.3(b)(ii)). The prepared material is heated under controlled conditions. These are arranged in such a way as to reduce the moisture content of the granules to between 0·5 and 1 per cent and, subsequently, to expand them quickly at a temperature of 1000-1100°C in either a two-stage rotary kiln or a shaft kiln with an updraught varying in velocity from 1·5 to 9 m/s. The particles soften and expand by an evolution of steam, whereupon they are carried away to an air-separator for gradation into sizes.

The bulk density of the finished product lies normally within a range of 80-240 kg/m³, but it can be reduced to about 16 kg/m³ by changing the size of granules, the rate of temperature rise and the temperature of expansion. Perlite, which is to be used in gypsum plaster and insulating concrete, should comply with the requirements specified in AS 1467, BS 3797 or ASTM C35 and C332;[3-6] and its grading should fall within the limits given in Table 33.2. The material being glassy in nature, consideration should be given to possible reactivity and its effect when perlite is used with high-alkali portland cement under strong hydrothermal conditions.

TABLE 33.2   GRADING OF LIGHTWEIGHT AGGREGATE FOR
                       INSULATING CONCRETE

Aggregate	Percentage by Weight Passing Sieve Size:						
	9·5 mm	4·75 mm	2·36 mm	1·18 mm	0·60 mm	0·30 mm	0·15 mm
Perlite	—	100	85-100	40-85	20-60	5-25	0-10
Vermiculite (coarse)	100	98-100	60-100	40-85	5-45	2-20	0-10
Vermiculite (fine)	—	—	100	85-100	35-85	2-40	0-10

(b) Sedimentary aggregates.

(i) *Expanded clay and shale.* Certain clays and shales, on being heated to incipient fusion, expand or bloat by gas which is generated within them. The cellular structure thus formed is retained on cooling and, according to the details of manufacture, the final product is either a coated rounded aggregate or a crushed sintered one. The higher cost of the former is usually compensated by its superior physical properties as an aggregate for concrete.[7]

Many factors contribute to the successful bloating of clay and shale and no clear-cut correlation exists between their chemical composition, bloating characteristics and physical properties on being processed. Gas is produced by a reduction of iron oxide to the ferrous state and a liberation of sulphur dioxide, carbon monoxide and carbon dioxide (from compounds containing the requisite chemical elements to form these gases).[8] As bloatable materials differ widely in behaviour, each one should be studied individually with a close control of the temperature in producing a pyroplastic condition.

The easiest way of classifying raw materials into bloaters and nonbloaters is to test their bloating characteristics in a small stationary kiln, which simulates the heating schedule and kiln atmosphere that occur in practice. Most commercially used bloating clays and shales expand within a temperature range of 1100-1200°C and have a pH value above 5. Higher temperatures may be used with material that is either deficient in gas-forming substance or difficult to bloat.

Materials with poor bloating properties can be improved by blending them with good bloating clays, or by adding relatively cheap materials which contain the requisite gas-forming or fluxing compounds. These additives are based on compounds of iron, sulphur, carbon, alkali and alkaline earths. Those employed for fluxing should not only cause incipient fusion, but they should help to keep the mass highly viscous for the purpose of occluding gas at the bloating temperature. In this regard, fluxes containing sodium and potassium (e.g. feldspar), are better than those containing calcium and magnesium (e.g. limestone, dolomite and gypsum). The range of vitrification may be increased by a proportion of carbonaceous material, because the melting point of some minerals is lowered in an atmosphere of carbon dioxide arising from combustion. An excess of carbon, however, may act as a deterrent to bloating and prolonged heating may be required to reduce it. The objective, through these means, is to produce economically an aggregate with a uniform structure of small cells with tight, ceramic, shell-like walls.

In the manufacture of coated rounded aggregate, clayey material is partly dried, ground and pelletised by balling it with water in a revolving drum or preferably on an inclined rotating dish.[9] Pellets of different sizes are made by varying the rate of rotation of the equipment and the proportion and size of the drops of water sprayed into it. Unweathered shale of uniform composition, which is suitable for bloating, can be prepared readily for firing by crushing and screening.[10] The aggregate thus produced with a hard, vitreous coating is screened into regular sizes.[11] Outsize material, on being crushed, may be mixed with the rounded particles.[135]

As large particles require a longer heating period than small particles for 50 per cent or more volumetric expansion, the different sizes are fed separately into a rotary kiln for calcination. Bloating is controlled by varying the heating schedule for each size, and fine material with a higher fusion temperature is sometimes introduced to improve the surface coating on the particles of aggregate being produced.

Sintering is an economical way of making lightweight aggregate from either waste material such as colliery shale, or from clay which bloats at temperatures above that at which sticking takes place. The raw material is granulated by crushing and mixed with fine coal, and spread evenly to a depth of 200-300 mm on a moving grate. The charge passes under an ignition hood, where sintering is effected as burning proceeds under a forced-air draught that is induced by suction from below. The clinker is discharged onto a vibratory screen, where underburnt fines are separated and returned to the pelletiser. The sintered material, when cool, is crushed and screened into aggregate with a range of sizes.[11]

(*ii*) *Vermiculite* is an alteration product of biotite and other micas that contain entrapped moisture. The ore is dried to about 3 per cent moisture content, crushed, screened and winnowed by a horizontal stream of air into several size grades. These are dropped separately through an intense flame and flash-roasted at about 1100°C for 4-8 seconds, whereupon they exfoliate into a fluffy mass. The heating must be sufficient to cause a permanent change in crystal structure and thus prevent possible efflorescence through rehydration. The aggregate thus produced should comply with the requirements specified in BS 3797,[4] ASTM C35 and C332,[5, 6] and the grading limits given in Table 33.2.

Vermiculite has a bulk density of 65-190 kg/m³. It is relatively soft and expensive, and mixes that are made with it have low strength, high water absorption (unless nominally waterproofed) and high drying shrinkage. The hollow structure and pearly lustre of this aggregate make it eminently suitable for thermal insulation and fire protection.

(c) By-product aggregates.

(*i*) *Clinker* that is to be used as aggregate in concrete should consist of well-burnt furnace residues which have been fused or sintered into lumps. Its content of combustible matter should be kept low, as only a small percentage of certain types of coal is sufficient to cause unsoundness through expansion or increased drying shrinkage if incorporated in concrete. Clinker to be used should be checked for soundness by the method described in BS 1165[12] or by autoclaving mortar test bars at 177°C for 2·5 hours. In general, its loss on ignition should not exceed 10 per cent, although up to about twice this amount is permissible for concrete that is to be used internally and kept dry continuously.

As the combustible content of clinker consists mainly of small particles, it can be reduced by screening out and discarding the fines before the coarse material is crushed and graded. Alternatively, the unburnt fuel can be removed by such means as flotation, reburning the clinker on a sintering grate with forced-air draught, or stockpiling the clinker for several months before using it. Prolonged stockpiling causes spontaneous oxidation of the unburnt fuel and enables deleterious soluble substances to be removed by the percolation of rain water.

Wall surfacings may become stained if they are applied over clinker concrete that contains particles with a high content of iron. These particles can be removed by magnets at the crushing stage[13] or, alternatively, possible rust formation can be inhibited by treating the clinker with milk of lime at the rate of 6-9 kg of lime per m^3. A coating of plaster over clinker concrete may show signs of popping if the coated surface contains particles of certain material, such as hard-burnt lime and gypsum, which expand slowly on being wetted. One preventive measure is to keep the clinker moist for a few weeks before using it.

Although the sulphur compounds in clinker aggregate, when expressed as sulphur trioxide, are limited to 1 per cent by weight, the aggregate is unsuitable for use in reinforced concrete because of the corrosion potential of sulphur on steel reinforcement. Clinker aggregate is used widely in concrete masonry units for partition walls, ribbed-slab infill, insulation and fireproofing elements, some consideration being given to the foregoing recommendations.

(ii) Foamed slag is produced by bringing molten blastfurnace slag into contact with limited quantities of water and steam by one or several means. These include high-speed pelletisers, special rotors, tilting buckets and aerating tilting platforms, jets of steam and compressed air impinging on a stream of molten slag, and large shallow pits provided with jets of water and compressed air.

Depending on the system adopted, the slag either bloats into partially sealed, rounded particles with a mixed gradation and good workability potential, or it expands into a mass which, on cooling and crushing, produces particles with a granular, honeycombed structure. The aggregate, on being screened and stockpiled into coarse and fine grades (i.e. under 19·0 mm and 4·75 mm), has a related maximum bulk density (dry) that is limited to 880 and 1120 kg/m³, the combined limit being 1040 kg/m³.[14, 15, 83]

Foamed slag for concrete aggregate should comply with the requirements of BS 877, which limits the content of sulphate expressed as sulphur trioxide to 1.0 per cent.[16] In certain countries, it is used widely in lightweight concrete blocks, for *in situ* concrete used in the thermal insulation of roofs, and in reinforced concrete construction. If structures of the last-mentioned type are to be used under moist, highly humid or hydrothermal conditions, then precautions should be taken to inhibit the possible corrosion of steel reinforcement (see Section 22.3).

Suitable measures for this purpose include a low sulphur content in the slag, a concrete mix containing at least 8·5 40-kg bags of cement per m³, thicker cover to steel reinforcement than would be required in ordinary reinforced concrete (see Section 3.4), and a densification of the concrete generally. In this regard, an improvement in density and surface texture can be obtained by suitable mix proportions, grading of aggregate and procedures for compaction. It is facilitated by a partial replacement of the slag aggregate by fine sand, fly ash or finely ground foamed slag, the two latter components serving as a pozzolanic additive to long-term advantage (see Section 3.1.2(b) and Chapter 17).[135]

(iii) Sintered fly ash is made by moistening and balling black-coal fly ash on an inclined rotating dish, such as is used in making expanded clay aggregate, and firing the pellets thus formed at a temperature of about 1200°C. Combustible matter in the raw material is used as fuel and the finished product consists of hard nodules or pellets, which are graded subsequently with a maximum size in the region of 16·0 mm.[15, 17, 115, 116]

(d) Manufacture.
Concrete with a wide range of properties can be made with light-weight aggregate which, for uniformity and strength, should have regular characteristics of grading and a minimum content of interparticle voids. Typical gradings are given in AS 1467, BS 3797 and in Tables 33.2 and 33.3 (that are reproduced from ASTM C330, C331 and C332),[6, 18, 19] the maximum size of particle for masonry and insulating concrete being 10 mm. These gradings can be expressed in terms of absolute volume by dividing the observed

TABLE 33.3   GRADING OF LIGHTWEIGHT AGGREGATE FOR STRUCTURAL CONCRETE.

Aggregate	Percentage by Weight Passing Sieves:								
	26·5 mm	19·0 mm	13·2 mm	9·5 mm	4·75 mm	2·36 mm	1·18 mm	0·300 mm	0·150 mm
Fine									
4·75 mm to 0	—	—	—	100	85-100	—	40-80	10-35	5-25
Coarse									
26·5-13·2 mm	95-100	—	0-10	—	0-10	—	—	—	—
26·5-4·75 mm	95-100	—	25-60	—	0-10	—	—	—	—
19·0-4·75 mm	100	90-100	—	20-60	0-10	—	—	—	—
13·2-4·75 mm	—	100	90-100	40-80	0-20	0-10	—	—	—
9·5-2·36 mm	—	—	100	80-100	5-40	0-20	—	—	—
Combined									
13·2 mm to 0	—	100	95-100	—	50-80	—	—	5-20	2-15
9·5 mm to 0	—	—	100	90-100	65-90	35-65	—	10-25	5-15

weight of each fraction by its specific gravity. This, for expanded clay and shale, may range from about 1·2 for particles of 14 mm nominal size to 2·0 for those passing a 0·15-mm sieve.

Because of differences in the properties of various types of light-weight aggregate, separate mix designs are required for each kind of aggregate. Mixes for structural concrete frequently incorporate well-graded natural sand as the fine aggregate, with or without a pozzolan such as fly ash (see Chapter 17), for purposes of economy, workability and uniform high strength. They are proportioned on a cement-content basis for requisite strength characteristics. In this connection, a series of trial mixes are made and adjusted either empirically on a void-filling basis or by means of specific-gravity factors which were adopted by Committee 211.2 of the American Concrete Institute (see Reference 34 of Part 2). A pozzolan, when used with steam curing or prolonged moist curing, has a beneficial effect generally on the physical properties of the hardened concrete (see Section 9.4 and Chapter 17; also Section 36.4).[20, 40, 70, 122]

Lightweight mixes are frequently characterised by harshness and segregation. These effects can be brought under control by the use of coated, rounded aggregate, increased cement content, air-entrainment, an increased fine-aggregate/coarse-aggregate ratio, a pozzolanic additive as a partial replacement of aggregate, and by mechanical means of compaction. The amount of mixing water should be no more than that which would allow the concrete to be placed and consolidated satisfactorily. An excess amount of water causes segregation and settlement, increases shrinkage, reduces durability, and lowers the compressive, tensile and bond strengths of the concrete.

Owing to the absorptive nature of uncoated lightweight aggregate (AS 1467) and the difficulty of measuring absorbed and free water,[141] the water content of specific mixes may be controlled by reference to measurements of their workability. Should a slump test be used for this purpose, an allowance must be made in its indicative values of workability because of a reduced gravimetric effect with lightweight aggregate concrete. The slump value of structural lightweight mixes is usually below 65 mm, and should it be necessary to increase the slump of a particular mix, an increase in water content should be accompanied by a proportionate increase in cement content for uniform compressive strength.

In order to expedite quality control, the aggregate should have a uniform moisture content at the time of batching. Early dampening of certain aggregate, while not essential, minimises segregation of the aggregate during storage and transportation and it curtails early stiffening, which might otherwise occur with a lightweight concrete mix. For this purpose, a water spray may be located over a belt conveyor so as to wet the aggregate as it passes to storage hoppers.

In the manufacture of structural lightweight concrete, the aggregate (which is usually in a semisaturated condition) is mixed with two-thirds of the mixing water for 1-2 minutes. The cement, air-entraining agent, pozzolan (if any) and remaining water are then added, and the mixing is continued for an additional 2-4 minutes. Air-entrainment is kept within a range of 4-7 per cent, the amount being determined by a volumetric method as described in Section 16.3.5 and ASTM C173.[21]

With perlite and vermiculite aggregates, a suitable batching procedure is to mix them into a cementitious matrix, which is prepared with water, an air-entraining agent and cement added in that order. For insulating concrete with a bulk density (dry) of 320-640 kg/m^3, the proportions of cement and aggregate in the mix vary from 1 : 4 to 1 : 8 by volume, with a cement factor in the range of five to ten 40-kg bags/m^3 and a water requirement of 270-300 litres/m^3 for perlite concrete and 50 per cent more for vermiculite concrete.

For uniform quality, lightweight concrete must be mixed thoroughly because of the difference in the bulk density of its components. The mixing period is longer than that required with ordinary concrete, and batching is done preferably with a pan or paddle mixer. Where a drum mixer is used, its speed of rotation should be increased slightly when mixing this type of concrete. Truck mixers conveying structural concrete from batching plants should be rotated at high speed for 0·5 minute prior to discharging their load (see Sections 3.3.6(a), (b) and 7.1.2).

Truck mixers conveying perlite or vermiculite concrete should be rotated sparingly (if at all) during transit and at high speed on arrival at the site. For a uniform, freely flowing mix with requisite density, this mixing operation requires from 5 to 10 minutes for perlite concrete and up to 5 minutes for vermiculite concrete. For a maximum yield of ready-mixed insulating concrete, truck mixers need not be washed out until after the last load has been discharged.

The compressive strength of a suitably proportioned, structural, lightweight mix may be increased up to 50 per cent by high-frequency vibration (see Chapter 23), care being taken to minimise segregation because of a tendency of large particles to float if overvibrated. On slabs, perforated or mesh-covered, double-

wheeled rollers, or "jitterbug" tampers with a 6 mm wire-net base
are useful devices for pushing large particles of aggregate into
the screeded surface of concrete, and for bringing mortar to the
surface for finishing purposes. It is necessary to give special
attention to re-entrant angles under horizontal projections in
precasting moulds and, in finishing operations, a variety of surface
textures can be obtained by wood-floating, steel-trowelling or lightly
brushing an initially hardened surface with a soft-bristled broom.
Continuous control should be exercised in mixing, placing and
curing the concrete, which may be checked periodically for
consistence by weighing the fresh concrete.

Masonry units should be dried after curing to an equilibrium
moisture content corresponding with the humidity conditions of the
area in which they are expected to be used. This operation, in
conjunction with proper laying practice, is essential in masonry
construction if defects caused by drying shrinkage are to be avoided
(see Sections 27.2, 33.3.2(b) and 33.4.3). To this end, drying
chambers or mobile heaters (circulating air at $105 \pm 5°C$) can
be used to reduce the time required for effective drying to about 1
day. The units should be stacked under cover, so that air will
circulate freely beneath and through the stacks, either at the plant
or the site; care being taken to store them off the ground and to
cover them during haulage in rainy weather.

33.2.3 FOAMED CONCRETE AND CALCIUM
SILICATE HYDRATE

Foamed concrete is an aerated mortar or cementitious matrix made
by introducing air or gas into a prepared mortar or slurry, so that
small cells or interstices are entrapped in the mass. The bulk
density of ordinary foamed concrete ranges from 320 to 480 kg/m³
for insulation, to between 1300 and 1600 kg/m³ for structural
purposes. Developed since the beginning of this century, the
material has made great progress as a building component since
1929, when Axel Eriksson (a Swedish architect) improved its
properties by high-pressure steam curing. Through customary
usage and standards of production in different countries, it has
come to be known as foamed, aerated, cellular, gas or pore concrete,
lightweight *calcium silicate hydrate* or light silicate concrete with a
wide range of properties and bulk density. The present application
throughout the world exceeds about four million cubic metres per
year.

In certain countries, high-grade reinforced foamed concrete is being used extensively for roof, floor and wall slabs up to about 6 m long, the usual range of bulk density being from 400 to 800 kg/m³. These products are invariably precast and autoclaved at centrally situated plants. This procedure is essential in order to obtain a really lightweight material with a calcium silicate hydrate base, which possesses the requisite properties of strength and dimensional stability that are acceptable for structural purposes.[15, 22, 23] For a given bulk density, the compressive strength of the products is governed by the proportions of cement, lime and siliceous aggregate, the degree of fineness of the silica particles, and the curing cycle employed.[23-26]

Autoclaved precast units have advantages, not only in light weight and superior physical properties, but in speed of erection and high thermal insulation. Where the drainage fall of a roof deck is provided through the elevation of its supports, a screeded insulating layer can be eliminated; thereby effecting a saving in cost and the time that would be needed for its hardening and drying stages. These features are exemplified particularly in Swedish building practice, where wall, floor and roof slabs in dwelling houses, and some 70 per cent of industrial roofing requirements, are built in autoclaved foamed concrete (see Plate 27).

For economical production, the output from an automated plant should be some 120-150 m³ per day; and economies in manufacture may be implemented by the conversion of a sand-lime brick plant from dense to lightweight material and by a possible use of industrial by-products, such as mine tailings or fly ash. In contrast to these highly industrialised plants, mobile plants are employed where necessary at building sites, for the manufacture of sprayed, *in situ* or insulating grades of ordinary foamed concrete.[27]

(a) Raw materials.
A wide variety of cementitious and natural or artificially produced raw materials may be used. These include portland cement (grey and white varieties), high-calcium hydrated lime or quicklime, hydraulic lime, and a range of fine siliceous or aluminous-siliceous materials. The siliceous or aluminous-siliceous ingredients comprise one or other of the following materials: fine pit sand, silica flour, mine tailings, ground clinker or sand, black-coal fly ash, diatomite, glass grinding waste, ground cinders, pumice dust, ground granulated blastfurnace slag, milled burnt clay or shale, volcanic ash, and perlite or micaceous fines, with or without an aluminous admixture such as kaolin or bauxite.[118]

High-alumina cement can be used where early high strength is required, but special care must be taken in curing it with cold water and using it under suitable conditions (see Section 3.1.2(b)). Fibrous reinforcement (see Section 3.4.2), used in insulating grades of concrete, may consist of alkali-resistant fibreglass, [145, 146] asbestos, fibreglass-reinforced plastic, polypropylene, nylon, animal fibre or straw, some 2-6 per cent by volume being incorporated, depending on the quality of the product.

In the manufacture of foamed concrete, it should be noted that different cements (which must be used in a fresh condition) are likely to produce different results, and that mixes to be autoclaved may be prepared with a suitable content of lime and finely divided silica. Hydrated lime that is to be used for this purpose should have a low content of calcium carbonate and hard-burnt particles that would expand under hydrothermal conditions. To achieve a sound result, the lime should have been fully hydrated in a pressure hydrator or steam autoclave before being used. Otherwise, lime containing 3 or more per cent of unhydrated calcium and magnesium oxides should be incorporated with 2 or more per cent of calcium chloride, which is added to the gauging water (see Section 18.2.1).

When manufacturing autoclaved foamed concrete or lightweight calcium silicate hydrate, it is necessary to know the amount of free calcium oxide in hydrated lime that is being added to a mix. For control purposes, this component can be determined expeditiously by the sucrose extraction method. In this method, a sample not exceeding 0·6 g is shaken for 15 minutes in 100 ml of 10 per cent sucrose solution. The resulting extract is titrated at 23°C with 1 N hydrochloric acid, using methyl orange as indicator, and it is retitrated at boiling point so as to ensure complete hydrolysis of the sucrate.[28]

For high-quality autoclaved products, it is necessary to use finely divided siliceous material or silica flour with a specific surface area falling within a range of from 200 to 1000 m²/kg. Although the strength gradient of finished products is likely to increase with increasing subdivision of the siliceous component within this range, a suitable value of its fineness for economical production is approaching that of high-early-strength portland cement or about 375 m²/kg. For each degree of fineness of silica flour, there is an optimum lime/silica ratio which will contribute to a high strength/ density ratio in the matrix of the finished product (see Figure 33.1) and, as shown in Figure 33.2, a straight-line relationship exists between various optimum lime/silica ratios and diverse values of specific surface area of the siliceous component (see Section 33.2.3 (c)).[29]

If silica flour were used with a specific surface area of 375 m²/kg, the optimum lime to silica relationship for batching purposes would be 1 : 2 by weight for hydrated lime and 1 : 2·67 for quicklime. The specific surface area is determined usually by an air-permeability method (BS 3406), which is exemplified in a "Rigden" or a "Blaine" apparatus (see Plate 11) and, although a fairly uniform value is desirable for control purposes, the effect of variations in the raw material can be minimised by adjusting the mix proportions in accordance with the straight-line relationship described (BS 4359).

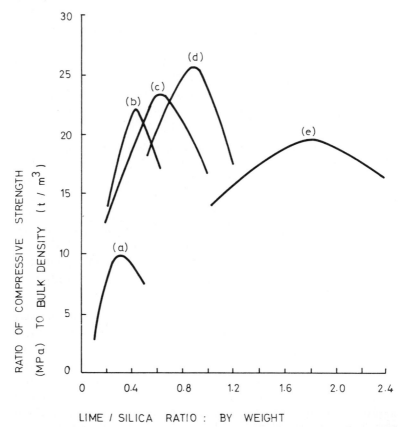

Fig. 33.1 Relationship between compressive-strength/bulk-density ratio and lime/silica ratio of calcium silicate hydrate. Specific surface area of silica flour in mix (m²/kg): (a) 65; (b) 280; (c) 427; (d) 830; (e) 1680. Lime: 92 per cent calcium hydroxide.

Note: Results apply to one set of curing conditions.

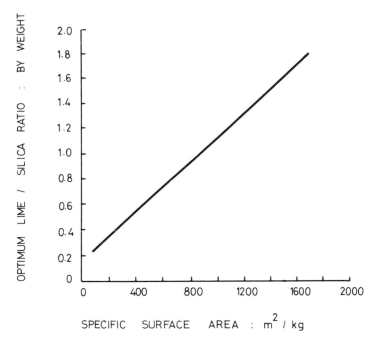

Fig. 33.2 Relationship between optimum lime/silica ratio and specific surface area of silica flour in calcium silicate hydrate. Lime: 92 per cent calcium hydroxide.

In the preparatory stages of manufacture, ground silica is produced by milling the requisite raw material in a dry or wet state for a predetermined period, its size-classification being effected by a pneumatic cyclone or hydraulic mechanical rake or spiral. Quicklime (if used) is ground to pass a 0·15-mm sieve, and the mix components are intimately blended by pan-milling or ball-milling, mechanical stirring or pneumatic agitation. A preliminary milling of hydrated lime will increase its efficiency for the manufacture of calcium silicate hydrate, care being taken to minimise atmospheric carbonation during the process.

Admixtures to be used consist of foaming agents (see Section 33.2.3(b)) and either a setting accelerator for foamed concrete containing cement or a hydration retarder for mixes containing ground quicklime in conjunction with hydrated lime and silica flour.[30] The rate of hardening of foamed concrete is increased by the use of high-early-strength portland cement or by using mixing water at a temperature of from 24°C to 66°C. It is doubled, more or less, by the use of one of the setting accelerators that are listed in Table 33.4.

TABLE 33.4   **SETTING ACCELERATORS**

Admixture	Amount (per cent by weight of cement)
Aluminium sulphate	1·0-2·0
Calcium chloride	2·0-3·0
Calcium formate	2·0-3·0
Sodium or potassium carbonate	0·75-1·5
Sodium hydroxide	0·5-1·0
Sodium silicate (20 per cent low-alkali solution)	4·0-6·0
Triethanolamine	0·1-0·125

The amount of setting accelerator required for early hardening is about twice that which would be used for ordinary concrete, and its use must be accompanied by an increased proportion of foaming agent. An excessive amount causes undesirable stiffening, increased shrinkage and reduced strength.

In lime-silica mixes a hydration-retarding agent (e.g. sucrose or ordinary sugar–see Section 10.2) is used in the proportion of 3-5 per cent of the quantity of quicklime introduced into the mix. Its purpose is to control the amount of heat developed during hydration of the quicklime and enable the mix to develop self-hardening characteristics by a partial interaction between its components, in addition to the immobilisation of some of the mixing water at an early age. Ground quicklime may be blended with hydrated lime for the purpose of inhibiting the possible ebullition of a large moulded mass and, as a variant of procedure, a slurry of hydrated lime and silica flour may be solidified (prior to autoclave treatment) by an admixture of set-retarded gypsum plaster.[31]

(b) Foaming agents.

The walls of entrapped air or gas cells must be sufficiently tough and stable in the initial stages to preserve homogeneity of the mass until it hardens. Suitable foaming agents are hydrolysed protein and keratin compounds,[84] sodium isopropyl naphthalene sulphate, petroleum naphthalene sulphate, sodium secondary alkyl sulphate, saponin, sodium alkyl aryl sulphate and highly stabilised saponified rosin and resin compounds. Gas-generating agents are flaky aluminium powder, calcium carbide and hydrogen peroxide combined with chloride of lime, which evolve hydrogen, acetylene and oxygen, respectively. The addition of a cement-dispersing or wetting agent to a mix improves the uniformity of size and distribution of the gas cells thus developed (see Sections 6.4 and 16.1).

The quantity of foaming or gas-generating agent required depends on a variety of factors. These include, for instance, the surface-active characteristics and concentration of the agent and the method of foaming; the operation of the particular equipment used and the length of mixing time; and the bulk density, proportions, consistence, temperature, constituents and volume of the mix. A polymer latex (see Section 21.2) may also be incorporated.

Preformed foam is made by entrapping air in a 2-5 per cent solution of foaming agent, which is of 15-35 per cent concentration, until its volume is increased about thirtyfold. This operation can be carried out with a 120-litre pneumatic foam ejector, which requires 5 litres of air per second at a pressure of 500-700 kPa. The foam can be generated, alternatively, by a whirlpool mixer provided with a whisking drum, a drive-and-idler gear homogenising pump, an immersed flotation cell or a perforated nozzle supplied with low-pressure compressed air. Foam from the last-mentioned generator is converted into a mass of uniform small bubbles, by being discharged through a fine screen or a tube housing a set of cutting blades. Entrained foam is developed in a mix by whipping air into it, by means of a bladed-cage whisk and by using sharp sand in the mix.

Foaming agents (e.g. "Aphrosol FC", "National-crete", "Phomene" and "Teepol 480") and equipment are obtainable from a number of proprietary sources. Where foaming agents contain organic compounds (e.g. keratin and protein), they should be stored out of contact with the atmosphere and used in a reasonably fresh condition. Although the direct preparation of foaming agents may be justified on certain occasions, it is preferable, wherever practicable, to use a suitable proprietary product. For the former purpose, the following formulations are given for guidance in the manufacture of saponified types of compound.

Saponified wood resin that is stabilised with animal glue can be prepared with the following ingredients by weight.

Wood resin: 2 parts.
Animal glue: 4 parts.
Sodium hydroxide in 20 per cent solution: 5 parts.
Water: 10 parts.

To prepare, soak the animal glue in the water, add the other ingredients, warm the mixture to 80°C, and agitate it until homogeneity is apparent. This compound is of 33 per cent concentration.

A resinous composition that is stabilised with casein can be prepared with a 15 per cent concentration as follows.

Resin soap (33 per cent concentration) mixed with an equal part of hot water: 3 parts.
Casein: 2 parts.
Potassium hydroxide in hot 2 per cent solution: 6 parts.
Water: 6 parts.

In this instance, the casein is soaked in the water, the potassium hydroxide solution is added with agitation, the resin soap is introduced and the mixture is stirred until homogeneity is apparent. During this operation, the temperature of the casein should not exceed 82°C. If the mixture is to be stored for several days, it should be preserved with an admixture of sodium pentachlorophenate, which is incorporated in the proportion of 3 per cent by weight of the casein.

Saponified "Vinsol" resin can be prepared readily as a 28 per cent solution with 100 parts by weight of pulverised resin (acid and saponification numbers of 95 and 125, respectively) and 9·5 parts of sodium hydroxide. The resin should pass a 0·30-mm sieve and it is added to 4 parts of 25 per cent sodium hydroxide solution that is diluted subsequently with 250 parts of water. The mixture is stirred for about 10 minutes and it is incorporated with 34 parts of 25 per cent sodium hydroxide solution, the stirring being continued for a further 20-30 minutes to effect complete saponification.

(c) Methods of foaming.
Air or gas can be incorporated in a mix in several ways: by mechanical (prefoaming or entraining), chemical, micropore or combined methods of operation. Each method produces concrete with specific physical and economic characteristics. For instance, products made by the first three methods are visibly cellular and those made by the fourth one have a capillary structure; while the cells in chemically foamed concrete are slightly ovoid in shape. Products of medium density are made simply by the entraining method, whereas low-density products are made more satisfactorily by a prefoaming or chemical foaming procedure.

Mixes containing a high proportion of finely divided or pozzolanic material are aerated more readily by a prefoaming than by an entraining procedure; while under ordinary conditions of manufacture, the amount of aeration introduced is controllable more easily by mechanical foaming than by the chemical method.

(i) Prefoaming. Products weighing 160-1100 kg/m³ (dry) can be made by gently mixing a prepared slurry or mortar with preformed foam in a mixer provided with steel-mesh paddles. Sufficient mixing water is used in the slurry or mortar to prevent nodules of fine material being formed during the mixing process, the requisite amount being about one-half of the weight of solid materials in the mix. The foam is injected continuously into the mixer and incorporated until the desired bulk density is obtained. Improved results are obtained by activating a slurry of suitable cementing material in a high-speed (colloidal) mixer for several minutes, transferring it to an ordinary concrete mixer and incorporating fine aggregate and preformed foam. Alternatively, a foaming solution may be aerated in a concrete mixer and then mixed with activated slurry and fine aggregate.

For high-grade autoclaved products with bulk densities (dry) ranging from 880 to below 400 kg/m³, typical mix proportions range from 1 part by weight of portland cement, 1·5 parts of hydrated lime and 3 parts of silica flour; to 1 part of portland cement, 1 part of hydrated lime and 2 parts of silica flour or the equivalent. In the foamed mix, the water/solids ratio amounts to 0·70 ± 0·15 and the agents/solids ratio lies within a range of 0·0003-0·003 by weight, the latter being based on the active constituent of the foaming agent with a hypothetical concentration of 100 per cent.

A schedule of approximate quantities per cubic metre of finished product is given in Table 33.5. The quantity of water thus indicated includes the amounts necessary to form a slurry and a 2·5 per cent solution with a foaming agent, which is formulated with a

TABLE 33.5   **QUANTITIES PER CUBIC METRE**

Bulk Density, Dry	Portland Cement	Hydrated Lime, 92 per cent Ca(OH)$_2$	Silica Flour, 375 m²/kg	Calcium Chloride	Water	Foaming Agent, 2 per cent concentration
(kg/m³)	(kg)	(kg)	(kg)	(kg)	(litres)	(kg)
1000	—	340	690	—	570	—
880	150	220	470	—	475	1·2
800	140	200	430	—	445	1·5
640	130	150	320	3·5	360	1·8
480	110	110	240	3·5	295	1·9
320	75	75	160	3·0	240	2·1
240	60	60	120	2·5	200	2·1

concentration of 25 per cent. The microcomposition of the finished product after being autoclaved has a high content of monocalcium silicate hydrate (see Section 31.1.2).[119]

In the preparation of lightweight calcium silicate hydrate products from a lime-silica mix, the various components of a batch are introduced into a mixer in the following sequence: cold water, hydration retarder, ground quicklime, hydrated lime, fine siliceous material and preformed foam. Alternatively, the dry materials may be vibratory-sieved and mixed into a preformed foam and retarder solution, the former containing 5 per cent of foaming agent with a 33 per cent concentration, and the latter some 5 per cent of sucrose by weight of quicklime in the mix. Approximate quantities per m^3 of autoclaved product, using a protein and keratin compound for foaming purposes, are indicated in Table 33.6.

TABLE 33.6   QUANTITIES PER CUBIC METRE

Bulk Density, Dry (kg/m³)	Ground Quicklime (kg)	Hydrated Lime (kg)	Silica Flour, 375 m²/kg (kg)	Sucrose (kg)	Water (kg)	Foam (kg)
640	100	70	420	5·0	310	18
480	75	50	310	3·9	240	21

(ii) Entraining. Products weighing 1200-1600 kg/m³ (dry) are made by whisking a prepared mix with a foaming agent. Agitation is continued for 3-5 minutes or until the desired bulk density (wet) is obtained. On no account must the mixing time exceed 20 minutes or the products will have low strength and high shrinkage. A typical mix for building blocks weighing 1450 kg/m³ (dry) is 1 part cement : 3 parts fine sharp sand (fineness modulus about 1·0) with a water/cement ratio of 0·6 and an agent/cement ratio of 0·001 (100 per cent active constituent) by weight.

(iii) Chemical. Products weighing 320-960 kg/m³ (dry) are made by causing a poured mix to swell by evolution of gas bubbles throughout the mass. The gas-generating agent that is used most commonly is flaky aluminium powder, incorporated with a froth-stabilising agent, in the proportion of 0·1-0·5 per cent by weight of cement or cement and hydrated lime in a lukewarm mix. Tricalcium aluminate and hydrogen are produced by chemical reaction between the powder and hydrated lime (including that which is hydrolysed from portland cement during setting) in the presence of water. The small bubbles of gas thus formed gradually increase in size and cause the mix to expand.

The reaction is controlled to effect the required expansion in about 0·5 hour or shortly before the initial hardening of the mix, care being taken to keep the surface intact while the material is in a plastic state. An excess of powder will produce an excess of gas and cause disruption of the surface. The rate of generation of gas is governed by the surface characteristics of the aluminium particles (which are usually below 50 μm in size and free from oil film) and the alkalinity and temperature of the mix. The hydrogen, on final dissipation, is replaced by air.

Best results are obtained with the use of activated cementing material, finely divided siliceous aggregate, hot mixing water, sodium hydroxide (added last to the mix), autoclaving, and carefully controlled conditions of production. Under optimum conditions, products weighing 640 kg/m^3 require about 0·5 kg of aluminium powder and 1-1·5 kg of sodium hydroxide per m^3 of concrete. This process is commonly used in large factories possessing highly mechanised means of manufacture.

(iv) Micropore. Products weighing 960-1100 kg/m^3 (dry) are made by mixing hydrated lime and finely divided siliceous material into a paste. This is poured into steel moulds provided with lids and steam cured afterwards at a pressure of 1000 kPa in an autoclave. The paste is proportioned with a lime/silica ratio of 0·14 + 0·0009 times the specific surface area (m^2/kg) of the siliceous component and with a water/solids ratio of 0·5-0·6 by weight. In this relationship, in comparison with Figure 33.2, the lime component consists of 100 per cent calcium hydroxide. A subsequent expulsion of excess water, which is turned into steam during the early stages of curing, develops a microporous structure which is retained in the hardened material.[29]

Products with a bulk density down to about 480 kg/m^3 (dry) can be made by increasing the water content of the mix, and by incorporating water-retaining materials or suspending agents in it so as to minimise segregation. Materials for this purpose include diatomaceous or volcanic earth, bentonite and carboxymethyl-cellulose (see Section 15.2.2(d)(vi)). Conversely, products with an increased bulk density can be made by reducing the water content of the mix and densifying the material by compressing it, either in a moulded or an extruded form, under various intensities of pressure. The pressed material, being strongly cohesive, can be autoclaved in a demoulded state.[33]

As the compactive pressure of the lime-silica mix is increased up to 140 MPa or more and the potential capillary structure is

decreased progressively to a negligible amount, so the bulk density
(dry) of the finshed product is increased up to about 85 per cent.[32]
This effect is illustrated in Figure 33.3, which shows that the
corresponding compressive strength can be increased up to sixfold
at an age of 1 day. Even with a moulding pressure of 20 MPa, an
autoclaved product can be made with a 1-day compressive strength
of approximately 140 MPa, a bulk density of only 1600 kg/m³, and
a strength/density ratio at 1 day which is ten times that of ordinary
concrete at 28 days.

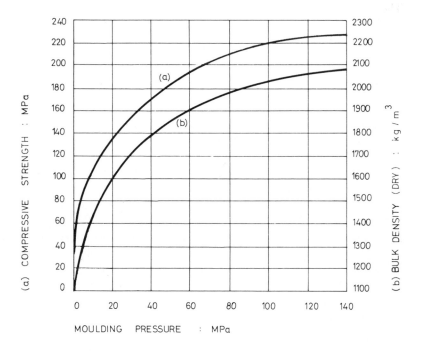

Fig. 33.3 Casting pressure, compressive strength and bulk density
relationship. Mix: 1 part hydrated lime : 2 parts silica flour (375 m²/kg) with
a water/solids ratio of 0.067-0.600 by weight. Curing cycle: 185°C for 8
hours. Lime: 92 per cent calcium hydroxide. Cubes: 25 mm in size and dry.
Age: 1 day.

These strength values would be reduced to some extent by an
increased proportion of early carbonation of the lime that is used in
the mix. A suitably proportioned lime-silica mix, on being blended
with portland cement and lightweight aggregate, may be used for
the manufacture of autoclaved, lightweight block masonry.

(*v*) *Combined*. Any of the preceding methods may be combined and a variety of mixes used in the manufacture of foamed concrete with bulk densities down to 160 kg/m³. Figure 33.4 illustrates a combination of the entraining and chemical methods of foaming, whereby a mortar mix is expanded 50 per cent mechanically before being expanded chemically in the moulds.

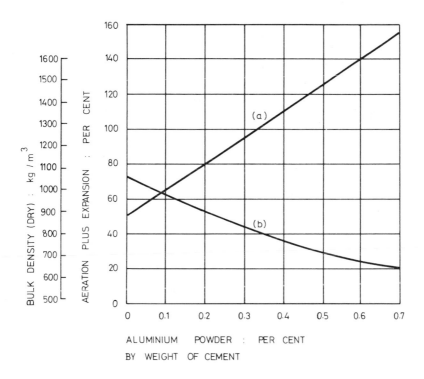

Fig. 33.4 Combined entraining and chemical foaming. Mix: 1 part portland cement : 1 part sand, water/cement ratio 0.6 by weight. (a) Aeration plus expansion $= 150X + 50$ per cent. (b) Bulk density (dry) $= \dfrac{320}{0.3X + 0.3}$ kg/m³. $X =$ aluminium powder, per cent by weight of cement.

(d) Manufacture.

(*i*) *Mixing*. Batching is controlled by weight for the solid components, by volume for liquid or slurry components with a consistent density, and by time for the incorporation of air. The mix should have a consistence approximating that of thick cream, this property being estimated by a slump-cylinder test or a spring.

loaded vane that is extended into moving material. In the prefoaming method of manufacture, preformed foam and slurry are mixed together slowly in a paddle mixer equipped with steelmesh blades, or otherwise in a drum or whirlpool mixer. In the entraining method, the mixing of ingredients is done in such equipment as a high-speed or colloidal mixer, an ordinary batch mixer geared to run at double or treble speed, a pan mixer or a bladed-cage whisk having a peripheral velocity of 7·5 m/s.

The requisite proportions of a mix and the order of charging its components into a mixer are indicated in paragraphs relating to the methods of foaming. In ordinary grades of foamed concrete where an appropriate compressive strength is to be attained, the proportions of cement to fine aggregate are varied inversely with the proposed bulk density of the product. When a setting accelerator is used in low-density mixes, it is incorporated in a prepared slurry before the foaming constituent is introduced in requisite amount.

(*ii*) *Casting*. The mixed material is conveyed in an agitated hopper, pumped through a pipeline, or discharged directly and continuously through a downpipe into ribbed steel formwork or moulds. These should be filled in such a way that an oil-based release agent does not enter the mix and cause striations during the filling process. Concrete that is made with very tough foam can be poured in lifts of storey-height without noticeable cell distortion or heterogeneity of structure. Big moulds may have a capacity of several cubic metres and they are filled, or partly filled for chemical foaming, with one batch of material. After being sprayed lightly with a paraffin-wax emulsion, they can be made liquid-tight at the joints by means of a lining of cellulose film, such as "Cellophane", which chars during high-pressure steam curing and thus serves as an additional release agent.

Moulded cementitious mixes which are to be autoclaved are allowed to solidify for about 6 hours, either normally without moisture loss or under conditions which may simulate those employed in low-pressure steam curing (see Section 9.4.2). The sides of big moulds are then opened, and the large partly hardened blocks of material thus formed are cut by tensioned wires into building blocks or planks. The material is remoulded by closing the sides of the moulds and it is moved on rail trolleys into cylindrical steel autoclaves for a substantial development of strength by high-pressure steam curing.

Reinforced flexural members are made by firmly fabricating plain or deformed high-yield reinforcement into welded grids, mats or cages with at least two mechanical anchorage attachments at each end. After being cleaned (e.g. by a two-stage acid-neutraliser dip), the grids are dipped twice in a cement slurry containing rubber latex, casein, ammonium hydroxide and possibly a corrosion inhibitor (see Section 22.3).[22, 36, 80] This slurry is formulated so as to harden substantially during a subsequent operation with high-pressure steam curing. After the coated grids have air dried, they are located accurately in the moulds by means of jigs.

In chemical and combined methods of foaming, the poured mix should expand until it is about 5 per cent above the top of the moulds, the excess material being sliced off and reused after initial solidification of the mass. Building blocks that are made in this manner are cast on end, so as to take advantage of differential strength characteristics that would accompany a possible formation of unidirectional ovoid cells. Edge grooves on planks for *in situ* jointing are formed before or after the curing operation by edge-tooling or milling appliances.

(*iii*) *Curing.* Ordinary foamed concrete on being demoulded or stripped of formwork is moist cured for at least 7 days before it is allowed to dry. Moulded material that is to be converted into a high-class product must be not only of suitable composition, but it must be cured in saturated steam in an autoclave which is purged of air and operated at a gauge pressure in the region of 1000 kPa (185°C) for an adequate period. The curing cycle is governed by a variety of factors, which include the composition, bulk density and mass of the material that is to be autoclaved, the economy and maximum temperature of operation, and the physical properties desired in the finished product.

The curing period for structural units ranges normally from 14 to 20 hours, which includes a period of about 4 hours to reach maximum temperature and a few hours for cooling.[79] The rate of heating at the start of the cycle must be gradual, in order to prevent the development of an excessive temperature gradient which could cause incipient fissures in the material. For pressed products of calcium silicate hydrate, the steam-curing period at a pressure of 1700 kPa (210°C) may be reduced to about 4 hours.

Some experimental test results obtained in the laboratory on small specimens of given size are shown in Figure 33.5. They indicate that for an appropriate steam-curing pressure and a particular product, there is an optimum curing period (8 hours in this instance) for the development of maximum compressive strength, and that there is no advantage in steam curing beyond this stage of development. The compressive strength of the specimens can be increased still further by increasing the steam-curing pressure to over 1400 kPa (200°C) for the optimum period, the increment being about 20 per cent at 2000 kPa (220°C).[32]

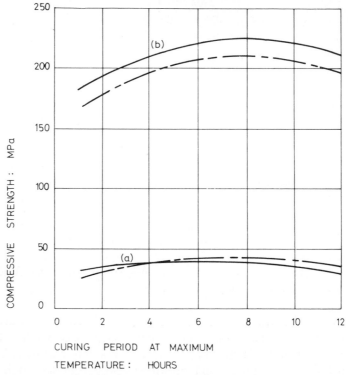

Fig. 33.5 Relationship between compressive strength and curing period of calcium silicate hydrate. (a) Plain mix of 1 part hydrated lime : 2 parts silica flour (375 m²/kg) with a water/solids ratio of 0.55 by weight and a bulk density (dry) of 1100 kg/m³ after high-pressure steam curing. (b) Pressed mix with similar proportions, but with a water/solids ratio of 0.067 by weight and a bulk density (dry) of 2100 kg/m³, the moulding pressure being 140 MPa. Steam temperature: — — — — —, 175°C and ——————, 185°C. Lime: 92 per cent calcium hydroxide. Cubes: 25 mm in size and dry. Age: 1 day.

High-pressure steam curing in industrial autoclaves is carried out automatically (as described in Section 9.4.6) with a longer effective cycle than is customary for dense concrete. Similar products should, amongst other things, be uniform in quality and dimensionally accurate. Means of achieving these ends include: an even intensity of heating throughout the length of autoclaves; the steam curing of material in closed or closely stacked steel moulds or when it is in an advanced state of hardness; the cyclical production of autoclaved units of regular composition and size; and the final trimming of products (where necessary) in high-speed milling machines.[117, 135]

(*iv*) *Drying.* The recommendations given in Section 33.2.2(d) for drying lightweight-aggregate concrete apply with equal emphasis to foamed concrete, particularly to the ordinary variety of material, in order to minimise its drying shrinkage. Demoulded products, on being autoclaved, may be dried partly under reduced pressure (see Section 9.4.6) and by being stacked in a large drying shed for several days. Here they are allowed to cool and dry to an equilibrium moisture content, which should approximate a value that would be expected after the units have been built into a structure (see Section 14.6).

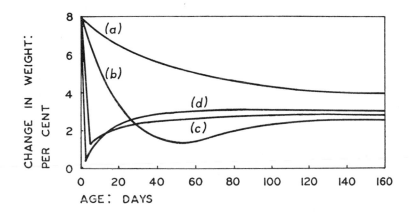

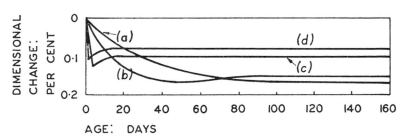

Fig. 33.6 Dimensional change of ordinary foamed concrete. Drying: (a) 23°C, 65 per cent relative humidity. (b) 23°C, 40 per cent relative humidity. (c) 60°C, 0 per cent relative humidity. (d) 105 ± 5°C, 0 per cent relative humidity. Storage: 23°C, 65 per cent relative humidity. Mix: 1 part cement : 3 parts sand with a water/cement ratio of 0.6 by weight. Specimens: 127 mm × 127 mm × 356 mm size with a bulk density (dry) of 1220 kg/m³. Age at drying: 28 days.

Masonry units of ordinary foamed concrete can be made dimensionally stable (under constant atmospheric conditions) by being dried below an anticipated equilibrium moisture content and then being allowed to return to it, as indicated in Figure 33.6. Circulating air at 105 ± 5°C for about 24 hours may be used for this purpose.

33.3 PROPERTIES OF LIGHTWEIGHT CONCRETE

33.3.1 NO-FINES CONCRETE

No-fines concrete made with 1 part by volume of cement and 8 parts of 20 mm nominal-size basalt aggregate has the following characteristics at 28 days.

Bulk density: 1760 kg/m^3.
Compressive strength (wet) of 150 mm × 300 mm cylinders: 5·5 MPa.
Ultimate torsional strength (wet): 1·4 MPa.
Bond strength between the concrete and reinforcement that has been coated with cement grout: 0·62 MPa.

This bulk density can be halved by replacing dense aggregate with 14 mm nominal-size expanded clay or shale aggregate. The drying shrinkage of no-fines concrete is approximately 0·02 per cent, which is about one-half that of ordinary concrete containing similar coarse aggregate. These two types of concrete should not be placed together, if cracking through differential shrinkage movements is to be avoided.

The thermal conductivity of no-fines concrete that is made with dense aggregate is about one-half that of ordinary concrete, the value of the latter being approximately 1·44 W/m·°C. With lightweight aggregate, the insulating value of a 200 mm thick no-fines concrete wall is equivalent to that of a 275 mm cavity wall in burnt clay brick. Moisture that penetrates a rendered no-fines concrete wall which is uniform in quality seldom migrates inwards more than twice the maximum gauge of the aggregate.

The ageing of the concrete causes but little increase in strength and a probable reduction in its elastic modulus (BS 1881).[34-36] A variant of procedure in making standard strength-test cylinders is outlined at the end of Section 14.4.1.

Lightweight-aggregate concrete shell roof at the TWA
Terminal, John F. Kennedy International Airport,
New York

Composite construction including
autoclaved foamed concrete
on a Swedish building estate

Prestressed Concrete Equipment
PLATE 28

"Vogt" tendon tension meter

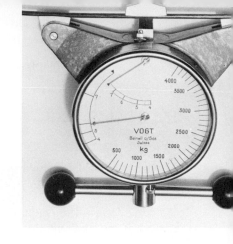

Typical post-tensioning anchorage

Typical post-tensioning
anchorage

"Freyssi" flat hydraulic jack

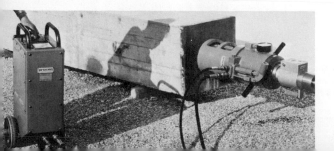

Typical
post-tensioning
hydraulic jack

33.3.2 LIGHTWEIGHT-AGGREGATE CONCRETE

(a) Strength.
The compressive strength of lightweight-aggregate concrete depends *inter alia* upon the physical and pozzolanic characteristics of the aggregate, the richness and water/cement ratio of the mix, the degree of compaction and type of curing, and the age and moisture content of the concrete. For each aggregate, under given conditions, a certain relationship exists between the compressive strength and bulk density of lightweight-aggregate concrete of known proportions, but only a general one exists for different kinds of lightweight aggregate considered as a whole.[38, 43]

For structural purposes, the compressive strength and bulk density of lightweight concrete can be increased in several ways: using coated, 16 mm lightweight aggregate with a one-half to two-thirds proportion of graded natural sand;[37] compacting effectively a properly designed mix with a maximum slump of 10-50 mm; using a cement factor of 8·5-11 40-kg bags per m³, and curing the concrete effectively. As the maximum strength attainable is limited ultimately by the strength of the aggregate, a stage in mix design can be reached when, for a particular lightweight aggregate, an increase in the cement factor may not cause an increase in strength of the concrete. A strongly coated, lightweight aggregate with medium-sized particles, therefore, may be a useful aid in the manufacture of high-strength concrete. The effect of an increased cement factor in this type of concrete, under moist-curing conditions, is a possible increase in the 7-day to 28-day strength relationship in comparison with that which applies normally with ordinary concrete (see Section 3.1.2(e)(vi)).

Typical values of the compressive strength and bulk density of various grades of lightweight concrete are indicated in Table 33.7, while some general data on various types of concrete are

TABLE 33.7   GRADES OF CONCRETE

Grade of Lightweight Concrete	Compressive Strength (Cylinder or Block) at 28 days (MPa)	Bulk Density, Dry (kg/m³)	
		Concrete	Aggregate
Structural	15-50	1400-1900	Coarse: 480-880
Fireproofing	5-15	1100-1600	Fine: 880-1100
Masonry	2-15	960-1600	Combined: 480-1000
Insulation	1-5	320-1100	Assorted: 65-1100

summarised broadly in Table 33.8.[15, 36-49, 70] For purposes of guidance in the proportioning of concrete with coated expanded clay or shale aggregate, a minimum cylinder compressive strength of 20 MPa at 28 days can be obtained with a mix of 1 part by volume of cement, 2·25 parts of graded natural sand and 3·5 parts of 16 mm gauge lightweight aggregate, the maximum slump being 50 mm and the bulk density (dry) of the concrete being 1600-1680 kg/m³. For a compressive strength of 35 MPa and a bulk density (dry) of 1680-1760 kg/m³, the mix proportions would be approximately 1 part of cement, 1·25 parts of sand and 2·75 parts of lightweight aggregate, the slump being limited to 10 mm for vibratory compaction (see Section 23.1).

TABLE 33.8   SUMMARISED DATA ON VARIOUS MIXES

Type of Concrete	Compressive Strength (Cylinder) at 28 Days (MPa)	Bulk Density, Dry (kg/m³)		Cement (bags per m³ of Concrete)
		Concrete	Aggregate	
Pumice	2-15	640-1400	480-880	4-12
Scoria	3-30	960-1800	640-1000	4-12
Perlite	1-6	400-960	80-240	4-11
Expanded clay and shale	4-50	960-1900	560-1000	4-12
Vermiculite	1-3	320-800	65-190	4-11
Clinker	3-15	1100-1600	720-1000	4-8
Foamed slag	3-35	960-1900	320-970	4-12
Sintered fly ash	4-40	960-1900	640-970	4-12

Procedures for the testing of lightweight-aggregate concrete are described in ASTM C330, C331, C495 and C513,[18, 19, 50] unless its strength is determined otherwise by ways appertaining to ordinary concrete (see Section 14.4). In this regard, the compressive strength of lightweight-aggregate concrete (on being moist cured for 7 days, dried at 23°C and 50 per cent relative humidity for 21 days and tested dry) is approximately the same as that of similar concrete (which is moist cured for 28 days and tested wet). The compressive strength of lightweight concrete that has been cured by the latter method, however, is increased some 15-20 per cent by drying unless limited by the strength of the aggregate.

With the former method of curing, test results indicate that the tensile splitting strength of lightweight concrete is about one-twelfth of its compressive strength and 0·6-0·8 of the tensile strength that would be obtained with 28-day moist curing. Results also show that it is a functional measure of unit shear strength (see Reference

88 of Part 3; also Section 34.1.4).[51, 52, 121, 125, 126] Steam curing with a maturity factor of 1200-1500°Ch ($-10°$C datum) reduces the 28-day compressive strength by approximately 10 per cent, but this effect is reduced appreciably at an age of 3 months (see Section 9.4.4).[43]

In connection with shear strength, tests tend to show less regular results with lightweight members, and diagonal cracking occurs usually at lower loads than for similar members made with ordinary concrete. General design practice, therefore, requires a reduction of 15-25 per cent in the permissible shear stress for lightweight-concrete slabs and subsidiary members, and the provision of appropriate reinforcement to resist all of the internal shear forces in main beams (AS1480 and BS CP110, CP114; see Table 3.25 and Appendix IV, 2).[15, 36, 51-53, 108, 109, 123, 140]

While pull-out tests indicate similar bond characteristics for lightweight and ordinary concretes, the bond strength of plain steel reinforcement in lightweight-concrete members that are cast horizontally may be down to one-half of the equivalent bond strength derived from these tests. This effect is due in general to a formation of layers of water and, consequently, shallow cavities beneath the horizontal steel reinforcement: these being formed by the bleeding of excess mixing water and some settlement of the concrete (see AS 1480, BS CP110, ACI Committee 318; also Table 3.25 and Section 3.4).[53] Except at early ages, a somewhat similar reduction in bond strength may be expected in steam-cured products containing either dense or lightweight aggregate but this effect, which is caused by differential hydration and strength development, is insignificant when high-tensile steel strand is used in dense, high-strength, ordinary concrete (see Sections 11.1.2(c), (d), 15.2.2(d)(vi), 23.1.3, 23.1.6 and 34.2.4; also Reference 64 of Part 4 and Reference 64 of Part 6, respectively).[78, 112, 113]

With plain reinforcement and moist curing, the bond strength is unaffected in vertically cast members, such as columns. With deformed high-yield reinforcement, a slip of 0·025 mm in lightweight concrete beams is attainable at a bond stress that is one-half to three-quarters of the stress needed to produce the same slip in ordinary concrete beams. The bond characteristics can be improved greatly with special mix design (including coated coarse and natural fine aggregates), controlled water content and air-entrainment, effective placement, compaction and moist curing and, in special instances, appropriate revibration of the newly placed mix (see Sections 16.1 and 23.2).[15, 36, 48, 58, 130, 131]

The modulus of elasticity of all-in lightweight-aggregate concrete increases with its bulk density and compressive strength, and it approximates 500 times the compressive strength of the concrete for short-duration loading within a range of working stress. This value for a given strength is about one-half that of ordinary concrete, which normally lies between 20×10^3 and 35×10^3 MPa (see Table 31.1), and it is increased up to 50 per cent when sand is used as the fine aggregate (see Section 34.1.4(c); also AS 1481). The reduced modulus of elasticity is of structural significance under working loads, because of its effect on prestress, deflection and the distribution of internal stresses across the section of flexural members (the neutral axis of beams being lowered), the proportionate load-carrying capacity of steel elements in reinforced concrete or composite columns, and the improved absorbence of shock loads by lightweight concrete members.[15, 36, 48, 54-58, 82, 108, 137, 148]

With a normal permissible stress in plain mild-steel reinforcement of 140 MPa, the width of cracks that form in lightweight concrete beams is not critical with regard to the possible corrosion of the reinforcement or the damage of surface finishes. With deformed high-yield steel reinforcement, a permissible tensile stress of 200 MPa (which may be used in lightweight concrete slabs) should be reduced by 10 per cent for the control of crack widths in beams (see Section 34.1.1).[15, 48, 58]

Provided that consideration is given to the relevant properties of the concrete and its component materials, as indicated in the foregoing remarks, there is little difference in the method of approach to the design of structures in either lightweight or ordinary reinforced concrete. Where in lightweight-slab construction prestressed dense-concrete pencils are used as reinforcement and laid directly on formwork, they should be designed with requisite concrete cover to the high-tensile wire tendons (see Section 34.1.3).[15] Bulk density is dealt with in ASTM C567 (Appendix I).

(b) Dimensional change.
The drying shrinkage of lightweight concrete usually exceeds that of ordinary concrete, its amount varying inversely (among other things) with the bulk density of the aggregate. A typical range of values, which are shown in Figure 33.7, was determined by the United States National Bureau of Standards from specimens that were stored at 21°C and 50 per cent relative humidity for 100 days.[61] The aggregate used should not contain a high proportion of

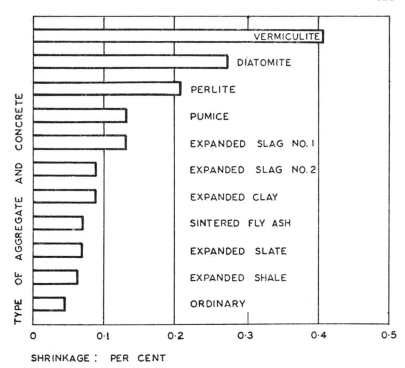

Fig. 33.7   Drying shrinkage of various types of concrete

dust, which would tend to segregate and cause crazing of exposed concrete surfaces.

The effective drying of concrete blocks for masonry construction completes their initial drying shrinkage, only a fraction of which is reversible by subsequent wetting and drying. For purposes of estimation, the average drying shrinkage of structural lightweight concrete may range from 0·05 to 0·07 per cent, the former value approximating that of ordinary concrete. It is possible, however, for high-quality lightweight-aggregate concrete to have a lower drying shrinkage than that of dense-aggregate concrete with an approximately equal cementitious matrix.[62, 63]

The drying shrinkage of sample specimens (taken from mature concrete blocks and dried to practically constant length from a saturated to an oven-dry condition) should not exceed 0·06-0·08 per cent (see Section 27.2; also AS 1500) or (by alternative drying test procedures) 0·04-0·09 per cent (BS CP121). The higher limit given in the British Standard is for lightweight blocks with bulk densities down to below 640 kg/m^3. In the Australian Standard, constant length is considered to have been reached when the change in length of test pieces, up to 300 mm in length, is less than 0·01 mm during drying for 44 hours at 50 ± 1°C and approximately 17 per cent relative humidity, and cooling for 4 hours to 23 ± 2°C in a desiccator containing calcium chloride. As these shrinkage limits do not necessarily ensure a crack-free masonry structure,[66] attention should be given to the proper preparation and use of building blocks in concrete masonry construction (see ASTM C427, Table 27.2; Sections 27.2 and 33.4).

In ASTM C330 and C331,[18, 19] the drying shrinkage of 50 mm × 50 mm × 285 mm specimens of lightweight concrete is limited to 0·10 per cent. The concrete is made with 1 part (by volume) of cement, 6 parts of combined aggregate (see Table 33.3) and sufficient water to produce a slump of 50-80 mm. The shrinkage is measured as a change in length between storage at 23 ± 2°C and 95 ± 2 per cent relative humidity, for 7 days, and storage at the same temperature, but at 50 ± 2 per cent relative humidity, to an age of 100 days. This test, which differs from that made on whole concrete blocks (ASTM C426 and Table 27.2), serves merely as a check on a concrete-making property of a lightweight aggregate.

In reinforced lightweight-concrete beams, the incidence of cracking may be a little more than in similar conventional concrete beams, but it is not necessarily proportional to the drying shrinkage of the concrete. This alleviation of effect is induced by the properties of a low elastic modulus, a reduced rate of drying, an increased capacity for tensile strain by creep (which may be up to about 1·5 times that of ordinary concrete), and a consequent relief of shrinkage stress to within the ultimate strength of the concrete in tension. High-temperature steam curing reduces the creep of the concrete by some 20 per cent, this proportion being more than trebled by means of high-pressure steam curing. Drying shrinkage can be curtailed by the use of expansive or nonshrinkage cement. In perlite concrete, the long-term shrinkage can be offset by an expansive effect, which may develop to an even greater extent when gypsum plaster is used instead of portland cement (see Sections 31.3.3 and 34.1.5).[43, 58]

Under similar loading conditions, the increased flexibility of lightweight concrete members causes beams to deflect between 10 and 30 per cent more than similar beams of ordinary concrete. This effect can be offset by a reduction in dead load and by providing an increased camber.[144] Excessive deflection should be prevented by increasing the requisite amount of steel reinforcement or by reducing the permissible span/depth ratios of beams, as laid down in concrete building codes (see Table 3.25), by 10-15 per cent. The increased creep and elastic characteristics of lightweight concrete should be taken into account in the design of long columns, prestressed elements and slender beams of large span, particularly those that are exposed to the weather (BS CP114).[15, 36, 47, 48, 58, 63-65, 82, 108] The coefficient of thermal expansion of the concrete is generally lower than that of ordinary concrete (see Sections 31.2 and 31.3).

(c) Thermal conductivity.

The thermal conductivity of lightweight-aggregate concrete is related to its unit weight and the mineralogical composition and proportion of lightweight aggregate in the concrete. The values for oven-dry material listed in Table 33.9 increase considerably as the moisture content increases,[38, 46, 59, 85] a comparable value for dry ordinary concrete being about 1·44 (Appendix IV, 5). Coated expanded clay or shale aggregate may be superficially impregnated with bitumen for improved insulation under damp conditions. Thermal diffusivity is referred to in Section 33.3.3(c).

TABLE 33.9 **THERMAL CONDUCTIVITY**

Thermal Conductivity, k (W/m·°C)	Bulk Density, Dry (kg/m³)
0·48-1·27	1900
0·35-0·81	1600
0·27-0·52	1300
0·18-0·36	1000
0·13-0·22	700
0·07-0·14	400

(d) Acoustic properties.

The sound transmission loss through airtight barriers may be estimated from Figure 33.8, which shows an approximate relationship between the average sound-reduction factor of single barriers and the logarithm of their mass per m² (see Reference 84 of Part 4).[85] It is evident from this relationship that the nature of the concrete used in partition walls is of little significance generally for sound insulation, provided that there are no air paths right through the material. Sound-insulation requirements for the walls and floors of dwellings are given in Section 33.3.3(d).[47, 60, 81, 110, 120]

A lean mix with a "popcorn" surface texture absorbs up to 60 per cent of sound waves or noise reverberation. This reduction is useful not only in dwellings, but also in auditoria, theatres and factories where sound control is expensive in conventional concrete structures.

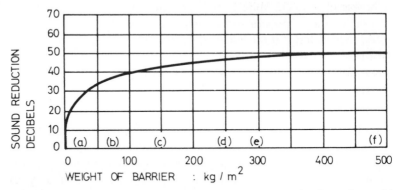

Fig. 33.8   Sound transmission. (a) 12 mm fibreboard, plate glass. (b) 100 mm concrete, 720 kg/m³. (c) 200 mm concrete, 720 kg/m³. 100 mm concrete, 1440 kg/m³. (d) 100 mm concrete, 2400 kg/m³. (e) 200 mm concrete, 1440 kg/m³. (f) 200 mm concrete, 2400 kg/m³.

(e) Water absorption.

The water absorption of lightweight-aggregate concrete varies not only with the type and grading of the aggregate and with time, but also with the quality of the cementitious matrix, the initial moisture content and the bulk density of the concrete. With lightweight aggregate, the important aspect of water absorption is not necessarily its total amount, but rather the amount which affects the workability of the concrete at an early age. For this reason, it is desirable that the aggregate should neither have too large a proportion of very fine or dust-like particles, nor have a maximum size in excess of 16·0 mm.[70, 71] Revibration can be of service (see Section 23.2).

Because of the reduced bulk density of lightweight-aggregate concrete, its water absorption is gauged better by volume than by weight. For purposes of comparison and estimation, a 24-hour absorption value of 8 per cent by weight (determined with 100 mm cubes of ordinary concrete) is equivalent to approximately 19 per cent by volume. Tests have shown that the volumetric absorption of structural lightweight-aggregate concrete can be kept within this amount, which may be regarded as an empirical limit for this concrete when exposed to the weather.[15, 58] Permissible values for unprotected masonry units are from 16 to 24 per cent (ASTM C55, C90 and C145).

Structural lightweight-aggregate concrete, nevertheless, has usually a greater 24-hour absorption than ordinary concrete of the same compressive strength at 28 days, but tests have shown that the permeability of the former is likely to compare favourably with that of the latter when estimated by the method described in AS 1757 and 1759, or BS 473 and 550 on concrete roofing tiles.[58, 72, 73] Structural lightweight-aggregate concrete, with an adequate cementitious matrix, depth of cover and degree of compaction, can be considered as being sufficiently impermeable to protect steel reinforcement against corrosion. Where this limit of volumetric absorption is likely to be reached or exceeded under external service conditions, protective measures may be introduced in the form of a suitable surface treatment or rendering on the concrete (see Section 33.4), or a coating of cement slurry on the steel reinforcement shortly before casting the concrete. For external exposure, the cover of reinforcement in cast *in situ* work is made not less than 50 mm (see Section 3.4).[15, 36, 46] A water-absorption test for lightweight aggregate is contained in AS 1467 and, for sandwich core materials, in ASTM C272.

(f) Durability.

Structural lightweight-aggregate concrete, properly designed and made, is as resistant against deterioration as dense ordinary concrete and it is capable of providing, with adequate cover, comparable protection against the corrosion of steel reinforcement (see Sections 15.2, 22.3 and 33.3.2(e)).[36, 46] Special cement (e.g. sulphate-resisting portland cement) and increased cement content with appropriate means for minimising crazing cracks should be used where the concrete must resist aggressive conditions (see Chapter 19). Tests for the soundness and durability of lightweight aggregate are incorporated in BS 3681 and ASTM C330 and C331,[18, 19] while those concerned with relevant properties of the concrete are described in Chapter 14.

Lightweight-aggregate concrete that is well compacted tends to carbonate to only a slightly greater extent than is customary for ordinary concrete, and, because of the vesicular nature of the aggregate, the concrete has a high resistance to frost action. The abrasion resistance of lightweight concrete, which is normally less than that of dense ordinary concrete, is quite sufficient for bridge decks.[67, 68]

The fire-protection value of lightweight-aggregate concrete is particularly advantageous in multistorey building construction.[110] As indicated in Figure 33.9, the fire resistance of a 100 mm

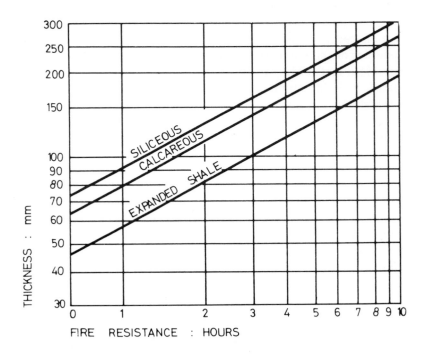

Fig. 33.9 Relationship between fire resistance and thickness of concrete

thickness of lightweight concrete containing expanded-shale or expanded-slag aggregate is approximately the same as up to 150 mm of ordinary concrete.[74, 101] Wire-mesh reinforced vermiculite or perlite concrete slabs, in a 1 : 4 volumetric mix of cement and aggregate, have a 2-hour–4-hour fire resistance in thicknesses of 25-57 mm.[75]

As reinforced lightweight-concrete members have a higher fire resistance than similar members of ordinary concrete, the structural stability of lightweight concrete buildings can be maintained readily during a 4-hour fire (see References 20-23 of Part 4).[46, 69, 76, 77] The fire-resistance performance of structural members can be improved by applying vermiculite or perlite surface finishes (see Section 34.1.3(b)). Siliceous aggregate (as indicated in Section 29.1) has a relatively low resistance to soaking heat (see Section 15.2.4). The spalling of concrete in fire is accelerated and accentuated by high moisture concrete.

33.3.3 FOAMED CONCRETE AND CALCIUM SILICATE HYDRATE

(a) Strength.

The compressive strength of foamed concrete is influenced by its bulk density, age and moisture content, the physical and chemical characteristics of component materials, the proportions of mix, and the methods of foaming, casting and curing. For any given combination of cementing material, aggregate and method of manufacture, a certain relationship exists between strength and bulk density, but a wide departure from this relationship can be obtained by varying the factors mentioned.

For instance, with ordinary foamed concrete, a change in the brand of portland cement can account for a variation of up to 50 per cent in the strength/weight ratio. Other things being equal, Figure 33.10 shows that the compressive strength of ordinary, insulating-grade material is higher when made by the prefoaming method than by the entraining method of foaming. When ordinary

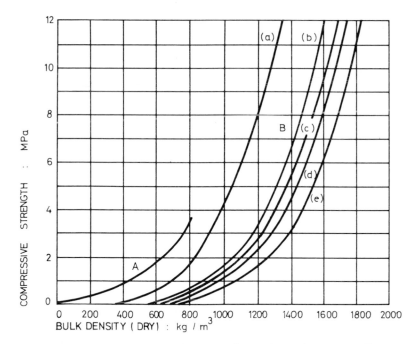

Fig. 33.10 Compressive strength of ordinary foamed concrete. Curves: A, prefoaming method; B, entraining method. Curing: 14 days in water. Cubes: 125 mm in size and dry. Age: 56 days.

Curve	Mix Proportions by Weight	
	Cement : Sand (FM 1.1)	Water/Cement Ratio
(a)	1 : 0	0·3
(b)	1 : 1	0·4
(c)	1 : 2	0·5
(d)	1 : 3	0·6
(e)	1 : 4	0·7

foamed concrete of medium unit weight is made by the entraining process, however, the compressive strength is not only relatively high, but it can vary within a range of 3-5 per cent for a 1 per cent change in bulk density.[127, 128]

The effect of fineness of sand on the compressive strength of ordinary foamed concrete is illustrated in Figure 33.11. It shows that a mature 1 : 4 foamed mix (with a bulk density (dry) of 1450 kg/m³) is 50 per cent stronger when made with fine sand than with coarse sand. This result occurs with an increased water/cement ratio for the finer grade of sand.[41]

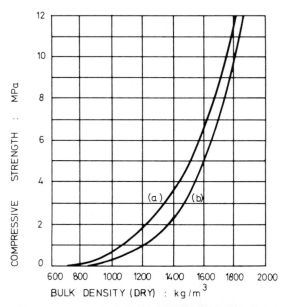

Fig. 33.11 Effect of fineness of sand on compressive strength of ordinary foamed concrete. Entraining method: (a) 1 part portland cement : 4 parts fine sand (fineness modulus = 1·1), water/cement ratio 0·7 by weight; (b) 1 part portland cement : 4 parts coarse sand (fineness modulus = 3·0), water/cement ratio 0·6 by weight. Curing: 14 days in water. Cubes: 125 mm in size and dry. Age: 56 days.

Ordinary foamed concrete can be increased in compressive strength by 20-30 per cent by vibrating the freshly cast mix at 25 Hz for 5 seconds. Effective moist curing of this type of concrete for 1-2 months increases its compressive strength up to about 100 per cent above that attained with air curing. A further increase with age (as indicated in Figure 9.1) is obtainable with continuous moist curing.

The compressive strength and other data given here on foamed concrete are based, unless stated otherwise, on *dry material*. This practice, which is used in Sweden,[15] has been employed in Australia for gauging the actual or comparable load-carrying properties of the material which, for best results, should be used in such a way as to maintain a low moisture content in service. Extensive tests have shown that the cylinder and cube strengths are, for practical purposes, the same; and the strength of water-saturated material is approximately 80 per cent of the strength of the same material after being oven dried and cooled, or allowed to reach a stable air-dry condition.[15]

As indicated in Figure 33.12, the compressive strength of ordinary foamed concrete, with a given bulk density, can be more than doubled by high-pressure steam curing in an autoclave. The increase in strength is associated with the formation of monocalcium silicate hydrate due to a reaction between hydrated lime and finely divided siliceous aggregate (see Sections 31.1.2, 33.2.3(a) and 33.2.3(c)). In autoclaved cement concrete, the lime required for this reaction with silica is hydrolysed from the cement.

In Table 33.10 are shown some typical compressive strengths (dry) of autoclaved chemically foamed concrete with a range of bulk density of from 400 to 800 kg/m³. The values thus indicated correspond with a direction of loading that is perpendicular to the direction of "rise" during manufacture rather than parallel to it. The modulus of elasticity of relevant structural grades varies from about 1400 to 3500 MPa, which is about one-tenth of the value pertaining to ordinary concrete. Other data[86] indicate that a minimum compressive strength (wet) of 2·75 MPa (which is specified in BS 2028 for lightweight-aggregate concrete blocks) can be met by autoclaved foamed concrete with a bulk density (dry) of 560 kg/m³ and over.

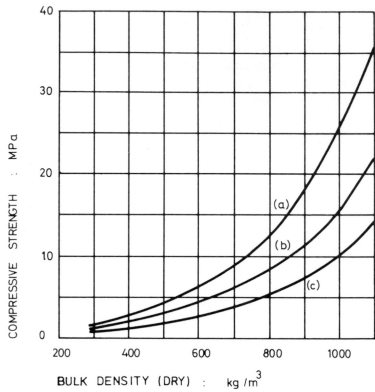

Fig. 33.12 Compressive strength of autoclaved foamed concrete, prefoaming and chemical foaming methods with autoclave curing. (a) Lime concrete. (b) Cement-lime concrete. (c) Cement concrete. Age: 1 day.

TABLE 33.10   AUTOCLAVED FOAMED CONCRETE

Grade	Compressive Strength, Air-dry (MPa)	Bulk Density, Dry (kg/m³)
Structural	5·2	800
	4·8	700
	3·9	600
	3·0	500
Insulation and light load	1·9	400

The tensile strength of autoclaved foamed concrete is about 0.25 ± 0.05 of its compressive strength with a strain at rupture of approximately 0·1 per cent. These characteristics have a favourable effect on the performance of structural slabs or planks, which are usually provided with compression reinforcement for safe handling purposes. Flexural tests on these units have indicated that a linear relationship exists between strain and its distance from the neutral axis, and cracking in the tensile zone does not occur until the nominal working load is increased by 50-100 per cent.[15]

The shear strength of structural slabs of autoclaved foamed concrete varies inversely with their thickness within a range of one-tenth to one-fifteenth of the compressive strength of the concrete. Shear reinforcement is seldom required in flooring and roofing units that are to be built into a monolithic system by means of jointing and surfacing procedures and protected against the weather. Secondary flexural members are so designed that they tend to fail as a result of yielding of the tension reinforcement and not as

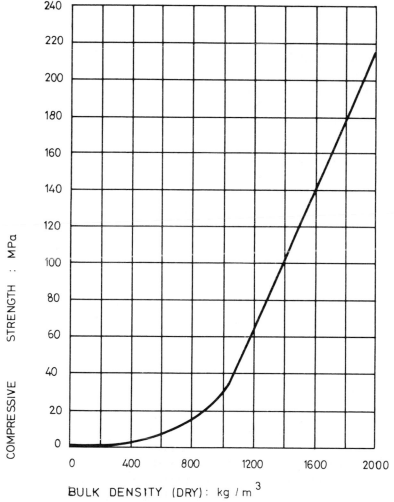

Fig. 33.13 Calcium silicate hydrate. Cubes: 25 mm in size and dry. Age: 1 day.

a result of shearing or crushing of the concrete. Increased depth or shear reinforcement should be used in lintels and trimmer beams which are to carry heavy loads, the load factor against failure being at least 2·5 in both bending and shear.[23, 93, 132]

Bond strength is provided by slight corrugations in dip-applied coatings on steel reinforcement, and by welded cross-elements which form a further means of mechanical anchorage. Cement-rubber-casein coated reinforcement has been found by test to have a bond strength in the region of 1·2 MPa in slabs of autoclaved foamed concrete with a bulk density of 500 kg/m³. The bond strength of reinforcement, with this particular coating, is little affected by its position in a casting mould during manufacture. An adequate bond strength is a significant feature with chemical foaming, which produces a variable amount of cavitation along the upper surface of top horizontal reinforcement of flexural members in their casting position. With bitumen-coated reinforcement, however, the bond strength of the coating is ignored for design purposes, and the anchorage of longitudinal reinforcement is assumed to be obtained by a sufficient quantity of welded cross-elements.[15, 26, 36]

Foamed concrete products that are stored in air at about 23°C and 60 per cent relative humidity increase in compressive strength by about 16 per cent during a period of 3 years. An early increase in strength is obtainable by storing cured varieties of the concrete in air enriched with carbon dioxide.[88, 89] Prolonged storage of lime-bonded foamed concrete in very humid air enriched with carbon dioxide, however, may cause strength to decrease by about 12 per cent or more.[22] When lime and silica-flour mixes are moulded under pressure and autoclaved under optimum conditions, the compressive strength of the resultant calcium silicate hydrate increases linearly with its bulk density above 1100 kg/m³ (see Figs 33.3 and 33.13), the elastic modulus in the 20-70 MPa range being 10×10^3-35×10^3 MPa (Plate 13).[32, 114, 132, 138, 139]

(b) Dimensional change.
The dimensional change of foamed concrete depends on its degree of aeration, mix proportions and methods of curing and drying; also the extent to which movement takes place depends on the composition, porosity, moisture changes and atmospheric carbonation of the concrete. Material with the greatest total drying shrinkage (about 0·5 per cent) is ordinary foamed neat cement with a high degree of aeration. By introducing sand into the mix and reducing the aeration, giving rise to a 1 part cement : 3 parts sand concrete weighing 1450 kg/m³, the drying shrinkage is reduced to

0·08 per cent. For wetting expansion, the corresponding values are 0·15 and 0·07 per cent.

Figure 33.6 shows that the drying shrinkage of ordinary foamed concrete can be brought under control by thoroughly drying moist-cured products in hot air. The more rapid the evaporation of moisture during the accelerated drying period, the lower the shrinkage and the shorter the time required to stabilise the material. The dimensional changes of foamed concrete products may be reduced by artificial carbonation[89] or by high-pressure steam curing, the latter bringing the drying shrinkage of the material to the lower half zone of the 0·04-0·09 per cent range indicated for lightweight concrete blocks in BS CP121 (see Sections 27.2.1 and 33.3.2(b)).

The stabilisation results very largely from the formation of hydrated calcium silicates in mixes containing lime and silica, and the conversion or partial conversion of amorphous calcium silicates in cementitious mixes into crystalline form.[119] High-pressure steam curing not only produces microcrystalline hydrated silicates of low basicity, however, but it forms (from silicates in the presence of aluminates) compounds called hydrogarnets. These compounds (containing lime, silica, alumina and water of crystallisation) are much more durable and stable than the amorphous form of cementing matrix in ordinary foamed concrete.[118] They are not exempt from mineralogical change and shrinkage-strain effects, however, which may be established in large or restrained units as a result of carbonation.[90, 91] Measurements have shown, nevertheless, that certain autoclaved lime and silica mixes may even increase slightly in length (within 0·02 per cent) with age.[25]

The creep and deflection effects of autoclaved structural units under working loads are similar to those obtained over a long period with dense concrete units. The theoretical deflection of precast flexural units under working load is based on a span/depth ratio of 30 for roof loadings and 24 for floor loadings. Its maximum value is intended to vary between 0·004 and 0·0025 of the span at working load, about one-half or more of the deflection being produced under test load and the remainder by plastic movement under sustained load. The theoretical deflection, which occurs without causing visible cracking, is never exceeded in secondary structural members. The compression steel is generally much more effective in increasing the stiffness of flexural members than the use of additional tension steel.[15, 36]

Autoclaved structural units possessing a high tensile strain at rupture have a high resistance to cracking. This characteristic is interconnected with the particle size of finely divided aggregate. It is augmented to some extent by thermal prestressing, which arises through the different coefficients of thermal contraction of the steel and concrete during the cooling stage of manufacture.

Ordinary products with a density gradient and low strength are likely to crack through differential drying and carbonation shrinkage. This tendency is lessened by limiting the depth of moulds or pours to about 750 mm, by autoclaving, by increasing the bulk density of the concrete, by using crack-distributing reinforcement and control joints, by inducing compression by partial prestressing, by using shrinkage-compensated cement, or by applying an appropriate surface coating.

The linear coefficient of thermal expansion of moist-cured foamed concrete containing sand is similar to that of ordinary concrete. Autoclaving reduces the value to about $8 \cdot 1 \times 10^{-6}$ per °C.

(c) Thermal conductivity.

Average thermal conductivity values of oven-dry foamed concrete are listed in Table 33.11, a comparable value for dry ordinary concrete being about 1·44. They increase very little with rising temperature but, through moisture absorption, they increase by approximately 10 per cent for each per cent increase in bulk density.[59, 85] Thermal diffusivity values (see Glossary, Appendix VI) indicate the facility with which concrete undergoes temperature change. They are lower for lightweight concrete than for ordinary concrete and clay bricks.[129, 142]

Test results show that the insulation value of a 275 mm cavity brick wall may be increased by about 23 per cent when the inner leaf of clay brick is replaced with foamed concrete of similar thickness and having a bulk density of 800 kg/m^3. Further improvement will accompany the use of concrete with a reduced bulk density.[86] Structural slabs (500 kg/m^3, 100 mm thick and 3 m span) will provide ample insulation for the roofs of industrial buildings by keeping their thermal transmittance value down to about 1·14 W/m^2·°C.[15, 132]

TABLE 33.11   THERMAL CONDUCTIVITY

Thermal Conductivity, k (W/m·°C)	Bulk Density, Dry (kg/m^3)
0·91	1900
0·63	1600
0·43	1300
0·27	1000
0·17	700
0·10	400
0·065	200

(d) Acoustic properties.

Foamed concrete is less effective than dense concrete in resisting the transmission of air-borne sound (see Figure 33.8 and Section 33.3.2(d)). The sound insulation of a foamed concrete wall or suspended floor is improved by plastering or applying a dense surface course, thereby increasing the weight per unit area and sealing any direct air paths (e.g. through partly filled joints) which freely transmit air-borne sound.[22, 60, 81, 85] The transmission of impact sound in apartments can be lessened by making suspended floor slabs discontinuous at dividing walls (due regard being given to its effect on structural stability) and by using a soft floor covering.

Uncoated foamed concrete has a sound-absorption coefficient of about 0·2-0·5 for low-frequency to high-frequency audible sound. This favourable reduction of reflected sound also applies to the interior surfaces of cavity walls; in partitions between apartments, these may be infilled for improved sound insulation.[132]

(e) Water absorption.

The water absorption of foamed concrete, because of its reduced bulk density, is gauged better by volume than by weight (see Section 33.3.2(e)). The capillary rise and water absorption values of autoclaved products are higher than those of comparable moist-cured units, which may not become fully saturated even after immersion in water for several weeks.[25, 41, 143] Although some grades of foamed concrete can resist rain penetration to the inner surface of a wall (under normal conditions of exposure) it is always desirable to render, coat or clad the material and keep it dry for maximum strength, thermal insulation, weatherproofness and dimensional stability.

Surface treatments should be selected or applied in general so as not to hinder the diffusion of water vapour from the concrete to the atmosphere. Under severe conditions of exposure, external walls may be clad with asbestos-cement tiles, concrete or brick tiles (nailed or hung), or other forms of decorative veneer. Rain-penetration tests on foamed-concrete block masonry rendered externally, and reinforced wall panels sprayed with a protective coating, have shown freedom from water staining on the back face for test periods exceeding 6 hours.[86, 127]

(f) Durability.

Foamed concrete takes up moisture and carbon dioxide from the atmosphere more readily than does dense concrete. The alkalinity of the concrete, which normally serves to passivate steel reinforcement against corrosion, is greatly reduced by high-pressure steam curing and by progressive carbonation. It is essential, therefore, that reinforcing steel be protectively coated against corrosion, particularly where structural units are to be used in moist and industrial atmospheres.

The protective coating required on reinforcing steel should have suitable properties to ensure satisfactory performance. These include imperviousness, adequate bond strength, ability to harden by high-pressure steam curing, and sufficient elasticity to follow the strains of the reinforcing steel and concrete when loaded. A cement-rubber-casein compound has been found to possess these characteristics.[80] A special filled grade of bituminous material, when applied thickly over a phosphate primer, has been found to comply with the requirements of the German Code of Practice for reinforced foamed concrete.[15, 23]

An effective protective coating on steel reinforcement in structural units makes the thickness of concrete cover of lesser importance for rust and spalling resistance. For an effective transfer of stress between the reinforcing steel and the foamed concrete in flexural members, however, the concrete cover should not be less than 15 mm.

On continuous roof decks, a possible cause of local stress and rupture in the protective waterproof membrane is avoided by stiffening the deck system over the supports. This is effected by embedding continuity rods with mortar in the grooves between adjacent roofing units, thereby flattening the slopes formed subsequently at the ends of abutting units. Insulated copper pipes carrying hot water can be kept free from possible stress corrosion by being suitably annealed, phosphorous-free, and embedded in foamed concrete that is free from ammonia-releasing protein.[87]

Foamed concrete is highly resistant to frost action provided that it is not allowed to become water-saturated. Where surface damage by impact or abrasion is likely, the material should be rendered, plastered, cladded or topped with fine dense concrete. Typical fire-resistance ratings of unplastered building units (made with 500 kg/m^3 material) are given in Table 33.12,[15, 92, 95, 128, 132] a 2-hour increase in performance being made possible by applying suitable surface finishes or using slabs with increased cover to the reinforcement (see Sections 33.3.2(f) and 34.1.3(b)).

TABLE 33.12   FIRE-RESISTANCE RATING

Unit	Thickness (mm)	Rating (h)
Nonloadbearing wall of blocks	65	2
Loadbearing wall of blocks	100	2
Reinforced roof slabs with 12 mm cover to reinforcement	125	2
Nonloadbearing wall of storey-height partition slabs	75	3
Nonloadbearing wall of horizontal reinforced wall slabs	150	6

33.4 USES OF LIGHTWEIGHT CONCRETE

33.4.1 GENERAL INFORMATION

The development of building materials is being increasingly directed towards those that are light in weight and best able to meet the needs of structural design and constructional productivity in a modern era. For this purpose, lightweight concrete has been developed in several forms possessing various combinations of structural strength, shock mitigation, weight reduction, heat insulation and fire resistance, nailability and durability, and ease of handling. It is not surprising, therefore, that versatile materials with these possibilities, when satisfactorily combined for economy or performance, are steadily gaining acceptance in regional building projects.

In contrast to no-fines concrete, which requires repetitional work generally for economical results, lightweight-aggregate concrete has wide scope of application because it can be utilised like ordinary concrete for precast and cast *in situ* construction. The use of this concrete in a twenty-storey office block results in an estimated saving of about 5 per cent in structural steelwork and, after allowing for the higher unit cost of the concrete, an overall economy in the project. In buildings such as warehouses, where the live load forms a higher proportion of the design load, the saving is less marked. Buildings only a few storeys high are not likely to show a saving in cost, but they may benefit if expensive piling is required in the foundation.[15, 94, 110]

Although foamed concrete is sometimes cast *in situ,* for the insulation of roofs, pipelines and cold stores and for the erection of tropical dwellings, it is mainly used in the form of autoclaved precast products. These include masonry blocks and reinforced units for roof, floor and cladding slabs, and for loadbearing and nonloadbearing partitions. In masonry work, lightweight building blocks can be laid within half the time or with less than half the labour required for ordinary brickwork. Reinforced roof panels are particularly advantageous for thermal insulation and structural economy, a 500 kg/m³ material (100-200 mm thick) being used for spans of 3-6 m.[135]

Reinforced foamed concrete floor slabs are useful where new or additional floors are to be put into an existing building which would otherwise require a strengthening of its structural support. They are advantageous in suspended ground floors where thermal insulation is desired, although with moderate to high floor loading they are more expensive to use than conventional concrete slabs. Wall panels with precision ground surfaces can be readily joined with an excellent finish, thereby enabling 75 mm partitions to be economically installed.

For the most favourable results in lightweight concrete construction, the design, testing and erection procedures employed should take into account the characteristic properties of the proposed materials to be used (see BS CP111, Table 3.25 and Reference 84 of Part 4).[23, 36, 95, 103-105, 132] To facilitate this in modern building, special equipment has been developed for the handling of reinforced lightweight products. This includes light-duty cranes and panel grabs, transporting and placing carriages, aligning levers, drilling and cutting tools, trowels and finishing devices.[95]

In common with other forms of construction, building during a rigorous winter may be done under a lightweight roof shelter that is raised by hydraulic jacks. For holding capacity in lightweight concrete, galvanised or alloy twisted spiral or cut nails are better than wire nails.[15, 95, 104-106] Correct methods must be observed when handling and using precast lightweight products (see Reference 66 of Part 1).[95, 111]

33.4.2 CAST *IN SITU* WORK

No-fines and lightweight-aggregate concretes, because of their relatively low hydrostatic pressure, require only light formwork, such as expanded metal or plywood on large stiffened frames. Ribbed panels of gypsum plaster, bituminised on one side, may be used as permanent formwork on the inside surface of no-fines concrete. Foamed concrete, because of its mobility, requires watertight formwork set upon hessian-based felt over a prepared base. Plastic linings may be used to produce a smooth surface finish.

Except for the need for sometimes prewetting the lightweight aggregate, longer mixing time and discretion in the use of vibrators during placing, lightweight-aggregate concrete is handled in much the same way as ordinary concrete. For instance, on large projects, mechanical equipment is required for handling both formwork and concrete. In multistorey construction, lightweight-aggregate concrete may be transported by truck mixers from the batching plant to an elevator tower, the transit and standing periods being used for mixing the concrete (see Plates 26 and 27).

Foamed concrete can be readily transported considerable distances, either horizontally or vertically, by pumping it under pressure through a flexible hose. Concrete (of all kinds) is placed in horizontal layers, proceeding continuously around a building to storey height, control joints being suitably placed in large areas of walling to take up drying-shrinkage movement. With reinforced lightweight-aggregate concrete, these joints may be located from 18 m to 30 m apart (see Section 31.2.1); whereas with ordinary foamed concrete, they are inserted at 3-3·5 m centres while the concrete is still plastic. Mesh reinforcement in foamed concrete walls is usually antirust treated before being used.

The external walls of single-storey buildings constructed with no-fines concrete are made 200 mm thick, the ground floor walls of two-storey and higher buildings being 250 mm and 300 mm thick, respectively. The partitions in these buildings are 100 mm thick when loadbearing and 75 mm when nonloadbearing. Walls of no-fines concrete should be made thicker if the cross-walls are located more than 5 m apart or if chases are more than 25 mm deep. Tall buildings of no-fines concrete (e.g. twenty storeys high) are possible if the concrete in the lower storeys is increased in strength. The external walls, which are of uniform thickness, are structurally stiffened by concrete floors, cross-walls and return walls, and designed with requisite load-carrying capacity (see AS 1170 and BS CP3—Chapter V, CP111).[35, 36]

Footings and walls below the ground-floor level of a no-fines concrete building are of conventional type and sufficiently strong to minimise movements due to differential settlement of the foundations. Damp-proof courses are situated at the base of walls, over lintels and (to allow for thermal movement) at the top of walls under flat concrete roofs. Wall ties are galvanised and steel reinforcement at wall openings and eaves level is coated with cement slurry for protection against corrosion. The slurry may contain in the mixing water 1-2 per cent of sodium benzoate or nitrite by weight of cement.

No-fines concrete walls are rendered externally with a composite mortar 15 mm thick, provided with weepholes at the base of walls and above lintels (see Appendix I). The mortar is gauged with 1 part portland cement : 1 part hydrated lime : 6 parts clean coarse-graded sand, or 1 part masonry cement : 3·5-4 parts of clean coarse-graded sand by volume. A decorative "Tyrolean" spatter finish may be applied with a similar mix containing coloured portland cement and an appropriate aggregate instead of sand. For internal plastering, a composite mortar backing coat can be finished with 1 part hydrated lime putty and 3 parts gypsum plaster by volume. No-fines concrete floors have increased insulation if made with lightweight aggregate (see Section 20.1.4(c)), and lightweight gypsum plaster is useful for air conditioning, fire resistance and condensation-free fabrication.

Floors laid directly on the ground (or on a 50 mm consolidated hardcore bed) are readily constructed of no-fines concrete either 150 mm thick with a 15-20 mm screeded mortar or fine-concrete surface, or 100 mm thick with 50 mm or more of high-quality dense concrete (see Sections 20.1.4(c) and 20.1.5(a)). In order to make the latter concrete surface vapourproof from below, it can be laid on a screeded mortar bed supporting an impermeable membrane, the membrane being lapped and sealed at the joins and made continuous with the damp-proof course in the walls. A slip-strip membrane between the floor and walls allows differential movement to take place. Lightweight-aggregate concrete floors are finished off with a perforated roller or "jitterbug" tamper, prior to wood-floating and steel-trowelling, or by applying a dense fine concrete compacted by a power float. A 40 mm pigmented mix of 1 part cement : 5 parts gravel (4·75-9·5 mm) can be used over 50 mm of rake-finished dense concrete for nonreflective road medians.

Insulating concrete can be used as a thermal barrier for walls, floors, roof decks (at least 40 mm applied thickness), hot-water or steam pipes, refrigerated chambers, skating rinks,[96] boiler and firebox installations, and the fire protection of prestressed concrete and structural steel (wrapped with wire mesh before the concrete is cast around it; see Sections 33.3.2(f) and 34.1.3(b)). Insulating concrete weighing 400-480 kg/m^3 is used in protected sections, but where there is likely to be impact from foot traffic, such as on roof decks, 800 kg/m^3 concrete, or a lighter one with a screeded surface course, is required. Exposed surfaces are cured with polyethylene sheets (see Section 9.3).

Low-density insulating concrete consists of a foamed neat-cement slurry or 1 part cement : 1 part sand mortar, or an air-entrained or plasticised mix of 1 part cement : 6-8 parts vermiculite or perlite, with 1-1·5 parts by volume of water. The thickness of insulating concrete required over a 100 mm concrete slab is usually 100 mm, falling to 50 mm; and an appropriate composition is selected to suit various factors of availability, economics and performance.[47] Expansion joints for perlite concrete should be made fairly wide (i.e. 15-20 mm per 30 m and 25 mm) at roof projections, the joints being filled with highly compressible material or designed as an air space.[133]

Roof-deck rendering consists of an air-entrained or plasticised 1 part cement : 4 parts clean coarse sand mortar, laid 15 mm thick in alternate bays not exceeding 10 m² in size, and given a drainage fall of not less than 1 in 60. Drain ducts, ventilators and joint-bridging facilities are provided, and the surface course is allowed to harden and dry before receiving a waterproof membrane, such as a three-layer bituminous felt (see Section 36.5; AS A98, A99, A120, A121 and BS 747, CP144.101; also ASTM D224, D226, D249, D250, D312, D371, D655, D1227, D1668).[95, 97, 98] A lightweight-aggregate no-fines bituminous concrete, if used in requisite thickness for insulation, can be sealed directly with asphalt.

Underground hot-water pipes to be insulated are protected with anticorrosive paint (see Section 22.3.2), prior to being wrapped with corrugated cardboard for movement differential. The lightweight concrete, on being cast, is enveloped with a watertight casing that will exclude water completely. The site must be well drained to ensure continuously dry insulation, using a no-fines concrete substrate and agricultural or perforated drainage pipes for the purpose.[99]

33.4.3 BLOCK MASONRY

Lightweight concrete blocks are used for such purposes as single-leaf and cavity walls, lining and curtain walls, nonloadbearing partitions and infill panels, soffit units in flat concrete slabs, cladding formwork, and the insulation of roof decks, floors and cold stores. Although blocks vary considerably in size, they are usually dimensionally coordinated on a 100 mm module containing a 10 mm allowance for joints (see Section 27.2.1, Table 27.2 and Appendix IX). Masonry blocks produced with different materials vary in bulk density from 400 to 1450 kg/m³, the total weight of each block being kept below 20 kg for ease in handling by one operative. In this regard, an autoclaved foamed concrete block 600 × 200 × 100 mm with a bulk density of 500 kg/m³ weighs about 6·3 kg.

All handling and building details should be arranged to ensure that lightweight masonry blocks remain as dry as possible. For instance, they should be stacked off the ground, sheeted over during transport, storage and construction in damp weather, and redried before use if allowed to become wet. Flashings and weepholes should direct water away from the inner leaf of cavity walls, and contraction joints should be provided at 6-9 m intervals in long walls.

These joints are not normally necessary in the inner leaf of cavity walls in small dwellings but, when required, they may be located at wall junctions with partitions (BS CP122).[47, 87] Contraction control joints in foamed concrete block walls may be formed either during construction or by cutting the concrete to a depth of about 40 mm. Ordinary cast *in situ* concrete may be clad with an insulating layer of foamed concrete. For this purpose, permanent precast formwork of foamed concrete may be fastened to the *in situ* concrete by bent wire ties, one end of the tie being driven into the top of each block as it is laid without mortar against a temporary skeletal frame.

Lightweight-aggregate concrete blocks can be cast with grooves. which are arranged for easy cutting at suitable spacings for the best assortment of sizes and potential use of offcuts.[100] Lightweight hollow and solid concrete blocks are used in much the same way as ordinary concrete masonry (see Section 27.2.2, AS 1475, 1640 and BS CP111, CP122, also Reference 84 of Part 4).[23, 36, 95, 103-105, 132] Lightweight block masonry should have solid blocks under floor joists, a mortar-free cavity in two-leaf walls, and bond beams of reinforced composite concrete design, and it should be kept in general above the damp-proof course level.

Autoclaved foamed concrete blocks have been used in a loadbearing capacity for dwellings up to five storeys high, the material having a bulk density of 640-800 kg/m^3 for the lower floors and 500 kg/m^3 for the upper sections.[99] These blocks (when manufactured with small slits and a tolerance of $\pm$ 2 mm) can be laid with dry joints and plastic locking discs (40 mm diameter $\times$ 3 mm thick) to form weathertight walls. During construction, a disc is placed above and below each vertical joint, so as to lock three blocks together.

Mortarless walls with special blocks can be quickly built and dismantled, when necessary, for future structural modifications.[102] Lintels over wall openings are of autoclaved reinforced foamed concrete, possessing the same bulk density as the block masonry, or of composite dense and lightweight concrete design. Advantages claimed for them are ease of handling by two workers, thermal compatibility with the walls, and ease of fixing accessories at windows.

The mortar for block laying and the two-coat rendering of single-leaf walls is gauged with 1 part portland cement : 2 parts hydrated lime : 9 parts clean coarse-graded sand, or 1 part masonry cement : 4-5 parts of clean, graded coarse to medium gauge sand by volume (see Appendix I). For cold weather work, severe conditions of exposure, or blocks of medium to high density, the mortar may be altered in composition to that of 1 part cement : 1 part lime : 6 parts sand, or a 1 : 3·5-4·5 masonry cement mortar, the mix in each instance being improved with activation by high-speed mixing. New walls should be allowed to dry and shrink to a state of equilibrium over a period of several weeks before surfacing. In wintry weather, heating and ventilation will hasten the drying operation.

The external surface of cavity walls can be weatherproofed by proprietary textured finishes or pigmented vapour-permeable treatments that are resistant to rain penetration, such as two or three coats of filled siliconate acrylic emulsion paint, cement-based paint with or without a finishing coat of silicone paint, or other suitable material, the first coats being applied liberally with vigorous brushing (see Sections 28.2, 28.3 and 28.4).[95] Walls low in suction are not surface dampened before being rendered or plastered. Those high in suction may be lightly sprayed with water to control the suction, and given a 1 part cement : 2·5 parts sand spatterdash a few hours before being rendered in two coats with a finished thickness of 15-20 mm. Alternatively, a dry internal surface can be given a coat of bonding agent before proceeding with the plastering, unless a special plaster mix is to be applied directly to the wall.

An undercoat for internal plastering may consist of a cement-lime sand or masonry cement mortar; a mix containing set-retarded gypsum plaster in the proportion of 1 part plaster : 2-3 parts sand; or a premixed lightweight plaster. The finishing operation is done in all instances with a wooden float. Alternative finishing coats include 1 part hydrated lime putty by volume : 3 parts gypsum plaster (AS CA27 and BS CP211) or a vermiculite or perlite gypsum plaster. Alkali-resistant paints of all types or oil paints over an alkali-resistant primer can be used internally, and wall tiles when required are laid on a mortar background with suitable adhesive.[47, 87, 95, 104]

In the construction of cold stores with a foamed concrete lining, it is necessary to exclude moisture that would reduce the efficiency of thermal insulation. For this reason the inner surface of the main wall and floor base of air-entrained concrete (which will be in contact with the blocks) are coated with bitumen. The blocks are laid in an air-entrained or lightweight mortar, and the floor is completed with a load-distributing layer of air-entrained mortar or concrete to receive an asphalt finish. Precast blocks or structural roof units can be used for the ceiling.

Weather tiles can be hung over a vapour barrier on external walls with galvanised or alloy cut nails (65 mm long), while 75-150 mm cut nails, wood screws or expansion shield anchorages can be used for attaching fittings to lightweight concrete masonry. Wire nails should not be used.

Note: For fixing purposes in 800 kg/m³ autoclaved foamed concrete, 75 mm and 100 mm cut nails have pull-out load values of about 1·2 kN and 1·9 kN, respectively. In similar 500 kg/m³ concrete, the pull-out values are 0·36 kN and 0·71-1·42 kN for 75 mm and 100-150 mm cut nails, 1·5 kN for 14-gauge 100 mm wood screws, and 2·25 kN for 100 mm screws in "Rawlplugs". The holding capacities are reduced by overdriving, and factors of safety required range from 5 to 10 for static loading, to 10 for shock or vibratory loading.[15, 95, 104-106, 132]

33.4.4 PRECAST REINFORCED MEMBERS

The physical and structural features of reinforced lightweight members are governed by the properties of the concrete used, the expected loading, span or slenderness ratio, and the ambient conditions in service. Precast wall panels of solid concrete or composite sandwich section are normally of single-storey height. They are made in varying widths that may incorporate window and door openings for application in single-storey to multistorey projects.

High-rise buildings in precast lightweight-aggregate concrete contain interconnected shear wall and flexural slabs, external wall panels, and infilled bolted joints of rigid or movement-control type, as governed by features in the structural design. Precast flexural members are of reinforced or prestressed concrete type, the latter being made with strong lightweight aggregate for spans of 9 m or more, provision being made during design for a possible increase in loss of prestress (see Section 34.1.5).[136]

While precast lightweight-aggregate concrete has, in general, a larger scope of application than autoclaved foamed concrete products, the latter can be accommodated in many ways in building construction. For instance, with 500 kg/m³ foamed concrete, flexural panels or planks some 508 mm or 610 ± 2 mm wide are capable of carrying uniformly distributed live loads of up to 4·8 kN/m² (4·8 kPa) over spans of up to 6·1 m, the appropriate thickness of the panels ranging from 75 mm to 300 mm. The factor of safety on total load is not less than 2·5 and the ultimate deflection at working load is kept below the span divided by two hundred and fifty (see Sections 33.3.2(b) and 33.3.3(b)).

Material with a higher bulk density and compressive strength can be used for structural purposes, due allowance being made for the weight of steel reinforcement and other fixtures in design.[95] Loadbearing wall units up to 1·5 m wide and 2·5-6·5 m long range from 75 mm to 300 mm thick, while nonloadbearing partitions are only 75-100 mm thick. Horizontal nonloadbearing wall units are used as panel infilling or cladding.

Vertical wall units are used for loadbearing purposes in dwellings up to three or more storeys high, or as cladding to framed structures. For slenderness ratios not exceeding 15, based on an effective wall height of one storey, standard panels 100-250 mm thick will support maximum uniformly distributed loads of 5700-14300 kg/m run. Load-reduction coefficients are used for greater slenderness ratios (AS 1480, BS CP111, CP114 and ACI in Table 3.25). Structural panels during erection can carry a point load of 160 kg before grouting, and flexural panels (unless specially reinforced) can be cantilever loaded for up to twice their thickness.[95, 132]

For solid external walls of autoclaved foamed concrete, the minimum thickness of panels used is 125 mm for industrial buildings and 200 mm for domestic and commercial buildings. Flexural units and vertical wall panels are held in position in a variety of ways. These include such accessories as anchored steel bands, 6 mm diameter continuity rods passing through anchored stirrups or slotted anchor plates, and the use of steel clips, double-clip plates, 50-175 mm cut nails, 225-300 mm spikes, and shear plates driven across joints at an angle of about 30° to the horizontal. Prefabricated frames with drilled holes at 450 mm centres can be fitted into walls with 150 mm cut nails.

With the exception of partitions, which are butt-jointed with a latex and cement adhesive, the panels mentioned are wetted along rebated edges and grooves, before being joined together with 1 part cement : 3 parts sand grout. Continuity reinforcement in one-third panel lengths is pressed into this material across the supports and after about 0·5 hour, the surplus grout is scraped off. The surface is protected with polyethylene sheet and closed to traffic for 2 days, the grouting operation being suspended when frost is present or the air temperature is below 2°C.

Foamed concrete roof units require a minimum bearing width of 45 mm on painted steelwork and 50 mm with a mortar bed on concrete, masonry, brickwork and timber. These widths are made 65 mm and 75 mm respectively, for floor units and, in each instance, a layer of bituminous felt is used on timber supports.[93, 95] Horizontal wall panels are laid dry above damp-proof course level with a 10 mm drip projection beyond the face of an underlying plinth. As indicated in Figure 33.14, the longitudinal joints are formed by tacking bituminous foamed plastic strips (e.g. polyurethane) or preformed mastic strips along the upper edge of each panel and then placing the next unit upon it.

The strips are fixed on at ground level and subsequently compressed under the weight of the panels, thereby forming a weathertight seal. The panels can be fixed to the external face of columns by using noncorrodible steel strips or wire clips and cut nails, or by bolting with the aid of noncorrodible cover strips over the vertical joints. With other means of fixing, the vertical joints must be sealed with a mastic to allow for thermal movement. Short panels at openings, where required, can be fixed together by steel dowels.

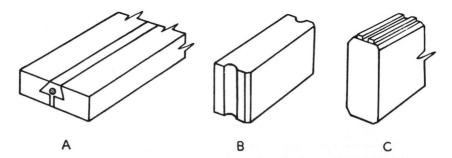

A   B   C

Fig. 33.14 Precast foamed concrete panels. A, floor and roof panels with continuity reinforcement over supports. B, vertical wall panel with joint-grouting grooves. C, horizontal wall panel with preformed jointing strips tacked onto it.

Control joints are placed across long buildings as in ordinary concrete construction and, on flat roofs, a bituminous covering can be applied directly to structural foamed concrete slabs. Main roofing joints are weatherproofed with crimped flashing and mastic or other appropriate means (BS CP144.101).[95, 98] If a solid roof slab is not thick enough to prevent condensation of moisture on the soffit under internal air conditions and external temperatures, then a ventilated-cavity roof should be installed.[36, 95] This design requirement applies particularly to industries having high internal temperatures and relative humidities.[132]

Foamed concrete floor slabs should be protected from construction traffic by the use of barrow runs, sand or other temporary covering. When a 40 mm fine concrete surface course is to be laid, the receiving surface should be well scored, cleaned, wetted and topped in bays not exceeding 10 m^2 in size. Alternatively, a 10 mm coat of mortar for tiling can be applied with an effective bonding medium, or 100×50 mm battens may be nailed to the concrete at 600 mm centres to take timber boarding, in conjunction with lightweight-aggregate infill.

Superficial damage to autoclaved foamed concrete units can be repaired on the site. Where a surface covering is to be applied, patching may be done with 1 part cement : 2 parts lime : 12 parts sand mortar or the equivalent; but where it is to be left untreated or painted, it is necessary to reproduce the basic colour and wire-cut texture. For this purpose, the patching mortar consists of a plasticised mixture of 1 part white portland cement (with the addition of 15 per cent of ordinary portland cement) and 3 parts of fine foamed concrete aggregate passing a 2·36-mm sieve and retained on a 0·30-mm sieve. The aggregate is soaked with water before use and patched areas, which are built up in layers not exceeding 10 mm thick, are finished with a wooden float or the toothed edge of a hacksaw blade. This type of material may, amongst others, be used for joint pointing in walls that are to be painted, skim-coated and painted, or size-coated and wallpapered or covered with hessian.

Foamed concrete panels should be brushed free of loose particles before being assembled or painted. The soffit of solid roof slabs and floor slabs can either be left untreated or be surface-finished with one or other of a variety of vapour-permeable materials. These include emulsion or latex paint with or without fine-sand filler, oil-free distemper, limewash, or a skim coat of finishing plaster, lightweight plaster or sand putty, the latter coats being made discontinuous over the soffit joints.

Surface finishes on lightweight-aggregate concrete members include a combination of broomed, smooth, textured, painted and decorative aggregate treatments. External paint finishes on lightweight concrete walls should be resistant to rain penetration and yet be vapour permeable. They may consist of filled siliconate acrylic emulsion paint, cement-based paint, with or without a finishing coat of silicone paint or other suitable material (see Chapter 28).[95] The soffit of a ventilated-cavity roof must be sealed with a vapour barrier, in order to maintain an optimum state of insulation and promote durability in the presence of moist aggressive atmospheres. Suitable materials for this purpose include chlorinated rubber lacquer, plastic or synthetic resin compounds, and bituminous paint.[132]

Autoclaved reinforced foamed concrete panels of storey height are used in the "Skarne" system of high-rise apartment blocks in Sweden, in an enterprising endeavour to save time, energy, weight, equipment and money in building construction. The basis of the system is a concrete tower, placed with sliding formwork. The tower is constructed first and constitutes a central core around which the rest of the building is erected. This core becomes a stabilising element for the building and contains all services, stairs and lifts (see Plate 27).

A crane is installed at the top, and the building is completed by erecting precast wall units and cast *in situ* concrete floors around the tower. The internal partition walls are, for the most part, loadbearing lightweight-aggregate concrete or dense concrete, while the remainder (plus the external walls) are of nonloadbearing foamed concrete design. External panels of 75 mm polystyrene foam between two 75 mm leaves of foamed concrete are used in conjunction with triple-glazed windows, where maximum insulation is required for efficient air conditioning in cold climates.[99, 107]

REFERENCES

1. Australian Experimental Building Station. "No-fines Concrete". *Notes on the Science of Building, NSB 8, NSB 12, NSB 15.*
2. NIEDHOFF, A. E. "Lightweight Pumice Concrete". *ASCE Proceedings*, 75, 6 (June 1949), pp. 743-57.
3. Standards Association of Australia. *Lightweight Aggregates*. AS 1467.
4. British Standards Institution. *Lightweight Aggregates for Concrete*. BS 3681, 3797.
5. American Society for Testing and Materials. *Inorganic Aggregates for Use in Gypsum Plaster*. Standard C35.
6. American Society for Testing and Materials. *Lightweight Aggregates for Insulating Concrete*. Standard C332.
7. DAVIS, R. E. and KELLY, J. W. "Lightweight Aggregates". *ASTM Special Technical Publication No. 83*, 1948.
8. HILL, R. D. and CROOK, D. N. "Some Causes of Bloating in Expanded Clay and Shale Aggregates". *Australian Journal of Applied Science*, 11, 3 (September 1960), pp. 374-84.
9. HODOROV, E. I. and NELIDOV, W. A. "Pelletization of Raw Materials". *Cement and Lime Manufacture*, 29, 5 (May 1956), pp. 55-59.
10. TRAUFFER, W. E. "Nytralite Barged Down Hudson to New York City Area". *Pit and Quarry*, 55, 8 (February 1963), pp. 86-95.
11. EVANS, R. H. and HARDWICK, T. R. "Lightweight Concrete with Sintered Clay Aggregate". *Reinforced Concrete Review*, 5, 6 (June 1960), pp. 369-400.
12. BSI. *Clinker Aggregate for Concrete*. BS 1165.
13. PECK, R. L. "Magnetic Devices Clean Cinder Aggregates". *Modern Concrete*, 26, 11 (March 1963), pp. 44-47, 83.
14. LEWIS, D. W. "Lightweight Concrete Made with Expanded Blast Furnace Slag". *ACI Journal*, 30, 5 (November 1958), pp. 619-33.
15. Reinforced Concrete Association. *One-day Symposium on Structural Lightweight Concrete*, Brighton, Britain, 1962.
16. BSI. *Foamed or Expanded Blastfurnace Slag Lightweight Aggregate for Concrete*. BS 877.
17. VALORE, R. and BOUX, J. F. "Enercon Ltd. Develops Fly Ash Process for Concrete Industry". *Concrete Products*, 72, 2 (February 1969), pp. 46-49, 69.
18. ASTM. *Lightweight Aggregates for Structural Concrete*. Standard C330.
19. ASTM. *Lightweight Aggregates for Concrete Masonry Units*. Standard C331.
20. ACI COMMITTEE 211.2. "Recommended Practice for Selecting Proportions for Structural Lightweight Concrete". *ACI Journal*, 65, 1 (January 1968), pp. 1-19; 74, 2 (February 1977), pp. 85-86.
21. ASTM. *Test for Air Content of Freshly Mixed Concrete by the Volumetric Method*. Standard C173.
22. "Symposium on Steam-cured Lightweight Concrete". *RILEM Bulletin*, New Series, 10 (March 1961), pp. 91-120.
23. German Industrial Standard DIN 4223. *Reinforced Roof and Floor Slabs of Steam-cured Aerated and Cellular Concrete*. Deutscher Ausschuss für Stahlbeton, Berlin, 1958.
24. TAYLOR, W. H. "Foamed Concrete". *Constructional Review*, 22, 6 (October 1949), pp. 11-18.
25. VALORE, Jr, R. C. "Cellular Concretes, Parts 1 and 2". *ACI Journal*, 25, 9 (May 1954), pp. 773-96; 10 (June 1954), pp. 817-36.
26. SHORT, A. and KINNIBURGH, W. "The Structural Use of Aerated Concrete". *Structural Engineer*, 39, 1 (January 1961), pp. 1-16; 11 (November 1961), pp. 364-77.
27. "Desert Sand Component for Concrete Mix in the East". *Building Digest*, 12, 12 (December 1952), pp. 410-11.
28. MOOREHEAD, D. R. and TAYLOR, W. H. "Sucrose Extraction Method of Determining Available Calcium Oxide in Hydrated Lime". *ASTM Bulletin*, 236 (February 1959), pp. 45-47.

29. TAYLOR, W. H. and MOOREHEAD, D. R. "Lightweight Calcium Silicate Hydrate: Some Mix and Strength Characteristics". *Magazine of Concrete Research,* 8, 24 (November 1956), pp. 145-50; 9, 26 (August 1957), p. 109.

30. TAYLOR, W. H. *Improvements in or Relating to the Manufacture of Products of Calcium Silicate Hydrate.* CSIRO, Australian Patent Specification No. 244,958.

31. "Cellular Concrete from Lime". *Concrete Building and Concrete Products,* 36, 9 (September 1961), pp. 379-80.

32. TAYLOR, W. H., COLE W. F. and MOOREHEAD, D. R. "High-strength Calcium Silicate Hydrate". *Nature,* 201, 4922 (29 February 1964), pp. 918-19; and *Society of Chemical Industry,* International Symposium on Autoclaved Calcium Silicate Building Products, London, 1965.

33. PECK, R. L. "Poreen. Buffalo Pilot Plant in Operation". *Modern Concrete,* 27, 3 (July 1963), pp. 36-41.

34. MALHOTRA, V. M. "No-fines Concrete—Its Properties and Applications". *ACI Journal,* 73, 11 (November 1976), pp. 629-644.

35. MACINTOSH, R. H., BOLTON, J. D. and MUIR, C. H. D. "No-fines Concrete as a Structural Material". *ICE Proceedings,* 5, 6 (November 1956), Part 1, pp. 677-94.

36. SHORT, A. and KINNIBURGH, W. *Lightweight Concrete.* Spon, London, 1968.

37. THOREK, S. W. "Technical Report on 28-storey Building in Melbourne for Consolidated Zinc Pty. Ltd." *IE Aust. Journal,* 34, 1-2 (January/February 1962), pp. 15-18.

38. Housing and Home Finance Agency. *Lightweight Aggregate Concretes.* United States Government Printing Office, Washington, 1949.

39. Housing and Home Finance Agency. "Design Data for Some Reinforced Lightweight Aggregate Concretes". *Housing Research Paper 26,* 1953.

40. ACI COMMITTEES 213 and 523. "Lightweight Concrete". *ACI Journal,* 64, 8 (August 1967), pp. 433-67; 65, 2 (February 1968), pp. 151-55; 68, 10 (October 1971), pp. 795-98; *ACI Publication SP-29,* 1971.

41. TAYLOR, W. H. "Lightweight Concrete". *Report Z4,* CSIRO Division of Building Research, 1954 (revised 1960).

42. SHORT, A. "Lightweight Aggregate Concrete". *Concrete* (London), 8, 7-9 (July/August 1974), Current Practice Sheets.

43. RAMOS, C. and SHAH, S. P. "Strength of Lightweight Aggregates and Concrete". *Journal of Materials,* 7, 3 (September 1972), pp. 380-87.

44. WELCH, G. B. and PATTEN, B. J. F. "Structural Lightweight-aggregate Concrete". *Constructional Review,* 37, 2 (February 1964), pp. 22-28.

45. LEWIS, R. K. and MATTISON, E. N. "Some Tests on Proportioning of Structural Lightweight Aggregate Concrete". *Report C3.3-1,* CSIRO Division of Building Research, 1961.

46. British Building Research Station. "CEB Recommendations and the Structural Use of Lightweight Concrete". *Current Paper 3/68,* BRS.

47. British Building Research Station. "Lightweight Aggregate Concretes". *BRS Digests (2nd Series) 111, 123.*

48. PFEIFER, D. W. "Sand Replacement in Structural Lightweight Concrete—Creep and Shrinkage Studies". *ACI Journal,* 65, 2 (February 1968), pp. 131-39.

49. SHIRAYAMA, K. "Estimation of the Strength of Concrete Made with Lightweight Aggregate". *Magazine of Concrete Research,* 13, 38 (July 1961), pp. 61-70.

50. ASTM. *Test for Compressive Strength of Lightweight Insulating Concrete.* Standard C495.

51. HANSON, J. A. "Tensile Strength and Diagonal Tension Resistance of Structural Lightweight Concrete". *ACI Journal,* 58, 1 (July 1961), pp. 1-39; 59, 3, 2 (March 1962), pp. 803-10.

52. ACI-ASCE COMMITTEE 326. "Shear and Diagonal Tension". *ACI Journal,* 59, 1 (January 1962), pp. 1-30; 2 (February 1962), pp. 277-333; 3 (March 1962), pp. 353-95; 9 (September 1962), pp. 1323-49.

53. ACI COMMITTEE 318. "Building Code Requirements for Reinforced Concrete". Appendix IV, 2(a)(i).
54. PAUW, A. "Static Modulus of Elasticity of Concrete as Affected by Density". *ACI Journal*, 32, 6 (December 1960), pp. 679-87.
55. PLOWMAN, J. M. "Young's Modulus and Poisson's Ratio of Concrete Cured at Various Humidities". *Magazine of Concrete Research*, 15, 44 (July 1963), pp. 77-82.
56. ALCOCK, D. G. and PAUW, A. "Controlled-deflection Design Method for Reinforced Concrete Beams and Slabs". *ACI Journal*, 59, 5 (May 1962), pp. 645-58.
57. ISAACS, D. V. and LANGLANDS, I. "Concrete Incorporating Lightweight Burnt-clay or Burnt-shale Aggregate". *Special Report No. 12*, Australian EBS, 1953.
58. EVANS, R. H. and DONGRE, A. V. "The Suitability of a Lightweight Aggregate (Aglite) for Structural Concrete". *Magazine of Concrete Research*, 15, 44 (July 1963), pp. 93-100.
59. VALORE, Jr, R. C. "Insulating Concretes". *ACI Journal*, 28, 5 (November 1956), pp. 509-32; 12 (June 1957), pp. 1249-56.
60. HARRIS, C. M. *Handbook of Noise Control*. McGraw-Hill, New York, 1957.
61. "Properties of Lightweight Aggregate Concretes". *Concrete*, 57, 7 (July 1949), pp. 22-25.
62. WASHA, G. W. "Properties of Lightweight Aggregates and Lightweight Concretes". *ACI Journal*, 53, 4 (October 1956), pp. 375-82.
63. PATTEN, B. J. F. "Drying Shrinkage and Creep of Expanded Lightweight Aggregate Concrete". *Constructional Review*, 33, 11 (November 1960), pp. 29-32.
64. McKEEN, R. G. and LEDBETTER, W. B. "Shrinkage-Cracking Characteristics of Structural Lightweight Concrete". *ACI Journal*, 67, 10 (October 1970), pp. 769-77.
65. BEST, C. H. and POLIVKA, M. "Creep of Lightweight Concrete". *Magazine of Concrete Research*, 11, 33 (November 1959), pp. 129-34.
66. MENZEL, C. A. "Fallacies in the Current Per Cent of Total Absorption Method for Determining and Limiting the Moisture Content of Concrete Block". *ASTM Proceedings*, 57 (1957), pp. 1057-76.
67. ANDREW, C. E. "Use of Expanded Shale Concrete in Bridge and Ship Construction". *Expanded Shale Concrete Facts*, 1, 1(1954).
68. ANDREW, C. E. "Structural Uses of Expanded Shale Concrete". *Expanded Shale Concrete Facts*, 1, 2 (1954).
69. Editorial, "Lightweight Aggregates for Concrete Block". *Concrete Products*, 66, 9 (September 1963), pp. 32-38.
70. Editorial, "Lightweight Aggregate Structural Concrete". *Concrete Products*, 66, 10 (October 1963), pp. 42-52.
71. HELMS, S. B. and BOWMAN, A. L. "Extension of Testing Techniques for Determining Absorption of Fine Lightweight Aggregate". *ASTM Proceedings*, 62 (1962), pp. 1041-53.
72. Standards Association of Australia. *Concrete Interlocking Roofing Tiles*. AS 1757 and 1759.
73. BSI. *Concrete Roofing Tiles and Fittings*. BS 473, BS 550.
74. THOMPSON, J. P. "Fire Resistance of Reinforced Concrete Floors". *ACI Journal*, 24, 7 (March 1953), pp. 677-80.
75. SNOW, F. S. *Lightweight Fire Protection for Structural Steelwork*. British Constructional Steelwork Association, London.
76. *Uniform Building Regulations, Victoria*. Victoria Government Gazette, Government Printer, Melbourne, 1974.
77. DAVEY, N. and ASHTON, L. A. "Investigation on Building Fires". *DSIR, National Building Studies, Research Paper 12*, HMSO, London, 1953.
78. BLAKEY, F. A. "Some Comparative Tests of the Bond of Plain and Indented Wires for Prestressed Concrete". *Constructional Review*, 28, 2 (June 1955), pp. 20-22; 4 (August 1955), p. 37.
79. GORYAINOV, K. E. and ZASEDATELEV, I. B. "Thermal Processes Taking Place in the Autoclave Treatment of Large Cellular Concrete Units". *CSIRO Translation 4644* from Beton i Zhelezobeton, 2 (1959), pp. 62-67.

80. International Siporex Ab. *Improvements in the Manufacture of Reinforced Light-weight Concrete.* British Patent Specification No. 829,051/60.
81. DSIR Britain. "Sound Insulation of Dwellings—I and II". *BRS Digests 88 and 89,* 1964.
82. ACI COMMITTEE 711. "Minimum Standard Requirements for Precast Concrete Floor and Roof Units". *ACI Journal,* 30, 1 (July 1958), pp. 83-94.
83. TIMMS, A. G. "Blast-furnace Slag as a Concrete Aggregate". *Modern Concrete,* 27, 6 (October 1963), pp. 29-32, 37; 7 (November 1963), pp. 29-33.
84. KOLOSEUS, E. J. and GEYER, B. G. "Cellular Concrete". *Industrial and Engineering Chemistry,* 52, 11 (November 1960), p. 28A.
85. ARNOLD, P. J. "Thermal Conductivity of Masonry Materials". *British BRS, Current Paper 1/70,* 1970.
86. DSIR, Britain. "Aerated Concrete—1: Manufacture and Properties". *BRS Digest (2nd Series) 16,* 1961.
87. DSIR, Britain. "Aerated Concrete—2: Uses". *BRS Digest (2nd Series) 17,* 1961.
88. KROONE, B. "Effect of Gaseous Carbon Dioxide on Concrete". *Indian Concrete Journal,* 37, 10 (October 1963), p. 379.
89. SLATANOFF, V. and DJABAROFF, N. "Co_2 Treatment of Foam-concrete in Order to Make Heat-insulating Sections". *CSIRO, Translation 4830,* 1961.
90. SILAENKOV, E. S. "Evaluation of Durability of Autoclave-cured Cellular Concrete Designated for Large-size Units". *Proceedings RILEM Symposium on Durability of Concrete,* Prague, 1961.
91. SILAENKOV, E. S. "Evaluation of the Durability of Large-sized Elements Made of Autoclaved Aerated Concrete". *DSIR BRS, Library Communication 1155,* 1962.
92. RYAN, J. V. and BENDER, E. W. "Fire Tests of Precast Cellular Concrete Floors and Roofs". *Monograph 45,* United States Department of Commerce, NBS, 1962.
93. " Autoclaved Aerated Concrete". British *BRE Digest 178;* BS CP116, Part 2.
94. WOOD, N. F. "The Trend to Lightweight Concrete". *Architecture and Arts,* 11, 7 (July 1963), pp. 21-22.
95. *Some Information and Technical Notes on Siporex.* International Siporex Ab, Costain Concrete, London.
96. HESPE, F. S. "Prince Alfred Park Ice Skating Rink". *Constructional Review,* 32, 6 (June 1959), pp. 19-23.
97. BALLANTYNE, E. R. and MARTIN, K. G. "Bituminous Roofs". *CSIRO DBR, Building Study No. 1,* 1960.
98. DSIR, "Built-up Felt Roofs". *BRS Digest (2nd Series) 8,* 1961.
99. DIAMANT, R. M. E. "Gas Concrete". *Architect and Building News,* 221, 16 (18 April 1962), pp. 567-72; 17 (25 April 1962), pp. 615-18.
100. CARHART-HARRIS, T. L. "The Development of Grooved Lightweight Concrete Blocks". *The Builder (London),* 196, 6063 (June 1959), pp. 1046-47.
101. GUSTAFERRO, A. H. "Fire Resistance of Architectural Precast Concrete". *PCI Journal,* 19, 5 (September/October 1974), pp. 18-37.
102. INTERNATIONAL YTONG Ab. "Plastic Locking Discs for Concrete Units". *The Builder (London),* 201, 6183 (November 1961), pp. 941-42.
103. MORLEY, C. R. "Factor of Safety and Permissible Stresses in Walls of Un-reinforced Blocks". *Proceedings RILEM Symposium on Steam-cured Lightweight Concrete,* Gothenburg, 1960.
104. *Thermalite Handbook and Data Sheets.* Thermalite Ytong, Warwickshire, Britain.
105. ADAMS, M. and PURVIS, J. F. "Autoclaved Aerated Concrete Building Blocks". *Technical Report TRCS2,* The Concrete Society, 1966.
106. DOVE, A. B. "Nailability of Concrete Blocks". *ACI Journal,* 32, 11 (May 1961), pp. 1509-12.
107. DIAMANT, R. M. E. "Ohlsson and Skarne Method". *The Architect and Building News,* 220, 14 (4 October 1961), pp. 514-16.

108. STIGTER, J. "Structural Lightweight Concrete". *Constructional Review*, 37, 5 (May 1964), pp. 25-30, 42.
109. HOGNESTAD, E., ELSTNER, R. C. and HANSON, J. A. "Shear Strength of Reinforced Structural Lightweight Aggregate Concrete Slabs". *ACI Journal*, 61, 6 (June 1964), pp. 643-56.
110. MACDONALD, A. J. and BARNETT, C. R. "Lightweight Building Construction". *New Zealand Engineering*, 19, 8 (August 1964), pp. 291-302.
111. MAKARICHEV, V. V. "Prestressed Lightweight and Cellular Concrete". *PCI Journal*, 9, 3 (June 1964), pp. 60-65.
112. FENWICK, J. R., GRIFFITHS, R. B. and BRETTLE, H. J. "Average Bond Stress Characteristics of ⅜ in. Diameter High Tensile Strength Steel Strand". *Constructional Review*, 35, 3 (March 1962), pp. 29-32.
113. KOLNER, V. M., KHOLMYANSKYI, M. M. and SEROVA, L. P. "Effect of Steamcuring on Bond Between Wire Reinforcement and Concrete". *RILEM International Conference on Problems of Accelerated Hardening of Concrete in Manufacturing Precast Reinforced Concrete Units*, Moscow, 1964.
114. COLE, W. F. and MOOREHEAD, D. R. "High Strength Calcium Silicate Hydrate: X-Ray, D.T.A., Chemical and Electron Microscope Results". *Society of Chemical Industry, International Symposium on Autoclaved Calcium Silicate Building Products*, London, 1965.
115. PEARSON, A. S. "Lightweight Aggregate from Fly Ash". *Civil Engineering*, 34, 9 (September 1964), pp. 50-53.
116. PFEIFER, D. W. "Fly Ash Aggregate Lightweight Concrete". *R/D Bulletin 1411*, United States Portland Cement Association, 1969.
117. MIRONOV, S. A., BARANOV, A. T., KRIVITSKII, M. Y. and ROZENFELD, L. M. "Aerated Concrete for Manufacture of Large Components". *DSIR, BRS Library Communication No. 1199*, 1964.
118. NOORLANDER, A. "Reducing the Shrinkage of Calcium-Silicate Bricks". *Society of Chemical Industry, International Symposium on Autoclaved Calcium Silicate Building Products*, London, 1965.
119. TAYLOR, H. F. W. "A Review of Autoclaved Calcium Silicates". *Society of Chemical Industry, International Symposium on Autoclaved Calcium Silicate Building Products*, London, 1965.
120. COPELAND, R. E. "Controlling Sound with C/M". *Concrete Products*, 68, 7 (July 1965), pp. 39-43.
121. LEWIS, R. K. and BLAKEY, F. A. "Moisture Conditions Influencing the Tensile Splitting Strength of Lightweight Concrete". *Constructional Review*, 38, 8 (August 1965), pp. 17-25.
122. LANDGREN, R., HANSON, J. A. and PFEIFER, D. W. "An Improved Procedure for Proportioning Mixes of Structural Lightweight Concrete". *PCA R & DL Journal*, 7, 2 (May 1965), pp. 47-65.
123. SHARPE, N. R. and NESBIT, J. K. "A Code for Structural Lightweight-aggregate Concrete". *Concrete and Constructional Engineering*, 60, 11 (November 1965), pp. 415-22.
124. JAIN, O. P. "Proportioning No-fines Concrete". *Indian Concrete Journal*, 40, 5 (May 1966), pp. 182-89.
125. LEDBETTER, W. B. and THOMPSON, J. N. "A Technique for Evaluation of Tensile and Volume Change Characteristics of Structural Lightweight Concrete". *ASTM Proceedings*, 65 (1965), pp. 712-28.
126. PFEIFER, D. W. "Sand Replacement in Structural Lightweight Concrete—Splitting Tensile Strength". *ACI Journal*, 64, 7 (July 1967), pp. 384-92; 65, 1 (January 1968), pp. 64-66.
127. "Lightweight Concrete—Aerated and Pumice Concretes". *Bulletins 86 and 88*, New Zealand Building Research Bureau, 1966.
128. ACI COMMITTEE 523. "Guide for Cellular Concretes Above 50 pcf, and for Aggregate Concretes Above 50 pcf with Compressive Strengths Less Than 2500 psi". *ACI Journal*, 72, 2 (February 1975), pp. 50-66.
129. PRATT, A. W. and LACY, R. E. "Measurement of the Thermal Diffusivities of Some Single-Layer Walls in Buildings". *Research Series 64*, Ministry of Technology, Building Research Current Papers, 1966.
130. NESBIT, J. K. *Structural Lightweight-aggregate Concrete*. Concrete Publications, London, 1967.

131. TEYCHENNE, D. C. "Structural Concrete made with Lightweight Aggregates". *Concrete,* 1, 4 (April 1967), pp. 111-22; 6 (June 1967), pp. 207-12.
132. BUEKETT, J. and JENNINGS, B. M. "Reinforced Autoclaved Aerated Concrete". *Technical Report TRCS3,* The Concrete Society, 1966.
133. ACI COMMITTEE 523. "Guide for Cast-in-Place Low Density Concrete". *ACI Journal,* 64, 9 (September 1967), pp. 529-35; 65, 3 (March 1968), pp. 228-29.
134. STIGTER, J. "Park Regis". *Constructional Review,* 41, 5 (May 1968), pp. 10-15.
135. "Lightweight Concrete Review. Parts 1 and 2: Production of Autoclaved Aerated Concrete, and Low-density Aggregates; Part 3: Some Applications". *Concrete,* (London), 2, 5 (May 1968), pp. 183-98.
136. ACI COMMITTEE 523. "Low Density Precast Concrete Floor, Roof and Wall Units". *ACI Journal,* 65, 7 (July 1968), pp. 507-12; 66, 1 (January 1969), p. 73.
137. SHORT, A. "CEB Recommendations and the Structural Use of Lightweight Concrete". *Concrete* (London), 1, 8 (August 1967), pp. 281-84.
138. Autoclaved Calcium Silicate Building Materials. *RILEM 2nd International Symposium,* Hanover, Germany, 1969.
139. GUNDLACH, V. H., HORSTER, E. and WULFRATH, G. R. "Influence of Pore Space in the Hydrothermal Synthesis of Calcium-hydro-silicates". *Tonind Zig (Clay Industry Journal),* 93, 3 (March 1969), pp. 107-14.
140. IVY, C. B. *et al.* "Shear Capacity of Lightweight Concrete Flat Slabs". *ACI Journal,* 66, 6 (June 1969), pp. 490-94; 12 (December 1969), pp. 1024-26.
141. LYDON, F. D. "The Problem of Water Absorption by Lightweight Aggregates". *Magazine of Concrete Research,* 21, 68 (September 1969), pp. 131-40.
142. HARMATHY, T. Z. and ALLEN, L. W. "Thermal Properties of Selected Masonry Unit Concretes". *ACI Journal,* 70, 2 (February 1973), pp. 132-42.
143. SCHWARZ, B. "Capillary Water Absorption of Building Materials". *BRE, Library Translation 1727,* 1972.
144. ACI COMMITTEE 435. "Deflections of Concrete Structures". *ACI Special Publication SP-43.*
145. WEST, J. M. *et al.* "Glass Fibre Reinforced Autoclaved Calcium Silicate Insulation Material". *BRE Current Paper CP 62/74,* 1974.
146. CORNELIUS, D. F. and RYDER, J. F. "New Fibrous Composites as Alternatives to Timber". *BRE Current Paper CP 66/74,* 1974.
147. SUSSMAN, V. "Lightweight Plastic-Aggregate Concrete". *ACI Journal,* 72, 7 (July 1975), pp. 321-23.
148. SWAMY, R. N. and BANDYOPADHYAY, A. K. "Elastic Properties of Structural Lightweight Concrete". *ICE Proceedings,* 59, 2 (September 1975), pp. 381-94.

PART 8

MODERN PROCEDURES AND PRODUCTS

CHAPTER 34

PRESTRESSED CONCRETE

34.1 DESIGN FACTORS

34.1.1 GENERAL INFORMATION

One of the disadvantages of ordinary reinforced concrete is that concrete is exceptionally weak in tension. The maximum possible elongation of concrete is about 0·01-0·02 per cent, whereas the elongation in the reinforcement when stressed to 124 MPa is about 0·06 per cent. A reinforced concrete member develops cracks on the tension side before the reinforcement is stressed to its design load (see Sections 3.4, 11.1.2(d) and 31.3).[134, 153]

The crack widths are proportional to the steel stress and, in the case of plain rods or early types of deformed bars, the diameter of the reinforcing elements. By using modern types of deformed reinforcing bar made of cold-worked, stress-relieved steel or hot-rolled, tempered alloy steel having a minimum yield strength of 410 MPa, the cracks thus formed can be controlled in width although increased in number (see Table 3.24 and Sections 33.3.2(a) and 34.1.4(b); also Glossary, Appendix VI). Their maximum width (which is about 1·5 times their average width) is influenced by the depth of concrete cover over reinforcement (see Section 22.3.2, AS 1480, Supplement and Commentary MP28, BS CP110 Part 1 (Appendix A), BS 5337 and ACI Committee 224).[11, 94, 103-108, 119, 135, 155, 156]

With a stress in modern deformed bars of 200 MPa, an effective reinforcement ratio of 0·02-0·20 and a concrete cover of 35 mm over well-distributed main steel, a typical average crack width lies within 0·075-0·100 mm. To ensure protection against corrosion, the maxi-

663

mum crack width permitted by the European Concrete Committee*
for deformed bars is 0·100 mm for members exposed to very
aggressive conditions, 0·200 mm for ordinary structural members
that are unprotected, and 0·300 mm for the latter members when
they are protected (see Sections 15.2.2(b) (i) (ii), 15.2.3 (g) and 31.2;
also CEB-FIP and NAASRA, Appendix IV, 2(a) (i)).[105, 139]

In ordinary reinforced concrete, the compressive strength in a
beam is distributed over that portion of the section which is above
the neutral axis. The remaining portion provides some shear
resistance, and a means of encasing and holding the reinforcement
at a particular depth, but otherwise it is used ineffectively for
carrying stresses due to bending. In general, therefore, the stress
and weight characteristics of ordinary reinforced concrete militate
against its use in long-span construction.

These effects are overcome in prestressed concrete (with or without the aid of fibre reinforcement (see Section 3.4.2)), in which
carefully located compressive forces are introduced into medium
to high-strength concrete for increased solidarity and flexural
load-carrying capacity. High-tensile steel tendons, tensioned for
the purpose, may be straight, draped, curved or parabolic in profile.
A permanent state of induced compression is thus established, and
this transforms concrete into an elastic material which will carry
forces that would otherwise cause critical tension or cracking.

M. Eugene Freyssinet conceived this principle in 1904 and
established it as a structural technique by applying it to the manufacture of power-transmission poles in 1928. The principle is
applicable in stages in one, two or three dimensions to slender,
hollow, and unusual forms of wide-span, spatial structures (see
Plates 1 and 30-32). Economies accrue from the use of a reduced
number of supporting members, the efficient use of materials, the
multiple use of easily stripped moulds, and a rapid rate of
construction with precast units containing pretensioned or post-
tensioned tendons. During manufacture or fabrication, the proof
testing of prestressed products and structures is automatic.

34.1.2 PRINCIPLES AND PERFORMANCE

Prestressing not only puts concrete into compression, so that little
or no tension will develop under maximum working load, but it also
enables structural forces to be balanced and guided most efficiently
to supporting columns or foundations. For instance, in a beam, thin

---

*CEB–Comité Européen du Béton, Paris.

shell or cantilever, the dead load and a proportion of the live load can be balanced by upwards components of force from curved or sloped tendons.

In spatial construction, structural efficiency can be enhanced by stiffening ribs along the lines of principal compression and by induced compression along the lines of principal tension. The outwards thrust of a dome can be counterbalanced by an applied hoop-prestress around the rim. The high impermeability and durability of prestressed concrete makes it particularly suitable for use in water-retaining structures (AS 1481 and BS 5337), piles, pontoons and uncovered roofs (see Reference 74 of Part 6).

The span of horizontal members of given weight is more than doubled by prestressing, and thin-webbed sections are possible because of greatly reduced diagonal tension. Webbed box-sections (e.g. 30 m long) may weigh less than 215 kg/m^2, the longitudinal tendons being located in the webs for fire protection and durability. The deflection of a flexural member under a particular set of loads can be eliminated by prestressing in such a way that the compressive stress in a section is distributed uniformly throughout its depth. With three-dimensional prestressing, the deformation and secondary stresses of a curved thin shell can be minimised, if not eliminated.

For a balanced load system in a uniformly loaded prestressed beam,[66] the profile of the tendon could be made parabolic with a sag value (h) derived from the formula

$$F = \frac{wl^2}{8h}$$

where F is the prestressing force, w is the unit load to be carried, and l is the span. The requisite profile for nominal zero deflection with other loads could be determined from a polygon of forces. For a parabolic tendon in a beam of length L

$$\text{Length of strand required} = L + \frac{8h^2}{3L} + \text{jacking length.}$$

The bearing face at the anchorage is sloped at an angle θ from the vertical, where

$$\tan\theta = \frac{4h.}{L}$$

The rise at a distance X from the lowest point of a tendon $= \dfrac{4X^2h.}{L^2}$

Deflections can be estimated from elastic-beam theory, taking into consideration such factors as creep, shrinkage, age, modulus of elasticity (which increases linearly with prestress),[69] ambient atmospheric and loading conditions, and variable bending moments due to a variable eccentricity of the tendons. If prestressed precast beams have a variable hog or upwards deflection, and they are to be placed adjacently, they should be laid to form a surface that has an imperceptible gradient or camber. Normally, the hog should not exceed span ÷ 300, if surface finishes are to be applied.

Prestressed concrete is practically immune from failure by fatigue under repeated loading within a normal working range, largely because the usual variation in steel stress is only of the order of a few per cent. If, through overloading, cracks were formed in a pre-stressed concrete member, they would disappear when the excess load was removed.

Admixtures containing calcium chloride should not be used in concrete or grout which could be brought into contact with prestressing steel. Model analysis and testing work should be done where necessary to facilitate design or check structural adequacy. Load tests are specified in codes on prestressing (see Section 34.1.4), and connection details are set out in Reference 54.

34.1.3 COVER

(a) Normal.
Concrete cover, exclusive of finishing coats, is the clear distance between the face of the concrete and the nearest surface of the tendon, duct or reinforcement (see Section 3.4). It should be sufficient not only for purposes of protection, but also to permit adequate compaction of the concrete. Tendons that are located outside structural concrete should be protected subsequently with dense concrete, the thickness of which is not less than the cover required for tendons inside the structural concrete under similar conditions.

In pretension building elements made under factory conditions, the concrete cover is not less than 20 mm for internal work, 25 mm for external work, 40 mm for work in contact with the ground, and 50 mm where conditions are particularly corrosive (e.g. exposure to saltwater or salt spray). In post-tension building elements, the concrete cover is not less than 25 mm for internal work, 40 mm for external work or work in contact with the ground, and 50 mm where conditions are particularly corrosive (e.g. exposure to saltwater or salt spray). Additional cover may be required where fire resistance must be considered.

In bridges and wharves, the concrete cover is not less than 50 mm to ducts and 40 mm to pretensioned tendons or to reinforcement. It may be reduced by 10 mm in webs and thin slabs, but for structures exposed to saltwater or sea spray, or in contact with the ground, it should be increased by 10 mm.

In liquid-retaining structures, the external concrete cover is not less than 40 mm, although it may be reduced to 25 mm where tendons placed around a tank wall are splash-coated with mortar at least 3 mm thick, and finally protected with a cover coat of pneumatic mortar, applied 25 mm thick when the tank is full (see Section 24.2). These cover requirements are increased by 10 mm and 7 mm, respectively, for adverse conditions of exposure. Each coat must be continuously moist cured from the time of initial set until 7 days after completion of the cover coat.

(b) Fire resistance.

Typical fire-resistance ratings to be incorporated in the design of structural building elements are given in Table 34.1 (AS 1481 and BS 4547).

TABLE 34.1 FIRE-RESISTANCE RATINGS

Classification	Period of Exposure (h)
Assembly buildings (i.e. public halls and theatres, but not drill halls, grandstands, libraries, places of worship, public lounges, schools and toilet rooms)	2
Boarding houses, guest houses	2
Clubs (residential), dormitories, hotels, motels	2
Domestic buildings (i.e. apartments and flats)	
One and two storey	Generally not required
Three storey and over	1·5
Factories, garages, laboratories, laundries, printing plants	Depending on circumstances, either 3 or 4
Office buildings, broadcasting studios, file rooms, television studios	2
Retail shops	3
Storage buildings (i.e. cold storage, vaults and strongrooms, warehouses)	4

To prevent a critical rise in the temperature of tendons when subjected to various periods of fire exposure, the concrete cover should be of sufficient thickness, as indicated in Table 34.2. Where this thickness exceeds 65 mm, some light steel-mesh reinforcement (e.g. 150 mm × 75 mm and 3·15 mm diameter or stronger), with a cover of not more than 25 mm, should be incorporated to retain the concrete in position around the tendons.

TABLE 34.2   CONCRETE COVER IN MILLIMETRES

Fire-resistance Period (h)	Dense-aggregate Concrete		Lightweight-aggregate Concrete*	
	Elements other than slabs	Slabs	Elements other than slabs	Slabs
0·5	25	20	25	20
1	40	25	35	25
1·5	50	35	45	35
2	65	40	55	40
3	80	55	70	50
4	100	65	90	55

* See Sections 33.2.2 and 33.3.2.

In composite construction, mechanical anchorages in the form of ties should be used to ensure an integral action of the component elements during a specified period of fire exposure. The fire resistance of a concrete element can be increased with an additional concrete topping or cover (see AS 1480 and Appendix B), a fire-barrier ceiling, or an applied insulation as indicated in Table 34.3 (see Section 15.2.4; also AS 1481 and BS 4547, CP299).

TABLE 34.3   APPLIED INSULATION IN MILLIMETRES

Dense-aggregate Concrete Cover (mm)	Added Insulating Cover to Increase Period of Fire Resistance to:					
	0·5 h	1 h	1·5 h	2 h	3 h	4 h
Elements other than slabs						
25	0	10	19	29	41	54
40	0	0	10	19	32	44
50	0	0	0	10	22	35
65	0	0	0	0	18	25
85	0	0	0	0	0	18
100	0	0	0	0	0	0
Slabs						
20	0	6	13	19	29	38
25	0	0	6	13	22	32
30	0	0	0	6	16	25
40	0	0	0	0	10	19
50	0	0	0	0	0	10
65	0	0	0	0	0	0

Alternative forms of insulation are as follows: mould-lining slabs of 1 part cement : 4 parts vermiculite or perlite concrete; either mould-lining slabs or a sprayed or trowelled cover of lightweight gypsum plaster, the mix being gauged with 50 kg of gypsum : 0·08 m³ of vermiculite or perlite and sprayed asbestos cover with a flow point not below 1120°C.

Lightweight-aggregate insulating concrete or plaster over 20 mm thick is lightly reinforced. The reinforcement used in mould-lining slabs is not less than 25 mm wire mesh of 1·0 mm diameter; and reinforcement used as a lath for plastering consists of expanded metal weighing at least 1·3 kg/m². The reinforcement is located not more than 20 mm from the outside surface of the added finish and is securely anchored without the use of explosion-driven fasteners, so as to prevent release at a temperature below 540°C (see Section 15.2.4).[20, 43]

34.1.4 PERMISSIBLE STRESSES

For comparative purposes between Australian and British Standard Codes, both cylinder and cube strengths (as specified) are quoted in Section 34.1.10, their values being converted where necessary from the data given in Section 4.4. Design methods for load-carrying, handling, transfer and constructional purposes are based generally on an accepted theory of elasticity (see Section 34.1.12) with permissible working stresses, such as those given in this Section; and a plastic theory whereby an evaluated design load (formulated with specified load factors) will be resisted by the ultimate strength of a member. The minimum loadings to be used in design are given in AS 1170 and BS CP3(v); and the appropriate statutes of building, road and railway authorities.

The bulk density of prestressed concrete generally may be taken as 1750-1950 kg/m³ when it is made with lightweight aggregate and 2500-2600 kg/m³ when made with dense aggregate, due allowance being made for the weight of steel components. The value for ordinary reinforced concrete, by comparison, is approximately 2400 kg/m³.

Rules for permissible stresses and other factors in the design of prestressed concrete are contained in several specification codes, such as AS 1481, BS CP110 Parts 1 and 3, CP115, CP116, CP2007; and ACI Committees 318, 344, 423; also FIP International and German Industrial DIN4227.[8, 9, 29, 41, 139, 167] The design of high-strength concrete mixes for prestressing, with or without steam curing, is dealt with in Sections 4.3.4 and 9.4. In this chapter, the *cylinder strengths* of concrete rather than its cube strengths are referred to, in order to accord with the rules given in AS 1481. An international code is given in Reference 139 of Part 8.

These codes contain working rules for the design of beam and slab sections, composite sections, shear connections, continuous and torsional systems, compression members, piles, liquid-storage tanks, and both statically indeterminate and special structures. In addition to specifying permissible stresses and load factors against failure, these codes contain formulae for the evaluation of design load and the design of members under ultimate-strength conditions. A load-balancing method for design purposes with curved or sloping tendons is illustrated in References 1-9, 66, 86, 87 and 122.

(a) Concrete.

(*i*) *Working stresses.* For design purposes, the permissible working compressive stress varies from $0 \cdot 3 \ F'_c$ plus one-half the working prestress (the total not exceeding $0 \cdot 4 \ F'_c$) for direct compression to $0 \cdot 4 \ F'_c$ for bending in bridges and $0 \cdot 45 \ F'_c$ for bending in buildings, where F'_c is the minimum compressive strength of the concrete at 28 days. For all uses, except in railway bridges, the last two values may be increased by $0 \cdot 05 \ F'_c$ for the support moments in continuous beams. For buildings, the permissible compressive stress may be increased by one-third, where the respective increase is due solely to stresses induced by wind or earthquake load.

The permissible initial compressive stress (for approximately uniform and triangular distributions of prestress) is $0 \cdot 5$ and $0 \cdot 6$ times, respectively, the minimum compressive strength at transfer. Where the initial stress approaches 20 MPa, the allowances for loss of prestress must be carefully considered and working stresses restricted to proper values accordingly. The initial prestress for general-purpose piles is approximately 7 MPa.

The permissible tensile stress in concrete due to bending is zero for bridges at maximum design load; whereas, for segmental members with unreinforced joints, a residual compressive stress is required to the extent of $0 \cdot 025 \ F'_c$ (see Section 34.2.6). A temporary overload of 40-50 per cent should not cause tensile stresses in excess of $7 \cdot 5 \ \sqrt{F'_c}$ for pretensioning or $6 \ \sqrt{F'_c}$ for post-tensioning; but, for post-tensioned segmental members with unreinforced joints, the latter limit is $2 \ \sqrt{F'_c}$.

The permissible tensile stress for buildings is $4 \ \sqrt{F_c}$ for pretensioning and $3 \ \sqrt{F'_c}$ for post-tensioning. An increase of 50 per cent is allowable for wind or earthquake loads. The bending tensile stress for buildings may be increased by up to 1·7 MPa, provided that it does not exceed 75 per cent of the tensile stress which would cause the first crack in a prototype performance test, the maximum concrete prestress is at least 10 MPa and steel is present in the tensile zone of the members.

The tensile stresses at transfer may vary from three to six times the square root of the minimum compressive strength of concrete at transfer. Nonprestressed reinforcement is used with the latter coefficient in design, or when the former coefficient is applicable to a segmental element, in order to ensure an uncracked section. Other aspects of design for shear and torsional strengths, principal tensile stresses, and bearing stresses at the supports and post-tensioned anchorages, should be dealt with as indicated in the codes cited here, and in References 109, 111, 116 and 120, also in other sources of information given at the end of this Part.[1-10, 19-34, 41-43, 46, 54, 66, 70-73, 78, 139-141, 148, 151, 157-163]

For lightweight-aggregate concrete, the coefficients for tensile stress or strength should be reduced in proportion to the ratio of the tensile strength of this concrete to that of dense-aggregate concrete having the same compressive strength F'_c. This ratio of tensile strengths will usually lie within a range of 0·70-0·85 (AS 1481). Handling and construction stresses should not exceed those permissible at transfer, due allowance being made for losses in prestress and an increase in strength up to the time of operation concerned.

In liquid-retaining structures, the walls of cylindrical tanks should be designed to resist bending moments and hoop stresses under all conditions of construction and loading. In this connection, the principal compressive stress in the concrete after all losses of prestress should not exceed 0·4 F'_c, and the residual compressive stress when a tank is full should not be less than 0·7 MPa horizontally and 0·35 MPa vertically. The tensile stress in the concrete when a tank is empty should not exceed 0·7 MPa although this limit may be increased to 1·4 MPa in reinforced concrete, the load factor against failure being not less than 1·8.[41, 72]

(*ii*) *Control tests.* Methods of sampling concrete for acceptance tests are described in Section 7.1.5. The Australian Code for Prestressed Concrete (AS 1481) requires that the average compressive strength of three representative samples of each section of work done generally should not be less than 1·33 times the specified minimum compressive strength at 28 days (F'_c) (see

Section 14.4 and Appendix I). In continuous work, where the amount of regular-quality concrete placed at any one time exceeds 20 m³ at time intervals not exceeding 2 weeks, the average value of six samples taken during two consecutive days of concreting should not be less than 1·20 F'_c, the lowest single value being 0·75 F'_c.

After thirty-six specimens have been tested, the number of representative samples may be reduced to three per section, provided that the average compressive strength of all samples taken from previous sections is at least 1·33 F'_c with no single test result below 0·75 F'_c; furthermore, the average value obtained in each successive section is at least 1·15 F'_c, with no single test result below 0·85 F'_c. In specially controlled work, the average value of three representative samples per section of work should be at least 1·10 F'_c with no single test result falling below 0·85 F'_c.

For the determination of the compressive strength of concrete at transfer, at least three specimens should be stored and cured under the same conditions as the concrete during production. Under steam-curing conditions, an adequate number of specimens are required to evaluate strength characteristics throughout the concrete. Non-destructive tests, core tests, load tests of prototypes at an age of 28 days, and structural-performance tests may be used for control and acceptance purposes. In flexural tests, the development of the first crack can be detected by a surface application of a volatile liquid (e.g. methylated spirits).

(b) Prestressing steel.

The tensile-stress-strain relationships for high-tensile steel and other types of steel are shown in Figure 34.1.

As the stress-strain relationship of high-tensile steel shows no definite yield point like that of mild steel and modern high-strength steel,[94, 104, 114] the concept of proof stress is used to indicate curvature in a graph of this relationship (Glossary, Appendix VI). Under conditions of load, an increase of stress in high-tensile steel causes the strain to follow a straight-line relationship, as shown by the line OA, the slope of which represents the modulus of elasticity of the steel. Subsequently the relationship becomes nonlinear, as shown at the point B (AS 1391 and BS 18).

If the stress were then reduced progressively, the stress-strain relationship would return linearly to zero, as shown by the line BC parallel to OA. A small residual strain OC thus remains permanently. If the strain represented by OC were 0·2 per cent on a 250-mm gauge length, the stress which produced it would be known as the 0·2 per cent proof stress, which is about 80-90 per cent of the ultimate tensile strength.

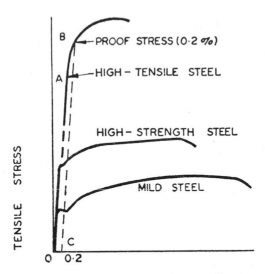

Fig. 34.1 Tensile stress-strain relationship

For jacking purposes, the tensioning stress should be kept within the 0·2 per cent proof stress. It should not exceed the following limits, where F'_s is the tensile strength of the prestressing steel.

0·80 F'_s for pretensioned tendons.
0·85 F'_s for stress-relieved post-tensioned tendons.
0·77 F'_s for post-tensioned tendons not stress-relieved.

The initial stress in prestressing tendons, immediately after transfer, should not exceed the permissible limits indicated here.

1. Stress-relieved, normal-relaxation or low-relaxation wire, and strand tendons and cold-worked, alloy-steel bars with rolled threads or wedge grips: 0·75 F'_s

2. Strand tendons not stress-relieved and hard-drawn wire tendons: 0·70 F'_s

For ground-anchorage tendons, the permissible design stress is limited to 0·60 F'_s.

Tables 34.7-34.9 list some physical characteristics of prestressing steel. The tendons should be fabricated and handled in an appropriate manner (see Section 34.1.10),[81] and the order of prestressing should ensure that the permissible stresses at any stage are not exceeded. Differences in force along any deflected tendon within a member should not exceed 5 per cent.

Specified requirements for hard-drawn, high-tensile and stress-relieved steel are given in AS 1302-1304, 1310-1313 and BS 2691, 3617, 4486, 4757; also ASTM A416, A421 and A431, hard-drawn spring wire being specified in ASTM A227. The ultimate strength of large, alloy steel bars (see Section 34.2.3) is nominally 1000-1100 MPa, and the permissible stress in ordinary load-distributing steel at anchorages is 140 MPa.

Prestressing steel should have a fully pearlitic structure (free of microsurface defects) by pretempering at above 500°C and it should not be heated for bending or by die drawing in order to preserve its properties. Care is needed to keep corrosion promoters, moist air, water, hydrogen sulphide, free carbon dioxide and the stray leakage electric currents of rail and tramway systems away from highly stressed tendons, and to keep the contact pH at 12-14 to avoid stress corrosion or hydrogen embrittlement. Surface flaws which weaken the metal may be removed by electrolytic polishing (see Sections 15:2.2(a), (c), 15.2.3(b), (d) and 19.2; also References 17-19 and 117 of Part 4).

In the pretensioning method, the bond length necessary for a gradual release and imposition of prestress varies from 50 diameters for units containing stress-relieved strand, to 100 diameters for units containing plain or indented wire, or wire with a small offset crimp (e.g. 0·300 mm at 40 mm pitch). Where the wire has a considerable crimp (e.g. 1·0 mm offset at 40 mm pitch), the bond length necessary would be about 70 diameters. These lengths are conditional upon the ends of the units being fully compacted and the strength of the concrete at transfer being not less than 30 MPa. With a lower strength at transfer (e.g. 25 MPa for factory-made precast elements), longer lengths are likely to be required. In each instance, some 80 per cent of the maximum stress is developed in 70 per cent of the bond length measured from the ends (see Sections 10.1 and 15.2.2(a)).

(c) Modulus of elasticity.

The elastic modulus of concrete, E_c (MPa), is calculable from established formulae (AS 1480), 1481 and ACI Committee 318; see Appendix IV, 2(a)(i)), such as

$$E_c = 0·043 \, D^{1.5} \sqrt{F'_c}$$

where D = bulk density, ranging from 1440 to 2600 kg/m³ (see Section 34.1.4), and F'_c = specified characteristic cylinder strength at 28 days or a nominated age.

Comparable data are given in Sections 14.9.4 and 31.3.2, also Reference 148 of Part 7 for lightweight concrete; and in BS CP110, CP114-CP116 for typical cube strengths (see Section 4.4), such as

E_c (10^3 MPa)	20	27	32	34	41	46
F'_c (MPa)		20	27	35	40	55	70

The elastic modulus of steel in tendons is

172×10^3 MPa for alloy steel bars of cold-worked, high-tensile type.

186×10^3 MPa for nineteen-wire steel strand as stranded.

193×10^3 MPa for seven-wire stress-relieved steel strand of regular and super-grade types; nineteen-wire steel strand of stress-relieved type; and steel wire of hard-drawn type.

200×10^3 MPa for steel wire of stress-relieved type.

The elastic modulus of nonprestressing steel is 200×10^3 MPa (see ACI Committee 318 and Table 3.25).

34.1.5 LOSS OF PRESTRESS

Because of immediate and deferred losses of prestress, the stresses induced initially are greater than the permissible working stresses. The cumulative losses, which are estimated by methods laid down in international and regional standard specifications, are adjusted where necessary to cater for new materials or systems and unusual conditions of exposure (e.g. temperatures above 40°C).

Statutory sources of information, data and procedures for evaluating effective stress include the following documents for reference: AS 1481; BS CP110 Parts 1 and 3, CP115, CP116 and 5337; ACI 318 *Building Code and Commentary,*[8] ACI-ASCE 423;[9, 29] ACI 435;[46] *ACI Manual of Concrete Practice,* Part 2; listings at the end of this Part; and in Appendix IV, 2(a) (i).[158, 160, 161]

Immediate losses of prestress are due to the following causes.

1. Elastic deformation of the concrete at transfer of load, the extent being governed by the values of elastic moduli given in Section 34.1.4(c). The loss is the product of the stress in the adjacent concrete and the modular ratio E_s/E_c for pretensioned tendons, and half the resulting product for members with post-tensioned tendons. These are assumed to be located usually at their centroid.

2. Friction in the duct, jack, anchorage and saddle bearers of external-type cables, which lead to stress variation along post-tensioned elements.

In post-tensioned members, friction loss varies with the type, length and curvature of the duct or sheath, and the method of construction. Cast-iron spacers are used with monostrand stressing of multistrand cables in order to separate the strands, prevent intertwining and minimise friction loss. Design-estimation values, derived from prestressing codes, company data texts and systems service catalogues (q.v.) are precautionarily test-verified during stressing operations.[1, 110]

3. Movement of the steel when a post-tensioning force is transferred from the tensioning equipment to the anchorage. Appropriate values for this and other associated forms of deformation are determined experimentally and checked on the site.

Deferred losses are due to the following causes.

1. Shrinkage of the concrete, the loss being the product of the modulus of elasticity for the steel and the anticipated shrinkage per unit length (see Section 34.1.6(a)).

The normal shrinkage and creep of concrete are reduced by high-temperature steam curing (see Sections 19.4.1, 31.2.1 and 31.3.3) and by using high-strength high-modulus concrete, such as that made with gap-graded coarse aggregate and a compatible gradation of coarse and fine aggregates (see Sections 4.5 and 23.1.5). The effect of high-temperature steam curing on deformation tends to lower the loss of prestress in precast units (see Reference 156 of Part 3, and References 60, 63 and 65 of Part 6). On the other hand, when prestressing is carried out by the long-line method of pretensioning, the tendency of such factors as high-temperature steam curing, partially exposed tendons, stress relaxation and an early immature development of bond strength is to increase the loss of prestress (see Section 33.3.2(a) and Reference 64 of Part 6).[24, 29, 78, 148, 149, 157, 158, 160, 161]

2. Creep of the concrete, the loss being the product of the modulus of elasticity for the steel and the anticipated creep per unit length (see Sections 31.3.3 and 34.1.6(b)).[24, 54, 78, 140, 141, 148, 149, 157, 158, 160, 161]

3. Steel relaxation, the resultant loss of prestress being of pure and modified varieties. Pure relaxation loss is determined

experimentally for the make, type and size of the prestressing tendon to be used in a project. It increases rapidly with temperature above the standard relaxation-test temperature of 20°C. Where the temperature of a structure during the first month after stressing is likely to exceed 40°C for continuous periods, the relaxation loss is increased by at least 25 per cent (see Sections 34.1.4 (b), (c) and 34.2.4).[129, 130]

With certain exceptions for design purposes, eventual losses of prestress may be taken as twice the percentage relaxation after 1000 hours, obtainable either from test data for the appropriate level of initial stress or from Table 34.8. Relaxation loss at reduced initial stress is assumed to vary linearly from 0, at 50 per cent, to twice the appropriate 1000-hour relaxation at 70 per cent of the minimum tensile strength. Otherwise, relaxation varies logarithmically with time.[158]

When curing is carried out by low-pressure steam curing, as typified in Section 9.4.3 and AS 1481, the losses determined for the time of transfer, and subsequently, are increased by 0·3 times the relaxation at 1000 hours × the ratio of the logarithm of the time of transfer in hours to the logarithm of 1000 hours. Furthermore, because of shrinkage and creep effects, the predetermined pure-relaxation loss in a tendon is reduced proportionately in accordance with the ratio of stress loss, caused by shrinkage and creep, to the initial stress in the tendon after transfer.

4. Miscellaneous deformations. These are connected with the effects of joints in segmental construction, frequently repeated loads, stress redistribution due to reinforcing-steel restraint, aggregate characteristics, time increments or differences in stressing stages, and the theoretical thickness of sections. In this category, deferred losses of prestress are estimated from test results.

Estimated cumulative movements, due to the foregoing inherent and ambient causes of deformation, are vital criteria for suitable bearing details in design.[54, 140, 141] The total loss of prestress in building practice is in the region of approximately 20 per cent for post-tensioning and 25 per cent for pretensioning methods of fabrication. Further increments of up to 5 or 10 per cent are feasible for dense-aggregate and lightweight-aggregate structural concretes, respectively (see Section 31.3 and Reference 36 of Part 7).

34.1.6 SHRINKAGE AND CREEP OF CONCRETE

The following subsections (a) and (b) are extensions of clauses 1 and 2 of deferred losses in Section 34.1.5.

(a) Shrinkage of concrete.
With empirical data published in AS 1481 or other literature,[139, 161] the shrinkage of high-strength concrete at any time may be predicted from the product of three determinable coefficients. These are related to environmental characteristics (k_1), the theoretical thickness of the member (k_2), and the development of shrinkage and creep with time (k_3).

For a tolerance no better than $\pm$ 40 per cent, using stable concrete aggregate, k_1 values may be chosen to suit the following conditions.

Environment:	k_1, Shrinkage (per cent)
Dry air and air-conditioned buildings	0·06
Open air generally, free from periods of prolonged high temperature or low humidity	0·04
Very humid air (e.g. over water)	0·02
Water immersion	0

For greater precision, k_1 values may be predicted from laboratory shrinkage measurements l_m after 8 weeks of drying (see Section 31.2.1 and Appendix I: e.g. AS 1012 Part 13 and 1481), as follows.

Environment:	k_1, Shrinkage
Dry air	$0·9\ l_m$
Open air generally	$0·6\ l_m$
Very humid air	$0·3\ l_m$

In predicting field shrinkage from measured laboratory shrinkage, a multiplying ratio of 100 to $[100 + 20\ p]$ may be used for estimating the effect of restraint due to p per cent of reinforcement. For estimation purposes, the theoretical thickness of members is frequently taken as the quotient of the area divided by the semi-perimeter in contact with the atmosphere. In connection with resultant values of 100, 200, 300, 400 and 500 mm, the k_2 coefficient mentioned previously is taken as 1·05, 0·80, 0·65, 0·55 and 0·50, respectively. The effect of shrinkage and creep with time, which is represented by k_3, may be rationalised approximately as shown in Table 34.4.

TABLE 34.4   **TIME EFFECT (k_3)**

Thickness (mm)	Time (days)	Coefficient (k_3)	Thickness (mm)	Time (days)	Coefficient (k_3)
100	3	0·10	400	20	0·05
	10	0·20		10^2	0·25
	10^2	0·60		10^3	0·75
	10^3	0·95		10^4	1·00
200	10	0·10	800	10^2	0·10
	10^2	0·40		10^3	0·50
	10^3	0·90		10^4	1·00

By comparison with the foregoing, BS CP115 has indicated that the ultimate shrinkage, with pretensioning, may be estimated normally as 0·03-0·04 per cent. With post-tensioning at between 2 and 3 weeks after concreting, the subsequent shrinkage may be estimated as 0·02 per cent, a higher value being used for an earlier age. For approximate purposes, one-half of the total shrinkage may be assumed to occur in the first month and three-quarters in the first 6 months after casting.[29, 78, 148, 149, 158, 160]

(b) Creep of concrete.
Subject to certain premises mentioned in subsection (a), the creep at any time may be predicted from the product of four coefficients, this being multiplied by the initial compressive stress in the concrete at the tendon at transfer (when the appropriate dead load is acting) and divided by the instantaneous modulus of elasticity of the concrete at 28 days (see Sections 34.1.4(c) and 34.1.12).

The coefficients are related to the environmental characteristics (k_1), the theoretical thickness of the member (k_2), the development of shrinkage and creep with time (k_3), and the degree of hardening of the concrete at the time of loading (k_4). For high-strength concrete, the value of k_1 varies as follows.

Environment:	k_1, Creep
Dry air and air-conditioned buildings	1·4
Open air generally (as in subsection (a))	1·1
Very humid air (e.g. over water)	0·8
Water immersion	0·5

The values of k_2 and k_3 indicated in Section 34.1.6(a) are supplemented with those of k_4 shown in Table 34.5, for the age of high-early-strength and ordinary portland cement concretes at the time of loading.

TABLE 34.5   TIME EFFECT (k_4)

Age at Loading (days)	Portland Cements	
	High-early-strength	Ordinary
	Coefficient k_4	Coefficient k_4
1	1·7	1·8
3	1·4	1·6
7	1·1	1·4
28	0·7	1·0
90	0·5	0·75
360	0·3	0·5

For prestress loss assessment, the creep may be assumed to be due to a stress that is the mean of the initial and ultimate stress values. The creep at temperatures between 20°C and 80°C is empirically equal to its predetermined value at 20°C times $[1 + k_5 (t - 20)]$, where t is the temperature in °C and k_5 is a coefficient dependent, as shown in Table 34.5, on the age of the concrete at the time of loading.

TABLE 34.6   TIME EFFECT (k_5)

Age at Loading	Coefficient k_5
7 days	0·015 per °C
28 days	0·030 per °C
6 months	0·045 per °C

Creep recovery may be assumed as being unaffected by temperature.

34.1.7 SPACING OF TENDONS AND DUCTS

Pretensioned tendons should have a clear spacing, horizontally and vertically at the ends of a member, of three times their diameter or 1·33 times the largest size of aggregate, whichever is greater. Ducts for post-tensioning should have a clear spacing, likewise, of 40 mm or 1·5 times the largest size of aggregate, whichever is greater. Deflected tendons or ducts may be bundled together in the middle third of a span, provided that they have a high rate of divergence towards the ends (see Section 34.1.2 and Fig. 34.3).

34.1.8 REINFORCEMENT IN MEMBERS

Reinforcement is required to resist the bursting and spalling forces induced by concentrated loads at the anchorages. It may take the form of a ligature grid, normal to the line of action of the tendons, or of spiral or fibre reinforcement (see Section 3.4.2).[19, 28, 117]

Web reinforcement is required in all beams in buildings, where the depth of the beam exceeds 600 mm and the depth of the web is more than four times the thickness. It is required in all bridge and crane girders and, wherever used, it should be spaced no further apart than the clear depth of the web.

The total cross-sectional area of mild-steel web reinforcement should be at least 0·15 per cent of the web area in plan. For railway bridges, the amount should be at least 0·25 per cent. Where high-tensile steel is used, the amount required may be proportionately reduced on a strength basis. Dynamically loaded beams should be provided with closed stirrups and longitudinal reinforcement of mild steel (see Section 3.4.1(a)).

Compression members containing longitudinal reinforcement are provided with lateral or helical reinforcement for requisite restraint against the buckling of each reinforcing rod. Those required to resist dynamic loading should contain closed stirrups or helical reinforcement as specified in the aforementioned prestressing codes. Torsional reinforcement, where required, should be in the form of closed hoops and longitudinal rods or, alternatively, of continuous helices (see Appendix IV, 2(a)).[111, 116, 120]

34.1.9 SLENDER BEAMS

Care should be taken in the handling of prestressed concrete members to ensure that they will not be subjected to undesigned stresses or distortion. Slender beams should be braced laterally during hoisting and erection, and until their permanent lateral support becomes effective. This precaution is important where the span/breadth ratio exceeds 60 or the depth/breadth ratio is about 4. If slender beams are to be inadequately supported or stiffened laterally under service conditions, the working stresses should be reduced in accordance with standard code requirements for design.

34.1.10 PROCEDURES AND PRODUCTS

Prestressing is carried out in the following ways.

1. Pretensioning of tendons in individual moulds, or on long beds or special tables.[74, 75]

2. Post-tensioning of a variety of tendons which are held by anchorages and fittings according to the following systems: BBRV, CCL, Dywidag,[85] Freyssinet-PSC, Macalloy, Preload and

VSL.[45, 71, 82, 157] Assemblies are illustrated in Reference 45 of Part 3.

3. Composite pretensioning and post-tensioning.

4. Chemical prestressing with expansive cement (see Section 3.1.2(a)).[88]

5. Electrothermic prestressing.[76]

6. Vacuum post-tensioning.[164]

In the *pretensioning method,* tendons of high-tensile wire (e.g. 5-8 mm diameter, see Table 34.7) or seven-wire, stress-relieved strand (7·9-18·0 mm diameter, see Table 34.8) are tensioned between end-grips or anchorages before the concrete is cast (AS 1314). The tension is applied usually by a single-strand hydraulic jack, the magnitude of the force being determined to within 1 per cent accuracy by a dynamometer, the elongation of the tendons, a tension meter or from gauges on the jack (see Plate 28). After the concrete has hardened, the tension is released and prestress is transferred by bond from the tendons to the concrete. Bond is increased at the ends by the elastic expansion of the steel there under reduced stress.[44, 80]

The method is very suitable for mass production with good factory control. When applied with individual moulds, the moulds should be sufficiently rigid to sustain the reaction to the prestressing load without distortion. They should avoid subsequent restraint to the elastic shortening of the concrete. Prestressing beds are normally worked on a 2-day cycle to permit the daily use of moulds. They consist of a concrete bed with massive abutments and movable steel bulkheads. The beds provide usually for a total prestress of 7000 kN at 600 mm above slab level for two-lane production or a single force of 6000 kN at 750 mm. Two gantry cranes are installed on long beds that contain inserts for fastening purposes.[137]

In long-span members, the prestressing force is made more eccentric at midspan than at the ends. Favourable stress conditions with straight tendons are obtained by using such techniques as hog-backed beams, deflected or draped tendons, or bond breaks for a desired distance from the ends. Tendon deflection by temporary saddles or locator yokes is most advantageous for spans of over 12 m. Bond prevention by a masking tube is an efficient means for varying the prestressing moment, because it enables the prestressing force to be varied in both magnitude and eccentricity in one operation.

The Australian Code for prestressed concrete (AS 1481) requires that the concrete should have a minimum characteristic (cylinder) compressive strength of 30 MPa at 28 days, the strength at transfer being above 25 MPa. Analogous requirements in the British Code (BS CP115) with standard cubes are 40 MPa and 27-25 MPa, respectively. The latter lower minimum strength is permitted for precast members, made under controlled factory conditions, where there is evidence of adequate anchorage of the tendons by bond (see Sections 4.3, 4.4 and Chapter 13; also Appendix II). Code requirements (see Section 34.1.4) should be observed for permissible variations in strength, and tensioning jacks should prevent strand unwinding under load. The ends of prestressed concrete units should be fully compacted and finally sealed, and the tendons released gradually for an efficient transmission of prestress. Limited prestressing may be used with lower strengths (BS CP110 Part 1).

In countries where labour costs are high, prestressed standardised units are used competitively with post-tensioned flat plates of *in situ* concrete, where structurally appropriate in building construction. *In situ* concrete is used in conjunction with preformed units in composite construction. Convenient units for wide-span buildings are of single-tee or double-tee shape and inverted channel sections (see Plate 30).

For this purpose, single-tee beams with straight or deflected tendons are often used as flooring units for up to 23 m or 30 m span, respectively, and as roofing units for up to 36 m span. They vary in width from 1200 to 2400 mm and in depth from 450 to 1050 mm, depending on loading and span requirements. Double-tee beams, 1200 or 1500 mm wide and 300-450 mm deep, are used for spans of up to 15 m, the flange edges of both these units being as thin as 40 mm. Inverted channel units (450-600 mm wide) and hollow-core units (450-1200 mm wide), each 100-200 mm deep, are used for spans of up to 7·6 m. Hollow-core slabs up to 2400 mm wide and 200-300 mm deep may be made of lightweight-aggregate concrete and used in spanning up to 30 m.

In the *post-tensioning method* (see Section 34.2.3) the concrete is usually cast with sheathed ducts and allowed to harden. The member may be cast in one unit or it may be made as an assembly of segmental precast units (see Section 34.2.6). It is then prestressed by tendons that are located usually within the ducts.[157]

Alternatively, in unbonded design, the tendons may consist of greased, plastic-coated, seven-wire strands. These are placed in position prior to casting the concrete, thereby obviating the need of preformed ducts and subsequent grouting in sections that are cast *in situ* for prestressing purposes. The post-tensioning is done by one or two hydraulic jacks, bearing against one or both ends, and the tendons are anchored by wedge-grips, cones or nuts which bear against thrust grids, spiral-ribbed sleeves and plates (see Plate 28).[84]

A variety of tendons are available for particular applications (see Tables 34.7-34.9). They include single high-tensile wire (e.g. 7 mm diameter), seven-wire, stress-relieved strand (7·9-18·0 mm diameter) in regular and super grades, nineteen-wire strand (18·0-31·8 mm diameter), and alloy steel bar (23-40 mm diameter) in regular and super grades that are provided with screwed couplers. Large tendons are made of multiwire or multistrand cables which may be diverged at the ends, simultaneously stressed and anchored in end blocks (see AS 1314, BS 4447 and CP110 Part 1).

Multiwire and multistrand cables are used, either singly or in groups, where large forces must be concentrated into small areas with a minimum loss due to friction. The relaxation of stress in these tendons by a standard test procedure (given in specifications for prestressing steel, Section 34.1.4(b)) is normally within 7 per cent and 12 per cent at, respectively, 0·7 and 0·8 times the minimum tensile strength after 1000 hours. The corresponding limits for low relaxation strand are 2·5 and 3·5, while the 0·7 strength limit for alloy steel bar is 4 per cent. Stressing jacks and site gauges are calibrated or checked for accuracy against a master gauge, and strands are severed preferably by a disc cutter or side-angle grinder than by an oxytorch.

Large cables are used as external tendons, with a polygonal profile, in long-span bridges. The tendons are deflected by saddles or dowels at transverse stiffeners, and they are encased finally with fine concrete that is bonded by stirrup reinforcement to the main concrete. In an area 150 mm square, for instance, a cable consisting of sixteen 28·6 mm strands can provide a final force of 7000 kN.

For post-tensioning work, the Australian Code for Prestressed Concrete (AS 1481) requires that the concrete should have a minimum characteristic (cylinder) compressive strength of 25 MPa at 28 days, the strength at transfer being not less than 22 MPa. Analogous requirements in the British Code (BS CP115) with standard cubes are 30 and 27 MPa, respectively (see Sections 4.3, 4.4, Chapter 13, Section 34.1.4 and Appendix II). End fixtures must be firmly embedded in fully compacted concrete and all steelwork must be protected against corrosion (AS 1314 and BS 4447, CP110 Part 1).

The tendon ducts are flushed with water containing hydrated lime (e.g. 12 g/litre) and subsequently blown out with oil-free compressed air. They are provided with tendon clearances of at least 3 mm and have vents at the crests of undulating profiles. After post-tensioning the tendons (preferably within a period of 2-3 weeks), the ducts are filled under vacuum[164] or injected with colloidal neat-cement grout (water/cement ratio, 0·45-0·50 by weight) under pressure (400-800 kPa). The grout is screened through a 1·18-mm sieve and introduced at an end or the lowest points of curved profiles through airtight, threaded connectors. Grout vents at summits, ends, and up to 50 m apart are subsequently locked-off.

The grout must be free from corrosion promoters (see Section 15.2.2(a)), water-gain segregation, and the possible use of excess aluminium powder. Evolvement of excess hydrogen gas from grout in a fresh cementitious mix is conducive to hydrogen embrittlement or a reduction of fatigue resistance of steel tendons (see Glossary, Appendix VI). A 1 part cement : 1·5 parts sand grout may be used in large ducts, the minimum (cylinder) compressive strength being 15 MPa at 7 days (see Sections 15.2.2(d)(vi) and 24.3.2).[81, 139]

The post-tensioning method is used, where economically suited, for the construction of long-span or large assembled projects. It is ideal for full, partial or secondary prestressing, where the upwards components of force in parabolic tendons can be used for the balancing of flexural loads and the control of deflection. Other structural advantages include the possibility of restressing and the use of external cables with cast *in situ* or precast concrete. These cables are characterised by a high concentration of force, low friction loss and loss of prestress, reduced thickness of concrete, and a flexible profile to suit stress and camber conditions.

Post-tensioning is applicable not only to simple beams and girders, but also to rigid frames, grid systems, trusses, continuous beams, cantilevers, water tanks, one-way and two-way slab systems, bond beams, thin shells, and ground anchors (AS 1481). In lift-slab building, the post-tensioning of internal tendons in two directions at ground level produces deflection-free slabs, with a minimum of structural depth and flexural creep. Post-tensioning equipment, with its low capital investment, is also used for prestressing concrete pavements.[10]

In prepost-tensioning (*combined method*) part of the prestress is applied by pretensioned straight or deflected tendons and the remainder by post-tensioned parabolic tendons. In a precast beam or slab, the primary prestressing moment resists the bending moment due to dead and erection loads, and possibly up to one-half of the anticipated live load. The secondary prestressing moment resists the residual moment due to live load. Because of the angle of incidence of nonlinear tendons, particularly those of parabolic shape, a large proportion of the end shear is resisted by the upwards components of prestressing force.

In practice, the pretensioning bed is released for early reuse and post-tensioning follows either at the factory or the site. The combined method is thus highly efficient, economical and effective. In composite construction with a roughened or tied interface, it provides a simple means for balancing dead load at various stages, maintaining uniform stress distribution in sections, and eliminating deflection or camber at all times.

34.1.11 FATIGUE STRENGTH

Fatigue tests on prestressed concrete beams conducted by the DSIR (Britain) have shown that several million repetitions of loading, within the normal working range, have no significant effect on the subsequent performance of beams. When pretensioned wires fail by fatigue, the ratio of the maximum load causing failure (after one million repetitions) to that causing failure under static conditions is as follows.

0·70-0·85 for plain wires,
0·65-0·75 for crimped or indented wires.

In beams with post-tensioned and grouted cables of plain wire, the ratio is about 0·8 for fatigue failure of either the steel or the concrete. Since the working load of beams is usually not greater than one-half of the ultimate static strength, the chance of failure by fatigue is small. Fatigue is dealt with in References 32, 33 and 67. Under static loading, there is little difference in behaviour between strand of 12·5 mm diameter and plain 5 mm diameter wire. Under repeated loading, the fatigue resistance of the former is adequate and comparable with that of deformed wire.

34.1.12 DESIGN THEORY

Consider a simple beam of area A with a moment of inertia I (Figure 34.2). By elastic theory, a bending moment M (that causes a stress f at a distance y^1 from the neutral axis) is equal to the moment of resistance, which is represented by the expression

$$M = \frac{fI}{y^1} = fZ$$

where Z = the section modulus at the point of stress concerned.

If the total prestressing force in the tendons is P, acting at a distance e below the neutral axis, the prestressing moment is Pe, and the stresses induced across a section of the beam would comprise

a direct stress $\dfrac{P}{A}$ and a bending stress $\dfrac{Pe\bar{y}^1}{I}$.

These stresses, on being combined, give a maximum compressive prestress at the bottom surface of

$$f_{pc} = \frac{P}{A} + \frac{Pe\bar{y}}{I}$$

and a tensile prestress at the top surface of

$$f_{pt} = \frac{P}{A} - \frac{Pe\bar{y}}{I}.$$

If $\dfrac{P}{A}$ were greater than $\dfrac{Pe\bar{y}}{I}$,

their difference would represent a compressive prestress at the top surface.

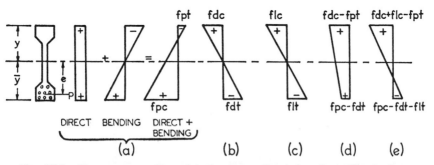

Fig. 34.2 Stress composition. (a) Prestress–direct; bending; direct plus bending (f_{pc}, f_{pt}). (b) Dead load stress (f_d, f_{dc}). (c) Live load stress (f_{lt}, f_{lc}). (d) Prestress plus dead load stress. (e) Prestress plus dead load stress plus live load stress.

The dead load of the beam induces a compressive stress f_{dc} at the top surface and a tensile stress f_{dt} at the bottom surface, while the live load to be carried induces the corresponding stresses f_{lc} and f_{lt} at these surfaces. The arithemetical summation of these stresses gives the final stresses that would occur at the outside surfaces of the beam. These should not exceed the permissible working stresses for design.

Principal tensile stresses in webs occur at points of maximum shear, and significant change of shear or change of section. The effect of shear on design becomes significant for thin-webbed members with heavy unbalanced loads on short spans. Nominal amounts of steel reinforcement are introduced into members, when necessary, for stress distribution and shear resistance (see Section 34.1.8).

Principal tensile stresses can be eliminated in prestressed concrete by using deflected or parabolic tendons, or by prestressing members both horizontally and vertically. In the member of Figure 34.3, if a draped tendon carries a force P^1 and has an angle of incidence θ, the shear force near the end, due to dead and live loads, would be reduced by the upwards component $P^1 \sin\theta$. The diagonal tension caused by shear in ordinary beams is reduced considerably by horizontal prestress.

Standard codes for design purposes that meet ultimate strength requirements are given in Section 34.1.4; shear provisions of AS 1481, BS CP110 and FIP are applied in References 1, 109, 139 and 162; torsional strength is considered in References 111, 116, 120 and 128; and differential shrinkage stresses in composite beams and *in situ* slabs are dealt with in the standard codes of Section 34.1.4, and in References 11-18, 79 and 158. Related subjects thus referenced include radiation shielding in Section 29.3; ACI Committee 435 data on deflection, and CEB-FIB international guidelines in Appendix IV, 2(a)(i).

Fig. 34.3   Deflected pretensioned tendons. (a) Temporary locator yokes.

Casting ribbed units

Precast units assembled
with post-tensioned tendons

Assembling
precast units

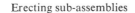

Erecting sub-assemblies

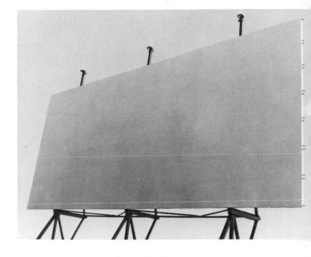

Completed screen
(34 m x 17 m)

Prestressed Concrete Construction
PLATE 30

Inverted channel floor units
in the Advertiser Building,
Adelaide

Art Gallery Road bridge
over Cahill Expressway,
Sydney

Post-tensioned
tendon bridge
at Jackadgery,
NSW

Parking station
in course of erection

34.2 APPLICATIONS AND USES

34.2.1 VARIETY

Prestressed concrete products should be standardised for economy in manufacture whenever possible. Typical products are

Bridge girder beams and segments.
Complete frames for prefabricated building construction.
Dams, tanks, pipes and 10 m long post-tensioned pipe assemblies.[34, 72, 150]
Floor beams, reinforcing elements, joist and box sections.[138]
Floor, road and aerodrome slabs.
Lighting standards, forms, flagpoles, kerbs, posts and fencing.
Planks, sheet piles and bearing piles (up to 67 m long).[47, 48, 72, 73, 147, 150]
Rail sleepers (BS 986).[22]
Roofing beams, trusses, purlins, lintels and spring-boards.
Telephone and electric power-transmission poles.[42, 150]

The cost of prestressed units, when made in quantities, is roughly comparable with ordinary reinforced concrete units for similar purposes; but, as they are more slender and lighter in weight, there is a saving in the cost of transport and handling. Large beams should, in general, be lifted at the ends, while the top of the beam is kept uppermost.

The use of prestressed precast units has considerable advantages in structures such as buildings and bridges, as such units avoid costly falsework.[70, 87] Prestressed sections of the pretensioned type may be mass produced and may be cut at any section without affecting the prestressing. For floors and roofs, prestressed precast units are placed at intervals and the space between these units is filled with precast slabs, the whole being converted into a monolithic structure by the introduction of a certain amount of concrete cast *in situ*. It is claimed that prestressed beams can be used as substitutes for steel joists and are only about twice as heavy as steel joists of equal strength and about 1·5 times as deep.

Prestressed precast units, which may be used in bridges to carry the heaviest loading, are placed side-by-side and cross-bolted or prestressed laterally together (Plates 30, 31, 32). Slabs prestressed in two directions are used in the method of erection in which entire floors are cast at ground level and jacked up to their final position, where they are connected to steel columns by means of steel plates cast in the slabs and welded to the columns. Connection details for prestressed precast concrete buildings are illustrated in Reference 54.

Prestressed planks for floors, roofs and lintels can be made of hollow-cored, high-strength, terracotta or concrete masonry. The units are laid end-to-end and held together by stressed steel wires, which are embedded in vibrated mortar along specially made grooves. Reviews on floor systems in which prestressing may be applied are given in the References.[35, 36]

Concrete pipes with longitudinal ducts may be post-tensioned into cylinders, a rubber ring (Durometer hardness of 55) being used at each joint. The capitalised maintenance of large prestressed concrete structures is low, because of the great durability of high-grade concrete in a state of compression. Prestressing has innumerable structural applications, all of which are not necessarily limited to concrete.

34.2.2 PRESTRESSED CONCRETE PIPES

A compressive stress can be induced circumferentially in a concrete pipe by expanding a freshly made pipe in its mould and allowing it to harden around an extended integral, high-tensile steel cage; or by winding high-tensile steel wire (under a controlled tension) around a composite concrete and steel cylinder or a concrete core-pipe. The wound wire is protected subsequently by a tenacious cover-coat of rich dense cement mortar or extruded epoxy resin mortar (see Chapter 21). The wire-wound concrete core-pipe process, which was developed in Australia by Rocla (see Section 23.3.2), is the most economical to implement (see Plate 18). Roll-on or entrapped rubber rings, or indented wedge-shaped, steel-sprung, rubber gaskets (see Section 34.2.3(a)) are used for jointing, and steel fittings with a suitable joint profile are used for bends and tees.

The wire-wound process of manufacture of spigot and socket prestressed (noncylinder) concrete pipes 5-7 m long, consists of the following steps.

1. High-tensile steel wires are placed longitudinally in a cylindrical steel mould and anchored at each end to heavy end-rings. The wires are fitted with end-anchors and tensioned by hydraulic jacks or a screwing device, the former aiding in subsequent transfer of load to the concrete.

2. A concrete core-pipe is cast, normally by the roller-suspension method, with a cement content of 15-25 per cent by weight of the total dry material. A richer and more plastic mix is used

in the socket than in the shaft of the pipe; and the pipe mould is designed to compensate for a slight differential diametrical movement in the stripped pipe, due to changes in thickness and mix design. The ends of each mould are covered immediately after filling for protection against draught to prevent early drying of the concrete.

3. The core-pipe is removed from the mould when, after low-pressure steam curing, the concrete has achieved sufficient strength to withstand the transfer of axial prestress from the mould to the core.

4. The core-pipe is further cured until sufficient strength is reached for wire-winding purposes. The circumferential precompression, which is then induced, is limited to 55 per cent of the compressive strength of the concrete at the time of winding.

5. The core-pipe is rotated in a wire-winding machine and spirally wound with high-tensile steel wire at the designed tension and pitch. The ends of the tensioned wire are anchored to the pipe and tensioning is effected by mechanical or hydraulic systems. Herein, a regenerative power control develops resistance to winding tension, which may reach 85 per cent of the ultimate tensile strength of stress-relieved wire.

6. The wound core-pipe is rotated in a mortar-coating machine, where a dense cement-rich mortar coat is applied mechanically at high velocity. The mortar, which is not leaner than 1 part cement to 2·5 parts aggregate, is premixed with a low water content (water/cement ratio below 0·35 by weight). The cementitious cover to the steel wire is not less than 20 mm.

7. The cement mortar coat is preceded and succeeded with coats of cement paste, in order to augment the passivation protection of an alkaline environment around the wound-wire tendons (see Section 22.3 and References 117, 133, 140 and 141 of Part 4). After further accelerated curing, under monitored hydrothermal conditions, the pipe is ready for testing.

8. For aggressive ground-water conditions, sulphate-resistant portland cement and an additional protective coating of coal-tar epoxy resin increase product-service durability.

9. Where ground salinity is likely to penetrate the external cover, the wound-tied tendons can be cathodically protected against corrosion by superimposing, by means of buried galvanic

anodes, a controlled electrical potential in the pipeline (see Sections 15.2.2(a) and 22.3.1; Reference 28, Section 4).

The same brand of portland cement is used in the mortar coat as in the concrete core, and calcium chloride or admixtures containing calcium chloride are avoided to prevent corrosion of the prestressing wire (see Sections 10.1, 15.2.2 and 22.3; also References 126 and 142 of Part 3). A longitudinal precompression (e.g. 1-2 MPa, depending on the circumferential prestress) is required to balance tensile stresses. These are induced by longitudinal strains and bending during the winding process, and by field handling and abnormal bedding conditions (AS CA33).[152]

The minimum (cylinder) compressive strengths of vibropressure-compacted specimens of core-pipe concrete and mortar coat are, for design purposes, 40 MPa at 28 days. Concrete strengths in practice, however, are often more than 70 MPa at 28 days (see Section 23.3.2). The minimum requirements for pipe-winding wire are specified in AS 1310 and BS 2691.

Prestressed concrete pipes are being made to 3000 mm diameter and tested to over 3050 kPa (305 m head). They are competitive with steel pipes and more economical than reinforced concrete pipes at high test pressures or under very high earth loads. Site factories, built near large pipeline contracts, reduce the cost of transport.

The test pressure applied to every pipe is often 1·5 times the sustained working pressure or, otherwise, it is the worst expected combination of internal pressure (including surge) and external loads on the pipeline. In pipe design, no tension is permitted in the concrete at sustained working pressure but, at test pressures, tensile stresses may reach 1·4 MPa in core-pipe walls and 2·8 MPa in precast concrete pressure pipes. In pipe design under combined loads, bending stresses (modulus of rupture) are divided by about 2·25 to obtain an equivalent direct tensile stress.

Experimental work and practical knowledge are essential adjuncts in design for best results. Rubber jointing rings should conform to the provisions of BS 2494. Standard specifications on concrete culverts and pipes are contained in Appendix I. They include AS 1392 and BS 4625, 5178, CP2010.5 on precast and prestressed concrete pressure pipes, respectively. Relevant information is contained in References 37-40 and the *Standard Specifications for Highway Materials and Methods of Sampling and Testing* (2 volumes), American Association of State Highway and Transportation Officials, Washington, 1974.[72, 150]

34.2.3 TYPICAL POST-TENSIONING SYSTEMS

Various post-tensioning systems, which are indicated in Section 34.1.10, are given a detailed descriptive coverage in technical catalogues and literature references, such as those cited at the end of this Part. For the sake of brevity, therefore, some comments will be confined to several of them, in alphabetical order, that are used significantly in Australian civil-engineering construction and building practice.[157]

(a) BBRV.

In the BBRV post-tensioning system, a direct form of anchorage bearing is obtained by "button-heading" the end of each high-tensile wire in tendons that incorporate up to 163 parallel wires. Each wire is usually 7 mm in diameter and of 1700 MPa minimum tensile strength, and the button heads, which prevent slippage during stressing operations, are as strong as the wires themselves. The method was developed in Switzerland during 1945-49 by Messrs Birkenmaier, Brandestini, Ros and Vogt, and it is well suited for segmental bridges and structures which may require couplers or special jacking units for internal use (see Section 34.2.6).[80, 159, 166]

The prestressing force, which is appropriate to the ultimate capacity of these grouped-wire tendons, ranges up to 10 000 kN or more. A typical anchor head at the jacking end of a tendon (requiring up to 1500 kN tensioning force) consists of a cylindrical block, which has concentric rings of holes for the wires and a central threaded hole for the draw bar that is used in tensioning. This block is threaded externally to accommodate a lock nut, which transmits the prestressing force to a bearing plate cast in the end of the structural element to be prestressed.

At the fixed end of the tendon, the cylindrical block and locking nut are often dispensed with and the wires (after being threaded through the bearing plate) are simply anchored by their button heads. The type and size of anchorage are varied to suit design requirements. When the tensioning force is to exceed 1500 kN, it is applied through a threaded tensioning socket that is introduced into the anchorage system. Shim-type anchorages are used with increasing forces (e.g. 3000-10 000 kN).

Tendon extensions are indicated on a graduated scale on the ram of hydraulic jacks, which range up to 12 000 kN capacity, and tensioning forces are measured directly by a load cell or dynamometer. Tensioning may be done in stages and initial losses in tendon stress are recovered before the ducts are grouted. Variants of the system are applied in:

1. Single or two-stage grouted ground anchorages (Section 5 of AS 1481).

2. Cable-stayed structures requiring special fatigue-resistant anchorages injected with epoxy resin.

3. The BBR-CONA method of stressing 12·5 or 15·2 mm diameter, seven-wire strand in flat-ducts and in single or multistrand tendons (see Section 30.1, also BBR Prestressing Manuals and Plates 25 and 28).[132, 139]

(b) Freyssinet-PSC.

The Freyssinet system, which has been in use since 1928, was evolved through the invention of prestressed concrete in France by M. Eugene Freyssinet. It was introduced into Australia by Pre-Stressed Concrete (Australia) in 1951. Although the underlying principle of operation remains unchanged, its scope of application has expanded very greatly through the introduction of new techniques and increased sizes of tendon. The anchorages in this system consist primarily of male and female cones made of heavily reinforced concrete or high-tensile steel for use with 12/7 mm diameter wires or 12/12·5-15·2 mm diameter, seven-wire strands with a load-carrying capacity of up to 1800 kN. The male cone has a central grout hole and a grooved surface to receive the individual wires or strands, which are wedged subsequently between the two cones.

Service versatility has increased since the late 1960s with the introduction of the PSC range of monostrand, monogrip, monogroup and slab systems for prestressing purposes. They cater normally for load-carrying capacities of up to 3850 kN, but can be extended to 10 000 kN for special structural purposes. Unit groups of 1-19/12·5-18·00 mm, seven-wire strands are accommodated at square, inclined and top recessed anchorages.

A characteristic feature in common is the seating of a forged-steel anchorage block on a cast-iron guide, this being coupled with a reinforcing spring for effective load distribution. Individual strands are anchored by three-piece jaw wedges or otherwise by a two-piece wedge-and-barrel fixture. This bears against a steel plate anchorage seated on a cast-iron guide.

In the slab system (see Section 34.2.3(d)), tendon ducting and support chairs are secured in place to stop possible flotation and dislodgement during concrete placement. Shrinkage cracking of concrete is controlled by jacking one strand per cable after 36 hours, the remainder being tensioned progressively as the strength of concrete increases during the pregrouting period.

(c) Macalloy.
The Macalloy system of using alloy steel bars was developed by Donovan H. Lee with McCall Macalloy of Britain in 1952. The properties of the bars are typified here.

> Minimum tensile strength: regular grade, 990 MPa; super grade, 1080 MPa.
> Proof stress (0·2 per cent): above 0·85 times the minimum tensile strength.
> Minimum elongation on a five-times-diameter gauge length: regular grade, 6 per cent; super grade, 5 per cent.
> Secant modulus at 650 MPa: 170×10^3 MPa.

The bars are available in lengths up to 20 m and in sizes of 23, 26, 29, 32, 35 and 40 mm diameters with stressing loads up to 1000 kN. The end-anchorage is formed by a bearing plate and high-strength nut on a rolled thread. Alternatively, the bearing plate may be threaded to receive the threaded bar at a dead-end anchorage. The rolled thread minimises the reduction in area at the root of the thread and gives a better physical structure than a cut thread.

Special high-tensile threaded couplers are used for joining lengths of bars into long tendons or for extending tendons in the progressive-stage construction of continuous structures. Bars of different diameters can be coupled together, and curvature in excess of self-weight deflection can be preformed into the bars where necessary. This versatile system, which is monitored by VSL Prestressing (Australia), finds many uses in flexural and vertical stressings, tanks, precast connections, stage-built projects, and ground anchorages for structures and dams where requisite tendon capacities may range up to 10 000 kN.[132, 139]

(d) VSL.
The VSL (Vorspann System Losinger abbreviated) multistrand and slab systems were stage-developed in the 1950s by Losinger, a civil engineering and contracting organisation in Switzerland. The multistrand system comprises the simultaneous stressing of a group of strands within a cable duct, each strand being not only wedge-secured, but also automatically engaged in common anchor blocks on release of the jacking load.

A tendon may be fashioned of 12·5 mm or 15·2 mm strands for jacking loads up to 10 000 kN, supergrade strand being normally used for economy. Dead-end anchoring is effected by an embedded bearing plate, upon which each strand is fanned out and secured by a swaged grip. Alternatively, the strands are looped around a U-shaped bearing plate for the same purpose. Shimming may be used to compensate for wedge draw-in where necessary, while overstressing and release-back of tendons can be carried out by the normally used centre-hole jack.

The slab system, which finds favour in building construction, uses sets of up to four 12·5 mm or 15·2 mm supergrade strands. These are located side-by-side in oval-shaped ducts, 75 mm × 19 mm deep, so as to obtain maximum drape or eccentricity in 150 mm shallow members or flat plates. The strands are stressed individually by light monojacks and the resultant tendon load, which is up to 800 kN, is sustained by elongated cast-steel anchorages. The ducts are subsequently pressure-grouted to form bonded elements as required. Supplementary facilities include bar tendons (see Section 34.2.3(d)), flat jacks (see Section 34.2.7) and ground-anchorage equipment.

34.2.4 PHYSICAL PROPERTIES OF STEEL TENDONS

Technical details of stress-relieved steel tendons are given in Tables 34.7-34.9, industrial catalogues, and the specification standards and data of Sections 34.1.4(b), (c), 34.1.10 and 34.2.3.[158]

34.2.5 SERVICE, EQUIPMENT AND MATERIALS

Technical services and equipment are available from qualified consultants and specialist representatives of established systems, materials and facilities. They are usually associated with the supply of tensioning appliances, high-tensile steel, and conduits for post-tensioning purposes. Gripping and anchorage efficiency is specified in AS 1314 and 1481.

34.2.6 SEGMENTAL CONSTRUCTION

When prestressed beams and columns are formed of a number of small precast units, contact surfaces that are ground flat provide an accurate bearing. With large units, unevenness of surface

TABLE 34.7   MINIMUM BREAKING LOAD

Tendon	Diameter (mm)	Minimum Breaking Load (kN)
Stress-relieved wire	8·0	77·9
	7·0	62·4-65·4
	5·0	33·4
	4·0	21·4
	3·0	13·1
Seven-wire strand		
regular grade	7·9	69
	9·3	94
	10·9	125
	12·5	165
	15·2	227
	18·0	311
super grade	7·9	74
	9·3	102
	10·9	138
	12·5	184
	15·2	250
	18·0	338
Nineteen-wire strand	18·0	370
	22·2	500
	25·4	660
	28·6	820
	31·8	980

Alloy steel bars (see Section 34.2.3(c)).

TABLE 34.8   MAXIMUM PERCENTAGE RELAXATION

Tendon	Limits after 1000 h at 20 ± 2°C from Minimum Initial Stress of:	
	0·70 Times Minimum Tensile Strength	0·80 Times Minimum Tensile Strength
Stress-relieved wire		
normal relaxation	6·5	8·5
low relaxation	2·0	3·0
Seven-wire strands (Regular and Super)		
normal relaxation	7·0	12·0
low relaxation	2·5	3·5
Nineteen-wire strands		
as-stranded	9·0	14·0
stress-relieved		
normal relaxation	7·0	12·0
low relaxation	2·5	3·5
Alloy steel bars	4·0	—

TABLE 34.9   MINIMUM PERCENTAGE ELONGATION

Tendon	Test Result
Stress-relieved wire	3·5 after fracture on a 250 mm gauge length.
Seven-wire strands	3·5 prior to wire fracture on a 600 mm gauge length.
Nineteen-wire strands	2·0 for as-stranded cable and 3·5 for stress-relieved cable prior to wire fracture on a 600 mm gauge length.
Alloy steel bars	As in Section 34.2.3(c) after fracture.

necessitates the use of a suitable gasket or jointing material. Joints up to 25 mm wide can be packed with portland cement mortar made with equal parts of cement and sand and with just sufficient water to give cohesion (water/cement ratio between 0·3 and 0·35). The compressive strength of this mortar is 30-40 MPa at 3 days at ordinary temperatures.

A stiff leaner mortar, where practicable, may be gauged 1 part cement with 2 parts sand by weight, and abutting clean surfaces may be indented to promote high shear and bond strengths. Jointing mortar is placed in layers with a depth less than three times their width, each layer being firmly caulked or rammed into place with the assistance of form backing. Wide joints, made 85-100 mm wide, are filled with concrete of designated segmental-unit strength, the concrete being compacted by ramming and vibration and subsequently moist cured (see Plates 1 and 29-32).

Vulcanised rubber bushes are sometimes used to form one or two ducts through the joints, so that cables can be easily pulled through them. Usually, 450 mm lengths of rubber tube are attached to long lengths of air hose of 6 mm internal diameter. They are drawn into position and inflated on bridging a joint. When the joint is completed, the tubes are deflated and drawn to the next joint. A steel lead is used to check clearances of the ducts. The joints can be fully stressed after 48 hours, when the strength of the jointing material is about 25 MPa. It is common to start prestressing large sections after 36 hours and to complete it after 48 hours.[125, 166]

In bridge building by the free-cantilevering method, the site work and erection time needed can be reduced by successively bonding precast girder segments together with a pretested, high-strength, epoxy resin adhesive, about 1 mm thick, the resin being cured under pressure (see Section 21.1). The segments may consist of hollow-box sections provided with shear keys. These units are progressively precast, face-matched, site assembled and prestressed as cantilevers, which advance symmetrically outwards from bridge piers without the use of falsework.

At the casting yard, the units are accurately cast successively against each other on continuous formwork, which is adjusted in profile to resemble the finished soffit. During this operation, a peel-off resin film is applied to induce subsequent separation. At the work site, each segment is held successively in place by a travelling jig, during both the application of resin and the subsequent prestressing. As the interfacial jointing quality is improved by transverse compression, the forward units are prestressed together immediately after applying the resin.[80, 159, 168]

34.2.7 FLAT HYDRAULIC JACK

The "Freyssi" hydraulic flat jack is a small, circular, elongated or rectangular steel bag with two edge nipples and a dished cross-section, as shown in Figure 34.4. It is capable of exerting extremely high forces that are sometimes needed in engineering construction. Circular jacks range from 70 mm to 900 mm diameter and exert forces ranging from 30 kN to over 8500 kN (see Plate 28).

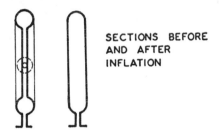

SECTIONS BEFORE
AND AFTER
INFLATION

Fig. 34.4   Flat jack

The injection of a fluid under pressure unfolds the edging and forces two plates apart, so as to obtain a movement of up to 25 mm under a pressure as high as 15 MPa. Batteries of jacks can be used in parallel or in series for increased force or stroke, respectively. On taking up the movement by wedging or packing with steel plates, the jack can be closed and recovered for limited reuse. Steel-leafed "Neoprene" and natural rubber pads, with a coefficient of friction of 0·022, are useful for bridge bearings (see Lindley, Appendix IV, 1).

If the jack cannot be removed, it may be left in place. In this instance, pressure is created with cement grout or synthetic resin and a suitable pump. The copper lead is sealed or shut off, the pump removed and cleaned, and the injected material is allowed to solidify. The spacing unit is finally encased with mortar or fine concrete (as for segmental construction).

Although the jack was originally devised for prestressing concrete beams between rigid abutments, it can be used for a variety of purposes. These include putting a thrust into arch ribs and causing them to lift from the formwork; compressing adjacent sections of concrete roads, runways, rafts and tunnels; underpinning buildings; preloading supporting structures,[68] measuring rock stresses; and testing the bearing capacity of piles.

34.2.8 PRESTRESSED CONCRETE CAST BY EXTRUSION

Prestressed concrete beams can be made in long lengths (e.g. 90 m) by means of slip moulds and the pretensioning system. The concrete is extruded from a travelling machine and, when cured, each beam is automatically fed to a saw table which cuts off the required lengths.

The beams are formed on flat concrete beds and the extruding machine, which is pneumatically driven, is suspended from an overhead gantry. It moves forwards in short steps, metering the correct quantity of concrete mix from a hopper, which is supplied from an overhead monorail tipper, and then vibrating it between side plates which, with the concrete bed, act as a form. The top of the beam is automatically levelled to any desired thickness between 100 mm and 200 mm by a moving screed. Hollows are formed by steel tubes which cannot be seen when the machine is in operation; rubber extensions to these tubes trail for some way behind the machine to prevent the hollows from caving-in.

While the machine travels at about 1·2 m per minute over one bed, a completed length of beam is cured on another at a temperature of 72-80°C. The heating method is economical, steam-pipes being embedded in the concrete beds which are insulated from the ground by vermiculite. The beams have insulated covers placed over them while being cured. Thermocouples relay information to an automatic control in the boiler house and any desired temperature cycle can be preset and implemented with precision.

On a third bed, a cured beam is rolled forwards to a saw table, where a 900 mm diameter diamond-tipped saw cuts off the required lengths. For this purpose, rollers at 2·5 m intervals along the bed are hydraulically raised, and the 90 m length of prestressed concrete is moved forwards by powered rolls, as in a steel mill. The tensioned steel wires are cut only on the saw table, so that as the beam moves along, wires for the next length to be extruded are automatically drawn from reels at the far end of the bed.

With beams of 330 mm × 100-200 mm cross-section and up to 9 m, the maximum output of the plant under these conditions is 0·5 kilometre per 8-hour day. Either dense concrete or lightweight-aggregate concrete may be used.[25, 77]

34.2.9 SAFETY PRECAUTIONS

If tensioned steel breaks or is accidentally released during the production of prestressed concrete, it whips out of the mould or the freshly placed concrete at high speed and with sufficient force to impale anyone nearby. Safety precautions[93] to prevent whipback include: temporary guards of stout steel stirrups, heavy bulkheads or concrete fence posts over open stressing beds; strongly constructed shields of steel or timber which are fitted behind each anchorage; the operation of jacks from the side of sections being made; and prohibition of access to areas on the stressing beds or behind the anchorages until tensioning operations are complete.

34.3 DATA ON PRESTRESSED CONCRETE BRIDGES

34.3.1 REQUIREMENTS OF MATERIALS

From Figure 34.5, it is seen that continuous beams with equal spans do not achieve significant economies over simply supported beams

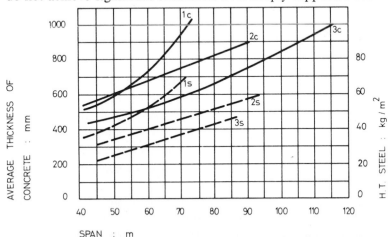

Fig. 34.5 Bridge data. Comparison of quantities of concrete and steel in terms of span. 1, simply supported beams. 2, continuous beams with equal spans. 3, continuous beams with unequal spans. c refers to concrete; s refers to steel.

for spans up to 55 m, but become effective in this regard at more than 60 m. The cost of foundations in each instance is about the same. Noticeable economies are obtainable by the use of continuous bridges with a central span and cantilever spans of different length. Their range of applicability depends upon the cantilever-span/central-span ratio (i.e. upon the degree of flexural fixity realised at the ends of the long span). When this ratio becomes as low as 0·3, the bridge may be considered to be 95 per cent fixed and the central span may be increased to about 120 m. For a high degree of fixity with short cantilever spans, an uplift of the outer piers may have to be resisted by rock anchorages or counterweights. The foundations, therefore, could become expensive.

The indicated spans for prestressed concrete bridges are based on concrete stresses of 14 MPa and an upper available limit of prestressing force of 7400 kN/m of cross-section. Although a maximum span of about 80 m can be used for simply supported bridges, there is no advantage in approaching this limit too closely and other solutions should be attempted for spans longer than 50-55 m. In fact, the specified headroom beneath the span sometimes militates against the use of simply supported beams for longer spans. By cantilevering with travelling forms, suspending a girder between cantilevers, or using an arch structure, bridges can be designed with main spans within a range of 150-300 m.[83, 112, 113, 118]

34.3.2 CROSS-SECTIONS

To meet the demands imposed by ordinary road traffic with minimum thicknesses of material, a deck-slab thickness should be at least 50 mm plus 10 mm per 300 mm of transverse span. For heavy traffic (multiwheeled axles or caterpillars), this estimated dimension should be increased by 25 mm or more, the slabs being double-haunched where structurally advantageous and economical.

The minimum practical thickness for a web, which is governed by cast *in situ* requirements, is approximately 100 mm plus 25 mm per 900 mm depth. The minimum thickness required for resistance to shear is the thickness for which the average shear stress across the web section is about 1·7 MPa. Prestressing in two directions may be supplemented with prestressing in the vertical plane (i.e. in the webs). Adequate bond is an essential requirement in prestressed construction.

34.3.3 SPACING MAIN MEMBERS

Increase of the transverse span results in a reduction in the number of main beams and, consequently (since webs are generally over-designed), of the dead weight. The extent to which the transverse span can be increased may be limited by the following practical considerations.

1. For a given longitudinal span, the spacing of the longitudinal beams has but little effect on the total number of longitudinal tendons. It follows that an excessive spacing of the main beams would necessitate many tendons being located in the bottom bulb of a beam cross-section. This concentration of tendons could cause some difficulty where they have to be transferred from the bottom to the top of the beam.

2. With prefabricated elements, a limit may be imposed by the available transport facilities and crane capacities.

There is a need for careful and proper positioning of tendons to ensure both ready concreting and adequate anchorage or bond. Bearings may be of metallic, rubber or Freyssinet hinge type.

34.3.4 DIAPHRAGMS

For the purpose of connecting longitudinal members and increasing their resistance to torque, diaphragms are important elements in the design of cross-sections. Their functions become more evident when exceptionally heavy loads have to be distributed, but they should not be placed too closely together. For general purposes, diaphragms at suitable intervals (e.g. four) may be structurally sufficient. The criterion of usage is that the local flexure under loads acting on an individual beam between two diaphragms, added to general flexure, must not exceed that which can be supported by this individual beam. Therefore, where beams are of variable depth, the spacing of diaphragms may be increased in the region of supports.

CHAPTER 35

SPATIAL STRUCTURES

35.1 FERROCEMENT FABRICATION OF BOATS AND BUILDINGS

Ferrocement fabrication involves the use of a very thin, resilient, highly reinforced slab (frequently corrugated or curved), which is made by pressing mortar (with a water/cement ratio of about 0·35 by weight) continuously through several layers of steel mesh and small diameter rods (see Section 24.2.6). These are first wired or clipped together, and faired up with a rubber or wooden mallet to form a section that is a little thinner than the thickness of the final unit.

The impregnating mortar is densely compacted within the reinforcing fabric by vibropressure, vibrovacuum or manual means. Because of the large specific surface area and volume of steel mesh used, there is a substantial improvement in the crack-resistant performance of the product. With up to 8 per cent of high-strength wire mesh, the tensile strength can be up to 35 MPa.

For a 35 mm thickness, eight layers of mesh weighing about 1·0 kg/m² are used, with an additional three layers of 6 mm diameter rods at 100 mm centres in both directions. The thickness may be reduced to 15 mm by using fewer layers of reinforcement, or it may be increased to 100 mm by using up to twelve layers of mesh, and one or more layers of 6 mm or 10 mm diameter rods between them. The high percentage (about 6 per cent) and distribution of reinforcement make the material act very much more like steel than ordinary reinforced concrete, and enable sections to be designed on the assumption that they are structurally homogeneous.

A typical mortar contains 40 kg of fresh portland cement and 4-5 kg of pozzolan per 70 kg of optimum-graded sand. Each batch is mixed in a paddle or pan mixer and, after several minutes (if practicable), remixed for improved workability and ease of subsequent application. A 3 mm cover has been found to be waterproof when high quality mortar and workmanship are used. The product has a low fire rating, because of its thin cover, but it can be used wherever steel framing is permissible in roof construction.

Ferrocement was developed and used by Pier Luigi Nervi in 1943, for the quick and simple construction of small ships of a size not exceeding 500 tonnes. The hulls of several motor vessels and sailing craft were designed on a basis of homogeneity and a working stress of 3·5 MPa. They include a 12 m ketch with a 16 mm thick hull that is strengthened with ribs at 1 m centres and finally sealed (see Sections 20.2(e), 21.1.6(c) and 28.3).

In boat hulls of this type, the ribs are formed of 25 mm diameter tubular-steel frames that are made continuous with the deck and other fixtures, securely braced and enclosed with 7 mm diameter longitudinal rods tied on at 50 mm centres. These are closely wrapped with seven layers of plain or weathered-galvanised light steel mesh (e.g. 1·4-0·7 mm by 12 mm size), the four outside layers and three inside layers being tied and laced together. The reinforcing mat thus made is impregnated with freshly mixed, slightly retarded, plasticised, pozzolanic or sulphate-resisting mortar (see Section 3.1.2(e); also Chapters 6, 15, 17, 19 and 31). It is forced through the fabric from one side and made good from the other.

Thorough moist curing for at least 21 days, using enclosed wet underfelt, is essential prior to ballasting in reinforced heavyweight concrete, buffing, etching with 1 part phosphoric acid : 3 parts water, rinsing, and final epoxy resin priming and sealing. A concrete hull, profiled longitudinally with 7 mm diameter high-tensile steel wires, can be converted into a shell having enormous strength by post-tensioning high-tensile steel strands, placed previously within skeletal, tubular-rib frames that shape the hull.[142-146]

Many spatial frames erected with ferrocement include the 95 m span corrugated vault of the Turin exhibition hall in Italy. It is claimed to be one of the most imposing edifices built during the last two centuries. Precast ceiling coffers of ferrocement, about 2·5 m wide, 1·5 m deep, 4·0 m long and 38 mm thick, were assembled on travelling scaffolding and joined together by reinforced concrete ribs that were cast *in situ* in the troughs and at the crests of the corrugations.

The coffers were cast on smooth oiled concrete moulds, which in turn were cast in plaster moulds. The visible underside of the units was cast in contact with the moulds. Each unit was closed at each end by stiffening diaphragms and joined to adjacent units in rows by a 38 mm thickness of strong mortar placed *in situ* between them. Reinforcing rods protruding from the tops and bottoms of the units were used to make them monolithic with reinforced concrete ribs that were poured subsequently and formed the main loadbearing elements. Glazed openings were provided for lighting purposes. Each four rows of units was supported by fan-shaped ferrocement units springing from inclined reinforced concrete buttresses. The dead weight of the roof was barely 120 kg/m^2. Similar vast, parabolic-vault canopies of precast trough units and glass were used in both the Olympic Sports and Vatican Audience Halls in Rome.

Corrugated ferrocement segments in a structure may be suitably enclosed to form a series of ceiling ducts that can be used for air-conditioning purposes. In this connection, thin precast concrete and cast *in situ* reinforced concrete decking may be placed externally and tied together with projecting dowels. Although the proportion of steel used in ferrocement work is high, a notable saving of dead weight in domes and vaults reduces the amount actually used. Protruding reinforcement at complex joints may be electrically welded for structural continuity.

Ferrocement moulds have been used as forming for cast *in situ* concrete slabs, on the lower side of which ribs were arranged in a tessellated pattern or in accordance with the isostatic lines of principal bending moments. Movable moulds, hardened on the surface, have been used in this way about fifty times each. Hollow-ribbed pans, arranged diagonally as a complete casing or in a diamond reticular pattern, have been used as permanent formwork to cast *in situ* reinforced work. The soffit and sides of great cast *in situ* concrete arches have also been formed with trough-shaped units.

Ferrocement fabrication is characterised by

Thin-shell structures with a high tensile-strength/span-weight ratio.

Intrinsic lightness and speed of assembly.

High resistance to shock and cracking, because of the extensive subdivision and all-over distribution of reinforcement.

Economy through a great simplification, or elimination, of formwork.

Economical adaptation to reticulate floor slabs and large ribbed roofs (vaults up to about 300 m span are projected), and to special structures, such as seacraft, pressure pipe lining, tanks and flexible pavements.

Ferrocement fabrication is an effective way of obtaining strength through form, with unlimited possibilities of application to new designs of spatial, reticular and lamella-type framework, including the curved tracery of reticulate vaults and domes, and the composite construction of giant corrugated ribbed arches (see Naaman and Shah, Rao and Gowder, Appendix IV, 4(b)).[49-53, 121, 165]

35.2 CTESIPHON STRUCTURES

Ctesiphon structures are corrugated-catenary, arched thin shells of large span (e.g. 50 mm thickness, 2400 mm pitch, 600 mm depth, 30 m span), which are supported on buttresses at internal valleys and on A-frames at end spans. The A-frames, which take lateral thrust, are tied at floor level by prestressed cables in ducts that pass across the width of the structure.

The name Ctesiphon is taken from the great Arch of Ctesiphon on the banks of the Tigris about 25 km from Baghdad. This structure is a catenary arch with a span of 27 m and a rise of 34 m. It was built of burnt mud bricks in about 550 AD. A Ctesiphon structure may have a rise above the springing level of only one-fifth the span, but it is always a corrugated catenary so as to achieve stiffness with great economy of material. It has the structural advantages of a doubly curved shell (such as occurs in nature), including pure compressive stress under its own dead load. Bending moments due to live loads are accommodated by the corrugations, which have additional reinforcement at the troughs and crests.

Fine-mix concrete for the shell has a minimum compressive strength of 20 MPa at 7 days. It is placed *in situ* on mobile formwork which is supported on travelling scaffolding. The latter consists of tied arched trusses mounted with coupled hydraulic jacks on rail bogies. Three corrugations are cast at a time, the formwork being subsequently jacked down temporarily about 1 m, so as to enable it to be moved forwards and raised into position for the next bay. The floor cables for each bay are post-tensioned, immediately before the formwork is moved, when the shell concrete is at a minimum age of 4 days.[121]

35.3 SHELL, FOLDED-PLATE AND TRANS-LUCENT CONCRETE STRUCTURES

Shell and folded-plate structures, which are of increasing interest and importance, fall into the categories of mathematical and model methods of analysis (see Appendix IV, 2(b)). Some of the factors to be considered in their investigation are indicated in References 55-65 of this Part and Reference 84 of Part 4, the calculus or matrix algebra involved being resolved where expedient with the aid of a computer. Useful applications of this implement, in experimental and developmental design projects, include the processing of logged data that are automatically recorded on punched paper tapes, and the determination of parametric-change effects in mathematical models that constitute an integral part of optimisation and probability studies (see Section 36.8.2 and Appendix IV, 2-4).

Precast concrete segments of large shell roof structures may be joined by means of epoxy resin and prestressing. They are lifted by tower cranes (up to 10 tonne capacity at 30 m radius) and supported temporarily by mobile scaffolding or articulated steel arches, which can be extended hydraulically to requisite profile and ensure limited full-load deflection. Tile lids, which can be bolted to concrete segments, have been developed by Pier Luigi Nervi (see Section 35.1) and may be incorporated in the design of curvilinear structures. They are made by laying tiles face downwards in a mould and backing them with 20 mm wire-reinforced concrete, the tiles being moulded so as to key strongly to the concrete.[81, 124, 136]

Shell structures offer much scope for the imaginative use of hanging roofs, hyperbolic paraboloids, conoids and other varieties of large spatial covering, in which strength resides in shape as well as in the component material. Hyperbolic paraboloid shapes are readily fabricated, because their surfaces are formed by a series of straight-line generators. These lines can be reproduced in the formwork or moulds, the joints of precast elements, and the post-tensioning ducts of units that are to be butted and tied together.[121, 126, 127, 133]

In suitable prestressed abutted assemblies, with a shell thickness of 40 mm, single-wire tendons can be anchored against steel scaffolding. This steelwork, which extends along the main joints, is cast subsequently into the shell with cast *in situ* concrete. Roof membranes or sealants are summarised in Section 36.5. An application of precast plate-units for roof construction is illustrated in Reference 89. Spatial structures of translucent concrete (see Glossary, Appendix VI) which are described in References 90-92 offer unique possibilities of light and shade in the construction of aesthetic canopies for public buildings. Strain-gauge stress analyses are used as a design check on sophisticated spatial structures, pilot-scale fabrications being used initially for design-refinement purposes (see Appendix IV, 4(b)).

CHAPTER 36

OPERATIONAL AND BUILDING TECHNIQUES

36.1 BUILDING ON MOVING GROUND

Wave-like surface movements resulting from subsidence of mine workings and, to a lesser extent, settlement on filled or faulty ground cause cracking, twisting, breakage and possibly dangerous reduction of the bearing length of beams in traditional construction. Differential subsidence causes a considerable stretch of the earth's surface, possibly as much as 100 mm over a length of 30 m, as well as vertical movements and tilt. A prediction of the magnitude of surface movements should be made. The more a building is wedded to the ground by fixtures, the more likely it is to suffer from horizontal movement of the ground.

In addition to lightness of weight and flexibility of structure, a building on moving ground should have a flat base over some form of material which will allow easy slipping, such as sand or poor binding gravel. The design must cope with two movements, horizontal and vertical, the former being dealt with by the ground slab and the latter by the superstructure.

For instance, a 125 mm slab is centrally reinforced continuously with enough steel fabric to keep it from parting company with itself at midportal joints that are painted with bitumen. The slab is like a sheet of paper, easily bent in all directions, but strong in tension. It can thus bend to follow vertical movements of the ground. A thin ground slab with no edge beams, held together by a single layer of tensional fabric reinforcement at mid-depth, is adequate for about 42 m of three-storey or 55 m of single-storey lightweight construction (see Fig. 36.1).

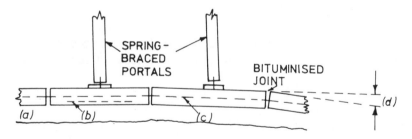

Fig. 36.1 Articulated slab. (a) Vapourproof membrane on granular bed. (b) Mat reinforcement under heavily loaded stanchion. (c) Continuous fabric reinforcement. (d) Change of slope up to 1 mm in 1 m.

An apron of paving slabs or other water-shedding surface around the building stops erosion under it by water. The total weight of a three-storey block of lightweight construction may be only 800 kg/m² below the ground slab, while a single-storey building may be only 560 kg/m². A bearing capacity in excess of these figures is ensured by compacting the prepared base with a vibrating roller.

A superstructure which can "caterpillar" necessitates the use of a loosely jointed steel framework of parallelogram portals. The columns are located on pins (grouted into the ground slab) and cross-braced at intervals by steel braces, which embody springs that will not move under ordinary design load conditions, but will "give" under abnormal loads due to subsidence. The springs and braces are so adjusted that they will control the building after subsidence waves have gone through the ground. The building, including drainage, is assembled so as to accommodate movements in the structural frame (see Fig. 36.2).[95-97]

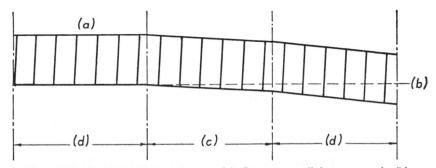

Fig. 36.2 Caterpillaring of frame. (a) Beams parallel to ground. (b) Horizontal level of ground before subsidence. (c) Rectangular portals. (d) Parallelogrammic portals.

Tests on a two-storey prototype structure have shown that the springs can pull a building back to vertical after being forced 100 mm out of upright. Shear on the foundation slab is about one-third of the weight of the building. Dry or weak jointing and ample tolerance in rebates are provided, and side-hung windows are bound only at the head and sill to accommodate lozenging effect. This example indicates how a building can be designed intelligently to suit circumstances (see BS 1377; and References 9 and 10 of Part 4).[169] Otherwise, a jacking system may be used for relevelling.

36.2 TERMITE-PROOF CONCRETE

Termites can enter a building through cracks in dense concrete or the voids in no-fines concrete, when these materials are used for floor slabs or wall construction. A powerful contact toxic effect, with considerable permanence toward termites, is provided by not less than 0·05 per cent of pure dieldrin (by weight) in the concrete. On the basis that a commercial emulsion of dieldrin contains 15 per cent of active constituent, ordinary concrete would require about 9 litres of this emulsion per m³. If the emulsion is used as a partial replacement of the mixing water, the toxicant does not affect the strength of the concrete. Soil poisoning by dieldrin or cyanide compound and the use of termite shields are alternative means of protection (AS CA43). To be termite resistant, plastic film should be thick, dense and suitably formulated with plasticiser, mineral filler and insecticide (e.g. aldrin or dieldrin; see AS 1694).

36.3 RESIDENTIAL CONCRETE FLOORS

36.3.1 DESIGN PROCEDURES

The design and behaviour of a residential raft-slab floor is strongly influenced by the potential mágnitude of adjacent differential soil movements or heave and sub-floor variations due to moisture changes at the site. Depending on geological origins, soil classifications, environmental and climatic conditions, the expected differential soil movement can range from one of insignificance to over 120 mm (see Reference 186, Part 5).

For guidance on raft slab design, working theories have been proposed by BRAB, City of Knox, Lytton, Fargher, Walsh, Fraser and Wardle, Holland and others (see References 33, 65, 82, 182, 188-190 of Part 4). Amongst these, Walsh has proposed a design procedure for determining a suitable interacting structural-element type of slab, which incorporates three basic profiles that are typified in Table 36.1 (see Reference 182 of Part 4). Professional engineering services for the third type should facilitate its usage in design, which is based on a simplified slab on mound or dish concept that gives rise to related interactive forms of flexural distortion.

Proven typical slabs, based on performance surveys, can be conducive (with technical adaptation) to further building usage in other suitable locations. Basically, they incorporate a membrane seal over a prepared foundation. They are designed with reinforced edge beams and internal edge-to-edge gridded beams of limited spacing, and with adequate slab top-steel and extra diagonal steel at re-entrant corners, in order to limit the effects of differential subgrade deformation (mainly doming) and potential surface cracking.

36.3.2 CONSTRUCTION

(a) Slab project.

Constructional considerations that complement the foregoing design factors are those concerning site drainage and water impermeability (see Sections 11.1.2 and 20.1.3); abrasion-resistance (see Section 12.1); alkali-aggregate reaction (see Section 15.2.3(a)); and durability against aggressive ground materials and moisture (see Section 19.1). Allied considerations are featured also in certain applied techniques. These include the construction of industrial concrete floors (see Sections 20.1 and 22.1); frost-heave prevention in cold stores (see Sections 16.1 and 20.1.4(c); also Reference 96 of Part 4); vacuum-dewatering (see Section 23.4) and moist curing of placed concrete (see Sections 9.1-9.3, 20.1.5(e) and 31.2.4); building methods (including flexible connections) on moving and termite ground areas (see Sections 36.1 and 36.2); and measures for enhancing surface serviceability generally (see Sections 20.2, 31.2.4, 32.2.2 and 36.4).

Towards an economical and rational design method for residential slabs, Holland and others are establishing, programming and monitoring ground-movement stations and experimental constructional projects, with financial sponsorship coming from supporting sources mainly through the Australian Engineering and Building Industries Research Association Limited (see References 188, 190 of Section 4, and subsequent progress reports). Primal slabs, used initially for the purpose, are designated in Figure 36.3 as Swinburne X and Swinburne Y.

**TABLE 36.1   SLAB TYPE SELECTIONS FOR RELEVANT
                                SITE CONDITIONS**

Soil Type	Typical Undersealed Slab Floor
Gravel, sands* and silts†	1. Thin slab (e.g. a 100 mm slab containing F72 [3·1 kg/m²] top fabric with 25 mm cover) and a single-splay edge beam, 300 mm or 400 mm total depth by 300 mm or 400 mm minimum width for a single or two-storey residence respectively. Trench reinforcement, with 50 mm cover, is a three-wire or four-wire layer respectively of F8TM (4·2 kg/m²) mesh. Internal slab strengthening is effected with a 1 m bottom strip of F82 (3·9 kg/m²) mesh under walls loaded at 7-14 kN/m; or with a monolithic grid of continuous double-splay beams, e.g. 250 mm total depth by 250 mm minimum width with three-wire F8TM mesh, located under more heavily loaded walls.
Clays and silty clays†	1. Thin slab as above for stable conditions, that have very little expected differential soil movement or heave due to moisture changes under a building since its construction.‡
	2. Light stiffened raft (e.g., 100 mm slab containing F82 top fabric with 25 mm cover, and a single-splay edge beam) 400 mm total depth by 300 mm or 400 mm minimum width for a single or two-storey residence respectively. Trench reinforcement with 50 mm cover, is respectively a three-wire or four-wire layer of either F11TM (8·3 kg/m²) mesh or F1118 (9·7 kg/m²) fabric. Double-splay, primary stiffening beams, 400 mm total depth by 300 mm minimum width, are made continuous from edge to edge and reinforced with three-wire F11TM mesh or F1118 fabric. They are placed under heavily loaded internal walls and at a maximum grid spacing of 4 metres. Secondary panel strengthening is effected as indicated in slab 1 above.‡
	3. Heavy stiffened raft for unstable conditions, i.e. a robust type 2 raft that is taken to 500 mm total depth or an adequate bearing medium.‡
Peats and low-strength soils	1. Suspended slab on deep foundation (e.g. a slab supported clear of the ground on walls and deep strip footings or on edge beams supported on deep piers).

*Loose sands should be compacted or a type 3 or 4 slab may be required.

†For moderately low-strength soils, for which the unconfined compressive strength is less than 7·5 times the average loading on the slab, a type 3 slab is required. For very low-strength soils a type 4 slab should be used.

‡CSIRO/AEBIRA. A report on the expansive behaviour of Melbourne soils for domestic construction is published by the C & CA, Australia, 1976.

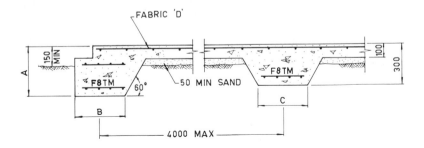

Fig. 36.3 Swinburne raft slab. Swinburne X: A (minimum) = 375 mm, B = 375 mm, C = 375 mm, D = F82 fabric. Swinburne Y: A (minimum) = 300 mm, B = 400 mm, C = 350 mm, D = F82 fabric. Slab underseal: 0·2 mm plastic membrane. All beams may have F8TM mesh also at the top.

The X-type is selected for use in Melbourne clay areas where the topsoil is stripped off, provided that the thickness of clay above the bedrock exceeds 0·6 m. The Y-type is used where the clay depth does not exceed 0·6 m and nonexpansive top layers above the clay are not stripped off. The beams should, in all cases, be of sufficient depth to ensure that they can be founded on firm clay or soil of suitable bearing capacity.

In sandy soil areas, the Swinburne Y slab can generally be used without internal beams. Continuous internal beams may be required, however, under or near all heavily loaded internal walls. These standard slabs are suggested minimum sections for one-storey and two-storey construction in clay soils, the greater loading of the latter having a damping effect on subslab soil movement.

In the preparatory stages of a project, borehole sampling, supporting capacity and settlement testing are factual determinants in variable soil profiles. These prerequisites are followed by drainage requirements, site formation, slab bedding and consolidation as indicated in Section 20.1.3, and coupled with a proviso that good-quality stripped topsoil is stockpiled apart for eventual topdressing purposes. On sloping ground, grillage raft construction is a useful medium through compacted soil fill, where it could be conveniently combined with the construction of a suspended slab.

For long-term concrete durability, the supporting subgrade and subslab granular bed (at least 50 mm thick) should be free from sulphates and sulphides (see Section 19.1.1); and slab-edge extensions of the subslab membrane are externally protected with suitable flashing material, asbestos cement, concrete masonry or pavement edging. In quaternary basalt clays, floater excavation can be avoided by thickening the subslab bedding, which, on very expansive clay, may consist of compact crushed rock some 200 mm thick. Building permits are required for these projects.

(b) Concrete.

Concrete of 20 MPa grade designation (see Appendix VIII, 5), with a slump of 50-100 mm, is commonly used in residential projects, although a higher grade with a reduced slump and limited air-entrainment (e.g. 2-3 per cent) is conducive to a superior surface finish without driers. Premixed concrete should be placed as near as practicable to its final position, with the likely aid of an extendable chute or a concrete pump. For early crack control, sheet-membrane curing and subsequent moist curing must be started immediately after surface-finishing the concrete, particularly in warm or windy weather.

An expeditious concreting programme is enhanced by using correct practical procedures. These include the proper support of slab reinforcements with base-plated bar chairs over a plastic membrane (see Section 3.4.1); the design of suitable mixes (see Sections 4.3 and 12.1); effective means of concrete placement, dewatering, compaction and finishing (see Sections 7.1, 7.2, 20.1.5(d), 23.1, 23.4 and 24.1); early sheet-membrane curing and adequate moist curing (see Sections 9.1-9.3, 20.1.5(e) and 31.2.4); local adjustments for extreme weather conditions (see Sections 32.1 and 32.2); means of controlling drying shrinkage and plastic shrinkage cracking (see Sections 15.2.2(b) and 31.2.1-31.2.4; also Appendix III); selective applications of surface finish (see Sections 20.1.5(d), 20.2, 36.4 and 36.6); and supervisory facilities of quality control (see Sections 13.1, 14.1 and Appendix II, 3). Building precautions against superstructural cracking are illustrated in Reference 186 of Part 5 and in Chen, Appendix IV, 1.[169]

(c) Heated floors.

Thermal space conditioning in cold climates can be carried out by direct or ducted means, or by heating concrete slabs to 27°C at the surface with either embedded off-peak electric cables or 20 mm bore polyethylene coils conveying water at 50°C. These elements are grouped into separate heating zones, which are thermostatically monitored or manually regulated on a temperature-time basis from a central terminal.

The slab (which is moisture protected by a membrane and located on a low-watertable site with a 150 mm hardcore bedding) is thickened to 125 mm for the inclusion of coils that are protected with at least 50 mm of superimposed concrete. The coils in each heating zone, which are of similar length, are omitted under walls and cupboards.

The conduits are positioned about 300 mm apart and tied at close intervals to a supporting layer of F52 square-mesh fabric ($1\cdot5$ kg/m²). Protected slab-edge insulation, reflective, foamed or fibrous insulants of walls and ceilings, and underside insulation of space-spanning slabs are factors which lower heat transmission losses and improve the thermal functional performance of heated concrete floors (see Section 20.1.4(a) and Reference 65 of Part 4).

36.4 TERRAZZO CONCRETE FINISHES

36.4.1 VARIETIES

Terrazzo concrete floor finishes (incorporating uniformly coloured portland cement, coloured marble and divider strips) combine beauty and ease of cleaning with long-wearing service at moderate cost. These finishes can be used in public, semipublic and residential buildings where fair slipperiness and the noise of footsteps are not likely to cause objection. The terrazzo concrete is site mixed or specially precast and laid on a screeded bed, which is either bonded to a concrete base course or separated from it. Glass aggregate, which could cause alkali-aggregate reaction, should be avoided (see Section 15.2.3(a)).

Bonded construction is commonly used for cast *in situ* finishes and unbonded construction is used where the finishing course (e.g. conductive flooring in hospitals) must be kept free from structural cracks that may occur undesignedly in the base course. It is used also where pressure-moulded precast tiles are to be employed over an unmatured concrete base.

36.4.2 SUPPORTING COURSES

A lightly reinforced base course laid on the ground should be 75 mm or more thick, provided with a drainage slope (e.g. 1 in

80-120) and given a green-cut or wood-floated finish for bonded and unbonded forms of construction, respectively (see Section 20.1.4). The mix proportions of the screeded bed depend on its thickness and bonding or nonbonding characteristics. They may consist of 1 part by volume of cement and either 1·5-2 parts of coarse-graded sand and 2·5-3 parts of 10 mm coarse aggregate, or 4-5 parts of coarse-graded sand, the slump after incorporating a workability agent (if any) being not more than 30 mm (see Sections 6.4, 16.2, 18.1 and 20.1.2(b)).

The thickness of the screeded bed varies generally from 13 mm to 32 mm, but it should be at least 40 mm for lightly reinforced unbonded construction with an *in situ* terrazzo finish (see Section 20.1.5(a)). The bed is laid in panels usually not exceeding 3·6 m square, or up to about 4·6 m square under favourable laying and curing conditions, and it is jointed coincidentally with joints in the base course. A 1 part cement : 1 part sand mortar is used on a clean prepared base course for bonding purposes (see Section 20.1.5(c)). Where a bond-breaking membrane is required, it may consist of polyethylene film, building paper or bituminous felt, with or without a 6 mm supporting layer of fine sand. Alongside walls, a break of bond should be provided for the full depth of ground floors.

36.4.3 FINISHING COURSES

A typical terrazzo mix consists of 1 part by volume of coloured portland cement (see Section 3.1.2(e)(ix)), 0-0·5 part of marble sand passing a 4·75-mm sieve and retained on a 0·30-mm sieve, and 2 parts of 10 mm to 14 mm nominal-size, coloured marble chips, the slump after incorporating a workability agent (if any) being not more than 30 mm (see Sections 6.4, 20.1.2(b) and 20.1.5(b)). The minimum thickness of the terrazzo finish is 12 mm for 10 mm nominal-size aggregate and 16 mm for 14 mm nominal-size chips. For precast veneer work, a thinner coating may be used with a 1 part cement : 2 parts marble mix containing a smaller size of aggregate and, in each instance, coloured cement mortar is used where necessary for bonding purposes (BS CP204).

Divider strips (2 mm thick) of brass, copper, zinc, coloured ebonite or coloured bakelite are installed so as to divide the surface into bays, generally not exceeding 1 m² in size, with a maximum dimension of 1200 mm. When used with precast tiles, the divider strips are located only at control joints in the screeded bed, because the possibility of surface cracking in this instance is small. They are halved at intersections and embedded in the newly laid screeded bed.

The top edges of the divider strips are kept uniformly 1 mm above the level of the finished surface so that they can be ground flush with it. The strips can be anchored to the terrazzo course by brass screws projecting alternately on each side at 150 mm centres. At construction or control joints, two divider angles or 3 mm strips are bedded in mortar on the base slab, and secured 6 mm apart with the aid of brass screws projecting vertically or horizontally at 150 mm centres into the screeded bed. The grooves thus formed are filled subsequently with a weak vermiculite mortar and a 6 mm depth of mastic.

Cast *in situ* terrazzo is laid as soon as the newly laid screeded bed has become sufficiently firm for surfacing purposes. After being screeded and provided with a small surcharge, it is compacted thoroughly by two or three rammings (at 15–20-minute intervals) or by rolling with a light roller (e.g. 25 kg) six times in each of two directions. The marble-packed surface is then floated, trowelled and cured (see Section 20.1.5(c),(d) and (e)). Where border or central decorations are required, they should be laid first with a similarly proportioned mix.

After 3 or 4 days of curing, the surface is wet ground, hosed and scrubbed clean, and treated with mortar that is similar in composition to the terrazzo matrix, so that any surface pittings or blemishes will be uniformly filled. After another 3 or 4 days of curing, the surface is more finely ground with successively finer grades of carborundum stone. It is then thoroughly cleaned, made good with matching paste where residual air holes occur, and wet ground again if necessary.

The floor, after the final grinding, is washed with hot water containing pure soft soap and, after drying, it can be slightly waxed to prevent dirt being absorbed by the surface. New terrazzo floors should be washed and waxed in this way every week for several months, while precast products, after they have thoroughly dried, may be sealed and polished directly. Of various preparations for the purpose, a typical composition comprises 1 litre of boiled linseed oil, 1 litre of turpentine, 0·5 litre of vinegar and 28 g of butter of antimony, the third component being omitted for use on a matured or neutralised surface (see Section 20.2(d); and References 49 and 111 of Part 4).

Pressure-moulded tiles for terrazzo finishes are backed with cement mortar or concrete, machine polished, and produced in sizes ranging from 150 mm square × 20 mm thick to 600 mm square × 40 mm thick (BS 4131). They are first brought to a damp condition by soaking them in water for about 15 minutes and draining them for a similar period. Cement paste of thick-paint consistence is then spread about 1 mm thick over the backs of the tiles just prior to laying them on the screeded bed, when they are tamped or tapped to the requisite level. Alternatively, several of the dampened tiles are laid at a time over a thin coating of cement paste, which has been applied over a relevant area of the screeded bed. The joints, which should be straight and 2-3 mm wide, are afterwards brushed clean and filled with a slurry of coloured cement and fine sand. The surface may be finally wet ground, cleaned and polished after 3 days. Precast terrazzo-faced concrete is specified in BS 4357.

36.4.4 MODIFICATIONS

Alternative terrazzo matrixes include fleximer compound and epoxy resin (see Sections 20.2(h) and 21.1.6(c)). Constructional modifications for conductive flooring (AS 1169 and BS 2050, CP204) include the following items.

An unbonded screeded bed provided with a vapour barrier and earthed reinforcement.
An admixture of acetylene carbon black introduced at the rate of 0·6 kg per 40-kg bag of cement in the screeded bed and 0·4 kg per 40-kg bag in the terrazzo and ancillary mortar mixes. Coloured ebonite or bakelite divider strips with dowelled anchorages of similar material.
Light mesh reinforcement in precast terrazzo tiles.[102, 123]

36.5 ROOF MEMBRANE SYSTEMS

Some typical roof membrane systems for weatherproofing concrete are as follows.

1. A three-layer bituminous felt is commonly used with a drainage fall of not less than 1 in 60 on flat roofs. Felts for the purpose are based on organic fibre, asbestos fibre, or glass fibre, the last-mentioned being characterised by dimensional stability in service performance. The first layer, which may have a gritted under-surface for vapour transmission to the verges, is partially bonded *in situ* with thirty-penetration blown bitumen having a softening

point of 85°C. A highly absorptive or somewhat dusting surface should first receive a bituminous primer, and a partial bonding of the underlay mentioned is facilitated by 12 mm diameter perforations in the underlay itself. The succeeding layers are fully bonded with bitumen, which is applied at a temperature of 200-230°C; some 14 kg of bitumen per 10 m² producing a layer a little less than 2 mm thick.

A roof-deck rendering, after being laid as specified in Section 33.4.2, should be protected initially with tarpaulins or polyethylene sheets in rainy weather. Provision should be made for proper drainage, drying, dimensional movements, ventilation and effective insulation requirements (see Sections 33.4.2 and 33.4.4). Shrinkage-line cuts in the rendering should be covered with strips of felt, 300-450 mm wide, these being bonded at both edges for a 50 mm width so that the overlying felt will stretch, rather than tear in response to minor deck movements. A flexible ridge formed by a rubber or plastic tube over a joint will accommodate larger movements but, at intervals of about 24 ± 6 m, a twin-kerb-and-cover control joint should be installed to take up major movements.

At parapet walls, fillets and felt upstands should be at least 150 mm high, the latter being covered with metal flashing that is made continuous with a damp-proof course. The built-up membrane is protected with a reflective surfacing, such as aluminium paint or light-coloured grit, or 25 mm concrete tiles laid in 9 m² bays that are separated by 10 mm joints. General sources of facts on this system are AS A98, A99, A120, A121, CA55 and BS 747, CP144.101; also ASTM D224, D226, D249, D250, D312, D371, D655, D1227, D1668; and References 95, 97 and 98 of Part 7.

2. For curved surfaces, a bituminous fibreglass mat approximately 5 mm thick can be formed *in situ*. It comprises a tack coat of clay-stabilised bituminous emulsion (see Section 20.2(g)) and two built-up surface coatings, each of which embodies one coat of fibreglass tissue and two dressings of clay-stabilised bituminous emulsion. A drying period is allowed after the second and fourth dressings of bituminous emulsion which, under slow drying conditions, should not become wetted by rain for 2 days, if possible. The hardened surface may be coated with bitumen-based aluminium paint, which is renewed after 2 years and then at 5-year intervals.

3. Four or five coats of synthetic rubber solution may be applied to a primed sound surface, each coat being 0·13 mm thick. A typical system contains two or three coats of polychloroprene (e.g. "Neoprene") and two coats of pigmented chlorosulphonated polyethylene (e.g. "Hypalon"). Surface priming can be done with a penetrating grade of chlorinated rubber lacquer or a polyamide-cured epoxy resin solution.

A silicone acrylic copolymer (see Sections 20.2, 20.2(e), 21.1.6(c), 28.3 and 28.4), may be used as an alternative surface primer and pigmented sealer (with or without filler), and strips of nylon tissue should be incorporated in dressings where they cross weakened-plane joints. A tough ethylene propylene terpolymer sheet (1·50 mm thick) may be laid with bituminous adhesive. Resilient bearings (5-7 mm thick; designed to carry up to 60 tonnes per m or ± 10 mm movements) are of rubber strip, slide or pads that are enclosed laterally with polystyrene or polyurethane sheet foam.

36.6 PRECAUTIONS AGAINST EFFLORESCENCE

Efflorescence (see Glossary, Appendix VI) may develop shortly after masonry walls have been built, because of the presence of certain soluble compounds (e.g. sodium, potassium and calcium compounds) and an abnormal moisture content. Relevant salts are deposited on surfaces as the moisture dries and atmospheric carbonation converts calcium hydroxide into calcium carbonate. The compounds referred to are inherent in the masonry materials used, and they may arise from jointing mortar, concrete building elements, back-up masonry and mortar, and ground-water contaminants (see Section 16.2). Initial dryness is required during construction (see Section 27.2.2(a)).

The amount (if any) of efflorescence produced annually decreases from year to year. The effect tends to be pronounced on masonry units that have, amongst other things, a high rate of water absorption. Before attempting to remove efflorescence, building faults which cause an abnormal wetting of walls should be corrected. This requires some attention to ground water conditions, damp-proof courses and facilities for drainage.

The blemish is removed usually by scrubbing the surface with water or, after being dampened, with 3 per cent hydrochloric acid solution and rinsing with water. Alternatively a dried surface may be wetted for 0·25 hour with 25 per cent diammonium citrate solution and scrubbed with water a few times. Otherwise, affected areas are wetted and plastered with moist papier-mâché, the resultant salt-absorbed layer being peeled off when dry and the process repeated.

Precautionary measures against efflorescence include the selection of materials with a low content of soluble salts, and building details that will protect masonry walls from excessive dampness. Concrete products should be stored suitably on job sites so as to avoid early drying and rewetting cycles, and contamination from salt-carrying ground water. Decorative facing slabs should be vented (see Section 28.12), and vapourproof membranes or water-shedding flashings should be used above substructure walls, bond beams or lintels, through parapets just above roof level, and below weep holes, copings and sills (see Section 27.2.2(d)).[98] Large eaves, splash shields, drip edges on projections and two coats of "breathing"-type paint (see Sections 28.2, 28.2.7 and 28.4) minimise wetting and drying effects in service (see Reference 184 of Part 5).

Efflorescence can be suppressed by retarding the initial evaporation of moisture from freshly screeded or reworked concrete surfaces, such as by spraying them with a film of dilute aliphatic (cetyl) alcohol prior to curing (see Sections 9.3, 31.2.4 and 32.2), and by spraying concrete products with a metallic silicofluoride, clear silicone acrylic or alkyd resin solution. A coating of curing compound emulsion or chlorinated rubber solution (see Section 9.3) on vented products can divert efflorescence, caused by early evaporation, onto surfaces where it does not matter.

Other factors for minimising possible efflorescence include

Low water/cement ratio and low rate of absorption: Sections 3.2 and 11.2.
Low-alkali and slow-hardening portland cements: Section 3.1.1.
Dense, uniform-quality concrete and masonry: Sections 6.4, 8.1 and 23.2.
Lime-saturated water, fog-spray or evaporation-free steam curing: Sections 9.2, 9.4 and 14.4.
Clean water and washed graded sand: Sections 3.2.1, 3.3.4(e) and 16.2.
Air-entraining or water-reducing agent: Sections 6.4 and 16.1.
Bleached ultramarine pigment, reactive with calcium hydroxide and producing high-early strength: Section 17.1.
Pozzolanic replacement of sand fines (e.g. 10-15 per cent by

weight of cement) and steam curing; portland blastfurnace cement; kaolin densifier in masonry: Section 17.1 and 18.3.[101] An admixture of paraffin wax emulsion, or butyl or ammonium stearate, where hardened water-repellent surfaces are not to be coated with cementitious material: Section 11.1.2(c).
Stain-suppressing release agent: Section 8.3.
Fired-glaze concrete masonry or dry mature units: Section 27.2.
Combustion-gas carbonation curing of medium-porosity concrete masonry after a 3-hour hydration period: Chapters 9 and 31.
High-pressure steam curing of mixes containing silica flour or fly ash: Section 9.4.6.

Concrete roofing tiles should have a waterproof surface and a low rate of water absorption, and they should be cured in a humid atmosphere at about 40°C for 24 hours. Care should be taken to prevent draughts, rapid evaporation and surface condensation through too high a curing-room temperature. A test for the perceptible efflorescence of clay bricks, when brought in contact with distilled water for several days, is described in AS 1225, BS 1257, 3921 and ASTM C67.

36.7 CUTTING THROUGH CONCRETE

36.7.1 DIAMOND-TIPPED TOOLS

Concrete saws and hollow drills are available from suppliers of specialist service for purposes falling outside the scope of pneumatic or electric hammers for severing concrete. When operated under a flow of water for cooling and dust-abatement purposes, concrete saws and core drills provide a useful means for cutting joints in concrete pavements (see Section 20.1.5(f)) between 8 and 24 hours after placement, and removing test specimens from hardened concrete (see Section 4.4 and BS 2064).[100, 131]
   Small thick sections of a heavily reinforced concrete wall (that have been purposely isolated by core-drilling) can be removed with the aid of a splitter. The procedure is useful where severance by an oxythermic lance (see Section 36.7.2) may initiate a fire hazard by causing a localised abnormal flow of molten steel within a business building. Operatives are advised to wear earmuffs when extensively using diamond-tipped tools, especially concrete saws.

36.7.2 OXYTHERMIC LANCING

A special oxygen-fired lance is used for cutting reinforced concrete. For instance, in making an opening in a wall, a number of perforations at about 150 mm centres are made around and across the area, which is to be isolated and finally knocked out with a sledge hammer. Cutting operations can be carried out alternatively by moving a carriage-mounted lance across a surface at the rate of about 50 mm/minute. The system is useful where noise, vibration and damage must be kept to a minimum in demolition or repair work.[99]

The lance consists of a black mild-steel pipe (e.g. 12 mm diameter by 3 m long) packed with mild steel wire (and sometimes aluminium wire for increased generation of heat) about 2·4 mm diameter. Alternatively, iron and aluminium powders may be fed pneumatically into the lance. The size of the lance is adapted to suit the size of individual holes required. The end is heated to full red heat (870°C) by means of an oxyacetylene welding blowpipe (1 m³/hour capacity), and subsequently ignited by oxygen supplied under pressure (400-900 kPa). It is normally used at 2500°C.

The oxygen is stored in 6·25 m³ cylinders and passed through a control valve, 10 mm diameter high-pressure rubber tubing, and a large-capacity, double-reduction regulator. The heated pipe and wire burn at a temperature sufficiently high to melt cement, sand and aggregate, and burn through steel reinforcement. A steady pressure is applied and subsequently accompanied by a swinging rotary movement (applied by two operators), so as progressively to enlarge the hole being formed (e.g. to 50–65 mm diameter). The rate of penetration with this lance is approximately 300 mm/minute. The consumption of oxygen and the lance is about 3·5 m³ and 4·5 m per m of hole, respectively. Operators are protected by leather clothing and an electric welder's helmet fitted with a light-shade lens. Strong-room security is enhanced by a lance-detector sonic system.

Miscellaneous applications of thermic lancing include the removal of slag from the gas passages in furnaces, cutting concrete for conduits or bolt fixtures, and the perforation of atomic reactor shields. While the procedure is not the most economical, it has the advantages of noise and vibration abatement, speed of operation and portability of equipment. Massive sections are readily cut with a powder lance, in which a mixture of aluminium and iron powder is burnt (in contact with pure oxygen) at a temperature of 3500°C. Thermic lancing is unsuitable for cutting limestone and dolomite, because of inherent carbon dioxide.

36.8 DEMOLITION WORK

36.8.1 ANALYSIS

Buildings require demolition when they exceed their span of economic usefulness. Demolition must be carried out in a safe and efficient way, according to the nature of each structure. In built-up areas, a concrete building is usually less risky than a brick one, from the standpoint of safety to the operatives, the public and adjoining buildings during demolition. The cost factor is greater, however, because of the time and equipment required, and the low salvage value of materials concerned.

A slow operation is expensive not only in man-hours, but in lost earning capacity while the site is unproductive. For a high rate of progress with low unit costs, it is necessary to use an assortment of mechanical tools on concrete structures. Salvaged items suitable for economic disposal include rolled-steel joists that have been stripped of concrete, and materials or fittings that can be recycled or reused (BS CP94).[154]

36.8.2 EQUIPMENT AND EXECUTION

For the demolition of large reinforced concrete structures, the mechanical equipment required includes such items as an air-compressor, pneumatic tools, a tower crane or crawler and mobile slewing cranes (e.g. 36 m or 55 m high), drop-balls and swing-balls, tractor-hammers, explosives, a pneumatic impact-breaker, cutting gear, tubular steel scaffolding, light front-end loaders and tip-trucks for the removal of debris. The particular type and capacity of equipment are governed by the size of the project.

Pneumatic tools vary in weight and size from 5 kg to over 2 tonnes, the heavy units being operated from a crane or possibly a pile-driving frame. The type and size of cranes are governed by structural considerations and site accessibility, large units requiring assembly and dismantlement. Cast-iron drop-balls and swing-balls vary in size from 0·5 to 1·5 tonnes or more, and they operate from heights of about 2·5-14 m. Drop-balls are of cylindrical shape with a blunt spear-pointed head, while swing-balls are pear-shaped for destructive efficiency with repeated blows. An hydraulic-impulse breaker or a pneumatic tractor-hammer is useful for dismembering concrete slabs or decks, using a quick succession of blows with a high rate of breakage.

Gladesville Bridge, Sydney
(Main span: 300 m)

Rip Channel Bridge, Brisbane Water, NSW

Approach viaduct of West Gate Bridge, Melbourne

Approach viaduct of West Gate Bridge, Melbourne

Explosives facilitate the fracture of large blocks of concrete, such as engine beds and column bases, or the progressive exposure of reinforcement in heavy sections. Delayed-action detonators are used for this purpose and to reduce the transmission of destructive shock waves to adjoining structures. Police and municipal permits must be obtained for the use of explosives, which require exacting safety precautions and noise control (AS CA23). In massive work, masonry drills, rock jacks, steel shims and wedges operated hydraulically are used for splitting purposes.

Various concrete-cutting facilities include oxythermic lancing, high-pressure water jetting (e.g. 100 MPa), laser or microwave technique, electron bombardment, diamond-tipped core drills and jig-guided saws. Reinforcing steel in concrete aids the burning efficacy of an oxythermic lance. Breaking-out work can be expedited along a line of overlapping holes or, where feasible, by expanding hydraulic or carbon dioxide-powered bursters in spaced holes.

Exposed steel rods up to 12 mm diameter can be sheared by bolt-cutters, thicker sections being cut readily by oxyacetylene or propane blowtorches. The simultaneous severance of steel-frame members can be carried out by remotely initiating and electrically igniting enclosed thermit powder: an aluminium and iron oxide mixture. Controlled collapse may be monitored by the pulling of attached steel hawsers with a crawler tractor.

Scaffolding, formwork, toms and shores are used with certain members that require dismantlement or support (e.g. spatial roofs, walls and timber floors). Structural principles should be observed when dealing with large prestressed concrete members. For instance, beams that have been stage-post-tensioned during construction may need partial destressing, in order to prevent their destructive hogging during the removal of upper storeys in tall buildings. Tendon-stress release can be induced by such means as concrete shortening, energy dissipation and electrical annealing of prestressed steel.

A general guide to organisation is to work on one part of a project at a time, and cut it into the largest sections that can be handled conveniently by equipment that could be used at the site. For example, sections weighing up to 2·5 tonnes could be handled by a 36 m crane and those up to 5 tonnes by a 55 m crane. With a multistorey building, the general procedure is to lift pneumatic tools, cutting appliances and a tractor-hammer by crane to the roof and top floors.

In the course of dismemberment, slabs are cut into sections up to 1·8 m wide, beams and columns are severed at their ends, and walls are cut into panels and undermined. Slabs and beams are supported in slings while the exposed reinforcement is being cut prior to their removal by crane to tip-trucks. Columns and wall panels are pulled over (onto a concrete floor), cut free and removed in segments by crane. When demolition reaches the levels of the lower floors, drop-balls and swing-balls may be used where flying fragments or falling sections will not be hazardous to the safety of personnel.

On sites where cranes cannot be used, the dismembered sections must be small enough in size to drop within the structure. Protective decking must be installed around buildings in built-up areas as a safety precaution. Defective, party-wall brickwork may be shored and strengthened with reinforced, surface-keyed, pneumatic mortar, in order to spread shore loads and serve as permanent formwork for subsequent reinforced concrete construction (see Section 24.2). Health and safety precautions require collective specialist services and demolishing techniques when handling the radioactive components of decommissioned nuclear-power stations.

36.8.3 RECOMMENDATIONS

Before building new concrete walls against those of existing buildings, the adjacent surfaces should be bituminised to ensure separation. A sound-insulating membrane, which would reduce intense vibration and noise transmission during demolition, should be secured against possible dislodgement when concreting is proceeding. A set of constructional drawings, supplemented with updated technical records of unusual features and structural changes, should be deposited in a cache in an accessible part of each new building, so as to facilitate future estimates and methods of demolition, including the installation of cranes. Structural records, otherwise, should be available for reference at Local Government Offices. Demolition work, which is construction in reverse, should be planned and proceeded with expeditiously.

36.9 PROJECT MANAGEMENT SYSTEMS

36.9.1 PRODUCTIVITY

For the effective execution of concrete production and construction projects, consideration must be given to the financial, personnel and industrial factors involved. These matters include such items as the business administration of working capital and organisational efficiency, cost/benefit evaluations, productivity and monetary indexes, and the best utilisation of resources to meet planned objectives (Appendix IV, 3). Achievement of the latter goal is expedited in some enterprises by elected craft-group representatives forming themselves into a works council which meets with management regularly for dialogue on emergent new ideas and disciplines, and their effective implementation to corporate and communal advantage. This mode of participation by consultation and co-determination is supplemented with incentive systems of motivation and vocational development, in order to activate creative service systems and keep executive personnel ahead of the implications of change.

On the industrial line, precast and processed products are being made in modular sizes in order to suit dimensional coordination and quick, economical worksite assembly (see Section 27.1). In consequence, there is being firmly established a strong demand for accurate programming in the precise manufacture, supply and fabrication of standardised components to predetermined schedules. At the same time, to minimise the time and effort spent on extensively repetitive and calculable work, logically arranged data are being processed by electronic computers (BS 3527).

To expedite the transformation of ideas into results, therefore, all procedures are being simplified, standardised, classified, codified, edited and automated for effective economic evaluation and control in large undertakings. Concurrently, current cost accounting, revelatory auditing, supporting services and security creditations are being used to assist management in evaluating performances and monitoring progress through productivity, with rapid means of viewing and directing factual events to beneficial effect.

36.9.2 METHODOLOGY

With progressive industrialisation, automated analytical techniques and data-management systems are being increasingly used as flexible tools to assist management in the task of solving practical problems and arriving at best decisions. By means of operational research in major industry, commerce and construction, for instance, it is often possible to deduce from past or simulated operations a more efficient scheduling system and way of achieving maximum results with the least expenditure of manpower, materials, energy, effort or money.

When operational research is used, it requires the collation of relevant data and the development of a mathematical or probabilistic model of the system that is under study. By manipulating the model and studying its behaviour, it is possible to predict the relevant effects of operational factors on the actual system thus simulated. One of the methods employed is to discover elements of constancy or stability in operations that have been repeated over a period of time; and, by identifying functions and evaluating their worth, to reduce the variables appearing in certain operations under examination (see Mallon, Appendix IV, 3; also Glossary, Appendix VI).

Technical methods of investigation, that are applicable to many industrial projects, involve such disciplines as linear, modular, dynamic and mixed integer programming, and the use of queuing, games, diversion and renewal theories, dimensional analysis, systems design and simulation technique. By sampling activities with work and method studies, videotape and variable-speed visual aids, operational systems can be planned to reflect improved organisation. These procedures can be used for such purposes as controlling inventories and costs, reducing the capital tied up in stocks, estimating manpower requirements, studying feasibility, optimising design, raising productivity, and revealing trends or probabilities that are in need of corrective action.

36.9.3 NETWORK SYSTEMS

On concreting and building projects, as well as in other forms of industry, the pictorial representation of a proposed working itinerary is an effective means of scheduling and controlling the flow of operations. For this purpose, a closed line-and-arrow diagram of a nodal-precedence diagram is prepared to serve as an operational model, which shows the interrelationships of all activities comprising a project. Appended information includes their duration, costs and requisite resources; and the earliest starting time, latest finishing time and the float or slack time are calculated for each activity (BS 4335).

The two principal techniques for analysing network systems are known as Critical Path Scheduling and PERT (Programme Evaluation and Revue Technique; see Appendix IV, 3). The former is applicable to most individual or multiple contracts, while the latter is more useful for projects where the activity times cannot be accurately estimated. Depending on the size or number of projects to be controlled and the amount of work involved, network systems may be analysed as a logical arithmetical exercise, either manually for a small project or by means of an electronic computer (AS X2).

The critical path of a network system is the particular sequence of activities that determines the duration of a project. It is that path of sequential activities that has no expendable float or slack time. When the critical path is known, other operations can be keyed into it and attempts made to shorten the total working time by concentrating on those critical activities that offer the greatest opportunities for improvement.

In comparison with bar chart schedules, a network analysis reveals not only the probable dates of key events, but also the effects on progress of any changes made in them. It enables an orderly assignment to be made of the available resources of equipment and manpower, first of all to critical activities and then to noncritical ones, with an order of priority that is suitably related to their amounts of slack time. The process can be used advantageously for estimating, scheduling work forces and material resources, comparing alternative methods of procedure, reviewing progress, evaluating deviations from schedule, and for recording costs and other data for future reference.

36.9.4 POTENTIAL

After rigorously processing all relevant data and evaluating new concepts by a computer, managerial decisions to take important courses of action must, in a team environment, be tempered with experience and sound intuitive judgment for best results.

In making recorded steady progress through the skilful use of operational systems, it is innate ingenuity, enthusiasm, concentration, thoroughness and persistent effort that enable people of diligence, destiny and determination to achieve and prevail. Hence, in reality

Man the shaper and maker,
Man the dreamer and doer,
Man the seeker and trier,
Man the arriver and answerer

gleans the fact that it is common sense and hard work, coupled with fitness, exertion and efficiency in performance that enable individuals to earn the right to quality of life in the human stream of cumulative creative services for beneficial purposes.

REFERENCES

1. COWAN, H. J. and SMITH, P. R. *Design of Prestressed Concrete in Accordance with the S.A.A. Code.* Angus & Robertson, Sydney, 1966.
2. BIRKELAND, H. W. "How to Design Prestressed Concrete Beams of Minimum Cross Section". *ACI Journal,* 71, 12 (December 1974), pp. 634-41.
3. *Prestressed Precast Concrete Design Handbook for Standard Products.* Concrete Materials, Charlotte N.C. 1968.
4. ANDERSON, A. R. and MOUSTAFA, S. E. "Ultimate Strength of Prestressed Concrete Piles and Columns". *ACI Journal,* 67, 8 (August 1970), pp. 620-35.
5. ABELES, P. W. *Introduction to Prestressed Concrete,* vols I and II. *Prestressed Concrete Designer's Handbook,* C & CA, Britain, 1971-76.
6. GUYON, Y. *Limit-State Design of Prestressed Concrete,* vols 1 and 2. Applied Science, Britain, 1974.
7. ANTILL, J. M. and NAGARAJAN, R. "The 1974 Australian Code for Prestressed Concrete". *IE Aust. CE Transactions,* CE17, 1 (June 1975), pp. 13-17.
8. ACI COMMITTEE 318. "Building Code Requirements for Reinforced Concrete, Chapter 18 Prestressed Concrete". *ACI Publication SP-10 and Commentary,* or *Manual of Concrete Practice,* Part 2; Appendix IV, 2 (a) (i).
9. ACI-ASCE COMMITTEE 423. "Concrete Members Prestressed With Unbonded Tendons". *ACI Journal,* 66, 2 (February 1969), pp. 81-87.
10. ACI COMMITTEE 325. "Design for Experimental Prestressed Pavement Slab". *ACI Journal,* 65, 4 (April 1968), pp. 249-65.
11. EDWARDS, A. D. and PICARDE, A. "Theory of Cracking in Concrete Members". *ASCE Proceedings,* 98, ST12 (December 1972), pp. 2687-2700.
12. "Composite Concrete Beams". *Bulletin 18,* Reinforced Concrete Research Council (PCA, Skokie, United States), 1968.
13. BARTLETT, R. J. "Shrinkage Stresses in Composite Floors". *Concrete and Constructional Engineering,* 53, 12 (December 1958), pp. 415-19; 54, 3 (March 1959), p. 119.
14. BRANSON, D. E. and OZELL, A. M. "A Report on Differential Shrinkage in Composite Prestressed Concrete Beams". *PCI Journal,* 4, 3 (December 1959), pp. 61-79.
15. SCHAFFER, H. R. "Computing Prestressed Concrete Composite Sections". *Civil Engineering,* 30, 6 (June 1960), pp. 72-73.
16. ACI-ASCE COMMITTEE 333. "Tentative Recommendations for Design of Composite Beams and Girders for Buildings". *ACI Journal,* 32, 6 (December 1960), pp. 609-28; 12 (June 1961), pp. 1659-64.
17. BIRKELAND, H. W. "Differential Shrinkage in Composite Beams". *ACI Journal,* 31, 11 (May 1960), pp. 1123-36; 32, 6, 2 (December 1960), pp. 1529-58.
18. KAJFASZ, S., SOMERVILLE, G. and ROWE, R. E. "An Investigation of the Behaviour of Composite Concrete Beams". *Research Report 15,* C & CA, Britain, 1963.
19. HAWKINS, N. M. "Behaviour and Design of End Blocks for Prestressed Concrete Beams". *IE Aust. CE Transactions,* CE8, 2 (October 1966), pp. 193-202; CE9, 1 (April 1967), pp. 169-70.
20. DAY, M. F., JENKINSON, E. A. and SMITH, A. I. "Effect of Elevated Temperatures on High-tensile-steel Wires for Prestressed Concrete". *ICE Proceedings,* 16 (May 1960), pp. 55-70.
21. WANG, Y.-L. "Design Chart for Computing Ultimate Flexural Strength of Prestressed Concrete Members". *ACI Journal,* 62, 9 (September 1965), pp. 1109-11.
22. HESPE, F. S. "Concrete Railway Sleepers". *Constructional Review,* 33, 1 (January 1960), pp. 30-38.
23. FREYERMUTH, C. L. "Design of Continuous Highway Bridges with Precast, Prestressed Concrete Girders". *PCI Journal,* 14, 2 (April 1969), pp. 14-39.

24. TADROS, M. K., GHALI, A. and DILGER, W. H. "Time-dependent Prestress Loss and Deflection in Prestressed Concrete Members". *PCI Journal*, 20, 3 (May/June 1975), pp. 86-98.
25. WASS, B. J. "Continuous Castings". *Concrete* (London), 4, 8 (August 1970), pp. 317-18.
26. RICE, L. "Computer Plotted Prestressed Concrete Design Charts". *ASCE Proceedings*, 97, ST3 (March 1971), pp. 863-79.
27. PRESTRESSED CONCRETE DEVELOPMENT GROUP. "Inverted T Beams for Spans from 25 to 55 ft. Box and I Section Beams for Spans from 40 to 85 ft. Box and I Section Beams for Spans from 85 to 120 ft." *Standard Beam Sections for Prestressed Concrete Bridges*, Parts 1-5, C & CA, Britain, 1963-1964.
28. GERGELY, P. and SOZEN, M. A. "Design of Anchorage-Zone Reinforcement in Prestressed Concrete Beams". *PCI Journal*, 12, 2 (April 1967), pp. 63-75.
29. ACI-ASCE COMMITTEE 423. "Prestressed Concrete Flat Plates". *ACI Journal*, 71, 2 (February 1974), pp. 61-71; 8 (August 1974), pp. 424-26.
30. GOBLE, G. G. and LAPAY, W. S. "Optimum Design of Prestressed Beams". *ACI Journal*, 68, 9 (September 1971), pp. 712-18.
31. GREEN, J. K. "Design of Bridges using Prestressed Concrete Beams of Standard Section". *Technical Paper PCS 7*, The Concrete Society, 1967.
32. REJCHA, C. "Simplified Bearing Plate Computations for Post-tensioning Anchorages". *PCI Journal*, 20, 4 (July/August 1975), pp. 102-111.
33. ACI COMMITTEE 215. "Symposium on Fatigue of Concrete". *ACI Journal*, 30, 2 (August 1958), pp. 191-259; *ACI SP-41*, 1975.
34. CREASEY, L. R. *Prestressed Concrete Cylindrical Tanks*. Contractors Record, London 1963; *ACI Journal*, 67, 9 (September 1970), pp. 657-72.
35. MAGUIRE, R. "Precast Prestressed Floors". *The Architectural Review*, 119, 710 (February 1956), pp. 142-46.
36. VOELCKER, J. "Floor Assemblies". *Architectural Design*, 26, 2 (February 1956), pp. 61-66.
37. SWANSON, H. V. "Design of Prestressed Concrete Pressure Pipe". *PCI Journal*, 10, 4 (August 1965), pp. 69-82.
38. CREPPS, R. B. "Wire-wound Prestressed Concrete Pressure Pipe". *ACI Journal*, 14, 6 (June 1943), pp. 545-55.
39. EVANS, R. H. "Application of Prestressed Concrete to Water Supply and Drainage". *ICE Proceedings*, 4, 3 (December 1955) Part 3, pp. 725-75.
40. CLARKE, N. W. B. and YOUNG, O. C. "Some Structural Aspects of the Design of Concrete Pipelines". *ICE Proceedings*, 14 (September 1959), pp. 67-96; 16 (May 1960), pp. 90-105.
41. ACI COMMITTEE 344. "Design and Construction of Circular Prestressed Concrete Structures". *ACI Manual of Concrete Practice*, Part 2, 1976.
42. THAKKAR, M. C. and BULSARI, B. S. "Optimal Design of Prestressed Concrete Poles". *ASCE Proceedings*, 98, ST1 (January 1972), pp. 61-74.
43. GUSTAFERRO, A. H. "Design of Prestressed Concrete for Fire Resistance". *PCI Journal*, 18, 6 (November/December 1973), pp. 102-16.
44. ARTHUR, P. D. and GANGULI, S. "Tests on End-zone Stresses in Pre-tensioned Concrete I Beams". *Prestress*, 15 (March 1966), pp. 4-22.
45. GODFREY, K. A., Jr. "Here's a Roundup of Post-tensioning Systems". *Concrete Products*, 64, 1 (January 1961), pp. 36-44; 3 (March 1961), pp. 42-70.
46. ACI COMMITTEE 435. "Deflections of Prestressed Concrete Members". *ACI Journal*, 60, 12 (December 1963), pp. 1697-1728.
47. LI, S.-T. and LIU, T. C.-Y. "Prestressed Concrete Piling—Contemporary Design Practice and Recommendations". *ACI Journal*, 67, 3 (March 1970), pp. 201-20.
48. WILLIAMS, N. S. "Transporting, Handling, Pitching and Driving of Prestressed Concrete Piles". *Structural Concrete*, 2, 5 (September/October 1964), pp. 212-33.

49. NERVI, P. L. *Structures.* F. W. Dodge Corporation, New York, 1956.
50. NERVI, P. L. *New Structures.* Architectural Press, London, 1965.
51. JOHNSTON, C. D. and MAFTAR, S. G. "Ferrocement—Behaviour in Tension and Compression". *ASCE Proceedings,* 102, ST5 (May 1976), pp. 875-89.
52. JOHNSON, C. D. and MOWAT, D. N. "Ferrocement—Material Behaviour in Flexure". *ASCE Journal,* 100 ST10 (October 1974), pp. 2053-69.
53. WALKUS, R. and KOWALSKI, T. G. "Ferrocement: A Survey". *Concrete* (London), 5, 2 (February 1971), pp. 48-52.
54. *Connection Details for Precast Prestressed Concrete,* C & CA, Aust., Sydney, 1966.
55. ACI COMMITTEE 334. "Concrete Shell Structures—Practice and Commentary". *ACI Journal,* 61, 9 (September 1964), pp. 1091-1108; 62, 2 (February 1965), p. NL35; 3, 2 (March 1965), pp. 1755-65; 6, 2, p. v; *Special Publication SP-28.*
56. "Symposium on Shell Structures". *Indian Concrete Journal,* 33, 12 (December 1959), pp. 401-80.
57. PARME, A. L. and CONNER, H. W. "Design Constants for Ribless Concrete Cylindrical Shells". *Bulletin No. 18,* International Association for Shell Structures, Madrid, 1964.
58. KALRA, M. L. "Formulas for Calculating Stresses in Long Cylindrical Single Shells". *Indian Concrete Journal,* 46, 2 (February 1972), pp. 70-75, 82.
59. BROTCHIE, J. F. "A Concept for the Direct Design of Structures". *IE Aust. Journal,* CE5, 2 (September 1963), pp. 61-66; CE6, 1 (March 1964), pp. 36-40.
60. CANDELA, F. "General Formulas for Membrane Stresses in Hyperbolic Paraboloidical Shells". *ACI Journal,* 32, 4 (October 1960), pp. 353-71.
61. AHM, P. and PERRY, E. J. "Design of the Dome Shell Roof for Smithfield Poultry Market". *ICE Proceedings,* 30 (January 1965), pp. 79-108; 35 (October 1966), pp. 316-35.
62. PAUW, A. and SANGSTER, W. M. "Characteristic Equation of Cylindrical Shells—A Simplified Method of Solution". *ACI Journal,* 59, 10 (October 1962), pp. 1505-12.
63. RAMASWAMY, G. S., TAMHANKAR, M. G. and NAITHANI, K. C. "Precast Pretensioned Folded Plate Roofs". *Indian Concrete Journal,* 40, 1 (January 1966), pp. 6-12.
64. SHAH, H. C. "Analysis of Continuous Folded Plate Roofs". *Indian Concrete Journal,* 41, 6 (June 1967), pp. 242-49.
65. FISCHER, R. E. *Architectural Engineering: New Structures.* McGraw-Hill, New York, 1964.
66. LIN, T. Y. *Design of Prestressed Concrete Structures.* Wiley, New York, 1964.
67. BENNETT, E. W. and MUIR, E. E. St J. "Some Fatigue Tests of High-strength Concrete in Axial Compression". *Magazine of Concrete Research,* 19, 59 (June 1967), pp. 113-17.
68. REED, J. J. and MANN, C. D. "Full Scale Testing Develops Efficient Preloaded Concrete Pillars". *ACI Journal,* 58, 5 (November 1961), pp. 625-38.
69. BRETTLE, H. J. "Effect of Variable Concrete Modulus of Elasticity on the Deflection of Prestressed Beams". *Australian Journal of Applied Science,* 11, 1 (March 1960), pp. 92-107.
70. RATHS, C. H. "Design Considerations for a Precast Prestressed Apartment Building". *PCI Journal,* 19, 1 (January/February 1974), pp. 12-61; 2 (March/April 1974), pp. 66-92; 6 (November/December 1974), pp. 16-27; 20, 1 (January/February 1975), pp. 74-98; 2 (March/April 1975), pp. 42-58.
71. TURNER, F. H. "Equipment for Prestressed Concrete". *Concrete and Constructional Engineering,* 57, 3 (March 1962), pp. 128-33; 4 (April 1962), pp. 152-59; 58, 5 (May 1963), pp. 219-22.
72. "Symposium on Concrete Construction in Aqueous Environments". *ACI Special Publication SP-8,* 1964.

73. BRUCE, R. N. and HEBERT, D. C. "Splicing of Precast Prestressed Concrete Piles". *PCI Journal,* 19, 5 (September/October 1974), pp. 70-97; 6 (November/December 1974), pp. 40-66.
74. GERWICK, B. C. "Principles of Mechanised Production of Prestressed Concrete". *PCI Journal,* 6, 3 (September 1961), pp. 14-21.
75. MIKHAILOV, V. V. "Recent Developments in Automatic Manufacture of Prestressed Members in the USSR". *PCI Journal,* 6, 3 (September 1961), pp. 34-46.
76. SKRAMTAEV, B. G. "Electrothermic Method of Pretensioning Bar Reinforcement of Precast Reinforced Concrete". *PCI Journal,* 6, 3 (September 1961), pp. 57-71.
77. EDWARDS, H. H. "Extrusion of Prestressed Concrete". *PCI Journal,* 6, 3 (September 1961), pp. 47-53.
78. HUANG, T. "Estimating Stress for a Prestressed Concrete Member". *PCI Journal,* 17, 1 (January/February 1972), pp. 29-34.
79. BRANSON, D. E. "Time-dependent Effects in Composite Concrete Beams". *ACI Journal,* 61, 2 (February 1964). pp. 213-30; 9 (September 1964), pp. 1207-9.
80. HUGENSCHMIDT, F. "Epoxy Adhesives in Precast Prestressed Concrete Construction". *PCI Journal,* 19, 2 (March/April 1974), pp. 112-24.
81. HOPKINS, G. "Prestressing Operations on the Sydney Opera House". *Constructional Review,* 38, 12 (December 1965), pp. 16-21; 45, 4 (November 1973), pp. 3-29.
82. ABELES, P. W. "Principal Post-tensioning Systems for Prestressed Concrete". *Concrete and Constructional Engineering,* 59, 2 (February 1964), pp. 79-82; 3 (March 1964), pp. 107-12.
83. BAXTER, J. W., GEE, A. F. and JAMES, H. B. "Gladesville Bridge". *ICE Proceedings,* 30 (March 1965), pp. 489-530; 34 (June 1966), pp. 244-57.
84. ZIELINSKI, J. and ROWE, R. E. "The Stress Distribution Associated with Groups of Anchorages in Post-tensioned Concrete Members". *Research Report 13,* C & CA, Britain, 1962.
85. JATANA, B. L. "Free Cantilever Method of Constructing Prestressed Concrete Bridges". *Indian Concrete Journal,* 37, 5 (May 1963), pp. 174-83.
86. LIN, T. Y. "Load-balancing Method for Design and Analysis of Prestressed Concrete Structures". *ACI Journal,* 60, 6 (June 1963), pp. 719-42; 12 (December 1963), pp. 1843-48.
87. LIN, T. Y. and KELLY, J. W. *Prestressed Concrete Buildings.* Gordon and Breach, New York, 1962.
88. ARONI, S., BERTERO, V. and POLIVKA, M. "Chemically Prestressed Concrete". *PCI Journal,* 13, 5 (October 1968), pp. 22-35.
89. ADLER, F. "Precast Plate-units for Roof Construction". *ICE Proceedings,* 23 (November 1962), pp. 321-36; 28 (June 1964), pp. 201-16.
90. NEWELL, H. W. "Translucent Concrete". *Reinforced Concrete Review,* 2, 1 (January 1950), pp. 49-84.
91. Glass/concrete Construction". *Concrete Construction,* 7, 2 (February 1962), pp. 33-36.
92. IRONMAN, R. "Take a Look at 'Glass Concrete' ". *Concrete Products,* 65, 7 (July 1962), pp. 26-31.
93. BRYSON, W. G. "How to Stay Safe in the Prestressing Yard". *Concrete Products,* 70, 7 (July 1967), pp. 28-32.
94. LUTZ, L. A., SHARMA, N. K. and GERGELY, P. "Increase in Crack Width in Reinforced Concrete Beams under Sustained Loading". *ACI Journal,* 64, 9 (September 1967), pp. 538-46.
95. HEATHCOTE, F. W. L. "Movement of Articulated Buildings on Subsidence Sites". *ICE Proceedings,* 30 (February 1965), pp. 347-68; 33 (March 1966), pp. 492-517.
96. LACEY, W. D. and SWAIN, H. T. "Design for Mining Subsidence". *The Architects' Journal,* 126, 3267 (October 1957), pp. 557-70; 132, 3417 (October 1960), p. 113.

97. SAMUELY, F. J. "Concrete-framed Building to Resist Mining Subsidence". *The Engineer (London)*, 209, 5433 (11 March 1960), p. 439.
98. AUSTRALIAN EXPERIMENTAL BUILDING STATION. "Dampness in Buildings". *Notes on the Science of Building, NSB No. 52.*
99. CASEY, E. T. "Thermic Boring—A Method for the Boring and Severing of Concrete". *Civil Engineering and Public Works Review*, 58, 679 (February 1963), pp. 217-19.
100. SINCLAIR, S. "Using the Diamond Drill for Drilling Through Reinforced Concrete". *Civil Engineering and Public Works Review*, 58, 678 (January 1963), pp. 67-71; 679 (February 1963), pp. 227-29.
101. MOLONY, B., HASHEM, M. A. and HOFFMANN, E. "Efflorescence on Concrete Masonry". *Constructional Review*, 35, 9 (September 1962), p. 35.
102. *Concrete Floor Finishes.* Concrete Association of India, Bombay, 1968.
103. GURALNICK, S. A. "High-strength Deformed Steel Bars for.Concrete Reinforcement". *ACI Journal*, 32, 3 (September 1960), pp. 241-82; 9 (March 1961), pp. 1193-1200.
104. NAWY, E. G. "Crack Control in Reinforced Concrete Structures". *ACI Journal*, 65, 10 (October 1968), pp. 825-36; 66, 4 (April 1969), pp. 308-11.
105. HOGNESTAD, E. "High Strength Bars as Concrete Reinforcement Part 2. Control of Flexural Cracking". *PCA R & DL Journal*, 4, 1 (January 1962), pp. 46-63.
106. KAAR, P. H. and MATTOCK, A. H. "High Strength Bars as Concrete Reinforcement Part 4. Control of Cracking". *PCA R & DL Journal*, 5, 1 (January 1963), pp. 15-38.
107. KAAR, P. H. "High Strength Bars as Concrete Reinforcement Part 8. Similitude in Flexural Cracking of T-beam Flanges". *PCA R & DL Journal* 8, 2 (May 1966), pp. 2-12.
108. GOTO, Y. "Cracks Formed in Concrete Around Deformed Tension Bars". *ACI Journal*, 68, 4 (April 1971), pp. 244-51; 10 (October 1971), pp. 798-99.
109. REYNOLDS, G. C., CLARK, J. L. and TAYLOR, H. P. J. *Shear Provisions for Prestressed Concrete in the Unified Code CP110: 1972.* C & CA, Britain, 1974.
110. WYATT, K. J. "Friction Losses in Post-tensioned Curved Prestressing Cables". *Constructional Review*, 37, 8 (August 1964), pp. 18-21.
111. BISHARA, A. "Prestressed Concrete Beams Under Combined Torsion, Bending, and Sheer". *ACI Journal*, 66, 7 (July 1969), pp. 525-38; 67, 1 (January 1970), pp. 61-63.
112. KERENSKY, O. E. and LITTLE, G. "Medway Bridge: Design". *ICE Proceedings*, 29 (September 1964), pp. 19-52; 31 (June 1965), pp. 160-204.
113. KIER, M., HANSEN, F. and DUNSTER, J. A. "Medway Bridge: Construction". *ICE Proceedings*, 29 (September 1964), pp. 53-100; 31 (June 1965), pp. 160-204.
114. Editorial. "Steel for Prestressing". Symposium, Madrid, 1968. *Indian Concrete Journal*, 44, 8 (August 1970), pp. 329-30.
115. "Corrosion Prevention for Post Tensioning Bars". *Contracting and Construction Equipment*, 18, 3 (November 1964), pp. 108-10.
116. SWAMY, M. "Behaviour and Ultimate Strength of Prestressed Concrete Hollow Beams Under Combined Bending and Torsion". *Magazine of Concrete Research*, 14, 40 (March 1962), pp. 13-24.
117. DOWRICK, D. J. "Anchorage Zone Reinforcement for Post-tension Concrete". *Civil Engineering and Public Works Review*, 59, 698 (September 1964), pp. 1101-7.
118. "Bendorf Bridge: A New World Record in Prestressed Concrete". *Engineering News-Record*, 173, 15 (8 October 1964), pp. 60-64.
119. DESAYI, P. "Determination of the Maximum Crack Width in Reinforced Concrete Members". *ACI Journal*, 73, 8 (August 1976), pp. 473-477.
120. COWAN, H. J. *Reinforced and Prestressed Concrete in Torsion.* Edward Arnold, London, 1965.

121. ACI COMMITTEE 334. "Concrete Thin Shells". *ACI Special Publication SP-28*, 1971.
122. ELLEN, P. E. "Ultimate Load Balanced Design of Prestressed Concrete". *Constructional Review*, 38, 4 (April 1965), pp. 17-25; 5 (May 1965), pp. 26-30; 10 (October 1965), pp. 19-25.
123. DSIR. "Granolithic Concrete, Concrete Tiles and Terrazzo Flooring". *Building Research Station Digest (Second Series) No. 47*, HMSO, London.
124. ARUP, O. N. and ZUNZ, G. J. "Sydney Opera House". *Structural Engineer*, 47, 3 (March 1969), pp. 99-132.
125. MULLER, J. N. *et al.* "Precast Segmental Construction". *PCI Journal*, 20, 1 (January/February 1975), pp. 28-61; 2 (March/April 1975), pp. 22-41.
126. BROTCHIE, J. F. *et al.* "A Post-tensioned Hyperbolic Paraboloid Shell". *Constructional Review*, 44, 1 (February 1971), pp. 68-74.
127. HAAS, A. M. *Thin Concrete Shells*. Wiley, London, 1967.
128. TAMBERG, K. G. "Elastic Torsional Stiffness of Prestressed Concrete AASHO Girders". *ACI Journal*, 62, 4 (April 1965), pp. 479-91.
129. MAGURA, D. D. *et al.* "A Study of Stress Relaxation in Prestressing Reinforcement". *PCI Journal*, 9, 2 (April 1964), pp. 13-57.
130. CARMICHAEL, A. J. "Some Stress Relaxation Properties of Prestressing Wires of Australian Origin". *IE Aust. CE Transactions*, CE7, 2 (October 1965), pp. 165-72.
131. GOKHALE, Y. C. "Experiments on the Sawing of Contraction Joints in a Concrete Pavement". *Indian Concrete Journal*, 39, 11 (November 1965), pp. 429-33.
132. EBERHARDT, A. and VELTROP, J. A. "1300-Ton-Capacity Prestressed Anchors Stabilize Dam". *PCI Journal*, 10, 4 (August 1965), pp. 18-43.
133. BILLINGTON, D. P. *Thin Shell Concrete Structures*. McGraw-Hill, New York, 1964.
134. "Cracking in Reinforced Concrete Members". *Bulletin 17*, ASCE Reinforced Concrete Research, 1967; *Indian Concrete Journal*, 40, 9 (September 1966), pp. 351-59, 371.
135. BLAKEY, F. A. and ANSON, M. "Design for Cracking". *Constructional Review*, 40, 7 (July 1967), pp. 19-22.
136. UTZON, J. "The Sydney Opera House". *Architecture in Australia*, 54, 4 (December 1965), pp. 79-92.
137. LAZLO, G. "Expanding a Prestressed-concrete Plant". *ASCE Civil Engineering*, 35, 4 (April 1965), pp. 42-44.
138. BURNS, N. H. "Development of Continuity Between Precast Prestressed Concrete Beams". *PCI Journal*, 11, 3 (June 1966), pp. 23-36.
139. FIP. *Prestressed Concrete Structures*, Provisional Edition, 1966; *Concrete Structures*, VI Congress, Prague, Recommendations and Appendixes, 1970; *Prestressed Concrete*, VII Congress, New York, 1974; *Ground Anchors*, Commission Report, New York, 1974 (C & CA, Britain); *Short to Medium Span Bridges and Prestressed Concrete in Buildings*, FIP/CIA Symposia, Sydney, 1976 (C & CA Australia).
140. BURTON, K. T., CORLEY, W. G. and HOGNESTAD, E. "Connections in Precast Concrete Structures—Effects of Restrained Creep and Shrinkage". *PCA Bulletin D117*, 1967.
141. GORDON, P. "Connections for Precast Members can be Trouble-free". *ASCE Civil Engineering*, 36, 7 (July 1966), pp. 62-64.
142. MAHAFFEY, P. I. "Ferro-Cement". *Technical Memorandum 8*, C & CA, Aust., 1970.
143. WALKER, K. "Lightweight Concrete Yachts". *Modern Boating*, 3, 10 (August 1968), pp. 36-39.
144. SAMSON, J. and WELLENS, G. *The Ferro-Cement Boat*. Samson Marine Design, Ladner, B. C., Canada, 1973.
145. HARTLEY, R. T. *Boat Building with Hartley*. Boughtwood, Takapuna North, New Zealand, 1970.
146. BENFORD, R. and HUSEN, H. *Practical Ferro-Cement Boatbuilding*. International Marine, Camden, Maine, United States, 1972.

147. GERWICK, B. C. "Prestressed Concrete Piles". *PCI Journal,* 13, 5 (October 1968), pp. 66-93.
148. PETERSEN, P. H. and WATSTEIN, D. "Shrinkage and Creep in Prestressed Concrete". *Building Science Series 13,* United States Department of Commerce, National Bureau of Standards, 1968.
149. KEENE, P. W. "Loss of Prestress due to High Temperature Curing". *Prestress,* 18 (March 1969), pp. 15-26.
150. *Mass-produced Prestressed Precast Elements.* Proceedings of Symposium, Madrid, 1968. C & CA, Britain.
151. GHALI, A. "Bending Moments in Prestressed Concrete Structures by Prestressing Moment Influence Coefficients". *ACI Journal,* 66, 6 (June 1969), pp. 494-97; 12 (December 1969), pp. 1027-28.
152. RITSON, R. C. "Design of Prestressed Concrete Pipes". *Technical Journal,* Rocla Concrete Pipes, 53 (December 1971), pp. 1-2.
153. GERGELY, P. "Distribution of Reinforcement for Crack Control". *ACI Journal,* 69, 5 (May 1972), pp. 275-77.
154. O'BRIEN, J. "Demolition of Concrete". *Contracting & Construction Engineer,* 26, 9 (September 1972), pp. 8-18.
155. WESTLAKE, B. J. and HAWKINS, N. M. "Deflection Considerations in Design of Reinforced Concrete Beams". *Constructional Review,* 42, 4 (November 1969), pp. 58-64.
156. DANILECKI, W. "Effect of Crack Width on Corrosion of Reinforcement in Reinforced Concrete". *Library Communication 1514,* British Building Research Station, 1969.
157. *PCI Design Handbook; Post-Tensioning Manual; Manual on Design of Connections for Precast Prestressed Concrete;* and *Architectural Prestressed Concrete Manual.* Prestressed Concrete Institute, Chicago, 1974.
158. GROUNI, H. N. "Prestressed Concrete—A Simplified Method for Loss Computation". *ACI Journal,* 70, 2 (February 1973), pp. 108-14; 6 (June 1973), p. N30.
159. MULLER, J. "Precast Segmental Construction". *PCI Journal,* 20, 1 (January/February 1975), pp. 28-61; 2 (March/April 1975), pp. 22-41; 4 (July/August 1975), pp. 34-42.
160. GHALI, A. and DILGER, W. H. "Rapid Accurate Evaluation of Prestress Losses". *ACI Journal,* 70, 11 (November 1973), pp. 759-63; 71, 5 (May 1974), p. 271.
161. PCI COMMITTEE. "Estimating Prestress Losses". *PCI Journal,* 20, 4 (July/August 1975), pp. 44-75.
162. SMITH, I. A. "Shear in Prestressed Concrete to CP110". *Concrete* (London), 8, 7 (July 1974), pp. 39-41.
163. PARME, A. L. "Prestressing Flat Plates". *PCI Journal,* 13, 6 (December 1968) pp. 14-32.
164. LANGE, H. "Post-tensioning by the Vacuum Method". *World Construction,* 27, 12 (December 1974), p. 17.
165. LOGAN, D. and SHAH, S. P. "Moment Capacity and Cracking Behaviour of Ferrocement in Flexure". *ACI Journal,* 70, 12 (December 1973), pp. 799-804.
166. GALLAWAY, T. M. "Precasting of Segmental Bridges". *ACI Journal,* 72, 10 (October 1975), pp. 566-72.
167. German Industrial Standard DIN 4227. *Prestressed Concrete Instructions for Design and Construction.* (Standards Association of Australia, 1976.)
168. SCHULTZ, R. J. "Epoxy Adhesives in Prestressed and Precast Concrete Bridge Construction". *ACI Journal,* 73, 3 (March 1976), pp. 155-59.
169. VENKATARAMANA, K. "Foundations on Treacherous Soils". *Indian Concrete Journal,* 50, 10 (October 1976), pp. 206-300, 319.

APPENDIXES

APPENDIX I

STANDARD SPECIFICATIONS

Standard specifications for tests and other data on concrete and related materials are tabulated. When using updated regional standards, variant details of procedure are taken into account in practice. Imperial standards have been used where their metric versions were unavailable at the time of publication.

TABLE OF STANDARDS*

Title	AS	BS	ASTM	ACE
Abrasion-resistance and scratch hardness	1141; 1465	812	C131; C235; C289; C418; C535	52; 58; 117; 130; 141; 145
Absorption of				
aggregate	1012.2; 1141; 1465; 1467	812; 882	C127; C128	107-109
concrete	1342; 1346; 1500; 1757; 1759	340; 368; 473; 556; 1881.5	C14; C76; C118; C140; C642	23; 32; 62
Accelerated weathering test of bituminous materials	—	—	D529	—
Admixtures				
air-entraining	1478; 1479	5075	C226; C233; C260; C595	12; 13; 71
chemical	1478; 1479; MP20.1; MP20.2	5075	C494	87
expanding	MP20.3; MP20.4	—	—	—

---

* Abbreviations are given in Appendix VII.

TABLE OF STANDARDS*—continued

Title	AS	BS	ASTM	ACE
fly ash (see Fly Ash)	—	—	—	—
permeability-reducing	MP20.1	—	—	—
pozzolanic (see Pozzolan)	—	—	—	—
set-controlling	1478; 1479	5075	—	—
thickening or suspending	MP20.2	—	—	—
water-reducing or cement-dispersing	1478; 1479	5075	—	—
workability or plasticising	1478; 1479	5075	—	—
Aggregates for concrete, mortar and grout	1141; 1465–1467	812; 882; 1198–1201	C29; C33; C35; C40; C88; C117; C123; C125; C131; C136; C142; C144; C235; C289; C294; C295; C330; C331; C332; C404; C702; D75	33; 100-128; 130; 131; 133-146; 564
Air content of freshly mixed concrete and mortar				
gravimetric method	—	—	C138†	7; 219
pressure method	1012.4	1881.2	C231	41; 75
volumetric method	1012.4	—	C173; C185	8
Air content of hardened concrete				
high-pressure procedure	—	—	—	83
micrometric procedure	—	—	C457	42
Air-entraining admixtures (see Admixtures, air-entraining)	—	—	—	—
Air-permeability test (see Fineness)	—	—	—	—
Alkalinity and free alkali in portland cement	1315	12	C114; C150	209
Alkali reactivity (potential) of cement-aggregate combinations	1141; 1465; 1466	—	C227; C289; C342; C441; C586	123; 128; 140
Analysis of freshly mixed concrete	—	1881.2	—	72
Analysis of hardened concrete	1012.15	1881.6	C85	30

---

* Abbreviations are given in Appendix VII.

† AASHTO T121.

TABLE OF STANDARDS*—continued

Title	AS	BS	ASTM	ACE
Artificial weathering tests, operation of light-exposure and water-exposure apparatus	—	—	E42	—
Asbestos-cement ducting	—	3954	—	—
Asbestos-cement pipes				
linings	—	—	C541	—
methods of testing	1711	486; 3656	C458; C500	—
nonpressure	1712	3656	C428; C644; C663	—
pressure, rubber rings for	1711 K138	486; 3506	C296; C500; D1869	—
Asbestos-cement decking	A179	3717; CP199	—	—
Asbestos-cement sheets and shingles	1611	690; 3536; 4036; 4476; 4624; CP143	C220- C223; C459; C460; C551	—
Autoclave expansion of portland cement	—	—	C151	224
Ball penetration in freshly mixed concrete	—	—	C360	46
Bitumen (further tests are specified in ASTM Standards, Part 15)	2008; A131	598; 3235; 3940	D4; D140; D1079; D2764	563; 564
Bitumen or asphalt emulsion	1160	434; 598; 2542; 3690	D244; D466; D1010; D1167; D2397; D2963	—
Bituminous macadam or paving and fibre pipe	A7	1621; 2040	D979; D1663; D1753; D1861; D2311; D2312; D2316; D2818	—
Bituminous damp-proof course	Int. 326; Int. 327	743; 1097; 1418	D41; D146; D147; D170; D171; D449; D450; D491; D1668	—

---

* Abbreviations are given in Appendix VII.

TABLE OF STANDARDS*—continued

Title	AS	BS	ASTM	ACE
Bituminous fabric, felt or membrane	A98; A99; A120; A121; CA55	747; CP144.101	D146; D224-D228; D249; D250; D312; D655; D751; D2178; D2521; D2626; D2643	—
Bleeding of cement paste, mortar and concrete	1012.6	—	C232; C243	9; 221; 245
Bond strength of				
mortar to masonry units	1475.1	—	E149	—
reinforcement	1480	CP114	C234†	24
Bulk density of				
aggregate	1467; 1480	812; 3681	C29; C127	106; 107
air-free mortar	1379	—	—	—
freshly mixed concrete	1012.5	1881.2	C138; C567	7
hardened concrete	1012.12; 1170.1 (Appendix C)	1881.6; 2028	C567	—
lightweight concrete	1480	—	C567	—
Cast stone	A22	1217	—	63
Cement analysis				
chemical	1315	12	C114	209
spectrophotometric	1378	3875	E275	—
Cement content of				
freshly mixed concrete	—	1881.2	—	72
hardened portland cement				
concrete	1012.15	1881.6	C85	30
soil-cement mixtures	—	1924	D806	—
Cement-water paint	—	—	—	565
Chemical admixtures (see Admixtures, chemical)	—	—	—	—
Chemical-resistant mortars	—	—	C287; C306; C307; C308; C321; C386; C395-C399; C413; C414	—

* Abbreviations are given in Appendix VII.

† ACI Committee 208 (Appendix IV, 4(b)).

TABLE OF STANDARDS*—continued

Title	AS	BS	ASTM	ACE
Chemical resistance of mortars	—	—	C267	254
Colorimetric test for organic impurities in sand	1141; 1465; 1466	812	C40	121
Colours and pigments	K54	381C; 1014; 2660	D1208; D1210; D1535; D1729; E308	—
Compacting-factor test	1012.3	1881.2	—	—
Compressive strength (incl. capping and curing) of				
ordinary concrete specimens	1012 Pts	1881.3;	C31; C39;	10; 11; 14;
(see also Ready-mixed concrete)	2, 8, 9, 14	1881.4	C42; C116; C192; C330; C495†; C617	15; 27; 29 —
hydraulic cement mortars	1315	12; PD572	C109; C349	227; 406
lightweight concrete	—	CP114	C330; C495	91; 92
masonry units	1500	1180; 2028	C140	
prepacked concrete	—	—	—	84
triaxial	—	—	—	93
Concrete blocks and bricks (see Masonry Units)	—	—	—	—
Concrete cores (drilled)				
measurement and testing	1012.14	1881.4	C42; C174; C513	28
Concrete culvert, drainage, sewer and water pipes	1342; 1392; 1597; CA33	556; 1194; 2494; 4625; CP2010.5	C14; C76; C118; C361; C412; C444; C477; C478; C479; C505; C506; C507	514
Concrete fence posts	N36	—	—	—
Concrete kerbs and channels	A175	340	—	—
Concrete tiles				
floor	—	1197	—	—
roofing	1757 1759	402; 473; 550	C222; C459	—
wall (see Wall tiling)	—	—	—	—
Creep and rupture testing of metals	—	3500	D1780; D2293; D2294; E139; E150	—

---

* Abbreviations are given in Appendix VII.
† AASHTO T22, T23, T24.

CONCRETE TECHNOLOGY

TABLE OF STANDARDS*—continued

Title	AS	BS	ASTM	ACE
Creep of concrete	1012.16	—	C512	54
Density of concrete specimens (see Bulk density)	1012.12; 1480	1881.5; 1881.6; 2028	C567	—
Distillation of petroleum products	—	—	D86	—
tars and tar products	—	—	D20	—
Drying shrinkage and moisture movement	1012.13; 1346; 1481; 1500	1180; 1217; 1881.5; 2028; 1364; 2908	C157; C341; C426; C490; C596	25; 56; 69; 73
Dynamic modulus of elasticity	—	1881.5; 4408	C215; C597; E317	18; 51
Epoxy resin (see also Chapter 21)	K172	3534; CP3003; 4045	D1652; D1763; D2393	590; 591
Explosives	CA23	—	—	—
False set of portland cement	—	—	C359	259
Fineness of hydraulic cement	1315	12; 146; 915; 1370; 3406; 4359	C115; C184; C204; C430	212; 214-218
Fire classification	1481	4547	—	—
Fire tests of building construction and materials	1530	476	D635; E119	—
Flexural strength of test specimens, including moduli of elasticity and rupture (see Young's Modulus)	1012.8; 1012.11	1881.4; 1881.5	C31; C42; C78; C192; C293; C348†; C469	10; 11; 16; 17; 19; 21; 27; 260
Floor coverings	—	4682; CP203	F141; F142	—
Flow cone	—	—	—	76

*Abbreviations are given in Appendix VII.
† AASHTO T23, T24, T97.

TABLE OF STANDARDS*—continued

Title	AS	BS	ASTM	ACE
Flow of heat from concrete with adiabatic temperature rise	—	—	—	34
Flow tables for concrete and mortar	—	—	C124; C230	210

Fly ash for use as
  an admixture in portland cement
    concrete
	1129; 1130;
	1317
		3892	C311; C441;
			C593; C595;
			C618
				255; 256;
				262; 263

a pozzolanic material with lime	—	—	C593	263
Footway paving slabs	—	368; CP2006	—	—
Freezing in air and thawing in water	—	—	C291	—
Freezing and thawing in water	—	—	C290	20; 114
Grout mixtures bleeding	1012.6	—	C232; C243	9; 221; 245
expansion	MP20.3-20.4	—	—	81; 588; 589
flow and slump	1316; 1475; MP20.4	—	—	79
fluidifier	—	—	—	88; 566
intrusion	—	—	—	85
setting time	1012.18	—	—	82
water retentivity	1316; 1640; A123	—	C91; C110; C270	80
Guarding machinery	CP3004	—	—	—
Gypsum boards and panels	A44; CA20	1230; 4022	C36; C318; C473; C630; C631	—
Gypsum building plasters	A43	1191	C28; C471-C475; C587; C588	—
Gypsum concrete	—	—	C35; C317	—
Heating in air and cooling in water	—	—	—	40
Heat of hydration of portland cement	1315	1370	C186	38; 229
Ice required to produce mixed concrete of a specified temperature	—	—	—	53

---

* Abbreviations are given in Appendix VII.

TABLE OF STANDARDS*—continued

Title	AS	BS	ASTM	ACE
Joint sealing compounds, waterstops and gaskets	1526; 1527	1878; 2494; 2499; 3712; 4254; 4255; 5212; 5292; 5337 CP110, Pt 1 (Appendix B)	B248; C443; C509; C510; C542; C603; C639; C679; D545; D994; D1190; D1191; D1752; D2828	503; 504; 507; 513; 527; 530; 532; 546; 567-577; 579-587
Lightweight aggregate	1467	3681; 3797; CP114; CP116	C35; C330-C332; C495; C641	91; 92; 134-136; 138
Lime quicklime, hydrated and hydraulic limes for building purposes and products	1672	890	C5; C25; C49; C110; C141; C207; C415	—
Liquid penetrant inspection	—	—	E165	—
Masonry units	1346; 1475; 1500; 1640	CP121; 1180; 2028; 1364 (see also Table 27.2)	C55; C67; C90; C129; C139; C140; C145; C426; C427	32; 62; 65-68
Masonry units chemical-resistant	—	—	C279	—
Mechanical mixing of hydraulic cement mortars of plastic consistence	—	—	C305	—
Membrane-curing materials	—	1521; 1763; 2739; 3177	C156; C171; C309; E96	300-302; 305; 308; 310-315
Microscopical analysis of the distribution of particles of subsieve size	—	—	E20; E175	—
Mixer performance	—	3963	—	55
Moduli of elasticity and rupture (see Flexural strength; Dynamic, Static Chord and Young's Moduli)	—	—	—	—

* Abbreviations are given in Appendix VII.

TABLE OF STANDARDS*—continued

Title	AS	BS	ASTM	ACE
Moisture condition of hardened concrete by relative humidity method	—	—	C427	76
Mortar air-free bulk density	1379	—	—	—
Mortars chemical-resistant resin type	—	—	C267; C306-C308; C321; C395; C399; C413; C531	—
Mortar lining of pipes	1281; 1516	—	—	—
Mortar-making properties of fine aggregate	—	—	C87	116
Mortar for masonry and plastering	A123; 1475; 1640	2028; CP211	C270; C398; C476	—
Mortar pneumatically placed; shotcrete	—	—	—	516
Packaged, dry combined materials for mortar and concrete	—	—	C387	74
Parquetry flooring	1261; 1262; CA31; 071	—	—	—
Permeability of concrete	—	CP2007	—	48
Petrographic examination of aggregate for concrete	—	—	C295	100; 127
Pigments	K54	314; 318; 1014	Standard, Pt 28	—
Plastering and rendering	CA27	CP211; CP221	—	—
Plastics design data	—	4618	—	—
fibreglass reinforced sheets	A66	3953; CP3003	C581; C582	—
polyethylene	1326	3012	D882; D1709; D2103; D2578; D3020	—
polyester	A66	3532; 3691; 3749	D1201; D2150	—

* Abbreviations are given in Appendix VII.

TABLE OF STANDARDS*—continued

Title	AS	BS	ASTM	ACE
polypropylene	—	—	D2342; D2530; D2445; D2578	—
polystyrene	A133; A134; CA28; K156	1493; 2552; 3126; 3241; 3290; 3837; 3932; 4041	C578; D703; D1463; D2125; D2362	—
polyvinyl chloride films and sheets	A180; A185; K124; K139	1763; 2739; 3757; 3878; 4203; CP3003	D1593; D1755; D1927; D2123	—
testing	1145; 1146; 1327; K94; K166	2782; CP3003	Standard, Pt 35	581
Plastic pipes				
polyethylene	1159; CA69; K125	1972; 1973; 3284; 3796; CP312, Pt 1	D2104; D2239; D2447; D2609; Standard, Pt 34	—
polypropylene				
pressure	—	4991	—	—
waste	—	5254; 5255	—	—
polyvinyl chloride	1254; 1260; 1273; 1415; 1462; 1464; 1477; A67; A68; A160; A185	3505; 3506; 4346; 4514; 4576; 4660; CP312, Pt 2	D1785; D2241; D2665; D2729; D2740	—
urethane foam (rigid for building)	—	4841.1; 4841.2; 5241	—	—
Poisson's ratio	1012.17	CP110.1	C469; E132	—
Portland cement (various kinds) (see Table 3.1)	—	—	—	—
Pozzolans	1129; 1130; 1317	3892	C311; C441; C593; C595; C618	255-257; 262; 263
Precast concrete	1480	CP116	Standard, Pt 14	—
Prestressed concrete	1481	CP115; CP116; CP2007	See Pt 8	—

---

* Abbreviations are given in Appendix VII.

TABLE OF STANDARDS*—continued

Title	AS	BS	ASTM	ACE
Rate of hardening of mortars sieved from concrete mixtures (by Proctor penetration resistance needles)	1012.18	—	C403	86
Ready-mixed concrete	1379	1926	C94	31; 50
Reinforced concrete	1315	CP114; CP116	†	—
Reinforcement for concrete and bond strength	1302-1304	785; 1221; 1144; 1478; 2691; 3617	A29; A82; A184; A185; A311; A331; A416; A421; A434; A496; A497; A615-7; C234	24; 500-502; 506; 510; 519; 520; 562
Relative humidity	—	—	E337	—
Rendering and plastering	CA27	5262; CP211	—	—
Roofing tiles	1757; 1759	402; 473; 550	C222; C459	—
Rubber	1683	903	Pts 37, 38	—
Rubber ring joints	1646; 1693	2494	C443; D1869	—
Sampling fresh concrete	1012.1	1881.1	C172	4
Sampling hydraulic cement	1315	12	C183	—
Scaffolding	1575; 1576; B231	CP97	—	—
Serviceability (see Table 3.25 and Appendix II, 2(g))	—	—	—	—
Shear strength	—	—	—	89; 90
Shrinkage (see Drying shrinkage)	—	—	—	—
Sieve analysis and grading	1141; 1465; 1466	812; 882; 1198-1201	C117; C136	103-105 264
Sieves for testing purposes	1152	410; 481	E11	101; 102

* Abbreviations are given in Appendix VII.
† ACI Committee 318 (Table 3.25; Appendix IV, 2(a)).

TABLE OF STANDARDS*—continued

Title	AS	BS	ASTM	ACE
Slump test	1012.3	1881.2	C143	5
Sodium and potassium oxides in portland cement by flame photometry	—	—	C114	209
Soundness of aggregate	1141; 1465; 1466	882	C88	114; 115; 137
Specific gravity of aggregate	1141; 1465-1467	812; 3681	C33; C127; C128	107; 108
concrete	—	—	C642	23
hydraulic cement	—	—	C188	228
Specific heat of aggregate and cement	—	—	—	124; 242
Static chord modulus (see also Young's modulus)	1012.17	—	—	—
Steel for prestressed concrete	1310-1313	2691; 3617; 4486; 4757	A416; A421; A431	—
Steel structures (see Tables 3.24 and 3.25)	—	—	—	—
Steels alloy	1050; 1444	PD6431	—	—
Sugar in cement, aggregate or concrete	1141; 1465-1467	—	—	213
Sulphate resistance of cement	—	4027	C150; C452	232
Sulphuric anhydride (sulphur trioxide) in portland cement mortar	1315	12	C265	253
Surface moisture in aggregate	1012.2	1881.5	C70	111-113
Temperature rise in concrete	—	—	—	38
Tensile splitting strength of concrete cylinders and cores	1012.10; 1012.14	1881.4	C330; C496†	77
Tensile strength of hydraulic cement mortars	—	—	C190	—
steel reinforcement and metals	1391; A46 (see Tables 3.24 and 3.25)	18; 485; 3228; 3500; 3688	E8; E21; E111; E139; E150; E151; E209, Pts 4 and 30	—

---

* Abbreviations are given in Appendix VII.
† ACI Committee 318 (Table 3.25; Appendix IV, 2(a)).

TABLE OF STANDARDS*—continued

Title	AS	BS	ASTM	ACE
Testing				
aggregate (see Aggregates; also Sections 3.3.4 and 33.2.2)	—	—	—	—
cement (see Section 3.1.1)	—	—	—	—
clay bricks	1225; 1226	3921; 2973	C67; C126; C216	—
concrete (see Chapter 14)	—	—	—	—
steel (see Section 3.4)	—	—	—	—
nondestructive	B258; B259, Pts 1-5; B260; B261	415; 1881.5; 2704; 3683; 3889	C215; C597	—
panels and trusses	—	—	E72; E73, Pt 18	—
soils and stabilised soils	1289	1377; 1924	D422; D558; D559; D698; D806; D915; D1195; D1241; D1557; D1632; D1633-5	—
Test requirements for members and structures	1012.11; 1480; 1481; MP28, Sect. C26 on 1480	CP111; CP114; CP115; 1881; 3500	A370, Pt 30	—
Thermal coefficient of expansion of				
coarse aggregate	—	—	—	125
concrete	—	—	—	39
mortar	—	—	—	126
Thermal conductivity	—	874	C177; C236	44; 45
Thermal diffusivity of concrete	—	—	—	35-37
Time of setting of hydraulic cement and concrete	1315	12	C191; C403	—
Unit weight (see Bulk density)	—	—	—	—
Verification of testing machines, extensometers, pressure gauges and strain gauges	B128; CA35	1610; 1780	E4; E74; E83	—

* Abbreviations are given in Appendix VII.

TABLE OF STANDARDS*—continued

Title	AS	BS	ASTM	ACE
Vitrified clay pipes	1741; CA56	1143	C12; C13; C200; C211; C301; C479	—
Voids in				
aggregate	—	812	C30	110
concrete	—	—	C642	—
Volume change of cement and concrete products	1346; 1500; A182	1180; 1217; 1364; 1881.5; 2028; 2908	C157; C341; C426; C490; C596	25; 56; 69; 73; 140
Wall tiling	A133; A134; CA28	1281; 2552; 3932; CP212	Pt 18	—
Water absorption (see Absorption)	—	—	—	—

Water
  analysis, hydrogen-ion
  concentration (pH) and
  requirements
	1871	1328; 1427; 2690; 3148	D512; D513; D516; D596; D857; D888; D933; D1068; D1125; D1126; D1255; D1293; D1428; D1889; D2688; D2776; D2777; D2778	400-406

Water retentivity of mortar	1316	—	C91; C110; C270	80
Waterstops	—	1878	B248	572
Water vapour transmission	—	3177	C355; E96	—
Weighing scales	—	1405; 1887; 2058	—	512
Weight per unit volume (see Bulk density)	—	—	—	—

* Abbreviations are given in Appendix VII.

TABLE OF STANDARDS*—continued

Title	AS	BS	ASTM	ACE
Welding				
electric for mild steel	1204; 1205; 1552-1554; 1586; 1588; 1674; 1796; B28; C97; C301; CC5; MP17; Z5; Z6	499; 638; 639; 709; 807; 938; 1140; 1389; 1453; 1719; 1856; 2630; 2642; 2901; 2910; 2937; 2996; 3065; 4515; 4870; 4871; 5135	A233; A316; A399	—
Workability				
remoulding test	—	—	—	6
Young's Modulus of Elasticity	1012.17; 1480; 1481	1881.5; CP110.1; CP115.2	C469; E111; E132†	19; 21

---

* Abbreviations are given in Appendix VII.
† ACI Committee 318 (Appendix IV, 2(a)(i)).

SITE SUPERVISION AND CONTROL

1. GENERAL SUPERVISION

The object of proper supervision is to check conditions at the site and see how they agree with those assumed for design purposes. It includes such functions as testing materials and fabricated units, keeping a field diary, preparing control charts and reports, taking photographs, reporting observations to higher authority and making warranted adjustments. Pocket beepers facilitate personal communication between the office and site supervisory staff. Items for selective use in check lists or daily reports are given here (see References 34, 167, 170, 194 and 207 of Part 3, also Reference 8 of Part 6).

(a) Identification.
Date; type, location, survey, plans, specifications and layout of work; critical path schedule (see Section 36.9.3); estimates of quantities and costs; shift; contractor's representative; inspector, design engineer and superintendent; a combined site inspection each month.

(b) Foundations.
Site investigations; meteorological and weather studies.
Composition and bearing-value of soil or rock; aggressive conditions.
Ground and drainage water; effects of seasonal variations.
Other observations (see Sections 18.3 and 26.5; also AS 1289, BS CP101; and Reference 9 of Part 4).

(c) Materials.

Kinds; sources; amounts received, used, wasted and on hand.

Field samples shipped to laboratory.

Field tests on aggregate; sieve analysis; per cent undersize; specific gravity; absorption; moisture content; bulk density and bulking (if on volume basis); deleterious substances; colorimetric test; fineness modulus.

(d) Proportions, batching and mixing (for each class of concrete).

Mix proportions; quantities per batch; computed yield (m^3 of concrete per bag of cement).

Period of operation of mixer; number and size of batches; computed total volume.

Consistence of concrete at mixer and forms.

(e) Parts of structure prepared for placing.

Excavation; piling; shoring; formwork; reinforcement; construction joints.

(f) Placing.

Weather (temperature; humidity; wind; sky; times of observation).

Adequacy of organisation and equipment.

Times of starting and stopping placement; delays; unusual batches.

Kinds and volumes of concrete placed; parts of structure completed.

(g) Tests on concrete.

Consistence at forms; bulk density; temperature.

Specimens for strength tests (identification of sample, moulds and specimens; mix data; curing; age of test; specimens sent to laboratory).

(h) Curing; removal of formwork (for various parts of structure).

Curing: date and age (in hours) at beginning; date of completion.

Formwork removed: sides, bottom of member (state age of concrete).

Condition of formed surfaces; defective areas repaired; defective sections replaced.

(i) Special work.

Grouting, sprayed mortar, ornamental concrete.

(j) Force-account work; extra work.

Labour, materials, equipment: usually covered by a separate daily report.

(k) General.
Condition of, or changes in, organisation, equipment or methods; safety requirements and systems; lighting for night work; regular maintenance and calibration of weighing and testing equipment; crane, hoist and scaffolding systems and regulations (Table 27.1); and unusual features.

2. INSPECTION OF CONCRETE

The following check list includes various items that might be covered by a detailed inspection of general concrete work (see References 77, 167 and 170 of Part 3). Items may be selected to suit the requirements of a particular project specification or building code.

(a) Preliminary.
Study of plans and specifications; building codes.
Division of duties between engineer's representatives.
Permissible tolerances of measurement.
Provision for records and reports.
Contractor's plant, equipment, organisation and methods.
Rights of way; interference with utilities or adjoining property.
Motorised-caravan or mobile testing laboratory (see Section 27.1.6).

(b) Materials.
General (applies to all materials).
   Identification; quantities; acceptability; uniformity; storage conditions; handling methods; waste.

Cement.
   Sampling for laboratory test.
   Protection from dampness.
   Maximum length of storage period.

Aggregate.
   Acceptability tests (e.g. gradation); silt; organic matter and other undesirable substances; soundness; resistance to abrasion; compressive strength in mortar or concrete; prewetting of lightweight aggregate.
   Control tests (e.g. moisture); absorption; specific gravity; bulk density; voids.

Water content and quality.

Admixtures.
   Category; proprietary information; composition; test data; dosage; interactive effects.

Reinforcing steel.
Size; bending; surface condition; standard specifications.

Prestressing steel.
Quality; compliance with code specifications; location; handling tendons; design and alignment of beds and moulds; stressing sequence; detensioning; retensioning; stress and elongation measurements; camber variations; tolerances; noncorrosive agencies or additives; suitably steam-cured and moist-cured structural concrete; safety requirements.

Accessories; equipment; gauges; jacks; moulds; applicators; fixtures.

Other materials (e.g. pozzolans); fibres; pigments; polymers; release agents; vapourproof membranes; joint sealants; surface coatings.

(c) Proportioning.
Control tests of aggregate.

Mix design and computations.
Grading of mixed aggregate; cement content; air-entrainment; batch quantities; yield.

(d) Before concreting.
Lines, grades and levels.

Excavation; foundations; subgrades.
Location, dimensions, shape; drainage; preparation of surfaces.

Formwork.
Location; alignment; provision for settlement.
Stability (bearing; shores; ties and spacers).
Inspection openings; preparation of surfaces; final clean-up.

Reinforcement in place.
Size (diameter; length; bending; end-anchorage; fabric).
Location (number of rods; minimum clear spacing; minimum coverage).
Splicing; connectors; welding.
Stability (wiring; chairs and spacers).
Cleanness (no loose rust; no oil, paint, dried mortar, frost).
Fixtures (location; stability; cleanness).

Openings not shown on plans.

Calibration of batching devices.

Condition of mixer; speed of operation.

Provision for continuous placement.

Adequate tools and workers for compaction, finishing and curing.

(e) Concreting.

Working conditions.
Weather; preparations completed; specified interval since previous placement; lighting for night work.

Batching.
Cement; aggregate; fibres; water; admixtures.
Check batching devices.

Mixing.
Minimum time; batches delayed in mixer; overloading.

Control of consistence.
Observation of concrete being placed; tests; adjustments in mix; air content at mixer and forms.

Conveying.
No segregation of materials; no excessive stiffening or drying out; time limits.

Placing.
Uniform and dense concrete; continuous operation; preparation of contact surfaces; mortar bedding; vertical drop; no dropping against forms or reinforcement; little or no flow after depositing; time and rate of placement in mass work; depth of layers; water gain; rock pockets; removal of temporary ties and spacers; disposal of rejected batches; placing concrete under water.

Compacting.
Thorough and uniform compaction; no overworking; design of mix for vibration; revibration; dewatering; pressing; rolling.

Construction joints.
Location; preparation of surface.

Control joints.
Joint material; location; alignment; stability; provision for movements.

Finishing of unformed surfaces.
Shallow surface layer of mortar; water gain; no overworking; first floating; alignment of surface; refloating; final hard-trowelling; plastic shrinkage cracking; weather protection; off-form finish.

Finishing of formed surfaces.
> Condition of surfaces upon removal of formwork (honeycomb, peeling, ragged tie holes, ragged form lines); repair of defects; surface treatment; no surface-drying.

(f) After concreting.

Protection from damage.
> Impact; overloading; marring of surfaces.

Time of removal of formwork.

Washout of mixers and trucks.

Curing.
> Surfaces continuously moist; time of beginning curing; length of curing period (see Section 9.1; Cold-weather and Hot-weather Concreting, Chapter 32).

(g) Tests on concrete.

Early-quality tests.
> Sampling; consistence or workability; air content; bleeding; weight per unit volume of freshly mixed and hardened concretes; proctor hardening; analysis of cement content and mix proportions of fresh and hardened concretes.

Strength tests (compressive, indirect, tensile, flexural, bond).
> Moulding specimens; storing specimens (standard conditions, field conditions); field tests; shipping specimens to laboratory; nondestructive tests; statistical analysis.
> Specimens of hardened concrete from structure (cores, prisms).

Serviceability tests.
> Design, testing and control for structural soundness and durability; surface finish and dimensional accuracy; dryness of delivered masonry; structural deformation or deflection and creep under load; drying shrinkage and crack widths under working conditions; impermeability and absorption; abrasion and erosion resistance; curing efficiency and corrosion inhibition (see Chapters 14, 15, 19, 22, 27 and 31, also Table 3.25, Appendix IV, 1 (Thorogood) and Appendixes IV, 2(a)(i) and IV, 4(b)).

(h) Records and reports.

Records.
> Materials; delivery dockets; mix computations; batching and mixing; placing and curing; special concrete work.

Reports.
> Daily diary, summary, photographs.

3. CONTROL OF CONCRETE

A summary of factors for the control of quality of concrete is as follows.

(i) Fresh cement is used from one cement works, and is carefully stored or drawn from a reserved silo. Preliminary sampling and testing of output are done for 3 months. Subsequent tests are made on samples if the cement characteristics are changed or the concrete strengths fall unduly. Where cement users have formed themselves into an association, tests can be carried out economically at an independent laboratory (see Section 3.1, Table 3.1 and Section 3.1.3(b)). Concrete silos are brush-sealed internally with two coats of coal-tar epoxy resin (see Section 21.2) and kept (yearly) free of cement encrustation.

(ii) Aggregate is obtained in several single sizes and stored separately in duplicate, flat-topped stockpiles. Each pile is refilled one day and used the next or thereafter, the bottom 600 mm serving as a drainage layer. Typical components are 40, 20 and 10 mm nominal sizes and two grades of sand. Multiple bins, adequate in capacity for controlled batching (see Section 3.3.5), are checked against wear in partitions at 2-month intervals.

(iii) The quality, grading and shape of aggregate are tested regularly and their proportions are varied if necessary (see Table 3.8).

(iv) All sampling at recorded intervals must be in accordance with the requirements of standard specifications (e.g. AS 1012, 1379, 1480 and MP28, Commentary C4 and BS CP110 Part 1, 1881, ASTM C172, C183, D75; see also Section 7.1.5).

(v) The specific gravity of calcium chloride solution (if any) is regularly checked by a hydrometer, and a mechanical admixture dispenser is adjusted to suit temperature change.

(vi) The materials are batched by weight, the weighing apparatus being regularly calibrated, cleaned daily (including knife-edges) and oiled.

(vii) The amount of water and weights of aggregate are frequently adjusted, according to the moisture content of the aggregate, so that correct proportions will be batched for uniformity of workability or strength.

(viii) The time of mixing and speed of mixer are checked

periodically, the number of revolutions being then determined; and mixer performance tests are made at half-yearly to yearly intervals for blade-checking purposes in the design and maintenance of major plant (see Sections 7.1.1 to 7.1.3).

(*ix*) Homogeneity and thorough compaction must be achieved and maintained by correct placing procedures (see Section 7.2) of slump-tested concrete.

(*x*) The air content of air-entrained concrete is checked at the site.

(*xi*) Weather-reporting services are considered daily; and newly placed concrete is protected from rain, wind and sun, and properly cured.

(*xii*) Tests are frequently made to ensure that concrete of the requisite quality is being produced. About a dozen test specimens a day for the first week or two may be reduced to three of a given kind per day when results are regularly satisfactory (see Sections 14.4 and 34.1.4(a)(ii), and Table 3.25, also AS 1012 Parts 1-18, 1379 and MP28 (Commentary C4 on 1480), BS CP110 Part 1 and 1881, ASTM C31, C39, C42, C116 and C192, References 72-78, 106, 107, 167 and 207 of Part 3).

The results obtained are graphed and analysed daily (see Chapter 13 and Section 14.1), so as to control penalty-rate delivery payment and facilitate adjustments in the cement and other contents of mixes for requisite properties at a suitable cost. Accelerated strength tests, when used, are of supplementary value only to standard 28-day compressive-strength tests.

In the manufacture of ready-mixed or premixed concrete, specimens in triplicate or duplicate are made from batches that have established mix proportions in common. The mean values of the strengths subsequently obtained at the control age are analysed in groups of at least twenty-five values per month. The resulting data are useful for evaluating quality, checking uniformity, and refining the design criteria of relevant mixes (see Sections 4.3, 12.1 and 13.2, also Reference 75 of Part 3).

Concrete suppliers and users are expected to heed the basic tenets of modern concrete technology and practice, and obtain specialist assistance in advance where necessary, for purposes of effective guidance in creative operations. Handling activities in batching, transporting, placing and curing are dealt with in Section 3.3.6 and in Chapters 7, 9, 23, 24, 32 and 33. Diligent executives can produce documentary evidence of factual performance to specified requirements.

SCHEMATI

HIGH STRENGTH		IMPERMEABILITY		MINIMUM OF CRACKS	
Reduction of mixing water, high-quality concrete	Extraction of part of mixing water	Vapour-proof membrane	Plastic workable mix with low water/cement ratio, increased cement content	Low water/cement ratio, coarse-graded aggregate, tensile strength	Control in casting, auto-claved or well-dried products, control joints
Mix design for vibration, vibratory compaction, pressure consolidation	Vacuum dewatering, porous form lining, remove concrete with high water gain	Watertight aggregate and joints	Air-entrainment (adjusted mix)	Revibra-tion, reworked surface, prepacked concrete	Vacuum dewatering, low heat of hydration
Dispersion of cement, wetting agents, water/cement ratio gradient	Electro-osmosis	Freedom from cracks and segregation, uniformity, low bleeding	Thorough mixing, proper placing, reduced water at top of lifts, revibration	Slow-hardening cement, expansive cement	Distribu-ting steel, induced com-pression
Autoclave or effective curing	Pneumatic and ferro-cement mortars	Expansive cement	Effective curing, surface treatment	Low-settlement, low-shrink mixes, early curing	Fibre and shrinkage reinforce-ment, polymer concrete
	Polymer concrete	Pozzolan, bentonite, adequate fines		Fibrous casing, service-ability	
	Control of quality	Polymer concrete			
		Prestressing			

IAGRAM

D PROCEDURES

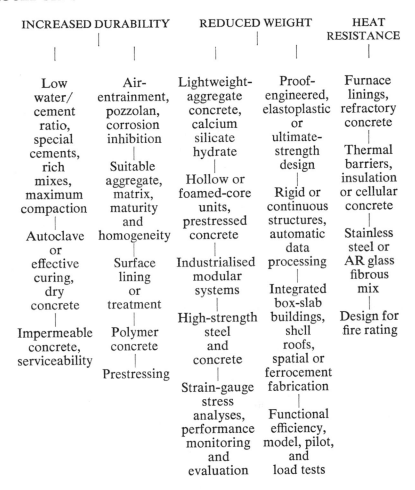

INCREASED DURABILITY		REDUCED WEIGHT		HEAT RESISTANCE
Low water/ cement ratio, special cements, rich mixes, maximum compaction	Air-entrainment, pozzolan, corrosion inhibition	Lightweight-aggregate concrete, calcium silicate hydrate	Proof-engineered, elastoplastic or ultimate-strength design	Furnace linings, refractory concrete
	Suitable aggregate, matrix, maturity and homogeneity	Hollow or foamed-core units, prestressed concrete	Rigid or continuous structures, automatic data processing	Thermal barriers, insulation or cellular concrete
Autoclave or effective curing, dry concrete	Surface lining or treatment	Industrialised modular systems	Integrated box-slab buildings, shell roofs, spatial or ferrocement fabrication	Stainless steel or AR glass fibrous mix
Impermeable concrete, serviceability	Polymer concrete	High-strength steel and concrete	Functional efficiency, model, pilot, and load tests	Design for fire rating
	Prestressing	Strain-gauge stress analyses, performance monitoring and evaluation		

BUILDING INFORMATION IN REFERENCE FORM

This consolidated reference list includes information on the design, management and execution of projects in building construction. Design considerations on prestressed concrete and spatial structures are contained in Part 8. References throughout the text to various textbooks and standards are intended to relate to their latest editions, which may be determined from technical institutions, publishers, retailers or libraries. Titles corresponding to the abbreviations used herein are given in Appendix VII.

The arrangement of the list is alphabetical by name of author.

1. CONSTRUCTION

ACI COMMITTEE 303. "Cast-in-Place Architectural Concrete Practice". *ACI Journal,* 71, 7 (July 1974), pp. 317-46.

ACI COMMITTEE 311. *ACI Manual of Concrete Inspection,* Publication SP-2, 1975.

ACI COMMITTEE 345. "Concrete Highway Bridge Deck Construction". *ACI Journal,* 70, 6 (June 1973), pp. 381-415.

ACI COMMITTEE 346. "Cast-in-Place Nonreinforced Concrete Pipe". *ACI Journal,* 66, 4 (April 1968), pp. 245-69.

ALLEN, P. W., LINDLEY, P. B. and PAYNE, A. R. *Use of Rubber in Engineering,* Maclaren, London, 1967.

ANTILL, J. M. *Civil Engineering Construction,* Angus and Robertson, Sydney, 1974.

ARNISON, J. H. *Roadwork Technology,* Vols 1-3, Iliffe, London, 1967.

BHATTACHARYYA, B. "Design of Swimming Pools". *Indian Concrete Journal,* 43, 12 (December 1969), pp. 482-87.

BLAHA, W. J. *et al.* "Housing Issue". *Concrete Products,* 71, 6 (June 1968), pp. 40-77.

BRITTON, H. W. "Prestressed Concrete Reservoirs of the Sydney Water Board at Mt Dorothy and Cecil Park". *IE Aust. CE Transactions, CE9,* 2 (October 1967), pp. 307-12.

BSI. *Building Drainage,* BS CP301; *Co-ordination of Dimensions in Building. Basic Sizes and Floor to Floor Heights,* BS 4011, 4176; *Foundations,* BS CP2004; *Foundations for Machinery,* BS CP2012; *Foundations and Substructures for Nonindustrial Buildings,* BS CP101; *Lightning Protection,* BS CP326; *Metal Scaffolding,* BS CP97; *Thermally Insulated Underground Piping Systems,* BS CP3009.

C & CA AUST., *Basic Guide to Concrete Construction;* and *Slab-on-Ground Floor Manual,* 1975.

CHEN, F. H. *Foundations on Expansive Soils,* Elsevier, Amsterdam, 1975. *Civil Engineering Standard Method of Measurement.* Institution of Civil Engineers, London, 1976.

Concrete Year Book. C & CA Annual Publication, Wexham Springs, Slough, Britain.

Construction Products Information Service. McGraw-Hill, New York.

Designer's Guide to Modern Steels. American Iron and Steel Institute, New York.

ERLIN, B. *et al.* "Repair of Fire Damage". *Concrete Construction,* 17, 3-5 (March-May 1972), pp. 98, 154, 225.

FERAHIAN, R. H. and HURST, W. D. "Vibration and Possible Building Damage Due to Operation of Construction Machinery". *DBR Research Paper 399,* NRC, Canada, 1968.

"Flat Roofs". *RIBA Journal,* 77, 5 (May 1970), pp. 217-23.

GAGE, M. *Guide to Exposed Concrete Finishes,* C & CA, Britain and Australia, 1970.

HAEGER, J. M. and SAPARIAN, S. S. "A New Concept of Storage Bin Construction". *ACI Journal,* 64, 9 (September 1967), pp. 575-79.

Handbook of Construction Techniques. McGraw-Hill, New York, 1976.

HASKER, J. "Slab on Ground Floors in Victoria". *Technical Report TR34,* C & CA Aust., 1972.

HIRSCH, T. J. "Recommended Practices for Driving Prestressed Concrete Piling". *PCI Journal,* 11, 4 (August 1966), pp. 18-27.

ICE ABSTRACTS. ICE Publications, 26-34 Old Street, London ECIV 9AD, January 1975 onwards.

ICE. "Ground Engineering". *Proceedings,* ICE Conference, London, 1970.

ICE. "Offshore Structures". *ICE New Civil Engineer,* Special Supplement, 26 September 1974.

ISAACS, D. V. "Excavation, Retaining Walls, and Foundations and Footings for Buildings". *Special Report No. 21,* Aust. EBS.

KORN, H. "Bridge Construction by Extrusion Sliding". *Concrete* (London), 9, 5 (May 1975), pp. 16-21.

LIN, T. Y. "Tall Buildings in Prestressed Concrete". *Architectural Record,* 12 (December 1965), pp. 165-70.

LINDLEY, P. B. and PAYNE, A. R. "Rubber Mountings for Large Structures". *Consulting Engineer,* 30, 9 (September 1966), pp. 40-42, 45.

Manual of Concrete Practice, Parts 1-3, American Concrete Institute, Detroit, 1976.

MARSH, P. *Concrete as a Visual Material,* C & CA Britain, 1974.

MAXWELL-COOK, J. C. *Structural Notes and Details,* C & CA Britain, 1975.

"Mechanical Fasteners for Concrete". *ACI Publication SP-22,* 1969.

MERRITT, F. S. *Building Construction Handbook,* McGraw-Hill, London, 1976.

"New Bentonite Techniques for Foundation Engineering". *International Construction,* 6, 12 (December 1967), pp. 26-33.

PERKINS, P. H. *Floors: Construction and Finishes.* C & CA Britain, 1973.

PERKINS, P. H. *Swimming Pools,* Elsevier, London, 1971; *Concrete Construction,* 17, 4 (April 1972), pp. 169-72.

"Piling, Diaphragm Wall and Ground Consolidation Feature". *International Construction,* 5, 3 (March 1966), pp. 2-79.

"Prestressed Swimming Pool". *Constructional Review,* 38, 6 (June 1965), pp. 15-18.

RATHBUN, I. T. *Building Construction Specifications,* McGraw-Hill, New York, 1972.

SAA, *Conditions of Contract: Supply and Erection of Plant and Machinery,* AS CB16, CZ20; *Civil Engineering Works,* AS CA24.1; *Conditions of Tendering and Form of Tender: Lump Sum Contract,* AS CA24.2; *Schedule of Rates Contract,* AS CA24.3; *Measurement of Civil Engineering Quantities,* AS A154; *Metric Conversion in Building and Construction,* AS MH1; *Metric Data for Building Design,* AS MH2, 1224; *Elastomeric Bearings for Use in Structures,* AS 1523; *Scaffolding Code,* AS 1576.

SALTER, W. L. *Floors and Floor Maintenance,* Elsevier Applied Science, Barking, Essex, 1974.

SMITH, R. C. *Materials of Construction,* McGraw-Hill, New York, 1973.

SORENSEN, C. P. "Solid Concrete Slab Floors". *Technical Study No. 38,* Aust. EBS.

Specification: Building Methods and Products. Architectural Press, London, 1976.

STIGTER, J. "Australia Square, Sydney. Construction and Structural Features of the 50-Storey, 600 Feet High Office Tower". *Constructional Review,* 39, 9 (September 1966), pp. 10-17.

STINSON, J. "Expo '67, Montreal". *Architects' Journal,* 145, 23 (June 7, 1967), pp. 1309-77.

Suppliers Index and Technical Indexes, Australia.

Sweet's Catalog Service. *Architectural Catalog File,* F. W. Dodge Corporation, New York, 1976.

TASKER, H. E. and SORENSEN, C. P. "Pier-and-Beam Footings for Single-storey Domestic Construction". *Technical Study No. 42,* Aust. EBS.

TAYLOR, D. C. "30-Storey Precast Concrete Flats, Park Street, South Melbourne". *Constructional Review,* 42, 2 (June 1969), pp. 42-51.

"Techniques: Concrete Domes in NSW Schools". *Constructional Review,* 49, 2 (May 1976), pp. 60-65.

THOROGOOD, R. P. "Accuracy of *In-situ* Concrete". BRE Britain, *Current Paper CP 99/75,* 1975.

TOMLINSON, M. J. *Foundation Design and Construction,* Pitman, London, 1969.

TROLLOPE, D. H. *et al.* "Symposium on Site Investigations". *IE Aust. CE Transactions,* CE 9, 1 (April 1967), pp. 1-106.

"Tunnelling Methods and Equipment". *International Construction* 14, 4 (April 1975), pp. 2-99.

WADDELL, J. J. *Concrete Construction Handbook.* McGraw-Hill, New York, 1974.

WASS, A. *Manual of Structural Details for Building Construction.* Spon, London, 1969.

WATSON, D. A. *Construction Materials and Processes.* McGraw-Hill, New York, 1972.

WEST, G. I. M. *et al.* "Tunnelling". *Civil Engineering and Public Works Review,* 63, 747 (October 1968), pp. 1087-1141.

WHEEN, R. J. "Positive Control of Multi-storey Construction Floor Loads". *Building Materials and Equipment,* 16, 3 (October/November 1973), pp. 59-64, 88.

ZIMMERMAN, L. G. "World of Fairs: 1851-1976". *Progressive Architecture,* 55, 8 (August 1974), pp. 64-77.

2. DESIGN AND MODELS

(a) Design.

(i) Strength and serviceability.

ACI COMMITTEE 209 *et al.* "Prediction of Creep, Shrinkage, and Temperature Effects in Concrete Structures". *ACI Publication SP-27,* 1971, pp. 51-93; "Flexural Crack Control in Structural Slab System". *Publication SP-30,* 1971, pp. 1-41; "Deflection". *Publication SP-43,* 1974, pp. 129-78, 515-45.

ACI COMMITTEE 215. "Design of Concrete Structures Subjected to Fatigue Loading". *ACI Journal,* 71, 3 (March 1974), pp. 97-121.

ACI COMMITTEE 224, "Cracking, Deflection, and Ultimate Load of Concrete Slab Systems". *ACI Publication SP-30,* 1971.

ACI COMMITTEE 301, "Specifications for Structural Concrete for Buildings". *ACI Manual of Concrete Practice,* Part 2, 1976.

ACI COMMITTEE 307. "Design and Construction of Reinforced Concrete Chimneys". *ACI Journal,* 65, 9 (September 1968), pp. 689-712; 66, 8 (August 1969), pp. 610-11; 67, 10 (October 1970), pp. 788-801.

ACI COMMITTEE 312. "Plain and Reinforced Concrete Arches". *ACI Journal,* 22, 9 (May 1951), pp. 681-91; 23, 4 (December 1951), pp. 692-711.

ACI COMMITTEE 313. "Design and Construction of Concrete Bins, Silos and Bunkers for Storing Granular Materials". *ACI Journal,* 72, 10 (October 1975), pp. 529-65.

ACI COMMITTEE 315. "New Developments in Detailing Practice". *ACI Journal,* 64, 5 (May 1967), pp. 234-39.

ACI COMMITTEE 317. "Reinforced Concrete Design Handbook—Working Stress Design". *ACI Publication SP-3; ACI Journal,* 65, 10 (October 1968), p. N16; 66, 2 (February 1969), pp. 87-91.

ACI COMMITTEE 318. "ACI Building Code Requirements and Commentary for Reinforced Concrete". *ACI 318-71 and 71C;* and *Annual Supplements,* 1974 onwards; *Manual of Concrete Practice,* Part 2; *ACI Journal,* 73, 1 (January 1976), pp. 15-25, onwards.

ACI COMMITTEE 320. "Safe Loadings for Existing Concrete Bridges". *ACI Journal,* 39 (January 1943), pp. 85-92 (1-11).

ACI COMMITTEE 322. "Structural Plain Concrete". *ACI Journal,* 68, 5 (May 1971), pp. 348-54; 11 (November 1971), pp. 870-72.

ACI COMMITTEE 324. "Minimum Requirements for Thin-section Precast Concrete Construction". *ACI Journal,* 59, 6 (June 1962), pp. 745-55; 60, 2 (February 1963), p. 171. (ACI Standard 525.)

ACI-ASCE COMMITTEE 326. "Shear and Diagonal Tension". *ACI Journal,* 59, 1 (January 1962), pp. 1-30; 2 (February 1962), pp. 277-333; 3 (March 1962), pp. 353-95; 9 (September 1962), pp. 1323-49.

ACI-ASCE COMMITTEE 333. "Design of Composite Beams and Girders for Buildings". *ACI Journal,* 32, 6 (December 1960), pp. 609-28; *ACI Manual of Concrete Practice,* Part 2, 1976.

ACI COMMITTEE 334. "Concrete Shell Structures—Practice and Commentary". *ACI Manual of Concrete Practice,* Part 2, 1976.

ACI COMMITTEE 336. "Design and Construction Procedures for Pier Foundations". *ACI Journal,* 69, 8 (August 1972), pp. 461-80; *ACI Manual of Concrete Practice,* Part 2, 1976.

ACI COMMITTEE 340. "Design Handbook". *Publications SP-17 (73)*, 1973; "Ultimate Strength Design Handbook, Vol. I". *ACI Publication SP-17A*, 1970; Errata, *ACI Journal*, 70, 10 (October 1973), pp. N35-N38; 71, 7 (July 1974), pp. N33-N35.

ACI COMMITTEE 344. "Design and Construction of Circular Prestressed Concrete Structures". *ACI Manual of Concrete Practice*, Part 2, Structural Design, Specifications and Analysis, 1976.

ACI COMMITTEE 345. "Concrete Highway Bridge Deck Construction". *ACI Journal*, 70, 6 (June 1973), pp. 381-415; 12 (December 1973), pp. 824-25.

ACI COMMITTEE 348. "Probabilistic Design of Reinforced Concrete Buildings". *ACI Publication SP-31*, 1972.

ACI COMMITTEE 423. "Concrete Members Prestressed With Unbonded Tendons". *ACI Manual of Concrete Practice*, Part 2, Structural Design, Specifications and Analysis, 1976 onwards.

ASCE-ACI COMMITTEE 426 "Shear Strength of Reinforced Concrete Members". *ASCE Journal*, 99, ST6 (June 1973), pp. 1091-1187; 100, ST8 (August 1974), pp. 1543-91; *ACI Manual of Concrete Practice*, Part 2, 1976.

ACI COMMITTEE 435. "Deflection of Concrete Structures". *ACI Publication SP-43; ACI Manual of Concrete Practice*, Part 2; *ACI Journal*, 63, 6 (June 1966), pp. 637-74; 12 (December 1966), pp. 1499-1510; 65, 6 (June 1968), pp. 433-44; 9 (September 1968). pp. 730-42; 12 (December 1968), pp. 1037-38; 69, 1 (January 1972), pp. 29-35; 70, 12 (December 1973), pp. 781-87; 71, 4 (April 1974), pp. 223-28.

ACI COMMITTEE 437. "Strength Evaluation of Existing Concrete Buildings". *ACI Manual of Concrete Practice*, Part 2, Structural Design, Specifications and Analysis, 1976.

ACI COMMITTEE 438. "Torsion of Structural Concrete". *ACI Publication SP-18, SP-35; ACI Manual of Concrete Practice*, Part 2; *ACI Journal*, 66, 4 (April 1969), pp. 314-54; 7 (July 1969), pp 576-88; 70, 6 (June 1973), p. N30; 71, 2 (February 1974), pp. 84-88.

ACI-ASCE COMMITTEE 441. "Reinforced Concrete Columns". *ACI Publication SP-50; ACI Journal*, 72, 8 (August 1975), pp. 431-35.

ACI COMMITTEE 442. "Design of Tall Buildings, including Tall Buildings in Seismic Areas". *ACI Publication SP-36; ACI Journal*, 70, 8 (August 1973), p. N29.

ACI COMMITTEE 442. "Response of Buildings to Lateral Forces". *ACI Journal*, 68, 2 (February 1971), pp. 81-106; 71, 2 (February 1974), p. 88; *ACI Publication SP-36* and *Manual of Concrete Practice*, Part 2, 1976.

ACI COMMITTEE 443. "Analysis and Design of Reinforced Concrete Bridge Structures". *ACI Journal*, 71, 4 (April 1974), pp. 171-200; 11 (November 1974), pp. 579-81; "Symposium on Concrete Bridge Design". *ACI Publication SP-23*, 1969.

ACI-ASCE COMMITTEE 512. "Recommended Practice for Manufactured Reinforced Concrete Floor and Roof Units". *ACI Journal*, 63, 6 (June 1966), pp. 625-36; 12 (December 1966), pp. 1495-97, NL29; 64, 4 (April 1967), p. 185. "Precast Structural Concrete in Buildings". *ACI Journal*, 71, 11 (November 1974), pp. 537-49.

ACI COMMITTEE 533. "Design, Fabrication, Handling, Erection and Testing of Precast Concrete Wall Panels". *ACI Manual of Concrete Practice*, Part 3, Products and Processes, 1976.

ACI COMMITTEE 543. "Design, Manufacture, and Installation of Concrete Piles". *ACI Journal*, 70, 8 (August 1973), pp. 509-44; 71, 10 (October 1974), pp. 477-92; 72, 4 (April 1975), p. 174.

ACI. *Manual of Concrete Practice; Manual of Concrete Inspection (Publication SP-2)*.

ACI Shear Committee. "Shear in Reinforced Concrete". *ACI Publication SP-42*, Parts 1 and 2, 1974.

ALLEN, A. H. *Reinforced Concrete Design to CP110—Simply Explained*. C & CA Britain, 1974.

ARE Association. *Manual of the American Railway Engineering Association*. Part 6 on "Crib Walls", 1970.

Australian Reinforced Concrete Design Handbook, SI Units. C & CA Aust., 1976.

BRESLER, B. *Reinforced Concrete Engineering*. Vols I, II and III, Wiley, New York, 1974.

BROOKS, D. S. "Design Procedures for Reinforced Concrete Members Subject to Combined Bending, Torsion and Shear". *Constructional Review*, 42, 2 (June 1969), pp. 58-64.

BSI. *Basic Data for Design of Buildings: Dead and Superimposed Loads*, BS CP3 V.1; *Wind Loads*, BS CP3 V.2; and *Durability*, CP3 IX; *Structural Recommendations for Loadbearing Walls*, BS CP111; *Structural Uses of: Concrete*, BS CP110 (three parts); *Precast Concrete*, BS CP116; *Prestressed Concrete*, BS CP115; *Reinforced Concrete in Building*, BS CP114; *Steel in Building*, BS 449; *Protection from Corrosion*, BS CP2008; and *Methods for Specifying Concrete*, BS 5328.

CALLENDER, J. H. *Time-Saver Standards for Architectural Design Data*. McGraw-Hill, New York, 1974.

CEB-FIP. *International Recommendations for the Design and Construction of Concrete Structures*. CEB-FIP, Paris, 1974.

CEFOLA, A. "Chart for the Spacing of Vertical Stirrups". *ACI Journal*, 62, 1 (January 1965), pp. 119-21.

CLARK, B. E. "Design Aids for Beams and Columns". *Concrete* (London), 7, 1 (January 1973), pp. 29-31.

CLOUGH, R. W. and PENZIEN, J. *Dynamics of Structures*. McGraw-Hill, New York, 1975.

Concrete Industries Yearbook. Pit and Quarry Publications, Chicago, 1976.

Concrete Pipe Design Manual. American Concrete Pipe Association, Arlington, Va., 1970.

Concrete Yearbook, The. Concrete Publications, London, 1976.

COWAN, H. J. and LYALIN, I. M. *Reinforced and Prestressed Concrete in Torsion*. Arnold, London, 1965.

COWAN, H. J. and SMITH, P. R. *Design of Reinforced Concrete*. Angus and Robertson, Sydney, 1968.

Criteria for Structural Adequacy of Buildings. I Struct. E, London, 1976. CRSI. *Design Handbook;* and *Placing Reinforcing Bars*. CRSI Publications, Chicago, 1976.

CUSENS, A. R. and KUANG, J. G. "A Simplified Method of Analysing Freestanding Stairs". *Concrete and Constructional Engineering*, 60, 5 (May 1965), pp. 167-72; 61, 1 (January 1966), p. 42.

DAVIES, J. D. "Bending Moments in Covered Square Concrete Tanks". *ICE Proceedings*, 37 (June 1967), pp. 317-23.

"Deflections of Concrete Structures", *ACI Publication SP-43*, 1974.

"Designing for Effects of Creep, Shrinkage, Temperature in Concrete". *ACI Publication SP-27; ACI Journal*, 69, 3 (March 1972), pp. 179-84.

DIVER, M. "Design of Reinforced Concrete Chimneys". *ACI Journal*, 67, 10 (October 1970), pp. 788-801.

EDITORIAL. " Progressive Strength Concrete Construction". *Constructional Review*, 45, 3 (August 1972), pp. 42-51.

EVANS, E. P. and HUGHES, B. P. "Shrinkage and Thermal Cracking in a Reinforced Retaining Wall". *ICE Proceedings*, 39 (January 1968), pp. 111-25; 40 (August 1968), pp. 539-68.

FARMER, L. E. and FERGUSON, P. M. "T-beams Under Combined Bending, Shear and Torsion". *ACI Journal*, 64, 11 (November 1967), pp. 757-66; 65, 5 (May 1968), pp. 417-21.

FERGUSON, P. M. *Reinforced Concrete Fundamentals*. Wiley, New York, 1973.

FERTIS, D. G. *Dynamics and Vibration of Structures*. Wiley-Interscience, New York, 1973.

FINTEL, M. *Handbook of Concrete Engineering*. Van Nostrand Reinhold, New York, 1974.

FINTEL, M. and KHAN, F. R. "Effects of Column Exposure in Tall Structures—Temperature Variations and Their Effects". *ACI Journal*, 62, 12 (December 1965), pp. 1533-56; 2, 63, 6 (June 1966), pp. 1833-35.

FINTEL, M. and KHAN, F. R. "Effects of Column Creep and Shrinkage in Tall Structures—Prediction of Inelastic Column Shortening". *ACI Journal*, 66, 12 (December 1969), pp. 957-67; 67, 6 (June 1970), pp. 473-75.

FIP. *Design and Construction of Concrete Structures*. C & CA Britain, 1970; see Reference 139 of Part 8.

GARDNER, N. J. and JACOBSON, E. R. "Structural Behaviour of Concrete Filled Steel Tubes". *ACI Journal*, 64, 7 (July 1967), pp. 404-13; 65, 1 (January 1968), pp. 66-69.

Geodex Systems and Digests. Geodex International, Sonoma, Calif.

GOODE, C. D. "Torsion, BS CP110 and ACI 318 Codes Compared". *Concrete* (London), 8, 3 (March 1974), pp. 36-40.

GOODE, C. D. and HELMY, M. A. "Design of Rectangular Beams Subjected to Combined Bending and Torsion". *Concrete* (London), 1, 7 (July 1967), pp. 241-44.

GOULD, P. L. "Interaction of Shear Wall-frame Systems in Multi-storey Buildings". *ACI Journal*, 62, 1 (January 1965), pp. 45-70; 9 (September 1965), pp. 1145-51.

GRAY, W. S. and MANNING, G. P. *"Concrete Water Towers, Bunkers, Silos and Other Elevated Structures*. C & CA Britain, 1973.

HAAS, A. M. *Design of Thin Concrete Shells*, vols 1 and 2, Wiley, New York, 1967.

HAGGLUND, A. "Building Design Principles". Paper at Conference of Fire Protection Associations Europe on Fire Safety in High-rise Buildings, Amsterdam, 1975.

HANCOCK, J. *Time-Saver Standards for Architectural Design Data*. McGraw-Hill, New York, 1976.

Handbook on the Unified Code for Structural Concrete: BS CP110. C & CA Britain, 1972.

HANSEN, R. J. *et al.* "Human Response to Wind-induced Motion of Buildings". *Proceedings ASCE, Structural Division*. 99, ST7 (July 1973), pp. 1589-1605.

Highway Bridge Design Specification and *Metric Addendum*. National Association of Australian State Road Authorities, Sydney, 1972-73.

IRISH, K. and WALKER, W. P. "Foundations for Reciprocating Machines". *Concrete* (London), 1, 10 (October 1967), pp. 327-37; 2, 2 (February 1968), p. 98. BSI *Foundations for Machinery*, BS CP2012.

JENSEN, A. and CHENOWETH, H. H. *Applied Strength of Materials*. McGraw-Hill, New York, 1975.

KANE, G. N. J. "Basic Facts Concerning Shear Failure". *ACI Journal*, 63, 6 (June 1966), pp. 675-92; 12 (December 1966), pp. 1511-28.

KEMPE, H. R. *Kempe's Engineers Year-Book*. Morgan, London, 1976.

KHAN, F. R. and FINTEL, M. "Effects of Column Exposure in Tall Structures—Design Considerations and Field Observations of Buildings". *ACI Journal*, 65, 2 (February 1968), pp. 99-110.

KOO, B. "Simplified Design for Torsion in Rectangular Reinforced Sections". *ACI Journal,* 70, 9 (September 1973), pp. 645-48.

KRIZ, L. B. and RATHS, C. H. "Connections in Precast Concrete Structures —Strength of Corbels". *PCI Journal,* 10, 1 (February 1965), pp. 62-75.

LEE, D. J. *Bearings and Expansion Joints for Bridges.* C & CA Britain, 1972.

LENCZNER, D. "Cohesive Arching of Bulk Materials in Bunkers and Silos". *Civil Engineering and Public Works Review,* 61, 724 (November 1966), pp. 1393-96.

LENCZNER, D. "Pressures in Containers of Granular Materials". *Concrete and Constructional Engineering,* 59, 5 (May 1964), pp. 165-72; 6 (June 1964), pp. 223-27.

LIN, T. H. *Theory of Inelastic Structures.* Wiley, New York, 1968.

LINDLEY, P. B. and PAYNE, A. R. "Rubber Mountings for Large Structures". *Consulting Engineer,* 30, 9 (September 1966), pp. 40-42, 45.

LONG, A. E. and BOND, D. "Punching Failure of Reinforced Concrete Slabs". *ICE Proceedings,* 37 (May 1967), pp. 109-35; 39 (January 1968), pp. 151-58.

MACPHERSON, J. D. "Cylindrical Reinforced Concrete Water Tanks with Thin Walls and High Hoop Stresses". *IE Aust. Journal,* 38, 4-5 (April/ May 1966), pp. 87-94; 9 (September 1966), pp. 249-50.

MAHOHAR, S. N. and DESAI, S. B. "Design of Circular Raft Foundations for Chimneys". *Civil Engineering and Public Works Review,* 62, 733 (August 1967), pp. 905-10.

MAISEL, B. I. and RELL, F. "Analysis and Design of Concrete Boxbeams with Side Cantilevers". C & CA Britain, TR42.494, 1974.

MALLICK, D. V. and SEVERN, R. T. "Behaviour of Infilled Frames Under Static Loading". *ICE Proceedings,* 38 (December 1967), pp. 639-56; 41 (September 1968), pp. 205-22.

MAUGH, L. C. and RUMMAN, W. S. "Dynamic Design of Reinforced Concrete Chimneys". *ACI Journal,* 64, 9 (September 1967), pp. 558-67; 65, 3 (March 1968), pp. 229-31.

MIRZA, M. S. and McCUTCHEON, J. O. "Behaviour of Reinforced Concrete Beams Under Combined Bending Shear and Torsion". *ACI Journal,* 66, 5 (May 1969), pp. 421-27.

MOORE, P. J. "A Review of Design Methods for Machinery Foundations". *IE Aust. Journal,* 39, 4-5 (April/May 1967), pp. 45-53.

MOWRER, R. D. and VANDERBILT, M. D. "Shear Strength of Lightweight Aggregate Reinforced Concrete Flat Plates". *ACI Journal,* 64, 11 (November 1967), pp. 722-29.

MULLER, J. F. G. "Reinforced Concrete Chimneys". *Concrete* (London), 1, 3 (March 1967). pp. 94-96; 2, 7 (July 1968), pp. 283-84.

NAASRA. *Highway Bridge Design Specification.* NAASRA, Melbourne, 1975.

NEWBERRY, V. W. and EATON, K. J. *Wind Loading Handbook* (BS CP3V.2). HMSO, London, 1974.

O'BRIEN, T. "Durability in Design". *Structural Engineer,* 45, 10 (October 1967), pp. 351-63; 46, 5 (May 1968), pp. 141-46.

PARK, R. and PAULAY, T. *Reinforced Concrete Structures.* Wiley, NY, 1975.

PHILLIPS, A. B. "Pressures in Silos". *Concrete and Constructional Engineering,* 60, 10 (October 1965), pp. 390-95.

PINFOLD, G. M. *Reinforced Concrete Chimneys and Towers.* C & CA, Britain, Viewpoint Publication 12.064, 1975.

Planning and Design of Tall Buildings. International Conference, Lehigh University, 1972; five volumes of Proceedings, ASCE, New York, 1973.

RAJAGOPALAN, K. S. "Design Aids for Torsion Design per ACI 318-71". *ACI Journal,* 70, 12 (December 1973), pp. 817-18.

RAMASWAMY, G. S. *Design and Construction of Concrete Shell Roofs.* McGraw-Hill, New York, 1968.

RANGAN, B. V. "Limit States Design of Slabs Using Lower Bound Approach". *ASCE Journal,* 100, ST2 (February 1974), pp. 373-89.

RANGAN, B. V. "Prediction of Long-Term Deflections of Flat Plates and Slabs". *UNICIV Report R-140,* 1975.

RANGAN, B. V. "Serviceability and Strength Design of Flat Plates and Flat Slabs". *IE Aust. National Conference Publication 75/6,* 1975, pp. 44-48.

REGAN, P. E. and YU, C. W. *Limit State Design of Structural Concrete.* Halsted, New York, 1973.

REYNOLDS, C. E. and STEEDMAN, J. C. *Reinforced Concrete Designer's Handbook.* C & CA, Britain, 1974.

RICE, P. F. and HOFFMAN, E. S. *Structural Design Guide to the ACI Building Code.* Van Nostrand Reinhold, New York, 1972.

RISH, R. F. "Forces in Cylindrical Chimneys due to Wind ". *ICE Proceedings,* 36 (April 1967), pp. 791-803; 38 (October 1967), p. 313.

ROGERS, P. *Reinforced Concrete Design for Buildings.* Van Nostrand Reinhold, New York, 1973.

SAA. *Minimum Design Loads on Structures,* AS 1170.1, 1170.2; *Concrete Structures Code,* AS 1480; *High-strength Structural Bolting,* AS 1511; *Metric Conversion in Building and Construction,* AS MH1; *Metric Data for Building Design,* AS MH2, 1224; *Prestressed Concrete Code,* AS 1481; *Steel Structures Code,* AS 1250, MA1.1-1.9; *Structural Steel Hollow Sections,* AS 1163; *Lightning Protection,* AS 1768; *Limit State Design,* AS 1973; *Earthquake-resistant Buildings,* AS DR761000.

SAFARIAN, S. S. "Design Pressure of Granular Materials in Silos". *ACI Journal,* 66, 8 (August 1969), pp. 647-55; 67, 7 (July 1970), pp. 539-47.

SCHMIDT, W. *Ultimate Strength Design Simplified—An Equivalent Stress Method.* Schmidt, Chicago, 1972.

"Seismic Design Seminar". *PCI Journal,* 17, 4 (July/August 1972), pp. 29-85.

Serviceability: design and tests. See Table 3.25 and Appendix II, 2*(g).*

"Shear in Reinforced Concrete". *ACI Publication SP-42,* 1974.

SMITH, J. W. and ZAR, M. "Chimney Foundations". *ACI Journal,* 61, 6 (June 1964), pp. 673-700.

STEFFENS, R. J. *Structural Vibration and Damage.* D of E, BRS Report, HMSO, London, 1974.

TAYLOR, R. "A New Method of Proportioning Stirrups in Reinforced Concrete Beams". *Magazine of Concrete Research,* 15, 45 (November 1963), pp. 177-81; 18, 57 (December 1966), pp. 221-30.

"Vibration Effects of Earthquakes on Soils and Foundations". *ASTM Special Technical Publication 450.*

"Vibrations in Buildings". Parts 1 and 2, *BRS Digests 117, 118,* Britain, 1970.

WADDELL, J. J. *Concrete Construction Handbook.* McGraw-Hill, NY, 1974.

WALLER, R. A. *Buildings on Springs.* Pergamon, London, 1969.

WALLER, R. A. "Design of Tall Structures with Particular Reference to Vibration". *ICE Proceedings,* 48 (February 1971), pp. 303-23.

WALSH, P. F., COLLINS, M. P. and ARCHER, F. E. "Design of Concrete Beams for Torsion". *IE Aust. CE Transactions,* CE9, 2 (October 1967), pp. 313-20; *Constructional Review,* 41, 2 (February 1968), pp. 20-25.

WALSH, P. F. "Reinforced Concrete Deflection Design". *CSIRO DBR Report 39,* 1975.

WEDGWOOD, R. J. L. "Design, Construction and Maintenance of Small Bridges". *ARRB 13th Regional Symposium.* Bega, NSW, 1976.

WINOKUR, A. and GLUCK, J. "Ultimate Strength Analysis of Coupled Shear Walls". *ACI Journal*, 65, 11 (December 1968), pp. 1029-36; 66, 6 (June 1969), pp. 500-501.

WINTER, G. and NILSON, A. H. *Design of Concrete Structures*. McGraw-Hill, New York, 1972.

WIRE REINFORCEMENT INSTITUTE. *Building Design Handbook*, WRI Publications, Washington, DC, 1976.

ZWEIG, A. "Design Aids for the Direct Design Method of Flat Slabs". *ACI Journal*, 70, 4 (April 1973), pp. 285-99; 10 (October 1973), p. 721; 71, 5 (May 1974), p. N33.

(ii) Computer.

ACI COMMITTEE 118. "Computer Graphics Applications"; and "Documentation for Computer Calculation Submittals to Building Officials". ACI *Manual of Concrete Practice*, Part 2.

ACI COMMITTEE 118. "Decision Logic Table Format for Building Code Requirements (ACI 318-71)". *ACI Journal*, 70, 12 (December 1973), pp. 788-92.

ALCOCK, D. G. "Computer Analyses of Reinforced Concrete Members and Foundations". *Structural Engineer*, 45, 9 (September 1967), pp. 303-11; 46, 5 (May 1968), pp. 139-40.

APERGHIS, G. G. "Slab Bridge Design and Drawing—An Automated Process". *ICE Proceedings*, 46 (May 1970), pp. 55-75.

ASCE. "Electronic Computation". *ASCE 6th National Conference*, Georgia Institute of Technology, 1974; *ASCE Journal*, 101, ST4 (April 1975), pp. 627-990.

BILLINGTON, D. P. "Computers and Thin Shell Analysis". *ACI Publication SP-16*, 1967, pp. 77-95.

BSI. *Data Processing*. BS 3174, 3480, 3527, 3635, 3658, 3732, 3880, 4058.

CARVER, D. K. *Introduction to Fortran II and IV Programming*. Wiley, NY, 1969.

CHRISTIE, I. F. "Use of Analogue Computers for Civil Engineering Problems". *ICE Proceedings*, 25 (July 1963), pp. 267-86; 27 (April 1964), pp. 831-34.

"Colour Computer Graphics for Architecture". *Architectural Record*, 154, 3 (September 1973), pp. 65-67.

Computer Aided Design Journal (Quarterly). IPC Science and Technology, Guildford, Britain.

CRANSTON, W. B. "A Computer Method for the Analysis of Restrained Columns". *Technical Report TRA402*, C & CA Britain, 1967.

DAVIES, J. D. and CHEUNG, Y. K. "Bending Moments in Long Walled Tanks". *ACI Journal*, 64, 10 (October 1967), pp. 685-90; 65, 4 (April 1968), pp. 347-48.

FENVES, S. J. *Computer Methods in Civil Engineering*, Prentice-Hall, Englewood Cliffs, N.J., 1967.

GAYLORD, C. J. "Computer Analysis and Design of Flat Slab Structures". *ACI Journal*, 64, 4 (April 1967), p. N25.

GIBSON, J. E. *Computing in Structural Engineering*. Applied Science, Britain 1975.

GOBLE, G. G. and MOSES, F. "Structural Optimization". *ASCE Journal*, 101, ST4 (April 1975), pp. 635-48.

HALL, A. S. and WOODHEAD, R. W. *Frame Analysis*. Wiley, New York, 1967. *Handbook of Mathematical Functions*. NBS Applied Mathematics Series 55, Government Printing Office, Washington, 1964.

HILL, L. A. "Automated Optimum Cost Building Design". *ASCE Proceedings*, 92, ST6 (December 1966), pp. 247-63.

HUDSON, W. R. and STELZER, C. F. "A Direct Computer Solution for Slabs on Foundation". *ACI Journal,* 65, 3 (March 1968), pp. 188-201.

MAJID, K. I. and WILLIAMSON, M. "Linear Analysis of Complete Structures by Computers". *ICE Proceedings,* 38 (October 1967), pp. 247-66; 40 (June 1968), pp. 205-21.

McCORMICK, M. M. "Optimum Design of Short Span Highway Bridges". *Contracting and Construction Engineer,* 25, 2 (February 1971), pp. 24-30.

McCRACKEN, D. D. *Fortran with Engineering Applications.* Wiley, New York, 1967.

NAYLOR, T. H. *et al. Computer Simulation Techniques.* Wiley, New York, 1967.

ORDWAY, F. "Matrix Algebra for Calculating Multicomponent Mixtures". *PCA Paper 74,* 1960.

PURUSHOTHAMAN, P. "Design of Reinforced Concrete Cantilever Retaining Walls on a Computer". *Indian Concrete Journal,* 40, 1 (January 1966), pp. 34-38.

ROBINSON, J. *Structural Matrix Analysis for the Engineer.* Wiley, New York, 1967.

ROGERS, P. *Reinforced Concrete Design for Buildings.* Van Nostrand Reinhold, Australia, 1973.

ROOS, D. *ICES System Design.* MIT Press, Massachusetts, 1966.

SAA. *Digital Data Transmission,* AS 1484, Parts 1-4; *Flowchart Symbols,* AS X2; *Programming Language: FORTRAN,* AS 1486.

SIMS, F. A. *et al.* "Computers and Bridges". *Concrete* (London), 6, 4 (April 1972), pp. 18-23.

SMITH, I. A. "Rectangular Columns to CP 110". *Concrete* (London), 10, 7 (July 1976), pp. 29-34.

SPINDEL, P. D. *Computer Applications in Civil Engineering.* Reinhold, New York, 1971.

Standardization of Input Information for Computer Programs in Structural Engineering. Institution of Structural Engineers, London.

STUBBS, I. R. "An Engineer Oriented Computer Language for the Design of Prestressed Concrete". *PCI Journal,* 13, 3 (June 1968), pp. 733-82.

TAMHANKAR, M. G., RAO, A. S. and KAPLA, M. S. "Standardization of Precast Prestressed Concrete Bridge Beams". *Indian Concrete Journal,* 42, 6 (June 1968), pp. 231-41, 255.

TORRES, G. G. B., BROTCHIE, J. F. and CORNELL, C. A. "A Program for the Optimum Design of Prestressed Concrete Highway Bridges". *PCI Journal,* 11, 3 (June 1966), pp. 63-71.

United States Standards Institute, *Fortran.* USS X3.9, X3.10.

Use of Digital Computers in Structural Engineering. Institution of Structural Engineers, London.

WEAVER, W. *Computer Programs for Structural Engineers.* Van Nostrand, London, 1967.

WEEKS, L. W. "Magic Memory Solves Raw Mix Problems". *Rock Products,* 63, 4 (April 1960), pp. 85-89.

WOODHEAD, R. W., UTTING, P. L. and HALL, A. S. "Computer Analysis of a 600-ft Three-dimensional Frame Building". *IE Aust. CE Transactions,* CE8, 2 (October, 1966), pp. 128-31; CE9, 1 (April 1967), p. 169.

YOUNG, W. M. "Staircase by Computer". *Concrete* (London), 9, 2 (February 1975), pp. 22-24.

(b) Models.

ALAMI, Z. Y. and FERGUSON, P. M. "Accuracy of Models used in Research on Reinforced Concrete". *ACI Journal,* 60, 11 (November 1963), pp. 1643-63; 2, 61, 6 (June 1964), pp. 2067-70.

AZEVEDO, M. C. and FERREIRA, M. J. E. "Construction of Models of Concrete Dams for Elastic Tests". *Technical Paper 232*, MPW, NCEL, Lisbon, 1964.

BALINT, P. S. and SHAW, F. S. "Structural Model of the 'Australia Square' Tower in Sydney". *Architectural Science Review*, 8, 4 (December 1965), pp. 136-49.

BASE, G. D. "Tests on Four Prototype Reinforced Concrete Hinges". *Research Report 17*, C & CA Britain, 1965.

BREEN, J. E. "Fabrication and Tests of Structural Models". *ASCE Proceedings*, 94, ST6 (June 1968), pp. 1339-52.

CHAPMAN, J. C. *et al.* "Structural Model Studies for the Hendrik Verwoerd Dam". *ICE Proceedings*, 2, 57, (September 1974), pp. 467-92.

CUSENS, A. R. and KUANG, J.-G. "Experimental Study of a Free-Standing Staircase". *ACI Journal*, 63, 5 (May 1966), pp. 587-604; 12 (December 1966), pp. 1487-94.

DESAYI, P. "A Model to Simulate the Strength and Deformation of Concrete in Compression". *Materials and Structures*, 1, 1 (January/February 1968), pp. 49-56.

ELMS, D. G. "Elastic Stress Distribution in a Coupled Shear Wall". *IE Aust. CE Transactions*, CE9, 2 (October 1967), pp. 195-200.

FRANCIS, F. *Computer Modelling and Simulation*. Wiley, New York, 1968.

HAHN, G. J. and SHAPIRO, S. S. *Statistical Models in Engineering*. Wiley, New York, 1967.

HEDGREN, A. W. and BILLINGTON, D. P. "Mortar Model Test on a Cylindrical Shell of Varying Curvature and Thickness". *ACI Journal*, 64, 2 (February 1967), pp. 73-83.

KORETSKY, A. V., COVINGTON, E. T. and ROBSON, R. J. "Gypsum Plaster, High Alumina Cement Mixes as a Material for Models of Concrete Structures". *Constructional Review*, 40, 12 (December 1967), pp. 26-33.

LENCZNER, D. "An Investigation into the Behaviour of Sand in a Model Silo". *Structural Engineer*, 12, 41 (December 1963), pp. 389-98.

LENCZNER, D. "Distribution of Pressure in a Model Silo Containing Cement". *Magazine of Concrete Research*, 15, 44 (July 1963), pp. 101-6.

LITLE, W.A. and PAPARONI, M. "Size Effect in Small-Scale Models of Reinforced Concrete Beams". *ACI Journal*, 63, 11 (November 1966), pp. 1191-204; 2, 64, 6 (June 1967), pp. 1571-82.

MAGURA, D. D. "Structural Model Testing—Reinforced and Prestressed Mortar Beams". *PCA R & DL Journal*, 9, 1 (January 1967), pp. 2-24.

MATTOCK, A. H. "Structural Model Testing—Theory and Applications". *PCA R & DL Journal*, 4, 3 (September 1962), pp. 12-23.

PAHL, P. J. and SOOSAAR, K. "Structural Models for Architectural and Engineering Education". *Research Report R64-03*, Massachusetts Institute of Technology, Cambridge, Mass., 1964.

PAIVA, De, H. A. R. and SIESS, C. P. "Strength and Behaviour of Deep Beams in Shear". *ASCE Proceedings*, 91, ST5 (October 1965), pp. 19-41.

PREECE, B. W. and DAVIES, J. D. *Models for Structural Concrete*. Contractors Record, London, 1964.

RANGANATHAM, B. V., RAO, K. S. S. and HENDRY, A. W. "Plaster Mortar for Small Scale Tests". *ACI Journal*, 64, 9 (September 1967), pp. 594-601.

RISH, R. F. "Forces in Cylindrical Chimneys due to Wind". *ICE Proceedings*. 36 (April 1967), pp. 791-803.

ROCHA, M. "Structural Model Techniques. Some Recent Developments". *Technical Paper 264*, MPW, NCEL, Lisbon, 1965.

ROCHA, M. and SILVEIRA, Da, A. F. "Use of Models to Determine Temperature Stresses in Concrete Arch Dams". *Technical Paper 230,* MPW, NCEL, Lisbon. 1964.

ROLL, F. "Materials for Structural Models". *ASCE Proceedings,* 94, ST6 (June 1968), pp. 1353-81.

ROWE, R. E. and BASE, G. D. "Model Analysis and Testing as a Design Tool". *ICE Proceedings,* 33 (February 1966), pp. 183-99; 36 (March 1967), pp. 677-95.

SABNIS, G. M. and WHITE, R. N. "A Gypsum Mortar for Small-scale Models". *ACI Journal,* 64, 11 (November 1967), pp. 767-74.

SERAFIM, J. L. "New Shapes for Arch Dams". *ASCE Civil Engineering,* 36, 11 (November 1966), pp. 38-43.

SHARPE, R. and CLYDE, D. H. "Rational Design of Reinforced Concrete Slabs". *IE Aust. CE Transactions,* CE9, 2 (October 1967), pp. 209-16.

SILVEIRA, Da, A. F., AZEVEDO, M. C. and FERREIRA, M. J. E. "Some Special Problems Relative to Concrete Dams Solved by Model Tests". *Technical Paper 228,* MPW, NCEL, Lisbon, 1964.

SOMERVILLE, G., ROLL, F. and CALDWELL, J. A. D. "Tests on a $1/_{12}$ Scale Model of the Mancunian Way". *Publication TRA394,* C & CA Britain, 1965.

STEPHENS, L. F. "A Simple Model Solution of the Suspension Bridge Problem". *ICE Proceedings.* 34 (July 1966), pp. 369-82; 36 (April 1967), pp. 883-86.

TAYLOR, J. R. *Model Building for Architects and Engineers.* McGraw-Hill, New York, 1971.

TURITZIN, A. M. "Dynamic Pressure of Granular Material in Deep Bins". *ASCE Proceedings (Structural Division),* 89, ST2 (April 1963), pp. 49-73.

WHITE, R. N. *et al.* "Models for Concrete Structures". *ACI Publication SP-24,* 1970.

ZIENKIEWICZ, O. C. and HOLISTER, G. S. *Stress Analysis.* Wiley, New York, 1965.

3. SYSTEMS; ECONOMICS; ENVIRONMENT

ANTILL, J. M. *Civil Engineering Management.* Angus & Robertson, Sydney, 1970.

ANTILL, J. M. and WOODHEAD, R. W. *Critical Path Methods in Construction Practice.* Wiley, New York, 1965.

ARCHIBALD, R. D. and VILLORIA, R. L. *Network-Based Management Systems (Pert/CPM).* Wiley, New York, 1967.

ARMSTRONG, K. G. "Economic Analysis of Engineering Projects". *ICE Proceedings,* 30 (January 1965), pp. 17-32; 34 (August 1966), pp. 512-41.

Association of Consulting Engineers. *Terms of Engagement.* ACE Federal and Chapter Offices.

BECKMAN, M. and KUNZI, H. P. *Lecture Notes in Operations Research and Mathematical Economics.* Springer-Verlag, New York, 1968 (forwards).

BSI. *Control Chart Technique When Manufacturing to a Specification,* BS 3507; *Management of Design for Economic Production,* BS PD6470.

BUTCHER, W. S. "Dynamic Programming for Project Cost-Time Curves". *ASCE Journal,* 93, CO1 (March 1967), pp. 59-73.

CAMPBELL, D. G. "Financial Control of a Concrete Company". *Contracting and Construction Engineer,* 22, 6 (February 1968), pp. 35-38.

CHANLETT, E. T. *Environmental Protection.* McGraw-Hill, New York, 1973.

CHESTNUT, H. *Systems Engineering Methods.* Wiley, New York, 1967.

COLE, L. J. R. "Optimization of Repetitive Building Work: Flowline Computer Program". *Chartered Builder,* 13 (May/June 1975), pp. 8-25.

"Computerized Automation Systems". *ASTM Standardization News,* 4, 5 (May 1976), pp. 8-29.

CORDELL NEWTON. *Building Cost Book and Estimating Guide.* Quarterly Publication, Australia.

Cost Control in Building Design. Research and Development Building Management Handbook, HMSO, London, 1968.

CSIRO. "Network Analysis Manuals". *DBR, Users Manuals T.A.67/1.0.* Highett, Victoria, 1968 (forwards).

DONOVAN, J. J. *Systems Programming.* McGraw-Hill, New York, 1972.

DUNHAM, C. W. *Contracts, Specifications and Law for Engineers.* McGraw-Hill, New York, 1971.

"Durable Concrete Structures and Products". Environmental design criteria are elaborated in Section Four of this textbook.

EARGLE, F. L. "Incentive Plan for a Ready-Mix Operation". *Modern Concrete,* 30, 11 (March 1967), pp. 72-74.

Engineering 1976-2001. IE Aust. Annual Engineering Conference, Townsville, Queensland, 1976.

ENRICK, N. L. *Management Operations Research.* Holt, Rinehart & Winston, London, 1965.

ENTWISTLE, A. and REINERS, W. J. "Incentives in the Building Industry". *DSIR Special Report No. 28,* National Building Studies.

FOSTER, N. *Construction Estimates From Take-Off To Bid.* McGraw-Hill, New York, 1972.

GATES, M. "Bidding Strategies and Probabilities". *ASCE Journal,* 93, CO1 (March 1967), pp. 75-107.

GEORGE, C. J. and McKINLEY, D. *Urban Ecology.* McGraw-Hill, New York, 1974.

GILL, W. H. *Law of Building Contracts.* Butterworths, London, 1969.

GOBOURNE, J. *Cost Control in the Construction Industry.* Newnes-Butterworths, London, 1973.

GRANT, E. L. and LEAVENWORTH, R. S. *Statistical Quality Control.* McGraw-Hill, New York, 1972.

GUTT, W. "Use of By-products in Concrete". BRE Britain, *Current Paper 53/74,* 1974.

HAHN, G. J. and SHAPIRO, S. S. *Statistical Models in Engineering.* Wiley. New York, 1967.

HANSSMAN, F. *Operations Research Techniques for Capital Investment.* Wiley, New York, 1968.

HARE, V. C. *Basic Programming.* Harcourt, New York, 1970.

KLERER, M. and KORN, G. A. *Digital Computer User's Handbook.* McGraw-Hill, New York, 1967.

"Legal Responsibilities in Concrete Construction". *ACI Symposium, Los Angeles Convention,* 1968.

LEGGET, R. F. *Cities and Geology.* McGraw-Hill, New York, 1973.

LINZEY, M. P. T. *et al.* "Economic Analysis of Tall Buildings—Total Building Design Using a Digital Computer". *Build International,* 7, 4 (September/October 1974), pp. 355-80.

LOPEZ, L. A. "Automated Data Management System". *ASCE Journal,* 101, ST4 (April 1975), pp. 661-76.

MALLON, T. R. "Adapting Computer Techniques for Ready-Mix Operations". *Concrete Products*, 72, 9 (September 1969), pp. 42-47.

McCREERY, T. I. "Improved Procedures for Calling Tenders and Administering Contracts". *IE Aust. Journal*, 38, 12 (December 1966), pp. 299-302; 39, 4-5 (April/May 1967), pp. 57-58; 6 (June 1967), pp. 81-82, N62.

MILLER, R. J. "Management in the Design Office". *Consulting Engineer*, 36, 7 (July 1972), pp. 29-31.

NANABHOY, R. "Simplified Costing for Contractors". *Indian Concrete Journal*, 32, 11 (November 1958), pp. 386-90; 33, 2 (February 1959), pp. 59-62.

"Network Analysis in Construction Design". MPBW, *Research and Development Building Management Handbook 3*, HMSO, London, 1967.

NEUFVILLE, de, R. and STAFFORD, J. H. *Systems Analysis For Engineers And Managers*. McGraw-Hill, New York, 1971,

O'BRIEN, J. J. *CPM in Construction Management*. McGraw-Hill, New York, 1971.

OUGHTON, F. *Value Analysis and Value Engineering*. Pitman, London, 1969.

PARKER, H. W. and OGLESBY, C. H. *Methods Improvement For Construction Managers*. McGraw-Hill, New York, 1972.

PECK, R. L. "How to Produce 33% More Concrete Blocks, Using 50% Fewer Employees!" *Modern Concrete*, 31, 7 (November 1967), pp 46-49.

REYNAUD, C. B. *The Critical Path Network Analysis Applied to Building*, Godwin, London, 1967.

RUDWICK, B. H. *Systems Analysis for Effective Planning*. Wiley, New York, 1969.

SANSOM, R. C. "Organization of Building Sites". *Special Report No. 29*, DSIR, National Building Studies, HMSO, London.

SEDDON, R. W. "Use of a Computer for Cost Accounting in Smaller Businesses". *Australasian Institute of Cost Accountants, Bulletin*, 10, 2 (April 1966), pp. 2-21.

SHULTS, E. C. "Application of a Management Information and Control System to the Ready-mix Concrete Industry in the USA". *Indian Concrete Journal*, 41, 6 (June 1967), pp. 238-41.

SLATER, D. "Planning for Least Cost". *Building*, 213, 6491 (October 1967), pp. 179-92.

SMITH, P. and TAMBERG, K. G. "System for Strength Evaluation of Concrete Bridges". *ACI Journal*, 67, 11 (November 1970), pp. 888-93.

SMITH, P. R. and JULIAN, W. G. *Building Services*. Applied Science, Britain, 1976.

TURK, A. *Ecology, Pollution, Environment*. Saunders, Philadelphia, 1972.

TURNER, B. T. *Management Training For Engineers*. Business Books, London, 1969.

"Utilization of Industrial By-products and Waste Materials". BRE, Britain, *Digest 255*, 1975.

WADE, D. H. "Critical Path Analysis and the Civil Engineering Industry". *ICE Proceedings*, 39 (February 1968), pp. 289-303.

WHITE, D. *et al. Operational Research Techniques*, Business Books, London 1969.

4. EQUIPMENT

(a) General.

BROUGHTON, H. F., EDEN, J. F. and VALLINGS, H. G. "Mobile Tower Cranes for Two- and Three-storey Buildings". *Special Report No. 31,* DSIR National Building Studies, HMSO, London.

"Concrete Brick Blocklaying Machines Aid C/M Systems Need". *Concrete Products,* 73, 1 (January 1970), pp. 54-55.

Construction Methods and Equipment. McGraw-Hill, New York (monthly issue).

Contractors' and Civil Engineers' Equipment Reference. Technical and Industrial Press, Sydney, 1969.

"Distributor for Concrete and Aggregate". *Concrete and Constructional Engineering,* 55, 6 (June 1960), p. 252.

DOUGLAS, J. *Construction Equipment Policy.* McGraw-Hill, New York, 1975.

EDEN, J. F. and BISHOP, D. "Mechanical Handling in Building Operations". *Structural Engineer,* 36, 2 (February 1958), pp. 61-72.

EN–R Field and Office Manual. McGraw-Hill, New York, 1976.

FEDERATION OF MANUFACTURERS OF CONSTRUCTION EQUIP-MENT, *British Construction Equipment,* FMCE, London, 1967.

FRYER, K. G. H. "Speeding up Building by Mechanization". *Note E1294,* DSIR, BRS.

ILLINGWORTH, J. R. *Movement and Distribution of Concrete.* McGraw-Hill, London, 1972.

Mine and Quarry Mechanisation (issued annually), Magazine Associates, Australia.

Specification. Architectural Press, London, 1976.

STANDARDS ON CRANES AND SCAFFOLDING. See Table 27.1.

SWEET'S CATALOG SERVICE. *Architectural Catalog File.* F. W. Dodge Corporation, New York, 1976.

VALLINGS, H. G. "Making Concrete Panels for Industrialised Building by Pressing". *National Builder,* 48, 1 (January 1967), pp. 24-29.

VALLINGS, H. G. *Mechanisation in Building.* Applied Science, Britain, 1976.

(b) Testing and monitoring.

ACI COMMITTEE 208. "Test Procedure to Determine Relative Bond Value of Reinforcing Bars". *ACI Journal,* 30, 1 (July 1958), pp. 1-16.

ACI COMMITTEE 437. "Strength Evaluation of Existing Concrete Structures". *ACI Manual of Concrete Practice,* Part 2, Structural Design, Specifications and Analysis.

"Analytical Techniques for Hydraulic Cement and Concrete". *ASTM Special Technical Publication No. 395,* 1966.

ARTHUR, J. R. F. and ROSCOE, K. H. "An Earth Pressure Cell for the Measurement of Normal and Shear Stresses". *Civil Engineering and Public Works Review,* 56, 659 (June 1961), pp. 765-70.

ASHWORTH, R. "Apparatus for Laboratory Freezing-and-thawing Tests on Concrete Specimens". *Magazine of Concrete Research,* 19, 58 (March 1967), pp. 45-48.

ASTM Proceedings. Annual Committee Reports. ASTM, Phil., Pa.

ASTM and BSI. *Strain Gauges for Concrete Investigations.* ASTM E251; BS 1881.

ATCHLEY, B. L. and FURR, H. L. "Strength and Energy Absorption Capabilities of Plain Concrete under Dynamic and Static Loadings". *ACI Journal,* 64, 11 (November 1967), pp. 745-56.

BAKOSS, S. L. "Low Cost Rig for Compression Creep Tests on Concrete". *Monograph CE 74/1,* NSW Institute of Technology, Sydney, 1974.

BALMER, G. G. and COLLEY, B. E. "Skid Resistance of Concrete". *ASTM Journal of Materials,* 1, 3 (September 1966), pp. 536-59.

BSI. *Design Requirements and Testing of Controlled-atmosphere Laboratories.* BS 4194; *Dial Gauges for Linear Measurement.* BS 907; *Measuring Instruments for Constructional Works.* BS 4484; *Non-destructive Methods of Test for Concrete.* BS 4408; *Thermal Properties and Insulation.* BS 874.

COLLEN, L. D. G. and KIRWAN, R. W. "Some Characteristics of Ferro-Cement". *Civil Engineering and Public Works Review,* 54, 632 (February 1959), pp. 195-96.

CONSIDINE, D. M. *Encyclopedia of Instrumentation and Control.* McGraw-Hill, New York, 1971.

CORSON, R. H. "A New Gauge for the Measurement of Internal Strains in Concrete". *Concrete and Constructional Engineering,* 60, 12 (December 1965), pp. 437-43.

"Cracking, Deflection, and Ultimate Load of Concrete Slab Systems". *ACI Publication SP-30.*

CROMARTY, R. E. and BRYDEN, J. G. "An Apparatus for Measuring Dimensional Changes in Fresh Cement Paste". *Magazine of Concrete Research,* 19, 61 (December 1967), pp. 239-42.

CRUZ, C. R. "Apparatus for Measuring Creep of Concrete at High Temperatures". *PCA R & DL Journal,* 10, 3 (September 1968), pp. 36-42.

CUEVAS, N. R. "Use of Creep Data to Obtain Properties of Materials". *Materials and Structures,* 2, 12 (November/December 1969), pp. 431-36.

DAVIS, H. E., TROXELL, G. E. and WISKOCH, C. T. *Testing and Inspection of Engineering Materials.* McGraw-Hill, New York, 1964.

EDITORIAL. "Factors in the Load Testing of Concrete Structures". *Indian Concrete Journal,* 41, 6 (June 1967), pp. 217-18.

ELVERY, R. H. and HAROUN, W. "A Direct Tensile Test for Concrete Under Long- or Short-term Loading". *Magazine of Concrete Research,* 20, 63 (June 1968), pp. 111-16.

ERLIN, B. "Methods Used in Petrographic Studies of Concrete". *PCA Research Bulletin 193,* 1966.

EVANS, R. H. and MARATHE, M. S. "Stress Distribution Around Holes in Concrete". *Materials and Structures,* 1, 1 (January/February 1968), pp. 57-60.

FERGUSON, P. M. and THOMPSON, J. N. "Development Length for Large High Strength Reinforcing Bars". *ACI Journal,* 62, 1 (January 1965), pp. 71-93.

FOOTE, P. "Spherical Seatings On Cube Testing Machines". *Concrete* (London), 8, 2 (February 1974), pp. 36-37.

FORRESTER, J. A. *et al.* "Apparatus for Rapid Analysis of Fresh Concrete". C & CA, Britain. *TR42.490,* 1974.

Fresh Concrete: Properties and Measurement. RILEM Seminar Proceedings, Leeds University, 1973.

GEYMAYER, H. B. "Strain Meters and Stress Meters for Embedment in Models of Mass Concrete Structures". *Technical Report 6-811,* United States Army Corps of Engineers, Vicksburg, 1968.

GEYMEYER, H. G. and COX, F. B. "Bamboo Reinforced Concrete". *ACI Journal,* 67, 10 (October 1970), pp. 841-46.

GOSBELL, K. B. and STEVENS, L. K. "Test Loading of a Full Scale Bridge". *Australian Road Research Board, Proceedings, Fourth Conference,* 1968, pp. 2018-41.

HANNANT, D. J. "Equipment for the Measurement of Creep of Concrete under Multiaxial Compressive Stress". *RILEM Bulletin,* 33 (December 1966), pp. 421-22.

HANSON, N. W. and KURVITS, O. A. "Instrumentation for Structural Testing". *PCA R & DL Journal,* 7, 2 (May 1965), pp. 24-38.

HEDSTROM, R. O. "Tensile Testing of Concrete Block and Wall Elements". *PCA R & DL Journal,* 8, 2 (May 1966), pp. 42-52.

HENDRY, A. W. and BRADSHAW, R. E. "A 600 Ton Multi-purpose Testing Frame". *Civil Engineering and Public Works Review,* 61, 724 (November 1966), pp. 1383-85.

HIME, W. G., MIVELAZ, W. F. and CONNOLLY, J. D. "Use of Infrared Spectrophotometry for the Detection and Identification of Organic Additions in Cement and Admixtures in Hardened Concrete". *PCA Research Bulletin 194,* 1966.

HOGBERG, E. "Mortar Bond". *Report 40,* National Swedish Institute for Building Research, Stockholm, 1967.

HONDROS, G. and MOORE, G. J. "Epoxy Resin Protection of Electrical Resistance Strain Gauges for General Use in Concrete". *Civil Engineering and Public Works Review,* 57, 671 (June 1962), pp. 756-58.

HSU, T. T. C. and MATTOCK, A. H. "A Torsion Test Rig". *PCA R & DL Journal,* 7, 1 (January 1965), pp. 2-9.

HUGHES, B. P. "Absorption of Concrete Aggregates". *Concrete* (London), 4, 9 (September 1970), pp. 367-68.

HUGHES, B. P. and CHAPMAN, G. P. "Complete Stress-Strain Curve for Concrete in Direct Tension". *RILEM Bulletin,* New Series, 30 (March 1966), pp. 95-97.

ISENBERG, J. "Study of Cracks in Concrete by X-radiography". *RILEM Bulletin,* New Series, 30 (March 1966), pp. 107-15.

KARPATI, K. K. and SEREDA, P. J. "Joint Movement in Precast Concrete Panel Cladding". *ASTM Journal of Testing and Evaluation,* 4, 2 (March 1976), pp. 151-56.

KEMP, E. L. *et al.* "Bond Characteristics of Deformed Reinforcing Bars". *ACI Journal,* 65, 9 (September 1968), pp. 743-56.

KEMPSTER, E. "Measuring Void Content: New Apparatus for Aggregates, Sands and Fillers". *BRS Current Paper 19/69,* Britain, 1969.

KINNEAR, R. G. "Formwork Pressure Balance". *C & CA Britain, Technical Report TRA/373,* 1963.

KOWALSKI, T. H. "Bamboo-reinforced Concrete". *Indian Concrete Journal,* 48, 4 (April 1974), pp. 119-21.

LEE, C. R. and BRYDEN-SMITH, D. W. "Experience with a Simple Moisture Gauge for Embedment in Concrete". *BRS Current Papers, Research Series 68,* Britain, 1968.

LERCHENTHAL, C. H. "Bonded Sheet Metal Reinforcement for Concrete Slabs". *RILEM Bulletin,* 37 (December 1967), pp. 263-69.

LIE, T. T. and ALLEN, D. E. "Fire Resistance of Reinforced Concrete Columns". *Proceedings IABSE Symposium,* Quebec, 1974, pp. 245-54; NRC Canada, *DBR Research Paper 622.*

LOUBSER, P. J. and BRYDEN, J. G. "Apparatus for Determining the Coefficient of Thermal Expansion of Rocks, Mortars and Concretes". *Magazine of Concrete Research,* 24, 72 (June 1972), pp. 97-100.

MAGURA, D. D. "Structural Model Testing—Reinforced and Prestressed Mortar Beams". *PCA R & DL Journal,* 9, 1 (January 1967), pp. 2-24.

MALHOTRA, V. M. "Concrete Rings for Determining Tensile Strength of Concrete". *ACI Journal,* 67, 4 (April 1970), pp. 354-57; 10 (October 1970), pp. 847-48.

MALHOTRA, V. M. and ZOLDNERS, N. G. "Comparison of Ring-Tensile Strength of Concrete with Compressive, Flexural and Splitting-Tensile Strengths". *Journal of Materials,* 2, 1 (March 1967), pp. 160-201.

MARSHALL, A. L. "Design and Construction of a Controlled-environment Chamber". *Magazine of Concrete Research,* 22, 72 (September 1970), pp. 185-86.

MARSHALL, W. T. *et al.* "Use of High-strength Friction-grip Bolts as Shear Connectors in Composite Beams". *Structural Engineer,* 49, 4 (April 1971), pp. 171-78.

McCOY, E. E. and ROSHORE, E. C. "Instruments for Measuring Pore Pressures in Concrete". *Technical Report 6-654,* United States Corps of Engineers, 1964.

McGONNAGLE, W. J. *Non-destructive Testing.* Gordon and Breach, New York, 1966.

MEULEN, G. J. R. VAN DER and DIJK, J. VAN. "A Permeability-testing Apparatus for Concrete". *Magazine of Concrete Research,* 21, 67 (June 1969), pp. 121-23.

MILLS, L. L. and ZIMMERMAN, R. M. "Compressive Strength of Plain Concrete Under Multiaxial Loading Conditions". *ACI Journal,* 67, 10 (October 1970), pp. 802-807.

MONFORE, G. E. "Electrical Resistivity of Concrete". *PCA R & DL Journal,* 10, 2 (May 1968), pp. 35-48; *Bulletin 224,* 1968.

MONFORE, G. E. and OST, B. "An 'Isothermal' Conduction Calorimeter for Study of the Early Hydration Reactions of Portland Cements". *PCA R & DL Journal,* 8, 2 (May 1966), pp. 13-20.

NAAMAN, A. E. and SHAH, S. P. "Tensile Tests of Ferrocement". *ACI Journal,* 68, 9 (September 1971), pp. 693-98.

NIELSEN, K. E. C. "Internal Stresses in Concrete". *RILEM Bulletin (Paris),* New Series 1 (March 1959), pp. 11-20.

NIELSEN, K. E. C. "Measurements of Water Vapour Pressure in Hardened Concrete". *Bulletin 35,* Swedish Cement and Concrete Research Institute, Stockholm, 1967.

OBERT, L., MERRILL, R. H. and MORGAN, T. A. "Borehole Deformation Gauge for Determining the Stress in Mine Rock". United States Bureau of Mines, *Investigation 5978,* Washington, 1962.

PATEL, S. B. and MEHTA, C. L. "Load-testing a Large-panel Precast System". *Indian Concrete Journal,* 42, 7 (July 1968), pp. 278-86.

"Performance of Existing Testing Machines". *Concrete* (London), 4, 9 (September 1970), pp. 347-51.

PERRY, E. S. and THOMPSON, J. N. "Bond Stress Distribution on Reinforcing Steel in Beams and Pullout Specimens". *ACI Journal,* 63, 8 (August 1966), pp. 865-75.

PINCUS, G. and GESUND, H. "Measuring Long-term Deformations in a Multistorey Reinforced Concrete Building". *Consulting Engineer,* 28, 2 (February 1967), pp. 102-9.

PIRTZ, D. and CARLSON, R. W. "Tests of Strain Meters and Stress Meters under Simulated Field Conditions". *ACI Publication SP-6, Paper 13,* 1963.

PLOWMAN, J. M. "Measurement of Stress in Concrete Beam Reinforcement". *ICE Proceedings,* 25 (June 1963), pp. 127-46.

PURKISS, J. A. and DOUGILL, J. W. "Compression Tests on Concrete at High Temperatures". *Magazine of Concrete Research,* 25, 83 (June 1973), pp. 102-8.

RAO, A. K. and GOWDER, C. S. K. "Study of Behaviour of Ferrocement in Direct Compression". *Cement and Concrete* (New Delhi), 10, 3 (October/December 1969), pp. 231-37.

"Rapid Analysis of Fresh Concrete". *Precast Concrete,* 4, 4 (April 1973), pp. 201-2.

RAWLINGS, B. "Experimental Equipment for Impulsive Testing of Structures". *IE Aust. Journal,* 39, 4-5 (April/May 1967), pp. 59-66.

ROCHA, M. "In Situ Strain and Stress Measurements". *Technical Paper 265,* MPW CEL, Lisbon, 1965.

ROCHA, M. and SILVEIRA, A. F. Da. "Assessment of Observation Techniques Used in Portuguese Concrete Dams". *Proceedings, Eighth Congress on Large Dams,* Edinburgh, 1964.

ROSENTHAL, I. and GLUCKLICH, J. "Strength of Plain Concrete Under Biaxial Stress". *ACI Journal,* 67, 11 (November 1970), pp. 903-14.

ROUSE, G. G. and BOUWKAMP, J. G. "Vibration Studies of Monticello Dam". *Research Report 9,* United States Bureau of Reclamation, 1967.

SAA. *Dial Gauges for Linear Measurement.* AS B80; *Thermostats and Energy Regulators.* AS C161; *Verification of Testing Machines.* AS B128.

SCHWIETE, H. "Testing at High Temperature". *Materials and Structures,* 2, 11 (September/October 1969), pp. 415-22.

SCOTT, I. G. "A Course in Resistance Strain Gauging". *Australian Journal of Instrument Technology,* 21, 4 (November 1965), pp. 134-39; 22, 1 (February 1966), pp. 23-32; 2 (May 1966), pp. 57-63; 3 (August 1966), pp. 92-97.

SIGVALDASON, O. T. "Influence of Testing Machine Characteristics upon the Cube and Cylinder Strength of Concrete". *Magazine of Concrete Research,* 18, 57 (December 1966), pp. 197-206.

SMEE, D. J. "Effect of Aggregate Size and Concrete Strength on the Failure of Concrete under Triaxial Compression". *IE Aust. CE Transactions,* CE9, 2 (October 1967), pp. 339-44.

SOMERVILLE, G. and BURHOUSE, P. "Tests on Joints Between Precast Concrete Members". *BRS Engineering Papers 45,* Britain, 1966.

"Structural Test Methods for Walls, Floors, Roofs and Complete Buildings—State-of-the-Art". *Building Science Series 58,* United States NBS, Washington, 1974.

"The Testing of Structures". Institution of Structural Engineers, *Report 40,* 1964.

THOMAS, H. S. H. "Measurement of Strain in Tunnel Linings Using the Vibrating-wire Technique". *Strain,* 2, 7 (July 1966), pp. 16-21.

TYLER, R. G. "Measurement of Strain and Stress in Concrete Bridge Structures". Road Research Laboratory, Britain, *Report LR189,* 1968.

WALLEY, F. "Some Special Applications of Prestressed Concrete to Civil Engineering Structures". *ICE Proceedings,* 33 (April 1966), pp. 567-97; 36 (April 1967), pp. 855-65.

WARD, W. H. and BURLAND, J. B. "Use of Ground Strain Measurements in Civil Engineering". British BRE, *Current Paper 13/73,* 1973.

WEBB, D. J. T. "Development of an Experimental Press for Large Concrete Units". BRS, *Current Papers, Construction Series 27,* Britain, 1966.

WEEKS. G. A. "Laboratory Testing of Large Structures". BRS *Current Paper 22/69,* Britain, 1969.

WOOD, R. H., NEEDHAM, F. H. and SMITH, R. F. "Test of a Multistorey Rigid Steel Frame". *Structural Engineer,* 46, 4 (April 1968), pp. 107-19.

5. INSULATION

ASTM. *Thermal Conductivity of Materials by Means of the Guarded Hot Plate.* ASTM C177.

ATKINSON, G. A. "Principles of Tropical Design". *Architectural Review,* 78, 761 (July 1960), pp. 81-92.

BARNED, J. R. and O'BRIEN, L. F. "Thermal Conductivity of Building Materials". *CSIRO DBR Report R2,* 1970.

BASTINGS, L. "Handbook on the Insulation and Heating of Buildings". *Information Series No. 18.* DSIR (NZ).

BAZLEY, E. N. *The Airborne Sound Insulation of Partitions.* Ministry of Technology, National Physical Laboratory, HMSO, London, 1966.

BERANEK, L. L. *et al. Noise And Vibration Control.* McGraw-Hill, New York, 1971.

BSI. *Sound Insulation and Noise Reduction,* BS CP3III; *Precautions against Fire,* BS CP3IV; *Heating and Thermal Insulation,* BS CP3VIII, 874, 1334, 1588, 2972, 2973, 3708, 3837, 3869, 3927, 3958.

Building Regulations. D of E BRE, Britain, 1974.

CAMPBELL-ALLEN, D. and THORNE, C. P. "Thermal Conductivity of Concrete". *Magazine of Concrete Research,* 15, 43 (March 1963), pp. 39-48; 16, 49 (December 1964), pp. 233-34.

CLOSE, P. D. *Sound Control and Thermal Insulation of Buildings.* Reinhold, New York, 1966.

"Condensation in Roofs". *BRE Digest 180,* Britain, 1975.

COPELAND, R. E. "Controlling Sound with Concrete Masonry". *Concrete Products.* 68, 7 (July 1965), pp. 39-43.

"Demolition and Construction Noise". *BRE Digest 184,* Britain, 1975.

Flat Compact Roofs. Building Research Bureau of New Zealand, Bulletins 111 and 112, 1968.

GALBREATH, M. "Fire Endurance of Concrete Assemblies". *DBR, Technical Paper 235,* NRC Canada, 1966.

HASSALL, D. *Reflective Insulation and the Control of Thermal Environments.* St. Regis-ACI, Sydney, 1974.

HOFTIEZER, A. B. "Practical Aspects of Noise Control in Crushed Stone Plants". *Pit and Quarry,* 47, 11 (May 1955), pp. 76-82.

"How Much Site Noise and Vibration?" *World Construction,* 27, 10 (October 1974), pp. 21-34.

Insulation Handbook. Lomax, Erskine & Co., London, 1969.

KOPATSCH, H. "Vented Cast Stone Facings for Buildings". *Technical Translation 1301,* NRC, Canada, 1967.

KORENKOV, J. "The Hygienic Properties of the 'Microclimate' of Flats". British *BRS, Library Communication 1169,* 1963.

LAWRENCE, A. *Architectural Acoustics.* Elsevier, Amsterdam, 1970.

National Building Research Institute, South Africa. "Condensation." *Information Sheet, DIS 88,* 1963.

NEW ZEALAND STANDARDS INSTITUTE. *Thermal Insulating Materials for Buildings.* NZSS 1340.

PARKIN, P. H. and HUMPHREYS, H. R. *Acoustics, Noise and Buildings.* Faber, London, 1969.

RETTINGER, M. *Acoustic Design and Noise Control.* Chemical Publishing, New York, 1973.

Standard U Values. D of E *BRE Digest 108,* Britain, 1975.

STEFFENS, R. J. "Structural Vibration and Damage". *D of E BRS Report,* HMSO, London, 1974.

Thermal Considerations in Roof Design. Canadian Building Digest, CBD70, 1965.

THIERY, P. *Fireproofing,* Elsevier, Amsterdam, 1970.

UNITED STATES DEPARTMENT OF COMMERCE. "Sound Insulation of Wall, Floor and Door Constructions". *NBS, Monograph 77; Building* ington.

(See also Sections 33.3.2(c), (d) and 33.3.3(c), (d).)

APPENDIX V

PETROGRAPHIC NUMBER OF AGGREGATE

For petrographic number determinations of aggregate (which are referred to in Section 3.3.4(k) of Part 1), rock-type classifications and related petrographic factors are tabulated for selective use.

PETROGRAPHIC DATA

Rock Type	Classification	Factor
Basalt—volcanic (slightly weathered)	Good	1
Carbonates (hard)	Good	1
Carbonates (sandy, hard)	Good	1
Gneiss (hard)	Good	1
Granite—diorite	Good	1
Greywacke—arkose	Good	1
Magnetite	Good	1
Pyrite (disseminated in trap)	Good	1
Quartzite (coarse grained)	Good	1
Quartzite (iron bearing)	Good	1
Sandstone (hard)	Good	1
Sedimentary conglomerates (hard)	Good	1
Trap	Good	1
Basalt—volcanic (soft)	Fair	3
Carbonates (slightly weathered)	Fair	3
Carbonates (sandy, medium hard)	Fair	3
Chert and cherty carbonates	Fair	3
Crystalline carbonates (hard)	Fair	3
Crystalline carbonates (slightly weathered)	Fair	3
Flints and jaspers	Fair	3
Gneiss (soft)	Fair	3
Granite (friable)	Fair	3

PETROGRAPHIC DATA—(Continued)

Rock Type	Classification	Factor
Pyrite (pure)	Fair	3
Sandstone (medium hard)	Fair	3
Carbonates (crystalline, very soft, porous)	Poor	6
Carbonates (deeply weathered)	Poor	6
Carbonates (ochreous)	Poor	6
Carbonates (shaly clay)	Poor	6
Carbonates (soft, sandy)	Poor	6
Carbonates (soft, slightly shaly)	Poor	6
Cementations	Poor	6
Chert and cherty carbonates (weathered)	Poor	6
Encrustations	Poor	6
Gneiss (friable)	Poor	6
Granite (friable)	Poor	6
Quartzite (fine grained)	Poor	6
Sandstone (soft, friable)	Poor	6
Schist (soft)	Poor	6
Clay	Deleterious	10
Iron formations (very soft)	Deleterious	10
Ochre	Deleterious	10
Shale	Deleterious	10
Sibley formation	Deleterious	10
Slates	Deleterious	10
Talc—gypsum	Deleterious	10
Volcanics (decomposed)	Deleterious	10

APPENDIX **VI**

GLOSSARY OF TERMS

This glossary includes only those terms which are not self-evident or are not fully defined where they occur in the text.

ACRYLIC: A transparent thermoplastic resin formed by polymerising the esters or amides of acrylic acid ($C_nH_{2n-2}O_2$).

ADHESION:

(a) The molecular attraction between bodies in intimate contact.

(b) The property by means of which a fluid or plastic substance sticks to the surface of a solid body.

ADIABATIC: A condition where heat neither enters nor leaves a system.

ADMIXTURE (ADDITIVE): A material (other than coarse or fine aggregate, cement or water) which is added in small quantities during the mixing of concrete, so as to produce some desired modification in one or more of its properties.

ADSORPTION: The attachment of a substance to the surface of a solid by virtue of forces arising from molecular attraction.

AGGREGATE: Uncrushed or crushed gravel, crushed stone or rock, sand or artificially produced inorganic materials, which form the major part of concrete.

Clinker: Well-burnt furnace residues which have been sintered into lumps (BS 1165).

Coarse: Aggregate which is retained on a 4·75-mm test sieve and contains only so much finer material as may be specified (see Section 3.3).

Crusher-run stone: Rock that has been broken in a mechanical crusher and which has not been subjected to any subsequent screening process.

792

Effective-size: The sieve opening which will just pass 10 per cent (by weight) of the particles of a grading.

Fine: Aggregate which passes a 4·75-mm test sieve and contains only so much coarser material as may be specified.

Gap-graded: Aggregate which contains particles of both large and small sizes, but in which particles of certain intermediate sizes are wholly or substantially absent (see Section 4.5).

Heavyweight: Aggregate with an unusually high specific gravity, such as barytes, iron ores, ilmenite, iron or steel, which is used for making high-density concrete.

Interlock: The projection of portions of aggregate particles from one side of a weakened plane joint or crack in concrete into recesses in the other side of same, thereby effecting load transfer in shear or compression and maintaining mutual alignment.

Lightweight: Aggregate of appreciably lower apparent specific gravity than that of ordinary aggregate (ASTM C35, C330-C332 and C495).

Saturated, surface-dry: Aggregate which has been immersed for 24 hours in water, and has subsequently had the surface moisture removed (see Sections 3.2.3 and 3.3.4(j)).

Ten-per-cent fines value: The crushing load which, when applied to a known weight of coarse aggregate, will produce fines amounting to 10 per cent of the original weight (see Section 3.3.4(f)(ii)).

Well-graded: Aggregate having a particle size distribution which will produce maximum density (i.e. minimum voids).

ALIPHATIC HYDROCARBONS: Solvents with open-chain molecular structure, derived from petroleum, and usually having low solvent power.

ALKALI-AGGREGATE REACTION: A chemical reaction, involving sodium or potassium ions and reactive silica, which generates expansive forces in hardened concrete that may be of sufficient magnitude to cause disruption. The sodium and potassium ions are derived from portland cement and sometimes from alkaline mixing water.

AMINE: Compounds prepared from ammonia by replacing hydrogen atoms with organic radicals.

AMPLITUDE: The maximum displacement from the mean position in connection with vibration.

ANODE: That part of a metal surface at which current enters an electrolyte and where corrosion occurs.

ANODE EFFICIENCY: The ratio of metal expended, in producing useful cathodic protection current, to total metal expended.

ANTIFOAMING OR AIR-DETRAINING AGENT: A product which greatly increases the surface tension of a liquid, and thereby reduces a tendency to foam during mixing or application.

AQUIFER: A layer of rock which holds water and allows water to percolate through it.

AROMATIC HYDROCARBONS: Strong solvents derived from coal tar with benzene ring molecular structure.

ATMOSPHERIC PRESSURE: The normal pressure of the air at sea-level, about 101·5 kPa. (See PASCAL.)

BACTERIAL CORROSION: The destruction of a material of construction (e.g. concrete, ferrous metal, copper and rubber) by chemical processes brought about by the activity of certain bacteria (e.g. *Sporovibrio* and *Desulphovibrio desulphuricans, Thiobacillus concretivorus* and *neopolitanus, Lactobacillus delbruckii, Proteus* and *Thiobacillus thio-oxidans*). These bacteria require sulphate, hydrogen, organic compounds, carbon dioxide, phosphate and humidity, sometimes without and at other times with oxygen, for their activity. They produce hydrogen sulphide, ammonia and sulphuric acid. These bacteria may be divided into two groups: autotrophic and heterotrophic. The former subsist on inorganic elements, whereas the latter require organic substances for growth (see Sections 15.2.3(d) and 19.1.1).

BASE PLATE (PAVEMENTS): A plate of metal or other suitable material placed under joints and the adjacent slab ends, so as to prevent an infiltration of soil and moisture from the sides or bottom of the joint opening.

BENEFICIATION: Improvement of the chemical or physical properties of a raw material or intermediate product by the removal of undesirable components or impurities.

BENTONITE: A clay composed principally of minerals of the montmorillonite group.

BLEEDING: The exudation of water from unhardened concrete.

BONDING LAYER: A layer of cement mortar (3-15 mm thick) which is spread on a moist and prepared, hardened concrete surface prior to placing fresh concrete.

BREEZE: An indefinite term which usually means clinker, but it may refer to coke-breeze.

BRICK: A manufactured rectangular prismatic unit, not exceeding 4000 cm^3 in gross volume, that is intended for use in bonded masonry construction.

BRITTLE FRACTURE: Fracture unaccompanied by visible plastic deformation.

BULK DENSITY: The weight (mass; including voids) per unit volume of a material on being compacted in a defined way (see Section 3.3.4(b) and Appendix I).

BULKING: The difference in volume of a given mass of sand or other fine material in moist and dry conditions; it is expressed as a percentage of the volume in a dry condition (see Section 3.3.6(b)).

BUSH-HAMMER: A hammer having a serrated face comprising rows of pyramidal points.

CABLE: A single concentration of steel wires intended for prestressing.

CALIFORNIA BEARING RATIO (CBR): The percentage ratio between the applied load required to cause a plunger of standard dimension to penetrate for a given distance and at a specified rate into a soil specimen, prepared under specified conditions, and an arbitrarily defined standard load (AS 1289; Ministry of Transport Road Note 29; ASTM D1883).

CAPILLARITY: The movement of a liquid in the interstices of a porous material due to surface tension.

CARBOXYL RADICAL: The univalent radical, $-COOH$, that is present in and characteristic of the formulas of all organic acids.

CATALYST: A substance that is added in small quantity to promote a chemical reaction, but which remains itself chemically unchanged at the end of the reaction.

CATHODE: That part of a metal surface at which current leaves an electrolyte and enters the metal.

CATHODIC PROTECTION: Partial or complete protection of a metal from corrosion by making it a cathode, using either a galvanic or impressed current.

CEMENT: A chemical binder, in general, or a finely ground powder which, in the presence of an appropriate quantity of water, hardens and adheres to suitable aggregate, thus binding it into a hard agglomeration that is known as concrete or mortar.

Antibacterial: A cement which contains a bacteria-inhibiting agent (e.g. 0·7 per cent pentachlorophenol) for hygienic, moist-concrete floors (see Sections 15.2.3(d) and 28.7).

Hydraulic: A cement which is capable of binding suitable aggregate into a concrete that can set and harden under water.

Low-alkali: A portland cement which has a low content of sodium and potassium oxides or sodium oxide equivalent (see Section 15.2.3(a)).

Oil-well: A hydraulic cement, which differs from ordinary portland cement since its setting time is very much longer at the high temperatures that occur in deep wells (see Section 10.2).

Ordinary and special: See Section 3.1.2.

CEMENT FACTOR: The weight of cement per cubic metre of hardened concrete.

CEMENT GEL: The colloidal material which makes up the major portion of the porous mass that constitutes mature hydrated cement paste.

CHARACTERISTIC STRENGTH: That value of concrete strength, as assessed by standard tests, which is exceeded by the strengths at 28 days of at least 95 per cent of the concrete (see Section 4.3.1 and Appendix VIII, 5).

CLAY: A fine-graded, hygroscopic silicate of alumina (usually mixed with impurities) which is plastic when moist.

COHESION: The ability of a material to resist by means of internal forces the separation of its constituent particles.

COLD WORKING: Steel processing by cold rolling, bending, drawing, punching, shearing, notching or upsetting. Cold working leads to reduced ductility and shock resistance, or increased brittleness (particularly at low temperatures) unless the metal is subsequently annealed or stress relieved by proper heat treatment (ASTM A143). (See STRAIN AGEING.)

COLLOIDAL GROUT: A grout which has artificially induced cohesiveness or ability to remain in suspension. It is mixed at high speed.

COLLOIDS: Noncrystalline substances which, when in solution, are unable to pass through an animal membrane. The constituent particles range in size from 0·2 to 0·005 μm approximately.

COLORIMETRIC VALUE: An indication of the amount of organic compounds present in fine aggregate (see Section 3.3.4(e)(vi)).

COMET: Concrete operations model evaluation technique, which is used as a managerial aid in computerised concrete production and control.

COMPACTING FACTOR: The ratio obtained by dividing the observed weight of concrete (which fills a container of standard size and shape when allowed to fall into it under standard conditions of test) by the weight of fully compacted concrete which fills the same container (see Section 6.2.4).

COMPACTION: The process of inducing a closer packing of the aggregate particles in concrete by the reduction of voids.

COMPOSITE CONSTRUCTION: A type of construction made up of different materials (e.g. concrete and structural steel) or of members produced under different conditions (e.g. *in situ* concrete and precast concrete).

COMPUTER: For automatic data processing, a device that is capable of automatically accepting data, applying a sequence of processes to the data and supplying the results of these processes (BS 3527, 4505, 4636 and CP95).

CONCRETE: A suitably proportioned mixture of coarse and fine aggregate, cement and water. Basically, the term means "to grow together".

Aerated: See FOAMED CONCRETE.

Cohesive: A workable concrete mix which does not segregate.

Colloidal: Concrete of which the aggregate is bound by colloidal grout.

Dense: Concrete containing a minimum of voids.

Dry-packed: A mix sufficiently dry to be consolidated only by heavy ramming.

Ecological: Concrete in which waste material (e.g. hardened concrete) is used in place of conventional aggregate.

Fat: A concrete containing a large proportion of mortar.

Foamed: Lightweight concrete in which the lightness is obtained by the formation in the plastic mix of bubbles of air or gas which are retained on setting and hardening.

Granolithic: Concrete suitable for use as a wearing surface finish to floors, which is made with specially selected aggregate of a suitable hardness, surface texture and particle shape (see Section 20.1.5).

Green: Concrete which has set but not appreciably hardened.

Heavy: Concrete made with specially selected heavy aggregate for radiation screening purposes (see Section 29.3).

In situ: Concrete which is deposited in the place where it is required to harden as part of the structure, as opposed to precast concrete.

Lean: Concrete with a high aggregate/cement ratio.

Lightweight: Concrete of appreciably lower bulk density than one made from gravel or crushed stone (see Chapter 33).

Mass: Plain concrete (i.e. concrete without reinforcement) in comparatively large volumes (see Chapter 30).

Micro: Concrete for model analysis in which the particle and steel sizes are scaled down to suit the scale of the project (see Section 4.5 and Appendix IV, 2(b)).

Monolithic: Concrete cast without joints, other than construction joints.

No-fines: Concrete which contains little or no fine aggregate (see Section 33.2.1).

Polymer: A composite prepared with a monomer or polymer: either by impregnating hardened concrete (PIC), or by mixing with aggregate (PC) or fresh cement concrete (PCC), the resin being subsequently polymerised, cured or hardened (see Section 21.4).

Precast: Concrete which is cast in moulds before being placed in position (see Chapter 27).

Prepacked: See Section 24.4.

Prestressed: Concrete in which effective internal stresses are induced deliberately, usually by means of tensioned steel, prior to loading the structure (see Chapter 34).

Ready-mixed: Concrete delivered at the site in a plastic condition and requiring no further treatment before being placed in the position in which it is to set (see Section 7.1.2).

Refractory: Concrete made with high-alumina or with calcium-aluminate cement and with refractory aggregate to withstand very high temperature (see Section 29.1).

Reinforced: Concrete in which rods, bars or fabric (usually of steel) are embedded in such a manner that the two materials act together in resisting loads on the concrete.

Stage-mixed: Ready-mixed concrete which is mixed partially in a central-plant stationary mixer and finally in a truck mixer en route to the site (see Section 7.1.2).

Spun: Concrete compacted by centrifugal action (see Section 23.3).

Terrazzo: Marble-aggregate concrete that is cast *in situ* or precast and ground smooth for decorative surfacing purposes on floors and walls.

Translucent: A combination of glass and concrete blended together in precast or prestressed panels.

Vacuum: Concrete from which water is extracted by a vacuum process before hardening occurs (see Section 23.4).

Vibrated: Concrete compacted by vibration during placing (see Chapter 23).

CONSISTENCE (CONSISTENCY): The capacity of fresh concrete, mortar or cement paste to resist deformation or flow (see Chapter 6).

CONTAINMENT GROUTING: See PERIMETER GROUTING.

CONTINUOUS GRADING: A particle-size distribution in which all intermediate size fractions are present, as opposed to gap-grading (see Sections 3.3.4(d), 4.3, 20.1.2 and 33.2.2).

CONTINUOUSLY REINFORCED PAVEMENT: A pavement without transverse joints other than tied construction joints, which are placed between successive day's concreting, and containing sufficient longitudinal reinforcement with lapped joins to ensure that transverse cracks will be held tightly closed. The central portion is fully restrained between 120-150 m lengths at the ends.

CONTROL FACTOR: The ratio between the minimum compressive strength and the average compressive strength (see Table 4.4).

CONTROL JOINTS: Provision for the dimensional change of different parts of a structure due to shrinkage, expansion, temperature variation or other causes, so as to avoid the development of high stresses. (See EXPANSION JOINTS.)

CORROSION: Disintegration or deterioration of concrete and metals by chemical agents from external sources.

COVER BLOCKS (SPACERS): Small precast mortar blocks used inside formwork or shuttering to ensure the correct cover to reinforcement (see Section 3.4).

CRAZING: The cracking of a surface layer into small, irregularly shaped, contiguous areas.

CREEP: A slow, inelastic or plastic deformation of concrete or steel under continued constant stress (see Section 31.3). (See SHRINKAGE for complementary deformation.)

CRITICAL PATH SCHEDULE: A methodical and graphical means of programming work with the aid of a line diagram, on which are shown (by one line per activity) the duration, cost and interrelationships of all activities comprising a project (see Section 36.9.3 and Appendix IV, 3).

CRITICAL STRUCTURAL SECTION: A section where a high rate of test sampling of concrete is necessary because of the importance of a structural element, or of difficulty in checking the quality of concrete should initial test specimens be suspect. Examples are heavily loaded columns and inaccessible footings.

CRYOGEN: A fluid having a boiling point below −150°C at atmospheric pressure.

CUMULATIVE BATCHING: Measuring more than one ingredient of a concrete batch in the same container by bringing the batcher scale into balance at successive total weights as each different ingredient is accumulated in the container.

CURING: The process adopted to ensure the hardening of concrete by preventing excessive evaporation or extremes of temperature (see Chapter 9).

CURTAIN GROUTING: Injection of grout into a subsurface formation in such a way as to create a zone of grouted material transverse to the direction of anticipated water flow.

CURTAIN WALL: A thin wall used as a shield or protection.

CYBERNETICS: The means of controlling a set of activities so as to keep them directed towards a particular goal. It is the study of control and communication between man and machine.

DAMP-PROOFING: The prevention of transmission of water vapour by capillary forces.

DENSITY (DRY): The weight (Mass; dry) of a substance per unit volume at a stated temperature (see Section 3.3.4(b); DENSITY AND MASONRY, Appendix I).

DEW POINT: The temperature at which the condensation of moisture would occur if the existing atmosphere were cooled without a change in vapour pressure.

DIATOMITE: A friable earthy material composed of nearly pure hydrous amorphous silica (opal) and consisting essentially of the frustules of the microscopic plants called diatoms.

DISPERSION: A two-phase system in which one phase, called the disperse phase, is permanently distributed as small particles through the second phase, called the continuous phase.

DRY MIX: A mix containing little water in relation to its other components.

DRYING SHRINKAGE: The shrinkage of concrete caused by evaporation. More precisely, it is the difference between the length of a specimen cut from concrete (which has been matured and subsequently saturated) and its length when dried to constant length, the result being expressed as a percentage (under specified conditions) of the dry length (see Section 31.2).

DURABILITY: The ability of concrete to resist weathering action, chemical attack and abrasion.

DYNAMIC LOADING: Loading from units (particularly machinery) which, by virtue of their movement or vibration, impose stresses in excess of those imposed by their dead load.

EFFECTIVE DEPTH: The distance in a beam or slab between the centre of the tensile reinforcement and the extreme surface in compression.

EFFECTIVE SPAN: The distance between the centres of supports, or the clear distance between supports plus the effective depth of the beam or slab, the lesser value being taken.

EFFLORESCENCE: Disfiguring deposits of soluble or insoluble salts that may form on the surface of cementitious, stone, brick, slab-faced or rendered external walls, as a result of evaporation of the water in which relevant materials were dissolved. Permanent staining or spalling of the surface may take place (see Section 36.6).

ELASTIC DESIGN: A method of analysis where the design of a member is based on a linear stress-strain relationship and corresponding limiting elastic properties of the material.

ELASTIC LIMIT: The limit of stress beyond which the strain is not wholly recoverable. (See ELASTICITY and PROPORTIONAL LIMIT.)

ELASTICITY: That property of a body by virtue of which it tends to recover its original size and shape after deformation.

ELASTIC MODULUS: See MODULUS OF ELASTICITY.

ELECTROLYTE: A conducting substance, normally a solution, in which an electric current flows by virtue of ionisation.

ELECTRO-OSMOSIS: A filtering of liquid conductors, under the influence of electric current, through porous or semipermeable partitions with a speed that is independent of their thickness but varying with their nature and section (see Section 23.5 and OSMOTIC PRESSURE).

ELUTRIATION: Purification by washing and straining or decantation; or separation by removing particle sizes or forms with varying densities through the use of air pressure.

EMULSION: The dispersion of a liquid, organic binder or synthetic resin in a normally immiscible liquid vehicle (e.g. water).

END BLOCK: An enlarged end section of a member, which is sometimes necessary to reduce anchorage stresses to allowable values.

ENDOTHERMIC: Pertaining to a reaction which occurs with the absorption of heat.

ENTRAINED AIR: Microscopic air bubbles intentionally incorporated in concrete or mortar.

ENTRAPPED AIR: Air in concrete which is not purposely entrained.

EPOXY: See Section 21.1.1.

ERGONOMICS: The design of work so that the best use is made of human abilities without exceeding human limitations (AS 1837; BS 4467).

EROSION: Deterioration brought about by the abrasive action of fluids or solids in motion.

EXOTHERMIC: Pertaining to a reaction which occurs with the evolution of heat.

EXPANSION JOINTS: Permanent joints between different parts of the work, formed to allow small relative movements normal to the joint to occur without the development of serious stresses. (See CONTROL JOINTS.)

FALSEWORK: The temporary structure erected to support work in the process of construction; comprising props, base plates, stringers, lateral bracing, shores and formwork for concrete beams and slabs.

FATIGUE: The weakening of a material caused by repeated or alternating loads.

FIBRE ASPECT RATIO: In fibre-reinforced concrete, the ratio of fibre length to fibre diameter.

FINENESS: A measure of the specific surface area or particle-size distribution.

FINENESS MODULUS: The sum of the cumulative percentage of aggregate retained on each of a series of sieves, each having a clear opening which is one-half that of the preceding one, the total being divided by 100 (see Section 3.3.4(c)).

FLASH SET: The setting of cement during or immediately after mixing.

FLEXURAL RIGIDITY: A measure of stiffness of a member, which is indicated by the product of modulus of elasticity and moment of inertia divided by the length of the member.

FLEXURAL STRENGTH: See TRANSVERSE STRENGTH.

FLOW FACTOR: The consistence of grout, as indicated by the time of efflux of a predetermined volume of the grout from a flow cone containing a precisely sized orifice.

FLUIDISATION: Aeration of dry powder to make it behave as a fluid.

FLY ASH (PULVERISED FUEL-ASH): Finely divided material (which may have pozzolanic properties) from the precipitators near flues of power stations using pulverised coal (see Sections 17.2 and 33.2.2(c)).

FORCE: An influence which produces or tends to produce motion or change of motion. (See MASS, NEWTON, PASCAL, WEIGHT.)

FORTRAN: Formula translation.

FOUNDATION: The ground upon which a substructure is supported.

FREE MOISTURE: Moisture not retained or absorbed by aggregate.

GAP-GRADING: An aggregate grading in which certain intermediate sizes of particles are wholly or substantially absent (see Sections 3.3.4(d), 4.5 and 23.1.5).

GAUGE: The nominal size of an aggregate. It is the minimum size of sieve through which at least 95 per cent of an aggregate will pass.

GEL: An extremely finely divided substance that has a coherent structure.

GROUT: A fluid mixture of cement, sand and water, or of neat cement and water.

GYPSUM: A crystalline mineral (hydrated calcium sulphate, i.e. $CaSO_4 \cdot 2H_2O$) from which gypsum plaster is manufactured.

HARDENER: A cross-linking agent used to effect the hardening of a resin system. Alternatively, a surface-treatment liquid for concrete (see Section 20.2).

HARSH MIX: A concrete mix which is difficult to place and work because of the grading, shape and texture of the aggregate and the proportions of cement and water.

HERTZ (Hz): SI unit of 1 cycle per second in periodic frequency.

HOT CEMENT: Cement which retains some of the heat generated during grinding.

HYDRATE: A compound of water with another compound or an element.

HYDROGEN EMBRITTLEMENT: In the stress corrosion or acid pickling of metal, atomic hydrogen is liberated as a corrosion product and its diffusion into highly strained regions (ahead of crack tips) tends to cause embrittlement and propagation of crevice fracture. The hydrogen content may arise as a result of steel-making, case-hardening, cold-working and galvanising techniques, cathodic protection, corrosive waters or processes, corrosion promoters, moisture present in the flux covering of electrodes during welding, and a destructive change in structure of high-alumina cement having a sulphide constituent. Hydrogen sulphide in an aqueous solution of a few mg/litre can cause fracture of stressed tendons. Delayed fracture can occur in high-tensile-strength steel (greater than 1000 MPa) at only 50 per cent of the 0·2 per cent proof stress. See Sections 3.1.2(b) and 15.2.2(a), 34.1.4(b); also STRESS CORROSION and STRAIN AGEING.

ICES: An integrated civil engineering system for optimal applications in computer design.

INDUSTRIALISED BUILDING: A process whereby the components of a building are manufactured in a factory and subsequently assembled on the building site.

INHIBITIVE PIGMENT: A pigment (e.g. zinc chromate and red lead) which retards or prevents the corrosion of metals by chemical or electro-chemical means, rather than by purely barrier action.

INITIAL DRYING SHRINKAGE: The difference between the length of a specimen (moulded and cured under special conditions) and its length when dried to constant length, expressed as a percentage of the dry length.

INVAR: A nickel-iron alloy, containing 36 per cent nickel, having a very low coefficient of expansion at ambient temperatures.

ION: One of the electrically charged particles into which the atoms or molecules of certain chemicals (i.e. salts, acids, alkalies and bases) are dissociated by solution in water and which make such a solution a conductor of electricity.

ISOTROPY: The behaviour of a medium having the same properties in all directions.

KELVIN: The basic SI unit of thermodynamic temperature that is symbolised by °K. °K = °C + 273·15. Thus the temperature interval is 1°C = 1°K.

KILOGRAM: The mass of the International Prototype Kilogram that is in the custody of the Bureau International des Poids et Mésures, at Sèvres, France.

LAITANCE: A weak porous substance that may form on the surface of concrete, by the accumulation and hardening of a layer of cement and fine particles of aggregate with a high water/cement ratio.

LASER: Light amplification by stimulated emission radiation.

LATEX: Emulsions of natural rubber and various synthetic resins.

LIFT: The concrete placed between two consecutive horizontal construction joints, a lift usually consisting of several layers.

LIME: The product arising from the calcination of calcareous material, with or without subsequent hydration (see Chapter 18).

Air-slaked lime: Quicklime that is partially hydrated and carbonated through exposure to moist air.

Dolomitic lime: Magnesian lime with a very high content of magnesia.

High-calcium lime: Lime with a high content of calcium oxide. It is known also as fat lime.

Hydrated or slaked lime: The product that is formed by treating quicklime with water and which consists essentially of calcium hydroxide. It is marketed as a powder.

Hydraulic lime: Lime in which a high proportion of the calcium oxide is combined with silica, alumina and iron oxide. It is made from limestone or chalk (which contains calcareous matter) and it will set and harden under water.

Lime putty: The product that is obtained by slaking quicklime with an excess of water or soaking hydrated lime and, after settlement, siphoning off the excess liquor. Lime putty is highly plastic and its consistence varies with water content.

Limestone: A rock which consists essentially of calcium carbonate, with or without magnesium carbonate. Geological designations are Carboniferous, Permian, Oolitic, Jurassic and Cretaceous (see Section 3.3.3(c)).

Magnesian lime: Lime which contains a significant proportion (over 5 per cent) of magnesium oxide or magnesia.

Milk of lime: A suspension of lime in water.

Quicklime: The direct product of calcination of calcareous material, the principal constituent being calcium oxide.

Semihydraulic lime: Lime which is intermediate in composition and character between high-calcium lime and hydraulic lime.

LIMIT DESIGN: A method of proportioning reinforced concrete members based on calculations of their ultimate strength.

LIQUID LIMIT: The water content at which a pat of soil, cut by a groove of standard dimensions, will flow together for a distance of 13 mm under the impact of twenty-five blows in a standard liquid-limit apparatus (AS 1289, BS 1377 and ASTM D423).

LITRE: The volume of 1 kg of pure water (very nearly 1 cubic decimetre in capacity) at the temperature of maximum density (4°C) under a pressure of 760 mm of mercury.

LOAD FACTOR: The ratio of the collapse load to the working load on a structure or section.

MARL: Calcareous clay containing some 35-65 per cent of calcium carbonate. It is found in the floor of shallow lakes, swamps and extinct freshwater basins.

MASS: The measure of the quantity of matter in a body, to which its inertia is ascribed. The quotient of the weight of the body and the acceleration g due to gravity ($9 \cdot 81$ m/s^2). However, mass is commonly called "weight".

MATRIX: The medium that surrounds or unites the components of an aggregate. The term applies also to algebraic and charted information for computer programming techniques.

METHODOLOGY: Orderly arrangement of ideas or procedures in any branch of operational activity.

MICRON: One-millionth of 1 metre.

MIX (noun): A colloquial term for concrete mixture.

MODEL: An idealised representation of reality.

MODIFIED AASHO COMPACTION TEST: A test which consists of compacting a sample of material (passing a 19-mm sieve) at an optimum moisture content to give maximum density. The material is compacted in five equal layers in a metal cylinder of 100 mm diameter and 0·95 litre capacity. Each layer is struck twenty-five blows with a 4·5 kg hammer (which has a 50 mm diameter striking face and falls through a distance of 450 mm; see ASTM D1557).

MODULAR RATIO: The ratio of the modulus of elasticity of steel in tension to the assumed modulus of elasticity of concrete in compression.

MODULE: A common unit, particularly specified for dimensional coordination, whereby standard lengths and sizes of elements facilitate the economic processing and fabrication of components in industrialised productivity and systems building.

MODULUS OF ELASTICITY: The ratio of stress to strain, within the elastic range of a material under applied load. It is designated in tension, compression, shear and force per unit of area. For materials that do not conform to Hooke's Law throughout their elastic range, either the tangent modulus or the secant modulus is used. The former is the slope of the tangent to the stress-strain curve at a low stress, and the latter is the slope of the secant connecting two points on the stress-strain curve at prescribed stresses (see Sections 14.9.4, 31.3, 33.3.2 and 34.1.5; also Young's modulus of elasticity, Appendix I).

MODULUS OF RIGIDITY: The ratio of shear stress to the corresponding shear strain. Alternative terms are "shear modulus" and "modulus of elasticity in shear", symbolised by G. For a homogeneous and isotropic material

$$G = \frac{E}{2(1 + \sigma)}$$

where E = modulus of elasticity and σ = Poisson's ratio.

MODULUS OF RUPTURE: The estimated maximum tensile stress developed in a specimen, undergoing a beam test, at the moment of rupture.

MOISTURE MOVEMENT: The difference between the length of a specimen of concrete or mortar (when dried to constant length) and its length when subsequently saturated, expressed as a percentage of the dry length.

MONOMER: A molecule of low molecular weight capable of reacting with identical or indifferent monomers to form a polymer.

MORTAR: A suitably proportioned mixture of fine aggregate, cement and/or lime, and water.

NETWORK: A diagrammatic representation of a programme or project plan, which shows the sequence and interrelationship of significant finite events in the plan to achieve end objectives under planned resource applications and performance specifications (BS 4335).

NEWTON: That force which, acting on a mass of 1 kg, produces an acceleration of 1 m/s^2 in the direction of the force.

Note: $1 \text{ N} = 0.224\ 81 \text{ lbf}$ or $0.101\ 97 \text{ kgf}$ and $1 \text{ lbf} = 4.448\ 22 \text{ N}$; $1 \text{ kgf} = 9.806\ 65 \text{ N}$. Typical structural notations are kN for axial force and shear; kN for point loading, kN/m for line loading, and kN/m^2 or kPa for area loading, and kN·m for moment. (See PASCAL.)

NOMINAL MIX: A volumetric or gravimetric description of the proportions of dry materials in a concrete mix.

NYLON: A synthetic thermoplastic polyamide (made by interaction of a dicarboxylic acid with a diamine) capable of extrusion when molten into fibres and sheets of extreme toughness, strength and elasticity.

OPERATIONS RESEARCH: A mathematical and systematic approach to decision making and optimisation of function.

OSMOTIC PRESSURE: The pressure exerted on a membrane with very fine pores through which a solution of high density travels to a solution of lower density, even against high water pressure. (See ELECTRO-OSMOSIS.)

OXYGEN INDEX: The percentage oxygen in a mixture of oxygen and nitrogen that will support, in a flammability test, the combustion of a material for 3 minutes.

PASCAL: A unit of pressure equal to 1 N/m^2. (See NEWTON and Appendix VIII.)

(*i*) $1 \text{ MPa} = 1 \text{ N/mm}^2 = 145.038 \text{ lbf/in}^2$; $1 \text{ lbf/in}^2 = 6.894\ 76 \text{ kPa}$.

(*ii*) 1 kPa = 7·501 mm of mercury = 0·2953 in mercury = 102 mm water = 4·015 in water = 10 mb.

(*iii*) 1 atmos = 101·5 kPa.

(*iv*) Typical structural notations are kPa for pressure and MPa for cross-sectional stress and Young's Elastic Modulus (AS 1170, Part 1).

PEARLITE: Microeutectoid constituent of steel, consisting of alternate lamellae of ferrite and cementite and containing 0·9 per cent carbon.

PERIMETER GROUTING: Injection of grout around the periphery of an area, which is subsequently to be grouted at higher pressure within boundary limits.

PERMEABILITY-REDUCING ADMIXTURE: An agent that reduces the rate of transmission of moisture, either in a liquid or vapour form, through concrete.

PERT: Programme evaluation and review technique in operational network analysis. The activity times employed have a probability attached to them in terms of standard deviation or variance.

PLASTER: A material or mixture of materials that is applied in substantial thickness to surfaces for protective or decorative coating purposes. It is designated as gypsum, lime, gypsum-lime or cement-lime plaster, according to composition.

PLASTICISER: An additive to improve the workability of concrete (see Section 6.4).

PLASTICITY: That property of a body by virtue of which it tends to retain its deformation after reduction of the deforming stress to its yield stress (see Section 31.3).

PLASTICITY INDEX: The numerical difference between the liquid limit and the plastic limit of a soil (AS 1289, BS 1377 and ASTM D424). It is the range in water content through which a soil remains plastic.

PLASTIC LIMIT: The water content at which a soil will just begin to crumble when rolled into a thread approximately 3 mm diameter (AS 1289, BS 1377 and ASTM D424).

PLUMS: Large stones in mass concrete.

POISSON'S RATIO: The ratio of lateral strain to longitudinal strain. It is usually 0·15 for concrete and 0·25-0·30 for most metals (ASTM E6).

POLYAMIDE: Any polymer in which the units are linked by amide or thioamide groups.

POLYESTER: A synthetic polymer in which the structural units are linked by ester groups, formed by condensing carboxylic acids with alcohols.

POLYETHYLENE (POLYTHENE): A plastic polymer of ethylene (C_2H_4).

POLYMER: A substance consisting of long chain or very large molecules that are composed of small single molecules of the same type.

Note: Other terms used in the plastics industry are defined in BS 1755.

POLYMER CONCRETE: See CONCRETE, *Polymer.*

POLYPROPYLENE: A plastic polymer of propylene (C_3H_6), similar to polyethylene but of greater strength.

POLYURETHANE: A polymer of urethane ($NH_2COOC_2H_5$).

POWER-FLOAT: A motor-driven appliance with a rotating disc that smooths, flattens and compacts the surface course of concrete floors.

PRINCIPLES: Common rules generally proved valid by experience.

PRODUCTIVITY: The ratio, in economic terms, of an output of goods and services to an input of resources (i.e. raw materials, plant and man-hour facilities).

PROGRAMME: In automatic data processing, a set of instructions, expressions and any other necessary data for controlling a computer run (BS 3527); in operations research, a stated course of action.

PROGRESSIVE STRENGTH: A technically managed, pro-gressively staged, efficient system of reinforced concrete building construction, invented in Australia by Civil & Civic Pty Ltd. Basically, the primary and secondary elements are cast successively in strength-developing and self-supporting stages, with the aid of standard, prefabricated, steel reinforcing-rod trusses and unshored recoverable forms.

PROJECT: In operational procedures, any set of actions directed toward a specific aim or objective.

PROOF STRESS: Stress applied to a material sufficient to produce a specified permanent strain (see Section 34.1.4).

PROPORTIONAL LIMIT: The greatest stress which a material is capable of developing without any deviation from proportionality of stress to strain (Hooke's Law).

PUNNING: A form of light ramming.

QUARTERING: The reduction in quantity of a large sample of material by dividing a heap into four approximately equal parts by diameters at right-angles, then removing two diagonally opposite quarters and mixing the two remaining quarters intimately together, so as to obtain a truly representative half of the original mass. The process is repeated until a sample is obtained of the requisite size (see Section 3.3.4).

RADICAL: Atomic constituent of a molecule that remains unchanged and behaves as a unit in many reactions.

REFRACTORIES: Materials, usually nonmetallic, used to withstand high temperatures.

REFRACTORINESS: In refractories, the property of being resistant to softening or deformation at high temperatures.

REINFORCED EARTH REVETMENT: A progressive assemblage of consolidated granular soil and galvanised-metal restraining strips, onto which are attached precast concrete, fibre-reinforced cement or metal interlocking face panels. These are fitted with rear lugs or internal reinforcing rods that are used for fastening purposes.

REINFORCEMENT: Rods, bars or fabric, usually of steel, embedded in concrete for the purpose of resisting particular stresses (see Section 3.4). They are not used as tendons.

Cold-drawn wire: Wire which has been drawn through a die at normal temperature.

Cold-twisted bars: Rolled mild steel bars which have been twisted when cold to increase their strength.

Cold-worked steel: Steel bars or wires which have been rolled or drawn at normal temperatures.

Crack-control reinforcement: Reinforcement that is designed to control the width of cracks in concrete construction by uniformly distributing therein a greater number of minute fissures or microcracks.

Deformed bar: Reinforcing steel bar having recurrent deformations with the object of increasing the bond strength.

Distribution rods: Small-diameter rods, usually at right-angles to the main reinforcement, intended to spread a concentrated load on a slab and to prevent cracking.

Dowels: Short rods, extending approximately equally into two abutting concrete units or pavement slabs, to permit movement and transfer shear loads at the joint. They are suitably supported and aligned during concreting operations.

Galvanised: See Table 22.1.

Hard-drawn wire: Wire which is cold-drawn from mild steel, and which has a tensile strength of 570-650 MPa.

Heavy edge reinforcement: Wire fabric reinforcement, for use in highway pavement slabs, having one to four edge wires heavier than the other longitudinal wires.

High-strength reinforcement: Steel reinforcement which has a tensile strength of 600-700 (or more) MPa.

Mesh reinforcement: An arrangement of rods or wires normally in two directions at right-angles, and tied or welded at their intersections or interwoven. Alternatively, it is a diamond mesh of expanded metal.

Mild-steel reinforcement: Reinforcement made of steel containing approximately 0·12-0·25 per cent of carbon and having a tensile strength of 430-510 MPa.

Shear reinforcement: Reinforcement, usually in the form of stirrups and bent-up rods, designed to resist shear.

Starter rods (stub-rods): Rods left projecting from concrete, in order to locate and provide continuity with other reinforcement.

Tie bars: Deformed bars that hold two abutting concrete elements together, but which are not designed like a dowel to transfer shear loads at the joint.

Transverse reinforcement: Links or helical reinforcement for columns, or reinforcement at right-angles to the main reinforcement..

Twin twisted bars: Two bars of the same nominal diameter twisted together.

Twisted-steel fabric: Factory-made reinforcing fabric made from cold-twisted steel bars.

RELATIVE HUMIDITY: The ratio of the quantity of vapour actually present to the greatest amount possible in an atmosphere at a given temperature. It is expressed as a percentage.

RELAXATION: Loss of load on a test piece that is being maintained at constant length and constant temperature.

RENDERING: The application, by means of a trowel or float, of a coat of mortar (see Section 16.2). See SPATTERDASH.

RESEARCH: The search for knowledge.

RESILIENCE: The work done per unit volume of a material in producing strain.

RETEMPERING: The remixing of concrete or mortar which has started to harden (see Section 7.1.4).

SAND: Fine aggregate resulting from the natural disintegration of rock, or from the crushing of quarried rock or gravel (see Section 3.3).

SAND EQUIVALENT: A measure of the amount of clay contamination in fine aggregate (see Section 26.2; also AS 1289 and ASTM D2419).

SAPONIFICATION: The formation of a soap by the reaction between a fatty acid ester and an alkali.

SCREED: A timber or metal straight-edge which is moved over guides for the purpose of striking off or finishing a concrete or mortar surface.

SEGREGATION: The differential concentration of the components of mixed concrete, resulting in nonuniform proportions (e.g. coarse and fine aggregate) throughout the mass.

SEQUENCE STRESSING LOSS: In post-tensioning, the elastic loss in a stressed tendon resulting from the shortening of the member when additional tendons are stressed.

SERVICEABILITY: The performance of a structure or product, which relates to the design criteria and control of deformations or deflections and crack widths under working conditions, as opposed to ultimate conditions (see Chapters 14, 15, 19, 22, 27 and 31, also Table 3.25 and Appendixes II, 2(g), IV, 2 and IV, 4(b)).

SETTING TIME (SET): The time required by a freshly mixed paste of cement and water to acquire an arbitrary degree of stiffness, as determined by specific tests for initial and final set (see Section 3.1.1), in which the stiffness is determined by the penetration of a needle of specified dimensions and weight.

SHRINKAGE: The volumetric deformation that concrete undergoes when not subjected to load or restraint (see Section 31.2). See CREEP for complementary deformation.

SHRINKAGE CRACKS: Cracks due to restrained shrinkage.

SILEX LINING: Protective lining for tube mills made of a hard siliceous rock. It is used in the manufacture of white cement where iron-contamination must be avoided.

SILT: Mineral particles, passing a 0·075-mm sieve, naturally deposited as sediment in water (see Section 3.3.4(e)).

SIMULATION: The use of mathematical models to help solve problems.

SLENDERNESS RATIO: The ratio of effective length or height of a wall or pier to its effective thickness. The lesser value is used in assessing the stability of a masonry wall or concrete panel or column against lateral loading.

SLUMP: The vertical distance through which the top of a moulded mass of fresh concrete sinks on removal of the mould, under specified conditions of test (see Section 6.2).

SLURRY: A liquid mixture of cement or other finely divided material and water.

SOIL: A generic term for natural surface material which is not of a stony nature.

SOUNDNESS: The freedom of a solid from cracks, flaws, fissures or variations from an accepted standard. It is a measure of the freedom of cement from expansion on setting as measured by the Le Chatelier test (see Section 3.1.1).

SPATTERDASH: A rich mix of portland cement and coarse sand (such as 1 : 2 by volume) which is thrown onto a background by a trowel, scoop or other appliance, so as to form a thin, coarse-textured, continuous coating. As a preliminary treatment before rendering, it assists bond of the undercoat to the background, improves resistance to rain penetration and evens out the suction of variable backgrounds.

SPECIFIC GRAVITY: The ratio of the weight (mass) of any volume of a material to the weight (mass) of an equal volume of fresh water at a stated temperature (see Section 3.3.4(j)).

SPECIFIC HEAT: The amount of heat required per unit mass to cause a unit rise of temperature, over a small range of temperature. For ordinary concrete and steel, it is 0·92 and 0·50 kJ/kg·°C, respectively.

SPECIFIC SURFACE AREA: The ratio between the total surface area of a number of particles and their total weight, usually expressed in m^2/kg. (See "Fineness" in Appendix I and Index.)

STEEL: See REINFORCEMENT in Section 3.4.1 and Glossary; also Appendix I and international standards.

STRAIN: A measure (due to force) of the change in size or shape of a body with reference to its original size or shape. Linear strain is the change (due to force) per unit of length in an original linear dimension. Shear strain is the tangent of the angular change (due to force) between two lines originally perpendicular to each other.

STRAIN AGEING: Ageing which occurs subsequent to the cold working of an alloy. In steels, strain ageing is predominantly due to the presence of uncombined nitrogen, which is dependent on the steel-making process, and results in a marked decrease of ductility notch toughness and impact strength. This strength may be reduced by as much as 90 per cent. By stabilising susceptible steel with aluminium or titanium, the formation of stable nitrides can cut the potential loss of impact strength to only 15 per cent. (See BRITTLE FRACTURE, HYDROGEN EMBRITTLEMENT and STRESS CORROSION, also Table 22.1 for hot-dip galvanising, and steel manufacturers.)

STRAIN HARDENING: Increasing the hardness, proportional limit and tensile strength of metal by cold-plastic deformation, cold-rolling or stretching.

STRENGTH:

Compressive: The maximum compressive stress which a material is capable of developing, based on the original area of cross-section. (See CHARACTERISTIC STRENGTH.)

Creep: The stress that causes a given creep in a given time and at a specified temperature.

Fatigue: The greatest stress which can be sustained for a given number of stress cycles without fracture.

Shear: The maximum shearing stress which a material is capable of developing, based on the original area of cross-section.

Tensile: The maximum, nominal tensile stress which a material is capable of developing under prescribed conditions.

Yield: The stress at which a material exhibits a specified limiting deviation from the proportionality of stress to strain.

STRESS: The intensity at a point in a body of the internal forces or components of force which act on a given plane through the point. It is expressed in force per unit of area.

Note: The stress or component of stress which acts perpendicularly to a given plane is called the normal stress, and it may be either tensile or compressive. That which acts tangentially to the plane is called the shearing stress. Nominal stress (as used in product specifications) is calculated on the basis of an original cross-sectional area, whereas true stress is calculated on an instantaneous area.

STRESS CORROSION: Brittleness which may occur in steel when subjected simultaneously to high tensile stress and corrosive influences. Chlorides, sulphates, nitrates, ammonia and hydrogen sulphide are major corrodents in effecting the stress-corrosion cracking of steel (see References 17-19 and 117 of Part 4). (See HYDROGEN EMBRITTLEMENT.)

STRESS RELIEVING: Low-temperature heat treatment of metal, to reduce peak levels of internal stress and slightly increase ductility with negligible loss of hardness and strength. (See STRAIN AGEING.)

SUBGRADE: The soil prepared and compacted to support a structure or a pavement system.

SUBSTRATE: The directly underlying surface that is to be treated, primed or sealed.

SURFACE AREA FACTOR: See SPECIFIC SURFACE AREA and Reference 63 of Part 2.

SURFACE MOISTURE: Moisture on the surface of an aggregate, as distinct from the moisture contained in the material itself.

SYNTHETIC RESIN: Member of a heterogeneous group of compounds that are produced from simpler compounds by condensation or polymerisation.

SYSTEM: Related elements working together to complete a function or produce an integral unit.

TAMPING: The operation of compacting freshly placed concrete by repeated blows of a tamper. This may incorporate an open-mesh screen or roller, or a bar grid to force coarse aggregate below the surface and facilitate strike-off, prior to finishing operations by floating or trowelling.

TEMPLATE: A full-sized mould, pattern or frame, shaped to serve as a guide in forming or testing contour or shape.

TENDON: A stretched element used in a concrete member or structure to impart prestress to the concrete (see Chapter 34).

THERMAL CONDUCTIVITY: The quantity of heat which will pass through unit area of a material in unit time, when unit difference of temperature is established between the faces of a unit thickness of it.

THERMAL DIFFUSIVITY: Thermal conductivity divided by the product of specific heat times bulk density. It is an index of the facility with which concrete undergoes temperature change (see References 12 and 113 of Part 6).

THIXOTROPY: The property of a fluid that enables it to stiffen in a short period on standing, but to change its viscosity on mechanical agitation to that of a mobile liquid, the process being reversible.

TOLERANCE: The permitted variation from a given dimension or quantity.

TOUGHNESS: The property of matter which resists fracture by impact or shock.

TRADITIONAL METHODS OF DESIGN: Methods of design that are based on the expected behaviour of a structure or structural element under loads up to and including working load, in which the essential technique of design is to limit the stresses or forces at working load to specified proportions of the ultimate strengths of the constituent materials or parts of the structure or structural element.

TRANSFER: The action of transferring load in prestressing tendons to concrete (see Chapter 34).

TRANSVERSE STRENGTH: The strength of a specimen tested in transverse bending; normally synonymous with MODULUS OF RUPTURE, but also used to refer to breaking load.

TRASS: A volcanic deposit of a pozzolanic character (see Chapter 17).

TRIAXIAL COMPRESSION TEST: A test in which a cylindrical specimen (encased in an impervious membrane if necessary) is subjected to a confining hydrostatic pressure and then loaded axially to failure.

ULTIMATE-STRENGTH DESIGN: Design that is based on the expected behaviour of a structure or structural element under loads up to and including failure load, in which the essential technique is to design the structure to fail at a load which exceeds the working load in the specified ratio given by the load factor.

UNIFORMITY COEFFICIENT: In sieve analysis, a figure determined by dividing the size opening that will pass 60 per cent of a sample that is being screened, by the size that will just pass 10 per cent. These sizes may be estimated by plotting a summation curve.

VAPOUR PRESSURE: A component of atmospheric pressure which is caused by the presence of water vapour. It is expressed in mm of mercury.

VIBRATOR: An oscillating machine used to agitate fresh concrete, so as to eliminate gross voids and produce intimate contact with containing surfaces.

VISCOSITY: Friction within a liquid which is due to mutual adherence of its particles.

VOLUME CHANGE: The expansion and contraction of hardened concrete resulting from wetting, drying or temperature variations.

WATER/CEMENT RATIO: The ratio of the amount of water (excluding that absorbed within the surface of aggregate) to the amount of cement in a concrete or mortar mix, the value being preferably stated as a decimal by weight (see Sections 3.2.2 and 33.2.2(d)).

WATER GAIN: See BLEEDING.

WATERPROOFING: The prevention of transmission of liquid water under an external pressure.

WEATHERING: Changes in colour, texture or chemical composition at the surface of a natural or artificial material due to the action of the weather.

WEIGHT: The force which gravitation exerts upon a material body. It is equal to the mass times the acceleration g due to gravity ($9.806\ 65$ m/s^2 or 32.16 ft/s^2; 9.81 m/s^2 or 32.2 ft/s^2 to nearest decimal place).

Note: In Australia and Papua New Guinea, the measurement of g does not exceed 9.800 m/s^2. By weight signifies nominally the quantity of a substance according to weight measurement.

WETTING AGENTS: See PLASTICISERS.

WETTING EXPANSION: The difference between the length of a specimen when dried and its length when subsequently immersed in water, the result being expressed as a percentage (under specified conditions) of the dry length.

WORKABILITY: The ease with which a given set of materials can be mixed into concrete and subsequently handled, transported, and placed with a minimum loss of homogeneity. (See CONSISTENCE.)

WYTHE (LEAF): Each continuous vertical section of a wall, one masonry unit in thickness, that is tied to its adjacent vertical section or sections by metal ties, bonders or grout.

YIELD: Of concrete, the volume of compacted fresh concrete produced per 40-kg bag of cement.
Of lime, the volume of putty at a standard consistence that is obtained from a unit volume of quicklime.

YIELD POINT: In certain materials, the first stress (less than the maximum attainable stress) at which an increase in strain occurs without an increase in stress. Where no well-defined yield point occurs, the yield stress is deemed to have been reached when the total extension is observed to be 0.005 of the gauge length (AS 1391; BS 18).

YOUNG'S MODULUS OF ELASTICITY: See MODULUS OF ELASTICITY.

BIBLIOGRAPHIC ABBREVIATIONS

Text abbreviations are given here and listed further in the *Abbreviations Dictionary* by R. De Sola, published by Elsevier, New York, 1974.

Symbol	Title
AASHTO	American Association of State Highway and Transportation Officials.
ACE	United States Army Corps of Engineers.
ACI	American Concrete Institute.
AISC	American/Australian Institute of Steel Construction.
ARRB	Australian Road Research Board.
AS	Australian Standard.
ASCE	American Society of Civil Engineers.
ASTM	American Society for Testing and Materials.
AWS	American Welding Society.
BDRI	Brick Development Research Institute.
BRE	Building Research Establishment.
BRS	Building Research Station.
BS	British Standard.
BSI	British Standards Institution.
C & CA Aust.	Cement and Concrete Association of Australia.
C & CA Britain	Cement and Concrete Association of Britain.
CBRI	Central Building Research Institute.
CEB	Comité Européen du Béton.
CIA	Concrete Institute of Australia.
CIRIA	Construction Industry Research and Information Association.
CMA	Concrete Masonry Association.
CS	Concrete Society.
CSA	Canadian Standards Association.
CSIRO	Commonwealth Scientific and Industrial Research Organisation.

Symbol	Title
DBR	Division of Building Research.
D of E	Department of the Environment.
DEC	Department of the Environment and Conservation.
DH & C	Department of Housing and Construction.
DIN	Deutsche Industrie Norm (German Industrial Standard).
DSIR	Department of Scientific and Industrial Research.
EBS	Experimental Building Station.
EN-R	Engineering News-Record.
FIP	Federation Internationale de la Précontrainte.
HRB	Highway Research Board.
IABSE	International Association for Bridge and Structural Engineering.
ICE	Institution of Civil Engineers.
IE Aust.	Institution of Engineers, Australia.
ISO	International Organisation for Standardisation.
MPBW	Ministry of Public Building and Works.
MPW	Ministry of Public Works.
NAASRA	National Association of Australian State Road Authorities.
NAS	National Academy of Sciences.
NATA	National Association of Testing Authorities.
NBRI	National Building Research Institute.
NBS	National Bureau of Standards.
NCEL	National Civil Engineering Laboratories.
NCMA	National Concrete Masonry Association.
NRCC	National Research Council Canada.
NRMCA	National Ready-Mixed Concrete Association.
NSB	Notes on the Science of Building.
PCA	Portland Cement Association.
PCI	Prestressed Concrete Institute.
RCA	Reinforced Concrete Association.
R & DL	Research and Development Laboratories.
RILEM	Réunion Internationale des Laboratoires d'Essais et de Recherches sur les Matériaux et les Constructions. (International Union of Testing and Research Laboratories for Materials and Structures.)
RRB	Road Research Board.
SAA	Standards Association of Australia.
SI	Système Internationale.
TRRL	Transport and Road Research Laboratory.
TR & S	Technical Research and Services.

CONVERSION FACTORS

1. GENERALITIES

The fundamental unit of length in the metric system is the metre (m), and it is from this that the units of volume (m³), capacity (litre) and mass (kilogram, kg) are derived. These units are simply related in that 1 cubic decimetre (dm³, $\frac{m^3}{1000}$) equals 1 litre, 1 litre of water is the volume of 1 kilogram of water under prescribed conditions (Glossary, Appendix VI), and the bulk density of water is 1000 kg/m³ or 1 kg/litre (see Section 3.3.4(b)).

While the newton is the recommended metric unit for force, in the English-speaking language the kilogram is often synonymously used for the designation of both weight and mass. Following precedents set by the Standards Association and Building Regulation Authorities in Australia, the Cement and Concrete Association, British Standards Institution and the *ACI Manual of Concrete Practice,* limited literary ambiguities have been included in the text for communication of practical meaning to general readers.

The conversion of Imperial to Metric units and *vice versa* is facilitated by the following factors, which may be supplemented with further data that are documentally furnished in Appendix IX.

2. PREFIXES

Prefixes to be combined with designated SI units are scheduled here, those marked with an asterisk being limited in selected usage.

Prefix		Factor by which the unit is multiplied	Prefix		Factor by which the unit is multiplied
Name	Symbol		Name	Symbol	
tera	T	10^{12}	*deci	d	10^{-1}
giga	G	10^{9}	*centi	c	10^{-2}
mega	M	10^{6}	milli	m	10^{-3}
kilo	k	10^{3}	micro	μ	10^{-6}
*hecto	h	10^{2}	nano	n	10^{-9}
*deka	da	10	pico	p	10^{-12}
			femto	f	10^{-15}
			atto	a	10^{-18}

3. TABULATED FACTORS

Fairly exact conversion factors, scheduled here in alphabetical sequence, can be suitably rounded-off for approximate usage in the building, construction and allied industries.

AREA

1 km² = 0·386 10 square mile	1 square mile = 2·589 99 km²
1 ha = 2·471 05 acres	1 acre = 0·404 69 ha
1 m² = 1·195 99 yd²	1 yd² = 0·836 13 m²
= 10·7639 ft²	1 ft² = 0·092 90 m²
1 mm² = 0·001 55 in²	1 in² = 645·16 mm²
(1 ha = 10 000 m²)	

BULK DENSITY

1 kg/m³ = 1·685 56 lb/yd³	1 lb/yd³ = 0·5933 kg/m³
= 0·062 43 lb/ft³	1 lb/ft³ = 16·0185 kg/m³

CALORIFIC VALUE: MASS AND VOLUME BASIS

1 kJ/kg = 0·429 92 Btu/lb	1 Btu/lb = 2326·0 J/kg
1 kJ/m³ = 0·026 84 Btu/ft³	= 2·326 J/g
	1 Btu/ft³ = 37·2589 kJ/m³

CAPACITY

1 litre = 0·219 98 gal (Imp)	1 gal (Imp) = 4·546 09 litres
= 0·264 18 gal (US)	1 gal (US) = 3·785 31 litres
= 1·7598 pt (Imp)	1 pt (Imp) = 0·568 25 litre
= 2·113 44 pt (US)	1 pt (US) = 0·473 16 litre
= 61·0255 in³	1 fl oz (Imp) = 28·412 25 ml
1 ml = 0·035 19 fl oz (Imp)	1 fl oz (US) = 29·572 70 ml
= 0·033 82 fl oz (US)	1 in³ = 16·3871 ml
1 litre/s = 0·0353 ft³/s (cusec)	1 ft³/s (cusec) = 28·317 litre/s

DENSITY*

ENERGY (WORK, HEAT)

1 MJ = 0·277 78 kWh	1 kWh = 3·600 MJ
1 kJ = 0·947 82 Btu	1 Btu = 1·055 06 kJ
1 J = 0·737 56 ft lbf	1 ft lbf = 1·355 82 J
1 J = 1 N·m	1 cal = 4·1868 J
1 kW = 1·341 02 hp	1 hp = 0·7457 kW

FORCE†

1 kN = 0·100 36 tonf	1 tonf = 9·964 02 kN
1 N = 0·224 81 lbf	1 lbf = 4·448 22 N
1 N = 0·101 972 kgf	1 kgf = 9·806 65 N

HEAT‡

HEAT FLOW INTENSITY

1 W/m² = 0·316 99 Btu/ft² h	1 Btu/ft² h = 3·154 59 W/m²

LENGTH

1 km = 0·621 37 mile	1 mile = 1·609 34 km
1 m = 1·093 61 yd	1 yd = 0·9144 m
= 3·280 84 ft	1 ft = 0·3048 m
1 mm = 0·039 37 in	1 in = 25·4 mm

MASS (WEIGHT)§

1 tonne = 0·984 21 ton (Imp)	1 ton (Imp) = 1·016 05 tonne
= 1·102 31 ton (US)	1 ton (US) = 0·907 18 tonne
1 kg = 2·204 62 lb	1 cwt (112 lb) = 50·8023 kg
1 g = 0·035 27 oz	1 cwt (100 lb) = 45·359 24 kg
1 tonne = 1000 kg	1 lb = 0·453 59 kg
	= 453·592 37 g
	1 oz = 28·3495 g

MASS PER UNIT AREA

1 kg/m² = 1·843 34 lb/yd²	1 lb/yd² = 0·542 49 kg/m²
1 kg/m² = 0·204 82 lb/ft²	1 lb/ft² = 4·882 43 kg/m²
1 g/m² = 0·003 277 oz/ft²	1 oz/ft² = 305·152 g/m²

MASS PER UNIT LENGTH

1 kg/m = 0·671 97 lb/ft	1 lb/ft = 1·488 16 kg/m
	1 lb/in = 17·8580 kg/m

* See Bulk Density.
† See Newton in Glossary, Appendix VI.
‡ See Energy.
§ See Glossary, Appendix VI.

MOMENT OF FORCE, TORQUE

1 N.m = 0·737 56 lbf ft	1 lbf ft = 1·355 82 N·m
= 8·850 75 lbf in	1 lbf in = 0·112 98 N·m

MOMENT OF INERTIA*

POWER

1 kW = 1·341 02 hp	1 hp = 745·700 W
1 W = 0·737 56 ft lbf/s	1 ft lbf/s = 1·355 82 W

PRESSURE, STRENGTH AND STRESS†

1 MPa = 0·064 75 tonf/in²	1 tonf/in² = 15·4443 MPa
= 9·323 85 tonf/ft²	1 tonf/ft² = 0·107 25 MPa
= 145·038 lbf/in²	1 lbf/in² = 6·894 76 kPa
1 kPa = 20·8854 lbf/ft²	1 lbf/ft² = 47·8803 Pa
1 MPa = 10·197 16 kgf/cm²	1 kgf/cm² = 0·098 07 MPa
1 kgf/cm² = 14·2234 lbf/in²	1 lbf/in² = 0·070 31 kgf/cm²
1 MPa = 1 N/mm², 1 Pa = 1 N/m²	

SECOND MOMENT OF AREA

1 mm⁴ = 2·403 × 10⁻⁶ in⁴	1 in⁴ = 416·231 × 10³ mm⁴
1 cm⁴ = 0·024 03 in⁴	1 in⁴ = 41·6231 cm⁴

STRENGTH AND STRESS‡

TEMPERATURE

$$°C = \frac{5}{9}(°F - 32) \qquad °F = \frac{9}{5}(°C + 32)$$

THERMAL CAPACITY

1 kJ/kg·°C = 0·238 85 Btu/lb·°F   1 Btu/lb·°F = 4·1868 kJ/kg·°C

THERMAL CONDUCTANCE, TRANSMITTANCE

1 W/m²·°C = 0·176 11 Btu/ft²h°F   1 Btu/ft²h°F = 5·678 26 W/m²·°C

THERMAL CONDUCTIVITY

1 W/m·°C = 6·933 47 Btu in/ft²h°F   1 Btu in/ft²h°F = 0·144 23 W/m·°C

---

* See Second Moment of Area.

† See Pascal in Glossary, Appendix VI.

‡ See Pressure.

UNIT WEIGHT*

VELOCITY

1 km/h = 0·621 37 mile/h	1 mile/h = 1·609 34 km/h
= 0·911 34 ft/s	= 0·447 04 m/s
1 m/s = 3·280 84 ft/s	1 ft/s = 1·097 28 km/h
= 2·236 94 mile/h	= 0·3048 m/s

VOLUME

1 m³ = 1·307 95 yd³	1 yd³ = 0·764 55 m³
= 35·3147 ft³	1 ft³ = 0·028 32 m³
1 cm³ = 0·061 02 in³	1 in³ = 16·3871 cm³
1 litre = 61·0255 in³	= 16·3871 ml
(1 cm³ = 1000 mm³, 1 m³ = 1000 litres)	

VOLUME RATE OF FLOW

1 m³/s = 35·3147 ft³/s	1 ft³/s = 101·9405 m³/h
	(1 cusec) = 0·028 32 m³/s
1 litre/h = 0·219 98 gal/h	1 gal/h = 4·546 09 litre/h

WEIGHT†

WORK‡

4. SPECIAL FEATURES

Selected conversion factors for convenient usage in the concrete industry are given here.

1 (40-kg) bag of cement/m³	= 0·717 26 (94-lb) bags of cement/yd³
1 (42·6-kg) bag of cement/m³	= 0·763 88 (94-lb) bags of cement/yd³
1 (50-kg) bag of cement/m³	= 0·896 58 (94-lb) bags of cement/yd³
	= 0·752 48 (1-cwt) bags of cement/yd³
1 kg/(40-kg) bag of cement	= 2·350 lb/(94-lb) bag of cement
1 kg/(42·6-kg) bag of cement	= 2·207 lb/(94-lb) bag of cement
1 kg/(50-kg) bag of cement	= 1·880 lb/(94-lb) bag of cement
	= 2·240 lb/(1-cwt) bag of cement

---

* See Bulk Density.

† See Mass, and also Glossary, Appendix VI.

‡ See Energy.

1 g/(40-kg) bag of cement = 0·0376 oz/(94-lb) bag of cement

1 g/(42·6-kg) bag of cement = 0·0353 oz/(94-lb) bag of cement

1 g/(50-kg) bag of cement = 0·0301 oz/(94-lb) bag of cement
= 0·0358 oz/(1-cwt) bag of cement

1 litre/(40-kg) bag of cement = 0·234 47 gal (Imp)/(94-lb) bag of cement
= 0·281 60 gal (US)/(94-lb) bag of cement

1 litre/(42·6-kg) bag of cement = 0·220 16 gal (Imp)/(94-lb) bag of cement
= 0·264 41 gal (US)/(94-lb) bag of cement

1 litre/(50-kg) bag of cement = 0·187 58 gal (Imp)/(94-lb) bag of cement
= 0·223 50 gal (Imp)/(1-cwt) bag of cement
= 0·225 28 gal (US)/(94-lb) bag of cement
= 0·268 42 gal (US)/(1-cwt) bag of cement

Nominal 40-, 42·6- and 50-kg bags of cement have volumes of 0.0265, 0.0283 and 0·0331 m³, respectively, and a 94-lb bag has a volume of 1 ft³.

1 m²/litre = 5·437 06 yd²/gal (Imp)
= 4·527 11 yd²/gal (US)

1 kg/litre = 10·0225 lb/gal (Imp)
= 8·3451 lb/gal (US)

1 litre/ m³ = 0·168 18 gal (Imp)/yd³
= 0·201 98 gal (US)/yd³

5. GRADE-DESIGNATED CONCRETE

Specified characteristic (cylinder) strengths are given here. Each item of standardised site-delivered concrete is a strength value, as assessed by standard tests, which is exceeded by the strengths at 28 days of at least 95 per cent of the concrete (see Sections 4.3.1 and 13.2.1).

Characteristic Strength*

Compressive		Indirect Tensile		Flexural	
MPa	lbf/in²	MPa	lbf/in²	MPa	lbf/in²
15	2200	2·0	290	2·5	360
20	2900	2·5	360	3·5	510
25	3600	3·0	430	4·0	580
30	4400	3·5	500	4·5	650
40	5800				
45	6500				
50	7300				
55	8000				
60	8700				

*1 MPa = 145 lbf/in² and 1000 lbf/in² = 6.89 MPa, approximately.

MEMORANDA

British Standards Institution:
  Accuracy in Building: BS PD6440.
  Conversion Factors and Table: BS 350.
  Coordination of Dimensions in Building:
    BS 4011, DD22, PD6432, PD6444, PD6446.
  Schedule of Weights of Building Materials: BS 648.
  Specification for pH Scale: BS 1647.
  The International System of Units (SI): BS 3763.
  Universal Decimal Classification (UDC): BS 1000.
  Use of the Metric System in the Construction Industry:
    BS PD6031.
  Use of SI Units: BS PD5686.

Standards Association of Australia:
  Conversion Factors and Tables: AS 1376, 1377.
  Coordinated Preferred Dimensions in Building: AS 1224, 1233,
    1234.
  Metric Conversion in Building and Construction: AS MH1.
  Metric Data for Building Designers: AS MH2.
  Metric Units for Use in the Construction Industry: AS 1155.
  *Metric Units for Use in Water Supply, Sewerage and Drainage
    (including Pumping):* AS 1686.
  Metric Units in Welding: AS 1880.
  The International System of Units (SI) and Its Application:
    AS 1000.

Addendum:
  Kaye, G. W. C. *Tables of Physical and Chemical Constants, and
    Some Mathematical Functions.* Longman, London, 1973.
  Metrication in the Construction Industry. MPBW, Britain.
  Metric Practice Guide. ASTM.

INDEX

Index*

*NOTE: Pages numbered in heavy print may have a priority of reference value in the entries.

832

833

834

oil-well, 227, 457, 797
ordinary portland, 3, **7**, 11, **22**, 343, 527
paint, 500
portland, 3, **7**, 11, 19, **22**, **343**, 527
portland blastfurnace, 11, 19
portland fly ash, 11, 23, **330**, 542, 704
portland-pozzolan, 11, **23**, 306, **327**, **330**, 346, 542, 704, 723
pozzolan, 11, **23**, **327**, 346, 542
setting accelerator, **222**, 402, 554, 569, **608**
setting and hardening, **8**, 12, 15, 199, **206**, **222**, 227, 509, **547**, 565
setting retarder, **227**, 431, **509**, 542, 545, 576
setting time and strength, 12, 814
slag, 24
Sorel, 309
special, 14, **344**, 352, 527
standards, 10, 12, 746
storage, 28, 760
sulphate-resisting portland, 10, 11, **23**, 343
sulphoaluminate, 15, 341, **463**
superfine, 21, 137
supersulphate, 11, 25, 343
transportation, 27
watertightness, 232
white, 24
Cementation, **455**, 685
Cement-based paint, 500
Cement-bound macadam, 464
Cement-coated tape, 96
Cement-lime concrete, 334
Cement-penetration macadam, 464
Cement-treated crushed rock, 467
Cement factor, **149**, 150, 232, **342**, 529, **542**, 543, 569, 602, 621, **622**, 746, 797
Characteristic strength, 115, 484, **797**, **829**
Chemical admixtures, *see* Admixtures
Chemical-resistant mortars, 387
Chemical attack, 9, **11**, **305**, 307, **339**, 375
Chlorinated rubber lacquer, 368
Clay, **50**, 63, **67**, 529, 797, *see also* Aggregate
Cleaning abrasive blast, 93
Coefficient of expansion, 555, *see also* Thermal expansion
Cold-weather concreting, 206, **222**, 435, 543, **568**
Cold bituminous mastic mortar, 372
Cold stores, 358, 359, 648
Cold working, 90, 797
Colloidal concrete, 464
Colloidal grout, 304, 324, **455**, 457, **466**, 797

Combustion gas curing, 219
Compacting factor, 119, 132, 133, **157**, 160, 163, **798**
Compaction, **36**, 182, **186**, 193, 227, 303, 306, 362, **417**, **431**, 465, 469, 472, 531, 603, **719**, 798
Composite construction, 668, 688, 798
Compressive strength (non-standard specimens), 141, 252
Compressive strength (standard specimens), 264, *see also* Strength: compressive
Concrete, 5, 798
aerated, **603**, 631, 641, 798
air-entrained, **127**, 151, 234, 303, 304, **315**, 319, 324, 346, 358, **424**, 440, **448**, 529, 543, 568, 723, 803
block masonry, *see* Masonry
building blocks, *see* Masonry
bulk density, 126, 335
cement, 3, **7**, 342, 352, 796
cement-lime, 334
colloidal, 464, 798
coloured, 24, 363, 485, 502
control, **118**, 246, **252**, **478**, 575, 576, **671**, 763, **764**
cores, 141, 266, 724, **747**
cover, **90**, 303, 401, 530, **629**, 640, **663**, **666**, **668**, 705
counterweight, 540
dense, 130, **186**, **417**, 536, 663, 798
density, *see* bulk density
dry-packed, 798
durable, 11, **295**, **342**, **351**, 359, 397, 760, **764**, 768
ecological, 798
fat, 798
ferro-cement, 704
fibre-reinforced, **94**, 177, 319, **605**
floors, 351, 387, 397, 712, 717, *see also* Floors
foamed, **603**, 631, 641, 798
grade designated, 828
granolithic, 359
green, 799
grouted, **455**, 457, 462, **466**, **487**, 538
heat-resistant, 527, 768
heavyweight, 537
high-alumina-cement, **11**, 14, **16**, 124, **139**, 140, **203**, 365, **527**, 568
high-early-strength, **140**, **206**, 214, **222**, **568**, 607, **608**, 634, 767
high-strength, 37, **130**, 136, 138, **149**, 426, **427**, 621, 633, **635**, **683**, **684**
impermeable, *see* Impermeability
insulating, 357, 528, 533, **571**, **595**, 602, 621, **622**, 644, **669**
in-situ, 176, **182**, **315**, 352, **417**, 435, **443**, **541**, 799

Fineness modulus, **60,** 82, 432, 450, **485,** 530, 804
Finishing, **190,** 362, **366,** 379, 387, 397, **439,** 497, **500,** 560, 647, 652, **719,** 723
Fire resistance, **311,** 527, **630,** 641, **667,** 705
Flame spalling, 366, 511
Flash set, 7, 17, 804
Fleximer compound, 364, 373
Floors:
    anti-static, 311, **376,** 398, 720
    articulated, 710
    conductive, 311, **376,** 398, 720
    granolithic, 359, 798
    granular metal, 397
    heated, 356, 359, **716**
    industrial concrete, 351, 764
    magnesium oxychloride, 309, 392
    monolithic, 360, 689, 717
    precast, 473, 486, **648,** 651, **683,** 689, 690
    prestressed, 498, 664, **683,** 685, 689, 690
    residential, 712
    surface treatments, **365,** 387, 391, 397
    terrazzo concrete, 388, **717,** 800
    tiled, 374, 719
    vacuum-processed, 434
Flow cone, 457, 748
Fly ash, 327, **330,** 457, 599, 601, 604, 804
Foamed concrete, 603, 631, 641
Foaming agents, 460, 608
Folded-plate structures, 706
Form oil, 197, 476
Formwork, 183, 188, **190,** 193, **197,** 417, 425, 437, 473, 509, 531, **574**
Fortran, 778, 804
Foundation, 305, 337, **339,** 343, **354,** 465, **471, 710, 714,** 758, 804
Freemoisture, 35, 38, 80, 804
Freezing chambers, 358, 359, 648
Freezing and thawing, 240, **271, 306,** 318, 436
Frost action, 240, 271, 306, 318, 430, **568,** 629, 640
Fundamental characteristics, 7, 295, **547,** 589, **663**
Further procedures for durability, 397

Galvanised steel, 93, 403
Gap grading, **14,** 427, 804
Gas treatment, **219,** 307, 347, **398, 565, 636,** 639
Gaseous and bacterial corrosion, **307,** 339, 795
Gauge, 61, 62, 76, 804
General information, 7, **46, 107,** 154, **190, 199, 259, 315, 327,** 330, 336, 380, 389, **417,** 434, **443,** 450, 455,

473, 482, 561, 568, **641, 663, 793,** 823
General requirements, 5, 57, 108, **232, 295,** 343, 355, 359, 484, 527, **767, 768**
General sieve data, **76**
Grading of aggregate, **61, 62,** 76, **107, 120,** 142, **144, 353,** 427, 467, **485,** 529, 595, **600,** 800, 804
Granolithic paving, 359, 798
Granular metal surfacing, 397
Graphical proportioning of aggregate, 143
Grouped-frequency method, 253
Grout, 304, 324, **455,** 457, 462, **466, 487,** 538, **685, 749,** 797, 805, 810
Gunite, *see* Pneumatic mortar
Gypsum plaster, 309, 608, **647, 669,** 749, 805

Handling:
    aggregate, 82
    cement, 27
    operations, 175, **182, 642,** 671, **689,** 690
    stresses, 481, 670
Harmful materials, 63, 64
Harsh mix, 130, **354, 427,** 542, **601,** 805
Heat and radiation resistance, 527
Heat of hydration, **11, 13,** 17, 459, **541,** 554
Heated concrete floors, 357, 359, **716**
Heating of materials, 570, 574
Heat-resistant concrete, 527, 768
Heat-resistant paint, 534
Heavyweight concrete, 537
High-alumina cement concrete, **11,** 14, **16,** 124, **139,** 140, **203,** 365, **527,** 568
High-early-strength:
    cement, 11, **17, 20,** 140
    concrete, **140, 206,** 214, **222, 568,** 607, **608,** 634, 767
High-pressure steam curing, **214,** 306, 310, 347, **549, 617,** 724
High-strength concrete, 37, **130,** 136, 138, **149,** 426, **427,** 621, 633, **635, 683, 684,** 767
High-strength steel, **89,** 93, 300, **663, 672,** 813
High-temperature steam curing, **206,** 347, 549
Historical, 3
Hot cement, 576, 805
Hot-weather concreting, **178, 202,** 203, 224, 227, **559, 576**
Hydrated lime, 166, 323, **334,** 336, 450, **486,** 605, **611,** 612, **647,** 806
Hydration and hardening, 8, **199,** 206, 222, **227, 547,** 549, 559, 561, 565, **568, 576, 617**

Hydraulic:
  cement, 7, 15, 797
  lime, 335, 807
Hydro valve, 545
Hydrocarbons:
  aliphatic, 794
  aromatic, 368, 795
Hydrogen sulphide and embrittlement, 18, 93, **297**, 340, 342, 403, 404, **674**, 685, **805**
Hydrophobic cement, 21
Hydrothermal curing, 32, 138, 217

Ices, 805
Igneous rocks, 53
Ilmenite, 82
Impact resistance:
  aggregate, 72
  concrete, 273, 624
Imperial units, 824, 829
Impermeability, **226, 232, 302,** 315, 345, 355, 436, **455,** 462, 489, 497, 500, **512,** 542, **549,** 629, 639, 665, **704, 767**
Incompatibility of materials, 309, 401
Industrial concrete floors, 351
Industrialisation, 474, 616, 729
Infra-red curing, 218
Inhibitive pigment, 404, 805
Initial drying shrinkage, 806, see also Drying shrinkage
Inspection, **252,** 479, 575, **760**
Insulating values, 533, **572,** 573, **627, 638,** 668
Insulation, 357, **528,** 571, 595, 621, **627,** 630, **638,** 641, **644,** 668
Ion, 338, 548, **561,** 806
Iron granules, 397, 461

Jackblock construction, 499
Jointing compounds, **237, 364,** 374, 465, 489, 493, 512, 556, 750
Jointing details, **187, 237,** 302, 320, **356,** 361, **364,** 374, **465,** 470, 489, **491, 512,** 534, 543, **557, 646,** 696, **697,** 706, **718,** 800

Laboratory equipment, 259, 478, **784**
Laitance, 188, 237, 304, 315, **358,** 361, **806**
Laser beam, 198, 806
Latex, 373, 389, 806
Laying concrete blocks, 495, 646
Laying concrete floors, 355, **359,** 397, **559,** 710, **712, 717**
Leaching, 310, 349
Leak stoppage, 17, 226, 234
Liftslab system, **498,** 685, 689
Lightning, 92
Lightweight calcium silicate hydrate, 603, 612, 636

Lightweight concrete, **589,** 663, 704, 768, 799
Lightweight-aggregate concrete, 418, 421, 448, **591, 621, 642,** 669, 671, **683**
Lime, 166, 323, **334,** 336, 450, 487, 605, **611,** 644, 647, **806**
Lime soundness, 336, 605
Lime-soil stabilisation, 337
Limestone, 50, **56,** 217, 308, **336,** 345, 555, 630, 725, 807
Limewash, 197
Lining formwork, 191, 425, 435, **476**
Liquid limit, 468, 807
Load factor, 669, 807
Loading notation, see Newton
Loss of prestress, 624
Low-alkali sodium silicate, 314, **368,** 455
Low-heat cement, 11, 21, 542
Low-pressure steam curing, 206, 549
Low-shrinkage mixes, **15,** 459, **460,** 541, **550,** 767

Macadam:
  cement-bound, 464
  cement-penetration, 464
Magnesium oxychloride, 309, 392
Main components, 7
Maintenance, 295, **377,** 386, **452,** 497, 722
Making good concrete, **5,** 34, **107,** 175, 232, 345, 417, **601,** 671
Management, 246, **728,** 758, 781
Marine work, 296, 318, 341, **348,** 401
Masonry, 208, 268, 321, 425, **482, 486,** 495, 535, **558,** 620, **625,** 634, 637, **645,** 723
Masonry cement, **18,** 324, **482,** 486, 495
Mass, 807, 825
Mass concrete, 149, **541,** 799
Massive construction, 541
Mastic mortar, **364, 372,** 374, 493, 514, 556
Material finer than 0.075 mm sieve, 11, 64, **68,** 327, **331, 605**
Matrix, 365, 376, **426,** 429, 807
Maturity factor, 210, 218
Measurement of water and liquid admixtures, 45
Measuring cylinder, 41
Mechanical compaction, see Compaction; Vibrated concrete
Membrane curing, **203,** 245, 267, 355, **363,** 498, 560, 723
Memoranda, 829
Metal in concrete, 403
Metallic silico-fluorides, 314, 367
Methodology, 729, 808
Microbiological corrosion, see Bacterial corrosion
Mineralogy and petrography, 50

844

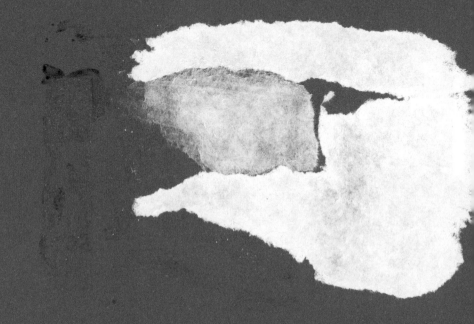